CARBONATE DEPOSITIONAL ENVIRONMENTS

CARBONATE DEPOSITIONAL ENVIRONMENTS

Edited by
Peter A. Scholle, Don G. Bebout, Clyde H. Moore

Published by
The American Association of Petroleum Geologists
Tulsa, Oklahoma 74101, U.S.A.

Association Editor: Myron K. Horn
Science Director: Edward A. Beaumont
Project Editors: Douglas A. White and Ronald L. Hart
Production and Design Editor: Gerri Winchell
Typographer: Carol Thompson

Table of Contents

Table of Contents

Table of Contents

Introduction

Peter A. Scholle
Don G. Bebout
Clyde H. Moore

*F*or more than 100 years geologists have been examining and describing modern sediments with an eye toward using characteristic features to aid in the interpretation of depositional settings of ancient strata. This field of interest developed particularly during the 1950s and 1960s with the creation of detailed models for modern carbonate deposition in Florida, the Bahamas, Cuba, the Persian Gulf, Belize, Pacific atolls, the Great Barrier Reef and other areas. An understanding of the depositional environments of these modern models, coupled with increased understanding of diagenetic effects, has led to vastly improved interpretations of ancient limestones. Such models also improved the "predictability" of many carbonate reservoir rocks.

In spite of the great strides made in our knowledge about carbonate depositional environments, their characteristic features have never been synthesized in a single work. Although excellent textbooks exist which describe some aspects of the interpretation of both ancient strata and modern sediments, systematic treatment of the entire subject is available only in the primary literature.

This book is an attempt to bring together this widely disseminated literature. The volume is specifically designed for use by the non-specialist—the petroleum geologist or field geologist—who needs to use carbonate depositional environments in facies reconstructions and environmental interpretations. Yet it is hoped that the book will also serve as a valuable reference for the specialist or advanced graduate student.

Toward that purpose, the book is extensively illustrated with color diagrams and photographs of sedimentary structures and facies assemblages. The text focuses on the recognition of depositional environments rather than on the hydrodynamic mechanisms of sediment movement. Assemblages of sedimentary structures and three-dimensional facies geometry are examined as criteria for recognition of depositional environments in surface and subsurface sections. Although individual sedimentary structures generally are not diagnostic criteria, frequently the entire suite of structures found in a rock sequence can be used to recognize carbonate depositional environments, particularly when used along with vertical and lateral facies sequences, rock-body geometry, grain size analysis, and other techniques. All of these criteria are stressed in various chapters of this

book. In addition, the presence or absence of diagnostic organisms (or trace fossils) as well as the ecological characteristics of such faunal assemblages are discussed throughout the book.

Early diagenetic processes are dealt with wherever they provide a clue to depositional settings. Rarely are diagenetic features absolute criteria for recognition of depositional environments but an understanding of such features may provide valuable additional evidence in such studies. In carbonate rocks, in particular, diagenetic processes start at the very moment of grain formation and deposition and the fabrics of diagenetic alteration (grain boring, leaching, cementation, etc.) are commonly environmentally sensitive.

Finally, an attempt has been made to provide some perspective on large-scale influences on carbonate deposition such as tectonic patterns, fluctuations of sea level, variations of climate, evolutionary patterns of organisms, and patterns of terrigenous sediment input. Carbonate sediments, because they are produced mainly by organisms (or through biochemical processes), are particularly susceptible to these influences. Although only a small number of different sedimentary environments exist, the factors mentioned above lead to a remarkable diversity of possible sedimentary patterns.

The ultimate purpose of this book is to improve exploration for oil, gas, and mineral deposits. Understanding of depositional environments and early diagenetic patterns are generally critical to the prediction of patterns of porosity and permeability. This is true both because depositional patterns commonly control patterns of water movement and diagenesis in carbonate rocks, and because a considerable amount of productive porosity in carbonate rocks is preserved from the depositional or early diagenetic setting. Thus, recognition of environments coupled with prediction of trends can lead to important exploration advantages as well as improvements in secondary recovery strategy.

The book is divided into individual chapters organized by environment of deposition. Although there is some variability between chapters due to the nature of available information for specific environments, most of the chapters are organized around a consistent outline. The introduction defines and describes the facies involved and provides a summary of the diagnostic criteria for the recognition of the environment. The second section provides information on relationships to adjacent facies, deposi-

tional setting, three-dimensional geometry of facies, and typical tectonic relationships along with sediment composition, texture and constituent organisms. A third section outlines the major sedimentary structures characteristic of the environment and is heavily illustrated to show physical and biological structures of primary or early secondary origin. In some chapters, information is also provided on typical sediment textures and log responses. A fourth section gives information on economic considerations such as porosity potential of the overall environment and subenvironments, common diagenetic effects, and overall petroleum- or minerals-trapping potential. Where available, ancient analogs are included at the end of each chapter.

The book is organized with chapters in a specific order from non-marine to deep-water settings. We have included environments in this book which are normally not considered to be particularly prospective for oil and gas. They have been included for a number of reasons. First, the recognition of depositional environments is not really based strictly on the potential of those environments for producing oil and gas. We would like to provide as complete a framework for recognition of en-

vironments as possible, and even units which are not necessarily producers will still be involved in facies interpretation. Second, some of the environments which are not currently considered to be economically prospective may contain, at some time in the future, "unconventional" reservoirs. We need only look back a few years to our views of lacustrine or deep-water sediments to see that environments which were previously completely written off as potential reservoirs have turned out, in some cases, to be very prolific ones. Lacustrine sediments are now major oil producers in the Uinta basin, in China, and in Brazil. Likewise, deep-water limestones have turned out to be productive in several areas within the North Sea basin and in North America.

This book describes only depositional environments of carbonate sediments. Equivalent sediments of clastic terrigenous origin have been discussed in AAPG *Memoir* 31, a separate book with parallel format.

Peter A. Scholle
Don G. Bebout
Clyde H. Moore

1

Subaerial Exposure Environment

Mateu Esteban
Colin F. Klappa

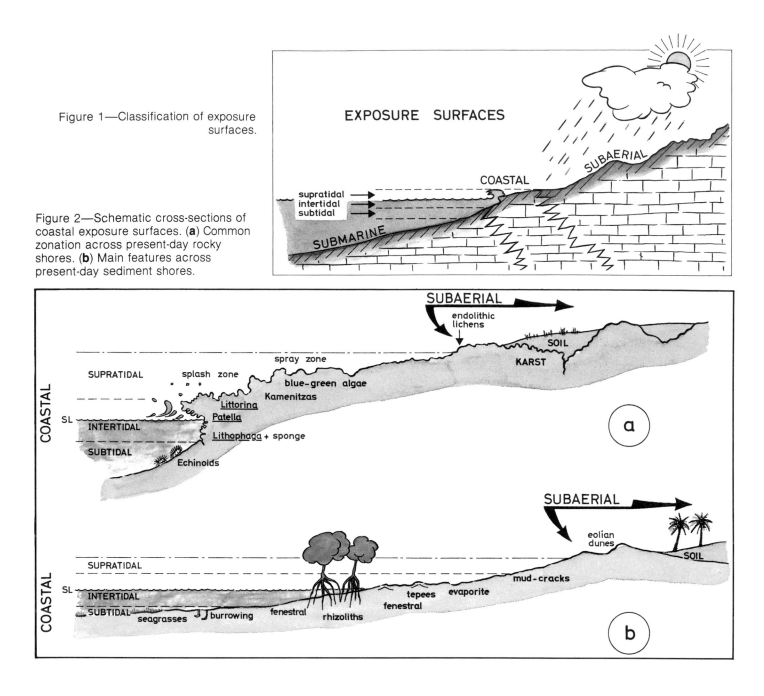

Figure 1—Classification of exposure surfaces.

Figure 2—Schematic cross-sections of coastal exposure surfaces. (**a**) Common zonation across present-day rocky shores. (**b**) Main features across present-day sediment shores.

*E*xposure surfaces occur on land and under the sea, but in this chapter we are concerned only with subaerial exposure surfaces. More specifically we are concerned with the effects of subaerial exposure on carbonate sequences. Subaerial exposure surfaces are areas where upper bounding surfaces of sediment or rock show the effects of being exposed at the Earth's surface. In order to recognize fossil subaerial exposure surfaces they need to be exposed long enough to allow subaerial diagenetic processes to modify or obliterate pre-existing fabrics. This will be recorded as a break in the sedimentary sequence. This usually means that significant periods of time have passed before exposed surfaces are buried by new deposits. What we mean by significant periods of time is a relative concept dependent on our limits of resolution and powers of observation. When considering absolutes of time we will offer only abstracts; our intention here is not to quantify absolutes of timing, duration or intensity of processes acting upon exposure surfaces, nor to examine critically the processes themselves. Rather we will document common and characteristic products of subaerial exposure, list criteria which aid in recognition of fossil subaerial exposure surfaces, and point out the significance and economic importance of subaerial exposure surfaces in ancient carbonate sequences.

Adhering to this outlined conceptual framework we can define a

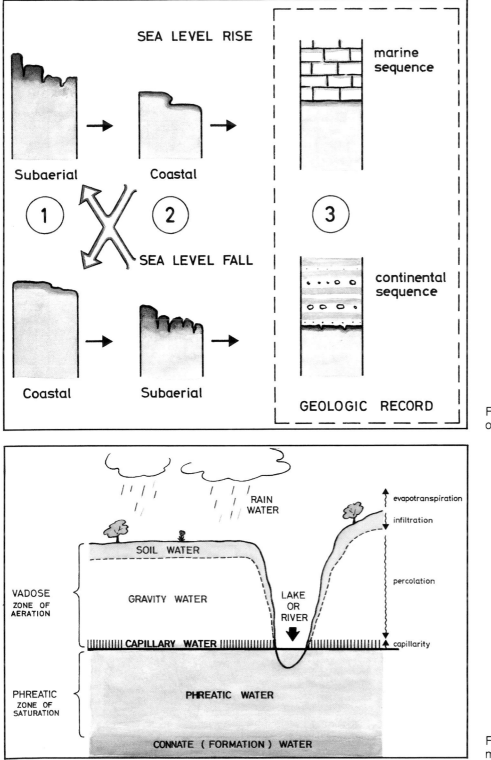

Figure 3—Major pathways of evolution of exposure surfaces.

Figure 4—Schematic representation of meteoric hydrologic zones (not to scale).

subaerial exposure surface as a distinct surface on land which indicates: (1) non-deposition and commonly erosion; and (2) a break in the sedimentary sequence. Regardless of cause or length, a subaerial exposure surface is a record of interruption in sedimentation. Other terms sometimes used to describe this condition include hiatus, diastem, break, disconformity, unconformity, hardground or discontinuity surface. Even though all these terms refer to exposure surfaces, they may be submarine, subaerial, or both submarine and subaerial in origin. A

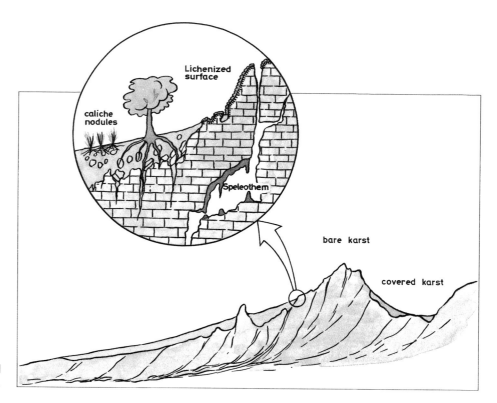

Figure 9—Co-existence of karst and caliche facies.

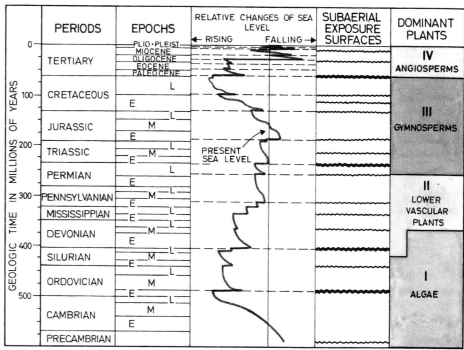

Figure 10—Relative global sea level changes, major known subaerial exposure surfaces and dominant plant groups during Phanerozoic time. (Adapted from Vail et al, 1977).

are subjected to identical subaerial diagenetic processes at their resulting exposed surfaces.

Given sufficient relative lowering of sea level, subaerial exposure surfaces may exist within shallow-water or deep-water marine carbonate sequences although their frequency should be greater in the former, assuming equal chances of preservation in the rock record. Eolian and lacustrine carbonates are but slight variations on the same theme. Once eolian carbonate sands are stabilized and lacustrine carbonates have dried out, the subaerial diagenetic processes which operate on these terrestrial carbonates will be the same as the processes which operate on subaerially exposed marine carbonate sequences. In other words, subaerial exposure surfaces obey no rules with respect to environment of formation of the subjected host carbonate. As long as

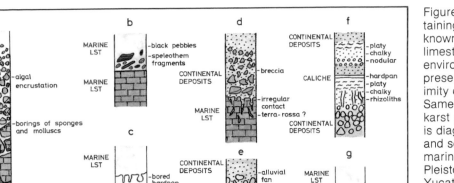

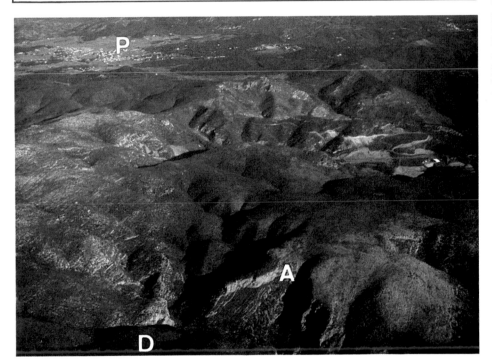

Figure 11—Generalized sequences containing exposure surfaces based on known examples. (**a**) Karstified Mesozoic limestone reworked in Miocene coastal environment. Most of the karst profile is preserved, with clear evidence of proximity of soil cover, Barcelona, Spain. (**b**) Same exposure surface as in (**a**) but karst profile is not preserved; sequence is diagnostic of close proximity of karst and soil. (**c**) Caliche profile in Tertiary marine limestone reworked in Plio-Pleistocene coastal environments, Yucatan, Mexico. (**d**) Sequence probably contains subaerial exposure but no diagnostic evidence is found, except for relics of possible terra-rossa in deep joints. Same exposure surface as (**a**) and (**b**). (**e**) Mesozoic limestone, intensively calichified, followed by Paleocene palustrine sediments and alluvial fans, Barcelona, Spain. (**f**) Calichified overbank deposits of Triassic alluvial fan complex, Barcelona, Spain. (**g**) Same exposure surface as (**a**), (**b**) and (**d**), but without traces of subaerial exposure facies. Only coastal exposure facies are recorded.

Figure 12—Karst surface landforms. Dolinas (**D**), polje (**P**) on a relic karst plateau. Relic alteration zone (**A**) outcrops as cliff on steeply dipping unaltered substrate. We believe that some of these relic karst features developed in earlier tropical climates. Present-day Mediterranean type of karst processes only produce slight modifications on the relic forms. Karst development from ?Lower Miocene (perhaps Lower Eocene?) to present on Cretaceous carbonates, Garraf Mountains, Barcelona, Spain.

there are: (1) subaerial conditions; (2) stabilization and non-deposition of sediment; and (3) sufficient time available for subaerial diagenetic processes to affect the exposed host carbonate, then a subaerial exposure surface and its underlying alteration zone will develop. Whether or not the original subaerial exposure surface and its underlying alteration zone will be preserved in the rock record depends largely on the presence or absence of any subsequent erosional phase.

Importance of Subaerial Exposure

As geologists, interested primarily in rocks, why should we concern ourselves with surfaces which represent missing pages in our story book of the Earth's history? There are at least three good reasons.

Firstly, subaerial exposure surfaces provide important information when faced with the task of trying to decipher the geologic history of a region. In many instances, it is just as important to know what is missing from a sedimentary sequence as it is to know what has been preserved. For example, when subaerial exposure

surfaces can be identified within marine carbonates, important deductions can be made regarding periods of regression or upbuilding of sediment packages above sea level. Secondly, subaerial exposure surfaces can be extremely useful horizon markers for outcrop or core correlations. Thirdly, and perhaps of greatest economic importance, subaerial exposure surfaces are sites where valuable natural resources can be concentrated, including: (1) oil, gas or water traps in which the sealing unit overlies the reservoir rock below the subaerial surface; and (2)

Figure 13—Tropical karst landforms. Tower and conical karst developed since the Pleistocene on Cretaceous carbonates. North side of Mayan Mountains, Belize.

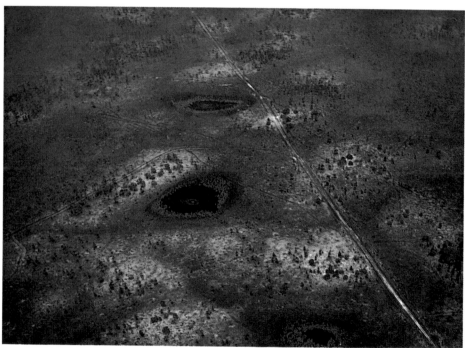

Figure 14—Dolinas. Aerial view of lowland plain with water-filled dolinas developed on Tertiary carbonates. These dolinas are believed to have formed in pre-Holocene temperate climates (Purdy, 1974). Location about 60 km (37 mi) north of Belize.

various accumulations and deposits of certain metals (lead, zinc, uranium, bauxite) related to soil processes or precipitation in karst cavities.

SETTING

Relationship to Lateral Facies

Subaerial exposure surfaces grade laterally seaward into coastal and submarine exposure surfaces (Fig. 1).

Exposure surfaces in general may be highly irregular, or smooth and planar. Thus, exposure surfaces are characterized by their topography and relative position to sea level. They may be developed on sediment or rock (Fig. 2).

Submarine exposure surfaces — Carbonate sediments or rocks are exposed on the sea floor below the lower limit of noticeable wave action.

This category includes submarine hardgrounds with or without their manganese, phosphatic or glauconite crusts and will not be considered further in this chapter.

Coastal exposure surfaces — Sediments or rocks are exposed in the peritidal environment above the lower limit of wave action and below the highest limit of the marine vadose salt spray zone. Some of the most com-

Figure 15—Fossilized rundkarren and solution pipe. Terra-rossa type soils fill depressions in karst landforms. Pleistocene, Bermuda.

Figure 16—Fossilized rundkarren. Intra-Triassic karst development on Lower Muschelkalk carbonates overlain by Upper Muschelkalk dolostones. Middle Muschelkalk red beds are represented as a condensed sequence of red soils covering the karst surface. Barcelona, Spain.

mon features of the coastal environment are shown in Figure 2a for rocky shores and in Figure 2b for sediment shores. Very irregular surfaces produced by wave action and bioerosion are characteristic of the rocky shore environment. Stabilization by plants is important in the sediment shore environment.

Of special interest for our purposes is the upper part of the coastal environment with rock pools infested by blue-green algae (Schneider, 1976), pelagosite crusts (Purser and Loreau, 1973) and the dissolution forms referred to as coastal marine karst. The coastal environment is of interest because of its juxtaposition to the subaerial meteoric environment and its high preservation potential in the geologic record. With changes of sea level, complicated patterns of

superimposed coastal and subaerial processes may act on the same exposure surface. Renewed sedimentation will preserve this polygenetic surface.

Subaerial exposure surfaces — Sediments and rocks are exposed to the atmosphere landward of any marine influence. Lake bottoms and deflation flats are excluded from this discussion. First evidence of subaerial

Figure 17—Sink-hole. Chaotic mass of red-stained brecciated blocks and sediments with marine Carboniferous fossils, developed as sink-hole fill in gray gently dipping Middle Ordovician limestones. Western Newfoundland, Canada.

exposure can be established by the appearance of endolithic lichens which represent the first stage of colonization and subsequent community succession in subaerial environments. This limit, established on the basis of lichen colonization, is not generally recognizable in the fossil record. A more practical limit for the geologist is the first appearance of karst and caliche diagenetic facies.

Succession and Modification of Exposure Surfaces

Sea level oscillations and variations in tectonic activity will produce important modifications and successions in exposure surfaces. With a relative sea level rise, areas subaerially exposed will eventually be overlain by a marine sediment (Fig. 3). During a relative fall of sea level, subaerial exposure surfaces will develop superimposed on the coastal and submarine exposure surfaces (Fig. 3) and eventually be overlain by a continental sequence (eolian sands, alluvial fans, lacustrine and palustrine sediments, fluvial sediments, porous and dense plant tufas). The last exposure surface prior to renewed sedimentation is the only one preserved on most occa-

sions; it is of no coincidence that the best evidence of subaerial exposure is found generally at the base of a sequence of continental deposits.

The profile of exposure surfaces may also change with time without relative sea level oscillations. Many exposure surfaces are accompanied by destructive processes and the general trend is toward lowering of elevation and planation of surfaces. Exposure surfaces may also be modified tectonically (fracturing, orogenic and epeirogenic uplifts) with the consequent reactivation of erosional and depositional processes.

Climate exerts an important control, both in time and space, on morphology and profile development of subaerial exposure surfaces. Spatially, a single extensive subaerial exposure surface can be subjected to different climatic regimes (for example, from a dry warm low plain to a wet cool mountainous region). With time, climatic belts may shift, subjecting a given point on a subaerial exposure surface to a new set (or new intensity) of climate-sensitive processes.

All these controls, position with respect to sea level, tectonic activity and climate, together with the inces-

sant evolution of diagenetic processes, will modify the topographic expression of exposure surfaces.

Subaerial Exposure Facies

Although caliche can develop in rock or sediment of any composition and karst may develop in evaporite bodies in addition to carbonate terrains, only karst and caliche developed in carbonate hosts are relevant to this discussion.

Generalities — Two major diagenetic carbonate facies are produced as a consequence of subaerial exposure, the karst facies and the caliche facies. Both diagenetic facies involve mobility of calcium carbonate. In the karst facies there is a net loss of calcium carbonate because of dissolution and removal of $CaCO_3$ solutions (supplemented by mechanical erosion). In the caliche facies there is a zero balance (local dissolution and reprecipitation of $CaCO_3$ without external sources) or a net gain (addition of $CaCO_3$) from elsewhere. Movement of $CaCO_3$-bearing solutions occurs both in the meteoric vadose and meteoric phreatic zones. These hydrologic zones are depicted in Figure 4, show-

Figure 18—Lapies. Irregular, jagged surface truncates Upper Miocene reef limestone overlain by Pliocene beach complex. Red-stained silt, probably derived from terra-rossa soils, covers karst surface. Mallorca, Spain.

ing the meteoric vadose zone subdivided into an upper dissolution zone of infiltration and a lower percolation zone of downward movement with negligible dissolution of calcium carbonate. Similarly the meteoric phreatic zone can be subdivided into an upper active lenticular zone (Jakucs, 1977) and a lower inactive deep zone (imbibition zone) which grades downward into connate or formation waters.

In the zone of infiltration, waters are active and may either (1) dissolve calcium carbonate because of aggressive atmospheric CO_2 and/or high levels of biogenic CO_2 derived from overlying soil cover, or (2) precipitate calcium carbonate if CO_2 levels are lowered by degassing or plant uptake. By the time water reaches the underlying percolation zone it is at $CaCO_3$-H_2CO_3 equilibrium and dissolution or precipitation of calcium carbonate is minimal. In contrast, in the uppermost part of the phreatic zone just below the water table, water movement is essentially horizontal and $CaCO_3$ is either removed by dissolution and mechanical erosion or precipitated. Continuing downward to the lenticular zone, whose position is determined by an underlying impermeable horizon or by the base

level of erosion, direction of water movement is controlled by hydraulic pressure rather than direct gravity. Limestone in this zone may undergo dissolution because of increased hydrostatic pressure and mixing corrosion, or precipitation may take place if mixed waters become supersaturated with respect to calcium carbonate. Water below the lenticular zone, even though under hydrostatic pressure, does not play a role in the hydrographic cycle of the overlying lenticular zone; it therefore attains equilibrium with its surroundings and becomes stagnant. This inactive zone grades downward through a transitional zone into connate waters whose properties and effects on calcium carbonate are poorly understood or unknown.

Karst Facies

The term karst has been used to designate specific landforms (including subterranean landforms) and a geographic region characterized by these landforms. Modern developments (Thrailkill, 1968, 1971; Sweeting, 1973; Jakucs, 1977) have stressed the concept of karst as a result of a complex set of processes (climatic, tectonic, edaphic, hydrologic, petrologic) with different evolutionary stages. On this basis, the

following definition is more useful to geologists: "Karst is a diagenetic facies, an overprint in subaerially exposed carbonate bodies, produced and controlled by dissolution and migration of calcium carbonate in meteoric waters, occurring in a wide variety of climatic and tectonic settings, and generating a recognizable landscape." Karst facies represent a net loss of calcium carbonate, although in some stages of karst evolution or in some parts of the profile, it is possible to have equilibrium or gain in the carbonate budget. A descriptive characterization of karst should be based on: (1) surface landforms (lapies, dolinas, poljes); (2) subterranean landforms (varieties of pores, caves, vugs, pipes); (3) speleothems (stalactites, stalagmites, flowstones, rimstones, globulites, cave pearls, lily pads, helictites, moon-milk); and (4) collapse structures due to removal of underlying carbonate.

Mineralogy of the karst facies is dominantly low magnesian calcite (LMC), although caves in dolostones show a wider variability in mineralogy with local deposits of aragonite, hydromagnesite, dolomite and nesquehonite. In this chapter we will refer generally to LMC deposits because they are more abundant and

Figure 19—Collapsed roof of cavern. Cavern formed in Upper Miocene reef limestone during time of formation of subaerial exposure surface of Figure 18. Mallorca, Spain.

cave deposits of other mineralogies show roughly the same gross morphologies.

Karst Profile — Karst presents a generalized facies pattern that can be summarized in a profile (Fig. 5) as follows:

(1) Infiltration (upper vadose) zone, characterized by surface landforms, with or without a developed soil cover (exposed carbonate usually has a lichen or algal cover, a protosoil) and by production of dominantly vertical caves (pecina, Figs. 20, 26). Most of the speleothems of this zone are the result of intense corrosion and degradation of carbonate host walls. These speleothems are fine grained deposits (moon-milk) made of random needle fibers of low magnesian calcite (LMC) and some types of globulites (popcorn) growing in degraded carbonate hosts. The final product is a characteristic accumulation in walls and cavern floors of white, fine-grained deposits of chalky or friable consistency, abundantly colonized by fungi and bacteria. The dominant processes in this zone are physicochemical dissolution and biological corrosion related to intense organic activity. Collapse breccia deposits can also be very abundant (Fig. 25) in zones up to 20 m (66 ft) thick in many European and North American karst profiles.

(2) Percolation (lower vadose) zone, characterized by net vertical water movement through pre-existing permeability pathways. In percolation zones dominated by vadose water seepage there is little dissolution, but localized areas of vadose flow (underlying major sinkholes, thick soil covers, open fractures) show active dissolution and hydraulic erosion. As a facies this zone is poorly defined, oscillating between 0 to 200 m (0 to 660 ft) thick, but the lower part near the phreatic water table is where speleothem formation is more abundant and more varied than elsewhere in the karst profile. The presence of a capillary fringe (up to 2 m or 6.6 ft above the water table) is likely related to this intense carbonate precipitation.

(3) Lenticular (upper phreatic) zone, characterized by intense formation of subhorizontal caves (jama) by hydraulic erosion and dissolution as a result of mixing corrosion and increasing hydrostatic pressure. Subvertical caves can also be formed but they are subsidiary to horizontal drainage patterns. Most karst cavern porosity is produced in this zone, especially just below the water table (Thrailkill, 1968) in allogenic karsts. Speleothem formation in this zone can also be important in senile stages of evolution, mainly at and a few centimeters below the water table (floating carbonate flakes or rafts, isopachous coatings, cave pearls, lily pads). Collapse breccias and water laid deposits are very abundant locally. In karst facies it is common to have perched water tables at various levels, thus introducing more complexity to the profile.

The lower limit of the karst profile is difficult to establish. Lenticular zones can be up to 100 m (330 ft) thick or more. The increase in hydrostatic pressure at the base of the lenticular zone may favor dissolution but, at a deeper level, waters will become stagnant and a diagenetic overprint related to subaerial exposure is no longer recognizable. On the other hand, the lenticular zone could be the ideal setting for pervasive phreatic cementation, a process accepted by many carbonate petrologists. Intergranular cements and speleothem deposits of the phreatic lenticular zone may be different products of the same cementation process; the differences being controlled by two factors — pore geometry and water flow pattern.

Evolution — The karst profile of Figure 5 could represent a mature stage of evolution of a karst facies. A dynamic view of the karst facies of-

Figure 20—Speleothem. Plan view of vertical solution pipe in Upper Miocene reef dolostone. Solution pipe wall is delineated by iron-rich crust which is overlain by banded calcite and minor aragonite flowstones and globulite (popcorn). Karst event is same as in Figures 18 and 19, but located here in Alicante, Spain.

Figure 21—Kamenitza. Recent solution rock pools developed on Tertiary carbonates in tropical climate, Yucatan, Mexico.

fers more complexity. To start with, in an absolutely massive, pure and homogeneous carbonate host, early stages of karst evolution would be confined to surface landforms (for example the experimental studies of Purdy, 1974). Heterogeneities such as fractures, joints, bedding planes, impurities in the carbonate host and vegetation favor the development of subterranean landforms. With a stable phreatic base level, the evolution of the karst profile (Fig. 6) implies the lowering of the surface and the downward shift of the infiltration zone. The surface evolves into complex landforms such as dolinas, uvalas, poljes and karst valleys and plains (Fig. 12). Insoluble residues of the carbonate host play a major role in karst facies evolution; they are the basic contributors to the formation of soil cover (with the consequent increase of biogenic CO_2 taken up partly by meteoric waters). Deposition of insoluble residues in subterranean pores (Fig. 33) and conduits controls the hydrologic patterns of karst. With the exception of tropical climates, the tendency is toward a limestone plateau where the infiltration and lenticular zones merge. The lenticular zone is drastically reduced because of clogging by insoluble residues. Tropical karst, on the other hand,

differs from this model. Accentuation of surface morphology, because of the relatively higher soil permeability and aggressiveness of waters in the lower depressed areas occupied by soils, produces conical and tower karst (Figs. 6, 13).

Profiles of karst facies are usually complicated by repeated changes in position of the phreatic water level (variations in tectonic and climatic

setting, sea level oscillations). For example, products of the phreatic zone can be subjected to processes in the upper vadose zone and become coated, corroded, dissolved or remodelled. Similarly, former vadose products can be drowned and modified in the phreatic zone (Fig. 6). In general, the karst profile is more active during periods following orogenic episodes because of: (1) the

Figure 22—Lapies. Enlarged open joint (kluftkarren) and root lapies modified by incipient rillenkarren. Kluftkarren and root lapies are possibly relic karst forms of wetter climate than present day, semiarid Mediterranean climate. Miocene limestones, Almeria, Spain.

creation of new joints and fractures in the host rock which allows percolation of more meteoric water, (2) the increase in topographic irregularities, with deepening of the water table and easy removal of soils on unstable slopes; and (3) the formation of topographic highs which can control local climatic patterns favoring orographic rains. The repetition of these changes strongly alters the host rock, with recognition of these multicyclic episodes and modifications extremely difficult but possible with careful geomorphological, petrological and paleontological studies. In the fossil record, only parts of the karst facies and profiles are likely to be preserved and the study of karst zones and their inferred evolution will be generally impossible.

The profile summarized here corresponds to authigenic karst (holokarst) where meteoric waters in the profile are collected in the infiltration zone of the same profile. Other possible types of profile occur where meteoric waters arrive into the profile from adjacent or suprajacent nonkarstic terrains (allogenic karst of Jakucs, 1977). Most karst profiles in nature are combinations of authigenic and allogenic types. Allogenic karst is very important in mature and senile stages of authigenic karst, typically

under thick soil cover or below noncarbonate formations. The allogenic karst facies results from valley sculpture by linear stream bed erosion and differs from authigenic karst mainly because in allogenic karst: (1) the drainage pattern has a marked polarity which clearly resembles an underground river, (2) deposition of speleothems and cave sediments is more abundant, and (3) scour caverns, typically with flat roofs, are abundant. For geologists working in the fossil record, the distinction between the authigenic and allogenic karst is extremely difficult; our discussion will refer to both types together.

Another important aspect of karst evolution is the intimate relationship with the soil cover. The definition of the term karst does not include soil-forming processes, although a variety of soils are common on different types of karst. These soils are an important control and consequence of karstification. As a control, soils are sites of biogenic CO_2 production and meteoric water storage. As a consequence, soils are formed from insoluble residues as a by-product of karstification. In general, early stages of karst evolution show poorly developed soil cover while later stages usually imply thicker and more extensive soil cover. The climatic setting,

together with lithological and topographic patterns of the carbonate host, are major factors which control development of the soil cover.

Residual soil deposits, such as terra-rossa and laterites, may be characteristic of specific climatic types of karst (Mediterranean, tropical) and can be recognized within fossil carbonate sequences (Figs. 18, 27, 36, 37). The soil cover of karst may also evolve in the direction of carbonate-rich soils of the caliche type (Fig. 38), particularly in mature and late stages of karst evolution. Because of their high lithification potential, caliche soils are one of the most useful criteria for recognition of fossil subaerial exposure surfaces and deserve careful attention.

Caliche Facies

Caliche (calcrete) is commonly defined as a fine-grained, chalky to well-cemented, low magnesian calcite deposit that formed as a soil in or on pre-existing sediments, soils or rocks in semiarid environments (Bretz and Horberg, 1949; Brown, 1956; Swineford et al, 1958; Durand, 1963; Blank and Tynes, 1965; Gile et al, 1966; Ruellan, 1967; Aristarain, 1970; Reeves, 1970; James, 1972; Esteban, 1972, 1974; Read, 1974). This essentially genetic definition is of limited

Figure 23—Rillenkarren and small kamenitza. Modern surface karst features developed on fine-grained limestone of Cretaceous age. Castello de la Plana, Spain.

Figure 24—Solution pipe and caliche profile. Solution pipe, developed in red-brown Pleistocene calcareous silts, lined with laminar micritic crust. Overlain by hardpan and nodular caliche of later Pleistocene age. Ibiza, Spain.

use to geologists looking for caliche in the fossil record. It is necessary to have a more descriptive definition which includes regional setting, lithologies and their sequences, structures, textures and fabrics. A descriptive definition has been proposed by Esteban (1976) and is adapted here, with some modifications, as follows: "Caliche is a vertically zoned, subhorizontal to horizontal carbonate deposit, developed normally with four rock types: (1) massive-chalky, (2) nodular-crumbly, (3) platy or sheetlike, and (4) compact crust or hardpan. The position and development of these rock types in a vertical sequence (profile) and laterally is highly variable. The only rather consistent relation is that the massive-chalky rock grades downward into the original rock or sediment through a transition zone, with strong evidence for both in-place alteration and replacement of the original rocks or sediment. Colors are commonly white and light brown, but red and black may be important. The predominant caliche fabric is a clotted, peloidal micrite with microspar channels and cracks. Accessory fabrics are rhizoliths, glaebules (pisoliths, ooliths, nodules, peloids), poorly laminated micrite and karst products. Microspar areas usually show evidence of

replacement of relic grains and of other primary and earlier diagenetic microfabrics." Caliche is an informal term used by geologists. Probably any carbonate-rich soil, with $CaCO_3$ concentrated in the form of glaebules, can be fossilized to form different types of caliche rocks.

Caliche Profiles — An idealized caliche profile is shown in Figure 7, and though only one of many possible variations it is the most common of modern western Mediterranean and Texas caliches. Other common profiles are shown in Figure 8. The term caliche profile refers to the complete vertical succession of morphologically distinct layers or horizons. Boundaries between horizons tend to show gradual transitions rather than abrupt changes. The main features of each horizon are summarized in Figure 7 and Table 1 and described as follows:

Hardpan — Being well indurated and lacking visible porosity, the hardpan is generally more resistant to weathering than underlying horizons and, thus, commonly stands out as a

prominent feature. Vertical thicknesses vary from 1 mm (0.04 in.) laminar layers (Fig. 75) to massive horizons, 1.5 m (5 ft) thick. The horizon is made up dominantly of well cemented microcrystalline or cryptocrystalline calcite. The thicker hardpans commonly show evidence of fracturing, non-tectonic brecciation (Figs. 41, 44, 60, 63, 68, 72), dissolution and recementation, and may contain glaebules (pisoliths) and rhizoliths. Colors are generally white or cream, although pale orange to brown are not uncommon. The hardpan may be macroscopically structureless or massive, laminated (Fig. 58), brecciated (Fig. 60) or nodular (Fig. 74).

Platy caliche — Platy caliche invariably occurs immediately below the hardpan horizon (Fig. 43) or, in profiles lacking a hardpan, occurs as the uppermost calcareous horizon or below a more recent soil cover. Platy caliche is distinguished from hardpan caliche by its horizontal to subhorizontal, platy, wavy or thinly bedded habit (Fig. 44), its planar fracture

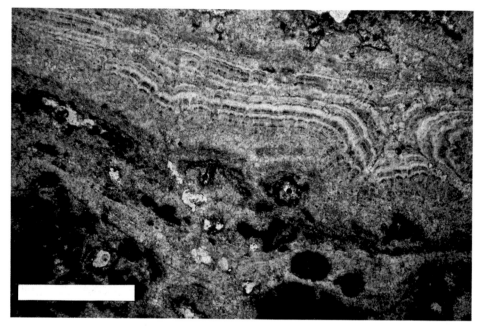

Figure 25—Collapse breccia. Part of fill in solution pipe formed in Pleistocene carbonates. Angular blocks of carbonate host coated with flowstone. Flowstone forms a geopetal fill within interblock porosity. Jamaica.

Figure 26—Speleothem. Laminar flowstone of Pleistocene age. (Field view of flowstone and host rock shown in Fig. 29). Ibiza, Spain. Thin section, plane polarized light. Scale bar = 1.0 mm.

porosity and greater friability, and the abundance of alveolar textures, rhizoliths and needle fiber fabrics. Recognizable transitional stages suggest that hardpans represent advanced stages of platy caliche development. Maximum and average thicknesses of platy caliche horizons are greater than those of hardpans, the maximum thickness recorded for platy caliche from Spain being 3.1 m (10 ft). Platy caliche generally grades downward into nodular caliche.

Nodular caliche — Nodular caliche is made up of glaebules which consist of discrete, powdery to indurated concentrations of calcium carbonate embedded in a less carbonate-rich matrix (Figs. 39, 40, 47). Individual glaebules range from silt-sized to pebble-sized particles. They may be spherical or subspherical, irregular or cylindrical in shape. Most cylindrical forms are vertically elongate (Fig. 43); some, however, show branching patterns. Glaebules occur either as isolated particles or as coalesced masses. In addition to greater concentrations of calcium carbonate and greater cohesion, glaebules usually can be distinguished from the surrounding matrix by color differences. Glaebules are commonly white to cream in color, whereas the matrix tends to be red or red-brown (Fig. 40) because of higher concentrations of acid-insoluble residues such as layer lattice minerals and ferric hydroxides. However, some glaebules, especially sand-sized glaebules with internal concentric fabrics, are dark red-brown whereas the matrix is pale brown to cream. Iron salt impregnation within these darker glaebules is usually responsible for this color difference. Nodular horizons invariably show diffuse upper and lower boundaries. In many profiles, the boundary between nodular and chalky-powdery horizons is so indistinct that this "boundary" can commonly be treated as a separated horizon, that is, the nodular-chalky horizon.

Chalky caliche — The chalky caliche is characterized by white to cream, unconsolidated silt-sized calcite grains. Cementation between grains is absent so that the material has the consistency of a powder. This horizon tends to be homogeneous structurally and texturally, although scattered, isolated glaebules are locally present (Figs. 39, 46). Areas around plant root systems appear to be favored sites for CaCO₃ accumulation, and this phenomenon gives rise to incipient nodular development. Maximum thicknesses of this horizon rarely exceed 1 m (3.3 ft). Chalky horizons are poorly developed or absent in most profiles where the host material has high interparticle porosity (for example, eolianites). The chalky horizon generally grades upward into nodular caliche and downward into the transitional horizon.

Transitional horizon — The term transitional horizon refers to the zone between unaltered host material and overlying caliche horizons that lack macroscopically discernible features inherited from the host material. The transitional horizon itself contains macroscopically discernible evidence of in-place alteration and partial replacement of the original host material (Fig. 51). This evidence includes: (1) relic sedimentary structures such as bedding (Fig. 46), (2) in-place relic fossils embedded in otherwise calichified host material, (3) in-place relic siliciclastic grains with distribu-

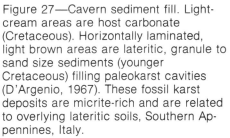

Figure 27—Cavern sediment fill. Light-cream areas are host carbonate (Cretaceous). Horizontally laminated, light brown areas are lateritic, granule to sand size sediments (younger Cretaceous) filling paleokarst cavities (D'Argenio, 1967). These fossil karst deposits are micrite-rich and are related to overlying lateritic soils, Southern Appennines, Italy.

Figure 28—Cave pearls and laminated cavern sediment. Same locality as in Figure 27. Cave pearls are also micrite-rich and occur in layers with characteristic coarsening upward grain size arrangements.

tion patterns inherited from the host material, and (4) relic mineral veins traceable from the underlying host material into the caliche profile without deviation or disruption. In profiles developed within bedded sedimentary deposits, dipping beds of unaltered host material can be traced upward into transitional and chalky horizons. Alteration takes place preferentially along bedding and joint

planes (Fig. 68). These planes of access allow water to move through host material more easily and are, thus, sites susceptible to diagenetic alteration and to penetration by roots. This horizon is essentially an in-place weathered zone consisting of partially degraded host material, making it difficult to fix the lower boundary. The transitional horizon may grade into the upper vadose zone

of karst and may show thicknesses of several meters (feet). In other profiles the transition horizon can be minimal or apparently absent.

Host material — The host material may be of any composition, texture, age and origin. The only significant factor of host material that influences caliche development is its mechanical stability; development of a caliche profile requires a stable substrate suf-

Figure 29—Speloethem and breccia. Brecciated (karst collapse breccia?) and recemented (and calichified) Kimmeridgian dolostone and laminar flowstone of Pleistocene age. Thin section photomicrograph of flowstone shown in Figure 26. Ibiza, Spain.

Figure 30—Endolithic lichens. Rock surface (Cretaceous limestone) shows an early development of rillenkarren which is smoothed out by endolithic lichens. Sharp, irregular contacts (**L**) between different lichen communities correspond to areas of intense frugal penetration into the rock. Garraf Mountains, Barcelona, Spain.

ficiently long for pedogenetic and diagenetic processes to operate. Even so, other factors of the host material such as permeability and calcium carbonate content can influence the rate of caliche development. The host material is distinguished from its overlying caliche profile by the absence of typical features which characterize caliche. Original structures, textures and fabrics of the host material are clearly recognizable; they have not been modified or obliterated by subsequent calichification in contrast to the transitional, chalky, nodular, platy and hardpan horizons which do show such alterations with increasing intensity upward and away from the host material.

Evolution — Unconsolidated caliches may be active, relic or fossil CaCO3-rich soils. Indurated caliches are lithified fossil CaCO3-rich soils. Soils are products of weathering and also the natural medium in which plants grow. The evolution, from a subaerially exposed marine carbonate to an indurated caliche rock, can be delineated simplistically into five stages:

Stage 1: Preparation of host material; weathering. Mechanical, physicochemical and biological dis-

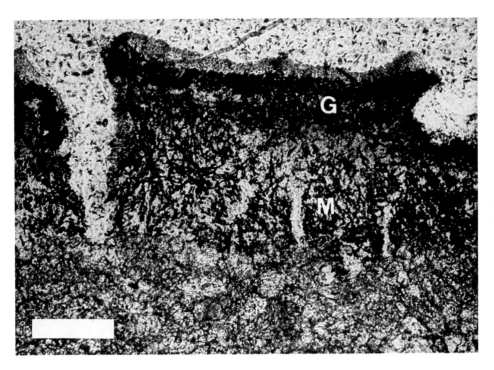

Figure 31—Endolithic lichens. Profile of modern endolithic lichen colonization in Jurassic dolostone. Green cavities below the surface are the algal components of the lichen symbiosis (gonidial zone, **G**). Black filaments correspond to the fungal hyphae (medulla zone, **M**). The lower part contains unaltered host rock. Spheroidal depressions and cracks in the surface are made by former reproductive bodies of the lichen, Mallorca, Spain. Photograph courtesy of L. Pomar. Thin section, plane polarized light. Scale bar = 0.5 mm.

integration generate a regolith or accumulation of weathered detritus. Soil development implies rates of accumulation exceeding rates of removal.

Stage 2: Soil development; pedogenesis. Unconsolidated sediment or weathered detritus is developed into a soil by changes produced by the action of organisms and by movement of water through the sediment.

Stage 3: Accumulation of calcium carbonate and differentiation into horizons. In the early stages of caliche development, the profile is composed of weathered materials with high porosities and permeabilities. Vertical movements of meteoric vadose water can take place relatively easily, and insufficient water is retained to supply the requirements of the vegetation. Some plants have to extend taproots vertically downward to the proximity of local water tables. Roots extend downward into fractures and joints within host material, modifying original structures and contributing to the disintegration of the substrate. Biological and physicochemical alterations of the host material culminate in the formation of a transitional horizon. Precipitation of calcium carbonate, without significant cementation because of mechanical and biochem-

ical instability of the profile, forms most of the chalky horizon. Pedoturbation (physical, chemical and biological disturbance of soil materials) precludes the formation of indurated layers.

As the accumulation of calcium carbonates increases, porosity and permeability of the profile decrease. Biological constituents of the soil may become calcified, thus forming biogenetic carbonate structures such as rhizoliths, calcified filaments, calcified fecal pellets, calcified coccoons and *Microcodium* aggregates. Wetting and drying of the soil favors development of shrinkage cracks and, later, precipitation of $CaCO_3$ within voids. In earlier stages of profile development, vertical water movements and vertical taproots tend to form vertically oriented elongate carbonate nodules. But at some point of profile development, it becomes easier for soil water to move horizontally rather than vertically. By this stage most plants form lateral root systems, their corresponding rhizoliths form the bulk of the platy caliche horizon. Vertical rhizoliths in lower parts of the profile tend to be large (5 to 20 cm; 2 to 8 in) and isolated; platy horizon rhizoliths are smaller (0.5 to 2 mm; 0.02 to 0.08 in), extremely abundant, branching and horizontally

oriented. Development of the platy caliche horizon from the nodular horizon may be a reflection of plant succession in a developing soil.

Stage 4: Lithification, cementation and fossilization. As accumulation of calcium carbonate increases, a point will be reached when soil organisms can no longer maintain viability. The intensity of soil-forming processes diminish and eventually cease to be important. Diagenetic processes, mainly cementation by low magnesian calcite, lead to the lithification and fossilization of the soil profile and formation of a hardpan.

Stage 5: Reworking, brecciation, weathering (new Stage 1). The lithified caliche profile, if it remains at the land surface, is subjected to further processes which will alter or destroy the profile. Lower plant (lichens, algae, fungi, bacteria) activities will form a protosoil, a pioneer stage in plant community succession. Eventually the developing soil profile is able to support higher plants. The root systems of these plants penetrate, dissolve and fracture the indurated hardpan. Disturbance of the caliche profile by vegetation may form tepee structures and rhizobreccias. Further pedoturbation, with carbonate dissolution and reprecipitation, leads to the formation of a

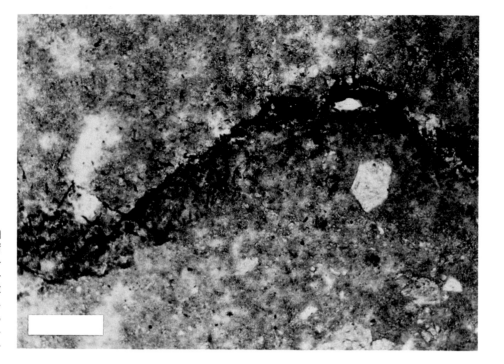

Figure 32—Fossilized lichen-colonized surface. Iron stained fungal hyphae of endolithic lichen overlain by dark iron-stained and organic-rich layer. Lichen-synthesized calcite prism shown on right both within and below organic-rich layer. Pleistocene caliche hardpan. Ibiza, Spain. Thin section, plane polarized light. Scale bar = 1.25 μm.

reworked, recemented, breccia-conglomeratic caliche hardpan. To complicate matters, the reworked recemented profile may be later subjected to a karst over-print.

Spatial and Time Relationships Between Karst and Caliche Facies

Processes operating in both karst and caliche diagenetic environments are not mutually exclusive. Different stages of karst and caliche evolution may co-exist at any one time and overlap in any one area (Fig. 9). In addition, with time the caliche facies may be subjected to karstification and renewed soil formation and, likewise, karstified surfaces may become calichified. Caliche and karst facies are dynamic systems involving complex processes controlled by many factors and including climatic variations, organic activity, and characteristics of the host substrate. Moreover, both facies influence their own evolution.

Some important differences between the karst and caliche facies do exist, however. One difference is that caliche is attributed to a specific climatic regimen (Goudie, 1973), namely a semiarid zone, whereas karst facies develop in all climates with water. In addition, while the caliche facies is restricted essentially

to the surface and near-surface subaerial vadose environment, the karst facies is a three-dimensional unit which occurs both at the surface of the Earth and extends downward through the vadose and into the meteoric phreatic environment. The karst and caliche facies thus show differences in magnitude and hydrologic regimen. Nevertheless, as will be seen later, some caliche products are similar to those which form in the upper vadose zone of the karst facies. The subaerial vadose zone is the location where lithosphere, atmosphere and biosphere interact in a complex manner to produce features common to both karst and caliche facies.

Subaerial Exposure Surfaces in the Geologic Record

Subaerial exposure surfaces from Silurian-Devonian times onward are well documented (see Bibliography). Earlier occurrences are scanty (Geldsetzer, 1976) and present more problems of interpretation. The weathering processes inferred above are heavily dependent on the presence and evolution of higher plants, mainly because of the production of biogenic CO_2 which increases the aggressiveness of meteoric waters. Before the Silurian, ? caliche and ? karst processes would necessarily be

governed by CO_2 obtained from lichens, algae, fungi and, possibly, from higher concentrations of atmospheric CO_2. We suspect that karst and caliche facies before the Silurian, without higher plant influence, had different features than the ones described in this paper. However, available references are not detailed enough for conclusive results in this respect.

It is well known that some intervals or episodes of the geologic record contain, simultaneously world-wide, abundant mineral deposits of lead-zinc and bauxites in karst terrains (Bernard, 1976; Geldsetzer, 1976; Padalino et al, 1976; Valeton, 1972 for general references). The references to possible fossil caliches (see Bibliography) also occur at times of mineralized karst; the same appears to be true for oil reservoirs in karst caverns. Some of these periods of generalized subaerial exposure have been frequently reported in the literature, either because of their economic importance or because of their intensive and widespread development, or both. As shown in Figure 10, these intervals of more pronounced or more studied subaerial exposure follow (by 10 to 15 m.y.) major stages or milestones in plant domination. This apparent coin-

Figure 33—Karst sediment fill. View looking downward onto upper surface (parallel to bedding). Reddish brown terra-rossa fill between coral heads of Pleistocene limestone, Florida.

cidence can be explained in two interrelated ways: (1) major episodes of generalized subaerial exposure may favor mass extinction of established terrestrial plants and expansion of new types, and (2) onset of domination stages of terrestrial plants may be reflected in the type and intensity of subaerial exposure facies. During extended periods of subaerial exposure, land colonized by expanding new types of plants could produce most intense diagenetic modifications in the host rocks, resulting in easily recognizable subaerial exposure facies and in most intense accumulation of mineral deposits.

As should be expected, major subaerial exposure facies occur during the global cycles of sea level fall (Fig. 10) established by Vail and others (1977). Global cycles of relative sea level oscillation are thought to be controlled by tectonic plate dynamics (Sloss, 1979; Vail et al, 1977). In this context, links between plate tectonics, global relative sea level oscillations, generalized subaerial exposure, plant evolution and the cycle of CO_2 deserve much more attention than can be given here. However, it is also important to remember that minor subaerial exposure surfaces will be produced independently of global episodes in parts of specific sediment packages subject to shallowing and shoaling upward sedimentation (sand shoals, reefs, deltas, tidal flats). These local subaerial exposures may or may not be sites for development of economically significant mineral deposits but, either way, they will still be of interest for deciphering regional geology.

The effects of Pleistocene sea level fluctuations in the Caribbean are recorded by relic yet distinctive morphological features in the present day landscape. Emergent shorelines, terraces and reef tracts are common around the perimeter of many Caribbean islands. On Barbados, where gradual but continued tectonic uplift has combined with the effects of glacio-eustatic sea level changes, a sequential record of emergent coral reef terraces has been produced. On the coastal plains of Belize and parts of Florida, dolinas and cenotes pepper the landscape and testify to a Pleistocene or earlier period of karsting. In the reef tract of Belize, drowned cenotes (blue holes) have been found with their floors 120 m (400 ft) below present sea level. Such observations record evidence of karsting during a glacial, low sea level stand followed by a rise during an interglacial. Karsting during interglacial stages of the Pleistocene has significantly increased macroporosity of exposed carbonate terrains in the Caribbean. Similar effects, with economic benefit, are seen further back in the geologic record, for example the Cretaceous El Abra of Mexico.

RECOGNITION OF SUBAERIAL EXPOSURE

General Procedure

The identification of subaerial exposure surfaces in the geologic record should be based on the following considerations: (1) understanding of the regional geology, including stratigraphy, structure and tectonic evolution, sea level oscillations, location of the regional and local paleotopographic highs, depositional facies and paleoenvironments, (2) detailed study of rock sequences, especially below marked discontinuity surfaces and within or at the base of continental deposits (e.g., alluvial fan, lake, swamp, glacial till, plant tufa); our aim will be eventual identification of a sequence assignable to a karst or a soil profile (Fig. 11), and (3) study of rock types, both in the field and in the laboratory, to identify characteristic macro- and microfeatures (Table 2). Any conclusion should be supported with the interpretation of

Figure 34—Karst breccia. Clasts of brecciated and corroded white limestone embedded in red terra-rossa sediment. Note laminar rind on clasts. Tubular voids are root molds. Pleistocene, Florida.

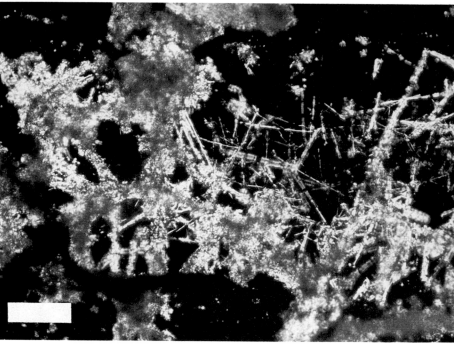

Figure 35—Moon-milk. Located in an open joint of the upper vadose zone of karst profile. Random needle fibers of low magnesian calcite (right) are progressively neomorphosed into microsparite mottles. Some of these mottles also contain relics of corroded carbonate host, Garraf Mountains, Barcelona, Spain. Photograph courtesy of L. Pomar. Thin section, XN. Scale bar = 0.1 mm.

lithologic sequences by integrating diagnostic and non-diagnostic but commonly present features of subaerial exposure facies (as in the examples of Figs. 11, 18, 48, 52). Of these three procedure stages, we will concentrate on the last, namely the study of the rock types, but without denying equal importance of the other two.

Different levels of security in the recognition of evidence of subaerial exposure can be established from relatively diagnostic features to commonly present but not indicative ones. Features listed below, taken in isolation, occur in distinctive but completely unrelated sedimentary and diagenetic environments. But, as stated so succinctly by Bathurst (1975, p. 417), ". . .safety is in numbers and a satisfactory decision can be reached if several criteria are combined." At the same time, it should always be remembered that many subaerial exposure surfaces may not contain characteristic features; but absence of these features does not necessarily mean that the deposit is not a subaerial exposure facies.

Figure 36—Surficial karstification. Irregular, V-shaped lapies filled with red-brown soil and clasts of host wall rock. Note modern root penetration. Karstification of Miocene limestone under a modern soil cover. Ibiza, Spain.

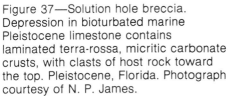

Figure 37—Solution hole breccia. Depression in bioturbated marine Pleistocene limestone contains laminated terra-rossa, micritic carbonate crusts, with clasts of host rock toward the top. Pleistocene, Florida. Photograph courtesy of N. P. James.

Diagnostic Features

Diagnostic indicators of subaerial exposure include recognition of karst, soils and duricrusts (Goudie, 1973). Here we will examine only karst and caliche features, and refer only to those features of potential preservation in the geologic record. Positive identification of the following features, provided they are in-place, is considered highly indicative of subaerial exposure.

Indicative of Subaerial Karst Facies — The following features are diagnostic of karst development away from the influence of supratidal processes (sea water capillary fringe and spray zone). The number of diagnostic landforms is increased when considering also coastal karst.

Surface landforms — Of the wide variety of surface landforms (see Sweeting, 1973, for example) few are of diagnostic value for subaerial (meteoric water) karst facies in the fossil record. Most useful to geologists are rundkarren forms in general, that is the rounded, smooth channels, troughs, pinnacles, 10 cm to about 1 m (0.33 to 3.3 ft) in dimension, formed under a humus cover (Figs. 15, 16). In exceptionally good outcrops, dolinas could also be

Figure 38—Transition zone. Calichified cracks (white) associated with terra-rossa and root penetration (**R**) in Miocene carbonates. Tarragona, Spain.

recognized. Dolinas (Figs. 12, 14) are circular to irregular, enclosed hollows of bowl, funnel or cylindrical shapes, 2 to 100 m (6.6 to 330 ft) deep and 1,000 m (3,281 ft) in diameter, formed by preferential dissolution along vertical joints or by roof collapse of underlying caves. In general, any landform will be more diagnostic of subaerial exposure if covered by relic soil.

Karst collapse breccia — Surface and subsurface accumulations of discrete bodies of carbonate breccia formed by foundering of cave roofs. Essential to the diagnostic value of karst breccia is the identification of collapse resulting from removal (dissolution or hydraulic erosion) of underlying carbonate rather than tectonism or sediment gravity flows. Common features to these breccias are the mixing of various types of cave sediments (sand, silt, clay), speleothem fragments and soils (Figs. 29, 34), and several generations of in-place speleothem coatings. Beds overlying the breccia may show upward attenuating V-shaped deformation (Fig. 19). The presence of organic borings of coastal organisms reduces the diagnostic value of collapse breccia since bioerosion and wave action at the base of coastal cliffs also produce collapse breccias.

Figure 39—Caliche profile. Well developed nodular horizon overlain by platy caliche and a thin surface hardpan at the top. This profile is developed in reddish Pleistocene loessic silts, exposed at the base of the profile. Tarragona, Spain.

Figure 40—Caliche glaebules. Nodular caliche horizon in the same locality as Figure 39. Glaebules (nodules) are soft and chalky in consistency, show tendency toward vertical elongation, and grew inside a red silty soil by the interaction of vertical water movement and possible roots.

Cavern porosity — According to Choquette and Pray (1970), "a pore system characterized by large openings, or caverns. Although much cavern porosity is of solution origin, the term is descriptive and not genetic..... A practical lower size limit of 'cavern' for outcrop studies is about the smallest opening an adult person can enter. Where the rock unit

Figure 41—Caliche hardpan breccia. A well developed hardpan (lower half) at the top of a Pleistocene caliche profile was destroyed as part of the weathering processes formed a new soil cover (upper half). Carlsbad, New Mexico.

Figure 42—Fossilized caliche profile. White hardpan at the top of a Pleistocene caliche profile shows a gentle undulatory morphology overlain by recent soil. Tarragona, Spain.

Figure 43—Caliche profile. Basal nodular horizon with white, chalky, vertically elongate glaebules (nodules) separated by red silt, grading upward into smaller, subspherical glaebules of lower part of platy caliche horizon. Thin (3 to 4 cm) hardpan crust on upper surface. Pleistocene caliche, Tarragona, Spain.

is known only from drilling, a practical lower size limit is that large enough to cause an easily recognizable drop of the drilling bit (a half meter or so)." Large cavern porosity can be concentrated in a subhorizontal plane, but commonly will be distributed in several interconnected horizons (variations of the phreatic water table). One of the reasons for the artificial lower limit of the term cavern is the possibility that smaller pores ("vugs") could form in other diagenetic environments without substantial proximity to subaerial exposure surfaces.

Speleothems — Although not abundant, speleothems (stalactites, stalagmites, flowstone, cave pearls) do occur in the fossil record (Figs. 20, 24, 28, 29), but are usually deeply

corroded and partly replaced by argillaceous micrite. These alterations are thought to result from repeated cycles of exposure in the upper vadose zone of the karst profile. In addition to in-place speleothems, reworked speleothem fragments within collapse breccias are also indicators of karst whereas clasts of speleothems lying on a discontinuity surface are valid indicators only of proximity to karst facies.

Indicative of Caliche Facies

Rhizoliths — Organosedimentary structures produced in roots by accumulation and/or cementation around, cementation within, or replacement of, higher plant roots by mineral matter (Klappa, 1980a). Five basic types of rhizoliths exist: (1) root

molds and/or borings which are simply cylindrical pores left after root decay (Figs. 56, 62, 72, 90); (2) root casts which are sediment- or cement-filled root molds (Figs. 57, 62); (3) root tubules which are cemented cylinders around root molds (Figs. 53, 59); (4) rhizocretions *s.s.* which are concretionary mineral accumulations around living or decaying roots (Fig. 54); and (5) root petrifications which are mineral encrustations, impregnations or replacements of organic materials whereby anatomical root features are partly or totally preserved (Figs. 88, 89). Most commonly, rhizoliths are formed by low magnesian calcite (LMC) accumulations of fine-grained, white- to cream-colored deposits (Calvet et al, 1975). Rhizoliths are generally millimeters to

Figure 44—Incipient platy caliche horizon. Horizontal to subhorizontal root mats penetrate and alter host calcarenite of Miocene age. Brecciated hardpan in upper part of the profile. Tarragona, Spain.

Figure 45—Horizontal root-mat with incipient glaebular and subsurface laminar hardpan development. Millimeter-sized glaebules (caliche ooliths) resemble marine ooliths. Recent caliche profile, Granada, Spain.

centimeters (fractions of an inch to inches) in diameter and centimeters to meters (inches to feet) long. They may be abundant or absent, but when present, they are reliable indicators of subaerial exposure since evidence of marine plant roots, such as those of *Thalassia* grasses, is rarely if ever preserved. Diagnostic reliability of rhizoliths is further increased by their location just below breaks in the sedimentary sequence (Fig. 55). However, rhizoliths may be confused with animal burrows; great care should be taken to separate these two different organosedimentary products (Klappa, 1980a). Apart from rhizoliths themselves, roots are primarily responsible for, or contribute to, the formation of platy caliche horizons, some brecciation textures (Klappa, 1980b), channel and root-moldic porosity, and alveolar texture. Roots, together with symbiotic fungi, are also responsible for the hitherto enigmatic structure of *Microcodium* (Klappa, 1978).

Alveolar texture (term introduced by Esteban, 1974) — Cylindrical to irregular pores, which may or may not be filled with calcite cement, separated by a network of anastomosing micrite walls (Figs. 65, 66). Pore

Figure 46—Transition and nodular zone of caliche profile. Altered host rock (Paleocene carbonates) still preserves traces of bedding in a transition zone of a Pleistocene caliche with incipient glaebule development. This profile is truncated by present day soil cover. Tarragona, Spain.

Figure 47—Fossil caliche. Caliche profile formed by a nodular horizon coalescing into an incipient platy caliche horizon. This profile developed on fluvial overbank silts and lutites of a Lower Triassic (Buntsandstein) alluvial fan complex. Barcelona, Spain.

diameters commonly range from 100 to 500 μm, but a few may reach 1.5 mm (0.06 in). In recent caliche, micritic walls are composed of banded LMC fibers, 0.5 to 2.0 μm wide and from 3 to 120 μm in length (Fig. 67). These banded needle fiber walls are commonly encrusted by equant microcrystalline or cryptocrystalline calcite in fossil caliche (Fig. 66). Alveolar texture may be abundant to absent but when present is found mostly within platy and hardpan horizons. Without specifying a name, Steinen (1974) described and illustrated alveolar texture and suggested that it may represent discrete channelways within sediment which had been penetrated by rootlets. Harrison (1977) described the same texture as root moldic porosity. In essence, alveolar texture is the product of coalesced millimeter-sized rhizoliths.

Microcodium Gluck — Elongate, petal-shaped calcite prisms or ellipsoids, 1 mm (0.04 in) or less in long dimension and grouped in spherical, sheet or bell-like clusters (Esteban, 1974). *Microcodium,* long thought to be algal, inorganic, bacterial or actinomycete in origin, is now believed (Klappa, 1978) to be the result of

Figure 48—Relic traces of fossil caliche below erosion surface. Sharp surface truncating the layering of a fenestral dolostone (lower part) is overlain by skeletal packstones. Thin section of the iron-stained fenestral rock just below the erosion surface shows alveolar texture, calcified filaments, root molds, glaebules and corroded quartz grains. Permian, Guadalupe Mountains, New Mexico.

Figure 49—Calichified planar subaerial exposure surface. Near vertically bedded Kimmeridgian limestones and dolostones overlain unconformably by interstratified colluvial calcareous silts and eolianites of Pleistocene age. Eolianites are calichified. Ibiza, Spain.

calcification of mycorrhizae, symbiotic associations between soil fungi and cortical cells of higher plant roots (Figs. 69, 70).

Tangential LMC needle fibers (term introduced by James, 1972) — The acicular or needle-shaped calcite crystals are arranged in a band with long axes of needles parallel and tangential to the band (modified from Bal, 1975). Sizes and shapes of these needle crystals are the same as those forming alveolar textures, and, in many instances, tangential LMC needle fibers form alveolar textures (Fig. 67).

Calcified cocoons — Ovoid to spherical, 1.0 to 3.0 cm (0.4 to 1.2 in) diameter cases of calcified puparia of soil-dwelling insects (Fig. 64). Recorded by Read (1974) and Ward (1975) in Australia and Yucatan respectively, these organosedimentary structures are also conspicuous in some Mediterranean paleosols. Although not abundant, calcified cocoons are reliable indicators of soils and, therefore, of subaerial exposure.

Caliche glaebules (including pisoliths and nodules) — A glaebule is a three-dimensional unit within the matrix of soil material, usually equant, prolate to irregular in shape,

Figure 50—Brecciated caliche hardpan. Upper surface of Figure 49. Extensive horizontal root networks of garriga vegetation are causing brecciation of Pleistocene hardpan and leading to soil development. Ibiza, Spain.

Figure 51—Transitional horizon. Relic in-place "boulders" of Kimmeridgian dolostone host rock wrapped in fine-grained micritic caliche matrix of lower-most Pleistocene caliche profile of Figure 49. Ibiza, Spain.

and distinguished from the enclosing matrix by a greater concentration of some constituent and/or difference in fabric (modified from Brewer, 1964). With respect to caliche profiles, glaebules consist of discrete, powdery to indurated concentrations of LMC micrite. Caliche glaebules are silt to pebble in size and may occur either as isolated particles (Fig. 73) or coalesced masses (Fig. 40). Many sand to granule size glaebules show poorly developed concentrical lamination (Fig. 81). The term caliche glaebule embraces various terms used to describe undifferentiated to concentric structures commonly recorded in caliche facies including caliche ooids, pseudooids, ooliths, peloids, pellets, pelletoids, coated particles, nodules, concentric structures and concretions.

Clay cutans — Brewer (1964) defined a cutan as "a modification of the texture, structure or fabric at natural surfaces in soil materials due to the concentration of soil constituents or in-place modifications of the plasma (relatively unstable soil matrix)." Brewer recognized several groups of cutans, distinguishable by the characteristics of the surfaces affected, mineralogy of the cutanic material and internal fabric of the

Figure 52—Sequence indicative of subaerial exposure. Underlying Miocene carbonates (**M**) show a flat upper surface without positive evidence of subaerial exposure. Overlying deposits contain diagnostic evidence of subaerial exposure such as pebbles of Miocene rock with lichen borings (**L**), reddish soils with vertical rhizoliths (**V**), and eolianite with rhizoliths along cross bedding (**E**). Mallorca, Spain.

Figure 53—Rhizoliths. Vertically oriented taproot rhizoliths and subhorizontal second order rhizoliths of Figure 52.

cutans. Cutans as defined by Brewer are pedological features and, thus, products of soil forming processes. While many pedological features such as glaebules, fecal pellets and animal burrows are not uncommon in sedimentary rocks, Teruggi and Andreis (1971) believed that cutans were exclusive to soils. However, we feel that calcite cements in carbonate rocks conform to Brewer's definition of cutans and, thus, cutans in general are not exclusive to soils. What we do believe is that some cutans are diagnostic of soils, specifically clay cutans. Although relatively uncommon in caliche, clay cutans probably are reliable indicators of soil formation (Fig. 79).

Circum-granular cracking (Swineford et al, 1958; Ward, 1975) — Irregular to globular masses separated by non-tectonic fractures and produced by alternate shrinkage and expansion are called circum-granular cracking (Figs. 82, 83). A common feature in soils in general and caliche in particular, circum-granular cracking could also occur in intertidal sediments but, as yet, has not been clearly recorded.

Figure 54—Rhizolith. Horizontal section of a vertical rhizolith. Area formerly occupied by the root corresponds to the brown central part of the rhizolith, and still there are left few remains of root moldic porosity. Photograph courtesy of F. Calvet, Pleistocene eolianite, Mallorca, Spain.

Figure 55—Caliche profile. Laminated caliche hardpan developed in a rhizolith-rich horizon in Pleistocene eolianite (white sediments in lower part). Note small sink-hole indicating later karst event. Yucatan, Mexico.

Common to both
Karst and Caliche Facies

The following features, diagnostic of subaerial exposure, are present in caliche and in the upper vadose zone of karst profiles.

Lichen structures — Subaerially exposed carbonates are commonly covered by lichens. Lichens cause textural and fabric changes of the uppermost (up to 2 cm; 0.8 in) colonized substrate (Fig. 31), commonly producing spongy, microsparitic or micritic layers containing organic-rich and organic-poor millimeter laminae, fungal borings, algal filaments and spheres, fungal hyphae and oily cells of fungi, reproductive structures and silt-sized to sand-sized calcite grains (Fig. 32) arranged in surface parallel layers (Pomar et al, 1975; Klappa, 1979). Endolithic lichens are more ac- tive in semi-arid climates whereas epilithic lichens dominate temperate to arctic climates. Lichen structures fossilize as borings into the rock or as calcified organosedimentary structures. When lichen structures show a well developed lamination they are referred to as lichen stromatolites (Klappa, 1979). Lichen structures, including lichen stromatolites are common on exposed upper surfaces of

Figure 56—Root molds. The top of a thick bed of a shallow marine Miocene limestone was colonized by vertical roots. Relics of alveolar texture seen in thin section support the interpretation as root molds. Mallorca, Spain.

Figure 57—Rhizoliths. Reddish fluvial silts show vertical tracks of grayish silts considered as rhizoliths. Paleocene, Barcelona, Spain.

both karst and caliche facies (Figs. 30, 71). We consider the presence of an extensive endolithic lichen cover (a protosoil) as the first evidence of subaerial exposure away from supratidal influence.

Random needle fiber low magnesian calcite (LMC) — This term, introduced by James (1972), is a synonym of "lublinite" and "pseudomycelia" in soil science. To produce a random needle fiber fabric, the basic distribution of acicular or needle-shaped LMC crystals is arranged in a random, loosely woven pattern within voids (modified from Bal, 1975). Ramifying, self-supporting but loosely woven LMC needles are common in caliche profiles and upper parts of karst profiles (Figs. 35, 84). In caliche, this fabric is common in voids within hardpan and platy horizons. Random needle fibers also form soft, powdery nodules within the chalky and nodular horizons. Size and morphology of these needles are similar to those which form alveolar texture and banded needle fiber fabrics. In karst facies, needle fiber fabrics occur along near-surface joints and are major contributors to moon-milk deposits of the upper vadose zone (Fig. 35).

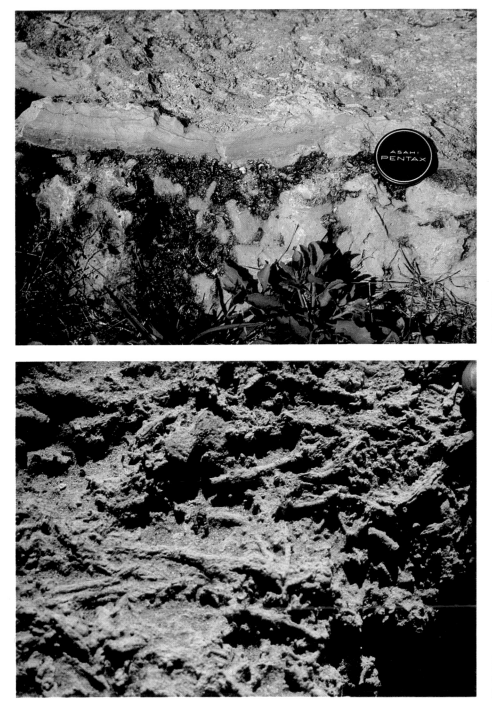

Figure 58—Laminar caliche hardpan. Laminar micrite hardpan of ?Pleistocene age capping reef core facies of Pliocene reef complex. Everglades, Florida.

Figure 59—Rhizoliths. Planar view showing millimeter-sized horizontal rhizoliths occurring just below the laminar hardpan shown in Figure 58.

Commonly Present in Subaerial Karst Facies but Non-diagnostic

The following morphologies are common in subaerial karst but also well developed in supratidal coastal settings. Many authors refer to "marine" karst as a specific type of karst formed under the influence of both sea water (salt spray and splash zones, sea water capillary fringe) and meteoric waters. Features developed in carbonate hosts in marine intertidal and subtidal settings are not considered as karst. Nevertheless, in special conditions (confined aquifers), subterranean karst can be active below sea level. In our discussion, these special conditions are not relevant; we will refer only to the distinction between supratidal coastal karst and subaerial karst. Because of the common juxtaposition and superposition of subaerial karst and coastal karst, the separation of these karst types may seem superfluous. However, this distinction is necessary to be consistent with our concept of terrestrial soil and subaerial exposure. For example, extensive areas (tens of kilometers; tens of miles) of barely emergent carbonate sand shoals in the supratidal zone could produce thick sedimentary sequences without con-

Figure 60—Caliche rhizobreccia. Brecciated dolomite clasts of Jurassic dolostone floating in a caliche matrix of Pleistocene age. Inset shows thin section photomicrograph of xylem vessel from a plant root which was extracted from dissolved caliche matrix between dolostone fragments. Ibiza, Spain. Scale bar = 100 μm.

Figure 61—Incipient caliche development. Pleistocene eolianite (**P**) overlain by active modern soil (**S**). Upper decalcified part of soil horizon contains abundant roots. Top of this calcareous-rich soil horizon (**arrow b**) shows irregular dissolution surface. Subsurface laminar caliche hardpan and horizontal root mat (**arrow**) overlies unaltered Pleistocene eolianite (**P**). Excavator is responsible for smooth vertical scrape marks (i.e., artifacts). Ibiza, Spain.

taining recognizable subaerial exposure surfaces.

Phytokarst (term introduced by Folk et al, 1973) — "Black-coated, jagged pinnacles marked by delicate, lacy dissection that lacks any gravitational orientation." This razor-sharp spongework of ragged pinnacles is formed by carbonate dissolution caused primarily by plants. Other authors refer to very similar features as root-lapies (deckenkarren), specifying the basic role of root penetration and presence of circular tubular sections (3 to 20 cm; 1.2 to 8 in wide) of tortuous branching channels. Jakucs (1977) noted that these spectacular lapies can reach down 25 m (81 ft) into the rock, producing 75% porosity, and notes further that, under tropical conditions, their evolution is extremely fast (may develop even in a single vegetal cycle of 4 to 10 years). Phytokarst (or root lapies) is very intense in tropical karst, but is also present in Mediterranean karst and in supratidal coastal environments, especially in substrates with high intercrystalline and/or interparticle porosity. In these coastal settings salt spray weathering contributes to the formation of phytokarst forms. In all these environments, phytokarst is

Figure 62—Root corrosion. Solutional voids at top of calcareous soil horizon shown in Figure 61 and arrowed (**b**). Tubular voids (**T**) are soil-filled root molds. Arrow (**R**) points to plant root.

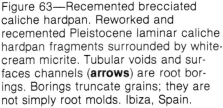

Figure 63—Recemented brecciated caliche hardpan. Reworked and recemented Pleistocene laminar caliche hardpan fragments surrounded by white-cream micrite. Tubular voids and surfaces channels (**arrows**) are root borings. Borings truncate grains; they are not simply root molds. Ibiza, Spain.

associated with kamenitzas (see below) and abundant blue-green algal borings. Root penetration (terrestrial plants, mangroves, seagrasses) and bioerosion in the low supratidal-intertidal zone (*Litorina, Patella,* chitons, *Lithophaga*) are probably contributing to rock disintegration and development of both phytokarst and kamenitzas.

Kamenitzas (solution pans, rock pools) — Small, flat-bottom depressions (up to 3 m or 10 ft in diameter and up to 1 m or 3.3 ft deep) in level calcareous land surfaces (Figs. 21,

23). Sides of these depressions can be steep, vertical or overhanging. These depressions start by vertical dissolution of water collecting between slight undulations (modified from Sweeting, 1973). Sides are typically infested by endolithic and chasmolithic algae. Common on subaerial karst surfaces, kamenitzas also occur in supratidal environments.

Rillenkarren and rinnenkarren (solution runnels) — Razor-sharp, finely chiselled, solution runnels from 1 to 2 cm (0.4 to 0.8 in) wide and deep, 20 cm (8 in) long, up to 50 cm

(20 in) wide and deep and 20 m (66 ft) or more long (based on Sweeting, 1973), forming downward oriented (gravitational) patterns along bare rock slopes (Fig. 23). These lapies subtypes rapidly (a few months) model other exhumed lapies (Fig. 22) and may evolve into pinnacle morphologies (spitzkarren; Fig. 18). Dimensions of runnels appear to be proportional to annual rainfall. These features develop preferentially on well cemented, hard and compact carbonates. Sharp dissolution runnels may also develop on carbonates ex-

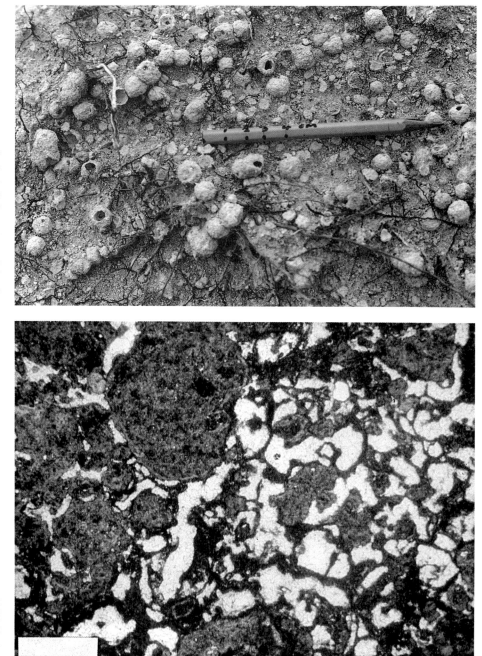

Figure 64—Calcified cocoons. Poorly developed caliche within Holocene wind-deposited littoral carbonate sands contains abundant spherical carbonate structures. Fractured structures show smooth inner and rough outer surfaces. Note walls of intact structures have holes which lead into internal chambers; interpreted as calcified cocoons (puparia) of soil-dwelling insects. Arrow points to terrestrial gastropod. Ibiza, Spain.

Figure 65—Alevolar texture. Complex network of micrite walls, partially filled by blocky calcite cements, between caliche glaebules. Pleistocene caliche, Tarragona, Spain. Scale bar = 0.5 mm.

posed in the coastal supratidal zone. Nevertheless, it seems that well developed rillenkarren and rinnenkarren should be strongly suggestive of subaerial exposure.

Commonly Present in Caliche Facies but Non-diagnostic

Black pebbles — Black limestone pebbles are a noticeable component in some caliche hardpans (Fig. 72). Ward and others (1970), after examining blackened eolianite and caliche rubble, attributed the blackening to the preservation of trapped organic matter within calcite crystals. They suggested that layers of fragmented dark-colored limestone may indicate subaerial exposure adjacent to hypersaline water. Klappa (unpub. data) has found that metallic oxides precipitated on fungal hyphae also produce blackening of limestone. In addition, black limestone may also form in the deep sea and in centers of large lakes where it may be tens to hundreds of kilometers from land. Once reworked, black pebbles derived from these limestones will be difficult to tell apart from those described by Ward and others (1970); study of lithofacies associations and fabric characteristics may help resolve this difficulty.

Floating texture and corroded carbonate grains — Embedded, silt-sized and sand-sized grains showing no point contacts in a micrite matrix give an arrangement that is commonly

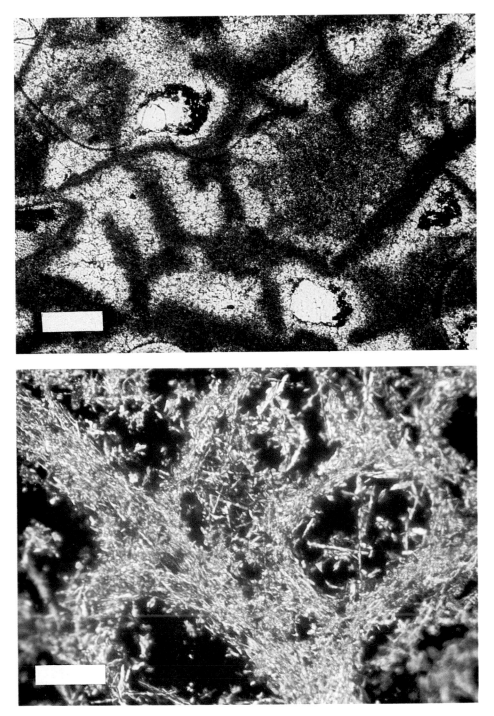

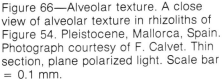

Figure 66—Alveolar texture. A close view of alveolar texture in rhizoliths of Figure 54. Pleistocene, Mallorca, Spain. Photograph courtesy of F. Calvet. Thin section, plane polarized light. Scale bar = 0.1 mm.

Figure 67—Tangential low magnesian calcite needle fibers. Parallel alignment of many needle fibers form interconnected bridge-like bands across voids. Pleistocene caliche, Barbados. Photograph courtesy of N. P. James. Thin section, XN. Scale bar = 50 μm.

referred to as floating texture (Brown, 1956). Floating texture has been explained in one of five ways: (1) normal sedimentary processes, (2) partial replacement of some grains by calcite, (3) partial or total replacement of some grains by calcite with other more resistant grains of different mineralogy being unaffected, (4) expansion as a result of penetration and dislocation of organisms by

weathering and disaggregation of bedrock or sediment, and (5) expansion as a result of displacive crystallization of calcite.

Corrosion of non-carbonate grains and partial replacement of grains by calcite (Fig. 86) has been widely documented in caliche and is probably the most important mechanism for producing floating texture in caliche; however, displacive

crystallization of calcite is favored by some workers (Assereto and Kendall, 1977; Watts, 1977, 1978) but this phenomenon has yet to be proved. Floating textures, although proposed by earlier workers as a criterion for recognition of caliche, can be produced in many environments by a wide variety of mechanisms; it is characteristic but not diagnostic of caliche facies.

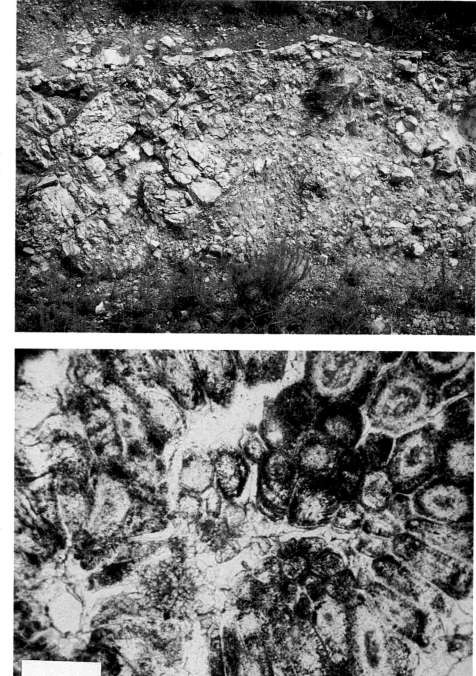

Figure 68—Transition zone. Intensely altered host rock (Cretaceous limestone) preserves some evidence of original bedding (inclined to the left). Thin caliche hardpan crust (Pleistocene) at the top is being destroyed under present soil cover. The conglomeratic appearance of the transition zone (developed during Paleocene) does not imply significant transport, but is produced largely by root penetration; dense *Microcodium* colonies form most of the matrix between the Cretaceous blocks. Barcelona, Spain.

Figure 69—*Microcodium*. Elongate calcite prisms with hexagonal basal section forming globular aggregates in the matrix of the "conglomerate" of Figure 68. Thin section, plane polarized light. Scale bar = 0.5 mm.

Vesicles, vesicular texture and vesicular porosity — Vesicles are spherical pores formed by rearrangement of sediment or crystallization of solutions around fluid bubbles in an unconsolidated material. Vesicles are larger than grain-supported voids and differ from both vug and fenestral porosity principally in their walls which are smooth, simple curves with circular outlines (modified from Brewer, 1964). In thin section, vesicular texture and alveolar texture are not readily differentiated, except that the latter may show tubular longitudinal sections whereas vesicular textures have only circular outlines. Vesicles are common in soils and beach sands.

Vermicular texture — Dense networks of micritic tubes and rods set in a micritic matrix give vermicular or "spaghetti-like" textures when viewed in thin section (Fig. 85). Vermicular texture forms by calcification of organic filaments (algae, fungi, actinomycetes). These calcified filaments are known from marine and meteoric settings and, thus, although common in caliche are non-diagnostic features.

Tepee or pseudo-anticline structures — Caliche pseudo-anticlines are

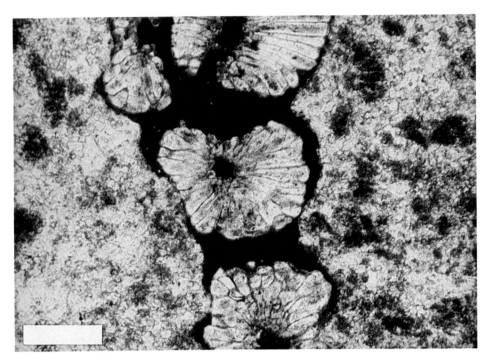

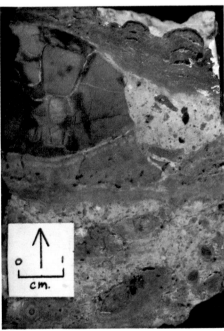

Figure 70—*Microcodium.* Vertical joint in Jurassic dolostone was enlarged by corrosion activity of *Microcodium* in tubular colonies. Black material in opened joint is dense, opaque argillaceus terra-rossa. This joint is located in the upper vadose zone of a karst profile developed during Lower Tertiary. Barcelona, Spain. Thin section, plane polarized light. Scale bar = 0.5 mm.

Figure 71—Caliche hardpan. Multi-generation hardpan of Pleistocene age showing glaebules (pisoliths), recemented hardpan clasts, exotic Mesozoic dolostone clast, and uppermost surface laminar micritic crust. Note inter-laminar voids within red-brown laminar crust and black rim to top layer. Laminae formed by calcification of lichen surfaces (lichen stromatolite). Ibiza, Spain.

Figure 72—Recemented brecciated caliche hardpan. Pink caliche matrix with black pebble limestone clasts. Irregular brown stained anastomosing channels and tubular voids are recent root borings. Root borings through black pebbles shown in upper left are filled with white-cream micrite. Pleistocene, Tarragona, Spain.

common in some caliche facies, but tepees or pseudo-anticlines in general can occur in marine and coastal environments (see Assereto and Kendall, 1977, for a review on tepees).

Commonly Present in Karst and Caliche Facies but Non-diagnostic
Irregular to smooth sharp upper surfaces, commonly truncating underlying sedimentary structures, may demarcate upper bounding sur-faces of karst and caliche facies (Figs. 18, 42, 48, 49).

Shallowing upward sequences culminating in non-specified exposure surfaces, missing paleontological zones, or in-place non-tectonic fracturing and brecciation are also commonly present.

Other features common to nondiagnostic karst and caliche facies include: microscopic features such as clotted micrite (structure grumeleuse; Figs. 80, 91), microspar, meniscus cements (Fig. 77), gravitational cements (Figs. 77, 78), calcified filaments (Fig. 85), calcified fecal pellets (Fig. 88) and microborings (Fig. 76); leached and vuggy porosity; lithoclasts in beds directly above sedimentary breaks (Fig. 50); laminar micritic crusts; chalky, friable to powdery white deposits (more diagnostic of subaerial exposure when below sedimentary break); iron oxide

Figure 73—Glaebules (pisoliths). Cretaceous limestone (lower part) is sharply eroded and overlain by a pisolith grainstone (Paleocene). Glaebules (pisoliths) originated in hypothetical near-by caliche profile but were transported laterally and resedimented on sharp erosion surface. Sedimentary transport and deposition generated grain size sorting in coarsening upward sequences. Tarragona, Spain.

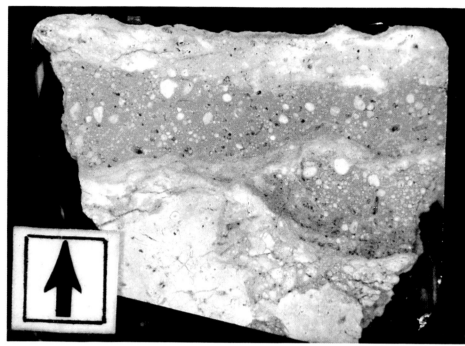

Figure 74—Caliche hardpan. Irregular, subhorizontal crust laminae are underlain by brecciated Pleistocene limestone and separated by darker carbonate with abundant white glaebules. Barbados. Polished slab, scale bar = 2 cm. (From James, 1972; his Fig. 4a).

staining (Figs. 48, 73); and crystal silt (Dunham, 1979a) which consists of geopetal fillings in secondary porosity. Crystal silt can be generated from the chalky horizon of caliche profiles or from moon-milk deposits of the upper vadose zone of karst profiles. This fabric is present in subaerially exposed carbonates but may also be produced in submarine environments (e.g. chips from boring organisms).

Features Related to Meteoric Diagenesis

Since the late 1950s, fresh water diagenesis has been considered of utmost importance by carbonate geologists, but perhaps has been overemphasized or misunderstood. It is pertinent here to remember that vadose diagenesis, leaching and moldic porosity are not necessarily related genetically and do not imply the development of subaerial exposure surfaces (although they can occur close together). Negative

isotope ratios are widely accepted as evidence of meteoric water influence. But many criteria considered as diagnostic in the late 1960s are now known to occur in very different environments. For instance, although pervasive calcite cementation, calcitization of dolomite ("dedolomitization") and aragonite, leaching of carbonate, syndepositional brecciation, growth of poikilotopic calcite, development of vuggy porosity, production of 'crystal silt' and dolomitization by mixing waters do occur under the influence of meteoric fresh water, many of these processes are also known in submarine environments (Lindstrom, 1963; Shinn, 1969; Schlager and James, 1978). Besides, the emphasis of our discussion has been the characterization of surfaces of subaerial exposure and their recognition in the geologic record, the processes listed above are not necessarily related to a surface of subaerial exposure.

Caution should be exercised in the use of these criteria as evidence of

subaerial exposure. Recent studies show increasing uncertainty in the absolute distinction between meteoric and marine diagenesis (Schlager and James, 1978; Klappa and James, 1979). Phreatic diagenesis, in particular, is an obscure and poorly known world.

CLARIFICATIONS

Palustrine Carbonates

One confusion zone in carbonates is the marginal areas of lakes. There is a continuous spectrum between lacustrine carbonates, palustrine (swamp, marsh) carbonates and caliche. Freytet (1973) pointed out marked similarities between fabrics of palustrine carbonates and fabrics of caliche. Palustrine environments are areas where lacustrine carbonates (or relic subaerial soils) are subjected to incipient soil processes (colonization by land plants, formation of glaebules, circum-granular cracking) but still saturated with lake water (for this reason they are also termed hydromorphic soils). We recognize the problem of distinction between palustrine carbonates and caliche but this problem can perhaps be resolved by studying vertical profiles and their lateral facies relationships, sup-

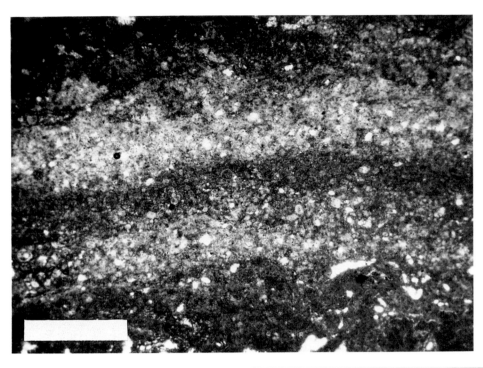

Figure 75—Laminar caliche hardpan. Subhorizontal micritic to microsparitic carbonate laminae; laminae in center contains abundant microglaebules. White specks are silt-sized quartz grains. Quaternary caliche hardpan, Barbados. Photograph courtesy of N. P. James. Thin section, plane polarized light. Scale bar = 0.5 mm.

Figure 76—Caliche hardpan. Clotted micrite overlain by laminar micrite. Spar-filled voids are root molds. Metallic oxide encrusted filamentous microorganisms in vertically oriented root mold. Poorly developed alveolar texture just left of center. Pleistocene. Ibiza, Spain. Thin section, plane polarized light. Scale bar = 100 μm.

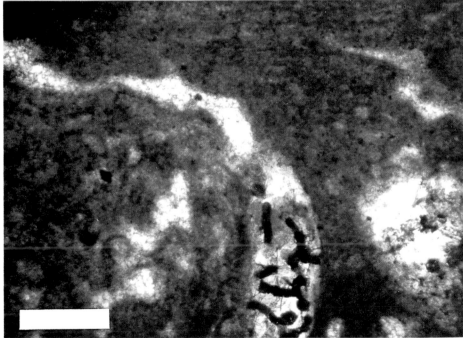

plemented by geochemical analyses, rather than studying fabrics alone.

Superficial Similarities

Laminated carbonate crusts — Algal stromatolites, lichen stromatolites (Fig. 71), "case-hardened" carbonate crusts, flowstones, laminar tufas (Figs. 92, 93), surface and subsurface laminar caliche hardpans (Figs. 45, 61) all show some similarities. However, differences do exist. These differences may represent formation in very different environments and can be recognized, usually in thin section and, with fortune or experience, in hand specimen.

Many references exist to laminated micritic crusts, in particular to the laminated crusts of Florida (Multer and Hoffmeister, 1968) as support for the soil origin of some rocks. These Floridian crusts show a non-laminated microcrystalline rind coated by porous and non-porous laminated micrite. The microcrystalline rind is probably a relic part of a pre-existing caliche profile. But the laminated micrite is formed by repeated calcite precipitation in layers of plant debris (plant litter) and cannot be considered as a true soil (as pointed out by Multer and Hoffmeister, 1968). Laminated crusts of the Floridian

type can be present in caliche hard-pans and on karst landforms; they are indicators of subaerial exposure. In essence, these crusts can be considered as micritic plant tufa, but they do not necessarily imply the presence of developed soil profiles or prolonged subaerial exposure; these laminated crusts can occur in barely emergent shoals.

The intention here is not to cover

petrographic characteristics of each crust type but, rather, to advise deliberation before reaching a conclusion on origin and significance of a particular carbonate crust.

Carbonate Pisoliths — Pisoliths, concentrically laminated spheroidal bodies larger than 2 mm (0.08 in), form in many environments. They are not restricted to caliche facies but occur also in shallow marine, deep

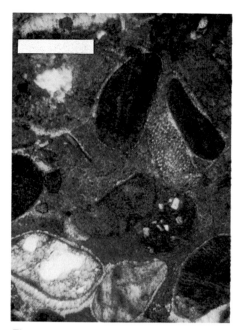

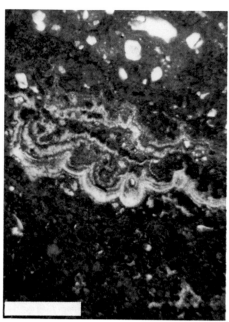

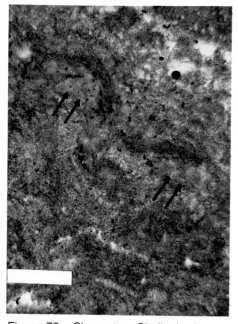

Figure 77—Gravitational and meniscus vadose cements. Calichified eolianite of Pleistocene age. Well-rounded sand-sized carbonate grains in Alizarine red-S stained caliche micrite. Sparry calcite between grains in lower part show meniscus outlines. Asymmetrical gravitational or pendulant cement beneath grain in lower left. Ibiza, Spain. Thin section, plane polarized light. Scale bar = 0.5 mm.

Figure 78—Microstalactitic vadose cement. Microstalactitic or pendulant fibrous calcite on former roof of quartz-rich caliche hardpan. Lower part of the photomicrograph shows microglaebular fabric. Pleistocene, Ibiza, Spain. Thin section, plane polarized light. Scale bar = 0.5 mm.

Figure 79—Clay cutan. Chalky horizon of caliche profile composed of silt-sized calcite crystals (microspar). Former wall of void now lined with red ferric-rich non-carbonate clay cutan (**arrows**). Pleistocene, Ibiza, Spain. Thin section, plane polarized light. Scale bar = 100 μm.

marine, peritidal, cave and lake environments. In some of the literature of the 1970s, pisoliths have been considered as diagnostic of subaerial exposure, mainly those showing polygonal fitting, downward elongations and perched sediment inclusions, occurring in beds with coarsening-upward graded bedding and affected by syndepositional, nontectonic fracturing. Although some caliche hardpans and speleothems can show similar features, Esteban (1976) questioned their value as evidence of either subaerial exposure or development of caliche soils. Many of these features are simply the result of in-place fabric evolution under some influence of hydrologic vadose regimes.

The environment of formation of the different types of pisoliths can be learned only after detailed study. Only then will pisoliths be of diagnostic value.

Isolating the Coastal Environment

Coastal environments represent belts between submarine and terrestrial realms and show attributes of both marine and subaerial influences. The coastal environment is also a confusion zone containing products commonly referred to as coniatolites ("marine caliche") and coastal marine karst. The term coastal marine karst is most adequate, but we have already mentioned the reasons for its separation from the subaerial karst. Here we will refer briefly to the considerable discussion brought about by different interpretations of coniatolites and their possible fossil analogs.

The terms coniatolite (Purser and Loreau, 1973) and "marine caliche" (Scholle and Kinsman, 1974) refer to aragonite and high magnesian calcite encrustations with a variety of morphologies (laminated crusts, dripstones, pisoliths) occurring in the intertidal and supratidal zone (Fig. 94). These products are well developed

along the Persian Gulf shorelines. Most authors referring to coniatolites (or "marine caliche") mention superficial similarities with caliche hardpans or speleothems and also possible partial analogies with the widely known "vadose pisolite" facies of the Permian of west Texas and New Mexico (Dunham, 1969b), but their features are not diagnostic of subaerial exposure or soil development (see Esteban, 1976, for a discussion).

As a conclusion, we believe that the isolation of the coastal supratidal and intertidal environments from the concept of subaerial exposure *s.s.* will benefit the accuracy of paleoenvironmental interpretations.

ECONOMIC CONSIDERATIONS

Oil and Gas Exploration

It is well established that many oil fields are related to unconformities. Chenoweth (1972) notes: "In view of this close association of petroleum with unconformities, one would

TABLE 1

| Horizon | Average Thickness (cm) | Boundary | Type | Porosity | | Cementation | | |
				% Field Estimate	% Point Counting	Cementing Agent	Continuity	Strength
0. Soil	32.5	Very abrupt to clear	Channel	30		Minor calcite	discontinuous	weak
1. Hardpan	20.6	clear to diffuse	—	—	0-13	calcite ± silica	continuous	indurated
2. Platy caliche	62.6	diffuse	fracture; moldic	10	3-17	calcite	continuous	strong
3. Nodular caliche	118.2	diffuse	interparticle	10-15	3-23	calcite	discontinuous	strong to weak
3. Nodular 4. chalky horizon	89.0	gradual to diffuse	interparticle	15-20		calcite	discontinuous	weak
4. Chalky horizon	54.5	gradual to diffuse	interparticle	15-20	9-35	calcite	discontinuous	weak
5. Transitional horizon	142.0		fracture; vug	15-25	3-31	calcite	discontinuous	weak to strong

Table 1. Physical properties of Quaternary caliche profiles (based on 326 measured profiles from Spain).

assume that concentrated studies of unconformities would occupy much of the time and talents of explorationists. Paradoxically, this is not so, and the unconformity. . .remains imperfectly understood.'' One possible reason, among many, for the paucity of detailed studies of the types, causes, and effects of unconformities is their non-diagnostic response in well-logs and seismic profiles. Without extremely refined techniques, now in development, different types of unconformities or exposure surfaces will be recorded as non-diagnostic kicks and reflectors.

Many, if not most, of these unconformities have suffered effects of subaerial exposure. During periods of subaerial exposure, carbonate sediments or rocks are susceptible to diagenetic changes which may either create or occlude porosity. Integrated regional studies, together with recognition of features outlined above, may help in understanding processes operating during subaerial exposure and, thus, in predicting patterns of porosity distribution.

Porosity potential is usually high in a paleokarst facies. A good example occurs in the Tampico region of Mexico. The Golden Lane trend follows

an elevated atoll-like rim of Albian-Cenomanian reef limestones (El Abra). This elevated platform edge was subjected to post-Cenomanian karstification and subaerial erosion before being buried by sealing rocks (shales) of Upper Cretaceous and Paleocene age. A similar age of karstification affects the Mesozoic sequence in many Mediterranean countries, locally prolonged up to the Lower Miocene, producing relatively important fields in the area (for example, Amposta field). A different example is the Arbuckle Limestone of the Central Kansas uplift. During both Late Ordovician and Late Pennsylvanian times, a karst landscape developed on the Arbuckle surface. Leached residuum, 1 to 10 m (2 to 30 ft) thick, formed a mantle during these karst episodes and acted as the Arbuckle reservoir rock (Walters, 1946). In general, seals for karst reservoir traps rely on subsequent sedimentation above the subaerial exposure surface.

Caliche facies, on the other hand, generally have low porosity potential. Caliche facies are not by themselves targets for oil exploration because their volumes and reservoir qualities are generally below economic viabili-

ty; in addition, their sealing potential, even of the hardpan, can be reduced dramatically because of fracturing and brecciation (tectonic, collapse, root penetration). Nevertheless, the discovery of caliche facies provides an optimistic signpost to possible underlying or juxtaposed karst reservoirs.

Other Mineral Resources

Lead-zinc deposits — Stratabound Mississippi Valley-type lead-zinc deposits (Brown, 1970) are common in some thick sequences of carbonate rocks, particularly dolostones, which have been subjected to major episodes of karstification. Whether introduction of metal-bearing fluids and the cavity forming processes of karstification are related genetically or not, is a debatable point (Bernard, 1976). Some authors use the term "thermokarst" for cavity forming processes by hot waters irrespective of their origin and assume a hydrothermal origin for these metallic deposits. For other authors, the hydrothermal metal-bearing fluid is thought to precipitate in a subaerial karst facies. Finally, other authors emphasize the role of vadose meteoric diagenesis and evolution of the deep phreatic

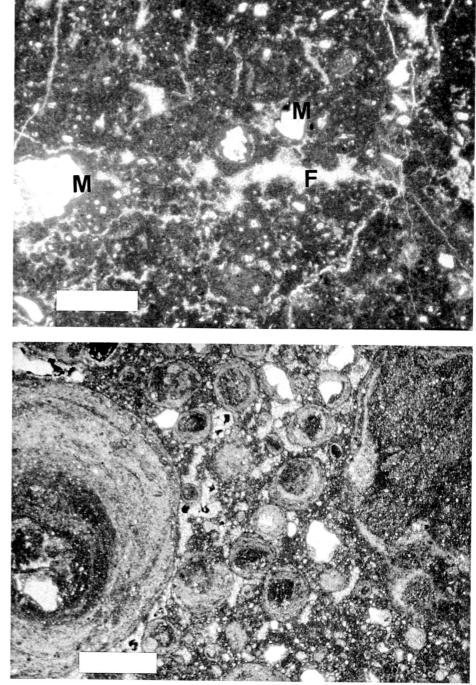

Figure 80—Clotted micrite. Clotted micrite fabric is formed by dense micrite glaebules (peloids) showing some incipient crumbly fracturing (**F**). Rock is now hard and completely cemented, showing rigid fracturing. *Microcodium* fragment (**M**). Paleocene, Barcelona, Spain. Thin section, plane polarized light. Scale bar = 0.55 mm.

Figure 81—Caliche glaebules (pisoliths). Poorly laminated, concentric micritic grains in Pleistocene caliche hardpan. White specks are detrital quartz grains. Tarragona, Spain. Thin section, plane polarized light. Scale bar = 0.5 mm.

(imbibition) zone as the adequate steps in the concentration process of formerly dispersed mineral fractions. But most interpretations need a physicochemical trap of metal-bearing fluids by impermeable fine-grained sediments. Therefore, karst development produces both reservoir and trap for this type of mineral precipitation.

Bauxite deposits — Surface and subterranean karst landforms are well known containers of the majority of world bauxite reserves (Valeton, 1972; Nicolas and Bildgen, 1979). Lateritic soils and terra-rossa may evolve in-place into bauxite; karstification may take place before, during and after bauxite mineral formation. Bauxites are subject to important resedimentation processes by fluvial systems during different episodes of their diagenetic evolution. Bauxites may also form in an allochthonous cover (eolian, fluvial) on karst from detrital components eroded from adjacent igneous rock bodies. Bauxites are known in the Late Proterozoic and Paleozoic, but major ores occur in Mesozoic and Cenozoic subaerial exposure facies, mainly in those related to major global sea level changes in the basal Eocene and Cretaceous.

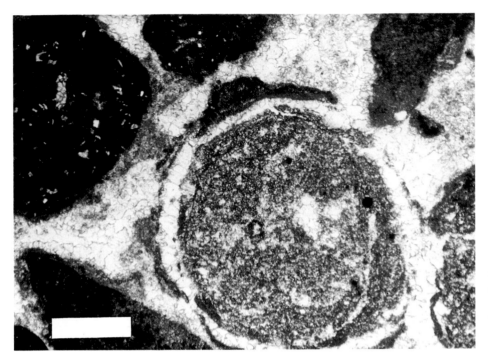

Figure 82—Circumgranular cracking. Caliche hardpan with non-laminated glaebules. Altered central grain shows partial leaching and cracking around the outer margin. Paleocene, Barcelona, Spain. Thin section, plane polarized light. Scale bar = 0.5 mm.

Uranium deposits — One of the most important types of settings for uranium accumulations are lacustrine, palustrine and caliche carbonates. In recent years, caliche profiles as uranium traps have received much attention. The mechanism of uranium accumulation implies lateral groundwater transport of uranium-rich solutions originated in extensive outcrops of granite and regolith. Precipitation to form carbonate-carnotite ore bodies occurs within the capillary fringe and the soil moisture zone. Evaporative concentrations, loss of CO_2, common-ion precipitation and oxidation are thought to be controlling factors of precipitation.

Water — Economic considerations on subaerial exposure surfaces should not neglect that many areas of the world, especially those in semiarid and arid climates, rely on karst systems for their vital water supply.

CONCLUDING REMARKS

Subaerial exposure surfaces in carbonate sequences are recognizable by a break in the record and by specific diagenetic facies superimposed upon exposed carbonates underlying these surfaces. These diagenetic facies are karst and soils. According to Jakucs (1977), karst is the formal imprint

upon the soluble parent rock of chemical and biological evolution of the overlying soil cover (including protosoils). When considering soils, emphasis here has been on caliche soils because of their high preservation potential in the geologic record.

Sediment texture of both karst and caliche facies will be governed by original substrate characteristics, intensity of subaerial diagenetic processes, and duration and direction of profile development. No generalizations can be made except perhaps that there is an overall trend toward obliteration of precursor facies and generation of fine-grained, commonly friable, low magnesian calcite products in both karst and caliche facies.

Modifications of post-Silurian subaerially exposed carbonates is controlled largely by direct or indirect biological processes which are expressed by the accumulation of biogenic products in the upper vadose zone. Before the Silurian, on the other hand, subaerial exposure facies are expected to differ from those described in this paper. Karst and caliche facies, as known today, will not be represented before the advent of substantial land plant colonization toward the end of the Silurian.

The interpretation of subaerial ex-

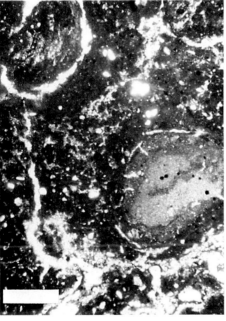

Figure 83—Circum-granular and intragranular cracking. Caliche glaebules increase their differentiation from the caliche matrix by complex patterns of cracking around and within grains. Cracking probably originated by repeated wetting and drying. White specks are quartz grains. Pleistocene, Tarragona, Spain. Thin section, plane polarized light. Scale bar = 0.5 mm.

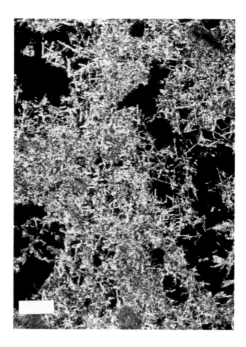

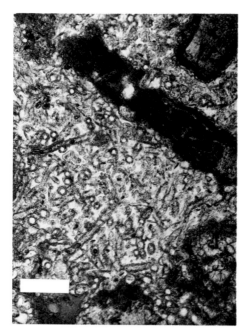

Figure 84—Random low magnesian calcite needle fibers. Random needle fibers of calcite form a loosely woven network within primary and secondary voids. Quaternary caliche hardpan. Barbados. Photograph courtesy of N. P. James. Thin section, XN. Scale bar = 100 µm.

Figure 85—Calcified filaments. Network of intertwined calcite tubes, each with a coating of micrite, possibly sloughed-off calcified root hairs. Pleistocene caliche. Barbados. Thin section, plane polarized light. Scale bar = 100 µm. (From James, 1972; his Fig. 6d).

Figure 86—Corroded quartz and floating texture. Irregular sand- and silt-sized quartz grains floating in caliche microspar. Caliche hardpan of Pleistocene age. Tarragona, Spain. Thin section, XN with gypsum plate. Scale bar = 100 µm.

posure facies presents many difficulties: (1) morphologies and products of both karst and caliche facies are not necessarily in equilibrium with the subaerial environment. Relic features are very common and may represent geologically long time spans (tens of millions of years); (2) particularly frustrating is the fact that many subaerial exposure facies undergo erosion and/or coastal to marine exposure and any evidence of subaerial weathering can be wiped out before being buried by new deposits; (3) duration of subaerial exposure cannot be measured by the thickness of the weathering profile or by the intensity of diagenetic transformation; (4) the intensity of diagenetic processes in the weathering zone is not necessarily related to proximity of the subaerial exposure surface. Both thickness of subaerial exposure facies and intensity of diagenetic processes are also controlled by hydrologic patterns, host lithology and climate; and (5) because subaerial exposure surfaces are commonly accompanied by erosion, in many ancient examples we have left only the lower part of weathering profiles. This may give the impression of incipient or poorly developed subaerial exposure diagenesis (weathering).

The main purpose of this paper has been to provide adequate criteria for

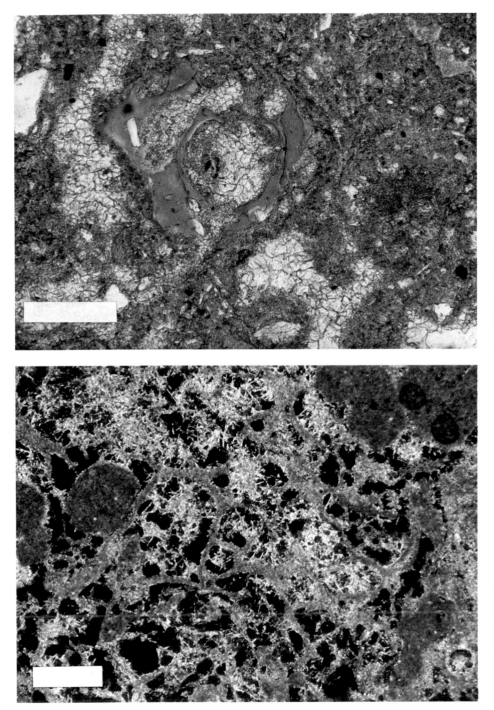

Figure 87—Corroded carbonate grain. Corroded and bored (by root hairs) miliolid foram in clotted micrite matrix with spar filled channel voids. Transitional horizon of Pleistocene caliche overlying mixed terra-rossa and eolianite also of Pleistocene age. Ibiza, Spain. Thin section, plane polarized light. Scale bar = 100 μm.

Figure 88—Rhizolith and calcified fecal pellets. Cellular pattern is calcite-encrusted plant cell walls of root. Subspherical sand-sized micritic particles are calcified fecal pellets produced by former root-feeding organisms. Pleistocene caliche hardpan. Tarragona, Spain. Thin section, XN. Scale bar = 0.5 mm.

the recognition of subaerial exposure surfaces in the geologic record. Our emphasis has been on rock facies rather than processes operating in these facies. We have offered a set of lithologic features with different diagnostic values from highly diagnostic to commonly present but non-diagnostic (Table 2). We believe that a careful use of these features, in conjunction with larger scale stratigraphic studies, will provide an adequate base for the identification of subaerial exposure facies.

As a final comment, we would like to give perspective to this contribution. This study does not pretend to be exhaustive or definitive; rather, it represents the first attempted marriage between partners hitherto thought mutually exclusive, the karst facies and the caliche facies. We have tried to confine our discussion to geologically relevant topics but, in so doing, have neglected important classical approaches of other disciplines. Despite these shortcomings, we hope this contribution will be a spring-board toward greater refinement.

Figure 89—Rhizolith (root petrifaction). Calcite-filled cells of plant root. Spar-filled tubular voids (**V**) shown in transverse section as circular structures are positions of the vascular system. Nodular horizon of Pleistocene caliche developed in Pleistocene eolianite. Ibiza, Spain. Thin section, plane polarized light. Scale bar = 125 μm.

ACKNOWLEDGMENTS

Ideas nurtured in this compilation were born in Catalunya and the Balearic Islands (eastern Spain) and reared in discussion and collaboration with L. Pomar and F. Calvet (both from the University of Barcelona) whose contributions, documented in the form of doctoral theses, sharpened our appreciation of subaerial diagenesis. During field studies, at various times and in different places, L. Pomar, F. Calvet, L. C. Pray, R. Salas and J. F. Meeder provided support, refinement and logic to our observations.

We acknowledge with thanks N. P. James, L. Pomar and F. Calvet for kindly lending photographs and R. G. C. Bathurst whose constant encouragement and enthusiasm made our struggles worthwhile and enjoyable.

Esteban acknowledges partial support from Consejo Superior de Investigaciones Cientificas (Institut Jaume Almera) and Memorial University of Newfoundland. Klappa acknowledges receipt of a Natural Environment Research Council Studentship while at the University of Liverpool (1975-1978) and a Memorial University of Newfoundland Post-Doctoral Fellowship (1978-1980).

SELECTED REFERENCES

General References on Exposure Surfaces

Allan, J. R., and R. K. Matthews, 1977, Carbon and oxygen isotopes as diagenetic and stratigraphic tools; surface and subsurface data, Barbados, West Indies: Geology, v. 5, p. 16-20.

Chafetz, H. S., 1972, Surface diagenesis of limestone: Jour. Sed. Petrology, v. 42, p. 325-329.

Goudie, A., 1973, Duricrusts in tropical and subtropical landscapes: Oxford, Clarendon, 174 p.

Jaanussan, V., 1961, Discontinuity surfaces in limestones: Geol. Inst. Univ. Uppsala Bull., v. 40, p. 221-241.

Klappa, C. F., 1979, Lichen stromatolites; criterion for subaerial exposure and a mechanism for the formation of laminar calcretes (caliche): Jour. Sed. Petrology, v. 49, p. 387-400.

_____ and N. P. James, 1979, Biologically induced diagenesis at submarine and subaerial carbonate discontinuity surfaces (abs.), *in* Recent advances in carbonate sedimentology in Canada, a C.S.P.G. symposium: Calgary, Canadian Soc. Petroleum Geols., p. 16-17.

Martin, R., 1966, Paleogeomorphology and its application to exploration for oil and gas (with examples from western Canada): AAPG Bull., v. 50, p. 2277-2311.

Multer, H. G., and J. E. Hoffmeister, 1968, Subaerial laminated crusts of the Florida Keys: Geol. Soc. America Bull., v. 79, p. 183-192.

Newell, N. D., 1967, Paraconformities, *in* C. Teichert and E. L. Yochelson, eds., Essays in paleontology and stratigraphy: Lawrence, Kansas, Univ. Kansas Press, p. 349-368.

Pomar, L., et al, 1975, Accion de liquenes, algas y hongos en la telodiagenesis de las rocas carbonatadas de la zona litoral prelitoral Catalana: Univ. Barcelona, Inst. Inv. Geol., v. 30, p. 83-117.

Rose, P. R., 1970, Stratigraphic interpretation of submarine versus subaerial discontinuity surfaces; an example from the Cretaceous of Texas: Geol. Soc. America Bull., v. 81, p. 2787-2798.

Schneider, J., 1976, Biological and inorganic factors in the destruction of limestone coasts: Contr. Sedimentology, 6, 112 p.

Semenuik, V., 1971, Subaerial leaching in the limestones of the Bowan Park Group (Ordovician) of central western New South Wales: Jour. Sed. Petrology, v. 41, p. 939-950.

Videtich, P. E., and R. K. Matthews, 1980, Origin of discontinuity surfaces in limestones; isotopic and petrographic data, Pleistocene of Barbados, West Indies: Jour. Sed. Petrology, v. 50, p. 971-980.

TABLE 2

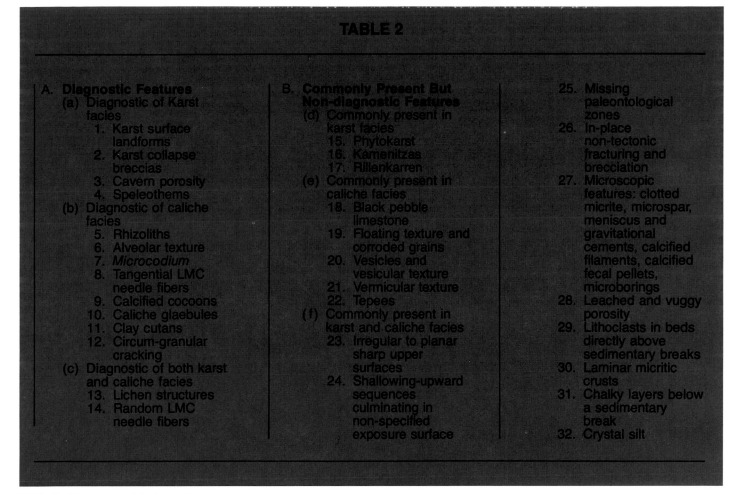

A. Diagnostic Features	B. Commonly Present But Non-diagnostic Features	
(a) Diagnostic of Karst facies 　　1. Karst surface landforms 　　2. Karst collapse breccias 　　3. Cavern porosity 　　4. Speleothems (b) Diagnostic of caliche facies 　　5. Rhizoliths 　　6. Alveolar texture 　　7. *Microcodium* 　　8. Tangential LMC needle fibers 　　9. Calcified cocoons 　10. Caliche glaebules 　11. Clay cutans 　12. Circum-granular cracking (c) Diagnostic of both karst and caliche facies 　13. Lichen structures 　14. Random LMC needle fibers	(d) Commonly present in karst facies 　15. Phytokarst 　16. Kamenitzas 　17. Rillenkarren (e) Commonly present in caliche facies 　18. Black pebble limestone 　19. Floating texture and corroded grains 　20. Vesicles and vesicular texture 　21. Vermicular texture 　22. Tepees (f) Commonly present in karst and caliche facies 　23. Irregular to planar sharp upper surfaces 　24. Shallowing-upward sequences culminating in non-specified exposure surface	25. Missing paleontological zones 26. In-place non-tectonic fracturing and brecciation 27. Microscopic features: clotted micrite, microspar, meniscus and gravitational cements, calcified filaments, calcified fecal pellets, microborings 28. Leached and vuggy porosity 29. Lithoclasts in beds directly above sedimentary breaks 30. Laminar micritic crusts 31. Chalky layers below a sedimentary break 32. Crystal silt

Table 2. Summary of features in subaerial exposure facies.

Karst Facies

Bogli, A., 1980, Karst hydrology and physical speleology: Berlin, Springer-Verlag Pub., 284 p.

Donahue, J. D., 1969, Genesis of oolite and pisolite grains; an energy index: Jour. Sed. Petrology, v. 39, p. 1399-1411.

Folk, R. L., and R. Assereto, 1976, Comparative fabrics of length-slow and length-fast calcite and calcitized aragonite in a Holocene speleothem, Carlsbad Caverns, New Mexico: Jour. Sed. Petrology, v. 46, p. 486-496.

_____ H. H. Roberts, and C. H. Moore, 1973, Black phytokarst from Hell, Cayman Islands: Geol. Soc. America Bull., v. 87, p. 2351-2360.

Jakucs, L., 1977, Morphogenetics of karst regions: New York, John Wiley and Sons, 284 p.

Kendall, A. C., and P. L. Broughton, 1978, Origin of fabrics in speleothems composed of columnar calcite crystals:

Jour. Sed. Petrology, v. 48, p. 519-538.

Legrand, H. E., and V. T. Stringfield, 1973, Karst hydrology — a review: Jour. Hydrology, v. 20, p. 97-120.

Purdy, E. G., 1974, Reef configurations; cause and effect, *in* L. F. Laporte, ed., Reefs in time and space: SEPM Spec. Pub. 18, p. 9-76.

Quinlan, J. F., 1972, Karst-related mineral deposits and possible criteria for the recognition of paleokarsts; a review of preservable characteristics of Holocene and older karst terranes: Montreal, 24th Internat. Geol. Cong., Sec. 6, p. 156-168.

Sweeting, M. M., 1973, Karst landforms: London, Macmillan Pub., 362 p.

Thrailkill, J., 1968, Chemical and hydrologic factors in the excavation of limestone caves: Geol. Soc. America Bull., v. 79, p. 19-46.

_____, 1976, Speleothems, *in* M. R. Walker, ed., Stromatolites, developments in sedimentology 20:

Amsterdam, Elsevier Sci. Pub., p. 73-86.

Caliche Facies and Soils in General

Aristarain, L. F., 1970, Chemical analysis of caliche profiles from the High Plains, New Mexico: Jour. Geology, v. 78, p. 201-212.

Bal, L., 1975, Carbonate in soil; a theoretical consideration on, and proposal for its fabric analysis. I. crystic, calcic and fibrous plasmic fabric: Netherlands Jour. Agr. Sci., v. 23, p. 18-35.

Blank, H. R., and E. W. Tynes, 1965, Formation of caliche in situ: Geol. Soc. America Bull., v. 76, p. 1387-1392.

Braithwaite, C. J. R., 1975, Petrology of palaeosols and other terrestrial sediments on Aldabra, western Indian Ocean: Philos. Trans. Royal Soc. London, v. 273, p. 1-32.

Bretz, J. H., and L. Horberg, 1949, Caliche in southeastern New Mexico: Jour. Geology, v. 57, p. 491-511.

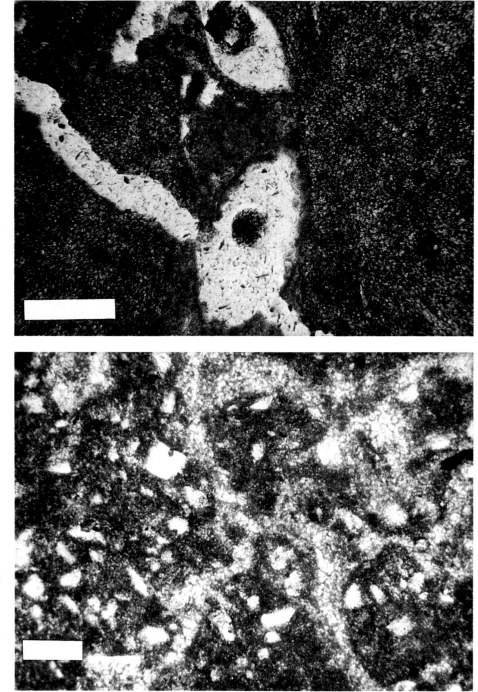

Figure 90—Root molds. Oblique section through root molds within now cemented chalky (microspar or silt-sized carbonate) horizon of Pleistocene caliche. Open fracture running NW-SE post-dates transected root mold. Ibiza, Spain. Thin section, plane polarized light. Scale bar = 100 μm.

Figure 91—Incipient glaebules and channels. Matrix in caliche hardpan with abundant silt-sized fragments of skeletal grains and detrital quartz, showing incipient differentiation into micrite glaebules by microspar channels. Pleistocene, Tarragona, Spain. Thin section, plane polarized light. Scale bar = 0.1 mm.

Brewer, R., 1964, Fabric and mineral analysis of soils: New York, John Wiley and Sons, 470 p.

Brown, C. N., 1956, The origin of caliche on the northeastern Llano Estacado, Texas: Jour. Geology, v. 64, p. 433-457.

Calvet, F., L. Pomar, and M. Esteban, 1975, Las rizocreciones del Pleistoceno de Mallorca: Univ. Barcelona, Inst. Invest. Geol., v. 30, p. 35-60.

Durand, J. H., 1963, Les croutes calcaires et gypseuses en Algerie; formation et age: Soc. Geol. France Bull., v. 7, p. 959-968.

Esteban, M., 1974, Caliche textures and *Microcodium*: Soc. Geol. Italiana Bull. (supp.), v. 92, p. 105-125.

_____, 1976, Vadose pisolite and caliche: AAPG Bull., v. 60, p. 2048-2057.

Gile, L. H., F. F. Peterson, and R. B. Grossman, 1966, Morphological and genetic sequence of carbonate accumulation in desert soils: Soil Sci., v. 101, p. 347-360.

Harrison, R. S., 1977, Caliche profiles, indicators of near-surface subaerial diagenesis, Barbados, West Indies: Canadian Petroleum Geol. Bull., v. 25, p. 123-173.

James, N. P., 1972, Holocene and Pleistocene calcareous crust (caliche) profiles; criteria for subaerial exposure:

Figure 92—Plant tufa (Travertino). Ornamental facing stone of One Shell Plaza Building, Houston. Note porous sponge texture. Origin and age of facing stone unknown.

Jour. Sed. Petrology, v. 42, p. 817-836.

Klappa, C. F., 1978, Biolithogenesis of Microcodium; elucidation: Sedimentology, v. 25, p. 489-522.

———, 1979, Calcified filaments in Quaternary calcretes; organo-mineral interactions in the subaerial vadose environment: Jour. Sed. Petrology, v. 49, p. 955-968.

———, 1980a, Brecciation textures and tepee structures in Quaternary calcrete (caliche) profiles from eastern Spain; the plant factor in their formation: Geol. Jour., v. 15, p. 81-89.

———, 1980b, Rhizoliths in terrestrial carbonates; classification, recognition, genesis and significance: Sedimentology, v. 27, p. 613-629.

———, in press, A process-response model for the formation of pedogenic calcretes: Jour. Geol. Soc. London.

Knox, G. J., 1977, Caliche profile formation, Salkhana Bay (South Africa): Sedimentology, v. 24, p. 657-674.

Krumbein, W. E., 1968, Geomicrobiology and geochemistry of the "Nari-Lime-Crust" (Israel), in G. Muller and G. M. Friedman, eds., Recent developments in carbonate sedimentology in central Europe: Berlin, Springer-Verlag Pub., p. 138-147.

——— and C. Giele, 1979, Calcification in a coccoid cyanobacterium associated with the formation of desert stromatolites: Sedimentology, v. 26, p. 593-604.

Perkins, R. D., 1977, Depositional frame-work of Pleistocene rocks in South Florida: Geol. Soc. America Mem. 147, p. 131-198.

Price, W. A., 1925, Caliche and pseudo-anticlines: AAPG Bull., v. 9, p. 1009-1017.

Read, J. F., 1974, Calcrete deposits and Quaternary sediments, Edel Province, Shark Bay, Western Australia, in Evolution and diagenesis of Quaternary carbonate sequences, Shark Bay, Western Australia: AAPG Mem. 22, p. 250-280.

———, 1976, Calcretes and their distinction from stromatolites, in M. R. Walter, ed., Stromatolites, developments in sedimentology 20: Amsterdam, Elsevier Sci. Pub., p. 55-71.

Reeves, C. C., 1970, Origin, classification and geologic history of caliche on the southern High Plains, Texas and eastern New Mexico: Jour. Geology, v. 78, p. 352-362.

———, 1976, Caliche; origin, classification, morphology and uses: Lubbock, Texas, Estacado Brooks, 233 p.

Ruellan, A., 1967, Individualisation et accumulation du calcaire dans les sols et les depots quaternaires du Maroc: Cah. Off. Rech. Sci. Tech., Outre-Mer, Ser. Pedol., v. 5, p. 421-460.

Swineford, A., A. B. Leonard, and J. C. Frye, 1958, Petrology of the Pliocene pisolitic limestone in the Great Plains: Kansas Geol. Survey Bull., v. 130, p. 97-116.

Teruggi, M. E., and R. R. Andreiss, 1971, Micromorphological recognition of paleosolic features in sediments and sedimentary rocks, in D. H. Yaalon, ed., Paleopedology; origin, nature and dating of paleosols: Jerusalem, Int. Soc. Soil Sci. and Israel Univ. Press, p. 161-172.

Ward, W. C., 1975, Petrology and diagenesis of carbonate eolianites of northeastern Yucatan Peninsula, Mexico, in K. F. Wantland and W. C. Pusey, eds., Belize shelf-carbonate sediments, clastic sediments, and ecology: AAPG Stud. in Geology No. 2, p. 500-571.

——— R. L. Folk, and J. L. Wilson, 1970, Blackening of eolianite and caliche adjacent to saline lakes, Isla Mujeres, Quintana Roo, Mexico: Jour. Sed. Petrology, v. 40, p. 548-555.

Watts, N. L., 1977, Pseudo-anticlines and other structures in some calcretes of Botswana and South Africa: Earth Surface Processes, v. 2, p. 63-74.

———, 1978, Displacive calcite; evidence from recent and ancient calcretes: Geology, v. 6, p. 699-703.

———, 1980, Quaternary pedogenic calcretes from the Kalahari (southern Africa); mineralogy, genesis and diagenesis: Sedimentology, v. 27, p. 661-686.

Yaalon, D. H., ed., 1971, Paleopedology; origin, nature and dating of paleosols: Jerusalem, Int. Soc. Soil Sci. and Israel Univ. Press, 350 p.

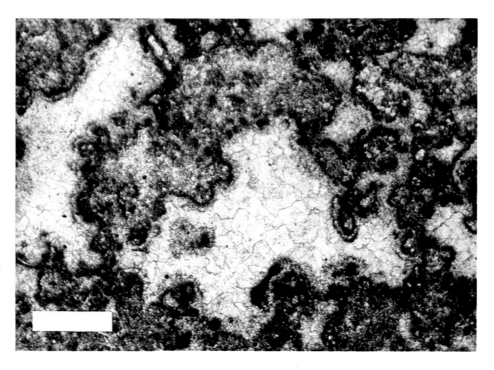

Figure 93—Plant tufa. Section of micrite peloids and tubular coatings with abundant blocky calcite cementation. Located above caliche hardpan. Paleocene, Barcelona, Spain. Thin section, plane polarized light. Scale bar = 0.5 mm.

Examples of Possible
Fossil Karst Facies

Bignot, G., 1974, Le paleokarst eocene d'lstrie (Italie et Yougoslavie) et son influence sur la sedimentation ancienne: Memoires et Documents, 1974 nouvelle scrie, v. 15, Phenomenes Karstiques, tome II, p. 177-185.

Carannante, G., V. Ferreri, and L. Simone, 1974, Le cavita paleocarsiche cretaciche di Dragoni (Campania): Boll. Soc. Nat. Napoli, v. 83, p. 1-11.

D'Argenio, B., 1967, Geologia del gruppo Taburno-Camposauro (Appennino Campano): Soc. Naz. Sci. Fis. Mat., (3), v. 6, p. 1-218.

Gilewska, S., 1964, Fossil karst in Poland: Erdkunde, v. 18, p. 124-135.

Jordan, G. F., 1954, Large sinkholes in Straits of Florida: AAPG Bull., v. 38, p. 1810-1817.

Kobluk, D. R., et al, 1977, The Silurian-Devonian disconformity in southern Ontario: Canadian Petroleum Geol. Bull., v. 25, p. 1157-1186.

Malloy, R. J., and R. J. Hurley, 1970, Geomorphology and geologic structure; Straits of Florida: Geol. Soc. America Bull., v. 81, p. 1947-1972.

Mullins, H. T., and A. C. Neumann, 1979, Geology of the Miami Terrace and its paleo-oceanographic implications: Marine Geology, v. 30, p. 205-232.

Pirlet, H., 1970, L'influence d'un karst sous-jacent sur la sedimentation calcaire et l'interet de l'etude des paleokarsts:

Ann. Soc. Geol. Belg., v. 93, p. 247-254.

Read, J. F., and G. A. Grover, Jr., 1977, Scalloped and planar erosion surfaces, Middle Ordovician limestone, Virginia; analogues of Holocene exposed karst or tidal rock platforms: Jour. Sed. Petrology, v. 47, p. 956-972.

Somerville, I. D., 1979, A cyclicity in the early Brigantian (D2) limestones east of the Clwydian Range, North Wales and its use in correlation: Geol. Jour., v. 14, p. 69-86.

Van der Lingen, G. J., D. Smale, and D. W. Lewis, 1978, Alteration of a pelagic chalk below a paleokarst surface, Oxford, South Island, New Zealand: Sed. Geol., v. 21, p. 45-66.

Walken, G. M., 1974, Palaeokarstic surfaces in Upper Visean (Carboniferous) limestones of the Derbyshire Block, England: Jour. Sed. Petrology, v. 44, p. 1232-1247.

Examples of Possible Fossil
Caliche Facies

Allen, J. R. L., 1974, Sedimentology of the Old Red Sandstone (Siluro-Devonian) in the Clee Hills area, Shropshire, England: Sed. Geology, v. 12, p. 73-167.

Burgess, I. C., 1961, Fossil soils of the upper Old Red Sandstone of south Ayrshire: Trans. Geol. Soc. Glasgow, v. 24, p. 138-153.

Cocozza, T., and A. Gandin, 1976, Eta' e significato ambientale delle facies

detritico-carbonatiche dell'altopiano di Campumari (Sardegna sud-occidentale): Soc. Geol. Italia Boll., v. 95, p. 1521-1540.

Folk, R. L., and E. F. McBride, 1976, Possible pedogenic origin of Ligurian ophicalcite; a Mesozoic calichified serpentinite: Geology, v. 4, p. 327-332.

Harrison, R. S., and R. P. Steinen, 1978, Subaerial crusts, caliche profiles, and breccia horizons; comparison of some Holocene and Mississippian exposure surfaces, Barbados and Kentucky: Geol. Soc. America Bull., v. 89, p. 385-396.

Hubert, J. F., 1978, Paleosol caliche in the New Haven Arkose, Newark Group, Connecticut: Palaeogeography, Palaeoclimatology, Palaeoecology, v. 24, p. 151-168.

Marzo, M., M. Esteban, and L. Pomar, 1974, Presencia de caliche fosil en el Buntsandstein del valle del Congost (provincia de Barcelona): Acta Geol. Hispanica, v. 9, p. 33-36.

Nagtegaal, P. J. C., 1969, Microtexture in recent and fossil caliche: Leidse Geol. Meded., v. 42, p. 131-142.

Roper, H. -P., and P. Rothe, 1975, Petrology of a fossil duricrust; the "Kerneoldolomit-Horizont", Permian, S. W. Germany (abs.): Nice, France, 9th Int. Sediment Cong., v. 2, p. 10.

Steel, R. J., 1974, Cornstone (fossil caliche) — its origin, stratigraphic, and sedimentological importance in the New Red Sandstone, western Scotland: Jour.

Figure 94—Pelagosite (coniatolite). Aragonite laminated encrustation on Jurassic carbonates. Encrustations in this locality are up to 1 cm (0.4 in) thick and occur in salt spray and splash zones of the Tarragona coastline, Spain.

Geology, v. 82, p. 351-369.

Walls, R. A., W. B. Harris, and W. E. Nunan, 1975, Calcareous crust (caliche) profiles and early subaerial exposure of Carboniferous carbonates, northeastern Kentucky: Sedimentology, v. 22, p. 417-440.

Winchester, P. O., 1972, Caliche-like limestones in the Lower Permian Laborcita Formation of the Sacramento Mountains, New Mexico (abs.): Geol. Soc. America Abs. with Programs, v. 4, p. 707.

Economic Considerations and Case Histories

Bernard, A. J., 1976, Metallogenic processes of intra-karstic sedimentation, *in* G. C. Amstutz and A. J. Bernard, eds., Ores in sediments: Berlin, Springer-Verlag Pub., p. 43-57.

Brown, J. S., 1970, Mississippi valley-type lead-zinc ores: Mineral. Deposita, v. 5, p. 103-119.

Carlisle, D., 1978, Characteristics and origins of uranium-bearing calcretes in Western Australia and South West Africa (abs.): Jerusalem, Israel, 10th Int. Sediment. Cong., v. 1, p. 119.

Chenoweth, P. A., 1972, Unconformity traps, *in* R. E. King, ed., Stratigraphic oil and gas fields — classification, exploration methods, and case histories: AAPG Mem. 16, p. 42-46.

Collins, J. A., and L. Smith, 1975, Zinc deposits related to diagenesis and intrakarstic sedimentation in the Lower

Ordovician St. George Formation, western Newfoundland: Canadian Petroleum Geol. Bull., v. 23, p. 393-427.

Enos, P., 1977, Tamabra limestone of the Poza Rica trend, Cretaceous, Mexico, *in* H. E. Cook and P. Enos, eds., Deep-water carbonate environments: SEPM Spec. Pub. 25, p. 273-314.

Geldsetzer, H., 1976, Syngenetic dolomitization and sulphide mineralization, *in* G. C. Amstutz and A. J. Bernard, eds., Ores in sediments: Berlin, Springer-Verlag Pub., p. 115-127.

Nicolas, J., and P. Bildgen, 1979, Relations between the location of the karst bauxites in the northern Hemisphere, the global tectonics and the climatic variations during geological time: Palaeogeography, Palaeoclimatology, Palaeoecology, v. 28, p. 205-239.

Padalino, G., et al, 1976, Ore deposition in karst formations with examples from Sardinia, *in* G. C. Amstutz and A. J. Bernard, eds., Ores in sediments: Berlin, Springer-Verlag Pub., p. 209-220.

Valeton, I., 1972, Bauxites — developments in soil science I: Amsterdam, Elsevier Sci. Pub., 226 p.

Walters, R. F., 1946, Buried pre-Cambrian hills in northeastern Barton County, central Kansas: AAPG Bull., v. 30, p. 660-710.

Other References Cited in Text

Assereto, R. L. A. M., and C. G. St. C. Kendall, 1977, Nature, origin and classification of peritidal tepee structures and related breccias: Sedimentology, v. 24, p. 153-210.

Bathurst, R. G. C., 1975, Carbonate sediments and their diagenesis; developments in sedimentology 12: Amsterdam, Elsevier Sci. Pub., 658 p.

Choquette, P. W., and L. C. Pray, 1970, Geologic nomenclature and classification of porosity in sedimentary carbonates: AAPG Bull., v. 54, p. 207-250.

Dunham, R. J., 1969a, Early vadose silt in Townsend mound (reef), New Mexico, *in* G. M. Friedman, ed., Depositional environments in carbonate rocks; a symposium: SEPM Spec. Pub. 14, p. 139-181.

————, 1969b, Vadose pisolite in the the Capitan Reef (Permian), New Mexico and Texas, *in* G. M. Friedman, ed., Depositional environments in carbonate rocks; a symposium: SEPM Spec. Pub. 14, p. 182-191.

Freytet, P., 1973, Petrography and paleoenvironment of continental carbonate deposits with particular reference to the Upper Cretaceous and Lower Eocene of Languedoc (southern France): Sed. Geol., v. 10, p. 25-60.

Lindstrom, M., 1963, Sedimentary folds and the development of limestone in an Early Ordovician sea: Sedimentology, v. 2, p. 243-275.

Purser, B. H., and J. P. Loreau, 1973, Aragonitic, supratidal encrustation on the Trucial Coast, Persian Gulf, *in* B. H. Purser, ed., The Persian Gulf: New York, Springer-Verlag Pub., p. 343-376.

Schlager, W., and N. P. James, 1978, Low-magnesian calcite limestones forming at the deep-sea floor, Tongue of the Ocean, Bahamas: Sedimentology, v. 25, p. 675-702.

Scholle, P. A., and D. J. J. Kinsman, 1974, Aragonitic and high-magnesian calcite caliche from the Persian Gulf — a modern analog for the Permian of Texas and New Mexico: Jour. Sed. Petrology, v. 44, p. 904-916.

Shinn, E. A., 1969, Submarine lithification of Holocene carbonate sediments in the Persian Gulf: Sedimentology, v. 12, p. 109-144.

Sloss, L. L., 1979, Global sea level change; a view from the craton, *in* J. S. Watkins, L. Montadert, and P. W. Dickerson, eds., Geological and geophysical investigations of continental margins: AAPG Mem. 29, p. 461-467.

Steinen, R. P., 1974, Phreatic and vadose diagenetic modification of Pleistocene limestone; petrographic observations from subsurface of Barbados, West Indies: AAPG Bull., v. 58, p. 1008-1024.

Vail, P. R., R. M. Mitchum, Jr., and S. Thompson, III, 1977, Global cycles of relative changes of sea level, *in* C. E. Payton, ed., Seismic stratigraphy — applications to hydrocarbon exploration: AAPG Mem. 26, p. 49-212.

Lead-Zinc Deposits of Bleiberg-Kreuth

Thilo Bechstadt
Barbara Dohler-Hirner

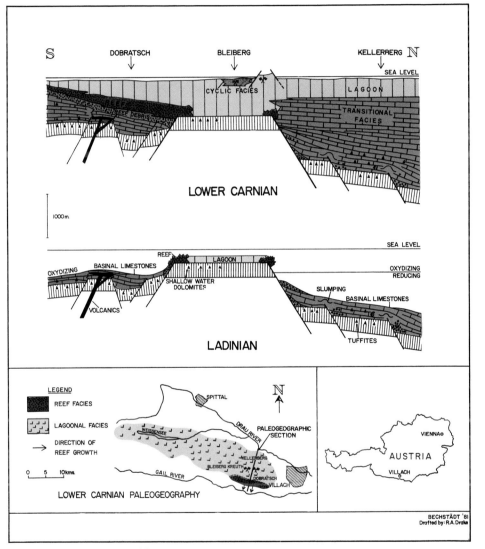

Figure 1—Paleogeographic cross sections and map of the Bleiberg-Kreuth area.

*L*ead and zinc ores of the Alpine Triassic occur mainly within shallow water sediments of lower Carnian age; these sequences underwent emersion periods and were affected by weathering and meteoric karstification (Lagny, 1975; Bechstadt, 1975a, b, 1979; Assereto et al, 1976).

The Bleiberg-Kreuth mineralization within the Wetterstein limestone is controlled by at least four geologic factors: (1) it occurs on lagoonal platforms, situated some distance from the mainland to the north, which might have been the original source of the metal; (2) on these platforms it is localized within areas where an extensive cavity network had been formed by karstic weathering; (3) the mineralization is bound to areas close to or with peritidal cyclic and evaporitic sedimentation ("special facies" of Schneider, 1964); and (4) it occurs below sealing shales (Raibl beds).

Only the subaerial exposure facies is described in more detail, whereas the conflicting theories concerning lead-zinc mineralization are mentioned only briefly (Figs. 1, 2).

Sediment Types

The different facies types can be partly related to water depths in peritidal areas. The first order sediment type is an often mud-supported, yellowish, Megalodont-bearing limestone. These lamellibranchs, together with locally abundant dasyclad algae show the original subtidal deposition of the rock. Solution fabrics, together with vadose cements, however, indicate postsedimentary changes in water depth.

Arenitic-ruditic limestone, often with small laminated fenestrae, con-taining abundant tepee-structures and tepee-breccias (Assereto and Kendall, 1977) are placed in the intertidal environment.

Milky-white dolomitic limestones resp. Dolomites with frequent laminated fenestrae, prism and sheet cracks, flat pebble conglomerates and various vadose cements (Bechstadt, 1974) hint to environments near and above the mean sea level (inter- to supratidal).

Black breccias and greenish marly carbonates (often called "green marls") are interpreted as subaerial exposure facies (Fig. 3). The black to brown breccia components can either be found within the green matrix or within brownish mudstones with fenestral fabrics (inter- to supratidal), the fenestral fabrics usually on top of the mudstones. The flabby figure and small cavities within the components point to carbonate solution which took place under a meteoric regime, as indicated by the vadose cements within the cavities. The black color is caused by pyrite and bitumen impregnation of the components. They are discolored normal Wetterstein limestones of mainly the subtidal type (Fig. 4). Associated inter- to supratidal "milky" components are not discolored, possibly due to early lithification. Because of the common fragile appenditures of the black breccia components they could not have been transported a greater distance in their present condition.

The green marly layers frequently show erosional features at their base, cut and fill structures, or are restricted to cavities and fissures or dikes within the underlying, mostly subtidal rock. Carbonate solution is indicated by the irregular size of the cavities, up to half a meter or even more in size. Megalodont-shells were dissolved as well, the former shells partly filled by the green matrix.

Statistical investigations (Bechstadt, 1975b) of the transition frequencies between the a.m. sediment types show a first order association of the green layers with inter- to supratidal sediments (with breccia components) on their top, followed frequently by thicker inter- resp. supratidal rocks. Because subtidal rocks are the dominant facies in the Bleiberg cyclothems, a tuffitic origin (assumed by some authors) is unlikely as no coincidence between inter- to supratidal conditions and volcanism can be expected.

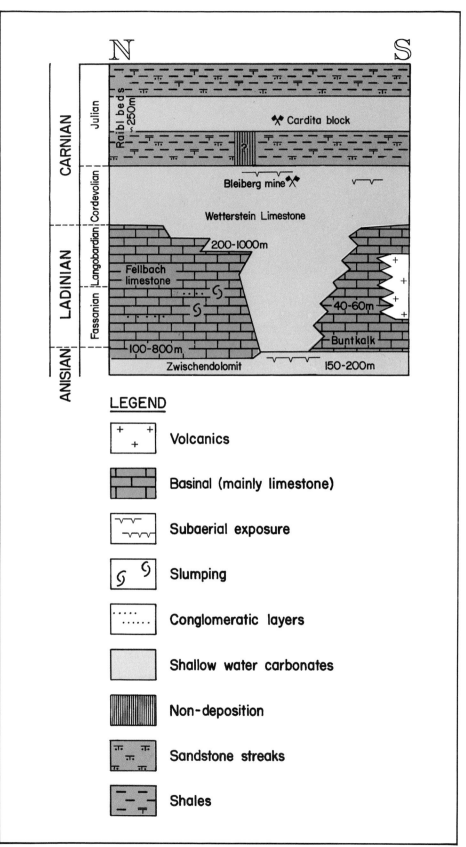

LEGEND

+ + +	Volcanics
	Basinal (mainly limestone)
	Subaerial exposure
ʂ ʂ	Slumping
	Conglomeratic layers
	Shallow water carbonates
	Non-deposition
	Sandstone streaks
	Shales

Figure 2—North-south stratigraphic section in the Bleiberg-Kreuth area (location is the same as the cross section in Fig. 1).

Figure 3—Green "marls" containing a dolomitic (but not discolored) component, overlain by inter-supratidal, milky dolomite with laminated fenestrae.

Figure 4—Green sediment with black breccia components. Note their irregular components.

The same association of sediment types is known from other stratigraphic levels inside and outside the Alps—for example, the Lofer cyclothems of the Norian Dachstein limestone. Cyclic emersions of lagoonal platforms are assumed here as well (see Fischer, 1964; Zankl, 1971). Barthel and Seyfried (in press) describe other occurrences from the stratigraphic record as well as recent black breccias (Figs. 5, 6).

Early Diagenesis

Meniscus cement and vadose pisolites are indicative of a vadose influence. Dripstone cement, indicative of inter- to supratidal resp. vadose environments can be found within cavities of subtidal rocks (Fig. 7).

This indicates a relative sea-level fall shortly after sedimentation. This type of cement is also common within the black breccia-components and the inter- to supratidal rock types themselves.

All of these cements are frequently dolomitized, partly together with internal sediments. This dolomitization of the former aragonite is thought to

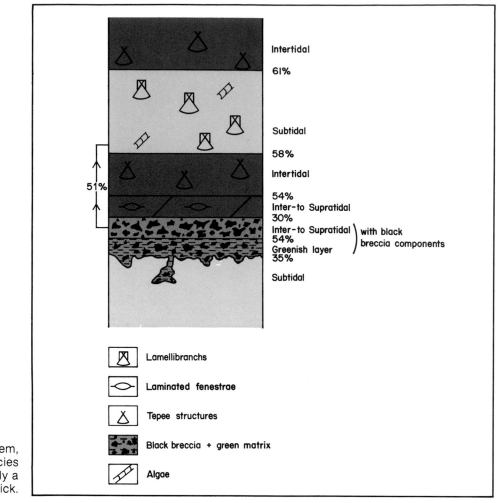

Figure 5—Idealized Bleiberg cyclothem, established from transition frequencies of facies types. One cycle is usually a few meters thick.

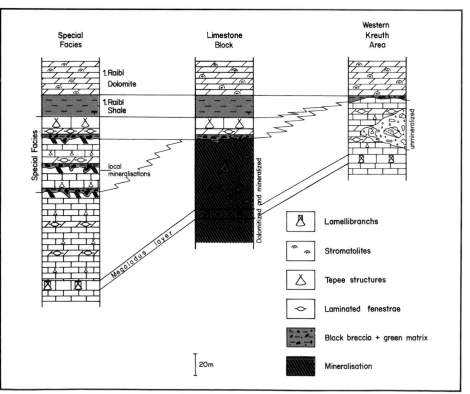

Figure 6—Schematic profiles within the Bleiberg-Kreuth mine. Only a few of the (many) cycles within the "special facies" are drafted.

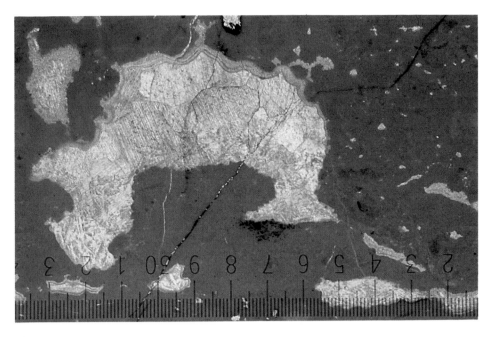

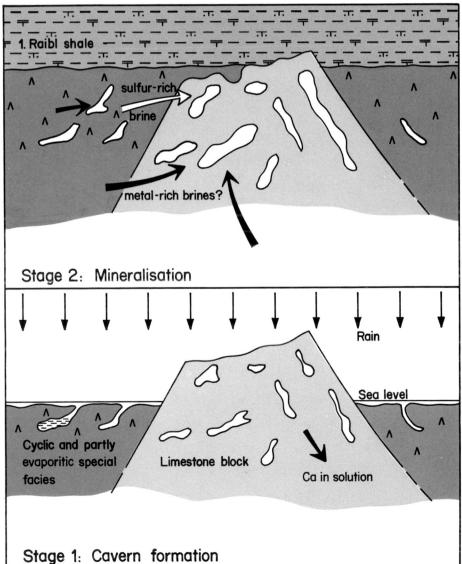

1. Raibl shale

sulfur-rich

brine

metal-rich brines?

Stage 2: Mineralisation

Rain

Sea level

Cyclic and partly
evaporitic special
facies

Limestone block

Ca in solution

Stage 1: Cavern formation

Figure 7—Dripstone cement on top of a cavity within inter-supratidal dolomite.

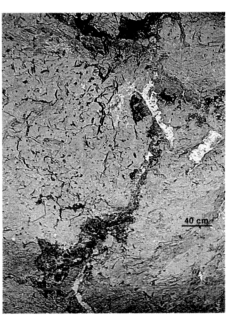

40 cm

Figure 8—Fissure within subtidal rock, filled with green sediment. Small cavities at the left of the big fissure are partly burrows and partly dissolved megalodont shells filled with green ''marls.''

Figure 9—Hypothetical model of events leading to mineralization. Stage 1 is of lower Carnian age, whereas Stage 2 is probably post-Carnian.

TABLE 1		1	2	3	4
Na$_2$O	(%)	0.35	0.33	0.23	0.41
MgO	(%)	1.09	1.82	6.52	9.6
Al$_2$O$_3$	(%)	1.02	1.01	1.06	2.4
SiO$_2$	(%)	1.35	1.42	1.19	4.93
P$_2$O$_5$	(%)	0.11	0.09	0.07	0.12
K$_2$O	(%)	0.21	0.23	0.18	0.64
CaO	(%)	53.31	51.06	46.15	40.37
TiO$_2$	(%)	0.06	0.04	0.02	0.14
MnO	(%)	0.13	0.13	0.09	0.11
Fe$_2$O$_3$	(%)	0.21	0.21	0.25	0.30
number (n)		58	39	9	8
V	(ppm)	9.51	12.64	22.14	31.7
Rb	(ppm)	16.00	18.02	18.78	36.2
Zr	(ppm)	3.01	5.03	12.57	18.0
Ba	(ppm)	63.49	55.64	73.45	111.1
Zn	(ppm)	88.96	72.55	30.83	129.25
Pb	(ppm)	10.14	10.84	9.83	10.01
Cr	(ppm)	14.44	14.22	14.35	30.13
Cu	(ppm)	6.02	7.73	5.23	8.58
Co	(ppm)	10.58	11.23	10.97	10.6
Ni	(ppm)	2.60	2.43	0.66	15.8
Sr	(ppm)	137.47	142.43	220.92	179.0
number (n)		55	47	10	10

Table 1. Mean content of selected elements within the subtidal (1), intertidal (2), inter- to supratidal (3), and subaerial exposure facies (marly carbonates) (4).

Figure 10—Small cavities below layer of green cement, seen at the upper right of the picture. The cavities are partly sediment-filled and mostly mineralized. Arrows of Figures 10 and 11 point to the same cavity.

be indicative of a relative sea-level fall as well. A recent analog was described from Quaternary limestones of Fuerteventura, Canary Islands (Muller and Tietz, 1971).

Megalodont shells are commonly dissolved. Geopetal green sediment, radiaxial fibrous calcite, and drusy sparite (but no vadose cements) can be observed within the former shell. The country rock must have been cemented at the time of the shell dissolution because there is almost no alteration of the former shell outline (Fig. 8).

Geochemistry

The geochemical data are based on

Figure 11—Detail of Figure 10. Arrow is the same one as in Figure 10, and points to a sediment-filled cavity, and then to a small sphalerite mineralization at the top. The bigger cavity shows a first generation of mostly chemically precipitated sphalerite, followed by galena and cement as second and third generations.

TABLE 2										
Element	Ca	Na	P	Ti	Zn	Ni	Mg	Sr	V	Zr
Subtidal	+	+	+	+	+	+	−	−	−	−
Inter- to Supratidal	−	−	−	−	−	−	+	+	+	+

two sections within the Bleiberg mine. Zinc, lead, copper, chromium, cobalt and nickel were measured by AAS, while the other elements of Table 1 were measured by x-ray fluorescence (Dohler-Hirner, in prep.). A comparison of the subtidal type with the inter- to supratidal type shows positive/negative correlations of some elements, with the intertidal values placed in between (Table 2).

Most of these results are confirmed by investigations of four other sections within Drau Range and the Northern Limestone Alps. The high strontium values of the supratidal type are unique to the Bleiberg section.

After the results of Ferguson and others (1975), zinc-enrichments should be expected at first sight within the supratidal type. The positive correlation with the subtidal sediments might be due to diagenetic

mobilizations, responsible for the zinc sulfides. However, extreme zinc anomalies, not seen in the averages of Table 1, are usually found within either the inter- to supratidal type or the green layers.

The green, marly carbonates show high values of most elements, being partly constituent of or fixed to clay minerals. These clays consist (nearly entirely) of illite (Krumm, personal commun., 1980). Only one sample contained traces of montmorillonite.

Depositional Environment

A restricted, partly evaporitic environment is indicated by the local occurrence of anhydrite and the limited fossil content. The dasyclad algae *Poikiloporella duplicata* and *Clypeina besici* are indicative of protected lagoonal environments some distance from a central reef (Ott, 1967, 1972).

The black breccia components and the cavity filling green layers, interpreted as reworked paleosoils, are

Figure 12—Cavity, interpreted as karstic, within "Limestone block" of Kreuth area. The cavity is filled with dolomitic components of surrounding Wetterstein Formation, embedded in a matrix of mostly sphalerite. Arrows of Figures 12 and 13 point to the same component.

related to oscillations in sea level. Within such peritidal lagoonal areas, sea-level falls of a small scale will result in emersion areas. The eustatic control of the cycles is favored because of the excellent correlation of the sections within the Bleiberg mine in an east-west distance of at least 7 km (43 mi). However, the facies to the north and south differ.

Most of the beds deposited immediately before the subaerial exposure (especially inter- to supratidal beds) were stripped away, reworked, and partly dissolved. Meteoric water filled small depressions and caused vadose diagenesis. The dolomitization of the vadose cements is thought to have taken place later, when the sea level started to rise again (evaporative pumping model). This was also the time of anhydrite formation.

H_2S and organic-rich layers were formed in small pools below the surface. They impregnated many of the breccia components, mainly along small fissures, causing a black color.

The result of a further sea-level rise is a relatively uniform distribution of the former paleosoils. The green matrix often fills only cavities and small dikes within the underlying rock, whereas the black breccia components can also be found several centimeters above, within inter- to

supratidal sediments. This redeposition of the former paleosoils is thought to be the explanation for their intercalation within the subtidal rock.

Other parts of the mine underwent longer periods of erosion, karstification and dolomitization, due to tectonic events. Karst cavities in these areas can be highly mineralized if they are covered by sealing shales. Other areas underwent karstification even during the following Raibl time. In the westernmost part of the mine an area of unmineralized dikes within the Wetterstein limestone is partly filled with Raibl marls and dark Raibl dolomites.

Mineralization

Locally present depositional fabrics within the lead-zinc ores, originally thought to represent external sedimentary deposition (Schneider, 1964; Schulz, 1964) are now explained as internal cavity-fillings. Dzulynski and Sass-Gustkiewicz (1977, 1980) believe in ore-bearing solutions causing the synchronous formation of (hydrothermal) karst structures and the emplacement of the ores. The mineralized cavities are assumed to be preferentially located nearly beneath ancient land surfaces, some being earlier meteoric karst cavities (Fig. 9).

Bechstadt (1975a, 1979) thinks of a meteoric karst as caused partly by sea level changes, and partly by tectonism. In this model the relation between mineralization and "special facies" is easier to understand. The metals might originate from an erosional area, assumed to be in the north. Lead isotopes show an age of about 350 m.y. (Carboniferous) for most Alpine lead-zinc deposits (Koppel and Schroll, 1978). An ascending transport of leached metals from beneath is unlikely, as the Carboniferous in most Alpine regions is represented by an unconformity.

The enrichment process might be due to the evaporitic conditions existing on parts of the lagoonal platforms. Lead-zinc concentrations of a factor of 200 to 300 within unstable mineral phases are known from experiments by Ferguson and others (1975). Even today the zinc maxima occur within the inter- and supratidal rock and the paleosoils. Lead and zinc freed from the unstable minerals, might have been transported by (hot or cold) pore fluids, using karst cavities as an aquifer. Sulfides may have been formed where metal-rich and organic sulfur-rich brines were mixing (Figs. 10, 11, 12, 13).

Figure 13—Detail of Figure 12.

SELECTED REFERENCES

Assereto, R., et al, 1976, Italian ore/mineral deposits related to emersion surfaces-a summary: Mineral. Deposita (Berl.), v. 11, p. 170-179.

_____, and C. G. Kendall, 1977, Nature, origin and classification of peritidal tepee structures and related breccias: Sedimentology, v. 24, p. 153-210.

Bechstadt, T., 1974, Sind stromatactis und radiaxial-fibroser Calcit Faziesindikatoren?: N. Jb. Geol. Palaont. Mh., v. 1974, p. 643-663.

_____, 1975a, Lead-zinc ores dependent on cyclic sedimentation (Wetterstein limestone of Bleiberg-Kreuth, Carinthia, Austria): Mineral. Deposita (Berl.), v. 10, p. 234-248.

_____, 1975b, Zyklische sedimentation im erzfuhrenden Wettersteinkalk von Bleiberg-Kreuth (Karnten, Osterreich): N. Jb. Geol. Palaont. Abh., v. 149, p. 73-95.

_____, 1979, The lead-zinc deposit of Bleiberg-Kreuth (Carinthia, Austria); palinspastic situation, paleogeography and ore mineralzation: Verh. Geol. B.-A., v. 1978/3, p. 221-235 (Proceed. 3rd ISMIDA, p. 47-61).

Dohler-Hirner, B., in prep., Untersuchungen zur mikrofazies und geochemie des oberen Wettersteinkalkes (Nordliche Kalkalpen, Drauzuc): Univ. Munich, West Germany, Doctoral thesis.

Dzulynski, S., and M. Sass-Gustkiewicz, 1977, Comments on the genesis of the eastern Alpine Zn-Pb deposits: Mineral. Deposita (Berl.), v. 12, p. 219-233.

_____ and _____, 1980, Dominant ore-forming processes in the Cracow-Silesian and eastern Alpine zinc-lead deposits: Proc. 5th Quadrennial IAGOD Symp., p. 415-429.

Ferguson, J., B. Bubela, and P.-J. Davies, 1975, Simulation of sedimentary ore-forming processes; concentration of Pb and Zn from brines into organic and Fe-bearing carbonate sediments: Geol. Resch., v. 64/3, p. 767-782.

Fischer, A. G., 1974, The Lofer-cyclothems of the Alpine Triassic, *in* D. F. Merriam, ed., Symposium on cyclic sedimentation: Kansas Geol. Surv. Bull., v. 169, p. 107-149.

Koppel, U., and E. Schroll, 1978, Bleiisotopenzusammensetzung von Bleierzen aus dem Mesozoikum der Ostalpen: Verh. Geol. B.-A., v. 1978, p. 403-409.

Lagny, P., 1975, Le gisement plombo-zincifere de Salafossa (Alpes italiennes orientales); remplissage d'un paleokarst triasique par des sediments sulfures: Mineral. Deposita (Berl.), v. 10, p. 345-361.

Muller, G., and G. Tietz, 1971, Dolomites replacing "cement A" in biocalcarenites from Fuerteventura, Canary Islands, Spain, *in* O. P. Bricker, ed., Carbonate cements: Johns Hopkins Univ. Stud. Geology, v. 19, p. 327-329.

Ott, E., 1967, Segmentierte Kalkschwamme (Sphinctozoa) nus der alpinen Mitteltrias und ihre Bedeutung als Riffbildner im Wettersteinkalk: Bayer, Akad. Wiss., math.-naturw. Kl., Abh., N.F. 131, 96 p.

_____, 1972, Die Kalkalgen-Chronologie der alpinen Mitteltrias in Angleichung an die Ammoniten-Chronologie: N. Jb. Geol. Palaont. Abh., v. 141, p. 81-115.

Schneider, H. -J., 1964, Facies differentiation and controlling factors for the depositional lead-zinc concentration in the Ladinian geosyncline of the Eastern Alps: Devel. Sedimentology, v. 2, p. 29-45.

Schulz, O., 1964, Lead-zinc deposits in the Calcareous Alps as an example of submarine hydrothermal formation of mineral deposits: Devel. Sedimentology, v. 2, p. 47-52.

Zankl, H., 1971, Upper Triassic carbonate facies in the Northern Limestone Alps, *in* G. Muller, ed., Sedimentology of parts of central Europe, p. 147-185.

Travertines

Ramon Julia

*T*ravertines are accumulations of calcium carbonate in springs (karstic, hydrothermal), small rivers, and swamps, formed mainly by incrustation (cement precipitation and/or biochemical precipitation).

The term travertine has a local origin from Tivertino, the old Roman name of Tivoli in Italy where travertine forms an extensive deposit. It has already been used by Lyell, 1863; Cohn, 1864; Weed, 1889; and Howe, 1932.

These deposits have also been reported as tufa, calc tufa, calcareous tufa, plant-tufa, carbonate concretions, petrified moss, Vaucheria tufa, Chironomid tufa, spring-sinter, calcic-sinter, sinter crust, and others, and by local names. The term tufa refers to highly porous, spongy deposits, and was already used by Plinius (thophus) for incrustation on vegetal remains and porous volcanic rocks. According to Pia (1933), the term sinter should be restricted to those deposits of an abiotic origin that are typically more dense and compact than tufa; most probably sinter-crust includes flowstones and other speleothems.

In order to avoid semantic confusion, the present trend is to designate as travertine all the carbonate incrustation on plant remains (in place and debris) without reference to the pore volume or density. As a matter of fact, many fossil travertines are very dense because of complete cementation of all the originally abundant cavities.

Travertines can form on higher and

Figure 1—Hand sample from a dam showing different morphological calcium carbonate surfaces according to vegetal communities. m = mosses (*Cratoneuron commu. tatum*). H = hepaticae (*Pellia fabbroniana*). c = cyanoficeae.

lower plants, but most commonly on algae (blue-green, green), mosses, hepatics (Fig. 1) and on insect larva caparace (Chironomids) (Fig. 2), and in some cases they may be considered as stromatolites.

Published works on recent and subrecent travertines are numerous, dating from the 19th and 20th centuries. Among the most remarkable research works were those carried out by the German school; Pia (1933-34), Wallner (1933-35), Stirn (1964), Irion and Muller (1968), and others. They point out the linkage between the different vegetal genera and the formation of travertines which shows a morphology according to the incrusted or incrusting vegetation.

Recently, more specific problems have been studied. From a petrologic point of view Malesani and Vannucci (1975) studied the precipitation of calcite-aragonite and the diagenetic transformations. Couteaux (1969), Marker (1973), Adolphe and Rofes (1973), Wiefel and Wiefel (1974), Schnitzer (1974), Jacobon and Usdoski (1975), and Geurts (1976) envisaged the genetic perspective and studied the correlation between the physico-chemical and biochemical parameters of algae and organisms involved in the processes of travertine formation. These works show two main trends: (1) The predominance of physico-chemical processes over the biochemical ones; the main

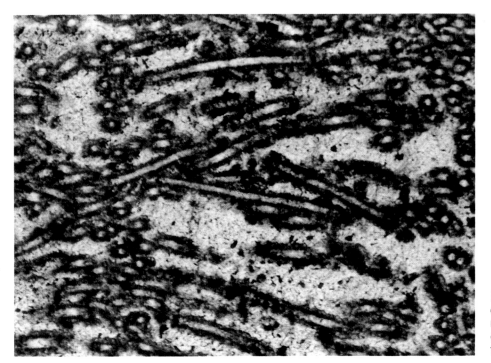

Figure 2—Travertine built by *chironomidae* tubes. These tubes are made up by external pseudoradial calcite crystals (50 μm) over an intern tube of microcrystalline calcite.

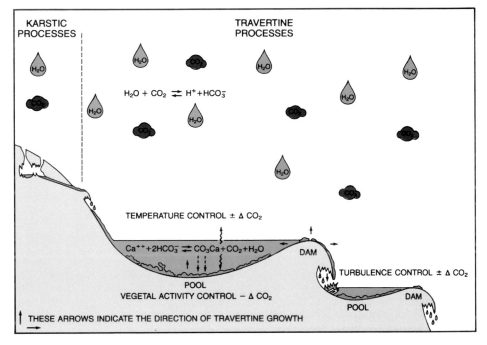

Figure 3—Diagram depicting concentration of calcium ions and carbon dioxide favorable to the formation of travertines. Note that factors affecting carbon dioxide concentration include vegetal growth and turbulence in the water body.

parameters being the evaporation, the turbulence and the temperature; and (2) the predominance of biochemical processes over the physicochemical ones. The vegetal activity (photosynthesis) regulates the rate of CO_2 in the water and hence indirectly the calcite precipitation. Moreover the vegetal activity may precipitate calcite within the living vegetal cells (Geurts, 1976).

Geological Setting

Precipitation of calcium carbonate requires that the water reach favorable concentrations of CO_2 and Ca^{++}. The element most subject to change is CO_2, which is controlled physically by Henry's Law (relation of CO_2 with temperature and pressure) and biochemically controlled throughout photosynthesis (Fig. 3).

In order to have precipitation of calcium carbonate it is necessary to have a decrease of CO_2 in water.

The processes of physico-chemical and/or biochemical incrustation occur mainly: (a) in karstic (Fig. 4) or thermomineral springs (Fig. 5), where the vegetation and/or temperature changes (mainly cooling of thermal water reaching the atmosphere) favored the calcium carbonate

Figure 4—Vertical sheets mainly built by incrustation of mosses in a hanged spring (Collegats, Lleida, Spain).

Figure 5—Travertine accumulation forming pools and dams mainly due to algal activity in Mammoth Hot Spring (Yellowstone). One can see different growth styles: horizontal sheets in pools, vertical to subvertical sheets in dams and waved inclined sheets when some little cascades are covered.

precipitation. For that reason the travertines have some characteristics inherent in the karst (for example long prismatic crystals resembling many speleothems) due to physico-chemical precipitation of calcium carbonate; (b) in small stream channels that usually show a characteristic pattern; they are shaped by several natural dams and pools (rimstone dams and rimstone pools) (Figs. 6, 7).

The dams are mainly formed by incrustation of vegetal communities (Fig. 1) (related with water level) and sometimes begin by interruption of the flow due to leaf, stem and trunk fall (Fig. 8). In the pools (always with lower and higher plants) detritic sedimentation processes coexist with incrustation processes. The clasts are mainly formed by fragments of dams destroyed by floods and by oncoliths;

and (c) in swamps and in littoral environments of some lakes where the vegetation (lower and higher plants) favors the precipitation processes forming travertines. In this case the travertines are interstratified with fluvial or lacustrine sediments.

Figure 6—Dams and pools built by travertine in a small stream related with different vegetal communities. (Dosquers stream, Girona, Spain, during dry season). The rectangle indicates the hand sample of Figure 1.

Figure 7—Travertine dams and pools formed by hydrothermal overflowing water (Turkey).

GEOMETRIC DISPOSITION

Generally it is possible to distinguish two main forms of occurrence; one in vertical, subvertical and overhanging laminae associated with the incrustation of the vegetation located in the waterfalls and travertine dams (hanged springs, cascades, etc.) (Fig. 4), and the other in horizontal or subhorizontal laminae associated with the existence of stream channels (shaped by pool and dams) and paludal environments (Figs. 5, 9).

In the first case there exists a marked overlap of incrusted laminae and in the second case it is common to find the horizontal superposition of detrital (clastic travertines or not) and bioconstructred travertine beds.

CLASSIFICATION

The classification of the different recent and subrecent travertine deposits has been based on several criteria.

One criterion is the physical properties, for instance Eisenstuck (1951) classified travertines according to the degree of coherence distinguishing hard travertines and incoherent

Figure 8— Hand sample of travertine showing different vegetal remains (leaves **l** in the top and stems **s** in the bottom) incrusted by calcium carbonate, due to algal activity (mainly cyanoficeae). Recent travertinization processes from Dosquers stream, Girona, Spain.

Figure 9—Pleistocene travertine deposit, with horizontal detrital sheets (**d**) and vegetal incrusted remains (**v**). Banyoles, Spain.

travertines.

Another criterion, this one used by the botanic German school (Wallner, 1934; Stirn, 1964) utilizes as a main indicator the type and nature of the plant vegetation involved in the formation of travertine. This trend has also been used later by Irion and Muller (1968).

Other classifications are based on the geomorphological position in which the travertines are formed. In this way Symoens (1951) distinguishes spring travertines (denominated after the Belgian term "cron") formed at the foot of hanged springs, and the travertines developed in river beds.

The most useful criterion is classification of the travertine as in-place and clastic; although commonly there are multiple intergradations.

In-place (autochton) travertines are calcium carbonate accumulations on plants in life-position. Clastic traver-tines are calcium carbonate accumulations: (a) on debris of higher and lower plants; (b) on clasts constituted by fragments of in-place travertines; (c) on other clasts (lithoclasts, bones, etc.); and (d) on nodular grains, on-cholithes, etc.

HAND SAMPLES

The most common features of recent and subrecent hand samples are

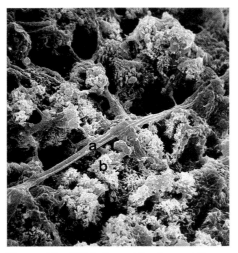

Figure 10—Travertine formed by leaves covered by calcium carbonate incrustations due to algal activity (Oscillatoriaceae).

Figure 11—Travertine dams and pools of centimetric size formed by hydrothermal overflowing water and algae colonization (Turkey).

Figure 12—SEM view of broken portion of incrusted leaf. Note the algal filaments (**a**) and the anhedral microcriptocrystals of LMC (**b**); x 1950.

Figure 13—SEM view of a broken portion from submerged recent algal travertine. The algal filaments (**a**) form a dense network and trap detrital grains (Q = quartz, d = diatoms, t = travertine, etc.); x 950.

the high porosity and morphological variety of the surface covered by calcium carbonate (Figs. 7, 8, 10). These features are more strongly developed in travertine dam facies and karstic springs. In dams, the water flow accumulates a great amount of vegetal debris rapidly affected by algal colonization (mainly cyanophyceae), incrusted and, for that reason, immediately incorporated to the dam (Figs. 8, 9). The vegetal colonization over the dams is quite

different depending on the level reached by the water. In the areas permanently submerged, the algae (cyanophyceae and clorophyceae) predominate (Fig. 7c); in areas affected by a water level oscillation, the hepaticae (Fig. 7h) are more developed, and lastly the mosses (Fig. 7m) prevail in the high parts of the dam, commonly recibing the water splash.

In the inner part of the pools the clastic facies predominate although

near the dam edges the accumulation of vegetal remains may be important, mainly small stems and leaves. The edges of the pools are colonized by the same vegetation that colonizes the dam (mosses, hepaticae and algae) and vegetation (Miriophilaceae, Charophyte, etc.) develops over the clastic facies of the bottom.

In some thermomineral springs where an important algae colonization exists, the hand samples show a typical morphology in steps (Fig. 11)

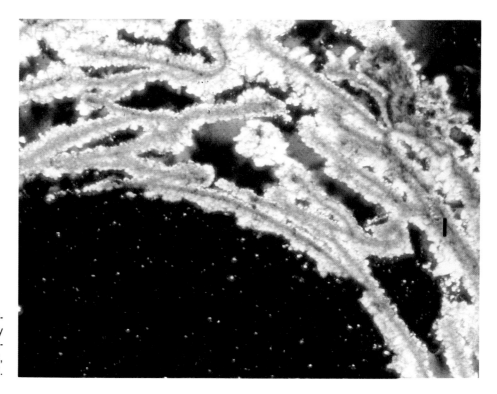

Figure 14—Thin section from a travertine formed by leaves (**1**) covered by fibrous calcite due to algal activity. Recent travertine of Dosquers stream, Girona, Spain; x 70 N.C.

with small pools and dams of centimetric size.

Mineralogy

At all times the prevailing mineral is calcite with a low Mg content (LMC) occasionally accompanied by aragonite, mainly in thermomineral water flows (Malesani and Vannucci, 1977) and with near 5% content of silica (coming from diatoms, frustules and rhizopoda caparaces) or detritic quartz.

FABRICS AND MICROTEXTURES

There are many fabrics related with the different organisms involved in the processes of travertine formation. The vegetation operates in different ways: decaying CO_2, precipitating biochemical calcite (Fig. 12) including metabolic calcite formed in the cells, trapping (Fig. 13) the particles carried by the water flow, and finally serving as substrate to predominant physico-chemical precipitation (Fig. 14).

From a practical point of view it is possible to distinguish two fabrics. One is directly related with algal activity and is characterized by the formation of micro-cryptocrystalline anhedral crystals of LMC and microfibers (Fig. 12). The second fabric is characterized by the deposition of euhedral crystals of LMC over moss stalks, hepatics, and Miriophilaceae, which reach the millimetric range (Fig. 15).

The different algal families may show small fabric variations owing to their rhythmic seasonal growth. This seasonal growth, mainly in the Oscillatoriacea, makes up a laminated stratification of millimetric to sub-millimetric range (Fig. 16). The spring lamina is the thinner and darker of the two.

The micro-cryptocrystalline anhedral crystals as well as the euhedral ones may experience intense neomorphic processes. The long euhedral crystals of LMC can be affected by degrading processes and can be turned into crypto-microcrystalline anhedral crystals (bio and chemical corrosions). On the other hand the crypto-microcrystalline anhedral crystals covering the algal filaments can be reorganized in such a manner as to form larger crystals of pseudoradial calcite.

These neomorphic processes occur very rapidly and this is attributed to the capability of intense water circula-tion through the highly porous and permeable travertine fabrics. For the same reason, travertine cavities are rapidly cemented and filled up by equant calcite mosaics (Fig. 17).

RECOGNITION IN THE FOSSIL RECORD

Some criteria have been used for recognition of travertine fossils records, the most important are supported by related plants and geometric disposition.

The travertine bodies are laterally discontinuous and may show subaerial exposure (caliche crusts, karst, etc.) or inter-stratification with fluvial and paludal deposits (lignite beds, marls, siltstones) (Fig. 9).

The travertine fossil records also show laminated botryoidal, radial and arborescent carbonate fabrics with an abundance of calcite in mosaics of large crystals (Fig. 12).

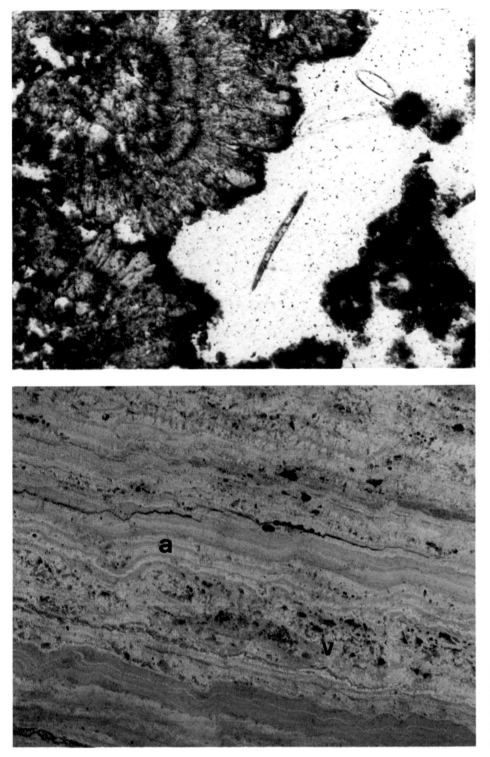

Figure 15—Thin section from a travertine made up by cyanoficeae algae. Fibrous calcite with long axes of crystals oriented perpendicular to vegetal nucleus; x 70.

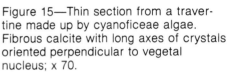

Figure 16—Travertine formed by algal activity (**a** = *Oscillatoriaceae*) showing the seasonal rhythmic growth (dark and white laminae), and by detrital vegetal remains also colonized by algae (**v**). (Upper Pleistocene, Banyoles, Spain); x 0.8.

SELECTED REFERENCES

Adolphe, J. P., and G. Rofes, 1973, Les concretionements calcaires de la Levriere (Eure): Bull. A. F. E.Q., v. 2, p. 79-87.

Cohn, F., 1864, Uber die entstehung des travertines in den Wasserfallen von Tivoli: N. Jb. Min. Geol. Palaont., v. 32, p. 580-610.

Coteaux, M., 1969, Formation et chronologie palynologique des tufs calcaires du Luxembourg belgo-grand ducal: Bull. A.F.E.Q., v. 3, p. 167-183.

Eisenstuck, M., 1951, Die kalktuffe und ihre molluskenfauna bei schmeiechen nahe Blaubeuren (Schwabische Alb): N. Jb. Geol. Palaont. Abh., v. 93, p. 247-276.

Geurts, M. A., 1976, Genese et stratigraphie des travertins de fond de vallee en Belgique: Acta Geograph. Lovainesa, v. 16, 66 p.

Howe, M. A., 1932, The geologic impor-

Figure 17—SEM view of a tube in a recent Hepaticae travertine showing the cement that occludes the cavity. These cavities are produced by degrading organic matter; x 205.

tance of the lime-secreting algae with a description of a new travertine-forming organism: U.S. Geol. Survey Prof. Paper, 170-E, p. 19-23.

Irion, G., and G. Muller, 1968, Mineralogy, petrology and chemical composition of some calcareous tufa from the Schwabische Alb Germany, *in* G. Muller and G. M. Friedman, eds., Recent developments in carbonate sedimentology in central Europe: Springer Verlag Pub., p. 157-171.

Jacobson, R. L., and E. Usdowski, 1975, Geochemical controls on a calcite precipitating spring: Contrib. Mineral. Petrol., v. 51, p. 65-74.

Malesani, P., and S. Vannucci, 1977, Precipitazione di calcite e di aragonite dalla acque termominerali in relazione alla genesi e all'evoluzione dei travertini: Atti Acad. Naz. Lincei, R.C. Cl. Sci. F, M.N. Ital, v. 58, p. 761-776.

Manfra, L., et al, 1974, Effetti isotopici nella diagenesi dei travertini: Geol. Romana, v. 13, p. 147-155.

Marker, M. E., 1971, Waterfall tufas; a facet of karst geomorphology: Zeist. Geomorph. SupBand 12, p. 138-152.

———, 1973, Tufa formation in the Transvaal, South Africa: Z. Geomorph., 17-4, p. 460-473.

Pia, J., 1933, Die rezenten Kalksteine: Tschemarks Min. Petr. Mitt. Erg. Bd.

Schnitzer, U. A., 1974, Kalkinkrustationen und kalksinterknollen in Lias-Quellwassern bei Elsenberg (Bl. Erlangen-Nord): Geol. B. Nordost-Bayen, v. 24, p. 188-191.

Scoll, D. W., and W. H. Taft, 1964, Algae contributions to the formation of calcareous tufa, Moro Lake, California: Jour. Sed. Petrology, v. 34, p. 309-319.

Stirn, A., 1964, Kalktuffworkommen und kalktufftypen der Schwabischen Alb.: Abh. Karst Hohlenkde, E, H.1, 91 p.

Symoens, J. J., et al, 1951, Apercu sur la vegetation des tufs calcaires de la Belgique: Bull. Soc. Royale Bot. Belg., v. 83, p. 239-352.

Wallner, J., 1934, Beitrag zur kenntnis der vaucheriatuffe: Zbl. Bakteriol. Parasitenk. Infek., 2. Abt., v. 90, p. 150-154.

———, 1934, Uber die bedeutung der sog. chironomidentuffe fur die messung der jahrlichen kalkproduktion durch algen: Hedwigia, v. 74, p. 176-180.

Weed, W. H., 1889, Formation of travertine and siliceous sinter by vegetation of hot springs: U.S. Geol. Survey, 9th Ann. Rept., p. 619-676.

Wiefel, H., and J. Wiefel, 1974, Zusammenhange zwischen verkarstung und travertinbildung un gebiet von weimar: Abh. Zentr. Geol. Inst. Berlin, no. 21, p. 61-75.

Economic Aspects of Subaerial Carbonates

J. Richard Kyle

*A*n important aspect of subaerial exposure facies is that karstification and associated processes are often responsible for the creation of secondary porosity and other features that may become sites for the accumulation of hydrocarbon or mineral resources, and in the case of some ore deposits, may be responsible for the creation of the valuable commodity. It is widely acknowledged that subaerial exposure may be an important aspect in the development of secondary porosity for hydrocarbon reservoirs (Moore, 1979). However, the recognition of ancient subaerial exposure surfaces is often difficult, particularly in the subsurface, due to their generally nondiagnostic response to standard geophysical techniques used in the petroleum industry. Generally the effects of exposure, including porosity development or destruction, are more readily apparent than the exposure surface.

The producing strata in many major oil fields are closely associated with unconformities. However, the details of this association often are not well established. The relatively wide spacing of exploration and production wells, i.e. data points, is not generally conducive to reconstruction of the nature of the ancient subaerial exposure surface. The close spacing of minerals, exploration cores and the three-dimensional view provided by mining exposures permit a far more detailed evaluation of the nature of the exposure surface and the subsurface diagenetic effects. For this reason, two case histories documen-

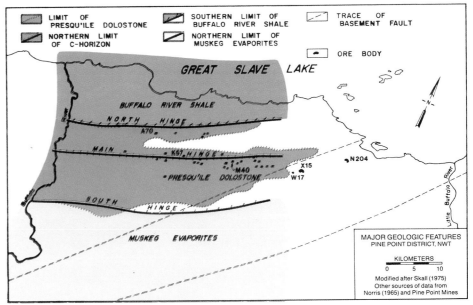

Figure 1—Major geologic features of the Pine Point district. (Reproduced with modifications from Mining Engineering, 1980, v. 32, p. 1617-1626.)

ting the economic aspects of subaerial exposure of carbonates were chosen from major mining districts. These ore-hosting carbonate strata are important oil-hosting reservoirs elsewhere, and the relationship of subaerial exposure to the development of both types of resource concentration is comparable.

MINERAL DEPOSITS ASSOCIATED WITH SUBAERIAL EXPOSURE SURFACES

Consideration of the effects of subaerial exposure and weathering on the formation of ore deposits is beyond the scope of this paper as it would include topics as far ranging as supergene sulfide enrichment (porphyry copper deposits), laterite formation (nickel silicate deposits), and placer concentrations (gold, gem stones, etc.). If the discussion is

restricted to the effects of karstification on sedimentary terranes, then the economic mineral deposits may be divided into four major types (Quinlan, 1972): (1) residual accumulations from the removal of soluble strata; (2) deposits preserved in solution-subsidence structures; (3) subsurface sulfur deposits related to bacterial reactions; and (4) hydrothermal precipitates in solution-produced cavities.

Residual accumulations commonly result from the solutional removal of predominantly carbonate strata and include both simple physical concentrations of pre-existing dispersed minerals and chemical reorganization of the constituents of the strata into new minerals. Economic concentrations of antimony, barite, bauxite, copper, fluorite, gold, iron, lead,

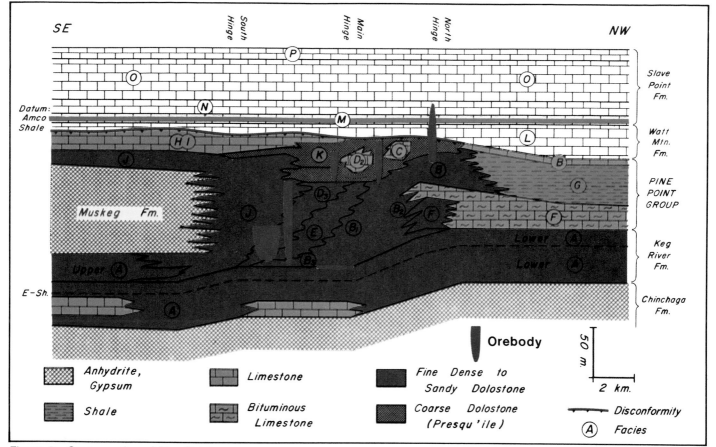

Figure 2—Cross-section of Pine Point barrier complex with schematic stratigraphic positions of selected orebodies. Note vertical exaggeration of 40X. (Reproduced with modifications from K. H. Wolf, ed., Handbook of Stratabound and Stratiform Ore Deposits: Amsterdam, Elsevier Sci. Pub., v. 9, ch. 11, p. 643-741.)

manganese, phosphate, tin, and zinc have been formed by these mechanisms. Deposits of clay, sand, diamonds, coal, and base and precious metals have been preserved in subsidence structures related to removal of soluble strata. Karst-related sulfur deposits are associated with evaporitic strata in which sulfate-reducing bacteria act on hydrocarbons and gypsum or anhydrite to produce hydrogen sulfide, calcite, and water. The reduced sulfur is oxidized to elemental sulfur by meteoric water.

A significant effect of karstification is the preparation of strata for mineralization through the development of zones of secondary porosity

that may serve as conduits for fluid migration and sites for ore precipitation. Deposits of base and precious metals, barite, fluorite, manganese, mercury, uranium, and vanadium have been localized by such features (Quinlan, 1972). Most investigators feel that the effects of subaerial exposure are largely separate from the mineralization event, but Bernard (1973) and Zuffardi (1976) propose that mineralization commonly occurs as a mature stage of karstification and that the ore constituents may have been released during earlier stages of rock solution. However, evidence for the nature of the ore-forming fluids for many deposits is not compatible with derivation from meteoric water at near surface temperatures (Roedder, 1976). Conversely, some authors suggest that solution brecciation is due to the corrosive action of the hydrothermal solutions that formed the ore deposits.

Karstification appears to have been an important process in the localiza-

tion of some of the world's major deposits of lead, zinc, and other metals and nonmetallic commodities that occur in carbonate strata of Phanerozoic age. This type of deposit is particularly well developed in the Paleozoic carbonate strata of the midcontinent area of the United States and is often referred to as "Mississippi Valley type" lead-zinc deposits. Similar deposits occur in the Appalachian, Cordilleran, and Arctic Archipelago regions of North America, as well as in Great Britain, central Europe, and northwest Africa (Brown, 1970). Although there are significant differences among districts, they have a number of unifying general characteristics (Ohle, 1959, 1980; Sangster, 1976). Most orebodies occur as stratabound concentrations in shelf carbonate sequences, usually dolostones, on the flanks of major intracratonic basins. The districts commonly occur at facies boundaries of depositional or diagenetic origin, and the mineralized stratigraphic sequence is typically

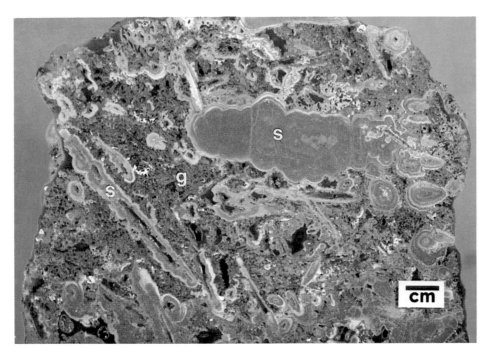

Figure 3—Hand specimen of high grade ore consisting of broken colloform sphalerite (**s**) with skeletal galena (**g**).

Figure 4—White baroque dolomite in upper Facies K adjacent to K57 prismatic orebody. Height of exposure is 1.5 m.

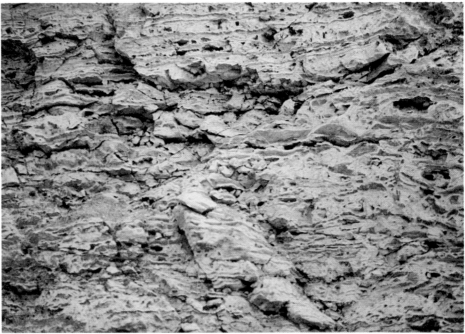

associated with an unconformity. Mineralogy of the deposits is relatively simple, generally consisting of large, well-formed crystals of sphalerite, galena, marcasite, pyrite, barite, fluorite, dolomite, calcite, and quartz that are the result of growth in secondary porosity in the carbonate strata. Fluid inclusion data indicate that the mineralizing fluids were extremely saline, with temperatures ranging from 60 to 175°C (Roedder, 1976); these mineralizing fluids are similar to the subsurface brines present in many sedimentary basins, for example the Gulf Coast (Carpenter et al, 1975; Land and Prezbindowski, 1981). The ore galenas generally have a wide range in lead isotope composition and are often enriched in the radiogenic isotopes (J-type), thus indicating a complex history.

Similarities between oil field brines and inclusion fluids and the spatial association between some hydrocarbon and sulfide accumulations have prompted some to suggest genetic affiliations between hydrocarbons and carbonate-hosted lead-zinc deposits (Anderson, 1978). Although there are different opinions concerning metal sources and causes of mineral precipitation, it is now widely accepted that these carbonate-hosted ore deposits, like petroleum and

natural gas, are the result of a normal sequence of events in an evolving sedimentary basin (Macqueen, 1979). This "basinal evolution" model for mineralization was first proposed for the Pine Point lead-zinc district in Canada by Beales and Jackson (1966).

The Pine Point district in Middle Devonian strata and the Tennessee districts in Lower Ordovician strata provide excellent examples of the ef-

fects of subaerial exposure and concomitant diagenesis on the development of ore-hosting porosity. Furthermore, the ore-hosting stratigraphic intervals and structures in these districts are important hydrocarbon reservoirs elsewhere.

Pine Point Lead-Zinc District

The Pine Point mining district is located on the south shore of the Great Slave Lake in the Northwest

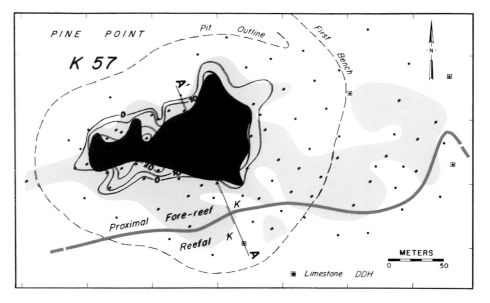

Figure 5—General geologic features of the K57 area. Dots indicate location of diamond drill holes; limestone drill holes indicate areas of upper Pine Point Group limestone preserved within Facies K. Isopach is detritus thickness in meters. Yellow area represents greater than 1% total sulfides over a 53.3 m interval; red area is greater than 10% total sulfides. (Reproduced with modifications from K. H. Wolf, ed., Handbook of Stratabound and Stratiform Ore Deposits: Amsterdam, Elsevier Sci. Pub., v. 9, ch. 11, p. 643-741.)

Territories of Canada (Fig. 1). The district has been a major base metal producer since 1964 and has yielded more than 50 million tons of ore averaging 9% combined lead-zinc; a similar tonnage of ore grade material remains. Orebodies vary considerably in size, geometry, metal percentages and ratios, sulfide textures, and host rock relationships. Individual sulfide concentrations are stratabound in relatively narrow intervals, but orebodies occur in several positions in a 200 m stratigraphic section (Fig. 2). The sulfide bodies are composed almost exclusively of variable amounts of sphalerite, galena, pyrite, and marcasite and associated dolomite and calcite. Individual deposits are zoned with a high grade core of galena and sphalerite (Fig. 3) that grades outward into lower grade material with increasing amounts of iron sulfides toward the perimeter of the sulfide body (Kyle, 1980). Transition between high total sulfide material and barren host rock is commonly abrupt (Fig. 4). About 50 sulfide bodies are present within a narrow Middle Devonian carbonate reefal complex which developed in shallow water along a zone of prolonged tectonic instability (Fig. 1; Skall, 1975). The Pine Point barrier complex was the dominant control of Middle Devonian (Givetian) sedimentation over a vast area and separated a deep water shale basin (Mackenzie

Basin) from an extensive evaporite basin (Elk Point Basin). The Pine Point barrier now dips gently to the southwest from the outcrop belt and hosts several major oil fields in northwestern Alberta and scattered smaller gas fields in Alberta, British Columbia, and adjacent Northwest Territories (De Wit et al, 1973).

Stratigraphy

A highly variable sequence of early Paleozoic evaporitic carbonate and siliciclastic strata up to 100 m thick overlies the Precambrian basement. The early Middle Devonian (Eifelian) Chinchaga formation unconformably overlies the early Paleozoic units and dominantly consists of an evaporite sequence up to 110 m thick (Norris, 1965). Strata of Givetian age host all known ore deposits in the district. Skall (1975) redefined the Givetian stratigraphy of the Pine Point barrier complex, and facies designated A through P replaced previous formational nomenclature (Fig. 2). The early Givetian Keg River Formation (Facies A) conformably overlies the Chinchaga Formation. The Pine Point Group includes the middle Givetian strata (Facies B through K) that form integral parts of the barrier complex (Skall, 1975). The lower part of the barrier (Fig. 2) consists largely of the fine-crystalline dolostones of Facies B (Off-reef), D (Organic Barrier), E (Fore-reef), and J (Back-reef)

and of the bituminous limestones of Facies F (Marine); the evaporite strata of the Muskeg Formation are, in part, lateral equivalents of these units. Dendroid, tabular, and massive stromataporoids, dendroid and massive corals, and brachiopods dominate the framework of the Organic Barrier. The upper part of the barrier consists of limestones of Facies B, C (Fore-reef), D, H and I (Back-reef), the coarse-crystalline Presqu'ile dolostone (Facies K) diagenetically developed from these units, and the Buffalo River Shale of Facies G (Marine). A partial disconformity separates the upper barrier from the late Givetian Watt Mountain Formation (Facies L). The Slave Point Formation (Facies M through P) conformably overlies Facies L and is overlain by the Upper Devonian (Frasnian) Hay River Shale. Skall (1975) documented that subtle tectonic adjustments along three N65°E trending "hinge zones" were responsible for the abrupt shifts in depositional environments resulting in the complex facies relationships.

Effects of Subaerial Exposure

It appears that karstification and dolomitization were major aspects of carbonate diagenesis related to subaerial exposure which were of great importance in the preparation of fluid-transporting and sulfide-hosting structures at Pine Point.

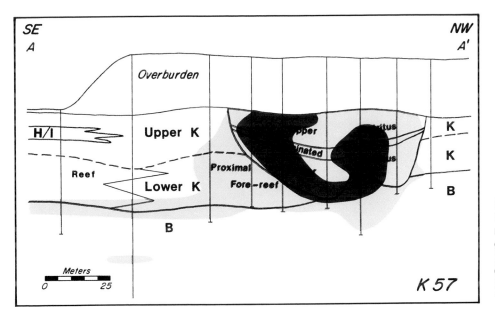

Figure 6—Geologic cross-section of the K57 area. Vertical lines represent diamond drill holes. Yellow area contains greater than 2% total sulfides; red area is greater than 30% total sulfides. (Reproduced with modifications from Mining Engineering, 1980, v. 32, p. 1617-1626.)

Karstification — The post-middle Givetian, pre-late Givetian partial disconformity is present in the southern and central part of the Pine Point area (Figs. 1, 2) where the basal late Givetian Watt Mountain Formation, Facies L, overlies the coarse-crystalline dolostone of the Presqu'ile Facies K or the undolomitized upper barrier limestones of Facies C, D-2, H and I. North of the coarse-crystalline dolostone development, a considerably thicker Facies L directly overlies Facies B without apparent disconformity (Fig. 2; Skall, 1975). The disconformable contact commonly consists of a zone of green clay and carbonate rock fragments overlying the upper Pine Point Group strata which may contain major green clay in vugs, fractures, and bedding planes for some distance below the rubble contact. Thickness of the rubble zone is highly variable, apparently depending on the local topography on the subaerial exposure surface, but generally is less than 5 m.

Major solution features related to this period of subaerial exposure appear to control the distribution of sulfides and to account for the geometry of the "prismatic" orebodies with restricted horizontal dimensions relative to vertical extent and of the "tabular" orebodies with restricted vertical extent relative to the horizontal dimensions (Kyle, 1980, 1981). Within many prismatic

orebodies in the upper Pine Point barrier are strata which are not present in standard Facies K sections, even immediately adjacent to the mineralized zones. Generally this material is light gray, fine- to coarse-crystalline, friable dolostone, commonly with a detrital texture. This detritus may be laminated and may contain fragments of green clay and fine-crystalline tidal flat dolostone, particularly in the upper portion. These fragments are generally only a few centimeters in the longest dimension but are occasionally as much as a meter.

The K57 area illustrates well the nature of the detritus and its relationship to adjacent strata and to sulfide concentrations of both the prismatic and tabular types. The K57 orebody and the geologically similar K62 orebody occur within the coarse-crystalline Presqu'ile dolostone along the Main Hinge Zone in the western part of the district. In this area, Facies K has a maximum preserved thickness of about 45 m and may be divided into upper and lower units. The lower unit consists of tan to light gray, coarse-crystalline dolostone which contains megafossil and uniform granular lithologic types. These rocks are interpreted to represent reefal and proximal fore-reef depositional environments, respectively; a minor amount of this unit is preserved as limestone (Facies D-2).

An irregular, interfingering facies boundary between the two lithologic types of the lower Facies K is present just to the south of the mineralized zone (Figs. 5, 6). Direct evidence concerning the original nature of the upper Facies K unit has been largely obliterated by the pervasive introduction of white dolomite. Isolated limestone remnants lateral to this unit suggest that this material was originally the lagoonal Facies H/I.

Within the zone of coarse-crystalline dolostone is an irregularly elliptical area containing light gray, fine- to coarse-crystalline, friable dolostone detritus (Figs. 5, 6). The detritus zone is about 190 m long and 100 m wide and has a maximum detritus thickness of about 35 m in two areas. Although the top of most of the detritus zone is truncated at its subcrop, Facies L units are present in four drill holes, thus suggesting that the present detritus thickness represents most of the original material. The thickest section of detritus overlies Facies B (Fig. 6). Near the detritus zone, the contact between Facies B and Facies K is indistinct because the upper part of Facies B is medium- to coarse-crystalline with poorly preserved faunal components. Three stratigraphic units have been recognized in the K57 detritus zone (Fig. 6; Kyle, 1980). The stratigraphically lowest unit consists of relatively clean, fine- to coarse-crystalline dolostone which contains

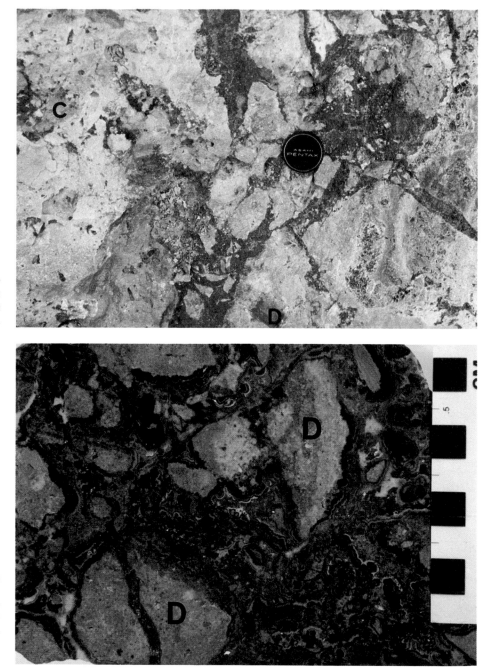

Figure 7—Detritus (doline fill) with scattered tidal flat dolostone (**D**) and green clay (**C**) fragments. K62 orebody; scale is 54 mm in diameter.

Figure 8—Hand specimen of detritus "fragments" (**D**) cemented by colloform sphalerite and galena. Note isolated colloform sphalerite crusts without substrates and galena veinlet transecting detritus fragment and its colloform sphalerite rim.

blocks of coarse-crystalline Facies K dolostone, particularly near the base. A thinner unit of laminated detritus, consisting of alternating bands of light and medium gray dolostone with rare green clay blebs overlies the lower detritus unit. The upper unit consists of nonlaminated detritus with common small fragments of green clay and fine-crystalline dolostone (Figs. 7, 8). Similar dolostone detritus has been observed in at least 10 other prismatic orebodies. Tabular ore-

bodies in the upper barrier complex lack major amounts of associated detritus. Although there are small areas of massive sulfide concentration in these zones, most of the ore consists of stratabound sulfides confined to a relatively narrow stratigraphic interval. Tabular sulfide zones are common in the lower part of the coarse-crystalline Facies K dolostone in some areas; many of the prismatic orebodies have contiguous tabular sulfide concentrations.

M40 is a tabular orebody located in the central portion of the Main Hinge Zone within a cluster of prismatic orebodies. The sulfide concentration in M40 is stratabound in the lower part of the coarse-crystalline Facies K dolostone and averages about 3 m in thickness, although it reaches a maximum of 15 m. Most of the ore occurs as open-space filling of vuggy porosity (Fig. 9), and local collapse breccias also host sulfides (Fig. 10). Although some massive sulfide con-

Figure 9—Coarse crystals of sphalerite and galena in tabular porosity zones in the lower Facies K dolostones. M40 orebody; height of exposure is 1 m.

Figure 10—Collapse breccia in lower Facies K dolostone cemented by sphalerite and galena. M40 orebody; height of exposure is 2.5 m.

centrations are present, sulfide textures indicate growth from the walls and suggest that these zones were large voids prior to mineralization. Some of these zones are thin and tabular and contain gravity-controlled stalactitic sulfide forms.

Solution features associated with sulfide concentrations within the coarse-crystalline dolostone of the upper barrier differ somewhat from the sulfide-hosting features within the fine-crystalline dolostones of the lower barrier. The latter sulfide-bearing zones occur northeast of the present-day erosional limit of the Presqu'ile (Fig. 1) but presumably were once covered by the coarse-crystalline dolostone. Sulfide concentrations may be either massive without lithic components or within breccias and intercrystalline porosity in reefal (Facies D-1) and back-reef (Facies J) lithologies. N204 is the most northeasterly and the lowest stratigraphically of the presently known mineralized zones (Fig. 1). Sulfides occur in the upper part of a laterally extensive zone of vuggy and intercrystalline porosity and incipient fracturing in the fine-crystalline Facies B dolostones just above the Keg River Formation (Fig. 2).

Solution features in the Pine Point barrier complex were of considerable importance in the localization of sulfide concentrations. The detritus zones appear to have been the loci for the deposition of the prismatic

Figure 11—Vadose and phreatic diagenetic zones in the upper Pine Point barrier during post-Middle Givetian erosional period. (Reproduced with modifications from Mining Engineering, 1980, v. 32, p. 1617-1626.)

Figure 12 A-D—Development sequence for sulfide-bearing detritus-filled dolines at Pine Point. (Reproduced with modifications from K. H. Wolf, ed., Handbook of Stratabound and Stratiform Ore Deposits: Amsterdam, Elsevier Sci. Pub., v. 9, ch. 11, p. 643-741.

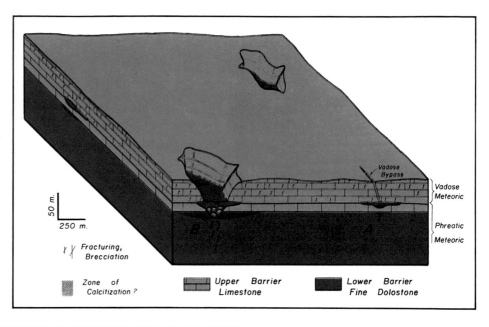

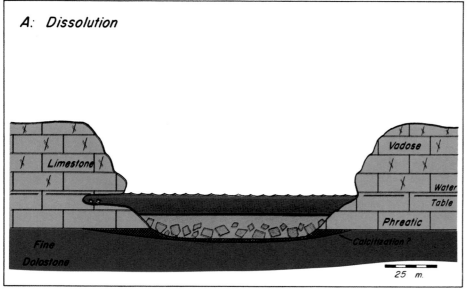

A: Dissolution

orebodies and contiguous tabular sulfide concentrations in the upper barrier. This association can be demonstrated readily by comparing the distribution of detritus with the total volumetric amount of sulfides.

K57 has a maximum total sulfide concentration of about 50 volume percent over a 53 m interval. The greatest sulfide concentration is within the detritus zone (Fig. 5), but an irregular stratabound zone of lower sulfide content is present to the east of the massive zone (Figs. 5, 6). Most of the sulfides in the stratabound zone occur as pore-filling of the uniform granular lithology (proximal

fore-reef) in the lower unit of the coarse-crystalline Facies K dolostone. Sulfide minerals within the detritus zone fill fractures and breccia interstices, but large intervals are massive. Despite extensive vuggy porosity within the pervasive white dolomite of the upper Facies K, the boundary of the prismatic orebody is sharp, and the upper Presqu'ile does not contain megascopic sulfides. Minor sulfides occur in intercrystalline porosity in the Facies B dolostone which underlies the Presqu'ile.

Solution features associated with sulfide concentrations in the coarse-crystalline Facies K dolostone of the

upper barrier are either large detritus-filled depressions open to the disconformable surface or macropores and stratabound zones of increased porosity in the lower Presqu'ile without apparent direct connection with the disconformable surface. These features are thought to have formed during prolonged subaerial exposure with attendant karstification and carbonate diagenesis during post-middle Givetian emergence (Kyle, 1980, 1981).

The sulfide-filled macropores in the lower part of Facies K (for example M40) that lack apparent direct connection with the disconformity are interpreted as former caves which developed in the upper part of the phreatic zone during the period of the stable water table (Fig. 11, situation A). Local collapse in the cave system created breccias which now also host sulfide minerals. Extensive solution of limestone is readily accomplished by meteoric water which has not become saturated with respect to calcite due to CO_2 loss during seepage of rainwater through the vadose zone to the water table (Thrailkill, 1968). Therefore, macropores and caves developed in the limestone of the upper Pine Point barrier where a bypass permitted vadose flow instead of vadose seepage to the water table (Fig. 11). The vadose bypasses did not remain open enough to permit

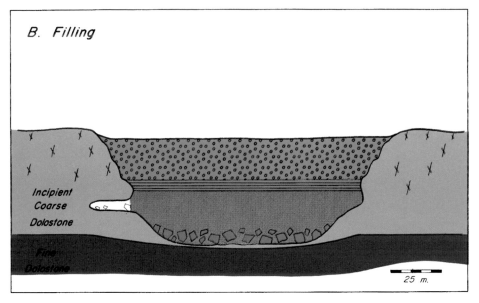

B. Filling

Incipient
Coarse
Dolostone

Fine
Dolostone

25 m.

complete filling of the caves by detritus during the close of the erosional period.

Horizons of increased porosity and permeability were apparently created in the upper part of the phreatic zone adjacent to the zones of major solution. This process was responsible for creation of most of the hosting porosity for the tabular sulfide concentrations which are common in the lower Presqu'ile particularly along the Main Hinge. The development of horizons of increased permeability was probably controlled by the local supply of meteoric water with the chemical capacity to produce increased porosity. Once meteoric water became saturated with respect to calcite, effective porosity enlargement ceased. This restriction may account for the apparent limited lateral extent of these horizons in the lower Presqu'ile rather than development of a continuous zone of increased permeability along the position of the upper part of the paleo-phreatic zone.

The detritus zones are interpreted as filled compound dolines which developed largely by solution activity in the limestone Facies C, D-2, H, and I of the upper barrier (Fig. 11, situation B). Original stratigraphic facies was not the major factor controlling development of the dolines because they occur in back-reef, reefal, and intercalated facies. The

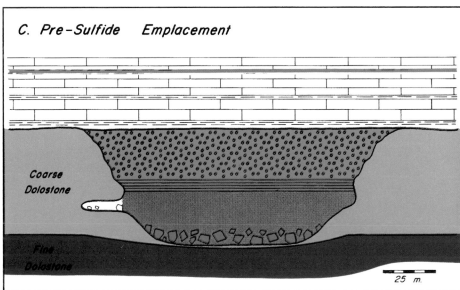

C. Pre-Sulfide   Emplacement

Coarse
Dolostone

Fine
Dolostone

25 m.

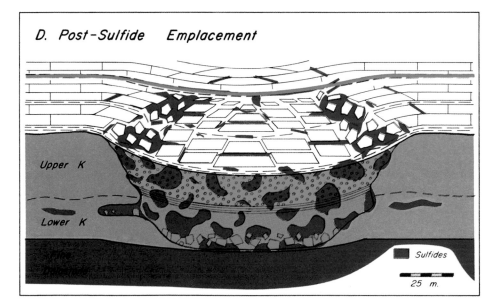

D. Post-Sulfide   Emplacement

Upper K

Lower K

Fine
Dolostone

Sulfides

25 m.

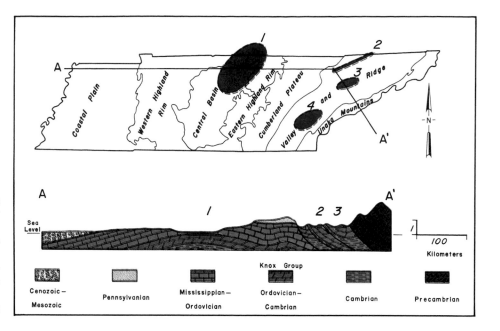

Figure 13—Location of mineral districts in the Knox Group relative to major physiographic and geologic provinces in Tennessee. 1 = Central Tennessee-Southern Kentucky; 2 = Copper Ridge; 3 = Mascot-Jefferson City; 4 = Sweetwater, (Reproduced with modifications from Economic Geology, 1976, v. 71, p. 892-903.)

dolines are elongate in a northeasterly direction parallel to the Hinge Zones, suggesting that karstification was controlled by the same structural factors that governed carbonate sedimentation. The dolines are as much as 400 m in length and have length to width ratios ranging from about 2 to 3.5. Doline walls were relatively steep and irregular as suggested by the abrupt transition from detritus to wall rock, and depths of at least 35 m are indicated. In addition to detritus-filled dolines, green clay and carbonate detritus occur in solution-enlarged joints, bedding planes, and vugs near the disconformity.

The developmental sequence envisaged for the formation of the sulfide-bearing, detritus-filled dolines is shown in Figure 12. During subaerial exposure, the upper barrier limestones were subject to chemical attack by rainwater, which is undersaturated with respect to calcite (Thrailkill, 1968). Solution was concentrated at joint intersections, generally along the dominant northeasterly trend and proceeded until the entire limestone sequence was penetrated (Fig. 12A). Additional solution was slowed at this stage, not only by the underlying chemically more resistant fine-crystalline dolostone, but also by hydrologic factors. The position of the meteoric water table may have been controlled

indirectly by sea level on the fore-reef side of the exposed barrier. Since most carbonate solution takes place in the upper part of the meteoric phreatic zone (Thrailkill, 1968) which apparently was confined within the upper barrier, extensive solution of the fine-crystalline dolostones by meteoric water was greatly inhibited. The water table eventually became relatively stable at a level within the lower part of the limestone units of the upper barrier (Fig. 12A). Lateral solution of limestone occurred near the water table adjacent to the dolines, and macropores were developed. Slumping of oversteepened doline walls resulted in a rubble pile of large limestone blocks and finer debris on the doline floor. Slumping of the doline walls may have effectively sealed off part of the macropores so that they were not completely filled with detritus.

Following the post-middle Givetian erosional period and development of the karstified barrier, marine transgression first affected the low-lying "nearshore" areas, and local tidal flat conditions were established. Periodically, probably during major storms, fine carbonate detritus was swept into the standing body of water in the dolines and sedimented in thin layers. Inundation of the erosional surface was sporadic with several periods of transgression and minor

regression. During the final closing of the erosional period, the regolith developed on the karstified barrier was swept into depressions on the surface, the largest of which were the dolines (Fig. 12B). This erosional material consisted of fine carbonate detritus with some green clay and tidal flat dolostone fragments eroded from the ephemeral tidal flat and lacustrine deposits of initial transgression. Marine transgression appears to have been effective in filling the dolines, and little compaction of detritus appears to have occurred during late Givetian sedimentation (Fig. 12C). Collapse of late Givetian strata overlying the detritus zones occurred at least in part as the result of volume reduction during sulfide emplacement (Fig. 12D).

The relationship of the post-middle Givetian episode of karstification to the sulfide-hosting structures in the fine-crystalline dolostones of the lower barrier complex is unclear. It seems unlikely that the zone of most intensive carbonate solution, the upper phreatic, could have extended into the lower levels of the barrier complex during the post-middle Givetian erosional period. However, carbonate solution may have occurred in the zone of mixing of fresh water and sea water which existed within the lower barrier during this period. Another possibility is that the sulfide-hosting

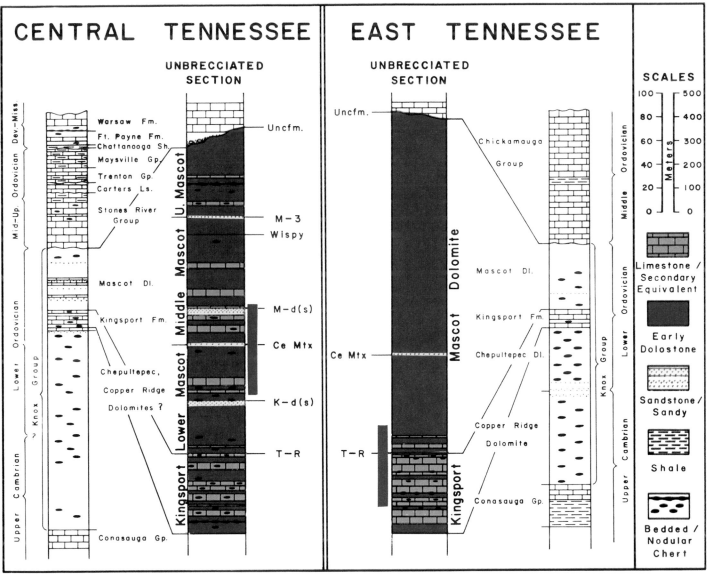

Figure 14—Generalized stratigraphic sections for the east and central Tennessee zinc districts. Red bars delineate the stratigraphic interval of major zinc concentrations. (Reproduced with modifications from Economic Geology, 1976, v. 71, p. 892-903.)

structures in the fine-crystalline dolostones resulted from an earlier period of subaerial exposure and carbonate diagenesis, as Maiklem (1971) has proposed for the middle Givetian of the Pine Point barrier complex.

Dolomitization — All of the Pine Point orebodies occur in dolostones of several types in several depositional facies (Fig. 2). The "fine, dense to sandy" dolostones contain two distinct types. Some units consist

of uniform, very fine-crystalline (generally less than 20μ), dense rocks with sedimentary structures and stratigraphic relationships that indicate deposition in a supratidal to intertidal environment. Most of the lower Pine Point Group consists of fine-crystalline (20 to 50μ), dense to sandy dolostones for which stratigraphic, petrologic, and faunal evidence indicates deposition as calcium carbonate sediments in a normal marine to slightly restricted subtidal environment (Fig. 2). Because of the extensive evaporite strata coeval with these dolostones, two major mechanisms for dolomitization have been proposed — seepage reflux of evaporative brines through the barrier

(Skall, 1975), and evaporative pumping of seawater through the barrier (Kyle, 1980).

The coarse-crystalline dolostone that occurs between the lower barrier units and the Watt Mountain Formation is defined as Facies K (Skall, 1975) (Fig. 2). The distribution of the coarse-crystalline dolostone is further restricted to the area between the Hinge Zones and below the post-middle Givetian disconformable surface (Figs. 1, 2). This lithology consists of dolomite crystals greater than 200μ in size, and the unit is a diagenetic facies superimposed on back reef, reef, and fore-reef strata (Skall, 1975). In contrast to the ubiquitous fine-crystalline dolostones of

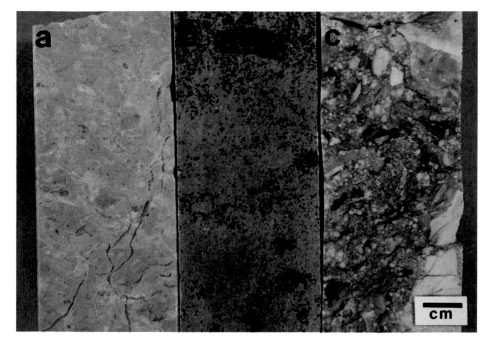

Figure 15—Cores representing the development of limestone "equivalents:" intraclastic pelsparite limestone (**a**) from the Mascot Dolomite; coarse crystalline dolostone (**b**) with about 10% relict calcite (stained with Alizarin red-S in dilute HCl); and coarse rock matrix breccia (**c**) consisting of fragments of chert and shale in a matrix of coarse crystalline dolostone.

the lower barrier, the original limestone lithologies are preserved as isolated remnants within the coarse-crystalline dolostone and as extensive back-reef strata of Facies H and I (Fig. 2). Contacts between coarse-crystalline dolostone and limestone commonly transect bedding, are irregular, and are relatively sharp. Relic depositional textures are preserved locally in the coarse-crystalline dolostone, particularly in the lower units. White dolomite is a common accessory, and its pervasive introduction has obliterated much of the original nature of the upper part of Facies K in some areas (Fig. 4). The contact of the coarse-crystalline dolostone with the underlying fine-crystalline dolostones of the lower barrier is generally abrupt but locally is gradational over a few meters.

Four mechanisms have been proposed for the origin of the coarse-crystalline Facies K dolostone: (1) dolomitization of coarse-grained reefal limestones (Norris, 1965); (2) recrystallization of fine-crystalline dolostones (Campbell, 1967); (3) dolomitization of barrier limestones by pre-sulfide heated brines circulating along the Hinge Zones during post-Givetian times (Skall, 1975); and (4) dolomitization of barrier limestone related to the mixing of

meteoric water with sea water during the post-middle Givetian erosional period (Kyle, 1980). The coarse-crystalline nature of Facies K does not reflect an original coarse-grained sediment because it is not restricted to the reefal facies and because the coarse-grained reefal lithologies of the lower barrier have not been converted to coarse-crystalline dolostone. The extensive recrystallization of fine-crystalline dolostone into Facies K is unlikely because of the thin beds of fine-crystalline Facies J dolostone preserved within the coarse-crystalline dolostone. The coarse-crystalline dolostone is not an alteration effect directly related to sulfide mineralization because it is much more extensive than sulfide mineralization, is not associated with the orebodies in the lower barrier, and is not present above the disconformity in those orebodies that extend into the Watt Mountain Formation.

Both the reflux and evaporative pumping mechanisms can be eliminated as possible methods of generating the coarse-crystalline dolostone because the evaporitic conditions responsible for the deposition of the Muskeg evaporites and tidal flat lithologies of Facies J had ceased to exist in the Pine Point area by the time of upper barrier sedimentation

(Skall, 1975). Instead, predominantly lagoonal sedimentation resulted in Facies H and I. The strongest evidence concerning the origin of the coarse-crystalline Facies K dolostone is its distribution below the Watt Mountain disconformity and immediately "landward" of the most "seaward" extent of the partial erosional surface (Fig. 2). The disconformity is not exposed in the Pine Point area, but the abundant drill hole information indicates that the paleo-erosional surface is one of relatively low relief with a characteristic karst topography which gently slopes "seaward," generally at less than 1 m/km. Thus, the situation provides a classic model for the dynamic mixing of meteoric water with normal marine water in the subsurface during the post-middle Givetian exposure of the barrier complex. The low relief and lack of major surface drainage on the erosional surface and the apparent lack of stratigraphic restrictions to vadose water flow in the upper barrier suggest that the water table was not much above sea level. A representative figure of 1 m above mean sea level approximately 10 km from the point of termination of the partial disconformable surface (that is, the middle Givetian "shore") is reasonable in comparison with the

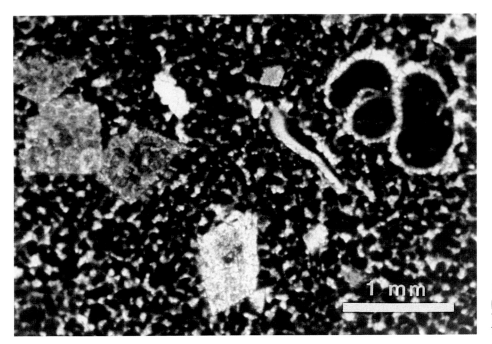

Figure 16—Photomicrograph of pelsparite with gastropod and trilobite fragment; the large dolomite rhombs retain pellet ghosts.

analogous modern hydrologic system of the northern Yucatan Peninsula (Back and Hanshaw, 1970). Assuming that the permeability of the upper Pine Point barrier complex was relatively homogeneous during the erosional period, the Ghyben-Herzberg principle can be applied (Back and Hanshaw, 1970). This principle states that for every unit of fresh water above the mean sea level, the thickness of the fresh water lens floating on salt water of ocean water density is about 40 units. With a hydraulic head of 1 m, the fresh water lens would have extended within the upper barrier to a depth of about 40 m below the mean level of the post-middle Givetian area. A zone of meteoric water and sea water mixing would have existed below this depth and could account for the position of the maximum thickness of coarse-crystalline Facies K dolostone about 10 km "landward" from "shore" and sloping gently upward to mean sea level.

Kyle (1980) has suggested that the development of the coarse-crystalline Facies K dolostone was related to the dynamic mixing of meteoric water with sea water as an integral part of barrier diagenesis during the post-middle Givetian erosional period. This mixing model has been proposed

for dolomitization in several varied carbonate environments (Folk and Land, 1975); at Pine Point it is consistent with evidence of paleo-geography and other effects of subaerial exposure. The mixing zone was probably rather thin at any particular time, but short and long term fluctuations in sea level resulted in migration of the mixing zone and concomitant dolomitization of the entire area now represented by Facies K (Fig. 2).

Post-middle Givetian subaerial exposure of the Pine Point barrier complex resulted in the formation of major solution structures and the coarse-crystalline Facies K dolostone. These solution features were aquifers for mineralizing fluids and loci for sulfide deposition in the Facies K dolostone. Dolines host prismatic orebodies, whereas caves and tabular permeable zones contain tabular orebodies. Origin of the sulfide-hosting breccias in the fine-crystalline dolostones of the lower barrier is less apparent, but these are also believed to be solution features. High-grade sulfide concentrations are localized in dolines and breccia zones because these transgressive features were the bypasses between different aquifers and acted as natural mixing sites for fluids of different character, one of

which contained metals and the other reduced sulfur. Fluid inclusion evidence indicates that the Pine Point sulfides were deposited by highly saline brines at temperatures of 50 to 100°C (Roedder, 1968). These brines appear to have originated within the sedimentary sequence, but the immediate metal source cannot be defined by present data. Composition of the ores indicates only abundant Pb, Zn, and Fe components in a 2:5:3 ratio. The basinal evolution model for Pine Point (Beales and Jackson, 1966) states that the metals now present in the orebodies were released by weathering of continental rocks and concentrated into the Mackenzie Basin shales in the form of absorbed metal ions on clay minerals. During compaction and diagenesis of these water-rich sediments, the metals were released to interstitial fluids as soluble metal chloride and organic complexes. Metal-bearing brines migrated laterally into the permeable reef complex and travelled in the direction of least hydraulic gradient. Reduction of evaporite sulfates in the Elk Point Basin resulted in concentration of hydrogen sulfide in parts of the carbonate complex. Mixing of metal-bearing brines with reduced sulfur resulted in rapid precipitation of metal sulfides. The basinal evolution

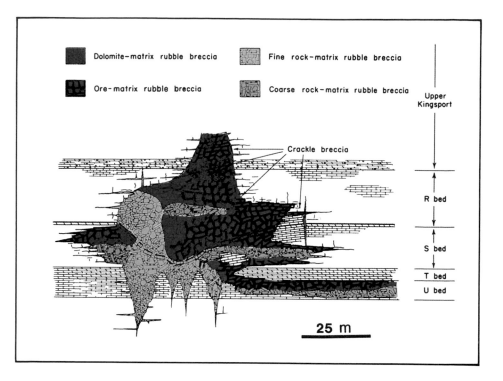

Figure 17—Generalized relationships among breccia types, Mascot-Jefferson City district. (Reproduced with modifications from J. D. Ridge, ed., Ore Deposits of the United States 1933-1967: Am. Inst. Mining, Metallurgy, and Petroleum Engineers, v. 1, p. 242-256.)

model is very appealing for the geologic setting at Pine Point, and variations of the model have been used to explain the origin of some other carbonate-hosted districts.

Tennessee Zinc Districts

Four major mining districts in Tennessee contain economic mineral concentrations in the upper part of the Knox Group, a 1,000 m thick, dominantly shallow water carbonate sequence of Cambro-Ordovician age (Figs. 13, 14). Ore deposits occur as open-space filling of irregular, stratabound breccia bodies in the Lower Ordovician (Beekmantown) Kingsport and Mascot formations. These strata dip from 10 to 40° to the southeast from surface exposures in the northeast-trending imbricate thrust fault belts of the Valley and Ridge Province in east Tennessee. The Knox Group carbonates in central Tennessee and southern Kentucky are covered by a minimum of 90 m of nearly horizontal younger strata. The Mascot-Jefferson City and Copper Ridge districts contain orebodies with breccias tightly cemented only by pale yellow sphalerite and dolomite with varying amounts of iron sulfides. The Sweetwater district has long been a significant producer of residual barite

developed from recent weathering of mineralized breccias, but recent subsurface exploration in the area has concentrated on the fluorite potential with byproduct zinc and barite. Reddish-brown sphalerite is the only economic product from the central Tennessee deposits, but the breccias contain local concentrations of galena, barite, fluorite, calcite, dolomite, and quartz. The reader is referred to a special 1971 issue of *Economic Geology* (v. 66, no. 5) for details on the Tennessee ore deposits.

Lower Ordovician carbonate strata in the Appalachians contain numerous areas with similar mineralized breccias, including economic deposits at Friedensville, Pennsylvania (Callahan, 1968), and Daniels Harbour, Newfoundland (Collins and Smith, 1975). Furthermore, the abundant open space still remaining in the breccia systems in some areas of central Tennessee and southern Kentucky results in the upper Knox Group being an important aquifer. The brecciated upper Mascot Formation where capped by relatively impermeable Middle Ordovician strata in a favorable structural position forms a reservoir for hydrocarbon production (Perkins, 1972).

Stratigraphy

The Knox Group carbonates in central Tennessee are underlain by a Cambrian sequence of carbonate and siliciclastic strata up to 500 m thick overlying the Precambrian basement. Although the depositional relationships are obscured by displacement and crustal shortening in the Valley and Ridge Province, it appears that the pre-Knox sedimentary sequence is as much as 1,000 m thick in the area of the mineral districts and increases to the east as the Upper Precambrian – Lower Cambrian siliciclastic wedge thickens into the Appalachian continental margin basin. The Knox Group is comprised of the Upper Cambrian Copper Ridge Dolomite and the Lower Ordovician Chepultepec Dolomite, Kingsport Formation, and Mascot Dolomite (Fig. 14). These formations are largely dolostone with some limestone, sandstone, and chert interbeds.

Because of the role of stratigraphy in localizing mineralization, the Kingsport and Mascot formations are of particular interest. These formations are dominated by sequences of very fine to fine crystalline (less than 50μ) dense dolostone beds, many of which are remarkably persistent over thousands of square kilometers and

Figure 18—Recent karst collapse breccia in Mississippian limestones, Putnam County, Tennessee. Height of exposure is 10 m.

have been widely used as marker beds for detailed stratigraphic correlations and structural interpretations. These dolostones seldom have any recognizable faunal constituents and exhibit sedimentary features that are indicative of shallow water depositional environments. These "early dolostones" are interbedded with varying amounts of limestone or "limestone equivalents" where these intervals have been diagenetically converted to coarse crystalline (greater than 200μ) dolostones (Fig. 15). Upper Knox limestones are pelmicrites and pelsparites with a sparse fauna dominated by gastropods but also including ostracod, trilobite, brachiopod, and pelmatazoan fragments (Fig. 16). Primary sedimentary structures such as shale partings, sand layers, and chert nodules zones can be traced laterally in single exposures from limestone into coarse crystalline dolostone. Economic mineral concentrations often occur in breccias within the transition zone between limestone and coarse crystalline dolostone, hence the importance of the "limestone edge" in minerals exploration. In east Tennessee the Kingsport Formation consists dominantly of limestone (or equivalents), whereas in central Tennessee approximately equal amounts of limestone and fine crystalline dolostone occur (Fig. 14).

The Mascot Dolomite in east Tennessee is largely fine crystalline dolostone; in central Tennessee, numerous limestone interbeds occur within the Mascot (Fig. 14).

The top of the Knox Group is a major regional unconformity over much of eastern North America (Harris, 1971). The post-Beekmantown section is dominantly argillaceous limestones, except for the Devonian-Mississippian Chattanooga Shale. A thick carbonate sequence of Mississippian age forms the Highland Rim provinces, and the Cumberland Plateau consists of Pennsylvanian deltaic sedimentary units (Fig. 13).

Effects of Subaerial Exposure

Unlike the Pine Point district where many of the orebodies are directly associated with the disconformity which represents the period of subaerial exposure that created the ore-hosting zones, the mineralized strata in the Tennessee districts occur considerably below the post-Knox unconformity. The principal ore-bearing zone in the Mascot-Jefferson City, Copper Ridge, and Sweetwater districts is the lower 20 m of the Mascot Dolomite and the upper 50 m of the Kingsport Formation (Fig. 14). In central Tennessee the most intensely mineralized interval is a 60 m section in the middle Mascot, although

significant zinc concentrations occur above and below this interval.

The post-Knox regional unconformity represents a major period of subaerial exposure. Local relief on the unconformity may be as much as 40 m, and considerable regional beveling has taken place with younger Knox strata eroded to the north and northeast of the mineralized areas. Middle Ordovician strata are thickest where they overlie paleotopographic depressions, and material similar to that immediately above the unconformity is present in fractures and breccias several hundred feet below the top of the Knox.

Breccias in the upper Knox Group may be divided into early breccia systems which are generally unmineralized and late breccia systems which host sphalerite and related minerals. Geometry of these collapse breccias consists of a core of rubble breccia surrounded by mosaic and crackle breccia passing outward into undisturbed wall rock (Fig. 17; Hoagland et al, 1965; McCormick et al, 1971; Hoagland, 1976; Kyle, 1976). Both early and late breccias occur in these general forms although early breccias are generally smaller and less well developed. Some individual breccia bodies affect tens of meters of section, but most are tabular and confined to thinner

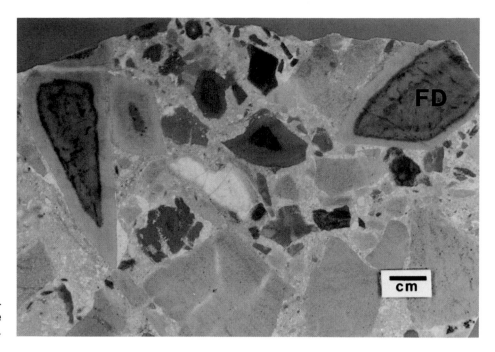

Figure 19—Early, fine rock matrix breccia with bleached borders on very fine crystalline dolostone (**FD**) fragments.

stratigraphic intervals; these geometric relationships can be seen in collapse breccias associated with recent karstification (Fig. 18).

Early, fine rock-matrix breccia contains fragments of very finely crystalline dolostone in a matrix of pale green, and finely crystalline dolostone with varying amounts of tripolitic chert (Fig. 19). In gross appearance some early breccia is similar to the material that immediately overlies the post-Knox unconformity. Fine rock-matrix breccia is present at various intervals within the Mascot Dolomite but is best developed in the upper 150 m, that is, near the post-Knox unconformity. Early breccia fragments often have bleached rims and single or multiple dark concentric bands or reaction fronts adjacent to unbleached cores (Fig. 19). Reaction rims have been observed in scattered breccia intervals from the top of the Knox to as much as 200 m lower (Kyle, 1976). Extensive brecciation is not necessary for good development of the rims; otherwise insignificant fractures are often made obvious by bleached borders.

Reaction rims are similar to bleaching patterns in iron sulfide mottled dolostones as a result of present-day surface oxidation. This analogy is further supported by the

presence of bleached borders on some fragments of Knox dolostone enclosed in the basal Middle Ordovician strata. Evidence for formation of reaction rims and early breccias during the post-Knox erosional interval is provided by these associations and by the predominant occurrence of these features in the upper part of the Knox near the unconformity.

Early breccias are unmineralized except where transected by later fractures. This late fracturing often took place without preference to matrix or breccia blocks indicating that early breccias were fairly well lithified by the time of late brecciation. However, much of this subsequent fracturing occurred along early fractures because of the relative weakness of the breccia matrix. It has been suggested that early breccias commonly form the framework upon which the late, ore-hosting breccia systems developed in east and central Tennessee (Fig. 17; Hoagland et al, 1965; Hoagland, 1976).

Late breccia systems in central Tennessee occupy intervals from above the Knox unconformity to below the top of the Kingsport Formation. The most consistently brecciated and intensely mineralized interval is the 60 m zone in the middle Mascot. This section is particularly favorable for

formation of tabular collapse breccias related to solution of the thick limestone interbeds. Other intervals may also contain significant intersections of mineralized breccia. In east Tennessee, late breccia systems are best developed in the lower 20 m of the Mascot Dolomite and upper 50 m of the Kingsport Formation.

Late breccia systems have two types of matrix or cement. The lower part contains coarse rock-matrix breccia which consists of medium and coarse dolomite rhombs that are matrix to differing amounts of very fine to coarse crystalline dolostone blocks, chert fragments, argillaceous material, quartz sand, and silt (Figs. 20, 21, 22). Vague boundaries often occur between coarse crystalline dolostone fragments and coarse matrix, and all variations between clean secondary coarse crystalline dolostone and detrital coarse rock-matrix breccia are present in the same stratigraphic position (Fig. 15).

Secondary dolomitization and solution of limestone appear to be complementary processes (Kyle, 1976). Volume-for-volume dolomitization of upper Knox limestone has occurred in some areas of Tennessee. Little collapse of overlying strata has taken place, and any mineral occurrences are largely restricted to vugs and minor

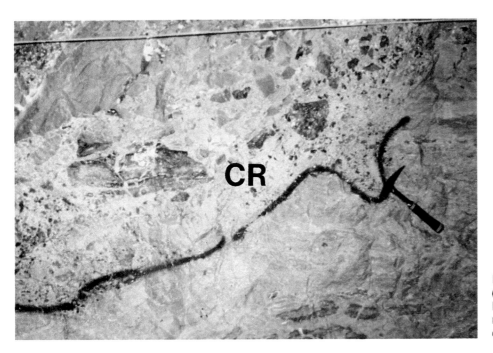

Figure 20—Coarse rock matrix breccia (**CR**; solution residue) in lower portion of late breccia body in central Tennessee; note reddish brown sphalerite cementing overlying breccia.

fractures. In other areas, solution was apparently the dominant process, and varying amounts of coarse rock-matrix breccia represent the insoluble components of the limestone intervals. In these areas the amount of collapse of overlying units is considerable with either tabular breccia zones developed in selected intervals or occasional "break-through" breccia bodies which affect most of the Mascot section. Displacement of breccia blocks is always downward and may be as much as 15 m. In some drill holes in central Tennessee mineralized late breccia systems continue without discontinuity as much as 60 m across the Knox unconformity (Hoagland, 1976).

The upper part of the late breccia systems may host zinc ore deposits. Sphalerite, galena, fluorite, barite, marcasite, pyrite, calcite, dolomite, and quartz form varying proportions of the cement of the mineral-matrix breccia (Figs. 23, 24). The relative amount of these cementing minerals may change greatly over short distances. All late breccias do not contain economic accumulations of sphalerite. In fact, most are cemented largely by carbonate gangue or still contain abundant open space between fragments.

Late breccia fragments do not have

bleached rims like early breccia blocks (Fig. 24), and late breccia blocks with broken reaction rims are occasionally present. These features indicate that bleaching is not directly related to solutions responsible for late brecciation or mineralization and that formation of late breccias took place after cessation of the period of oxidation.

The mechanism commonly proposed for the formation of the late breccia systems in east Tennessee is enlargement of early breccia zones as the result of limestone solution by mineralizing fluids (Hoagland et al, 1965). However, in central Tennessee, sulfides were deposited in equilibrium with early calcite crystals suggesting that the sulfide-depositing fluids could not have been responsible for limestone solution (Kyle, 1976). It is suggested that the early, fine rock matrix breccia systems represent detritus filled solution structures directly related to the post-Knox erosional period. Further, solution and dolomitization of limestone beds may be indirectly related to the erosional period. The regional tilting of the Knox strata accompanying uplift and erosion would have resulted in recharge areas capable of supplying meteoric water along permeable zones to strata well below the top of the

Knox. Subsurface solution of limestone would take place during this period. With the termination of the erosional period by marine transgression, conditions would exist for the establishment of irregular sea water/meteoric water mixing zones. That is, regional topographic highs would remain as recharge areas for the subsurface, while low-lying areas would be inundated by the marine water. With increased meteoric water/marine water mixing (Folk and Land, 1975), dolomitization would become the dominant process, and dolomitized solution residue (coarse rock matrix breccia) and coarse crystalline secondary dolostones would result. Much of the collapse of the fine crystalline dolostone beds overlying the voids due to solution of limestone probably did not take place until considerably later accompanying loading of the upper Knox Group by Middle Ordovician (and younger?) sediments. Therefore, the ore-hosting collapse breccias may not have developed in their present form until long after the termination of the erosional period. However, there is increasing evidence that ephemeral periods of subaerial exposure related to depositional cycles in the upper Knox strata may have played an important role in the development of

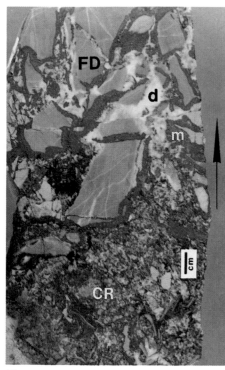

Figure 21—Hand specimen representing a late breccia system from the Copper Ridge district. Coarse rock matrix breccia (**CR**) overlain by a collapse breccia of fine crystalline dolostone fragments (**FD**) cemented by marcasite (**m**) and dolomite (**d**).

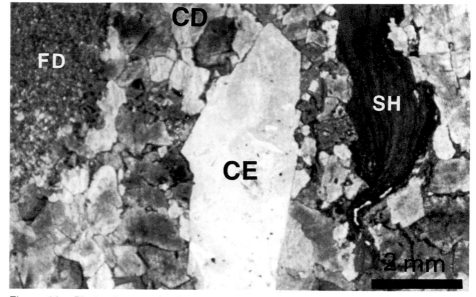

Figure 22—Photomicrograph of coarse rock matrix breccia with fine crystalline dolostone (**FD**), chert (**CE**), and shale (**SH**) fragments in a coarse crystalline dolostone matrix.

the later breccia systems.

Fluid inclusion data indicate that the mineralizing fluids were concentrated brines ranging from 80 to 150°C (Roedder, 1971). Hoagland (1976) has proposed that these fluids were derived by sediment compaction in the Appalachian Basin and were transported through the regional Knox paleo-aquifer to the mineralized areas. Cementation of the upper part of the late breccia systems by sphalerite and related minerals may have occurred over a considerable period of time. Although difficult to determine exactly, mineralization may have taken place as early as Middle Ordovician in east Tennessee, whereas a late Paleozoic age for mineralization can be supported for central Tennessee.

CONCLUSIONS

The Middle Devonian strata of the Pine Point district and the Lower Ordovician strata of the Tennessee districts provide excellent examples of the effects of subaerial exposure on the development of ore-hosting secondary porosity in carbonate terranes. The effects of exposure differ somewhat between the areas because of the carbonate depositional environments and the nature of subaerial exposure surfaces. Nevertheless, many of the ore-hosting features such as strata-bound porosity zones in secondary dolostones and collapse breccia bodies are similar. The consistent relationship of carbonate-hosted ore deposits with unconformities and subtle ephemeral exposure surfaces suggests that diagenetic processes related to subaerial exposure may be more important than generally recognized in the localization of this important type of ore deposit. Similar processes also produce secondary porosity for important hydrocarbon reservoirs.

ACKNOWLEDGMENTS

I am grateful for the opportunity to have been involved in exploration and research in the Tennessee and Pine Point districts and am indebted to many individuals and mining companies in these areas for numerous courtesies. Partial support for this publication was provided by the Geology Foundation of The University of Texas at Austin.

SELECTED REFERENCES

Karstification and Subaerial Exposure
Back, W., and B. B. Hanshaw, 1970, Comparison of chemical hydrogeology of the carbonate peninsulas of Florida and Yucatan: Jour. Hydrology, v. 10, p. 330-368.
Esteban, M., and C. F. Klappa, 1982, Subaerial exposure surfaces, in P. A. Scholle, ed., Carbonate depositional environments: AAPG Mem. 33, this volume.
Folk, R. L., and L. S. Land, 1975, Mg/Ca ratio and salinity; two controls over crystallization of dolomite: AAPG Bull., v. 59, p. 60-68.
Mathews, R. K., 1974, A process approach to diagenesis of reefs and reef associated limestones, in L. F. Laporte, ed., Reefs in time and space: SEPM Spec. Pub. 18, p. 234-256.

Figure 23—Collapse breccia consisting of fine crystalline dolostone blocks cemented by yellow sphalerite and dolomite. Note the different dolostone lithologies represented by the fragments, indicating stratigraphic disruption accompanying collapse. Mascot-Jefferson City district; height of exposure is 4 m.

Moore, C. F., Jr., 1979, Porosity in carbonate rock sequences, *in* Geology of carbonate porosity: AAPG Continuing Education Course Note Series 11, p. A1-A124.

Quinlan, J. F., 1972, Karst-related mineral deposits and possible criteria for the recognition of paleokarsts; a review of preservable characteristics of Holocene and older karst terranes: Sec. 6, 24th Internat. Geol. Cong., p. 156-168.

Thrailkill, J. V., 1968, Chemical and hydrologic factors in the excavation of limestone caves: Geol. Soc. America Bull., v. 79, p. 19-46.

Zuffardi, P., 1976, Karsts and economic mineral deposits, *in* K. H. Wolf., ed., Handbook of strata-bound and stratiform ore deposits: Amsterdam, Elsevier Sci. Pub., v. 3, p. 175-212.

Carbonate-Hosted Ore Deposits

Anderson, G. M., 1978, Basinal brines and Mississippi Valley-type ore deposits: Episodes, v. 2, p. 15-19.

Bernard, A. J., 1973, Metallogenic processes in intra-karstic sedimentation, *in* G. C. Amstutz and A. J. Bernard, eds., Ores in sediments: New York, Springer Verlag Pub., p. 43-57.

Brown, J. S., 1970, Mississippi Valley type lead-zinc ores: Mineral. Deposita, v. 5, p. 103-119.

Carpenter, A. B., M. L. Trout, and E. E. Pickett, 1974, Preliminary report on the origin and chemical evolution of lead- and zinc-rich oil field brines in central Mississippi: Econ. Geology, v. 69, p.

1191-1206.

Jackson, S. A., and F. W. Beales, 1967, An aspect of sedimentary basin evolution; the concentration of Mississippi Valley-type ores during late stages of diagenesis: Canadian Petroleum Geols. Bull., v. 15, p. 383-433.

Land, L. S., and D. R. Prezbindowski, 1981, The origin and evolution of saline formation water, Lower Cretaceous carbonates, south-central Texas, U.S.A.: Jour. Hydrology, v. 54, p. 51-74.

Macqueen, R. W., 1979, Base metal deposits in sedimentary rocks; some approaches: Geosci. Canada, v. 6, p. 3-9.

Ohle, E. L., 1959, Some considerations in determining the origin of ore deposits of the Mississippi Valley type: Econ. Geology, v. 54, p. 769-789.

———, 1980, Some considerations in determining the origin of ore deposits of the Mississippi Valley type; part II: Econ. Geology, v. 75, p. 161-172.

Roedder, E., 1976, Fluid inclusion evidence on the genesis of ores in sedimentary and volcanic rocks, *in* K. H. Wolf, ed., Handbook of strata-bound and stratiform ore deposits: Amsterdam, Elsevier Sci. Pub., v. 2, p. 67-110.

Sangster, D. F., 1976, Carbonate-hosted lead-zinc deposits, *in* K. H. Wolf, ed., Handbook of strata-bound and stratiform ore deposits: Amsterdam, Elsevier Sci. Pub., v. 6, p. 447-465.

Pine Point Lead-Zinc District

Beales, F. W., and S. A. Jackson, 1966, Precipitation of lead-zinc ores in car-

bonate reservoirs as illustrated by Pine Point ore field, Canada: Trans. Inst. Mining Metallurgy, v. 75, p. B278-B285.

Campbell, N., 1967, Tectonics, reefs, and stratiform lead-zinc deposits of the Pine Point area, Canada, *in* J. S. Brown, ed., Genesis of stratiform lead-zinc-barite-fluorite deposits in carbonate rocks: Econ. Geology Mon. 3, p. 59-70.

De Wit, R., et al, 1973, Tathlina area, District of Mackenzie, *in* The future petroleum provinces of Canada: Canadian Soc. Petroleum Geols. Mem. 1, p. 187-212.

Kyle, J. R., 1980, Controls of lead-zinc mineralization, Pine Point district, Northwest-Territories, Canada: Mining Engineer, v. 32, p. 1617-1626.

———, 1981, Geology of the Pine Point lead-zinc district, *in* K. H. Wolf, ed., Handbook of strata-bound and stratiform ore deposits: Amsterdam, Elsevier Sci. Pub., v. 9, p. 643-741.

Maiklem, W. R., 1971, Evaporative drawdown—a mechanism for water level-lowering and diagenesis in the Elk Point basin: Canadian Petroleum Geols. Bull., v. 17, p. 194-233.

Norris, A. W., 1965, Stratigraphy of Middle Devonian and older Paleozoic rocks of the Great Slave Region, Northwest Territories: Geol. Survey Canada, Mem. 322.

Roedder, E., 1968, Temperature, salinity, and origin of the ore-forming fluids at Pine Point, Northwest Territories, Canada, from fluid inclusion studies:

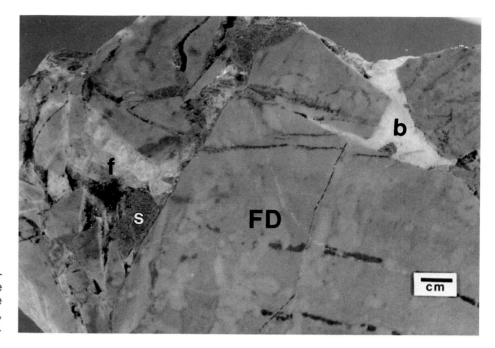

Figure 24—Hand specimen of a collapse breccia from central Tennessee consisting of fine crystalline dolostone blocks (**FD**) cemented by sphalerite (**s**), fluorite (**f**), and barite (**b**).

Econ. Geology, v. 63, p. 439-450.

Skall, H., 1975, The paleoenvironment of the Pine Point lead-zinc district: Econ. Geology, v. 70, p. 22-45.

Tennessee Zinc Districts

Callahan, W. H., 1968, Geology of the Friedensville zinc mine, Lehigh County, Pennsylvania, *in* J. D. Ridge, ed., Ore deposits of the United States, 1933-1967: New York, Am. Inst. Mining, Metall. Petroleum Engineers, p. 95-107.

Collins, J. A., and L. Smith, 1975, Zinc deposits related to diagenesis and intrakarstic sedimentation in the Lower Ordovician St. George Formation, Western Newfoundland: Canadian Petroleum Geols. Bull., v. 23, p. 393-427.

Crawford, J., and A. D. Hoagland, 1968,

The Mascot-Jefferson City zinc district, Tennessee, *in* J. D. Ridge, ed., Ore deposits of the United States, 1933-1967: New York, Am. Inst. Mining, Metall. Petroleum Engineers, p. 242-256.

Harris, L. D., 1971, A lower Paleozoic paleoaquifer-the Kingsport Formation and Mascot Dolomite of Tennessee and southwest Virginia: Econ. Geology, v. 66, p. 735-743.

Hoagland, A. D., 1976, Appalachian zinc-lead deposits, *in* K. H. Wolf, ed., Handbook of strata-bound and stratiform ore deposits: Amsterdam, Elsevier Sci. Pub., v. 6, p. 495-534.

_____, W. T. Hill, and R. E. Fulweiler, 1965, Genesis of the Ordovician zinc deposits in East Tennessee: Econ. Geology, v. 60, p. 693-714.

Kyle, J. R., 1976, Brecciation, alteration,

and mineralization in the Central Tennesse zinc district: Econ. Geology, v. 71, p. 892-903.

McCormick, J. E., et al, 1971, Environment of the zinc deposits of the Mascot-Jefferson City district, Tennessee: Econ. Geology, v. 66, p. 757-762.

Perkins, J. H., 1972, Geology and economics of Knox dolomite oil production in Gradyville East Field, Adair County, Kentucky, *in* Proceedings of the Technical Sessions, Kentucky Oil and Gas Association Annual Meetings, 1970-71: Geol. Survey, Spec. Pub. 21, p. 10-25.

Roedder, E., 1971, Fluid-inclusion evidence on the environment of formation of mineral deposits of the Southern Appalachian Valley: Econ. Geology, v. 66, p. 777-791.

Paleokarst Development

W. Martinez del Olmo
Mateu Esteban

*A*long the margins of the western Mediterranean, Mesozoic carbonates in structurally controlled paleotopographic highs present a well developed paleokarst of local economic importance with significant oil fields. In the offshore of northeastern Spain, different horizons of the Jurassic-Cretaceous sequences are overlain by Miocene marine sediments, although some low areas contain a basal Paleogene of continental facies. Mesozoic series in this area are dominated by carbonate rocks with important dolomitic intervals. Subaerial exposure and karstification covers a variable time span (40 to 60 m.y.) depending on the location of the paleotopographic highs. Onlapping Miocene sediments (carbonates, shales and sands) date from Lower to Upper Miocene.

The Mesozoic-Miocene unconformity is marked by up to 100 m of Mesozoic breccia and it is usually difficult to establish a sharp limit between Mesozoic and Miocene units. The breccia clasts are highly irregular in size and shape (from large boulder to pebbles) and are cemented by complex generations of calcite cements and terra-rossa. At the upper part, coastal marine reworking of the breccia during the Miocene is commonly present. Cavern porosity is well developed as well as calcitization (dedolomitization) and leaching of dolostones. This set of features is interpreted as indicative of the presence of a subaerial exposure surface. All

Figure 1—Unconformity Mesozoic-Miocene (green horizon) interpreted as subaerial exposure surface, offshore northeastern Spain. Yellow reflectors correspond to the Upper Miocene, the limit Lower-Middle Miocene is established at the orange reflector. Note hogback morphology of the erosion surface.

Figure 2—Four examples of paleokarst development in Mesozoic carbonates, offshore northeastern Spain, with characteristic gamma ray and sonic log response. The subaerial exposure surface corresponds to the same event represented in Figure 1.

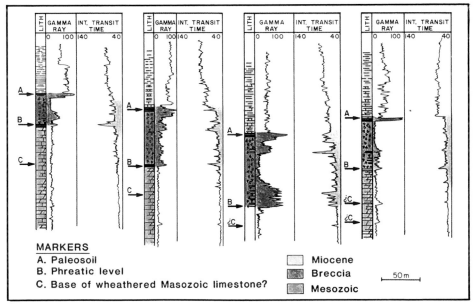

MARKERS
A. Paleosoil
B. Phreatic level
C. Base of wheathered Masozoic limestone?

☐ Miocene
▓ Breccia
▒ Mesozoic

50 m

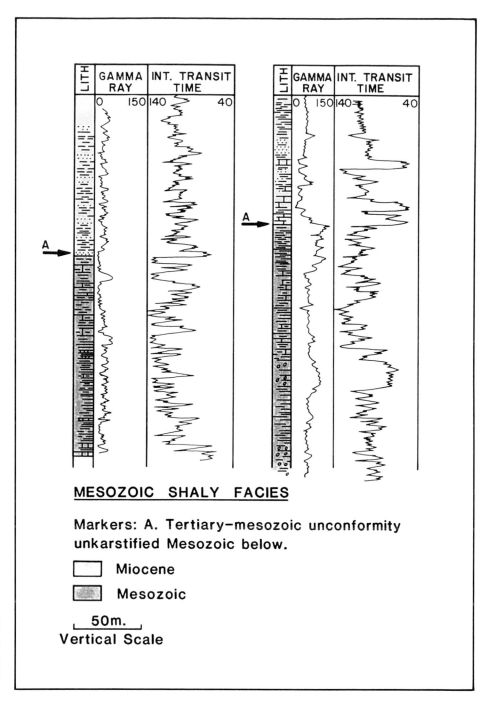

MESOZOIC SHALY FACIES

Markers: A. Tertiary-mesozoic unconformity
unkarstified Mesozoic below.

☐ Miocene

▨ Mesozoic

⌐ 50m. ⌐
Vertical Scale

Figure 3—Two examples of the same subaerial exposure surface shown on Figures 1 and 2, but affecting unkarstified, shaly Mesozoic basement. Logs show no particular response at the unconformity.

seismic profiles in the area (Fig. 1) show a marked reflector at the base of the Miocene onlap, locally in angular discordance with the Mesozoic basement. This surface can be easily correlated to nearby land exposures, where additional evidence of subaerial exposure can be found (*Microcodium*, rhizoliths, caliche profiles; see Figs. 11a, b, d and g, 68, and 70).

Although non-diagnostic, log response is quite characteristic and useful in local correlations. In the more typical situations (Fig. 2), the gamma-ray log shows a marked increase at the brecciated horizon. We can distinguish two sharp peaks at the top and the bottom of the brecciated interval (A and B in Fig. 2), characteristics of clean shale lithology. Usually A is more intense than B,

with other minor peaks irregularly distributed along the brecciated interval. The sonic log shows a progressive decrease in Δt from the top to the bottom of the brecciated interval (A-B in Fig. 2), mostly noticeable at the upper part (first 40 to 50 m). The sonic response also suggests the presence of fractures at and below the breccia. The lower part of the brecciated interval is characterized by a lower but still non-stabilized Δt. Toward deeper levels, Δt progressively disminishes up to a point C (deeper than B) where the response is completely stabilized and corresponds to the characteristic of the unaltered carbonate rock formation.

This log response can be explained in light of the ideal karst profile described in the text. The top of the profile (A) shows a highly radioactive peak, interpreted as a paleosoil and zone of infiltration. The horizon B is probably produced by the top of the phreatic level and its oscillations. This is the zone of major cavern formation and deposition of cave sediments, corroborated by the occurrence of total loss of circulation in the well at this point. The interpretation of the deep interval B-C is less satisfactory, but obviously represents the transition to the unaltered rock; perhaps it coincides with the transition zone of the ideal karst profile described in the text. Where the subaerial exposure surface occurs on a non-carbonate substrate, log response is quite different in character (Fig. 3), and presents difficulties in the location of Mesozoic-Miocene unconformity.

2

Lacustrine Environment

Walter E. Dean
Thomas D. Fouch

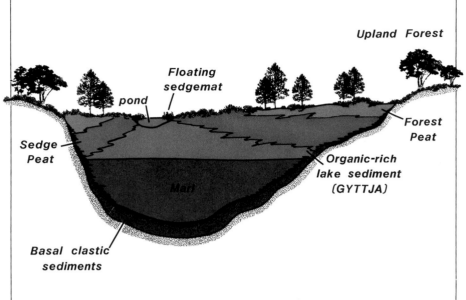

Figures 1, 2—Typical, moderately deep, north temperate zone lake in late stage of development. Open lake water is bounded by a floating mat of sedges and reeds forming a sedge-reed peat (**a**). A bog forest (**b**) accumulating forest peat surrounds and eventually replaces the sedge mat and is in turn surrounded by dry land and upland forest (**c**). As lake productivity increases, the open lacustrine sediment becomes richer in organic matter and forms an olive to black, fine-grained sediment called gyttja that may or may not contain carbonate depending upon the chemistry of the lake system. Such a lake system can produce and preserve sediments that contain lipid-rich, herbaceous, and woody organic matter.

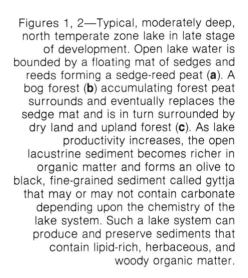

Sedimentary rocks deposited in lacustrine systems are known from much of the world, but relatively few have been the focus of exploration for oil and/or gas. Most large oil and gas fields are related to, or found in, marine strata that are the source or reservoir rocks for hydrocarbons. Large accumulations of oil and gas trapped in rocks believed to have formed in ancient lake systems, however, are known from the western part of the United States, parts of South America and Africa, Indonesia, Russia, and from much of China. In addition, shows of hydrocarbons and small oil and gas fields developed in lacustrine strata are known from several other areas.

The purpose of this chapter is to illustrate some of the most important features in lacustrine carbonate rocks, and particularly those that formed in lake basins and have the potential to produce oil and gas. Hydrocarbon source, reservoir, and trap units of Chinese, African, South American and United States fields were deposited in large lacustrine basins comparable in size to inland seas. Some of these lakes existed for

several millions of years, a life span not generally characteristic of most present-day lakes. The ancient lake basins of China contain the principal oil- and gas-bearing lacustrine beds in the world. The Uinta Basin of northeastern Utah is another large lake basin that contains large volumes of hydrocarbons. Unlike the lacustrine strata of China that consist mainly of siliciclastic rocks, those of the Uinta Basin consist of large proportions of both siliciclastic and carbonate rocks. The primary reservoirs for hydrocarbons in the Uinta Basin are fractured sandstone and siltstone, but the hydrocarbon source rocks are carbonate mudstones.

Lakes are frequently thought of as freshwater systems, but are commonly alkaline and/or saline. It is also evident from analysis of physical, chemical, and biological data that most ancient lakes were dynamic, and the depositional facies reflect frequent changes in their chemistry and bathymetry. As a result, characteristic features used to recognize a particular depositional phase of a lake's history may differ significantly from those of a different lake phase. Thus, the distribution of sedimentary structures and fossils in lacustrine facies can vary significantly from one phase to another as physical and chemical characteristics of the hydrologic basin change. Because of the extreme variability it is very difficult to pro-

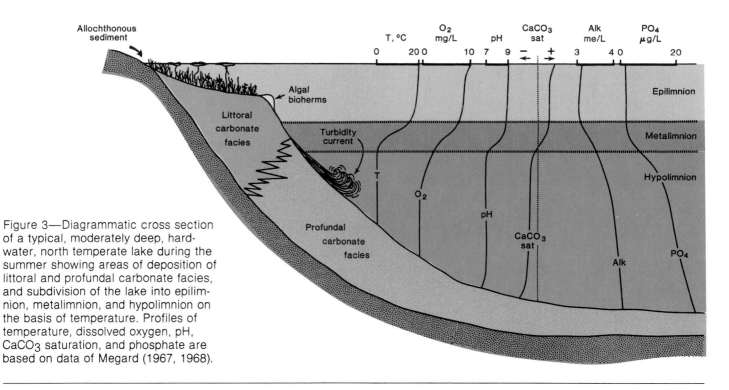

Figure 3—Diagrammatic cross section of a typical, moderately deep, hardwater, north temperate lake during the summer showing areas of deposition of littoral and profundal carbonate facies, and subdivision of the lake into epilimnion, metalimnion, and hypolimnion on the basis of temperature. Profiles of temperature, dissolved oxygen, pH, $CaCO_3$ saturation, and phosphate are based on data of Megard (1967, 1968).

pose a unique set of the physical, chemical, and biological criteria that specifically characterize lacustrine depositional environments.

Many physical, chemical, and biological features of lacustrine rocks are not unique to this depositional environment. Some illustrations in this chapter are of rocks that formed in large lacustrine depositional systems and both the lacustrine and associated peripheral fluvial facies are illustrated. Much of the oil and gas in the Uinta Basin and China is recovered from lacustrine rocks, but much is also recovered from rocks that formed in environments at the fluctuating margins of the lake or in environments well removed from the lakes. Hydrocarbons in nonlacustrine beds are believed to have been derived from lacustrine rocks and to have migrated into peripheral facies.

Many descriptions and discussions of a great variety of ancient and modern lacustrine depositional systems are available in the literature, and many of these papers are listed in the reference section of this chapter. However, it is evident that only a few of these were of sufficient size, or produced and preserved sufficient organic matter, to form oil and/or

gas in significant quantities. Ancient lake beds found in the western part of the United States and specifically in Utah, Colorado, and Wyoming have been the focus of many studies of the depositional histories of the rocks. For this reason, and because of the authors' familiarity with these units and of comparable lakes, many of the materials used in this chapter are taken from studies of these rocks by ourselves and colleagues.

DEPOSITIONAL SETTINGS

Figure 1 is a diagrammatic illustration of the distribution of sedimentary facies resulting from the postglacial development of a typical north temperate lake, and Figure 2 is a photograph of the surface of such a lake. North temperate lakes commonly contain both carbonate and siliciclastic sediments. The physical, chemical, and biological processes in such a lake are similar in many ways to those inferred for some ancient lakes. Because much is known of the interrelationships between these processes in modern lakes, it is helpful to review this information in order to understand and correctly interpret facies relations in ancient lacustrine

sequences.

The four most important components in the sediments of a typical moderately deep (about 25 m), hardwater[1], north temperate lake are: (1) detrital material, (2) biogenic silica, (3) organic matter, and (4) carbonate minerals. The relative importance of each of these components changes as the lake evolves from a newly filled basin to dry land over a period of about 10,000 years.

Basal sediments are commonly sand, silt, and clay derived from erosion of bedrock in the drainage basin, or transported into the drainage basin as glacial drift. These contain very little organic matter because the organic productivity of the lake at this stage is very low (oligotrophic) and there is little vegetation on the surrounding land. If the drainage basin contains some limestone and/or dolomite, or calcareous glacial drift, leaching of this material produces an accumulation of dissolved calcium carbonate within the lake. Once the lake water

---

[1]Hardwater lakes contain substantial concentrations (usually more than 1.0 milliequivalents per liter) of total dissolved alkaline-earth cations, mostly calcium and magnesium, and surface waters are usually saturated with respect to $CaCO_3$ at least during the summer months.

Figure 4—Typical crystals of low-Mg calcite. Photograph is of a sediment-trap sample from Green Lake, Fayetteville, New York. Bar = 10μm.

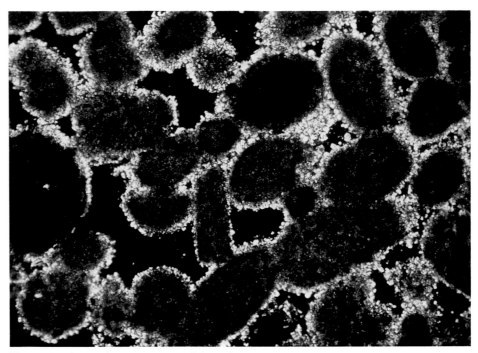

Figure 5—Aragonite cement precipitated subaqueously on micritic aragonite fecal pellets from the brine shrimp *Artemia salina* from Lake Urmia, Iran. The aragonite in the fecal pellets is precipitated in the lake during seasonal chemical events. Photograph by Kerry Kelts.

is saturated with respect to calcium carbonate, one or more carbonate minerals may precipitate and form a carbonate-rich sediment (marl) with varying proportions of clastic material.

At the same time that carbonate is leached from the surrounding areas, nutrients are also leached and begin to accumulate in the lake. Organic productivity of the lake increases and more organic debris either is incorporated into the sediment or decays, releasing more nutrients for organic growth. In addition, organic matter is also contributed to the lake from vegetation transported from the drainage basin. The organic matter that accumulates in the lake is a mixture of both allochthonous material (pollen grains, leaves, needles, seeds, woody material, and other organic detritus) from the drainage basin, and autochthonous material (largely debris from planktonic algae and aquatic macrophytes). Allochthonous organic matter tends to be enriched in humic, high-molecular-weight, carbon-rich compounds, whereas autochthonous phytoplankton debris is enriched in low-molecular-weight, nitrogen-rich lipids. Studies of sedimentary plant pigments in recent lake sediments indicate that most of the organic matter in moderately to highly productive lakes is autochthonous. Throughout the open-lake history of the lake, therefore, the most important source of organic matter is from floating microscopic algae or phytoplankton. The sediment

at this highly productive (eutrophic) stage of the lake's development is likely to be a brown or black, lipid-rich organic sediment called gyttja. The organic content is usually greater than 20% (about 10% organic carbon) and may be as high as 50% (about 25% organic carbon). If the lake is highly eutrophic with a hypolimnion that is anoxic throughout much of the year, or if the lake is permanently stratified (*meromictic*) with permanently anoxic bottom waters (*monimolimnion*) that contains hydrogen sulfide and/or methane, the sediment is likely to be a black, foul-smelling ooze called sapropel that contains abundant lipid-rich organic matter and ferrous sulfides.

As the lake fills and becomes shallower, the littoral zone of rooted aquatic vegetation becomes wider, and a floating mat of sedges begins to grow outward into the lake. The algal gyttja around the margins of the lake is replaced by sedge peat (Fig. 1).

Figure 2 is a photograph of the surface of a lake at this stage of development. The sedge peat eventually covers the gyttja as the lake continues to become shallower and smaller in area. Once the sedge peat provides a stable substrate, a bog forest can develop and eventually the sedge peat will be covered with a layer of bog-forest peat. The resulting relations among lacustrine sedimentary facies (Fig. 1) would consist of: (1) basal clastic sediments grading upward into marl as the sediments become increasingly rich in carbonate, (2) marl grading upward into algal gyttja recording increased productivity of the lake, (3) gyttja grading into sedge peat, and (4) sedge peat grading into bog-forest peat as algal organic matter is replaced by herbaceous organic detritus from aquatic and forest vegetation. Such a lake system has the capacity to generate and preserve lipid-rich, herbaceous, and woody organic materials that are potential sources of both oil and natural gas.

Most temperate lakes are dimictic, characterized by semi-annual over-turns, one in the spring and one in the fall. Overturn occurs because water has a maximum density at about 4°C. Therefore, surface water

Figure 6—Crystals of trona ($NaHCO_3.Na_2CO_3.2H_2O$) from the surface deposit of Lake Magadi, Kenya, a saline, sodium-carbonate-bicarbonate lake in the Rift Valley of East Africa. The lake is underlain by a deposit of bedded trona up to 50 m thick that has accumulated over the past 9,000 years (Eugster, 1970; Eugster and Hardie, 1978). Lake Magadi is fed by alkaline hot springs that create perennial water bodies at the lake margins even during the dry season. Note hammer for scale. Photograph by Hans P. Eugster.

Figure 7—Nahcolite ($NaHCO_3$) nodules consisting of aggregates of bladed crystals in laminated dolomitic marlstone (typical oil shale) of the Green River Formation, Piceance Creek Basin, Colorado (Dyni, 1974). Note that laminae of marlstone are deformed around the nahcolite nodules, but some bladed crystals have grown into the enclosing marlstone. Photograph by John R. Dyni.

will sink to the bottom when cooled to 4°C in the fall and when warmed to 4°C in the spring. These overturns are the heartbeat of a lake and, indeed, are important in maintaining life in the lake. Overturn is the main way that oxygen is supplied to the bottom waters of a lake. Similarly, during overturns the surface waters are restocked with nutrients that accumulate in the bottom waters during summer or winter stagnation.

At fall overturn, the entire water column is isothermal at 4°C. Eventually the temperature of the surface water reaches 0°C and the lake freezes. The lake can no longer receive oxygen from the atmosphere, and oxygen production by photosynthesis is at a minimum. Oxygen consumption by respiration and decay continues and eventually the lake waters may become depleted in dissolved oxygen. Oxygen depletion occurs first in the bottom waters and moves upward through the winter. If oxygen depletion occurs rapidly, or if ice remains on the lake for a long period of time, the entire water column may become deoxygenated, resulting in a winter kill when much of the fish population suffocates.

Ice finally melts and the surface water is heated to 4°C. At this time, the lake is isothermal, and the water can be completely mixed by wind, carrying oxygenated surface waters to the bottom and returning nutrients stored in the bottom waters to the surface. During the spring and summer, the surface water is continually warmed, and by June or July a temperature profile of the lake will be similar to that in Figure 3. At this time, the lake water is divided into three water masses by temperature-related density differences. The upper water mass or *epilimnion* may be isothermal as a result of wind mixing. The thickness of the epilimnion is therefore determined largely by the depth of wind mixing. The middle

Figure 8—Laminated kerogenous carbonate mudstone (oil shale) that has been deformed. Man's finger points to cavity caused by leaching of water-soluble nahcolite nodule to form the bird's nest zone in the saline facies of the Parachute Creek Member of the Green River Formation along Evacuation Creek, eastern Uinta Basin, Utah. Beds are exposed along mined trace of gilsonite vein in the Green River Formation (Eocene) near Bonanza, Utah.

Figure 9—Interbedded halite (deeply eroded layers) and fine-grained to microcrystalline nahcolite from the same horizon in two cores of the Green River Formation separated by about 5 km, Piceance Creek Basin, Colorado. Each halite-nahcolite couplet may be a seasonal unit of sedimentation (a varve; Dyni et al, 1970). Unlike diagenetic nodules of nahcolite (Fig. 7), the nahcolite laminae probably precipitated directly from lacustrine brine (Dyni, 1974). Photograph by John R. Dyni.

water mass (*metalimnion*) is characterized by a rapid change in the temperature-depth curve (this region of the curve is called the thermocline). The lower water mass or *hypolimnion* is usually characterized by a gradual decrease in temperature to the bottom. Depending upon the abundance of plant life in the lake and the extent of wind mixing, the epilimnion may be saturated or even supersaturated with oxygen from the atmosphere and from photosynthesis. In the hypolimnion, however, oxygen may be completely eliminated by respiration and decay. The depth at which oxygen production by photosynthesis equals oxygen consumption by respiration and decay is called the compensation depth. The compensation depth varies daily but is usually in the metalimnion. The part of the lake bottom that is populated by rooted aquatic vegetation is called the littoral zone (Fig. 3). The part of the lake bottom under the hypolimnion is called the profundal zone. There is usually a transition zone called the sublittoral zone between the littoral and profun-

Figure 10—Crystal of ankerite from the water-insoluble residue of bedded nahcolite and halite in the Green River Formation, Piceance Creek Basin, Colorado. Bar = 10 μm. Photograph by John R. Dyni.

Figure 11—Fossils of the prosobranch gastropod *Goniobasis tenera* (Hall) preserved in a packstone of the Green River Formation near Soldier Summit, Utah. These gastropods lived in shallow, fresh, oxygenated water at the edge of Lake Uinta in the early Eocene. Scale is in centimeters.

dal zone that is inhabited mostly by algae and mosses.

DEPOSITION OF CARBONATE IN LAKES

For simplicity, the sediments of a marl lake can be considered a four-component system of carbonate, organic matter, biogenic silica (mostly diatom frustules), and detrital siliciclastics. An increase in the relative abundance of any one of the four components will result in decreases in the relative abundances of the other three. As a result of dilution by organic matter, clastic material, and biogenic silica, sediments of most marl lakes, even those known to precipitate large quantities of $CaCO_3$, usually contain less than 50% $CaCO_3$.

The major sources of carbonate in lake sediments are: (1) inorganically precipitated carbonate, (2) photosynthesis-induced, inorganically precipitated carbonate (hereafter called bio-induced carbonate), (3) biogenic carbonate consisting of debris from calcareous plants and animals, and (4) allochthonous (detrital) material derived from carbonate rocks in the drainage basin. Unlike marine sediments in which most carbonate is derived from remains of calcareous organisms, most of the carbonate in lake sediments is inorganic or bio-induced; however, ostracode- and mollusk-rich layers do

occur. Bio-induced precipitation of carbonate is obvious in the littoral zone of a hard-water lake where rooted aquatic vegetation may be covered with a crust of carbonate that may exceed the weight of the plant (Wetzel, 1975). In the central, open-water (pelagic) part of the lake, bio-induced precipitation of carbonate is important, but less obvious because the plants are microscopic algae.

Mineralogy

The most abundant carbonate mineral in both profundal and littoral sediments of marl lakes is low-Mg calcite, that is, calcite containing less than 5 mole-percent Mg (Fig. 4). Small amounts of aragonite occur, but usually as biogenic aragonite from mollusks. According to Muller and others (1972) primary aragonite can usually form in lakes if the Mg-to-Ca ratio in the water is greater than about 12. Ratios of Mg-to-Ca

this high are rare in temperate marl lakes but may occur in alkaline saline lakes (Fig. 5). The Mg-to-Ca ratio of the water also determines whether primary high-Mg calcite (>5 mole-percent Mg) and diagenetic dolomite will form. Observations by Muller and others (1972) indicate that primary high-Mg calcite precipitates when the Mg-to-Ca ratio in the water is between 2 and about 12, and diagenetic dolomite occurs in lakes with high-Mg calcite as a primary carbonate mineral and a Mg-to-Ca ratio between 7 and about 12. Dean and Gorham (1976) reported high-Mg calcite and dolomite in surface sediments of marl lakes in the semiarid prairie regions of western and southern Minnesota. Of the 46 lakes they studied, the lake with the most dolomite in its profundal sediments (Elk Lake, Grant County, Minnesota) was the only one with a high Mg-to-Ca ratio (7.7). Dean and Gorham also found that the Mg-to-Ca ratio in precipitated calcite increases with increasing Mg-to-Ca ratio in the lake water. Muller (1970) and Muller and Wagner (1978) also found that the Mg-to-Ca ratio of lake water determines the Mg content of

Figure 12—Laminated kerogenous carbonate that has been desiccated to form polygonal cracks. Cracks are filled with mollusk skeletal fragments including the shells of the gastropod *Goniobasis* sp. These beds are part of the Green River Formation near Soldier Summit in the Uinta Basin. Kerogenous units were formed and preserved in anoxic water at the fluctuating margin of Lake Uinta. The mollusks probably lived in oxygenated water near the shoreline and skeletal debris filled the cracks during shoreline fluctuations. Scale is in inches and centimeters.

Figure 13—Fossil unionid bivalves (*Plesielliptio* sp.) preserved in packstone composed of bivalve and gastropod skeletal debris in the Green River Formation near Soldier Summit, Utah. These strata were deposited in a shallow, nearshore lacustrine environment with aquatic plants and fresh, oxygenated water (Fouch et al, 1976). Divisions on scale are in millimeters.

high-Mg calcite precipitated by extensive growth of phytoplankton in Lake Balaton, Hungary.

Sodium carbonate and bicarbonate minerals such as trona ($NaHCO_3 \cdot Na_2CO_3 \cdot 2H_2O$; Fig. 6), nahcolite ($NaHCO_3$; Figs. 7 to 9), and natron ($Na_2CO_3 \cdot 10H_2O$) are common in sediments of ancient saline alkaline lake systems (Culbertson, 1966; Hite and Dyni, 1967; Bradley and Eugster, 1969; Dyni, 1974; Robb and Smith, 1974; and Eugster and Hardie, 1975) and modern saline lakes (Eugster and Hardie, 1978; Smith, 1979). Precipitation of these minerals requires extreme brine concentrations of about 300,000 ppm total dissolved solids (Eugster and Hardie, 1978; Smith, 1979), which, although rare, have been reached in some large modern and ancient lake systems. For example, saturation with respect to trona in Lake Magadi, Kenya, is reached only after lake waters have been concentrated about 250 times beyond the point at which all alkaline-earth carbonates have been precipitated (Jones and others, 1977). Iron carbonates

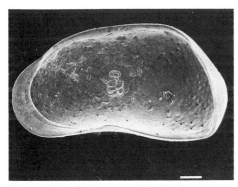

Figure 14—Calcareous bivalved carapace of a Pliocene ostracode. Bar = 100 μm. Photograph by Richard M. Forester.

Figure 15—Living charophyte (*Chara* sp.) containing about 50% dry weight CaCO₃ (left hand) and carbonate mud that consists mostly of low-Mg calcite derived from charophytes (right hand), Green Lake, Fayetteville, New York.

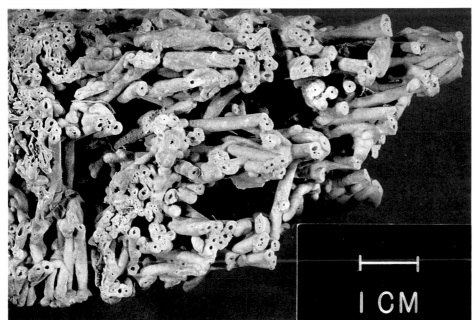

I CM

Figure 16—Encrustations of low-Mg calcite around stems of charophytes, Blue Hole, Ohio.

siderite (FeCO₃) and ankerite (Ca[Mg$_x$ Fe$_{1-x}$]CO₃) are common in some ancient lacustrine strate such as the lower Tertiary Green River Formation (Fig. 10; Smith and Robb, 1966; Robb and Smith, 1974; Desborough and Pitman, 1974; Desborough, 1978), but with one exception they have not been reported from modern lacustrine sediments. The exception is a possible occurrence of siderite reported by Anthony (1977) from varved sediments of Lake of the Clouds, an iron-meromictic lake in northern Minnesota. He found X-ray diffraction lines that correspond approximately to the six strongest lines of siderite. Formation of siderite requires a reducing environment and an extremely low sulfide concentration so that ferrous carbonate can form instead of ferrous sulfide (Kelts and Hsu, 1978). Desborough (1978) reported a large variety of Ca-Mg-Fe carbonates in oil shale from the Green River Formation. He recognized abundant substitution of Mg for Fe in siderite to form Mg-siderite, and substitution of Fe for Mg to form ankerite and Ca-ankerite. Smith and Robb (1966) concluded that the dominant carbonate mineral in the oil-rich Mahogany zone of the Green River Formation in the Piceance Creek Basin, Colorado, was ankerite with

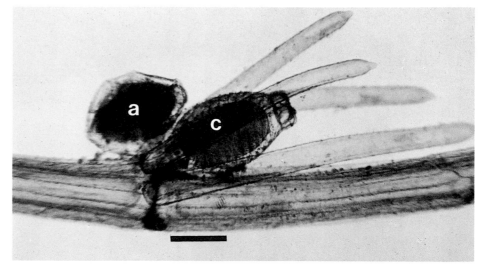

Figure 17—Antheridium (**a**) and oogonium (**c**), the male and female reproductive structures of *Chara* sp. The outer covering of the oogonium (gyrogonite) calcifies and is a distinctive, common calcareous fossil in lacustrine carbonate sediments and rocks. Bar = 0.2 mm.

an average composition of Ca(Mg.85Fe.15)CO3 and not stoichiometric dolomite. Cole and Picard (1978) found that dolomite and ankerite are most abundant in the distal, open-lacustrine oil shales in the Parachute Creek Member of the Green River Formation in the Piceance Creek Basin, Colorado, and eastern Unita Basin, Utah. Manganese carbonate (rhodochrosite) has been reported associated with ferromanganese deposits in Green Bay, Lake Michigan, by Callender and others (1973).

In the discussions that follow, we generally use the term carbonate to mean collective carbonate minerals in a sediment with a predominance of low-Mg calcite, but with the understanding that some aragonite, high-Mg calcite, and/or dolomite may also be present. We use CaCO3 in some places, usually in the context of a precipitate in the water, with the implication that the mineral is calcite but possibly has a variable Mg content. Additional recent discussions of lacustrine carbonate mineralogy are in Jones and Bowser (1978) and Kelts and Hsu (1978).

Profundal Carbonate

Most of the carbonate that accumulates in the sediments of a typical marl lake is generated within the lake. Most detrital sediment that enters a lake is deposited in the littoral zone. Some detrital material re-

Figure 18—Calcareous outer cover (gyrogonite) of the female reproductive structure (oogonium) of a late Pleistocene charophyte. Bar = 100 μm. Photograph by Richard M. Forester.

mains in suspension and is deposited as pelagic sediment in the profundal zone, but these contributions are usually minor. Redeposited littoral carbonate may be a major source of profundal carbonate in some lakes, but rarely is it the dominant source. Examples of what we consider to be extreme redeposition of littoral sediment occurs in Green and Round

Figure 19—*Phacotus* sp., a planktonic, lacustrine green alga from Elk Lake, Clearwater County, Minnesota. This is one of the few calcareous planktonic organisms that contributes carbonate to lacustrine sediments. Bar = 1 μm.

Lakes, Fayetteville, New York (Figs. 22 to 24). Ludlam (1974) estimated that about 50% of the profundal sediment in Green Lake consists of redeposited littoral sediment transported by turbidity currents (Fig. 22). The sides of Green and Round Lakes are unusually steep (about 30°), however, and much of the littoral zone is rimmed with overhanging algal bioherms (Figs. 37, 38; Dean and Eggleston, 1975; Eggleston and Dean, 1976). Slumping is also common on the steep slopes of these two lakes (Figs. 23, 24). Littoral sediment transported into the profundal zone by turbidity currents is usually easy to

Figure 20—Varve laminations from sediments in Elk Lake, Clearwater County, Minnesota. Light layers consist mostly of fine-grained low-Mg calcite and diatoms (Fig. 21); dark layers consist mostly of diatoms and clay. Marks on scale are in millimeters.

Figure 21—Fine-grained low-Mg calcite and diatoms (*Stephanodiscus niagarae*) from a varve lamina in sediments from Elk Lake, Minnesota (Fig. 20). Bar = 100 μm.

recognize because of differences in color, grain size, textures, bed forms, and presence of littoral plant and animal remains. Fine-grained detrital and resuspended littoral carbonate transported to the center of the lake in suspension and deposited as profundal sediment would be difficult to distinguish from authigenic carbonate formed in the epilimnion, but contributions from these two sources are likely to be very small.

The main distinction between pro-fundal and littoral carbonates is that littoral carbonates contain a larger proportion of biogenic $CaCO_3$, most-ly debris from mollusks (Figs. 11 to 13), ostracodes (Fig. 14), and charophytes (Figs. 15 to 17); profun-dal carbonates rarely contain calcareous organic remains except for small contributions from resuspended littoral material. There are a few freshwater planktonic calcareous organisms such as the coccolith *Hymenomonas* (Hutchinson, 1967) and the green alga *Phacotus* (Fig. 19; Kelts and Hsu, 1978), but, in general, the amount of carbonate contributed to profundal sediments by calcareous organisms is small. This is also the most important distinction between profundal lacustrine sediments and pelagic deep-sea sediments.

If detrital and biogenic contribu-tions to carbonate in the profundal zone are minor, inorganic and bio-induced precipitation of $CaCO_3$ by removal of CO_2 becomes the main

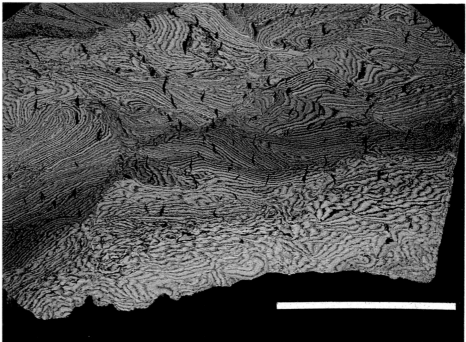

Figure 22—Varve laminae and turbidite layers (**t**) in sediments from Green Lake, Fayetteville, New York. The light lamina in each varve couplet consists mostly of low-Mg calcite crystals precipitated in the surface waters (Fig. 4), and the dark lamina consists mostly of low-Mg calcite, clay and organic matter (Ludlam, 1969). The light laminae contain an average of about 80% calcite and less than 1% organic carbon, whereas the dark laminae contain an average of about 55% calcite and more than 2% organic carbon. Most of the calcite is precipitated between May and October (Brunskill, 1969). Turbidite layers intercalated with normal varve laminations account for about 50% of the sediment that accumulates on the floor of the main basin of the lake (Ludlam, 1974). Bar = 5 cm. Photograph by Stuart D. Ludlam.

Figure 23—Natural slumping of varved sediments in Green Lake, Fayetteville, New York, near the foot of the steep basin slope. Sediments of different ages are of different hues. The pale laminae in the most recent sediments (post 1950) are cream-colored, and the varves are relatively thick. The oldest varves (before about 1850) have olive hues of gray, tan, and brown, and the varves are relatively thin. Varves of intermediate age are gray and of intermediate thickness. The cracks and holes are an artifact of rapid drying. Bar = 5 cm. Photograph by Stuart D. Ludlam.

source of profundal carbonate. Precipitation of $CaCO_3$ can theoretically occur as soon as saturation is reached, but metastable supersaturation of up to several times theoretical saturation without precipitation is the rule rather than the exception. The most important factor controlling the CO_2 budget in most moderately to highly productive hard-water lakes is the balance between CO_2 consumption by photosynthesis and CO_2 production by respiration and decay. A simplified photosynthesis-respiration equation is:

$$energy + CO_2 + H_2O \leftrightharpoons CH_2O + O_2.$$

Rates of CO_2 removal by phyto-

plankton photosynthesis can be very rapid (more than 1.0 g $C/m^2/day$ in very eutrophic lakes) and can increase the pH to 9.0 or more (Wetzel, 1975; Megard, 1967, 1968). In extreme examples, CO_2 may be depleted faster than it can be replaced from the atmosphere, and this depletion favors a predominance of phytoplankton that can utilize carbon from bicarbonate as well as from CO_2 for photosynthesis (Wetzel, 1975). This increase in pH caused by photosynthetic removal of CO_2 can result in supersaturation and precipitation of $CaCO_3$ and is the basis of bio-induced carbonate precipitation. Bio-induced precipitation is difficult to prove, because without detailed photosynthesis experiments it is difficult to demonstrate that CO_2 removal is due to photosynthesis and not diurnal temperature fluctuations.

The most convincing evidence for bio-induced carbonate precipitation was provided by Megard (1967, 1968) who investigated primary productivity in six lakes in Minnesota and at the same time measured pH, carbonate saturation, alkalinity, and calcium-ion concentration. He observed that there was a linear relation between the rate at which carbon was fixed by

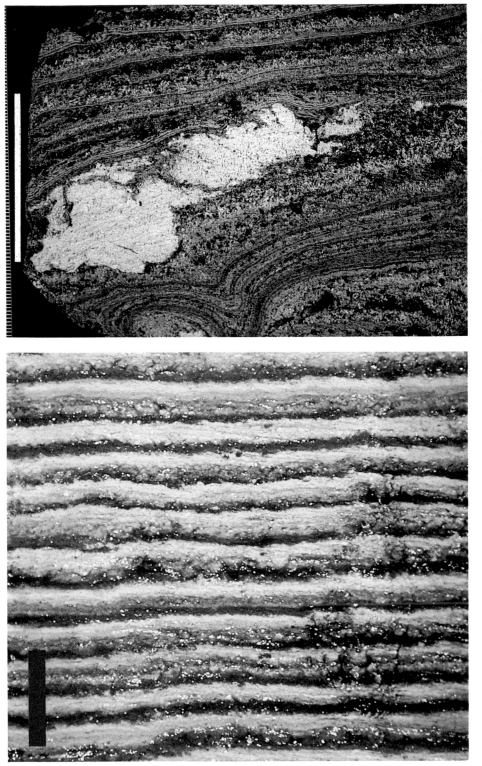

Figure 24—Varve(?) laminae and turbidite layers in sediments from Round Lake, Fayetteville, New York. It is not known if the dark and light laminae-couplets are strictly annual accumulations (varves). Most of the material in both laminated sediment and the irregular white mass is composed of discrete calcite crystals similar to those found in sediments from Green Lake (Fig. 4; Ludlam, 1969), but larger. The white mass appears to be derived from marl deposited around the periphery of the lake at a period of higher lake level and contains numerous gastropod shells. The turbidites consist mostly of debris from the littoral zone, particularly calcite casts of charophytes and organic fragments of other macrophytes. These sediments are difficult to sample without disturbance; as a result, some of the structure is an artifact of the sampling procedure. Bar = 5 cm. Photograph by Stuart D. Ludlam.

Figure 25—Varve laminae from the central plain of Lake Zurich, Switzerland. Varve couplets generally range between 2 and 5 mm in thickness with 80 to 90% water content. A typical varve cycle consists of a dark layer containing organic sludge with algal filaments, iron sulfides, and clay grading upward into a lacy network of diatom frustules and organic matter, overlain by a light layer containing diatom frustules and calcite at the base and almost pure calcite at the top (Kelts and Hsu, 1978). Varve couplets, episodically interrupted by turbidites, have been forming in Lake Zurich since 1895. Bar = 1 cm. Photograph by Kerry Kelts.

photosynthesis and the rate at which calcium and alkalinity were depleted from the water, presumably by precipitation of $CaCO_3$. His empirical results indicate that for these six lakes an average of four moles of carbon must be removed to precipitate one mole of $CaCO_3$. His empirical results indicate that for these six lakes an average of four moles of carbon must be removed to precipitate one mole of $CaCO_3$.

Some of the more important chemical manifestations of phytoplankton photosynthesis and

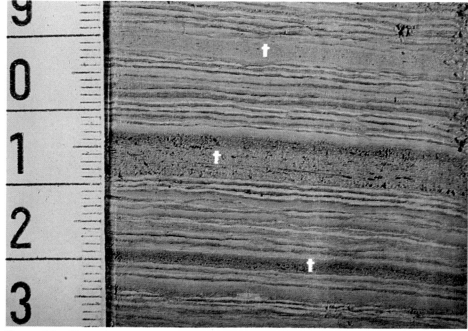

Figure 26—Rhythmic laminae (varves?) from the central part of Great Salt Lake, Utah, in about 3 m water depth. Scale is in millimeters. The lamination is suspected to be due to seasonal alternation of organic (algal) production and carbonate (mostly calcite) precipitation. These laminae formed in a more brackish stage of the lake prior to 12,000 years B.P. The sediments contain a few diatom frustules, ostracodes, and ash layers. Photograph by Kerry Kelts.

Figure 28—Varve laminae from the mid-Holocene in the deep-basin plain of Lake Zug, Switzerland. Light layers consist mostly of seasonally precipitated calcite; dark layers consist mostly of clay, organic detritus, and diatom frustules. Thin graded turbidites (t) of clastic carbonate episodically interrupt the varved sequence. Scale is in millimeters. Photograph by Kerry Kelts.

Figure 29—Varve laminae from the Rita Blanca Lake Beds of Anderson and Kirkland (1969; early Pleistocene), western panhandle of Texas. Light layers consist mostly of calcite, silt-size quartz grains, clay, and ostracodes; dark layers consist mostly of clay and organic matter (Anderson and Kirkland, 1969). Bar = 0.5 cm. Photograph by Douglas W. Kirkland.

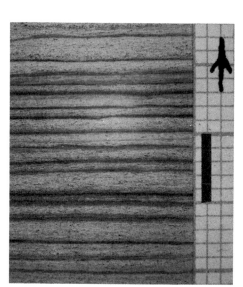

Figure 27—Rhythmic laminae (varves?) from Great Salt Lake, Utah. Bar = 1 cm. The thin reddish laminae consist mostly of filaments of red-brown algae. The light layers consist mostly of aragonite and clay. Photograph by Kerry Kelts.

carbonate precipitation in the open lake are illustrated by the profiles in Figure 3. Because of net photosynthesis in the epilimnion and net respiration in the hypolimnion, dissolved oxygen is usually saturated or supersaturated in the epilimnion during the day in summer and decreases markedly in the hypolimnion, commonly reaching zero (anoxic). A profile of dissolved CO_2 would essentially be a mirror image of the O_2 profile. Nutrients (illustrated by phosphate in Fig. 3) are depleted in the epilimnion by the phytoplankton and released in the hypolimnion on decay of the phytoplankton. Removal of CO_2 by photosynthesis in the epilimnion may increase the pH to 9.0 or more. As a result, the CO_3^{2-}-to-HCO_3^- ratio increases and the epilimnion becomes supersaturated with respect to $CaCO_3$. Precipitation of $CaCO_3$ is indicated by depletion of alkalinity (and Ca^{+2}) in the epilimnion.

Some, or in extreme examples, most of the $CaCO_3$ precipitated in the epilimnion may never reach the bottom because of dissolution during

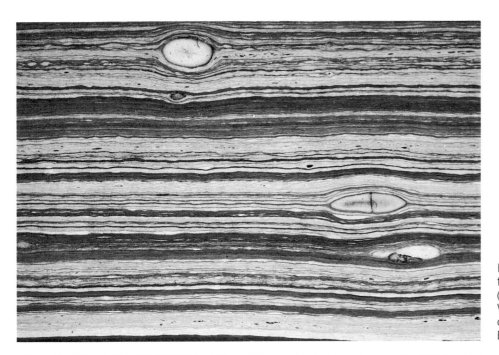

Figures 30, 31—Laminated limestone from the Green River Formation (Eocene), Fossil Syncline, west-central Wyoming. The large clasts are fish coprolites. Photographs by Hans P. Eugster.

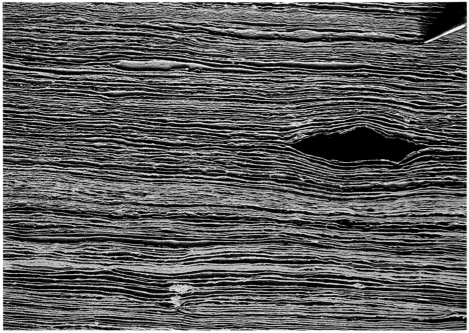

Figure 32—Photomicrograph of well-laminated, kerogenous, dolomitic marlstone (oil shale) typical of much of the open lacustrine facies of the Green River Formation, Piceance Creek Basin, Colorado. Light layers are composed mostly of altered tuffaceous material and dolomite; dark layers are rich in organic matter. Several groups of continuous laminae are interrupted by boudinage-like pull-apart structures called loop-bedding. Bar = 2 mm.

settling through undersaturated hypolimnetic waters. This dissolution is less than might be expected because organic coatings protect the $CaCO_3$ and reduce the rate of dissolution. Dissolution of carbonate is indicated by a decrease in pH to 7.0 or less and an increase in alkalinity (and Ca^{+2}) in the hypolimnion (Fig. 3).

As a result of inorganic and/or bio-induced precipitation in open epilimnetic waters, $CaCO_3$ in profun-dal sediments is often predominantly autochthonous. The evidence suggests that most of this autochthonous carbonate is precipitated by phytoplankton photosynthesis (Megard, 1967, 1968; Otsuki and Wetzel, 1974).

The greatest effects of diagenesis are dissolution of carbonate due to lowering of pH by oxidation of organic matter in the sediments and the conversion of high-Mg calcite to

Figure 33—Photomicrographs showing correlation of tuff-rich laminae between two cores from the upper part of the Parachute Creek Member of the Green River Formation, Piceance Creek Basin, Colorado. Distance between core locations is about 10 km. Bar = 1 cm.

Figure 34—Open lacustrine Pleistocene upper member of the Lisan Formation, north end of the Dead Sea, Israel. Laminated aragonite (light) and calcite marl (dark) formed in Lake Lisan, the Pleistocene precursor of the Dead Sea. Begin and others (1974) included these strata in their interfan facies. Sedimentary features include minor scour structures, thinning and thickening of laminae, and contorted bedding. Rare micro-cross laminations, ripple marks, and mud cracks, and common large-scale convolute bedding, are present in some beds of this facies (Begin et al, 1974). Photograph by Peter Scholle.

dolomite. There appears to be a delicate balance between the rate of carbonate precipitation by phytoplankton productivity and the complete destruction of carbonate in lakes that are highly productive. Dean and Gorham (1976) found that lakes in central Minnesota containing sedimentary organic matter in excess of about 39% contain no carbonate in profundal sediments, whereas lakes in the same area with similar water chemistry but lower organic contents contain abundant carbonate. Decomposition of organic matter in the hypolimnion and profundal sediments apparently generates enough CO_2 to dissolve any carbonate that may have formed in the epilimnion. Megard (1968) concluded that much of the $CaCO_3$ that is generated by phytoplankton photosynthesis in the epilimnion during the summer and survives dissolution in the undersaturated hypolimnion is dissolved in the sediments the following winter.

Most of the sediment that accumulates in the profundal zone contains components that vary in abundance throughout the year in response

Figure 35—Green (**G**) and Round (**R**) Lakes, Fayetteville, New York. Line of cross section in Figure 36 is shown by the white line. View is approximately to the north.

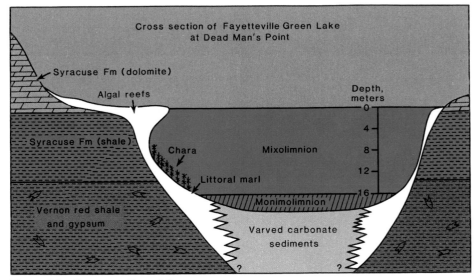

Figure 36—Cross section of Green Lake, Fayetteville, New York, showing relationships between carbonate facies, bedrock types, and boundary (chemocline) between the mixolimnion and anoxic monimolimnion in this meromictic lake. The cause of meromixis in Green Lake is the influx of springs below about 17 m that are more saline than lake surface waters because of dissolved gypsum from the Silurian Vernon Shale. The overlying Syracuse Formation is also of Silurian age. Location of line of cross section is shown in Figure 35.

to seasonal variations in rates of supply. The annual accumulations of sediment may consist of a simple two-component couplet (for example summer $CaCO_3$, winter clay) or complex sequences consisting of many organic and inorganic components, each of which may have a seasonal pulse in rate of supply. Usually the profundal zone is sufficiently oxic to support a benthic epifauna or infauna that will mix the sediments and destroy the delicate annual layers (varves); as a result, the profundal sediments may be structureless. However, if the profundal zone is sufficiently anoxic to exclude benthic organisms, for example if the lake is meromictic, the highly-seasonal components that accumulate on the bottom may result in the preservation of annual sediment laminae (Bradley, 1929a; McLeroy and Anderson, 1966; Anderson and Kirkland, 1969; Ludlam, 1969; Kelts and Hsu, 1978; Figs. 20 to 34). Varves are a powerful interpretive tool for the paleolimnologist because they provide high-resolution time calibration for determining rates and timing of lacustrine processes. In addition, the varves can provide precise time correlation between locations within the lake basin.

In summary, the bulk of carbonate

that accumulates as profundal sediment in marl lakes is the resultant of the rate at which $CaCO_3$ is produced in the epilimnion, and the rate at which it is dissolved in the under-saturated hypolimnion. Most of this carbonate is low-Mg calcite, but high-Mg calcite and dolomite may also form if the Mg-to-Ca ratio is sufficiently high. Although dolomite is commonly assumed to be a mineral that is associated with high-salinity environments (Eugster and Hardie, 1978; Eugster and Surdam, 1973), it can form in freshwater environments with a Mg-to-Ca ratio > 1.0.

Littoral Carbonate

Littoral sediments in marl lakes usually contain a higher percentage of carbonate than profundal sediments. This difference is in part due to a greater rate of carbonate production in the littoral zone and in part due to carbonate dissolution in the hypolim-

nion and on the lake bottom in the profundal zone.

The rate of inorganic precipitation of $CaCO_3$ is greater in the littoral zone because of higher diurnal and seasonal temperatures. The rate of bio-induced carbonate precipitation is also greater in the littoral zone because both algae and rooted aquatic vegetation are removing CO_2. Bio-induced precipitation of $CaCO_3$ is also more obvious in the littoral zone because $CaCO_3$ encrusting aquatic vegetation can easily be observed. Most detrital sediment is trapped in the littoral zone. This detrital sediment may dilute littoral carbonate if it consists mostly of siliciclastic material, or increase the carbonate content of the littoral sediment if it consists mostly of carbonate debris.

Biogenic debris may be a major if not dominant source of carbonate in littoral sediments. Remains of calcareous animals, especially mollusks and ostracodes, are common. Another source of biogenic carbonate is commonly from charophytes, calcareous green algae of the genus *Chara* (Figs. 15 to 18). The importance of charophytes as major contributors of $CaCO_3$ in marl lakes has been recognized for many years. Charophytes contain both internal and external $CaCO_3$. The external $CaCO_3$ is mostly precipitated by

Figures 37, 38—Algal bioherms, Green Lake, Fayetteville, New York. The main framework of the bioherms is provided by precipitated calcite that fills voids and cements sediment trapped and bound by blue-green algal filaments. The carbonate cement and sediment are both composed exclusively of low-Mg calcite. Growth of the bioherms begins below the water surface on solid objects such as tree branches (Fig. 39), cans, bottles, and jars (Figs. 40, 41). The bioherms grow upward to the water surface then outward to form lobate, overhanging ledges that protrude 2 to 8 m from the shore (Dean and Eggleston, 1975; Eggleston and Dean, 1976).

Figure 39—Beginning of algal bioherm growth on a tree branch from Green Lake, Fayetteville, New York. Scale is in centimeters.

photosynthetic removal of CO_2 and HCO_3^- in the same way that the other rooted plants in hard-water lakes are commonly encrusted with $CaCO_3$. Internal calcification within the algal cells may amount to more than half of the dry weight of the plant. Most of the internal and external $CaCO_3$ derived from *Chara* cannot be recognized as biological remains, although the calcareous cortication tubules are sometimes preserved in low-energy environments. In extreme situations, the stem-like charophyte thalli may serve as nucleae for additional inorganic or organic carbonate accumulation (Dean and Eggleston, in press). The protective outer cover of the female reproductive structure (oogonium) of *Chara* calcifies and provides a distinctive ornamented egg-shaped fossil

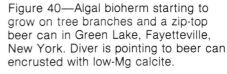

Figure 40—Algal bioherm starting to grow on tree branches and a zip-top beer can in Green Lake, Fayetteville, New York. Diver is pointing to beer can encrusted with low-Mg calcite.

Figure 41—Encrustations of algal-precipitated low-Mg calcite (incipient algal bioherms) on can, bottle, and jar from Green Lake, Fayetteville, New York.

Figure 42—Charophytes growing on the slopes of an algal bioherm, Green Lake, Fayetteville, New York. These calcareous algae are the dominant sediment producers in the littoral zone of Green Lake.

(Figs. 17, 18) that can be common in the sediments of both modern and ancient lakes.

Algal carbonates (stromatolites) are common in alkaline saline lakes of the western United States such as Great Salt Lake, Utah; Mono Lake, California (Fig. 54); Lake Winnemucca, Nevada (Figs, 52, 53); and Pyramid Lake, Nevada; but also occur in some less saline marl lakes in glaciated parts of the United States. Algal bioherms are particularly well developed in Green and Round Lakes, Fayetteville, New York (Figs. 35 to 42; Brunskill and Ludlam, 1969; Dean and Eggleston, 1975; Eggleston and Dean, 1976). Algal bioherms are

Figure 43—Algal bioherms, Eocene Wilkins Peak Member, Green River Formation, Sweetwater County, Wyoming (Bradley, 1929b).

Figure 44—Algal bioherms exposed in plan view on floatblock in the Laney Member of the Green River Formation (Eocene), Delaney Rim, Sweetwater County, Wyoming. Photograph by Emmett Evanoff.

also common in some ancient lake systems such as the Green River Formation (Bradley, 1929b; Surdam and Wray, 1976; Figs. 43, 44, 48 to 51) and the Ridge Basin Group (Pliocene) of California (Link and Osborne, 1978; Fig. 62). Lacustrine oncolites (cryptalgal pisolites) have been described from a number of modern and ancient lakes (Fig. 55, 56, 58 to 62). Although algal carbonates are not particularly common in marl lakes, they commonly constitute a major part of the littoral lithofacies in those lakes where they do occur.

Oolites composed of low-Mg calcite have been reported from the margins of Higgins Lake, Michigan, by Wilkinson and others (1980), but to our knowledge, this is the only reported occurrence of oolites and pisolites (other than oncolites) in freshwater lakes. They do occur, however, in streams, in some saline lakes such as Great Salt Lake, Utah (Fig. 65; Eardley, 1938, 1966), Pyramid Lake, Nevada (Jones, 1925) and hot-spring-fed pools (Figs. 63 and 64; Risacher and Eugster, 1979). Oolites and pisolites also occur in the lacustrine Green River Formation (Bradley, 1929b; Picard and High, 1972) and other ancient lacustrine

Figure 45—Large hemispherical stromatolite of Oligocene age in the Confusion Range, Utah. Recrystallization of the carbonate minerals has formed bladed crystals that may have destroyed fine laminations that may have existed. Photograph by R. E. Anderson.

Figure 46—Laminated lacustrine algal carbonate that formed around a well-rounded lithoclast (now removed). Lacustrine beds are part of an unnamed Oligocene unit in the Conger Range, Utah. Photograph by R. E. Anderson.

strata (Figs. 66 to 68).

In summary, a typical littoral marl is usually coarse-grained (coarse silt to sand), contains more than 60% carbonate, has a large component of calcareous plant and animal debris and a large component of carbonate precipitated by algal and macrophyte photosynthesis, and may have a large detrital component. This mass of carbonate sediment commonly builds a

platform or marl bench that progrades outward from shore with a steep drop-off at the lakeward margin. A good example of lacustrine carbonate facies, including oncolites, resulting from development of an extensive marl bench in a shallow marl lake in Michigan, is described by Murphy and Wilkinson (1980). They concluded that the progradational marl-bench sequence comprises the

bulk of temperate lacustrine deposits, although the lake they studied had a small profundal zone.

MODEL FOR COMPARISON

The distribution of the interpreted depositional environments of the western part of Lake Uinta, Utah, as it existed in the early Eocene is shown in Figure 72 (modified from Ryder et

Figure 47—Recrystallized stromatolitic limestone of the Pliocene and Miocene Horse Camp Formation in the Grant Range, Nevada. These beds formed at the shoreline of an alkaline lake in a graben. The beds grade laterally up depositional dip into conglomerate and down depositional dip into recrystallized thin-bedded limestone.

Figure 48—Stromatolitic algal boundstone in the Eocene Douglas Creek Member of the Green River Formation in the Tavaputs Plateau, Utah. Boundstone overlies fluvial and shallow lacustrine siliciclastic beds and was formed at the fluctuating shoreline of Lake Uinta in saline water (Fouch et al, 1976).

al, 1976). This diagram is most appropriate for one period of Lake Uinta's existence in northeastern Utah. Kerogen-rich lake beds of the Green River Formation as typically developed in the Green River Basin of Wyoming and the Piceance Creek Basin of Colorado are also present at several different horizons of the Green River Formation in the Uinta Basin, but represent a significant variation from the depositional condi-

tions that existed throughout most of the history of Lake Uinta in northeastern Utah. Although the grouping of the rocks shown in Figure 72 is not appropriate for all lacustrine rocks, it has served as a practical model for interpreting both surface and subsurface geochemical, biological, and physical rock data. In addition, it separates the rocks into beds that contain probable source, reservoir, and trap units for hydrocarbons. For

purposes of this paper, the facies relations diagrammed in Figure 72 serve as a guide for the illustration and interpretation of many of the physical and biological features found in terrigenous rocks of large ancient lakes and lake complexes of the world.

The continental sedimentary rocks of the early Eocene of the Uinta Basin can be divided into three major depositional facies: (1) alluvial, (2) marginal lacustrine, and (3) open lacustrine. Ryder and others (1976) indicated that each of these facies contains a suite of sedimentary structures and biologic constituents that collectively indicate the depositional environments of different stages of Lake Uinta's existence.

ECONOMIC CONSIDERATIONS

Lacustrine carbonate rocks are known to be intercalated with beds composed of saline minerals such as halite, trona, nahcolite, and dawsonite; some of these salts are the focus of mining activities. However, the lacustrine carbonate beds are perhaps most important as sources of petroleum, because they are commonly found in stratigraphic units that yield oil and gas in many parts of the

Figure 49—Douglas Creek Member of the Green River Formation, East Tavaputs Plateau, Utah. Stromatolitic algal boundstone (**A**) overlain by an ostracodal, curviplanar, laminated to small-scale cross-laminated, calcareous sandstone (**B**). The sandstone is overlain by a laterally-continuous, small-scale, cross-laminated, ostracodal, grain-supported limestone (**C**) as much as 1 m thick that is in turn overlain by stromatolitic carbonate rock. This depositional complex represents cyclic carbonate shoal deposits overlain by terrigenous strata formed in deltaic and lower delta-plain environments. Scale is in inches and centimeters.

Figure 50—Stromatolitic boundstone from east Tavaputs Plateau, Utah. This sample is from near the boundstone of the Green River Formation illustrated in Figure 49.

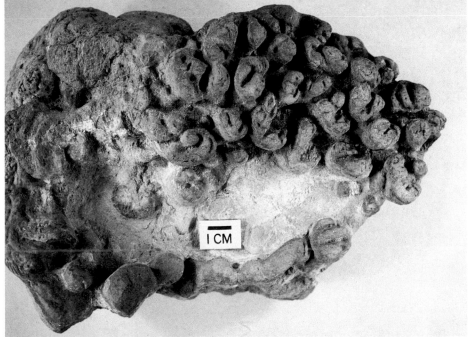

world, and they form the principal oil-shale deposits of the Green River Formation in the Rocky Mountains of the western part of the United States. Although lacustrine beds have been historically related to commodities of economic interest, published documentation of the composition and significance of the lacustrine carbonate rocks is generally sparse and of limited scope. However, several significant factors seem to indicate that an understanding of the sedimentology and geochemistry of carbonate rocks is particularly important to petroleum exploration efforts in ancient lacustrine depositional systems.

The principal petroleum-bearing lacustrine rocks in the world are sandstones, but much of the oil recovered from these units probably formed in kerogenous carbonate rocks and migrated into siliciclastic reservoirs. However, carbonate lacustrine rocks may prove to be important reservoirs in the Campos Basin of Brazil (de Castro and Azambuja, in press) and perhaps in other basins of South America and Africa (Brink, 1974; Brice and Pardo, 1980; Chignone and De Andrade, 1970). Nonmarine carbonate reservoir units in the Campos Basin were apparently

formed from bivalve shells and ostracode carapaces that were concentrated in shoal and beach deposits in shallow water at the margins of lakes during fresher water phases of their history. Some of these shallow-water deposits are redeposited as turbidites near the basin center. These bioclastic sediments were subsequently lithified to form grain-supported rocks.

Bivalve-bearing, grain-supported carbonate units also occur in some of the ancient lake beds of China, but their relation to oil and gas accumulations is uncertain. In the United States, such bivalve and ostracode grainstone beds are usually less than 1 m thick and are not known to be major petroleum reservoir rocks. However, some of these beds in

Figure 51—Algal bioherms, Green River Formation (Eocene), Nine-Mile Canyon, Utah. Photograph by John H. Hanley.

Figure 52—Dense Holocene stromatolitic algal columns along the western shore of (dry) Lake Winnemucca, Nevada. These structures, which grew to heights of 3 m or more, are one morphology of a wide spectrum of algal carbonates that develop in such semi-arid lacustrine systems. Photograph by Bruce H. Wilkinson.

lacustrine sequences in the central Rocky Mountains are bitumen-bearing on surface exposures.

Pisolites, oolites, and/or oncolites formed during more saline or alkaline water stages of the lakes of China, South America, Africa, and the United States. Although the role of these grain types in the development of nonmarine petroleum reservoir beds is poorly documented, oolite and pisolite grainstone locally may form relatively small reservoirs for oil in the subsurface in South America and

the United States. These grain types are locally bitumen-bearing on surface exposures in the Uinta Basin, Utah.

Carbonate minerals in lake beds seem to have played an important role in the formation of porosity in both carbonate and siliciclastic reservoir units. Much of the porosity in the fractured nonmarine Tertiary sandstone of the Uinta Basin is of secondary origin. Fouch and Pitman (1981) believe that most of the porosity developed from the leaching of

carbonate minerals, and the principle reservoirs for oil in much of the Green River Formation in the Uinta Basin are in fluvial and lacustrine sandstone units in which carbonate grains and cements have been dissolved (also see Fouch, 1975; Fouch, 1981). Secondary porosity may prove to be the principle porosity type found in both siliciclastic and carbonate rocks that formed in nonmarine aquatic environments that contained carbonate-rich water. In systems containing abundant $CaCO_3$

Figure 53—Porous lithoid algal pinnacles from western (dry) Lake Winnemucca, Nevada, These columns, like those in Figure 52, developed on coarse, nearshore, basaltic gravels and were not associated with spring emanations on the lake floor. Photograph by Bruce H. Wilkinson.

Figure 54—Algal pinnacles, Mono Lake, California. These large algal carbonate structures form in response to spring discharge directly into the lake. Photograph by Bruce H. Wilkinson.

Figure 55—Beach composed of size-sorted oncolites, Onondaga Lake, Syracuse, New York (Dean and Eggleston, in press).

in solution, parts of both terrigenous and carbonate sediments deposited in or near lakes may be quickly cemented with calcite prior to the formation of petroleum and prior to the extensive compaction of grains by burial. In contrast, those grains that were deposited in alluvial settings well removed from the lake may not contain water rich in $CaCO_3$ and may not be cemented with carbonate minerals prior to extensive compaction. For this reason, the alluvial sediments may be more compacted than marginal-lacustrine sediments, although their burial histories may be similar. Lacustrine and nonlacustrine rocks with little remaining primary porosity may be the end product of early carbonate-mineral cementation of sediments formed in or near lakes prior to significant compaction combined with the compaction of sediments deposited in peripheral alluvial settings away from lakes. However, dissolution of carbonate cements can produce abundant pores in those sediments deposited in or near lakes which underwent early cementation by carbonate minerals.

Perhaps the greatest economic significance of carbonate lake beds is as a source of petroleum. Fine-grained carbonate units form the so-called "oil shales" of the Green River Formation. Tissot and others (1978) indicated that the kerogenous carbonate units form the principle petroleum source rocks in Paleocene and Eocene Green River Formation in

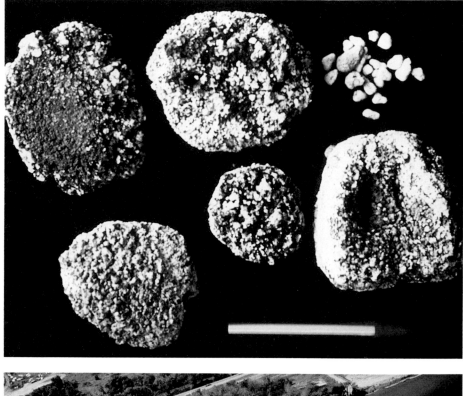

Figure 56—Larger oncolites from about 2-m water depth in Onondaga Lake, Syracuse, New York. Smaller oncolites in lower left are rounded, reworked oncolites from beach for comparison (Fig. 55). Most oncolites of all sizes contain one or more stems of charophytes as nucleae. Charophytes do not grow in Onondaga Lake today and were probably eliminated by a marked increase in salinity of the lake that resulted from the introduction of $CaCl_2$ as a biproduct of soda-ash manufacturing on the lake shore about 1885. This data suggests that growth of the oncolites began at least 100 years ago. At present, $CaCl_2$ enters the lake through Nine-Mile Creek, causing a marl delta to form at the mouth of the creek (Fig. 57).

Figure 57—Marl delta at the mouth of Nine-Mile Creek as it enters Onondaga Lake, Syracuse, New York. This example of massive precipitation of $CaCO_3$ is caused by mixing of $CaCl_2$ industrial waste waters with lake waters that are already saturated or nearly saturated with respect to $CaCO_3$. Sedimentation rates of $CaCO_3$ in Onondaga Lake may be as high as 5 to 10 cm per year (Dean and Eggleston, in press).

the large oil fields of the Uinta Basin, Utah. In addition, carbonate units contain abundant lipid-rich organic matter in Cretaceous and Paleogene lake beds in the Great Basin of the United States (Fouch et al, 1979). For example, the nonmarine Sheep Pass Formation is believed to be the principle source of oil recovered from Oligocene fractured welded tuffs and Paleogene lacustrine carbonate and siliciclastic beds of the Sheep Pass at the Eagle Springs oil field in Nevada (Claypool et al, 1979) (Fig. 73).

ACKNOWLEDGMENTS

We are grateful to J. P. Bradbury, J. H. Hanley, M. H. Link, and B. H. Wilkinson for helpful reviews of the manuscript and to Louise Reif who corrected its many drafts on the word processor.

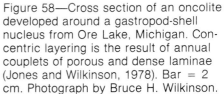

Figure 58—Cross section of an oncolite developed around a gastropod-shell nucleus from Ore Lake, Michigan. Concentric layering is the result of annual couplets of porous and dense laminae (Jones and Wilkinson, 1978). Bar = 2 cm. Photograph by Bruce H. Wilkinson.

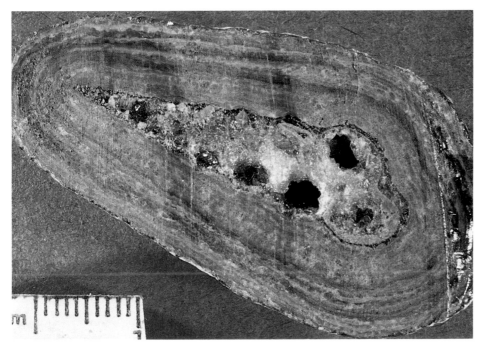

Figure 59—Cross section of an oncolite developed around a gastropod shell of the genus *Goniobasis,* Flagstaff Limestone (Paleocene-Eocene), Sevier County, Utah (Weiss, 1969). Marks on scale are in millimeters. Photograph by John H. Hanley.

SELECTED REFERENCES

Anderson, R. Y., and D. W. Kirkland, 1969, Paleoecology of an Early Pleistocene lake on the High Plains of Texas: Geol. Soc. America Mem. 113, 211 p.

Anthony, R. S., 1977, Iron-rich rhythmically laminated sediments in Lake of the Clouds, northeastern Minnesota: Limnology and Oceanography,
v. 22, p. 45-54.

Begin, Z. B., A. Ehrlich, and Y. Nathan, 1974, Lake Lisan, the Pleistocene precursor of the Dead Sea: Geol. Survey of Israel Bull. 63, 30 p.

Bradley, W. H., 1929a, The varves and climate of the Green River Epoch: U.S. Geol. Survey Prof. Paper 158, p. 87-110.

———— 1929b, Algae reef and oolites of the Green River Formation: U.S. Geol.
Survey Prof. Paper 158-A, p. 203-223.

———— and H. P. Eugster, 1969, Geochemistry and paleolimnology of the trona deposits and associated authigenic minerals of the Green River Formation of Wyoming: U.S. Geol. Survey Prof. Paper 496-B, 71 p.

Brice, S. E., and G. Pardo, 1980, Hydrocarbon occurrences in nonmarine, presalt sequences of Cabinda, Angola (abs.): AAPG Bull., v. 64, p. 681.

Figure 60—Man's hand rests on oncolite packstone intercalated with sandstone in the Flagstaff Member of the Green River Formation on the Wasatch Plateau, Utah. Some oncolites formed around the nonmarine gastropods *Goniobasis* sp. and *Viviparus* sp. (Fouch et al, 1976).

Figure 61—Oncolites in outcrop of the Flagstaff Member (Paleocene-Eocene), of the Green River Formation, near Red Narrows, Utah County, Utah. Light Area in center = Approx. 5 mm (Weiss, 1969).

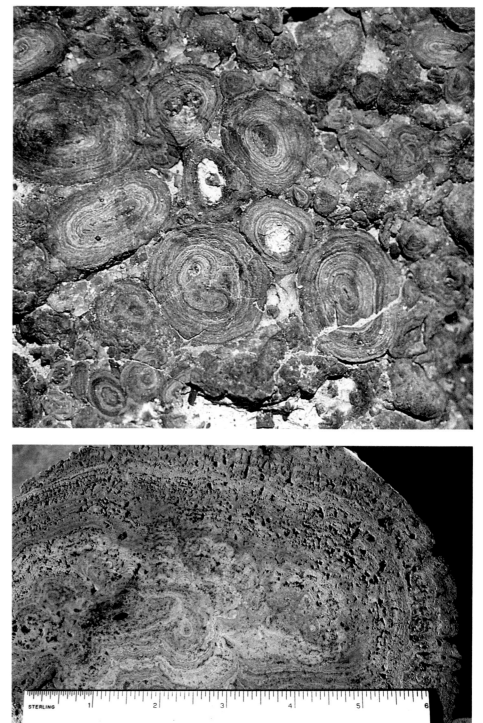

Figure 62—Oncolite from the nearshore lacustrine facies of the Ridge Basin Group (Pliocene), California (Link and Osborne, 1978). Oncolites are associated with remains of ostracodes, mollusks, and plants. Photograph by Martin H. Link.

Brink, A. H., 1974, Petroleum geology of Gabon Basin: AAPG Bull., v. 58, p. 216-235.

Brunskill, G. J., 1969, Fayetteville Green Lake, New York; II. Precipitation and sedimentation of calcite in a meromictic lake with laminated sediments: Limnology and Oceanography, v. 14, p. 830-847.

_____ and J. D. Ludlam, 1969, Fayetteville Green Lake, New York; I. Physical and chemical limnology: Limnology and Oceanography, v. 14, p. 817-829.

Callender, E., C. J. Bowser, and R. Rossmann, 1973, Geochemistry of ferro-manganese carbonate crusts from Green Bay, Lake Michigan (abs.):

Trans. Am. Geophys. Union, v. 54, no. 4, p. 340.

Claypool, G. E., T. D. Fouch, and F. G. Poole, 1979, Chemical correlation of oils and source rocks in Railroad Valley, Nevada (abs.): Geol. Soc. America Abs. with Programs, v. 11, no. 7, p. 403.

Cole, R. D., and M. D. Picard, 1978,

Figures 63, 64—Holocene pisoliths, 1 to 200 m in diameter, forming in shallow, hot-spring-fed stagnant pools of water that are supersaturated with respect to calcite because of loss of CO_2 by algal photosynthesis and by degassing, Andean Altiplano, Pastos Grandes, Bolivia (Risacher and Eugster, 1979). Photographs by Hans P. Eugster.

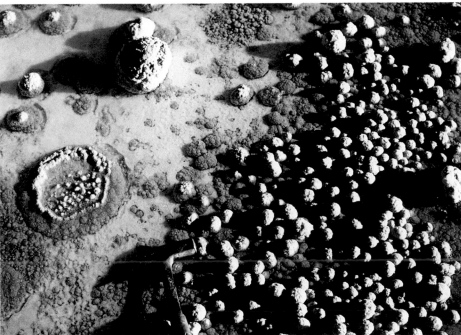

Comparative mineralogy of nearshore and offshore lacustrine lithofacies, Parachute Creek Member of the Green River Formation, Piceance Creek Basin, Colorado, and eastern Uinta Basin, Utah: Geol. Soc. America Bull., v. 89, p. 1441-1454.

Culbertson, W. C., 1966, Trona in the Wilkins Peak Member of the Green River Formation, southwestern Wyoming: U.S. Geol. Survey Prof. Paper 550-B, p. B159-164.

Dean, W. E., and J. R. Eggleston, 1975, Comparative anatomy of marine and freshwater algal reefs, Bermuda and central New York: Geol. Soc. America Bull., v. 86, p. 665-676.

_____ and _____, in press, Freshwater oncolites, Onondaga Lake, New York: Sedimentary Geology.

_____ and E. Gorham, 1976, Major chemical and mineral components of profundal surface sediments in Minnesota lakes: Limnology and Oceanography, v. 21, p. 259-284.

deCastro, J. C., and N. C. de Azambuja, F°, in press, Facues e Analise Estratigrafica da Formacao Lagoa Feia, Cretaceo Inferior de Bacia de Campos, Brasil: Actas del VIII° Congreso Geologico Argentino.

Desborough, G. A., 1978, A biogenic-chemical stratified lake model for the origin of oil shale of the Green River Formation; an alternative to the playa-lake model: Geol. Soc. America Bull., v. 89, p. 961-971.

_____ and J. K. Pitman, 1974, Signifi-cance of applied mineralogy to oil shale in the upper part of the Parachute Creek Member of the Green River Formation, Piceanace Creek basin, Colorado, in D. K. Murray, ed., Guidebook to the energy resources of the Piceance Creek basin, Colorado: 25th Field Conf. Rocky Mtn. Assoc. Geols., p. 81-89.

Dyni, J. R., 1974, Stratigraphy and nah-colite resources of the saline facies of the Green River Formation in north-west Colorado, in D. K. Murray, ed., Guidebook to the energy resources of the Piceance Creek basin, Colorado: 25th Field Conf., Rocky Mtn. Assoc. Geols., p. 111-122.

_____ R. J. Hite, and O. B. Raup, 1970, Lacustrine deposits of bromine-bearing halite, Green River Formation, north-western Colorado, in 3rd Symposium on Salt, vol. 1: Cleveland, Ohio, Northern Ohio Geol. Soc., p. 166-180.

Eardley, A. J., 1938, Sediment of Great

Figure 65—Bars composed of oolites at the shoreline of the Great Salt Lake, Utah. The bars are aligned parallel to the shoreline.

Figure 66—Marginal lacustrine facies, Douglas Creek Member of the Green River Formation, Nine-Mile Canyon, West Tavaputs Plateau, Utah. Pisolitic and oolitic ostracodal grainstone (**A**) (Fouch et al, 1976) is capped by stromatolitic algal boundstone (**B**). This carbonate complex was deposited in shoal waters in Lake Uinta. A channel-formed sandstone (**C**) overlies and locally scours the boundstone. The terrigenous beds represent fluvial deposition on a delta (Ryder et al, 1976).

Lake, Utah: AAPG Bull., v. 22, p. 1305-1411.

———, 1966, Sediments of Great Salt Lake: Utah Geol. Soc. Guidebook to Geology of Utah, no. 20, p. 105-120.

Eggleston, J. R., and W. E. Dean, 1976, Freshwater stromatolitic bioherms in Green Lake, New York, *in* M. R. Walter, ed., Stromatolites: Amsterdam, Elsevier Sci. Pub., p. 479-488.

Eugster, H. P., 1970, Chemistry and origin of the brines of Lake Magadi, Kenya: Spec. Paper No. 3, Mineralog. Soc. America, p. 215-235.

——— and L. A. Hardie, 1975, Sedimentation in an ancient playa-lake complex; the Wilkins Peak Member of the Green River Formation of Wyoming: Geol. Soc. America Bull., v. 86, p. 319-334.

——— and ———, 1978, Saline lakes, *in* A. Lerman, ed., Lakes–chemistry, geology, physics: New York, Springer-Verlag Pub., p. 237-293.

——— and R. C. Surdam, 1973, Depositional environment of the Green River Formation of Wyoming; a preliminary report: Geol. Soc. America Bull., v. 84, p. 1115-1120.

Feduccia, A., 1978, *Presbyornis* and the evolution of ducks and flamingos: Am. Scientist, v. 66, p. 298-304.

Fouch, T. D., 1975, Lithofacies and related hydrocarbon accumulations in Tertiary strata of the western and central Uinta Basin, Utah, *in* D. W. Bolyard, ed., Symposium on deep drilling frontiers in the central Rocky Mountains: Rocky Mtn. Assoc. Geols., p. 163-173.

———, 1979, Character and paleogeographic distribution of Upper Cretaceous (?) and Paleogene nonmarine sedimentary rocks in east-central Nevada, *in* J. M. Armentrout, M. R. Cole, and H. TerBest, eds., Cenozoic paleogeography of the western United States, Pacific coast paleogeography symposium 3: Pacific Sec., SEPM, p. 97-111.

———, 1981, Chart showing distribution of rock types, lithologic groups and

Figure 67—Pisolite grainstone in the Green River Formation of the Tavaputs Plateau, Utah. These beds form the uppermost part of a laterally continuous unit that grades upward from horizontally laminated, kerogenous, carbonate mudstone to small-scale cross-stratified grainstone at the top. This shoal cycle formed on a beach at the margin of Lake Uinta. This unit is bed A in Figure 66.

Figure 68—Pisolites exposed at the margin of a Miocene and/or Pliocene lake in the Horse Camp Formation, Grant Range, Nevada.

depositional environments for some lower Tertiary and Upper Cretaceous rocks from outcrops at Willow Creek–Indian Canyon through the subsurface of the Duchesne and Altmont oil fields, southwest to north-central parts of the Uinta Basin, Utah: U.S. Geol. Survey Oil and Gas Inv. Chart, OC-81, 2 sheets.

_____, et al, 1976, Field guide to lacustrine and related nonmarine depositional environments in Tertiary rocks, Uinta Basin, Utah, *in* R. C. Epis and R. J. Weimer, eds., Studies in Colorado field geology: Colo. School Mines Prof. Cont., no. 8, p. 358-385.

_____ J. H. Hanley, and R. M. Forester, 1979, Preliminary correlation of Cretaceous and Paleogene lacustrine and related sedimentary and volcanic rocks in parts of the eastern Great Basin of Nevada and Utah, *in* G. W. Newman and H. D. Goode, eds., Basin and Range symposium: Rocky Mtn. Assoc. Geols. and Utah Geol. Assoc., p. 305-312.

_____ and J. K. Pitman, 1981, Sedimentologic and mineralogic controls on reservoir characteristics of unconventional hydrocarbon-bearing Tertiary rocks, Uinta Basin, Utah: Western Gas Sands Project Status Rept., Oct.-Nov.-Dec., p. 7. (Avail. as Rept. DOE1BC110003-18, from Nat. Tech. Inf. Service, U.S. Dept. of Commerce, Springfield, Va., 22161.

Figure 69—Mudcracks in dolomitic mudstone in the playa mudflat facies of the Wilkins Peak Member, Green River Formation, Green River Basin, Wyoming. Cracks are filled with carbonate and siliciclastic sand (Eugster and Hardie, 1975). Photograph by Hans P. Eugster.

Figure 70—Lithoclastic packstone intercalated with laminated kerogenous mudstone (oil shale) that was periodically subaerially exposed and desiccated. Cracks are filled with ostracode grainstone. These beds are part of the Green River Formation in Nine-Mile Canyon, West Tavaputs Plateau, Utah.

Ghignone, J. I., and G. De Andrade, 1970, General geology and major oil fields of Reconcavo Basin, Brazil: AAPG Mem. 14, p. 337-358.

Hite, R. J., and J. R. Dyni, 1967, Potential resources of dawsonite and nahcolite in the Piceance Creek Basin, northwest Colorado, in 4th Symposium on Oil Shale: Colo. School Mines Quarterly, v. 62, p. 25-38.

Hutchinson, G. E., 1967, A treatise on limnology, v. 2; introduction to lake biology and the limnoplankton: New York, John Wiley and Sons, 1115 p.

Jones, J. C., 1925, The geologic history of Lake Lahontan: Carnegie Inst. Washington Pub. 325, p. 3-50.

Jones, B. F., H. P. Eugster, and S. L. Rettig, 1977, Hydrochemistry of the Lake Magadi Basin, Kenya: Geochim. et Cosmochim. Acta, v. 41, p. 53-72.

_____ and C. J. Bowser, 1978, The mineralogy and related chemistry of lake sediments, in A. Lerman, ed., Lakes–chemistry, geology, physics: New York, Springer-Verlag Pub., p. 179-235.

Jones, F. G., and B. H. Wilkinson, 1978, Structure and growth of lacustrine pisoliths from Recent Michigan marl lakes: Jour. Sed. Petrology, v. 48, p. 1103-1110.

Kelts, K., and K. J. Hsu, 1978, Freshwater carbonate sedimentation, in A. Lerman, ed., Lakes–chemistry, geology, physics: New York, Springer-Verlag Pub., p. 295-323.

Link, M. H., and R. H. Osborne, 1978, Lacustrine facies in the Pliocene Ridge Basin Group; Ridge Basin, California, in A. Matter and M. E. Tucker, eds., Modern and ancient lake sediments: Internat. Assoc. of Sedimentols., Spec. Pub. No. 2, p. 169-187.

Ludlam, S. D., 1969, Fayetteville Green Lake, New York; 3. The laminated sediments: Limnology and Oceanography, v. 14, p. 848-857.

_____, 1974, Fayetteville Green Lake, New York; 6. The role of turbidity currents in lake sedimentation: Limnology and Oceanography, v. 19, p. 656-664.

Figure 71—Tracks of a shore bird (note webbing marks) in the Green River Formation, Utah. Tracks were probably made by the flamingolike wader *Presbyornis* whose fossil remains are known from localities in the Wilkins Peak Member of the Green River Formation of Wyoming, Utah, and Colorado (Feduccia, 1978). Photograph by Hans P. Eugster.

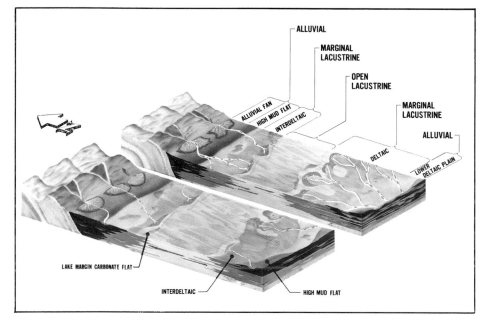

Figure 72—A diagrammatic block diagram (modified from Ryder et al, 1976), illustrating the distribution and interpreted depositional environments of open-lacustrine, marginal-lacustrine, and alluvial facies of the western part of Lake Uinta, Utah, as it existed during the early Eocene. Diagonally striped blue pattern = grain-supported carbonate rock; solid blue pattern = mud-supported carbonate rock; yellow pattern = sandstone; claystone units are shown in their natural colors. Some thin siliciclastic turbidites are present at delta fronts and in the open-lacustrine facies. Much of the open-lacustrine facies is composed of kerogen-rich carbonate units. The width of Lake Uinta in the diagram is about 40 km. Vertical exaggeration is between 15 and 20.

McLeroy, C. A., and R. Y. Anderson, 1966, Laminations of the Oligocene Florissant lake deposits, Colorado: Geol. Soc. America Bull., v. 77, p. 605-618.

Megard, R. O., 1967, Limnology, primary productivity, and carbonate sedimentation of Minnesota lakes: Univ. Minnesota, Limnology Research Center, Interim Rept. 1, 69 p.

_____, 1968, Planktonic photosynthesis and the environment of carbonate deposition in lakes: Univ. Minnesota, Limnology Research Center, Interim Rept. 2, 47 p.

Muller, G., 1970, High-magnesian calcite and protodolomite in Lake Balaton (Hungary) sediments: Nature, v. 226, p. 749-750.

_____ G. Irion, and U. Forstner, 1972, Formation and diagenesis of inorganic Ca-Mg carbonates in the lacustrine environment: Naturwissenschaften, v. 59, p. 158-164.

_____ and F. Wagner, 1978, Holocene carbonate evolution in Lake Balaton (Hungary); a response to climate and impact of man, *in* A. Matter and M. E. Tucker, eds., Modern and ancient lake sediments: Internat. Assoc. Sedimentols., Spec. Pub. No. 2, p. 57-81.

Murphy, D. M., and B. H. Wilkinson, 1980, Carbonate deposition and facies distribution in a central Michigan marl lake: Sedimentology, v. 27, p.

Otsuki, A., and R. G. Wetzel, 1974, Calcium and total alkalinity budgets and calcium carbonate precipitation of a small hard-water lake: Archiv fur Hydrobiologie, v. 73, p. 14-30.

Picard, H. D., and L. R. High, Jr., 1972, Criteria for recognizing lacustrine

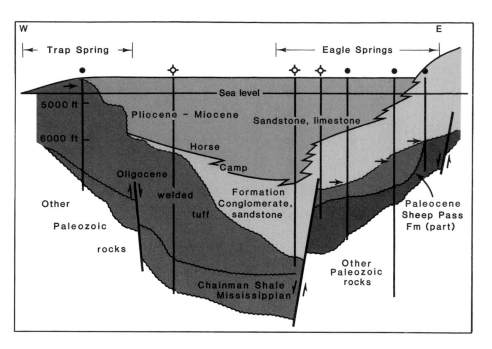

Figure 73—Cross section in Railroad Valley, Nevada, illustrating Tertiary sedimentary and volcanic rocks from the Eagle Springs oil field to the Trap Spring oil field. Oil is produced from fractured lacustrine carbonate rocks of the Sheep Pass Formation and from fractured welded tuff at Eagle Springs. Claypool and others (1979) indicated that the oil at Eagle Springs originated from the thermochemical maturation of organic matter in lacustrine rocks of the Sheep Pass Formation. The Ely Formation is of late Paleozoic age.

rocks, in J. K. Rigby and W. K. Hamblin, eds., Recognition of ancient sedimentary environments: SEPM Spec. Paper No. 16, p. 108-145.

Risacher, F., and H. P. Eugster, 1979, Holocene pisoliths and encrustations associated with spring-fed surface pools, Pasos Grandes, Bolivia: Sedimentology, v. 26, p. 253-270.

Robb, W. A., and J. W. Smith, 1974, Mineral profile of oil shales in Colorado core hole no. 1, Piceance Creek Basin, Colorado, in D. K. Murray, ed., Guidebook to the energy resources of the Piceance Creek Basin, Colorado: 25th Field Conf., Rocky Mtn. Assoc. Geols., p. 91-100.

Ryder, R. T., T. D. Fouch, and J. H.

Elison, 1976, Early Tertiary sedimentation in the western Uinta Basin, Utah: Geol. Soc. America Bull., v. 87, no. 4, p. 496-512.

Smith, G. I., 1979, Subsurface stratigraphy and geochemistry of late Quaternary evaporites, Searles Lake, California: U.S. Geol. Survey Prof. Paper 1043, 130 p.

Smith, J. W., and W. A. Robb, 1966, Ankerite in the Green River Formation's Mahogany Zone: Jour. Sed. Petrology, v. 36, p. 486-490.

Surdam, R. C., and J. L. Wray, 1976, Lacustrine stromatolites, Eocene Green River Formation, Wyoming, in M. R. Walter, ed., Stromatolites: Amsterdam, Elsevier Sci. Pub., p. 535-541.

Tissot, B., G. Deroo, and A. Hood, 1978, Geochemical study of the Uinta Basin; formation of petroleum from the Green River Formation: Geochim. et Cosmochim. Acta, v. 42, p. 1469-1485.

Weiss, M. P., 1969, Oncolites, paleoecology, and Laramide tectonics, central Utah: AAPG Bull., v. 53, p. 1105-1120.

Wetzel, R. G., 1975, Limnology: Philadelphia, W. B. Saunders Co., 743 p.

_____ and A. Otsuki, 1974, Allochthonous organic carbon of a marl lake: Archiv fur Hydrobiologie, v. 73, p. 31-56.

Wilkinson, B. H., B. N. Pope, and R. M. Owen, 1980, Nearshore ooid formation in a modern temperate region marl lake: Jour. Geology, v. 88, p. 697-704.

Eolian Environment

Edwin D. McKee
William C. Ward

*C*arbonate dune sands line the coasts in areas of carbonate sedimentation in many parts of the world. Although they form a common and conspicuous facies in Holocene and Pleistocene carbonate sediments, few eolian limestones are reported from ancient rocks. This is either because dune deposits are scarce in older carbonate rocks, or because few geologists have recognized wind-laid limestone. The latter may be the more likely case, inasmuch as recognition of carbonate eolianite is difficult, particularly in the subsurface, because of its similarity to clastic limestone deposited in high-energy, shallow-marine environments.

The widespread distribution of wind-deposited strata that consist largely or entirely of calcium carbonate was recognized at the start of this century by J. W. Evans (1900) who presented a comprehensive and fairly detailed description of mechanically formed limestone in Kathiawar, India. Since the report of Evans, and especially in recent years, a number of localized studies of calcareous eolianites have been made in various parts of the world — Bermuda, Bahamas, Yucatan, the Mediterranean coast and elsewhere. In most cases, these studies have served to confirm the original ideas concerning requisite conditions for development. These are: (1) a warm climate favorable for a source of CaCO$_3$ sediment; and (2) on-shore winds required to deliver coastal carbonate sand inland to areas of deposition.

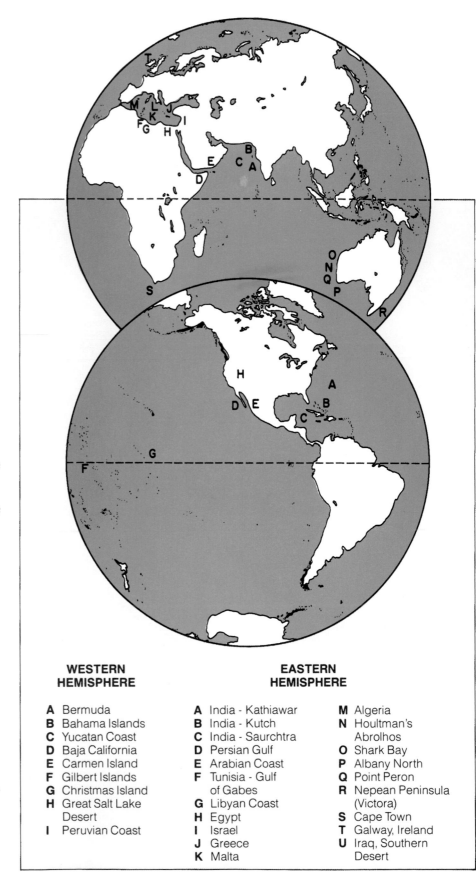

WESTERN HEMISPHERE

A Bermuda
B Bahama Islands
C Yucatan Coast
D Baja California
E Carmen Island
F Gilbert Islands
G Christmas Island
H Great Salt Lake Desert
I Peruvian Coast

EASTERN HEMISPHERE

A India - Kathiawar
B India - Kutch
C India - Saurchtra
D Persian Gulf
E Arabian Coast
F Tunisia - Gulf of Gabes
G Libyan Coast
H Egypt
I Israel
J Greece
K Malta
M Algeria
N Houltman's Abrolhos
O Shark Bay
P Albany North
Q Point Peron
R Nepean Peninsula (Victora)
S Cape Town
T Galway, Ireland
U Iraq, Southern Desert

Figure 1—Global distribution of principal areas of carbonate eolianites and wind-deposited limestones.

Figure 2—Air photo showing Pleistocene dune ridges which form the backbone of Isla Mujeres off the Yucatan Peninsula. Looking southeast down the trend of the middle Upper Pleistocene eolian ridge. Subaqueous remnant of western ridge (right) is seen as light stripe on right half of photo. Photo by W. C. Ward.

General Description

Most carbonate eolianites are deposited as coastal dunes adjacent to high-energy beaches in warm climates where abundant calcareous sand accumulates. Generally carbonate dune rocks consist of well-sorted, cross-stratified clastic limestones composed of sand-size carbonate grains blown inland from the beach. Constituent grains are ooids, pellets, and/or any of a wide variety of skeletal fragments, depending on what organisms are produced in the adjacent shallow-marine waters. In some localities the carbonate sand is mixed with quartz sand.

Calcareous dunes do not always remain as coastal fringes, and in some regions advance far inland. Most notable is the Thar Desert of India where the wind has transported large quantities of foraminifers inland as much as 400 km or more. The abundance of their skeletons or tests in the sand decreases only gradually (Goudie and Sperling, 1977). In the Wahiba sands of Saudi Arabia, the carbonate fraction in some areas is thought to have been blown for great distances from coastal sabkhas and marine terraces.

Interior desert winds may also build up carbonate dunes, although these dunes are relatively uncommon. In the Southern Desert of Iraq where the carbonate content is generally less than 30% and "is concentrated in the finest fraction," the main source of the eolian clastics is considered to be the Mesopotamian Plain (Skocek and Saadallah, 1972). Along the Persian Gulf in Saudi Arabia and Iran, fine sand and dust, some of it highly calcareous, has been transported long distances from the northwest by the prevailing wind (Emery, 1956). Other illustrations of interior source areas for carbonate clasts are in the Ica Desert of Peru where Newell and Boyd (1955) observed dunes formed largely of Eocene mollusk shell fragments, and in the Great Salt Lake Desert, Utah, where Jones (1953) recorded ooids, gypsum fragments, ostracod clasts and algal particles, all derived from marginal deposits of a former large lake.

Diagnostic Criteria

The depositional environment of calcareous eolianites is notably difficult to recognize with certainty in many places, as has been pointed out by Glennie (1970) and by Ward (1975). Few examples have been described from ancient rocks, both because of their similarity in composition to sands of other depositional environments and because alteration through diagenesis commonly makes recognition very difficult. The problem of determining with confidence the origin of these deposits is well illustrated by the Miliolite Limestone of western India and adjacent regions which has been referred to coastal marine deposition by numerous geologists and to eolian deposition by many others. Likewise, the Gargaresh and related limestones along the Mediterranean shores have been variously attributed to both marine and continental genesis (Hoque, 1975).

Because the clastic constituents of most calcareous eolianites are derived largely from the skeletons of marine organisms such as bivalves, corals and foraminifers, or from inorganic grains such as ooids, they are commonly identical to sand grains on nearby beaches or bars. Composition,

Figure 3—Holocene eolianite back of beach on Isla Blanca, Yucatan Peninsula. Exposure face is parallel to strandline. Maximum height of dune is 2.5 m. Photo by W. C. Ward.

therefore, cannot normally be considered a reliable means of distinguishing between such environments. The presence of marine organisms in eolian deposits doubtless has led to many misinterpretations as is indicated by numerous examples of facies assignments that were based entirely on the presence of certain specific animals.

Where the clasts in eolianites include fragments from the shells of terrestrial animals, such as land snails or from the bones of vertebrate animals, they mostly, but not always, represent nonmarine conditions, and when considered with other factors such as wind direction, may support an eolian interpretation.

Various features of texture seem to be characteristic of eolian deposits, but most also occur in other environments and cannot be considered proof of wind deposition. Among these features are roundness, polish, and abrasion of grains, all of which are likewise common attributes of

beach and some other sands (Folk, 1967). Because such features normally reflect the process of transportation and not of deposition, and may represent a second or third generation of sedimentation, they are not valid evidence of wind activity. Good sorting and the winnowing out of fine particles (dust) likewise are characteristic features of most calcareous eolianities, but are by no means restricted to sands of that environment.

A feature common to many eolian limestones and carbonate dunes, but seldom observed in beach or marine deposits, is the preservation of root casts and molds. These are recorded as common in Australian eolianites (Teichert, 1947) and are prominent in the Bahaman deposits (Ball, 1967), as well as in many other places.

Probably the most diagnostic criteria for the recognition of calcareous eolian deposits, as with quartz sand deposits, are the structures. Cross-strata constitute the

largest and principal type of structure, but unfortunately few geologists have given quantitative descriptions of the dip, length of foresets, dimensions of sand body, or other significant features. Of ten descriptions of calcareous eolianites recently reviewed, nine described their structures as "distinctive cross-strata," "dune-like cross-strata," "well-developed false bedding" or used other nonquantitative statements.

Most of the characteristics of cross-stratification attributed to dunes of quartz sand (McKee, ed., 1979) seem to apply also to calcareous dunes. Large-scale foresets — dipping surfaces that are greater than 20 ft in length and may be 60 or 70 ft in length — are common features. Degrees of dip of foresets are mostly high angle, commonly 30° and, in some deposits, 33 or 34°. The basal parts of foresets commonly have a tangential relation to the underlying surface -- a result of the large amount of suspension-load deposi-

Figure 4—Air photo showing multiple dune ridges (upper left) on Isla Blanca barrier island, Yucatan Peninsula. Dune, beach, and nearshore carbonate sands have thin oolitic coatings. On landward side (left beyond photo) is large lagoon of carbonate mud. Small isolated tidal pools (left side of photo) become hypersaline in some seasons. Photo by W. C. Ward.

tion.

Although most eolian dunes have cross-stratification that is either tabular-planar, or wedge-planar, some coastal dunes of carbonate type develop trough structures with a festoon pattern as described from the Bahamas (Ball, 1967) where they fill saddles between lobes. Also common on some calcareous dunes are the convex-upward foreset structures distinctive of many parabolic-type dunes. This structure, common in Bermuda, has been discussed by MacKenzie (1964a, 1964b) who stressed its unique character in this type of dune.

Some minor structures, though relatively scarce, are among the best criteria for the recognition of eolian deposition in carbonate deposits. Ripple marks are preserved uncommonly but, where found, constitute excellent evidence of wind activity. They are readily distinguished from water-formed ripple marks both by ripple index and by their distinctive orientation parallel to the dip of high-angle foresets (McKee, 1945). Raindrop pits formed in dry sand likewise may be readily recognized by their cup-like craters, commonly reoriented on a sloping sand surface as described and illustrated by McKee (1945).

Foresets of eolian sands may "exhibit crumpling of some laminae" as stated by Glennie (1970). Furthermore, distinctive types and combinations of contorted beds are characteristic of wind-deposited sand, and serve to differentiate between various degrees of cohesion as in dry, wet, saturated, and crusted sand.

Figure 5—Eolian limestone of Bahama Island. Ridge of eolianite **A** (top), **B** (bottom) near Eleuthera Island bordering northeast Providence Channel. Photographs by Peter A. Scholle.

Figure 6—Cross-strata **A** (right), **B** (below) in eolian limestones, Government House Lake, Rottnest Island, Western Australia. Photographs by Curt Teichert, 1946.

Such structures have been determined by laboratory experiments (McKee, Douglass, and Rittenhouse, 1971) and tested on coastal dunes of Brazil by McKee and Bigarella (1972).

Summary of Criteria[1]

The following summary of the properties of Quaternary eolianites may aid in the identification of eolian limestones:

(a) geometry — generally elongate bodies of carbonate grainstone trending parallel to the strandline; thickness highly variable within a single body;

(b) associated facies — interfingers with beach and near-shore sands of similar composition; in some settings interfingers landward with lagoonal mud or evaporite-pond sediments; (in prograding sequence, overlies beach sand and might underlie lagoonal deposits or caliche);

(c) composition — sand-sized skeletal fragments, ooids, pellets, and

[1]Prepared by W. C. Ward.

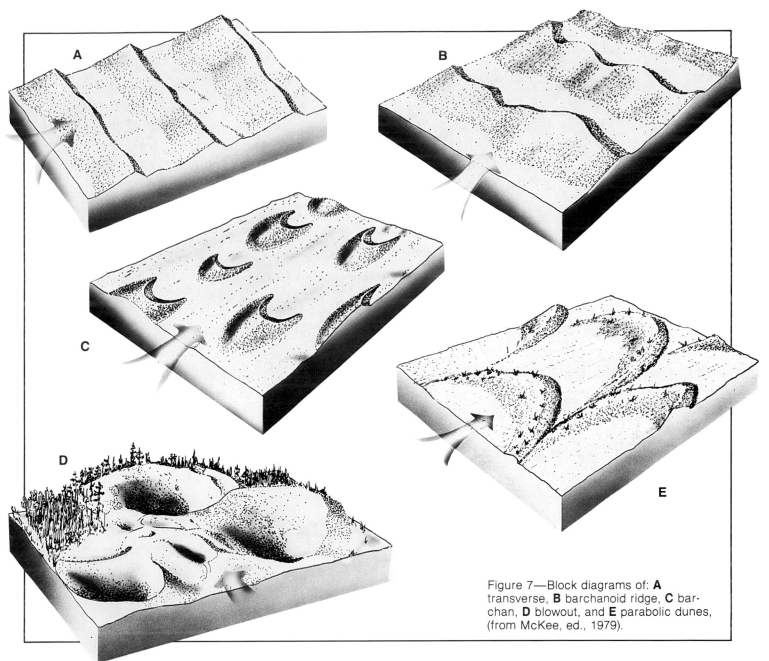

Figure 7—Block diagrams of: **A** transverse, **B** barchanoid ridge, **C** barchan, **D** blowout, and **E** parabolic dunes, (from McKee, ed., 1979).

peloids without larger marine shells such as commonly occur in beach and nearshore carbonate sands;

(d) grain size and sorting — fine- to medium-grained without the gravel-sized constituents which are common in beach deposits; mostly well-sorted;

(e) stratification — alternating fine and coarse laminae; generally highly cross-stratified on large scale, high-angle foresets dipping predominantly landward.

(f) early-stage cements — vadose cements such as meniscus, pendulous, and needle-fiber;

(g) trace fossils — rhizocretions, root-hair sheaths, and micro-borings associated with root systems of dune plants; burrows rare and;

(h) other evidence of subaerial deposition — paleosols; land snails and other terrestrial fossils; subaerial crusts (caliche layers); karst features.

Of these characteristics the most diagnostic for subsurface evaluation is probably the combination of b (vertical sequence) and d (well-sorted carbonate grainstone without coarser material). Where cores or dipmeter data are available, e (high-angle

cross-strata) and g (lack of burrows) should be added to the combination. Paleosols and terrestrial fossils are diagnostic, but are uncommon. Penecontemporaneous vadose cements, rhizocretions, and caliche can develop in marine grainstones which are deposited along prograding coastlines.

Recognition of carbonate dune rock in the geologic record is important for several reasons: (1) Identification of eolian limestones facilitates location of ancient strandlines and islands — this is im-

Figure 8—Cala Conta, Ibiza Island, Balearic Islands, Mediterranean. **A** (right), Pleistocene eolian limestone with high-angle cross-strata and caliche cap; red-brown colluvial silt below; **B** (below), Holocene eolian dune showing root system; scale in center is 20 cm. Photographs by C.F. Klappa.

Terminology

The sediments and their lithified equivalents that are treated in this chapter have two distinctive features — a composition that is dominantly calcium or magnesium carbonate and a genesis through eolian processes. Thus, it might be supposed that a name encompassing these two elements might readily be determined and receive general acceptance. A survey of the pertinent literature, however, shows that no one term has had exclusive adoption and that approximately 20 names or variants (Table 1) have been applied to the calcareous dune deposits and consolidated rocks derived from them.

As indicated in Table 1, the feature of $CaCO_3$ composition has been variously indicated in the sediment name or rock name by use of such terms as calcareous, carbonate, calcarenite, lime or limestone, or when the grains consist of a single distinctive type, by such terms as oolitic or foraminiferal. The agent of deposition is mostly indicated by the names eolian, wind-blown, dune sand or dune rock. In many descriptions, one or the other of these elements is

portant in stratigraphic analyses and in studies of early diagenesis; (2) Eolian ridges have profound effect on the distribution patterns of subsequent carbonate sediments. Recognition of dune-ridge limestone in the rock record, therefore, should be significant in predicting facies distribution in overlying limestones; and (3) Eolianites are potential oil and gas reservoirs because they are

cemented early, but not always pervasively, enhancing the chance that some primary and early-secondary porosity may be preserved after compaction during deep burial. In some carbonate sequences, eolianite is the most-updip facies and may interfinger with lagoonal-mud source rocks.

TABLE 1

Name	Locality	Reference
Calcareous dune rock	Australia	Fairbridge (1950)
Calcareous dune formations	Australia	Bird (1972)
Calcareous low dunes	India	Evans (1900)
Highly calcareous and cemented dune sands	Thar Desert, India	Goudie and Sperling (1977)
Wind-laid calcarenite	Yucatan	Ward and Brady (1973)
Wind-formed calcareous rock	India	Evans (1900)
Wind-blown lime sand	Bermuda	MacKenzie (1964)
Carbonate dunes	Bahamas; Libya, Saudi Arabia	Ball (1967); Glennie (1970)
Carbonate dune rock	Yucatan	Ward (1976)
Carbonate-sand eolian dunes	Yucatan	Ward and Brady (1973)
Wind drift shell sand	Bermuda	Verrill (1907)
Wind-blown calcareous deposits	India	Evans (1900)
Dune lime sand	Onotoa Atoll	Cloud (1952)
Dune limestone	Australia; India	Teichert (1947); Shrivastave (1968)
Aeolian calcareous rock	India	Evans (1900)
Eolian (aeolian) limestone	Bermuda; Yucatan Australia; Bermuda	Sayles (1931); Ward (1975) Teichert (1947); Verrill (1907)
Eolian carbonate sand	Bahamas	Ball (1967)
Eolian ridges, carbonate sand	Bahamas	Ball (1967)
Eolian calcarenites	Australia	Bird (1972)
Cross-stratified oolitic rocks, fossil dunes	Tunisia	Fabricius (1970)
Foraminiferal aeolian deposits	India	Goudie and Sperling (1977)
Gypsum-oolite dune complex	Utah	Jones (1953)
Gypsum-oolite dunes	Utah	Jones (1953)

Table 1. Terms applied to wind-deposited dunes, dominantly of CaCO3 grains, and their lithified equivalents.

implied but not specifically stated in discussions, so that such terms as dune sand, calcareous sand, or dune rock must be in context to be clearly understood.

Although modern dunes are mostly active, and become cemented and stabilized with time, there seems to be great variation in the amount of time required for lithification, so only in a very general way can active dune sands be assigned a Holocene age and cemented sands a Pleistocene or earlier age.

The name "eolianite" was proposed by Sayles in 1929 "for all consolidated sedimentary rocks deposited by the wind" (Sayles, 1931); where they were composed largely of calcium material, they were termed "calcareous eolianite." Whether any particular rock unit should be called a calcareous eolianite or an eolian limestone (stressing the process or the composition), however, seems relatively unimportant and is mostly a matter of desired emphasis.

Geography

Carbonate dunes and eolian limestones probably have a more world-wide distribution than is generally recognized (Fig. 1). However, they are largely restricted to regions of warm climate and they mostly form adjacent to coastal areas. They are uncommon in interior deserts, but may develop locally where alluvial fans furnish a source of carbonate sediment or where calcareous particles of an earlier generation have accumulated. A list of the principal coastal localities throughout the world where carbonate eolianites are recorded is presented in Table 2.

DEPOSITIONAL SETTING

Location and Development of Dunes

Typical calcareous eolianites may be mechanical deposits of pure CaCO3 or may be mixed in various proportions with quartz sand and other detrital grains. They characteristically have a simple texture, being formed of accumulations of fossil fragments (bioclasts), ooids, pelletoids, or microfossils such as foraminifers, or they may contain a mixture of these. Most carbonate dunes have a basic structure consisting of cross-strata of two principal

TABLE 2

Locality	References
Bermuda	Verrill, 1907; Sayles, 1931; MacKenzie, 1964a, b.
Bahama Islands	Newell and Rigby, 1947; Illing, 1954; Ball, 1967.
Yucatan Coast	Ward and Brady, 1973; Ward and Wilson, 1974; Harms, Choquette and Brady, 1974; Ward, 1975.
Baja California	Anderson, 1950; Phleger and Ewing, 1962.
Northwestern India	Chapman, 1900; Glennie, 1970; Lele, 1973.
Persian Gulf	Evans, 1900.
Tunisia	Fabricus, 1970.
Libyan coast	Glennie, 1970; Hoque, 1975.
Egypt	Selim, 1974.
Israel	Yaalon, 1967.
Greece; Malta	Hoque, 1975.
Algeria; Italy	Hoque, 1975.
West Australia	Teichert, 1947; Fairbridge, 1950; Malek Aslani, 1973.
Victoria, Australia	Bird, 1972.
South Australia	Von Der Borch, 1976.
Onotoa Atoll, Gilbert Islands	Cloud, 1952.
Christmas Islands	Fosberg, 1953.
Galway, Ireland	Evans, 1900.
Cape Town, South Africa	Evans, 1900.
Great Salt Lake Desert	Jones, 1953.
Thar Desert, India	Gaudie and Sperling, 1977.
Southern Desert, Iraq	Skocek and Saadallah, 1972.
Ica Desert, Peru	Newell and Boyd, 1955.
Oman	Glennie, 1970.

Table 2. List of principal coastal localities containing carbonate dunes and/or eolian limestones.

types — low-angle windward slopes and high-angle foresets. As with quartz sand dunes, the upwind strata are largely destroyed during dune advance and only the high-angle cross-strata of the lee side normally are preserved. Carbonate eolianites, in general, are greatly affected by diagenetic changes, especially cementation, early in their development.

The initial stages in dune-forming along a coastal area consist of the transfer of dry beach sand or bar sand by onshore winds to form one or more ridges parallel to the coast. In Yucatan (Figs. 2, 3, 4) such ridges border the Caribbean Sea and commonly separate lagoons from the sea (Ward and Brady, 1973), whereas in the Bahamas (Fig. 5A, B) long dune ridges form low islands (Illing, 1954). Along the southern ocean coast of Victoria, Australia, "prevailing westerly winds are driving sand inland and crests rise more than 200 feet above sealevel" (Bird, 1972). In western Australia (Fig. 6A, B) numerous parallel rows of beach ridges show "an element of dune ridge about them" (Fairbridge, 1950).

The multiple seaward ridges of the Bahamas (Ball, 1967) are said to compose the highlands of most islands. These ridges parallel the slope breaks, especially the beaches and bars. Largest dunes are upwind, on the eastern edge of the platform which is near the source, and the smallest dunes are downwind (Ball, 1967). In Baja California, dune fields trend parallel to the barriers along the coast (Phleger and Ewing, 1962).

Figure 9—Punta Chincho, Ibiza Island, Balearic Islands, Mediterranean. **A** (left), Pleistocene carbonate eolianite; upper part cross-stratified, lower part contains calcified root system; 20 cm scale. **B** (below), High-angle (30°) cross-strata in calcareous dune sand; nodular appearance from calcified roots and salt weathering; 20 cm scale. Photographs by C.F. Klappa.

Along the coast of western India, ridges of calcareous eolianite are common on sand bars, marking the present coast line and standing about 10 ft above sealevel (Verstappen, 1966). Earlier dune ridges are recorded to have been formed on old shorelines by the advancing Pleistocene sea (Shrivastava, 1968).

On atolls of the Pacific, carbonate dunes are formed on the lagoon side of gravel ramparts where sand, largely of foraminifers, is blown from the adjoining reef flat (Cloud, 1952). Figure 3 of Cloud's report shows dune locations with respect to the reef flat. Most atoll dunes are not more than 2 to 3 m above high tide according to Fosberg (1953).

Following the development of many calcareous dune ridges, the processes of cementation and stabilization begin almost at once, so dunes of early Holocene and Pleistocene age are commonly lithified.

Most calcareous eolianites clearly indicate by the directional vectors of their cross strata that they were formed by prevailing winds or by at least reasonably constant on-shore winds. This feature accounts for the normal transverse orientation of dune ridges and for the common development of such dune types as the transverse, parabolic, and blowout. On the Libyan coast, however, detailed measurements by Hoque (1975) showed evidence of bidirectional winds with azimuthal modes of northeast and southeast.

Causes or controls of the alternating sequences of calcareous eolianites and soil zones in Bermuda, India, Yucatan and some other areas have been considered by various geologists (Sayles, 1931) to be changes in sea level during the Pleistocene. Periodic drops in sea level, attributed to the forming of ice

Figure 10—Pleistocene eolianite, Isla Mujeres off the Yucatan Peninsula. Upper layers of weakly cemented eolianite with thicket of rhizocretions weathering in relief. Structureless layer (0.3 m thick) is paleosol which contains land snails and calcified cocoons of soil-dwelling weevil. Fewer rhizocretions penetrate lower unit. Photo by W. C. Ward.

sheets, are considered favorable to exposure of much sand for dune build-up and so, at least in Bermuda, a glacial-interglacial stratigraphic sequence has been recognized.

Dune Types

The classification of dune types that is used in this discussion of eolian limestones is that recommended in U.S. Geological Survey Professional Paper 1052 on Global Sand Seas (McKee, ed., 1979). This classification segregates into types the various dune forms and relates these types to specific environments of deposition. This classification is based on two principal attributes of dunes — the shape or form of the sand body and the position and number of slipfaces (steep lee-side surfaces) on it.

Although the recommended classification was developed largely from studies of quartz sand dunes of interior deserts, few if any differences between the largely quartz and largely carbonate dunes can be recognized in either the operative processes or in the resulting dune forms. Mostly the differences are in the proportions of each dune type represented and, of course, in their composition.

Figure 11—Caliche crust lining soil pocket formed as solution hole in Pleistocene carbonate dune rock. Eleuthera Island, Bahamas. Photo by E. A. Shinn.

Because nearly all extensive deposits of eolian limestones are formed on subaerial surfaces bordering the sea in warm temperate latitudes or in arid climates, most of the dunes of which these deposits consist are the same types as those commonly recognized in coastal areas of quartz sand accumulation. Most of

the dunes result from redeposition of clasts by unidirectional onshore winds that transport carbonate sand composed of skeletal fragments, foraminifers, ooids or other types, from nearshore marine sources, especially beaches.

The transverse dune (Fig. 7A) is the most commonly mentioned dune type in the literature on carbonate eolianites. The reason being that most relatively straight dune ridges, parallel to the coast, are normally referred to as transverse dunes. Such ridges with foresets dipping shoreward are a common feature in the Bahamas (Ball, 1967) and are found on the Yucatan coast, bordering beaches (Ward, 1975). Along the coast of Libya (Hoque, 1975), Pleistocene dunes developed where "carbonate beach materials were carried landward to form transverse sand dunes." The relationship of transverse to barchanoid forms (Fig. 7A, B, C) is illustrated in Bermuda where transverse types have been described (MacKenzie, 1964b) as having developed from the coalescing of small dunes to form ridges. In Great Salt Lake Desert, large lateral carbonate dunes are referred to (Jones, 1953) as a transverse type "bordering on the

Figure 12—Pleistocene eolianite on Isla Mujeres, Yucatan Peninsula, Mexico. **A** (left), Abundant rhizocretions; grains-skeletal. **B** (below), Dominantly skeletal grains. Photographs by M. Esteban.

barchane shape."

Barchan dunes (Fig. 7C), which are probably the most elemental type, are characteristically formed in areas of unidirectional wind and relatively small sand supply. In eolian sands of the Southern Desert in Iraq, they apparently are fairly numerous and occur in many places, but the carbonate content in this sand is relatively small, "reaching a maximum of 30 percent" (Skocek and Saakallah, 1972). These dunes are not coastal. On a barrier bordering tidal lagoons of Baja California, extensive barchan dunes, "many more than 100 feet high," are reported by Phleger and Ewing (1962), but here also the carbonate content of the sand is very low — in this area only about 1% foraminifera and shelf — so they cannot be classed as eolian carbonates. Other examples indicate that the barchan dune is not typical of carbonate eolianites.

Distinctive Dune Types

Certain dune types in the recommended classification form a distinctive evolutionary sequence under conditions of unidirectional wind, with a sand supply that is decreasing. This charactertistic sequence involves change from transverse to barchanoid ridges to barchan dunes. Such a sequence is typical of many interior deserts where vegetation is sparse (McKee, 1966). The tendency toward

development of a similar sequence, with parabolic dunes instead of barchan dunes and with rows of coalescing parabolics rather than barchanoid ridges, occurs in areas of considerable vegetation, including some coastal areas containing calcareous eolianites. Thus, in regions of partial dune stabilization, compound parabolic dunes that consist of U-shaped or V-shaped sand hills combined in rows to form ridges, seem to be the counterparts of desert barchanoid ridges.

Rows of compound parabolic dunes seem to be especially well

developed in the Thar Desert of India where they have been referred to (McKee, ed., 1979) as "rakelike" in form because of the resemblance between the nearly parallel trailing arms and the tines of a rake. The amount of carbonate sand in these dunes is not recorded, but their extensive development in northwestern India probably reflects the control of sand movement in that area by monsoon rains and by resulting vegetation. In the western Rajasthan desert near Bikaner such dunes are described as "well stabilized" and include "layers of calcium carbonate" (Saxena and

Figure 13A,B—Pre-Holocene and Holocene eolianites at Cancun, Yucatan Peninsula, Mexico. Most grains are marine ooids. Photographs by M. Esteban.

dune ridges formed from the combining of lobes seem to be characteristic. As described by MacKenzie (1964a) "the individual dunes appear to be predominantly lobate sand bodies which have coalesced to form irregularly defined transverse dune ridges." In these deposits the windward side is "markedly irregular; festooning, scour-and-fill and filled-blowout surfaces are common" (MacKenzie, 1964b). These ridges, therefore, represent a form of compound parabolic dune.

Blowouts (Fig. 7D) and parabolic dunes (Fig. 7E), as might be expected, constitute common forms in some coastal fields of carbonate sand, especially where vegetation is abundant and serves to partly stabilize the sand. In Bermuda, lobate sand units that have convex upward foresets and curved crest lines that are concave upwind are interpreted as "trace parabolic or U-shaped dunes" (MacKenzie, 1964a). In the Thar Desert of India, where dunes "consist mainly of clustered parabolic and obstacle" types, sands are described as "highly calcareous and cemented dune sands" and as "foraminiferal aeolian deposits" (Goudie and Sperling, 1977).

Interdunes

Because a large percentage of carbonate dunes seem to have formed as

Singh, 1976). They are said to be mostly coalesced and to "run in a chain for a distance of 1 to 2 km." The term "coalesced parabolic" dune is used by Singh (1977) for this dune type.

In the coastal areas of the Bahamas, lobes that seem to be comparable to the noses of parabolic dunes and are distributed along calcareous sand ridges, are described by Ball (1967) and referred to as "eolian spillover lobes." As illustrated by Ball (1967), some eolian ridges that form "the multiple seaward ridges. . .on most of the islands" are largely composed of these lobes. They have been partially stabilized by vegetation as indicated by root casts, and the cross-strata dips "range through an arc of about 180 degrees centered onto the platform," both of which features indicate their affinity to the parabolic dune type.

In Bermuda, as in the Bahamas,

Figure 14—Wedge-planar cross-strata of mixed quartz and carbonate sand; low angle strata overlying high angle; Persian Gulf coastal dune. Photo by E. A. Shinn.

Figure 15—Stratified carbonate eolianite, Rottnet Island, Western Australia. **A** (right), Phillip Point; foreset beds dip below sea level. **B** (below), Lake Bagdad. Photographs by Curt Teichert, 1938.

"beach ridges," the nature of interdune corridors between them should be predictable. The process of forming interdunes consists of the accumulation of fine sediments settling from suspension or introduced by waves or tides, together with unsorted debris and, locally, with the carbonaceous remains of plants. The extent and ultimate preservation of these deposits is largely dependent on the rate of dune migration across the interdune surface as described in McKee and Moiola (1975).

Unfortunately, few records of the textures and structures of interdunes between calcareous eolianites have been made. In Yucatan, where extensive dune deposits occur, interdune structures are described by Ward (1975) as "near-horizontal strata." In the Great Salt Lake Desert, Utah, carbonate eolian cross-strata (Jones, 1953) are said to be overlain by nearly horizontal layers of windblown material. These structures are expectable for interdunes. In some areas, however, similar characteristics are ponded surfaces or ephemeral lakes where muddy sediment accumulated and tidal flats developed. Interdune deposits along the Libyan Coast are described (Hoque, 1975) as red-brown units with high clay and silt content, but no cross-strata.

The thickness of an interdune deposit is clearly a function of the time available for accumulation before burial by an encroaching dune. Thus, where dunes tend to stabilize early because of cementation or other diagenetic processes, relatively large amounts of fine sediment may be deposited, and weathering processes may act upon the sediment to develop a soil. In Bermuda, where a series of five soils is recognized between dune deposits, such soils have been interpreted as the products of interglacial times when sea level was high (Sayles, 1931). Paleosols likewise are recognized within dune sequences in coastal

Figure 16A,B—Pleistocene eolianites on Mallorca, Balearic Islands. Abundant rhizocretions; fragments of mollusks, red algae, bryozoans, echinoids. Tabular-planar cross-strata. Photographs by M. Esteban.

primary or penecontemporaneous product formed "beneath interdune flats between stranded barriers" in the Coorong area of South Australia (Von Der Borch, 1976). The dolomite is associated with structural elements characteristic of shallow-water to subaerial origin such as desiccation cracks, intraclast breccias and bird's-eye fenestrae.

Height of Calcareous Dunes and Thickness of Eolian Limestones

Few reliable measurements are available on the size of modern calcareous dunes. In many places, such dunes occur on the tops of consolidated dune rock (Table 3), most of which is probably of Pleistocene age, but the older dunes are not always distinguishable from nonlithified or partly lithified Holocene dunes. On the basis of the sparse data, calcareous coastal dunes must be considered relatively small, as compared to many large quartz-sand dunes of interior deserts; commonly they are less than 100 ft high (Table 4).

Pleistocene eolian limestones which occur primarily in coastal areas of various parts of the world have been the subject of many studies, but, as a rule, these limestones have not been dated with respect to the time span represented, or even in many places,

Libya (Hoque, 1975), and "interbedded with dune sands" in southeastern Australia, as well as a few other areas.

Among the low dunes of Onotoa Atoll in the Gilbert Islands, what might be considered a dune-interdune sequence is described by Cloud (1952) as similar to modern dune sands "but with a humus layer weakly to moderately well developed."

Although sabkahs, ephemeral lakes and other water bodies — both tidal and fresh — may develop in inter-dune areas, not all such basins of deposition can be considered as interdunal. They may merely represent features of an environment that happened to be in the vicinity of eolian accumulation, but are developed quite independently of the eolian processes. The result of flooding between dune ridges by marine currents and the complicated distribution patterns in delta-like fashion is described for the Bahamas by Ball (1967).

Microcrystalline dolomite (less than 10 to 20 microns), is described as a

TABLE 3

1. Isla Mujeres, Yucatan (Ward, 1975, p. 511). Younger [Pleistocene] dunes blown on top of earlier dunes.
2. Isla Cancun, Yucatan (Ward and Brady, 1973, p. 233). Active dunes piled against and atop of Pleistocene dune rocks.
3. Victoria, southeast Australia (Bird, 1972, p. 349). Modern dunes are on older lithified dunes.
4. Point Peron, western Australia (Fairbridge, 1950, p. 39). Holocene dunes overlie older consolidated dune rocks.
5. Abrolhos Islands, western Australia (Teichert, 1947, p. 153). Partially cemented dunes overlie Pleistocene dunes.
6. Coastal terrace, southern Tunisia (Fabricius, 1970, p. 757). A rim near present-day shoreline is of Holocene age.
7. Libyan coast west of Tripoli (Glennie, 1970, p. 128). Uncemented Holocene dunes overlie lithified Pleistocene dunes.

Table 3. Relationships of some calcareous dunes to carbonate dune rocks.

Figure 17—Holocene eolianite on Isla Cancun, Yucatan Peninsula. Exposure face parallel to dip of leeside foreset laminae (landward is to the right). Photo by W. C. Ward.

the part of the Pleistocene in which they were formed. Figures for the thickness of principal eolianite limestone formations generally range from 100 to 200 ft (Table 5), although in some areas, such as Bermuda, the maximum thickness is considerably greater (Verrill, 1907). Here, the total thickness apparently represents the sum of several deposits of wind-accumulated sand, each con-trolled by changing base level, and separated by soils or other products of nondeposition.

Paleosols and Stabilization by Plants

Carbonate dune sands are stabilized not only by partial or complete cementation of particles, but also by vegetation, especially where roots are extensive (Fig. 8B). Root casts — rhizocretions of Kindle (1923) or dikaka of Glennie and Evamy (1968) — where well preserved, give evidence of partial stabilization in some places as in Bermuda (Verrill, 1907), on the Bahama Banks (Ball, 1967), and in Victoria, Australia (Bird, 1972). Calcified root systems are well illustrated in the Balearic Islands of the western Mediterranean (Figs. 9A, B).

Paleosols occur as layers within se-

TABLE 4

Locality	Height (in feet)	Reference
Western Australia	88	Fairbridge (1950, p. 39).
Baja California	100+	Phleger and Ewing (1962, p. 149).
Isla Blanca, Yucatan	12	Ward and Brady (1973, p. 233).
Isla Cancun, Yucatan	50	Ward (1981, pers. commun.)
Bermuda	<100	Verrill (1907, p. 49).
Joe's Hill, Christmas Island	45	Fosberg (1953, p. 2).
Great Salt Lake Desert, Utah	10-30	Jones (1953, p. 2530).

Table 4. Heights of characteristic calcareous eolianites.

Figure 18—Sea-cliff exposure of eolianite of Isla Mujeres of the Yucatan Peninsula. Large-scale cross beds dip steeply landward (left). Uppermost beds are almost flat-lying and are penetrated by numerous rhizocretions which weather in relief. Caliche crust caps the dune rock. Photo by W. C. Ward.

quences of eolian limestones (Fig. 10) or as pocket fillings (Fig. 11) in numerous regions. In dune deposits of southeastern Australia where paleosols separate two or more sequences of eolian cross-strata, they are described by Bird (1972) as differing from wind-deposited sand (1) in their color — red, brown or yellow of iron oxides, locally black from carbonaceous material, (2) in their texture — sandy silt or silty clay, and (3) in the common presence of an underlying irregular calcrete surface. Australian buried soil zones associated with carbonate dunes are mostly 1 to 2 ft thick. A comparable paleosol unit in the Thar Desert of India is stated by Goudie and others (1973) to "separate two phases of aridity and represent an interval of significant wet climate conditions of the region." In both places, numerous fossil root casts are preserved in the dunes below.

A considerable stratigraphic section including five calcareous eolianites and six soil units, each with a local name, is recognized in Bermuda (Sayles, 1931). These extensive dunes are attributed by Sayles to times of glaciation when sea level was low and land surfaces correspondingly large, storms were numerous and vegetation sparse; the ancient buried soils were considered to represent interglacial times. They are deep red and commonly a few feet thick.

TABLE 5

Locality	Formation or Age	Thickness (in feet)	Reference
Isla Mujeres, Yucatan	Pleistocene ridge	90+[1]	Ward (1975, p. 511).
West side, Isla Mujeres, Yucatan	Pleistocene ridge	18+[1]	Ward (1975, p. 511).
Isla Cancun, Yucatan	Quaternary	80	Ward and Brady (1973, p. 233).
Bermuda	Quaternary	"Few hundred"	Sayles (1931, p. 385).
Bermuda	Quaternary	200-268 Est. 350+	Verrill (1907, p. 49).
Victoria, Australia	Older lithified dunes	200+	Bird (1972, p. 349).
West Tripoli, Libya	Gargaresh calcarenite	90±	Hoque (1975, p. 395).
Libya	Late Quaternary	90	Fabricius (1970, p. 757).
Tunisia	Late Quaternary	9-30	Fabricius (1970, p. 757).
Abu Dhabi	Quaternary, "Unit 1"	20.1	Evans and others (1969, p. 148).
Kathiawar, India	Miliolite Formation	200	Biswas (1971, p. 149).
Kutch, India	Miliolite Formation	100	Biswas (1971, p. 153).
Kathiawar, India	Miliolite Formation	Coast 900; Veraval 138[2]	Lele (1973, p. 305).
Abrolhos Islands, Australia	Dune limestone	30	Teichert (1947, p. 152).

[1]Height above sealevel
[2]Borehole

Table 5. Thicknesses of some characteristic eolian limestone formations.

Relationship to Changes in Sea Level and Climate

Carbonate dune rocks on Bermuda were thought by Sayles (1931) to have been deposited during glacial periods, when lowered sea level exposed the inner-shelf sands to wind erosion. Five paleosol zones were recognized by him within the Pleistocene eolianite sequence on Bermuda, and he attributed each of these fossil soils to a period of weathering during times of warm interglacial climates. Dune rocks along the western Australia coast, however, were believed by Fairbridge and Teichert (1953) to have formed at different high stands of sea level during warm, arid interstadial phases. These geologists concluded that eolianites may build up during either regressive or transgressive stages. Eolianites on Bermuda, according to Bretz (1960) and Land and others (1967), accumulated during interglacial high stands of sea level. A study of dune rocks of Israel prompted Yaalon (1967) to caution that climate varies irregularly in space and time, so that neither eolianite nor buried soils by themselves can be used as criteria for specific climatic episodes of the Quaternary.

Most Quaternary eolianites are believed by Fairbridge (1971) to have built up rapidly at the beginning of the regressive phases, when colder temperatures brought increased aridity. Thus coastal eolianites are "fine indicators corresponding to the beginning of a cold period" (Fairbridge, 1971). Along the northeastern coast of the Yucatan Peninsula, eolianites accumulated during the early stages of the later Pleistocene regression of sea level and again during the maximum transgression of the Holocene sea level (Ward, 1970, 1975).

In conclusion, the causes or controls of the alternating sequences of calcareous eolianites and soil zones in Bermuda, India, Yucatan and some other areas have been considered by various geologists (for example, Sayles, 1931) to be changes in sea level during the Pleistocene. Periodic drops in sea level, attributed to the forming of ice sheets, are considered favorable to exposure of much sand for dune build-up and so, at least in Bermuda, a glacial-interglacial stratigraphic sequence has been recognized.

Figure 19—Large-scale, high angle foresets dipping landward, carbonate eolianite at Bannerman Point, Eleuthera Island, Bahamas. Photo by E. A. Shinn.

ORIGIN AND CHARACTER OF CARBONATE SAND

Sand Sources

The development and maintenance of a carbonate-sand dune field is clearly controlled by the adequacy of its calcareous sand source and by the presence of a prevailing wind to transport and deposit the sand. By far the greatest source of carbonate sand is along the margins of seas where marine bioclasts, ooids and foraminifers are forming and accumulating. Beaches are probably most important as the immediate source of wind-blown calcareous sand; the supply of sand on them is constantly being replenished by longshore currents and, as the sand dries, movement inland is accomplished by onshore winds.

Illustrations of carbonate dune sands that are derived from beaches or other shore deposits are numerous. In the Thar Desert of western India, the southwesterly monsoons seasonally carry inland much sand from the Rann of Cutch and adjacent seacoast (Wadia, 1953). Other records from India of calcareous eolian sands, derived from littoral deposits, are by Evans (1900) and by Qadri (1957). Along the Mediterranean coast of Africa, sand similar in textural attributes and skeletal composition to the present beach sand, suggests that

the extensive eolian limestones of that area were formed from Pleistocene beaches under a dominantly northwest wind (Hoque, 1975). Additional dune regions developed by beach-furnished sands are the Bahamas (Ball, 1967), and Bermuda (Sayles, 1931; MacKenzie, 1964b).

In some regions calcareous dunes along coastal fringes advance far inland. Most notable in this respect is the Thar Desert of India where the wind has transported large quantities of foraminifers inland as much as 400 km or more. The abundance of their skeletons or tests in the sand decreases only gradually (Goudie and Sperling, 1977). In the Wahiba sands of Saudi Arabia, the carbonate fraction is thought to have been blown from coastal sabkhas and marine terraces, in some places, for great distances.

Interior desert winds also may build up carbonate dunes, although such dunes are relatively uncommon. In the Southern Desert of Iraq, the main source of the eolian clastics is considered to be the Mesopotamian Plain (Skocek and Saadallah, 1972), where the carbonate content is generally less than 30% and "is concentrated in the finest fraction." Along the Persian Gulf in Saudi Arabia and Iran, fine sand and dust, some of it highly calcareous, is transported long distances from the

northwest by the prevailing wind (Emery, 1956). Other illustrations of interior source areas for carbonate clasts are in the Ica Desert of Peru where Newell and Boyd (1955) observed dunes formed largely of Eocene mollusk shell fragments, and in the Great Salt Lake Desert, Utah, where Jones (1953) recorded ooids, gypsum fragments, ostracod clasts, and algal particles all derived from marginal deposits of a former large lake.

On the island of Bermuda where Sayles (1931) made extensive studies of the eolianites, he noted that although they are composed chiefly of pelecypod shell fragments, they contain some foraminifers, fragments of crab shells and algae. They have an insoluble residue that includes eight principal minerals as determined by E. S. Larsen (Sayles, 1931). These minerals are of special interest because, although pyroxene, perovskite, magnetite, and glass were probably derived from the volcanic platform underlying the island, the quartz, orthoclase, zircon, rutile, and tourmaline are typically continental minerals lacking satisfactory explanation for their presence.

Character of Sand

The character of sand in carbonate dunes in most places closely resembles that of adjacent beaches or shallow

Figure 20—Low-angle wedge-planar cross beds, Pleistocene eolianite, Isla Contoy, Yucatan Peninsula. Photo by W. C. Ward.

marine areas that serve as principal sources of clastic sediment. This similarity of grains between wind-deposited sands and near-shore or beach sands is illustrated in Libya and elsewhere along the coast of north Africa as described by Glennie (1970). Similar comparisons are made by Ball (1967) for eolianites of the Bahamas, by Harms and others (1974) for dunes of coastal Yucatan, and by MacKenzie (1964b) for Bermuda dune limestones.

Among marine clastic sediments, the principal ingredients that contribute in various proportions to the forming of eolian deposits are: (1) the fragments of shells, usually referred to as skeletal material; (2) the tests of foraminifers which may be transported for considerable distances by the wind; (3) oolitic accumulations which, in favorable environments, form vast masses of pure carbonate sediment; and (4) peloidal grains developed from the erosion of earlier limestones. All four of these grain types may occur together in a calcareous eolianite, but the proportions of each varies widely from one locality to another.

Skeletal material — Fragments of skeletal material (bioclasts) form most of the eolianites in some regions (Fig. 12B). In coastal deposits of Libya, calcium carbonate which constitutes from 75 to 88% of the sediment is described as "coquinoid material" by Hoque (1975). In Bermuda, the dune sand in some places (Verrill, 1907) is said to consist "almost entirely of broken shells," and in Baja California, eolian deposits are reported to be composed of "rounded to subround-ed shell fragments" (Anderson, 1950). In some regions, mixtures of numerous other types of skeletal material occur with shell fragments as, for instance, on the west coast of Australia where fragments of echinoidea, corals, foraminifers and bryozoans are listed (Teichert, 1947; Fairbridge, 1950) as being associated with molluscan shells. In western India, the eolian sands are described (Chapman, 1900) as a mixture of oolites, foraminifers, excretory pellets and echinoderm spines, with clasts from mollusks. Also common in some deposits are fragments of algal crusts or stromatolites.

Foraminifera — Certain small marine animals, the shells of which commonly are worn but not broken, may form major parts of some carbonate rocks. Foraminifers are a

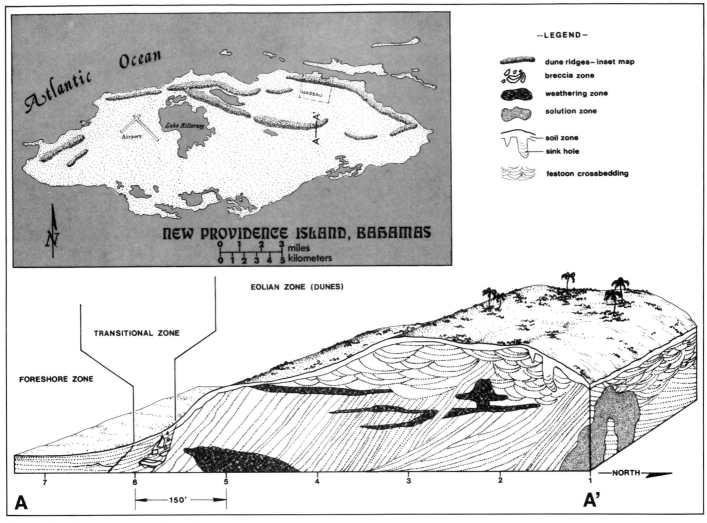

Figure 21—Generalized structures in a Pleistocene carbonate deposit along East Street road cut on New Providence Island, Bahama Islands. The foreshore carbonates on the left are younger and have transgressed over part of the calcareous dune. The entire sequence is now lithified. Tabular-planar cross-beds in lower part were formed by paleowinds from the north, cross-strata above were by paleowinds from the east or west. Field sketch by T. S. Ahlbrandt.

good example. In western India, the abundance of their tests in the Miliolite Limestone, which is considered eolian by many geologists, has been referred to by Shrivastava (1968) and Lele (1973); the name applied to that limestone refers to the abundance of miliolids included in it. In Baja California, 15 species of foraminifers of open-ocean types were recorded (Phleger and Ewing, 1962) in a single sandy calcareous eolianite. In Bermuda, 34 species of foraminifera were found in one specimen of fine-grained eolian sand-rock (Sayles, 1931).

In regard to the generally good preservation of foraminifers, Glennie (1970) stated that they are commonly better preserved among coastal wind-deposited sands than in deposits of marine environments. Furthermore, foraminifers are susceptible to long-distance transport by the wind, as shown by Goudie and Sperling (1977), who described the common occurrence of their tests in at least two locations in the Thar Desert of India about "400 km from the nearest sea" source. Also significant is the record of largely calcareous dunes, containing much foraminiferal material that was accumulated in a temperate climate as represented on the Galway Coast of Ireland (Evans, 1900).

Carbonate dunes composed dominantly of foraminifer tests are especially characteristic of wind-blown deposits on islets of the Pacific Atolls where beaches largely consist of several species of these animals. Examples of such dunes are cited by Cloud (1952). High proportions of foraminiferal sands are also recorded in various parts of limestones attributed to eolian origin near the coast of western India — Kathiawar, Kutch, and elsewhere — as described by Shrivastava (1968), Biswas (1971), and others. Numerous tests of foraminifers, mainly rotalids and miliolids, "mixed with quartzose sand" are reported far inland in the Thar Desert of India by Goudie and

Figure 22—Large-scale trough cross-strata in carbonate eolianite, Castle Rock, Bermuda. Photo by E. A. Shinn.

Sperling (1977). Some fusulinid tests occur in most other regions of carbonate eolianites including Australia, Bermuda, Baja California, and the Mediterranean coasts. The temperate-climate calcareous eolianites of Galway, Ireland, are recorded as "largely calcareous and contain much foraminiferal material" (Evans, 1900).

Other marine animals — Among marine animals, other than foraminifers, characteristically found in coastal eolian carbonates are gastropods and ostracods. Both of these groups are reported by Shrivastava from the Miliolite Limestone near the coast of Saurashtra in India (Shrivastava, 1968) and are said to be similar to animals now living in the nearby sea. The gastropod *Strombus bubonius* is recorded (Hoque, 1975) from eolian limestone on the Libyan coast and from Tunisia (Fabricius, 1970). This species is said to be "now endemic in tropical waters off the Senegal Coast and the Canary Islands, but not in the present Mediterranean."

Terrestrial animals — Carbonate dunes and eolian limestones commonly contain the skeletal remains of land animals. Terrestrial gastropods are especially numerous as noted on Bermuda (Sayles, 1931), the Bahamas (Ball, 1967), Great Salt Lake Desert

(Jones, 1953), and elsewhere. In "false-bedded blown sands" near Cape Town, South Africa, eolian deposits contain, in places, the bones of land animals among masses of comminuted marine shells (Rogers and Schwarz, *in* Evans, 1900). In Victoria, southeastern Australia, an extinct species of giant kangaroo is recorded by Bird (1972) as contributing evidence of the age of the eolian limestone containing it.

Ooids — Ooids are a most important source of carbonate material in many regions of dune formation. They commonly imply a nearby environment of deposition consisting of shallow, agitated waters, though they may have accumulated from the weathering of oolitic deposits of an earlier age. Areas in which ooids have been recorded as a principal component of eolianites are the coast of Yucatan (Fig. 13), where they form a belt of clean carbonate sand (Ward and Brady, 1973; Ward, 1975), the Mediterranean coast of Tunisia (Fabricius, 1970), where the "ooids are reworked from submarine and onshore oolites," and the coastal plain of Egypt, where ooids have probably been formed in agitated marine waters (Selim, 1974). Table 1 of Selim (1974) summarizes the constituent composition of the limestones of several ages represented in the Egyp-

tian area and shows that ooids form more than 50% of all constituents in the first two limestone ridges. Ooids are also important constituents in the eolianites of western India as discussed by Chapman (1900), by Shrivastava (1968), and by Biswas (1971). A most unusual occurrence of oolitic eolianite, however, is in the Great Salt Lake Desert where ooids, derived from lake waters of the former Lake Bonneville, are associated with gypsum crystals, fine quartz grains and some shell fragments in a cool-climate but desert environment (Jones, 1953).

Carbonate dust particles — Although in most desert regions dust particles are largely removed from dune deposits and carried away in the atmosphere, carbonate dunes and eolian limestones in some areas are reported to contain appreciable amounts of $CaCO_3$ in the cryptocrystalline form. A sample of dust carried over the Persian Gulf by winds from the northwest was recorded by Emery (*in* Graf, 1960) to contain "83 percent calcite." In the Southern Desert of Iraq, mainly along the western margin of the Euphrates River alluvial plain, dunes are formed where "travelling sandy materials are brought to a halt. . .due to higher air humidity" (Skocek and Saadallah, 1972).

Figure 23—Trough cross beds on flank of Holocene dune, Isla Cancun, Yucatan Peninsula. Photo by W. C. Ward.

Although the carbonate content in these deposits is mostly within the 10 to 15% range, the amount of carbonate "decreases with increasing grain size," suggesting that a major source is from desert soils and alluvial sediments of the Mesopotamian Plain.

Mixed grains — A mixture of quartz and other nonsoluble grains with carbonate grains occurs in various proportions in many eolianites (Fig. 6). In western India, Biswas (1971) recorded that coastal deposits are free from quartz, but that inland deposits contain 60 to 90% quartz. Quartz, plagioclase, igneous rock material and ferruginous casts, associated with carbonate material in India, are also described (Evans, 1900). Along the Arabian

coast, dune sands were determined by Evans and others (1969) to include 25 to 40% polished, well-rounded quartz and feldspar grains in sediment composed dominantly of calcium carbonate clasts. In southeastern Australia, carbonate dunes were calculated (Bird, 1972) to consist of 50 to 90% calcareous sand with the remainder mostly quartz.

TEXTURAL ATTRIBUTES

Principal Properties of Grains

The principal textural properties of carbonate sand in eolianites are grain size, sorting, degree of rounding, grain surface features, and orientation of grains. Mostly these properties are similar to those represented in nearby source areas because they

reflect the attributes of beaches or nearshore marine environments from which their sands were derived. Thus, most eolianite deposits are composed of fine to medium, well-sorted and moderately to well-rounded carbonate grains, but exceptions resulting from abnormal source materials, different distances of transport, and mixing of depositional agents, occur in numerous areas.

Grain Size

Fine and medium grain sizes are most commonly represented in eolian carbonates (Table 6). Considerably coarser grains, formed as lag concentrates under strong wind conditions, are described from the Ica Desert, Peru (Newell and Boyd, 1955). These lag deposits were derived largely from

TABLE 6

| | Size | | | | | |
Locality	Very Coarse	Coarse	Medium	Fine	Very Fine	Reference
Bermuda			X			MacKenzie (1964, p. 53b).
Bahamas				X		Ball, (1967 p. 573).
Yucatan			X	X		Ward and Brady (1973, p. 233).
Yucatan			X	X		Harms and others (1974, p. 144).
Baja California				X		Phleger and Ewing (1962, p. 162).
Onotoa Atoll			X	X		Cloud (1952, p. 58).
Libya			X			Hoque (1975, p. 395).
Abu Dhabi				X	X	Evans and others (1969, p. 148).
Kutch, India			X	X		Biswas (1971, p. 154).
Thar, India			X			Goudie and Sperling (1977, p. 630).
Victoria, Australia			X	X		Bird (1972, p. 350).

Table 6. Dominant or mean grain size, calcareous eolianites.

disintegrated mollusk shells of Eocene age. Other very coarse eolian carbonate sands are in the Great Salt Lake Desert, Utah, where 1/2- to 1-inch layers of granule-size algal fragments alternate "with thin layers of gypsum-arenite" sand (Jones, 1953). These beds of coarse grains were postulated to be indicators of "repeated intervals of maximum wind velocity."

At the other extreme, some wind-deposited sediments contain appreciable amounts of carbonate materials in the form of dust particles. Where wind is blowing toward the sea from an inland source of carbonate sediment such as a caliche surface or a calcareous desert fan, it may carry much calcareous dust. Prevailing northwest winds over the Persian Gulf, for instance, were recorded by Emery (*in* Graf, 1960) to carry dust containing 83% calcite. Analysis of a Persian Gulf dust sample (Emery, 1956) showed percentages of 1.6 fine sand, 78.9 silt, and 19.7 clay. Deposits of such carbonate dusts are recorded from Iran (Skocek and Saadallah, 1972) where a relatively widespread bimodal grain-size distribution in the dunes is attributed to "the infiltration of fine fractions transported by dust storms."

Grain Sorting

Virtually all reports on eolian carbonate sands agree on the excellence of their sorting. Calcareous dune sands in the Bahamas are described as "characteristically well sorted," (Ball, 1967). Along coastal Yucatan, determinations of 29 samples showed 28 of them to be very well to well sorted (Ward and Brady, 1973) and in carbonate dunes of Kutch, India, "a very high degree of sorting" is believed to be indicated by the uniformity

Figure 24—Trough cross-bedding formed in spillover deposits of carbonate
eolianite, Bannerman Points, Eleuthera
Island, Bahamas. Photo by E. A. Shinn.

Figure 25—Deformational structures in
avalanche sand, bevelled by plane of
slip-face; eolian carbonate rock; Bannerman Point, Eleuthera Island, Bahamas.
Photo by E. A. Shinn.

in shape and grain size (Biswas, 1971). Likewise, in Australia (Bird, 1972), along the Persian Gulf (Evans et al, 1969), and in numerous other localities, calcareous dune sands are recorded to be well sorted.

Roundness of Grains

Carbonate sand grains are eroded by wind action two to four times as fast as are quartz grains, according to Glennie (1970). Therefore, one might expect most carbonate eolianites, if relatively pure in CaCO3 composition and if derived from a source such as a beach where abrasion has been ac-

tive, to have well-rounded grains. A corollary to this generalization is that dunes of mixed carbonate and quartz sands may contain mixtures in grain roundness.

The validity of the foregoing generalization is supported by the studies of Evans and others (1969) in Abu Dhabi on the Persian Gulf where they found well-rounded grains of comminuted shells, but only moderately rounded grains of quartz and feldspar, and by the work of Shrivastava (1968) on the miliolite rocks of the Saurashtra coast in India which consist of quartz grains generally subangular to subrounded, whereas the associated shell fragments and ooids are rounded and polished. Roundness of some grains may have been inherited and actually represents the work of beach processes.

Surface Features

Because of their CaCO3 composi-
tion, grains in calcareous dunes or eolian limestones do not commonly develop conspicuous surface features such as high polish, frosting, or prominent scratches. Well-rounded, polished grains, formed from com-
minuted shells, are reported from Abu Dhabi by Evans and others (1969), but, in general, such features have not been mentioned for most areas. Tests of foraminifers and tiny gastropod and pelecypod shells in

western India are described (Biswas, 1971) as "worn-out" from having been rolled and abraded during long wind transport.

A detailed study of polish on car-
bonate grains of the Yucatan coast has been made by Folk (1967) who concluded that, like rounding, polish was largely the direct result of wave energy and, for that area, furnishes no evidence of chemical polish. Although dune deposits were not specifically mentioned, the beach sources of eolian sands on windward sides of the islands displayed highest polish and, on grains along the lee sides, polish was lacking. Two other factors seem to have controlled the amount of polish on the Yucatan sands:

(1) Grain size — There was essen-
tially no polish on grains "coarser than about 50 mm or finer than about 0.15 mm"; and

(2) Type of grain — "Dense black limestone polishes more easily than coral fragments and these polish more easily than *Halimeda* flakes.

Skewness

Little attention seems to have been given to the attribute of skewness in the sand of carbonate dunes, and probably little significant information can be deduced from this property. Skewness determinations were made for sands of the Libyan coast (Ho-

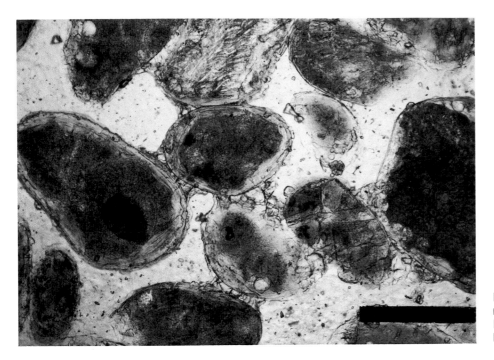

Figure 26—Photomicrograph of meniscus cement in Holocene eolianite, Isla Cancun, Yucatan Peninsula. Plane light. Bar scale is 200μ.

que, 1975); 16 samples showed 9 to be negative and 7 positive. Skewness values ranged from +0.21 to −0.34 with a mean skewness that was nearly symmetrical.

SEDIMENTARY STRUCTURES

Cross-Stratification

The composition and texture in clastic rocks give evidence concerning the source and the mode of transportation, respectively, but the sedimentary structures, because they are formed during accumulation of grains, best record the environment of deposition. In carbonate dunes and in the eolian limestones that develop from them, cross-stratification is the most characteristic and conspicuous structure. The principal types of cross-strata and strata that are recognized in quartz sand dunes are tabular- and wedge-planar, festoon trough-type, and horizontal types. These also occur in carbonate dunes but in different proportions.

Types of cross strata — A majority of carbonate dunes are formed along coasts where onshore winds develop ridges parallel to the strandline (Fig. 15). Thus, a planar structure with slipfaces dipping landward is a common form (Figs. 16, 17, 18, 19). On

the coast of Israel such structures are described (Yaalon, 1967) as forming straight parallel sets, several meters long. On the Mediterranean coast of Libya, mostly tabular-planar crossbeds occur (Hoque, 1975), and in the Bahamas, at numerous places a "single huge sand wave [may form] one very large set of foreset crossbeds," that probably consists of tabular-planar structures. They are well illustrated for Bermuda in several figures by Verrill (1907, Figs. 19, 22, 29) and by MacKenzie (1964a).

Wedge-planar structures are likewise well represented in the common transverse-type dunes on many coasts (Figs. 14, 20). In coastal deposits of western India "mostly wedge-type current bedding" is characteristic, according to Biswas (1971) and from the Yucatan coast "dominantly wedge-shaped rather than tabular" cross-strata are described by Harms and others (1974). Wedge-planar structures range in height from a few to 10 m or more, as seen in sea cliffs of Isla Mujeres. Even the inland dunes in the Great Salt Lake Desert apparently are remarkably uniform in orientation, and consist of wedge planar cross-strata as illustrated in Figure 3 of Jones (1953).

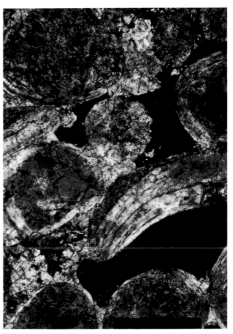

Figure 27—Photomicrograph showing grain-contact (meniscus) cement in Holocene eolianite, Isla Cancun, Yucatan Peninsula. Crossed nicols. Bar scale is 200μ.

A distinctive type of structure that has been recorded from numerous carbonate eolianites is the trough-type that forms a festoon pattern (Figs. 21, 22, 23). Such structures are uncommon in the dunes of inland

TABLE 7

Locality	Reference	Dip Foreset (degrees)	Dip Up-wind (degrees)
Salt Lake Desert, Utah	Jones (1953, p. 2532).	30-32 maximum	—
Bermuda	MacKenzie (1964a, p. 1449).	30-35	10-15
Kutch, Kathiawar, India	Biswas (1971, p. 155).	Very gentle (>5) to 20-30	—
Libya	Hoque (1975, p. 396).	Average for 24 localities 17-27	—
Western Australia	Fairbridge (1950, p. 41).	32	
Northeastern Yucatan	Ward (1975, p. 518).	28-39 maximum	<10

Table 7. Angles of dip recorded for foreset and for windward-side laminae in various dune fields.

deserts, yet they are reported from carbonate coastal dunes of Libya (Hoque, 1975), Yucatan (Ward, 1975), Bermuda (MacKenzie, 1964a) and the Bahamas (Ball, 1967). Unfortunately, little has been recorded about the size and character of these structures. They are "of various scales" and "occur within dune cores" according to Ward; they develop in saddles between "spillover lobes" (Fig. 24) in dune ridges as described by Ball and by MacKenzie. How they compare with the festoon-troughs of meandering streams (Frazier and Osanik, 1964; Bernard and Major, 1973) or in braided rivers (Smith, 1970) is not known.

Horizontal strata are recorded from some carbonate dune complexes. In north African coastal deposits, such strata are said by Hoque (1975) to be present, but not common; he did not indicate whether they were "windward topset beds" or interdune deposits.

Angles of foreset dips — Most foreset dip angles that have been recorded for carbonate eolianites show features similar to those of dunes formed from quartz sand. In general, foresets of the slipfaces are high angle (Figs. 15, 16, 18, 19), approaching the angle of repose for dry sand or about 34°, and those of the windward side are low angle or less than 15°. Measurements that have been recorded from some areas of detailed study are shown in Table 7.

The foreset dips for Libya given by Hoque (1975, Table 1) seem low for dune slipfaces, but these are only mean values; maximum values are not given. These values are based on 321 readings and include 55% that are high angle or greater than 20°, and 45% that are low angle, less than 20°.

Measurement of strata — Quantitative data on the dimensions of strata, and of sets or cosets of carbonate eolianites are scarce. The length of foresets in Bermuda eolian limestones is given as from 1 to 50 ft by MacKenzie (1964b); for coastal Yucatan they are merely referred to as "large scale" that "dip steeply" (Ward, 1975).

The thickness of sets has also received little attention, although this is one feature that can easily be compared with structures in ancient rocks. For Yucatan the thickness of sets is stated (Harms et al, 1974) to

range from a few to 10 m or more. The mean thickness of units (sets) on the Libyan coast is given as 148 cm and they range from 77 to 308 cm (Hoque, 1975). Sets and cosets on Bermuda are recorded as ranging from 1 to about 75 ft (MacKenzie, 1964b).

Traces of foresets — Considerable attention has been given in some cross-strata investigations to the profile shape and the orientation of foresets. A high-angle, sloping surface may meet the underlying plane at an acute angle, but more frequently it forms a tangent with the base (MacKenzie, 1964a; Ward, 1975). The cause of the tangential development is probably the same as in quartz sand deposits — either the sand is poorly sorted and the finer grains form bottomsets, or an appreciable amount of sand is deposited as fallout from suspension at the front of the avalanche sediment.

A second distinctive feature of the foreset trace is its curvature as seen on a vertical plane parallel to the wind direction. A large majority of all foresets, both eolian and aqueous, are concave upward or are nearly straight, because of the nature of set-

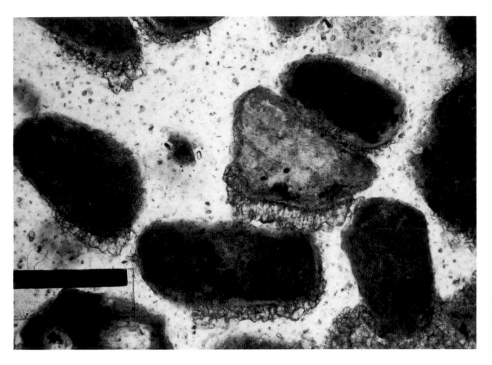

Figure 28—Photomicrograph showing pendulous cement in Holocene eolianite, Isla Blanca, Yucatan Peninsula. Plane light. Bar scale is 200μ.

tling at or near the angle of repose. In carbonate dunes, the occurrence of some foresets that are convex upward has been noted (Hoque, 1975; MacKenzie, 1964a). This upward convexity has been attributed by MacKenzie (1964a) to early stabilization of the dunes and, therefore, is considered a feature distinctive of calcareous sand. Similar structures, however, are found at White Sands, New Mexico, in gypsum dunes (McKee, 1966) and in the Killpecker field, Wyoming, in siliceous dunes (Ahlbrandt, 1975). In each place they occur only in parabolic dunes where the bases of projecting fronts or noses are oversteepened. Convex-upward foresets, therefore, are believed to be a function of undercutting by crosswinds on the noses of parabolic dunes and are not controlled by sand composition.

In Libya, traces of foreset beds, on horizontal planes cutting carbonate dunes, were noted by Hoque (1975) to be "strongly arcuate" and were interpreted by him to represent festoon or trough structures.

Azimuth patterns and paleocurrent directions — Mean dip-direction vectors and the amount of spread have been determined for cross-strata dips in a few areas of eolian carbonate distribution. Cross-stratified limestones near the coast of Libya were determined (Hoque, 1975) to have a mean azimuth of 102° and a standard deviation of 80°. The vectors had a bimodal distribution with modes in the northeast and southeast sectors. Similar studies on the Yucatan coast (Ward, 1975), based on 205 foreset measurements, showed a spread in dip directions mostly within arcs of about 90°. On the Bahamas, Ball (1967) determined that cross-bed dips in "spill-over" lobes, scattered along dune ridges, generally range through "an arc of about 180° centered onto the platform." All of these studies seem to confirm a fair consistency of sand transport direction in the dune fields.

Thickness of laminae and beds — The laminae in calcareous dunes seem to be similar in thickness, extent, and sorting to comparable siliceous dunes of similar environments. In general, sorting is excellent where appreciable sand has been transported. Experiments with colored sands (McKee, Douglass and Rittenhouse, 1971) show that on a dune slipface, sand will sort into laminae of distinct size grades in the distance of a few feet. On the other hand, because a majority of

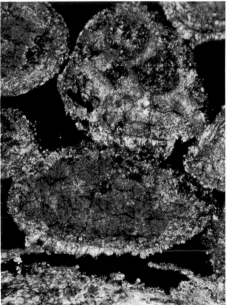

Figure 29—Photomicrograph showing eolianite grains cemented on the bottoms (pendulous cement) and partly dissolved on the tops. Pleistocene eolianite, Isla Mujeres off the Yucatan Peninsula. Crossed nicols. Bar scale in upper right is 200μ.

calcareous dunes are formed close to the source of material which commonly is a beach or other nearshore deposit, extraneous large particles such as unbroken shells, coral fragments, or beach trash may be in-

Figure 30—Photomicrograph showing needle-fiber cement in Pleistocene eolianite, Isla Mujeres off the Yucatan Peninsula. Crossed nicols. Bar scale is 100μ.

Figure 31—SEM photomicrograph of needle-fiber cement in Pleistocene eolianite from Isla Mujeres off the Yucatan Peninsula.

corporated.

An alternation of finer- and coarser-grained carbonate laminae is typical of eolian sands, according to Glennie (1970) and this seems to be borne out by the long, even cross-strata that are etched out in most dune peels. Dune strata in general are relatively thin. In Bermuda, they are described (MacKenzie, 1974a) as ranging from 2 mm to several cm. In the Great Salt Lake Desert, oolitic dune foresets are reported (Jones, 1953) to be between 2 and 8 mm, although granule-size fragments of algae locally form layers ranging from 1.3 to 2.6 cm in thickness, that alternate with the thin foresets.

Dune cross-strata in the Bahamas are recorded by Ball (1967) as being "paper thin." He pointed out that both this thiness of laminae and the large scale of the cross-beds are features that are unusual in marine carbonates of the area.

Minor Structures

Ripple marks — Ripple marks constitute a diagnostic, but relatively uncommon structure in eolian limestones. Although these structures frequently cover the windward side and, occasionally, the leeward side of modern calcareous dunes in the same manner as on active quartz sand dunes, the ripple marks are seldom preserved. Migration of dunes tends to destroy nearly all strata deposited upwind; only those relatively uncommon ripples to lee which are buried by avalanching sand have a good chance of preservation. This selective preservation of lee-slope sediments has been described by McKee (1945) with reference to the Permian Coconino Sandstone of Grand Canyon and seems to apply to most dune deposits.

Two characteristics of wind-formed ripple marks seem to be diagnostic features and serve to differentiate them from subaqueous ripple marks. One of these features, first noted by Kindle (1917), is the ripple index, or ratio of height to wave length. Wind-ripple marks are characteristically low and therefore have high indices. The degree of sorting in the sand controls the index, according to Bagnold (1943) who conducted experiments in this field and concluded that the ratio for very uniform sand grains "was usually 1 to 70 and never more than 1 to 30."

The second diagnostic feature of eolian ripple marks is their characteristic orientation — parallel to the dip directions of the high-angle foresets on the leeward side of dunes. This feature, which is seen as a series of parallel ridges and troughs on the surfaces of cross-strata, is distinctive and wherever preserved on dune rocks is a good indicator of genesis (McKee, 1945).

Relatively few ripple marks on carbonate eolianites seem to be recorded in geologic literature. Along the Yucatan coast, dune deposits are described as "abundantly cross-stratified and rarely ripple-marked" (Ward, 1975). In Bermuda, ripple marks have been noted, along with root casts and other minor structures (MacKenzie, 1964a). Coarse sand ripple marks (0.15 to 0.30 m high; 1.0 to 1.5 m apart) in Peru are described by Newell and Boyd (1955). These rippled sands are considered a product of weathering and deflation and finer fractions have been selectively winnowed; locally, they are formed of fragments of fossil mollusks of Eocene age (Newell and Boyd, 1955).

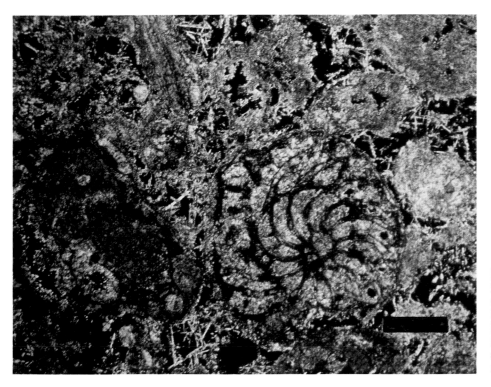

Figure 32—Photomicrograph showing microcrystalline-rind and needle-fiber cements in Pleistocene eolianite, Isla Mujeres off the Yucatan Peninsula. Peneroplid foraminifera (center) and *Halimeda* (upper right) are penetrated by small borings (rootlet borings?). Crossed nicols. Bar scale is 100μ.

On the coast of Israel, well-developed ripples with coarsest particles on crests are reported by Yaalon (1967).

Breccias and break-aparts — Sedimentary structures in dunes, commonly referred to as break-apart laminae and breccias, have been reproduced in the laboratory and studied experimentally (McKee, Douglass, and Rittenhouse, 1971). As shown in Table 1, column g, of that report, the development of such structures seems definitely to be a function of the degree of cohesiveness in the sand body. Wet sand is responsible for most breccias; wet crusts on dry sand commonly shatter into break-aparts. Both phenomena were found to be common in some quartz sand dunes on the wet Brazilian coast (McKee and Bigarella, 1972).

In carbonate dunes and eolian limestones, breccias and break-aparts have been noted in some places. This association is to be expected as many eolian carbonate deposits are accumulating in relatively humid coastal areas. Brecciation occurs within and between eolianite units in Bermuda (MacKenzie, 1964a), and breccia, composed of "altered fragments of eolianite," is said to occur at several horizons in sections on the Bahamas

(Ball, 1967). A caliche breccia is also reported in Yucatan and is believed to be related to the development of caliche surfaces (Harms et al, 1974). On the Bahamas it occurs locally at cliff bases (Fig. 18).

Contorted bedding — Contorted bedding of various types is a common feature of many dunes, both carbonate and quartz-sand types, especially in areas of high humidity and appreciable rainfall that are responsible for slumping on high-angle slipfaces (Fig. 25). Numerous varieties of contorted bedding have been recognized and described through laboratory experiments (McKee, Douglass, and Rittenhouse, 1971); most of these types were observed in wet coastal dunes of Brazil (Bigarella, Becker, and Duarte, 1969; McKee and Bigarella, 1972).

Contorted bedding in carbonate dune deposits has been recorded from Libya where it was described as "penecontemporaneous slump structure" (Hoque, 1975) and from Yucatan where it was attributed to "overload slumping on the dune slopes" (Ward, 1975). On the North African coast, the "foresets [of certain eolian limestones] exhibit crumpling of some laminae. . .very sug-

gestive of slumping seen on the flanks of recent dunes from nearby" (Glennie, 1970).

Rhizocretions or root traces — The term "rhizocretion" was proposed by Kindle (1923) and "dikaka" by Glennie and Evamy (1968) for root cast structures of dune vegetation, preserved in the form of caliche (Figs. 9, 12). The term "calcrete rhizoconcretion" has been used more recently, apparently for the same structures, by Sherman and Ikawa (1958) and by Ward (1975). These root systems are apparently common in calcareous eolianites, as well as in other dune deposits (Figs. 10, 16). They have been recorded from Bermuda (MacKenzie, 1964a), the coast of Yucatan (Harms et al, 1974; Ward, 1975), the coast of Libya (Glennie, 1970), and many other dune areas. They are described by Harms and others (1974) as zones "of thicket-like root casts," and by Glennie (1970) as "cemented plant root molds."

Root systems of eolianites in southeastern Australia are described by Bird (1972) as cylindrical calcrete forms and they commonly mark the presence of paleosols among eolian calcarenites. In Western Australia, root structures that are concentrically-

Figure 33—Rhizocretions weathering in relief from weakly cemented Upper Pleistocene eolianite on Isla Mujeres off the Yucatan Peninsula. Photo by W. C. Ward.

filled calcium carbonate casts are said to be extensive (Teichert, 1947). In Rajasthan, India, numerous root casts within the calcreted dunes are reported to be 56 to 68% calcium carbonate (Goudie et al, 1973).

The presence of a solid core of caliche "in most of the branching and anastomosing rhizocretions" in Pleistocene eolianites of Yucatan has been described in detail by Ward (1975). Root casts with hard, hollow sheaths are also present, but less common. In Holocene deposits, hard-core rhizocretions are rare, although a few horizontal and vertical root casts have formed where cementation has filled root holes.

Small tubules of microcrystalline calcite, also in Pleistocene eolianites of the Yucatan coast, have been studied with care and described by Ward (1975). These tubules are referred to as "root-hair sheaths" and have been variously attributed to three possible origins: algal filaments, extensions from the cells of rootlets, and fungal hyphae. Ward (1975) interpreted them as most likely the precipitates around root hairs of dune plants.

Bioturbation by animals — Bioturbation, defined as the disruption and mixing of sediments by organisms,

has been shown to be a common feature in many dune fields, especially those of inland deserts (Ahlbrandt, Andrews, and Gwynne, 1978). In carbonate dune fields, most of which are in coastal areas, bioturbation by plant root systems seems to be common in many regions. The root structures, as stated before, are referred to as rhizocretions (Kindle, 1923) or dikaka (Glennie and Evamy, 1968). Faunal bioturbation, in contrast, seems to be fairly uncommon or has not been recognized in many carbonate dunes.

Referring to the Bahama dune sands, Ball (1967) stated that "burrowing is rare in wind-blown carbonates," whereas root casts are quite common. In western Australia, "insect puparia" and fossil "cocoons" in solution pipes, together with abundant root casts, are recorded by Fairbridge (1950). Borings in carbonate sands are mentioned by Glennie (1970). In general, however, references to burrows in calcareous eolianites are rare, possibly because an actual scarcity occurs as a result of early diagenesis and development of a cemented crust making an unfavorable habitat. Possibly, also, the explanation is that "many geologists are reluctant to associate burrows with eolian deposits" as stated by

Ahlbrandt, Andrews, and Gwynne (1978) in connection with interior dunes.

Distinguishing between root molds (rhizocretions) and the burrows or borings of certain animals that cement or agglutinize the sand tubes with secretions is not always easy. Various features of shape and distinctions in color resulting from selective oxidation have proven useful, however, for differentiating between root molds and animal burrows as described by Ahlbrandt, Andrews, and Gwynne (1978), and by Ball (1967).

EARLY-DIAGENETIC FABRICS[1]

Eolianites are characterized by early-stage cements and other diagenetic fabrics that form in the vadose zone. Meteoric water percolating through carbonate dunes becomes supersaturated with respect to $CaCO_3$ through the partial dissolution of the aragonitic and Mg-calcitic grains. Much of the $CaCO_3$ taken into solution is reprecipitated as low-Mg-calcite cement in the pores of the dune sand. Therefore, carbonate

---

[1]"Early-Diagenetic Fabrics" was prepared by W. C. Ward.

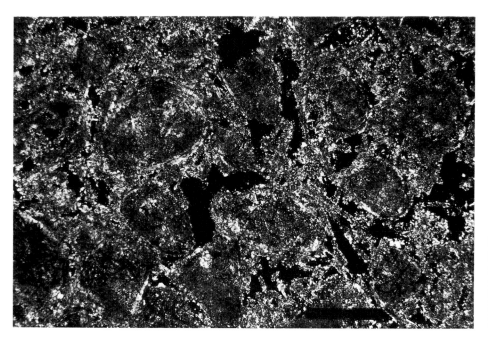

Figure 34—Photomicrograph of Pleistocene eolianite with microcrystalline-rind cement and microcrystalline-calcite root-hair sheaths in intergranular pores. Isla Mujeres off the Yucatan Peninsula. Crossed nicols. Bar scale is 200μ.

dunes may be lithified almost as soon as they accumulate. The early-diagenetic fabrics of eolianites are related, in large part, to rainfall, evaporation, and transpiration by dune plants, pene-contemporaneous with deposition.

Early-Stage Cements

The "vadose-character" of most early-stage cements in eolianites is easily recognized. One diagnostic characteristic is variability (Dunham, 1969). Cements in Quaternary eolianites tend to be highly variable in size, shape and distribution, even within an area of a few millimeters in a single rock. Fine-grained layers tend to be more completely cemented than coarse-grained layers. Furthermore, cement types among eolianites of different ages or locations may be different, particularly in crystal size (Ward, 1973, 1978). In addition to this typical variability, early cements in eolianities are characterized by the predominance of vadose types, such as meniscus, pendulous, and needle-fiber cements, and the absence of fibrous grain-rimming cements.

Grain-contact and meniscus cements — In general, cement in an eolianite is preferentially precipitated at grain contacts, where surface tension holds droplets of percolating meteoric water. The grain-contact ce-

ment may be microcrystalline, but is generally finely to coarsely crystalline calcite. The growth of grain-contact cement is restricted by the meniscus or outer limit of water held in the corners of pores (Dunham, 1971). As crystal growth proceeds, the pore face of the cement takes on the shape of the meniscus, resulting in rounding of the corners of the pores. Meniscus cement (Figs. 26, 27) is well developed in many eolianites.

Pendulous cement — Calcite cement preferentially precipitated on the bottoms of grains is common in some eolianites (Figs. 28, 29). This cement type, commonly called "pendulous cement," is widely accepted as evidence of vadose-zone cementation. In Quaternary eolianites of Yucatan the percentage of grains with pendulous cement is highly variable, ranging from zero in some layers to as much as 65% in other places.

Needle-fiber cement — Needle-like crystals of calcite cement (Fig. 30) are common in Pleistocene eolianites of Yucatan, particularly in proximity to ancient weathered surfaces and rhizocretions. This "needle-fiber cement" (Ward, 1970), "whisker crystals," (Supko, 1971), and "needle-fibers," (James, 1972), may be as indicative of vadose-zone cementation as are meniscus and pendulous cements.

This cement occurs both in intergranular pores and in intraparticle dissolution voids as unoriented, straight-fibers as much as 200 m long and usually less than 4 m wide. The needle-fibers are composite crystals composed of strings of imbricated flattened rhombs of calcite (Fig. 31).

Microcrystalline-rind cement — In some eolianites, grains are coated with irregular rinds of microcrystalline calcite (1 to 2 μm) that grades poreward into bladed and rhombic crystals of calcite less than 25 μm in greatest dimension (Fig. 32). This cement type in eolianites of Bermuda is called "grain-skin cement" (Land et al, 1967). In rocks cemented with finely crystalline calcite, few intergranular pores are completely filled.

Pore-occluding cements — In the Holocene eolianites of Yucatan, scattered pores may be totally filled with calcite cement, but adjacent pores remain cement free (Ward, 1975). Pore-occluding cement has a wide range of crystal sizes and shapes, including finely crystalline anhedral blocky spar, fine to coarse bladed druse, and large single crystals which take on the shape of an intergranular pore. In addition, echinoderm fragments take on large syntaxial overgrowths which occlude some pore space. Finer-grained layers and more tightly packed

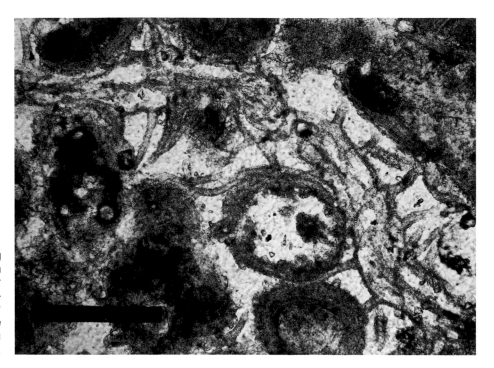

Figure 35—Photomicrograph showing root-hair sheaths extending from calcified rootlet. Smaller tubules appear to be nonseptate extensions of epidermal cells (root hairs) of the larger multicellar tube (rootlet). Pleistocene eolianite from Isla Cancun, Yucatan Peninsula. Plane light. Bar scale is 200μ.

patches tend to have a higher percentage of occluded pores. In a middle Pleistocene eolianite of Israel, pore space is entirely occluded by drusy calcite cement (Gavish and Friedman, 1969).

Diagenetic Fabrics Related to Roots

In some eolianites the cement types and other early-diagenetic features are associated with the root systems of dune plants. For example in Yucatan, eolianite needle-fiber cement and microcrystalline-rind cement are particularly abundant in the vicinity of root traces and in paleosols, but sparry-calcite cement is scarce (Ward, 1975). Other typical diagenetic fabrics result from precipitation of calcite around roots and root hairs.

Rhizocretions — Rhizocretions or root-casts (Fig. 33) may be recognized in cores as vertical or horizontal tubular structures of hard brown microcrystalline calcite. They range in diameter from a few millimeters to several centimeters. Generally the internal structure consists of wavy laminae of microcrystalline calcite (caliche-like), concentric about the long axis of the tubes. The core of the rhizocretions may crudely preserve some of the cell structure of the root.

Root-hair sheaths — Minute tubules of microcrystalline calcite are abundant in interstitial pores in some eolianites (Figs. 34, 35, 36). These hollow sheaths are 5 to 15 μm in diameter with a wall thickness of 1 to 2 μm. They are particularly abundant near rhizocretions and are probably calcified replicas of root hairs (Ward, 1975). Most sheaths are composed of finely crystalline calcite similar to the microcrystalline-rind cement (Fig. 34) and some are partly made up of needle fibers, the long axes of which lie tangent to the tubules.

Microcodium — Diagenetic structures composed of calcite prisms arranged in spherical, elliptical, or sheet-like bodies (Fig. 37) are formed in some Pleistocene eolianites (Ward, 1975; Calvert et al, 1975). These structures, which are widely reported from limestone and caliche, are referred to as *Microcodium*. Klappa (1978) found evidence that they are the produce of calcification of mycorrhizal (root plus fungus) associations.

Microscopic borings — Grains near rhizoconcretions may be riddled with tubular borings 15 to 25 m in diameter (Fig. 36). The holes are about the same size as root-hair sheaths (some of which are preserved in the holes) and may be the result of root-hair and rootlet penetration of the carbonate grains.

Many grains in association with rhizoconcretions and root-hair sheaths also contain long, straight fungal borings 1 to 2 m in diameter. Fungal hyphae grow in close association with root systems in dunes (Webley et al, 1952) and are filamentous plants that can exist in the shallow subsurface.

Subaerial crusts — Caliche or calcrete crusts (Figs. 38, 39) are developed on and within Quaternary eolianites in several localities (Johnson, 1967; Ward, 1970; Read, 1974; Semeniuk and Meagher, 1981). Criteria for the recognition of subaerial crusts are given in Chapter 1 of this volume.

Progressive Early Diagenesis

The progressive development of early-diagenetic fabric in eolianites has been studied in Bermuda (Land et al, 1967), Israel (Gavish and Friedman, 1969), and Yucatan (Ward, 1970, 1975).

Most of the Holocene dunes in Yucatan are at least weakly lithified. In general, more pore space is occupied by sparry-calcite cement in progressively older Holocene eolianite. Cement in young eolianites

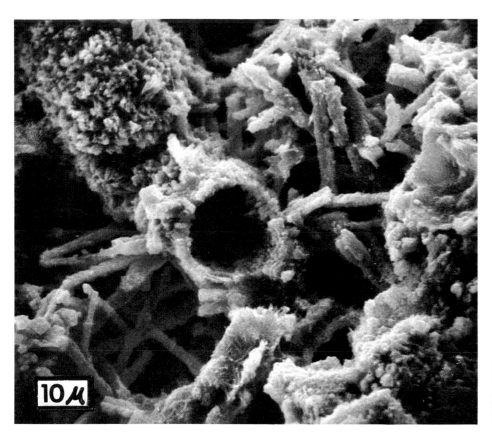

Figure 36—SEM photomicrograph showing microcrystalline-calcite root-hair sheath (center) surrounded by needle-fiber cement. Pleistocene eolianite, Isla Mujeres off the Yucatan Peninsula.

is not extensive enough to entirely occlude porosity, although intergranular porosity may be substantially reduced where the rock is cemented with coarsely crystalline spar. Porosity ranges from about 25 to 40% in the Holocene eolianites. Almost no dissolution of grains occurs in these calcarenites.

On Bermuda, the youngest Pleistocene eolianite is uncemented or is cemented at grain contacts only; porosity is about 20% where cemented. The middle Pleistocene dune rock is commonly cemented with grain-skin cement (microcrystalline-rind); porosity is 30 to 45%. The oldest Pleistocene eolianite is extensively cemented and aragonite grains are dissolved; porosity is 36%, the dissolution nearly balanced by cementation.

In Israel recent carbonate dunes are unconsolidated. Upper Pleistocene eolianite is cemented at grain-to-grain boundaries; much of the interparticle porosity is retained. Quartz and other silicate grains are partly replaced by calcite. In middle Pleistocene

eolianite, the intergranular pore space is entirely occluded by drusy calcite cement. Aragonitic skeletal fragments were dissolved out, forming molds now filled with calcite. The diagenetic fabric of the oldest Pleistocene eolianite is essentially the same as that of the middle Pleistocene dune rock.

Pleistocene eolianites of Yucatan are of at least three slightly different ages, but all are presumably late Pleistocene. The youngest Upper Pleistocene eolianite is cemented predominantly with microcrystalline-rind and needle-fiber cement with more sparry-calcite cement, and porosity is slightly less. The oldest of the Yucatan eolianites has more sparry cement and lower porosity (less than 25%) than do the others. Calvert (1979) and Calvert and others (1980) found that there are differences in the route and rate of diagenesis in Holocene and Pleistocene eolianites from different parts of Mallorca, depending on the original composition and the climate.

AGES OF CARBONATE EOLIANITES

Pleistocene Examples

Most carbonate dunes and eolian limestones are classed, according to age, as either Holocene or Pleistocene. No direct relationship between degree of lithification and age is apparent, although all or nearly all nonlithified, active dunes are of modern or Holocene age — not Pleistocene. Many Holocene dunes, however, are partly cemented. Pleistocene dunes may be either weakly or strongly cemented, but all of them are at least partially lithified (Fig. 12).

In many regions with extensive ridges or terraces of stabilized dune sands attributed to Pleistocene age, modern active carbonate dunes are also present. Dunes of these two age groups may occur one above the other (Table 3) as on the Libyan coast west of Tripoli where uncemented Holocene dunes are described as overlying lithified Pleistocene dunes (Glennie, 1970),

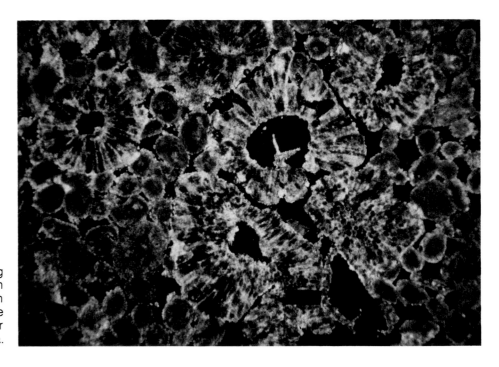

Figure 37—Photomicrograph showing *Microcodium* (globular bodies with radiating prisms and hollow centers) in Pleistocene eolianite, Isla Contoy, off the Yucatan Peninsula. Crossed nicols. Bar scale is 200μ.

and in Australia where loose Holocene dunes are on top of the consolidated dune rocks of the Coastal Limestone (Fairbridge, 1950).

In most regions of eolian carbonate sands, including western India, the Yucatan coast, and elsewhere, remnants of early dune limestones are located at higher levels or in different positions from the active coastal dunes of today. Bermuda, moreover, is one of the few localities of abundant Pleistocene eolianites in which modern carbonate dunes are uncommon. This lack of active dunes can be attributed to the islands which are now "heavily wooded and quite free from drifting sand" (Sayles, 1931). A partly submerged cliff of cross-stratified eolian limestone at Harrington Sound, Bermuda, is illustrated by Verrill (1907).

Generalized age designations such as "early Pleistocene," "late Pleistocene," "sub-Recent," "early Holocene," and "late Quaternary," are applied by various geologists in describing the carbonate eolianites of certain regions. Determinations based on radio-carbon dating have only been obtained in some places. Thirty-six age determinations of late Pleistocene and of Holocene eolian sands made in the Abu Dhabi area of the Persian Gulf (Evans et al, 1969)

with the radio-carbon method were used in interpreting events occurring over the last 7,000 years, including the eolian sand-dominated late Pleistocene and a later time of rising sea and mixing of sand with subaqueous carbonates.

The necessity of carefully selecting samples of carbonate sand for dating is illustrated by the discovery of late Pleistocene ooids in modern sands of coastal terraces in southern Tunisia. This modern oolitic sand, which was dated at 20,000 to 30,000 years before the present by carbon-14 measurement, has been shown (Fabricius, 1970) to represent reworked material; ooids of that type are not forming today in the region.

Pleistocene eolian limestones in regions of extensive source sediment have accounted for some tremendous accumulations. In Bermuda, for instance, 95% of exposed land is composed of this rock (MacKenzie 1964b). In some regions, dunes of Pleistocene age are now submerged, presumably due to changes in sea level. A notable example is at Shark Bay, Australia, where "dunes which were formed during the low stand of the sea level now separate Shark Bay from the Indian Ocean" (Malek-Aslani, 1973).

Ancient Eolianites

Relatively few carbonate rocks from the pre-Pleistocene have been interpreted as eolian limestones. A principal reason for this scarcity is the great difficulty in recognizing wind-blown carbonate deposits and in differentiating between them and other carbonate sands of the near-shore environment. Dune composition and dune texture are often essentially the same as those of adjacent beaches or bars, as one probably was derived from the other. This feature is strikingly illustrated in Bermuda (MacKenzie, 1964b), along shores of Bahama Islands (Ball, 1967), on the Mediterranean coast of Libya (Glennie, 1970) and in many other places.

One of the oldest carbonate rocks to be attributed to an eolian origin is the upper Precambrian Wilhite Formation of the Ocoee Supergroup in eastern Tennessee (Hanselman et al, 1974). This formation is said to contain an appreciable number of carbonate rocks that "compare favorably with modern shallow-water [types]" and are interpreted to include subaerial deposits. Elsewhere these carbonate deposits are described (Hanselman et al, 1974) as having evolved partly in a subaerial environment. Although many of their sedimentary textures have been

Figure 38—Caliche layers developed on Pleistocene eolianite, Isla Mujeres off the Yucatan Peninsula. Micritized eolianite is overlain by wavy, laminated crust, which is overlain by conglomeratic layer with crude reverse-graded bedding. Photo by W. C. Ward.

destroyed by diagenetic recrystallization, enough remain "to permit the identification of their depositional environments."

A few records from the Permian period have been interpreted as wind-deposited carbonate rocks. These records include the Magnesian Limestone from Scarcliffe, Derbyshire, England, described by Glennie (1970) as possibly "originally deposited as a carbonate dune on the desert shores of the Zechstein Sea, and was later dolomitized."

The Great Oolite Series of the Jurassic age in England was considered, many years ago (Evans, 1900) to be of eolian origin because cross-bedding and microscope sections showed a close resemblance to eolian limestones of Kathiawar, India, and to West Indian eolianites. Subsequent work, however, suggests that these rocks are not eolian, but probably represent subtidal bars and shoals (Klein, 1965). The strata that contain reptilian eggs, overlying the Jurassic oolite beds, were likewise believed to be of eolian origin (Evans, 1900).

ACKNOWLEDGMENTS

For assistance in compiling the extensive worldwide data and in preparing a synthesis of this material, appreciation is extended to Sarah Andrews. For drafting the several figures included, thanks are given to William Chesser and for much constructive criticism to T. S. Ahlbrandt.

SELECTED REFERENCES

Ahlbrandt, T. S., 1975, Comparison of textures and structures to distinguish eolian environments, Killpecker dune field, Wyoming: Mtn. Geologist, v. 12, no. 2, p. 61-63.
_____ S. Andrews, and D. T. Gwynne, 1978, Bioturbation in eolian deposits: Jour. Sed. Petrology, v. 48, no. 3, p. 839-848.
Anderson, G. A., 1950, 1940 E. W. Scripps cruise to the Gulf of California, Pt. I, Geology of islands and neighboring land areas: Geol. Soc. America Mem. 43, 53 p.
Auden, J. B., 1952, Some geological and chemical aspects of the Rajasthan salt problem: Proc. Symp. Rajasthan Desert, Natl. Inst. Science India Bull., v. 1, p. 53-67.
Bagnold, R. A., 1943, The physics of blown sand and desert dunes: New York, William Morrow and Co., 265 p.

Ball, M. M., 1967, Carbonate sand bodies of Florida and the Bahamas: Jour. Sed. Petrology, v. 37, p. 556-591.
Becher, J. W., and C. H. Moore, 1976, The Walker Creek Field; a Smackover diagenetic trap: Trans., Gulf Coast Assoc. Geol. Socs., v. 26, p. 34-56.
Bernard, H. A., and C. F. Major, Jr., 1963, Recent Meander Belt deposits of the Brazos River; an alluvial "sand" model: AAPG Bull., v. 47, p. 350.
Bigarella, J. J., R. D. Becker, and G. M. Duarte, 1969, Coastal dune structures from Parana [Brazil]: Marine Geology, v. 7, no. 1, p. 5-55.
Bird, E. C. F., 1972, Ancient soils at Diamond Bay, Victoria: Victorian Naturalist, v. 89, no. 12, p. 349-353.
Biswas, S. K., 1971, The miliolite rocks of Kutch and Kathiawar, western India: Sed. Geology, v. 5, p. 147-164.
Bramkamp, R. A., and R. W. Powers, 1958, Classification of Arabian carbonate rocks: Geol. Soc. America Bull., v. 69, p. 1305-1318.
Bretz, J. H., 1960, Bermuda; a partially drowned, late mature, Pleistocene karst: Geol. Soc. America Bull., v. 71, p. 1729-1754.
Calvert, F., 1979, Evolucio diagenetica en els sediments carbonatats del Pleistoceno Mallorqui: Univ. Barcelona, Ph.D. thesis, 273 p.
_____ L. Pomar, and M. Esteban, 1975, Las rhizocrecioues del Pleistoceno de Mallorca: Univ. de Barcelona, Inst.

Figure 39—Hard dense caliche crust which caps ridges of Pleistocene eolianite on Isla Mujeres off the Yucatan Peninsula. Caliche crust helps preserve the depositional topography of the dune rock. Also it is an aquiclude that protects underlying eolianite from the modern climate, thereby inhibiting diagenesis and preserving porosity and permeability. Photo by W. C. Ward.

Invest. Geology, v. 30, p. 35-60.
_____ F. Plana, and A. Traveria, 1980, La tendencia mineralogica de las eolianites del Pleistoceno de Mallorca, mediante la aplicacion del metodo de Chung: Acta Geologica Hispanica, v. 15, p. 39-44.

Carter, H. J., 1849, On foraminifera, their organization and their existence in a fossilized state in Arabia, Sindh, Kutch, and Khattyawar: Jour. Bombay Branch, Royal Asiatic Soc., v. 3, pt. 1, p. 158.

Chapman, F., 1900, Mechanically-formed limestones from Junagash (Kathiawar) and other localities: Geol. Soc. London Quart. Jour., v. 56, p. 559-583, 588-589.

Chimene, C. A., 1976, Upper Smackover Reservoirs, Walker Creek Field area, Lafayette and Columbia Counties, Arkansas, in North American oil and gas fields: AAPG Mem. 24, p. 177-204.

Cloud, P. E., Jr., 1952, Preliminary report on the geology and marine environments of Onotoa Atoll, Gilbert Islands: Pacific Science Board, Natl. Research Council, Atoll Research Bull., no. 12,p. 1-73.

Dunham, R. J., 1969, Early vadose salt in Townsend mound (reef), New Mexico, in G. M. Friedman, ed., Depositional environments in carbonate rocks–a symposium: SEPM Spec. Pub. 14, p. 139-181.
_____, 1971, Meniscus cement, in O. P. Bricker, ed., Carbonate cements: Johns

Hopkins Univ. Studies in Geology, no. 19, p. 197-300.

Emery, K. O., 1956, Sediments and water of Persian Gulf: AAPG Bull., v. 40, p. 2354-2383.

Erwin, C. E., D. E. Eby, and V. S. Whitesides, 1979, Clasticity index; a key to correlating depositional and diagenetic environments of Smackover reservoirs, Oaks Field, Claiborne Parish, Louisiana: Trans., Gulf Coast Assoc. Geol. Socs., v. 29, p. 52-62.

Evans, G. V., P. B. Schmidt, and H. Nelson, 1969, Stratigraphy and geologic history of the Sabkha, Abu Dhabi, Persian Gulf: Sedimentology, v. 12, p. 145-159.

Evans, J. W., 1900, Mechanically-formed limestones from Junagarh (Kathiawar) and other localities: Geol. Soc. London Quart. Jour., v. 56, p. 559-583, 588-589.

Fabricius, F. H., 1970, Early Holocene ooids in modern littoral sands reworked from a coastal terrace, southern Tunisia: Science, v. 169, p. 757-760.

Fairbridge, R. W., 1950, The geology and geomorphology of Point Peron, western Australia: Royal Soc. Western Australia Jour., v. 34, p. 35-72.
_____, 1971, Quaternary shoreline problems at INQUA, 1969: Quaternaria, v. 15, p. 1-18.
_____ and C. Teichert, 1953, Soil horizons and marine bands in the coastal limestones of Western Australia: Jour. and Proc. Royal Soc. N.S. Wales, v.

86, p. 68-87.

Folk, R. L., 1967, Carbonate sediments of Isla Mujeres, Quintana Roo, Mexico, and vicinity, in A. E. Weidie, ed., 2d ed., Field trip to peninsula of Yucatan guidebook: New Orleans Geol. Soc., p. 100-123.
_____ M. O. Hayes, and R. Shoji, 1962, Carbonate sediments of Isla Mujeres, Quintana Roo, Mexico and vicinity, in Yucatan field trip guidebook: New Orleans Geol. Soc., p. 85-100.

Fosberg, F. R., 1953, Vegetation of central Pacific atolls, a brief summary: Pacific Science Board, Natl. Research Council, Atoll Research Bull., no. 23, p. 1-26.

Frazier, D. E., and A. Osanick, 1961, Point-bar deposits, Old River locksite, Louisiana: Trans., Gulf Coast Assoc. Geol. Socs., v. 11, p. 121-137.

Gavish, E., and G. M. Friedman, 1969, Progressive diagenesis in Quaternary to late Tertiary carbonate sediments; sequence and time scale: Jour. Sed. Petrology, v. 39, p. 980-1006.

Glennie, K. W., 1970, Desert sedimentary environments, in Development in sedimentology 14: Amsterdam, London, New York, Elsevier Sci. Pub., 222 p.
_____ and B. D. Evamy, 1968, Dikaka—plants and plant-root structures associated with aeolian sand: Palaeogeography, Palaeoclimatology, Palaeoecology, v. 23, p. 77-87.

Goudie, A. S., B. Allchin, and K. T. M.

Hedge, 1973, The former extensions of the Great India Sand Desert: Geog. Jour., v. 139, pt. 2, p. 243-257.

_____ and C. H. B. Sperling, 1977, Long distance transport of foraminiferal tests by wind in the Thar Desert, northwest India: Jour. Sed. Petrology, v. 47, no. 2, p. 630-633.

Graf, D. L., 1960, Geochemistry of carbonate sediments and sedimentary carbonate rocks; part 1, carbonate mineralogy and carbonate sediments: Ill. State Geol. Survey, Circ. 297, 39 p.

Hanselman, D. H., J. R. Conolly, and J. C. Horne, 1974, Carbonate environments in the Wilhite Formation of central eastern Tennessee: Geol. Soc. America Bull., v. 85, p. 45-50.

Harms, J. C., P. W. Choquette, and M. J. Brady, 1974, Carbonate sand waves, Isla Mujeres, Yucatan, in Field seminar on water and carbonate rocks of the Yucatan Peninsula, Mexico; northeastern coast: Ann. Mtg. Field Trip Guidebook, Geol. Soc. America, p. 122-147.

Hoque, M., 1975, An analysis of cross-stratification of Gargaresh calcarenite (Tripoli, Libya) and Pleistocene paleowinds: Geol. Mag., v. 112, no. 4, p. 393-401.

Illing, L. V., 1954, Bahaman calcareous sands: AAPG Bull., v. 38, p. 1-95.

James, N. P., 1972, Holocene and Pleistocene calcareous crusts (caliche) profiles; criteria for subaerial exposure: Jour. Sed. Petrology, v. 42, p. 817-836.

Johnson, D. L., 1967, Caliche on the channel islands: Mineral Inf. Service, Calif. Div. Mines and Geology, v. 20, p. 151-158.

Jones, D. J., 1953, Gypsum-oolite dunes, Great Salt Lake Desert, Utah: AAPG Bull., v. 37, p. 2530-2538.

Kindle, E. M., 1917, Recent and fossil ripple mark: Canada Geol. Survey Museum Bull. 25, 121 p.

_____, 1923, Range and distribution of certain types of Canadian Pleistocene concretions: Geol. Soc. America Bull., v. 34, p. 609-648.

Klappa, C. F., 1978, Biolithogenesis of Microdocium; elucidation: Sedimentology, v. 25, p. 489-522.

Klein, G. deVries, 1965, Dynamic significance of primary structures in the Middle Jurassic Great Oolite Series, southern England: SEPM Spec. Pub. 12, p. 173-191.

Land, L. S., F. T. MacKenzie, and S. J. Gould, 1967, Pleistocene history of Bermuda: Geol. Soc. America Bull., v. 78, p. 993-1006.

Lele, V. S., 1973, The miliolite limestone of Saurashtra, western India: Sed. Geology, v. 10, p. 301-310.

MacKenzie, F. T., 1964a, Geometry of Bermuda calcareous dune cross-bedding: Science, v. 144, no. 3625, p. 1449-1450.

_____, 1964b, Bermuda Pleistocene eolianites and paleowinds: Sedimentology, v. 3, no. 1, p. 52-64.

Malek-Aslani, M., 1973, Environmental modeling; a useful exploration tool in carbonates: Trans., Gulf Coast Assoc. Geol. Socs., v. 23, p. 239-244.

McKee, E. D., 1945, Small-scale structures in the Coconino Sandstone of northern Arizona: Jour. Geology, v. 53, no. 5, p. 313-325.

_____, 1966, Structures of dunes at White Sands National Monument, New Mexico (and a comparison with structures of dunes from other selected areas): Sedimentology, v. 7, 69 p.

_____, ed., 1979, A study of global sand seas: U.S. Geol. Survey Prof. Paper 1052, 429 p.

_____ and J. J. Bigarella, 1972, Deformational structures in Brazilian coastal dunes: Jour. Sed. Petrology, v. 42, no. 3, p. 670-681.

_____ J. R. Douglass, and S. Rittenhouse, 1971, Deformation of lee-side laminae in eolian dunes: Geol. Soc. America Bull., v. 82, p. 359-378.

_____ and R. J. Moiola, 1975, Geometry and growth of the White Sands dune field, New Mexico: U.S. Geol. Survey Jour. Research, v. 3, no. 1, p. 59-66.

Newell, N. D., and D. W. Boyd, 1955, Extraordinarily coarse eolian sand of the Ica Desert, Peru: Jour. Sed. Petrology, v. 25, no. 3, p. 226-228.

_____ and J. K. Rigby, 1957, Geological studies on the Great Bahama Bank, in Le Blank and Breeding, eds., Regional aspects of carbonate deposition—a symposium: SEPM Spec. Pub., no. 5, p. 15-72.

Phleger, F. B., and G. C. Ewing, 1962, Sedimentology and oceanography of coastal lagoons in Baja California, Mexico: Geol. Soc. America Bull., v. 73, no. 2, p. 145-182.

Qadri, S. M. A., 1957, Wind erosion and its control in Thar; symposium on soil erosion and its control in the arid and semiarid zones: Karachi, Joint auspices of F.A.C.P. and U.N.E.S.C.O., p. 169-173.

Read, I. F., 1974, Calcrete deposits and Quaternary sediments, Edel Province, Shark Bay, Western Australia, in B. W. Logan, et al, eds., Evolution and diagenesis of Quaternary carbonate sequences, Shark Bay, Western Australia: AAPG Mem. 22, p. 250-282.

Saxena, S. K., and S. Singh, 1976, Some observations on the sand dunes and vegetation of Bikaner district in western Rajasthan: Annals of Arid Zone-15, p. 313-322.

Sayles, R. W., 1931, Bermuda during the Ice Age: Proc., Am. Acad. Arts Sciences, v. 66, p. 382-467.

Selim, A. A., 1974, Origin and lithification of the Pleistocene carbonates of the Salum area, western coastal plain of Egypt: Jour. Sed. Petrology, no. 44, p. 757-760.

Semeniuk, V., and T. D. Meagher, 1981, Calcrete in Quaternary coastal dunes in southwestern Australia; a capillary-rise phenomenon associated with plants: Jour. Sed. Petrology, v. 51, p. 47-68.

Seth, S. K., 1963, A review of evidence concerning changes of climate in India during the protohistorical and historical periods: Proc., Rome Symposium, UNESCO and World Meteorol. Organ., p. 443-450.

Sherman, G. D., and H. Ikawa, 1958, Calcareous concretions and sheets in soils near South Point, Hawaii: Pacific Science, v. 12, p. 255-257.

Shrivastava, P. K., 1968, Petrography and origin of Miliolite Limestone of western Saurashtra coast: Jour. Geol. Soc. India, v. 9, no. 1, p. 88-96.

Singh, S., 1977, Geomorphological investigation of the Rajasthan Desert: Jodhpur, Central Arid Zone Research Inst., Mono. 7, 44 p.

Skocek, V., and A. A. Saadallah, 1972, Grain-size distribution, carbonate content and heavy minerals in eolian sands, Southern Desert, Iraq: Sed. Geology, v. 8, p. 29-46.

Smith, N. D., 1970, The braided stream depositional environment; comparison of the Platte River with some Silurian clastic rocks, north-central Appalachians: Geol. Soc. America Bull., v. 81, p. 2995-3014.

Supko, P. R., 1971, "Whisker" crystal cement in a Bahamian rock, in O. P. Bricker, ed., Carbonate cements: Johns Hopkins Univ. Stud. in Geology, no. 19, p. 143-146.

Teichert, C., 1947, Contributions to the geology of Houtman's Abrolhos, Western Australia: Proc., Linnean Soc. South Wales, v. 71, p. 145-196.

Verrill, A. E., 1907, The Bermuda Islands; part IV-geology and paleontology: Trans., Conn. Acad. Arts and Sciences, v. 12, no. 145, p. 45-203.

Verstappen, H. T., 1966, Landforms, water, and land use west of the Indus Plain: Nature and Resources, v. 2, no. 3, p. 6-8.

Von Der Borch, C. C., 1976, Stratigraphy and formation of Holocene dolomitic carbonate deposits of the Evorong area, South Australia: Jour. Sed. Petrology, v. 46, no. 4, p. 952-966.

Wadia, D. N., 1953, Geology of India, 3rd ed.: London, MacMillan & Company, 531 p.

Ward, W. C., 1970, Diagenesis of Quaternary eolianites of N.E. Quintana Roo, Mexico: Houston, Rice Univ., Ph.D. dissert., 206 p.

_____, 1973, Influence of climate on the early diagenesis of carbonate eolianites: Geology, v. 1, p. 171-174.

_____, 1975, Petrology and diagenesis of carbonate eolianites of northwestern Yucatan Peninsula, Mexico, *in* K. F. Wantland and W. C. Pusey III, eds., Belize Shelf carbonate sediments, clastic sediments, and ecology: AAPG Stud. in Geology, no. 2, p. 500-571.

_____ 1978, Indicators of climate in carbonate dune rocks, *in* W. C. Ward and A. E. Weidie, eds., Geology and hydrogeology of northeastern Yucatan: New Orleans Geol. Soc., p. 191-208.

_____ and M. J. Brady, 1973, High-energy carbonates on the inner shelf, northeastern Yucatan Peninsula, Mexico: Trans., Gulf Coast Assoc. Geol. Socs., v. 23, p. 226-238.

Webley, D. M., D. J. Eastwood, and C. H. Gimingham, 1952, Development of a soil microflora in relation to plant succession on sand dunes, including the "rhizosphere" flora associated with colonizing species: Jour. Ecology, v. 40, p. 168-178.

Yaalon, D. H., 1967, Factors affecting the lithification of eolianite and interpretation of its environmental significance in the coastal plain of Israel: Jour. Sed. Petrology, v. 37, no. 4, p. 1189-1199.

4

Tidal Flat Environment

Eugene A. Shinn

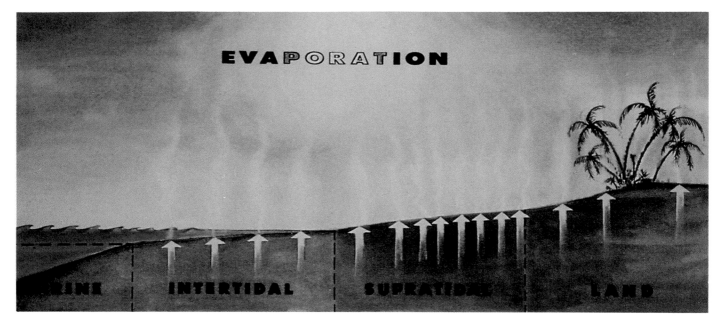

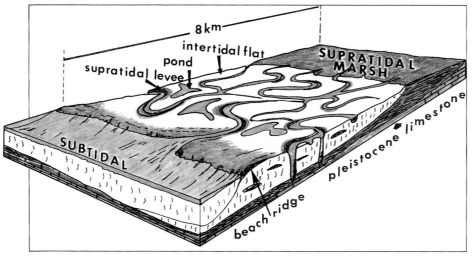

Figure 1—Schematic diagram showing sedimentary environments, their relationship to sea level and relative amounts of capillary evaporation. Facies changes are transitional with the change from supratidal to continental (land) being the most gradual and subtle of the three.

Figure 2—Block diagram schematically showing major facies on the Andros Island onlap (transgressive) tidal flat model. Supratidal zones are shown in brown, the major one being the supratidal marsh which, because of onlap sedimentation, underlies the entire tidal flat system, except where removed by tidal channel erosion.

Recognition of ancient tidal flats[1] in the geologic record, whether they be siliciclastic or carbonate, owes much to sedimentary and diagenetic investigations of modern analogues. The classic studies of northern European siliciclastic tidal flats by Van Straaten (1954), those of the English Wash area by Evans (1965), and finally the sophisticated refinements of the various sedimentary processes of this special environment by Reineck (1967) opened an important era of understanding. Additional refinement of clastic tidal flat processes was provided by Klein (1971, 1972). The extension of ideas

and concepts developed from siliciclastic investigations to areas of ancient carbonate tidal flat deposition soon furthered geologic understanding of these rocks. It was the recognition of similarities and differences between modern siliciclastic tidal flats and ancient carbonate counterparts that stimulated interest in modern carbonate tidal flats. Workers in carbonate counterparts owe much to the early investigations on modern stromatolites by Black (1933), whose work preceded realization of the economic importance of tidal flats. Later studies of modern sediments on the Bahama Banks by Illing (1954) and Newell and Rigby

(1957) helped focus attention on the Andros Island tidal flats; however, it was two "turning point" discoveries that gave tidal flats worldwide interest and more sharply focused the investigations that followed. The first discovery was that dolomite is forming on modern carbonate tidal flats and in areas where physical and chemical conditions are similar to those of large tidal flats; for example, Netherlands Antilles (Deffeyes et al, 1965); Persian Gulf (Wells, 1962); Bahamas (Shinn et al, 1965), and the Qatar Peninsula of the Persian Gulf (Illing et al, 1965). Until then, all forms of dolomite were considered an enigma, because there were no

Figure 3—Block diagram schematically showing major facies on the Persian Gulf Trucial Coast offlap (regressive) tidal flat model. Major supratidal zone is the sabkha, the arid climate equivalent of the supratidal marsh. Supratidal sabkha sediments have accreted over the gypsum, algal mat, burrowed intertidal and subtidal lagoonal sediments. In many places a cemented crust which forms in the upper subtidal or lower intertidal zone has been incorporated within the regressive sequence. The gypsum zone is laterally equivalent to and can be traced into a continuous anhydrite layer farther inland beneath the sabkha. Tidal deltas composed mainly of ooids form at tidal passes through the cross-bedded barrier island sands. The barrier island is composed of cerithid gastropod sands and ooids with scattered coral fragments. Coral reefs grow just seaward of the islands but are absent opposite the tidal deltas. Much of the tidal delta ooids has been cemented

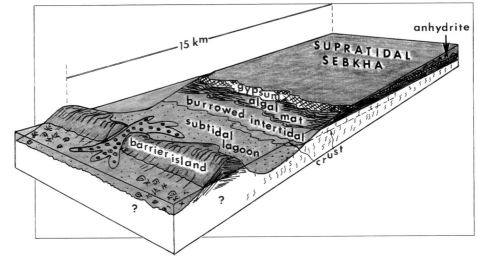

under submarine conditions, especially on the seaward side, and beachrock is abundant in the barrier island sands. Composition of facies underlying the barrier island, coral reefs and tidal deltas is not well known; however, a thin submarine cemented crust forms the substrate controlling coral growth.

modern analogues. The second discovery was the recognition and documentation of economic oil reserves in Paleozoic carbonate tidal flat accumulations by Roehl (1967). That study paved the way for recognition of many ancient oil-producing tidal flats.

Owing to the above breakthroughs, detailed investigations of both diagenetic and sedimentary processes proceeded rapidly. Studies were conducted on the Persian Gulf Trucial Coast by Kendall and Skipwith (1969), Kinsman (1964), Butler (1965), Evans (1966), Evans and others (1969), and Schneider (1975), and by the Royal Dutch Shell group on both the Trucial Coast and Qatar Peninsula (Purser and Evans, 1973; Shinn, 1973a). Additional studies in similar climatically arid settings in western Australia include those of Logan and others (1964), and Hagan and Logan (1975). Humid-climate tidal flats were studied at Andros Island in the Bahamas by Shinn and others (1969), followed eight years later by a strikingly illustrated book by Hardie (1977). Studies of arid environments in the Persian Gulf by the Royal Dutch Shell group and others were compiled in a comprehensive book edited by Purser (1973). The most recent compilation on tidal

flats, which includes both modern and ancient siliciclastics and carbonates, is a casebook edited by Ginsburg (1975).

For many years, geologists recognized mudcracks in the geologic record and realized that their presence suggested sedimentation at or just above sea level. Now, with more sophisticated knowledge, we know of other diagnostic indicators, and together with information from modern examples, we can determine paleogeographic settings, geometries, and stratigraphic traps associated with these ancient deposits. For example, tidal flats generally form belts parallel to land and may be bordered by open marine or basinal deposits that can serve as petroleum source areas. We also know that the various facies within a tidal flat system, such as intertidal, supratidal and channel deposits, can form reservoirs and seals with predictable distributions. The key to using this knowledge, however, is, as with other sedimentary rocks, accurate identification of the various sedimentary structures indicative of tidal flat accumulation.

The purpose of this chapter is to generalize our vast but diffuse knowledge of the various kinds of tidal flats, both modern and ancient, and to graphically describe the most

diagnostic sedimentary structures and sequences. Modern examples will be used frequently throughout this chapter, because they provide the fundamental basis of understanding. In addition, attempts will be made to speculate and pictorially show how stratigraphic traps may be produced in various parts of a tidal flat system.[2]

TIDAL FLAT ENVIRONMENTS AND SEDIMENTATION

Tidal flats are integrated systems. All tidal flat systems, except those in areas dominated by wind tides, are composed of three basic environments: supratidal, intertidal, and subtidal. Within these environments there are several subenvironments. The basic environmental zones shown in Figure 1 include the following: (1) supratidal sediments, those deposited above normal or mean high tide and exposed to subaerial conditions most of the time because they are flooded only by spring and storm tides; spring tides occur twice each month, and

---

[1]Tidal flats and their arid counterparts, called sabkhas, are interchangeable in this text.

[2]Highlights of this chapter are depicted in the film, "Stratigraphic Traps: the Tidal Flat Model," produced by AAPG.

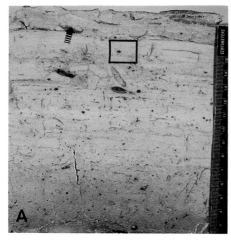

Figure 4A—Slab of plastic-impregnated core of supratidal levee laminations showing a 2-cm-thick lens of storm-deposited flat pebble breccia at top of core. Arrow shows imbricate laminated pebble. Note bird's-eye voids scattered here and there, which are distinct from plant roots. Also notice transition to less well laminated upper intertidal sediment at base of core. Outlined area shows location of thin section shown in Figure 4B.

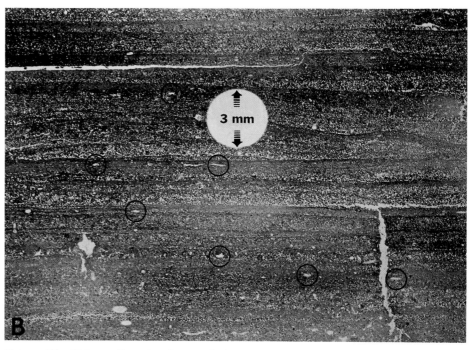

Figure 4B—Thin section showing graded pellets and lateral disconformity of supratidal levee laminations. Material removed from circle was analyzed by X-ray diffraction. Sediment contains approximately 5% modern dolomite. Small circles show bird's-eye voids which parallel laminations.

storm tides, the largest of all, occur sporadically during certain seasons and are less frequent; (2) intertidal sediments, which lie within normal high tide and normal low tide; they are exposed either once or twice daily depending on the tidal regime and local wind conditions; and (3) subtidal sediments, which are seldom, if ever, exposed to air. For the purpose of this paper, the term "subtidal" is restricted to those sediments seaward of a tidal flat system or within a system, such as in tidal channels.

The information presented in this chapter is based on two models, the humid, transgressive or onlap Andros Island model (Fig. 2) and the arid, regressive or offlap Persian Gulf model (Fig. 3). In both models the major supratidal flat (the supratidal marsh at Andros Island and the "sabkha" of the Persian Gulf), forms a belt paralleling the land. This supratidal belt invariably merges transitionally with continental deposits above the reach of even the highest storm tides. Minor supratidal environments occur as levees on the outer bends of meandering tidal channels within the intertidal zone, or on

the back side of low-lying beach ridges (Fig. 2). The important supratidal zone, the one that is most extensive and most likely to be preserved, is the supratidal marsh or its arid equivalent, the sabkha.

Because of the mechanics of supratidal sedimentation, there is a basic difference between sedimentary structures formed on levees and beach ridges and those deposited in the marsh or sabkha (Hardie, 1977). Levees and beach ridges are flooded frequently by spring tides that carry little sediment; thus, individual sedimentary laminae are thin (Fig. 4). Storm tides, even though they may carry large sediment loads, tend to deposit little in these environments, because soft, pelleted sediment does not easily settle out when water moves rapidly over these environments. In the supratidal marsh, however, highly sediment-charged water eventually comes to rest and layers exceeding 2 cm in thickness may be deposited in a few hours (Fig. 5; Shinn et al, 1969). During storms,

large quantities of sediment settle on intertidal and subtidal flats, but few laminations survive in the subtidal zone due to bioturbation. Some are preserved, however, in the intertidal zone of arid tidal flats because of extreme salinities.

Storm layers in the supratidal marsh are almost invariably sandwiched between layers of organic carbon-rich algal material, which proliferates during the long intervals between storms. The individual algal or storm layers can be traced laterally for tens of meters. Such thick algal material is generally lacking on levee and beach ridge environments, and individual thin laminations can seldom be traced laterally for more than a few centimeters (Fig. 4; Shinn et al, 1969).

SUPRATIDAL ZONE DIAGNOSTIC SEDIMENTARY STRUCTURES

Mudcracks

Probably no single sedimentary structure is better known or more indicative of a depositional environment than polygonally arranged mudcracks caused by shrinkage of carbonate mud (Figs. 6A, 6B, 7A, 7B). Although such shrinkage can occur in playa lakes far from the sea, they are

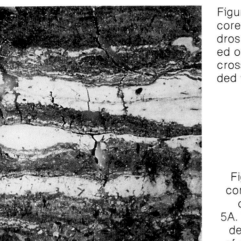

Figure 5A—Slab of paraffin-impregnated core from the supratidal marsh at Andros Island. Thick storm layers composed of pelleted mud containing micro-crossbedding. Storm layers are interbedded with dark algal-rich laminations.

Figure 5B—Portion of supratidal marsh core taken on a part of the marsh a few centimeters higher than that in Figure 5A. This core contains lightly lithified and desiccated layers and greater oxidation of organic-rich layers than that in Figure 5A. Large objects are mangrove and grass roots.

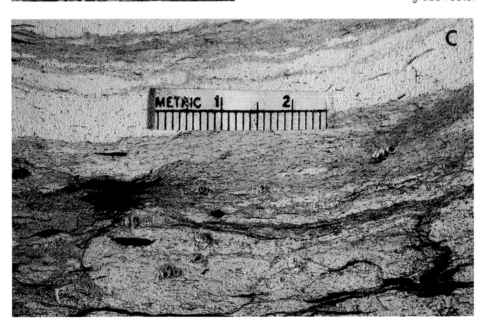

Figure 5C—Portion of a supratidal marsh core artificially compacted to a depth equivalent to 900 ft (274 m) of burial. Note lateral extension of organic matter and destruction of bird's-eye and root voids. Gastropod shells show no signs of crushing. Scale in cm.

more often associated with the supratidal environment of tidal flats that border the sea. Syneresis cracks, which may look similar to desiccation cracks, are known to form in subtidal clay mineral deposits. The phenomenon, however, has not been documented in carbonate sediments.

Shrinkage cracks are commonly observed in modern intertidal settings, but there is little evidence that they are preserved in this environment. The considerable information available suggests that mudcracks are best preserved in the supratidal and possibly the upper part of the intertidal zones.

Thick storm layers dry and shrink to produce large mudcracks and mud polygons, whereas thin layers shrink to form smaller cracks and polygons. When thick storm layers are deposited in areas lacking prolific algal mats, shrinkage cracks tend not to perpetuate themselves upward through subsequent layers. When algal mats are present, however, cracks tend to be widely spaced and perpetuate upward with the addition of each succeeding algal or sedimentary lamination, even when the layers are thin. Thus, where algae are abundant, the rule that thick layers result in large desiccation cracks and thin layers make small cracks generally does not hold true. Cracks in sediment dominated by algae may be very large (Figs. 8A, 8B) or small (Figs. 9A, 9B) or nonexistent (Fig. 10A). The reason for this difference is not understood but is probably related to the length of exposure (the exposure index of Hardie, 1977) and the thickness of individual storm layers within the mats. Intertidal flats are often devoid of algal mats or shrinkage cracks due to the activities of organisms (Fig. 10B).

The classical mudcrack is V-shaped. All mudcracks initially form in this shape, but in carbonates the "V" shape often changes. During exposure, the edges of the cracked mud polygons become rounded

Figure 6A—Depositional surface of supratidal lime mud on Cluett Key in Florida Bay showing healed mudcracks in a layer deposited by Hurricane Donna in 1960. Cracks are filled with porous and permeable material weathered from adjacent polygons and sediment deposited during subsequent storm floodings.

Figure 6B—Mudcracks of Ordovician age in Kentucky. Note similarity with modern example in Figure 6A. Photo courtesy of Earl Cressman (U.S. Geological Survey).

through weathering. The result can be a feature that is sausage-shaped in section, like the well known boudinage of ancient limestones (Figs. 11, 12). Because of this weathering, dense, impermeable mud polygons can become encased in the more porous and permeable sediment derived from the weathering process. An important aspect of this process is selective dolomitization. Shinn

(1968a) reported on selective dolomitization of permeable sediment surrounding lime mud polygons in the lower Florida Keys (Fig. 13). This form of early selective dolomitization is believed to be responsible for many limestone layers, lenses, and flat pebbles in ancient dolomite. These features are especially common in tidal flats of Paleozoic age (Fig. 12).

Laminations

Laminations in tidal flat sediments generally result from spring or storm tide deposition (Ball et al, 1963). As suggested earlier, they tend to be thicker in the more distant or landward portions of the system (the supratidal marsh or the sabkha; see Figs. 2, 3, 5A, 5B, 5C) and thinner in the more seaward part (Figs. 4A, 4B), although combinations of thick and

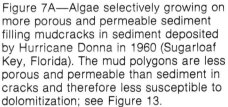

Figure 7A—Algae selectively growing on more porous and permeable sediment filling mudcracks in sediment deposited by Hurricane Donna in 1960 (Sugarloaf Key, Florida). The mud polygons are less porous and permeable than sediment in cracks and therefore less susceptible to dolomitization; see Figure 13.

Figure 7B—Shrinkage cracks in algal mat, which is not related to storm layer, Andros Island tidal flat. Compare with example in Figure 7A.

thin laminae can occur in any single location. Thick layers in the more landward part of arid sabkhas are often destroyed by desiccation, however.

Tidal flat sediment of both modern and ancient age is deposited as layers of sand-sized mud pellets, and although these pellets may be destroyed during dolomitization or compaction, they may even form micro-crossbedded laminae. Graded pellets are commonly present in supratidal storm layers (Fig. 4B). Rippling, a feature restricted to fine- to coarse-grained sizes and generally thought to be a submarine phenomenon, can be common in supratidal facies (Fig. 14).

Horizontal laminations, whether graded or ungraded, thick or thin, with or without crossbedding, are restricted to supratidal and upper intertidal conditions in modern tidal flats as well as to those of late Paleozoic through Cenozoic age. The presence of such laminations is indirectly controlled by burrowing organisms. In older Proterozoic or lower Paleozoic rocks, before the development of extensive infaunas or browsing organisms, sedimentary laminations may have been preserved

Figure 8A—Large mudcrack pattern in thick algal mat on edge of brine-filled pond at Inagua Island in the Bahamas. Gypsum is precipitating in cracks and on margins of upturned crack edges.

Figure 8B—Similar algal mats on tidal flats of the Trucial Coast in the Persian Gulf.

Figure 9A—Desiccated algal mat in lower supratidal zone in the Persian Gulf. Small mound next to lens cap in upper center of photo is made by algae populating pelleted sediment thrown up by burrowing worm. With their convex upward laminations and central canal, these mounds resemble the Precambrian algal feature, *Conophyton* (see Shinn, 1972).

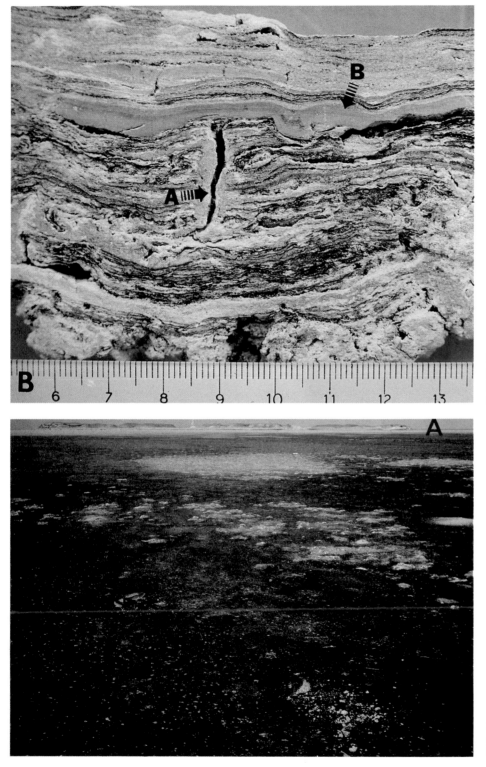

Figure 9B—Sliced section of algal sediment shown in Figure 9A. Note desiccation crack (**A**) covered by windblown sediment containing 65% dolomite (**B**).

Figure 10A—Localized algal mat without desiccation cracks in supratidal zone of the Persian Gulf on the Qatar Peninsula.

in all environments, as suggested by Garrett (1970) and Hofman (1969). Although storms may deposit layers in the subtidal and lower intertidal zones, abundant burrowing organisms that thrive in these environments quickly churn and homogenize nearly all primary laminations. Few organisms, on the other hand, can endure the rigors of exposure and fluctuating salinity in the supratidal environment; thus, there is little destruction of primary sedimentary features. Plant life is also severely restricted, especially in arid tidal flats or sabkhas; therefore, roots are seldom so abundant as to completely destroy sedimentary structures. Plant

Figure 10B—Burrowed intertidal sediment in the Persian Gulf. Small burrows were made by crabs and larger depressions by mud skippers, a small fish that stalks the flats at low tide.

Figure 11—Mudcracked supratidal dolomite from Cretaceous rocks in west Texas near San Angelo. Note bird's-eyes, rounded mudcrack polygons and large crack in thick storm layer with vertical tubes made either by worms or escaping gas bubbles. Similar burrow-like voids appeared in hurricane-deposited layers of similar thickness within a few weeks of deposition. Such mudcracked dolomites containing dinosaur tracks occur in Cretaceous rocks in central Texas. Specimen courtesy of Clyde Moore (Louisiana State University).

life and resulting root effects, however, do increase where the sedimentary surface is more elevated or in a landward direction, where supratidal sediments merge into continental deposits. Such a transition may also be associated with a facies change to a more siliciclastic environment. Such a transition from supratidal carbonates to siliciclastic redbeds is common in tidal flats of Permian age in west Texas and New Mexico.

Algal Structures
The environmental significance of algal mats and domes became apparent as a result of Black's (1933) work on Andros Island, Bahamas. Black's studies stimulated interest in and resulted in publications concerning algal structures, but it was not until the publication of Logan and others (1964) that the description and environmental significance of various forms became simplified enough for recognition and classification by nonspecialists. Logan and others (1964) recognized three basic forms: (1) flat laminated forms (generally referred to

Figure 12—Gray lenses of limestone floating in silty dolomite of Cambrian-Ordovician age from Maryland. The limestone lenses are thought to be rounded mudcrack polygons that escaped early dolomitization similar to the modern example in Figure 13. Specimen is from cyclical tidal flat limestone and dolomite sequence described by Matter (1967).

Figure 13—Modern cemented supratidal crust from Sugarloaf Key, Florida, showing a storm layer (arrow) composed of sun-dried pelleted lime mud. The surrounding cemented matrix contains up to 25% high-calcium dolomite. This style of selective supratidal dolomitization is thought to be the origin of the features in Figure 12.

Figure 14—Ripples formed in pelleted supratidal sediment on a channel levee at Andros Island. Locality is same as for core shown in Figures 4A and 4B. Rippling caused during storm and spring tide floodings probably explains laterally discontinuous laminae shown in Figures 4A and B.

as algal mats), (2) separate hemispherical or club-shaped domes, and (3) laterally-linked hemispherical shapes. Examples of these may be found in tidal flat accumulations ranging in age from Paleozoic to Holocene (see Figs. 15A, 15B, 16A, 16B). Such examples have been described by numerous authors, including Hofman (1969), Hoffman (1967), and James (1977).

Correct interpretation of tidal flat laminations can be complicated by the presence of algae. Some layers are induced entirely by algae, whereas others have a purely sedimentary origin. The origin of ancient laminations, where algae have long since rotted away, can therefore be extremely difficult to determine. One must find clues, such as antigravity structures (for example, sediment on the near vertical sides of domes), to determine if algae have assisted deposition. In Figure 17, a core from the Permian San Andres dolomite in west Texas shows oxidized supratidal laminations grading downward into

Figure 15A—Modern subtidal to intertidal columnar stromatolites at Hamlin Pool, Western Australia. Note that heads continue seaward into subtidal conditions. In a landward direction, the heads grade into laterally continuous layers on low supratidal flats associated with evaporitic minerals. Photo courtesy of R. N. Ginsburg (University of Miami Comparative Sedimentology Laboratory).

Figure 15B—Same locality as Figure 15A but showing rippled intertidal carbonate sands between cemented algal heads. Individual heads have long axes aligned perpendicular to shore and to trend of ripple crests. Photo courtesy of R. N. Ginsburg. For a more complete description of area, see Logan and others (1964), and Davies (1970).

gray reduced intertidal or subtidal sediment. The distinctly laminated domal structure, which is clearly algal in origin, strongly suggests that the more horizontal laminations with which it is associated were also of algal origin. Without the presence of this domal structure, it would be difficult indeed to determine if the laminations were physical or algal in origin (see Fig. 17). Algal tubules also provide clues to algal origin, but they

are unlikely to be preserved if the rocks have been dolomitized. Various forms of algal mats on modern tidal flats are shown in Figures 8, 9, and 10.

Observation on numerous occasions of algal mats in the Persian Gulf has shown that they act as sticky flypaper to trap windblown dust. Several times a year, northerly winds, called shamahls, cause "west Texas-type" dust storms that transport fine-

grained material across the Gulf from the Zagros Mountains of Iran. Within hours, this dust coats the mats, imparting a rusty-tan color. Within a day or two, algae recolonizes the surface, thus producing a distinct lamination of detrital material. The thick brown layer in Figure 9B was interpreted to be of eolian origin. If the layer in Figure 9B had been deposited from water, it would have been mainly aragonitic. Instead, it

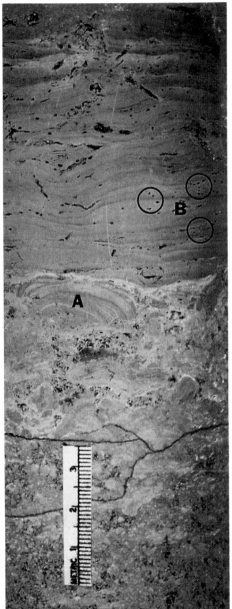

Figure 16A—Laterally-linked hemispheroids (LLH classification of Logan et al, 1964) formed by stromatolitic algae of Cambro-Ordovician age in Maryland. Photo from same locality as in Figure 12.

Figure 16B—False LLH stromatolites from the Florida Keys. This organic-rich stromatolite-like encrustation is actually a soilstone crust with little in common with true intertidal and supratidal algal stromatolites. Without knowing the sedimentary sequence, it may be impossible to differentiate such features from genuine algal structures in the geologic record. See Multer and Hoffmeister (1968) and Robbin and Stipp (1969) for genetic interpretation of soilstone crusts.

Figure 17—A core from the Permian San Andres dolomite in west Texas showing transition from oxidized algal-laminated supratidal dolomite with algal dome (**A**) and bird's-eye voids filled with anhydrite (**B**) and reduced burrowed intertidal dolomite below. Porosity is completely filled with anhydrite. At least 18 similar supratidal-intertidal transitions were observed in the core described in Figure 46. Scale in cm.

was composed of clays, quartz silt, and about 60% detrital dolomite. During several dust storms, vaseline-coated glass microscope slides were exposed to the wind and the adhered particles were examined petrographically. Typically, such dust contained about 60% detrital dolomite. The crystals measured about 30μm across, with quartz silt in the same size range. The remainder of the material was composed of clay minerals.

The obvious conclusion is that supratidal algal laminations in arid areas can result both from alternating flooding and deposition of marine sediment and algal mat growth or alternating eolian deposition and algal growth.

Bird's-eye Structures

Bird's-eyes, an old American term which was recently redefined and called fenestrae by Tebbutt and others (1965), are considered reliable indicators of supratidal deposition when they occur in predominantly muddy rocks. Bird's-eyes, or fenestrae, are considered identical to

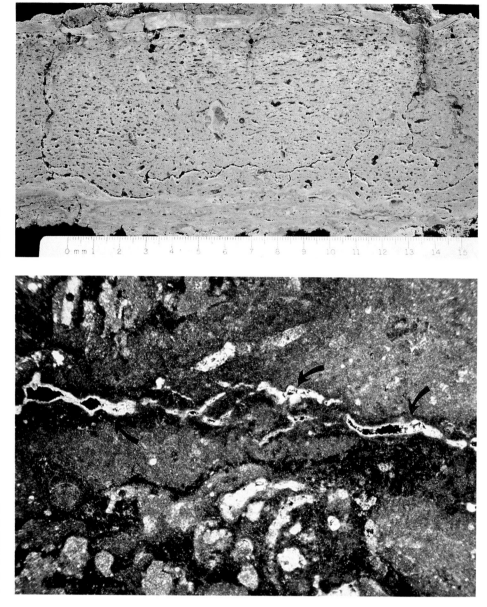

Figure 18—Semi-lithified modern supratidal crust from Sugarloaf Key, Florida, showing abundant planar-type bird's-eye voids (arrows).

Figure 19—Thin section photo of filled and partially filled bird's-eye voids in dolomitic supratidal crust from Andros Island. These early filled voids should resist obliteration during burial and compaction. The cement is either aragonite or magnesium calcite.

the "shrinkage pores" or "loferites" of Fischer (1964). These features are small millimeter-sized vugs that form in supratidal sediments as a result of shrinkage and expansion, gas bubble formation, air escape during flooding, or wrinkles in algal mats (Shinn, 1968b). In ancient limestones, bird's-eye vugs are generally filled with calcite or anhydrite. Often they contain geopetal internal sediment, as in the Ordovician rocks described by Grover and Read (1978). True supratidal bird's-eye structures should not be confused with other calcite-filled voids, which they often resemble. Shinn (1968b) found by ex-perimentation and examination of hundreds of modern sediment cores from Florida, the Bahamas, and the Persian Gulf, that bird's-eye vugs can be divided into two distinct types — planar-shaped, generally un-connected voids (Fig. 18), which tend to form parallel to individual laminae, and randomly distributed bubble-shaped voids. The latter almost always occur in supratidal sediments lacking sedimentary or algal laminations. Bird's-eye vugs may form in intertidal and even sub-tidal sediments, but evidence that they are preserved in these en-vironments is lacking. These features are preserved, however, in the upper part of intertidal sediments and in-crease in abundance through the tran-sition to supratidal sediments.

Supratidal bird's-eye vugs are preserved because they form in an ac-tive diagenetic environment, where early lithification is common and voids can be quickly filled with car-bonate cement, internal sediment, or evaporites. Figure 19 shows calcite-filled bird's-eyes in modern dolomitic crusts from Andros Island. The bird's-eyes in the Permian example in Figure 17 are filled with anhydrite that probably formed penecontem-poraneously with deposition. Some

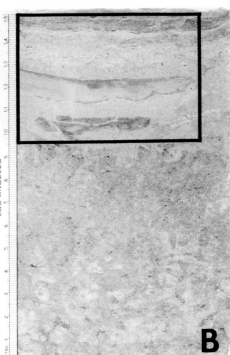

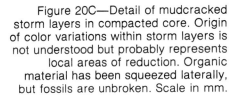

Figure 20A—A plastic-impregnated core from the supratidal zone of a mud island in Florida Bay. Note bird's-eye vugs and vertical worm or gas escape structures shown by arrows. Vertical voids are believed to be similar to those of Cretaceous age shown in Figure 11 and those of Ordovician age shown in Figure 22.

Figure 20B—Companion core to that shown in Figure 20A. Core has been artificially compacted to a burial depth of 10,000 ft (3,048 m). Note that bird's-eye voids and vertical tubular voids have been obliterated. Note mottled texture developed in lower portion of core which has been exaggerated by compaction. Detail of outlined area is shown in Figure 20C.

Figure 20C—Detail of mudcracked storm layers in compacted core. Origin of color variations within storm layers is not understood but probably represents local areas of reduction. Organic material has been squeezed laterally, but fossils are unbroken. Scale in mm.

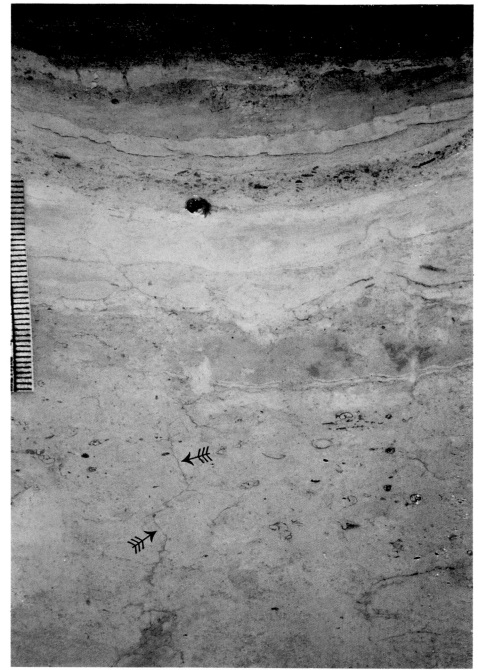

Figure 21—Portion of a compacted core of modern supratidal sediment containing algal mat (at surface) and thin algal layers and storm layers. Note subvertical irregular stylolite-like lines shown by arrows. The lines result from compaction of vertical voids that have a thin organic lining. Core from Crane Key in Florida Bay has been compacted to equivalent of 10,000 ft (3,048 m) of overburden. Scale in mm.

bird's-eye structures may even be formed by the development of supratidal evaporites (Illing, 1959). Contrary to an earlier conclusion by Shinn (1968b), bird's-eye vugs in uncemented sediment can be destroyed during compaction (Figs. 20A, 20B, 20C, 21), as shown by recent compaction experiments (Shinn et al, 1980). The presence of bird's-eye vugs and well-preserved pelletal texture in ancient limestones and dolomites is therefore considered evidence of both supratidal deposition and early cementation (Shinn et al, 1980).

Pseudo-Bird's-eye or Fenestral Structures

Some ancient tidal flat rocks contain "subvertical tubular fenestrae," as described by Read (1975) in rocks of Devonian age and by Cressman and Noger (1976) in rocks of Ordovician age in Kentucky. Although probably not of the same origin as the planar and bubble-like voids, these features are nevertheless common in both modern and ancient supratidal settings. Shinn (in press) has suggested that features identifiable as burrows or roots should not be included under the bird's-eye or fenestral classification even when they occur with true planar or bubble shaped fenestrae (bird's-eyes or loferites). Burrows and root tubes can form in other environments; thus, they alone may not be reliable indicators of tidal flat deposition. When they occur with other exposure

features, however, as is shown later, then they can be considered additional criteria for tidal flat deposition. Under these circumstances, the term "pseudo-bird's-eye," "pseudo-fenestra," or "pseudo-loferite" is preferred. Both pseudo- and true bird's-eyes or fenestrae also form in modern muddy subtidal environments, but because of compaction, they are not preserved. They are probably rare in ancient subtidal en-

vironments. Note the vertical voids indicated by arrows in Figure 20A. Although common, their origin has never been adequately explained. The author observed that these features formed within a few weeks in thick (up to 10 cm) lime mud layers deposited during Hurricane Donna in Florida in 1960. They measured 1 to 3 mm in diameter and had a dome-shaped upper termination. Unlike tubular voids made by burrowing

Figure 22—Calcite-filled vertical tubes in a storm layer of Ordovician age in Kentucky. Although often associated with supratidal deposition, such features should not be confused with true bird's-eye voids. Photo courtesy of Earl Cressman (U.S. Geological Survey).

worms, these voids seldom penetrate to the surface. These features, however, superficially resemble both worm burrows and plant root voids, including the subtidal root voids made by the turtle grass, *Thalassia*. Although these features resemble burrows and plant root holes, no worms or roots were observed in the voids formed in fresh storm-deposited sediment. Figure 22 shows pseudo-bird's-eye voids in Ordovician rocks that look identical to those observed in Hurricane Donna storm layers. A similar example of Cretaceous age is shown in what is clearly a thick storm layer in Figure 11. Notice in Figure 21 the dark vertical stylolite-like markings that resulted from compaction of vertical voids shown in Figure 20A.

Intraclasts

All sedimentary features described earlier are subject to erosion and redeposition to produce distinctive sedimentary structures. Intraclasts, probably eroded and redeposited mainly during storms, occur in two major areas — on the supratidal flats, and in subtidal channels, where they become part of the basal channel lag. Tidal channel deposition and its sedimentary products will be discussed later. Because hardening and cementation of muddy sediment on tidal flats is commonly restricted to supratidal conditions, this environment, therefore, provides most tidal flat intraclasts. Material eroded from muddy subtidal and intertidal environments, on the other hand, is generally in the form of grapestone, individual pellets, or mud-sized particles and fossils. Furthermore, because most clasts are derived from the supratidal environment, they therefore contain diagnostic sedimentary features, such as sedimentary

and algal laminations, in addition to bird's-eye vugs.

Mud polygons produced by desiccation are highly susceptible to erosion and redeposition. Before-and-after observations showed that the flat pebble conglomerate (Figs. 23A, 23B) was deposited almost instantaneously on a small supratidal flat in Florida Bay during Hurricane Donna in 1960. It was composed of redeposited mud polygons. Similar intraclasts can be seen in the top of the supratidal levee core shown in Figure 4A. In both examples, the intraclasts were deposited as sedimentary lenses 1 to 6 cm thick and less than 1 m across.

As shown in Figure 24, intraclasts can also occur as redeposited slabs of dolomitic crust. Those in Figure 24 were stranded in imbricate fashion along the shore of an almost-abandoned tidal channel west of Abaco Island in the Bahamas. Similar accumulations occur sporadically on the muddy beach ridge shoreline of the Andros Island tidal flats. Several

Figure 23A—Desiccation polygons redeposited on Cluett Key in Florida Bay during Hurricane Donna in 1960. Core taken with paint can is shown in Figure 23B.

Figure 24—Dolomitic supratidal crusts eroded and redeposited in imbricate fashion along edge of nearly abandoned tidal channel. Similar breccias are abundant in the geological record.

Figure 23B—Plastic-impregnated core (location shown in Figure 23A) showing imbricate mud intraclasts deposited during Hurricane Donna. Note how weathering has disrupted earlier storm layers and produced soil-like intraclasts. Thin algal layer separates Hurricane Donna lithoclasts from earlier storm layer. Note vertical voids in underlying sediment. Bird's-eye voids are abundant.

ancient examples of tidal flat intraclasts are shown in Figures 25 and 26.

Soil Clasts

A special kind of irregularly shaped intraclast forms where supratidal sediments are elevated high enough above the saline water table to support abundant plant life. In modern humid-to-semi-arid environments, the plants are generally grasses and palms. In these areas a combination of factors, such as alternate wetting and drying, shrinkage and contraction, disruption by roots and associated burrowers (earthworms), cause internal shifting of sediment components. Layers, for example, become disrupted and may appear as though transported by wind or water currents. Well known to pedologists, this soil process is probably an incipient stage of caliche formation. Examples of this form of intraclast formation are shown from both modern and Permian rocks in Figures 27 and 28.

INTERTIDAL ZONE

Intertidal sediments are deposited between normal high and low tide and generally accumulate as a belt seaward of the supratidal flats and landward of the subtidal zone. Many tidal flats, especially the onlap or transgressive accumulations at Andros Island (Fig. 2), are dissected by a complex system of tidal channels. Intertidal flats of the offlap or regressive accumulations in the Persian Gulf contain relatively fewer tidal channels.

Diagnostic sedimentary laminations and structures are generally lacking in both modern and ancient intertidal accumulations. Lack of laminations in this environment is attributed entirely to homogenization by burrowing organisms. Aside from a characteristic oxidized color and a generally lower fossil diversity, intertidal sediments in humid areas are practically indistinguishable from nearby subtidal sediments. They are distinguishable, however, from the subtidal deposits of tidal channels, which dissect intertidal sediments at regular intervals.

In the Persian Gulf (Fig. 3), where the climate is arid, the presence of evaporites in the upper part of the intertidal zone precludes the presence of most burrowing organisms. The mid-to-lower part of the intertidal zone (tidal range about 2 m) is dominated by burrowing fiddler crabs and a small fish of the gobie family that burrows and stalks the flats at low tide to catch crabs. As shown in Figure 3, algal-laminated sediments form in the upper intertidal zone and extend into the overlying supratidal

Figure 25—Limestone flat pebble breccia in dolomite supratidal laminations (probably algal) of Ordovician age in Kentucky. Photo courtesy of Earl Cressman (U.S. Geological Survey).

Figure 26—Limestone flat pebbles from a cyclical tidal flat sequence described by Matter (1967) in a matrix of fossil hash of Cambro-Ordivician age in Maryland. This accumulation may represent a basal channel deposit.

Figure 27—Soil processes in the upper part of the supratidal zone on Andros Island produced these clasts by disrupting individual storm layers. Note numerous bird's-eye voids, both bubble-like and planar. Scale in mm. Compare this plastic-impregnated modern example with the Permian example in Figure 28.

Figure 28—Anhydrite-filled supratidal "soil clast" dolomite from the Permian San Andres dolomite in west Texas. Subsurface depth in feet is shown on rock. Scale in mm.

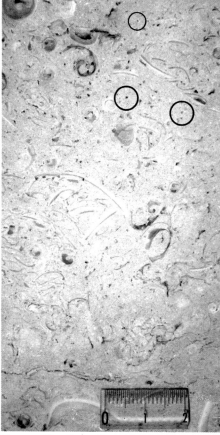

Figure 29—Transitional intertidal to supratidal facies from a modern tidal flat on the Qatar Peninsula in the Persian Gulf (see Shinn, 1973a) showing oxidized sediment with bird's-eye voids (in circles) and abundant shells. Abundance of shells is unusual and related to local conditions. Scale in cm.

zone. Because of high salinity and rapid burial by accreting supratidal sediments, reduction and buildup of H_2S is common. Only the upper part of the intertidal zone in these accretionary environments shows signs of oxidation. Bird's-eye vugs, generally the bubble-like variety, may occur in the upper intertidal zone in those areas where algal mats are lacking (Fig. 29). If algal mats are present, the vugs are predominantly of the planar variety.

Adjacent Marine Sediment
Except for tidal channels, the subtidal zone forms a belt seaward of the intertidal zone (Figs. 2, 3). This zone of generally muddy sediment is vital to the tidal flat system, because it is the source of sediment needed for accretionary growth. For the purpose of this paper, the outer limit of the muddy subtidal zone is 5 km from the intertidal zone. Depending on the geographic setting, the subtidal zone may extend seaward, in some cases for hundreds of kilometers, and includes accumulations such as ooid bars, coral reefs, "ramps" composed of both cemented and uncemented grainstone, and deep basinal deposits. Generally, however, the muddy adjacent marine sediments within 5 km are in shallow, low-energy conditions and are thus easily stirred into

Figure 30—Gray reduced burrow-mottled subtidal sediment from the same core as shown in Figure 29. Sediment does not contain bird's-eye voids. Scale in cm.

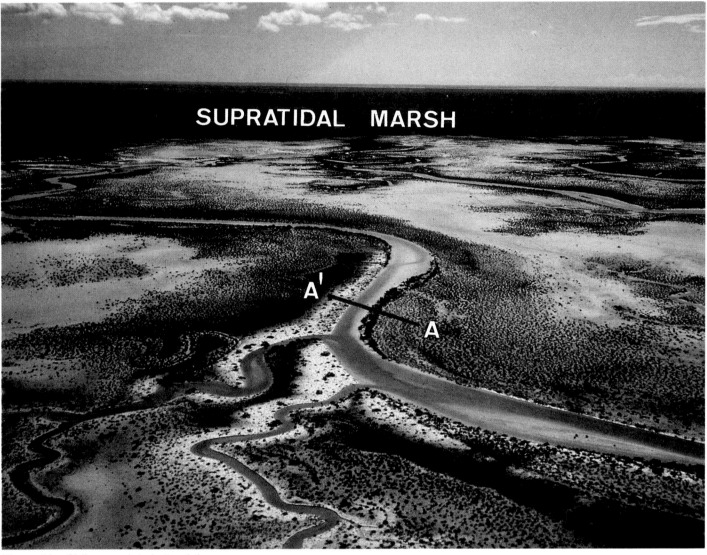

SUPRATIDAL MARSH

Figure 31—Oblique aerial photo of the Andros tidal flat belt showing location of core transect shown in Figure 32. Note light-colored levee sediment on outer bend near A' bordered by dark algal mat coating, in turn bordered by stunted red mangrove trees that peter out in subtidal pond area. Supratidal marsh, colored dark by dense algal mats, is shown in distance.

suspension and transported to the flats during storm-tide conditions.

Subtidal sediments are predominantly composed of pelleted muds, although they can vary in grain size, composition and hardness. These muds are almost always homogenized by burrowing and lack primary sedimentary structures. Never exposed to subaerial conditions, these sediments are also reduced and have a characteristic mottled-gray color (Fig. 30).

In some areas, such as in the lagoon seaward of the Persian Gulf Trucial Coast (Fig. 3) and at Qatar (Shinn, 1973a), winnowed grainstone layers in the upper subtidal or lower intertidal become cemented to form a resistant crust. Such crusts have low permeability; thus, when they become incorporated into an outwardly accreting tidal flat, fluids can be effectively retarded from vertical movement. Equally important, these resistant crusts often, but not always, prevent downward cutting of tidal channels, thereby restricting channel deposition to the intertidal zone. In addition to these thin, nearshore crusts, there are more thoroughly cemented crusts forming in subtidal sediments farther offshore, starting at a depth of about 3 to 4 m (Shinn, 1969). These deeper and thicker crusts with eroded and bored upper surfaces may also eventually underlie tidal flat sediments in a progradational setting in which the tidal flat builds seaward. Such crusts have been traced for dozens of kilometers seaward to depths of 15 to 20 m, and their low permeability of only a few millidarcies (Shinn, 1969) can also effectively impede vertical movement of fluids. The existence of similar layers in ancient tidal flats could have a great influence on oil or water movement.

Tidal Channels

Tidal channels comprise a special

Figure 32—Cross section based on three cores across channel bend shown in Figure 31. Scale in feet (8 ft = 2.4 m, 240 ft = 73.2 m). Note point bar-like sequence with channel lag grading upward into burrowed intertidal facies. Basal channel lag consists mainly of cerithid gastropods and locally contains abundant fragments of dolomite derived from erosion of levee as shown in next two figures. Note that supratidal marsh sediment at base of core through levee (A'), as well as entire sequence including channel levee, have undergone erosion as channel migrates from right to left. Given stable sea level and sediment supply and time, the entire intertidal zone could be reworked by tidal channels many, many times, thus concentrating dolomite which forms on levees, as well as converting a relatively impermeable intertidal flat to a more permeable blanket deposit. See Shinn and others (1969) for more detail.

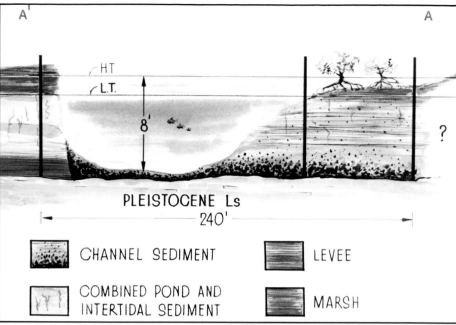

Figure 33—View of eroded channel levee on Andros tidal flat in the Bahamas. Dolomitic crust is undermined and works its way to bottom of channel during lateral migration.

Figure 34—A channel levee in the Persian Gulf (Qatar Peninsula; see Shinn, 1973a, for more detail). In this locality the levee is composed of cerithid gastropod sand and the cementation is by beachrock processes rather than dolomitization. Burrowing crabs have modified sediment in bottom left on the point bar, but some crossbedding may be preserved in the subtidal parts of the channel sequence.

Figure 35—Three cores of basal channel accumulations from the modern tidal flats on Andros Island, Bahamas. Cerithid gastropods and soritid foraminifera are the main sand-sized grains in cores A, B and C. Fragments of dolomite are mixed in, and upward grading is apparent. Note that inclined bedding in middle part of core C has escaped bioturbation. Dark objects are mangrove roots.

and dynamic subenvironment of the subtidal zone. Tidal channels range in depth from 0 to 15 m or more along the Persian Gulf Trucial Coast and 0 to 3 m on Andros Island in the Bahamas. All channels, whatever their maximum depth, will become shallower in a landward direction to the point where they disappear.

Coring and subsurface observations at Andros Island have shown that tidal channels migrate laterally in a manner similar to fluvial systems (Shinn et al, 1969). This similarity led to the concept that highly channeled tidal flats are analogous to river deltas turned wrong side out, that is, the sea, rather than the land, is the sediment source and channels with their distributaries branching in a landward direction provide the pathway for sediment delivery.

When tidal channels meander and migrate laterally, a "point bar" ac-

cumulation, similar to but different from that of fluvial channels, is laid down. With continued migration, a blanket of tidal channel sediment may be spread over a large area. Given the conditions of time and lack of seaward accretion or landward onlap, tidal channels could completely rework the intertidal zone many times over.

The major difference between sedimentary structures produced in

Figure 36—Evaporation and a continual supply of sea water from a fluctuating ground water table lead to concentration of brines near supratidal surface. Dolomitization and formation of evaporites result from these special conditions found only in the supratidal zone just centimeters above mean tide level. Such areas are also recharged at least twice a month during spring tide flooding. Under humid conditions at Andros Island, evaporites, such as gypsum, are periodically dissolved away by rain and morning dew.

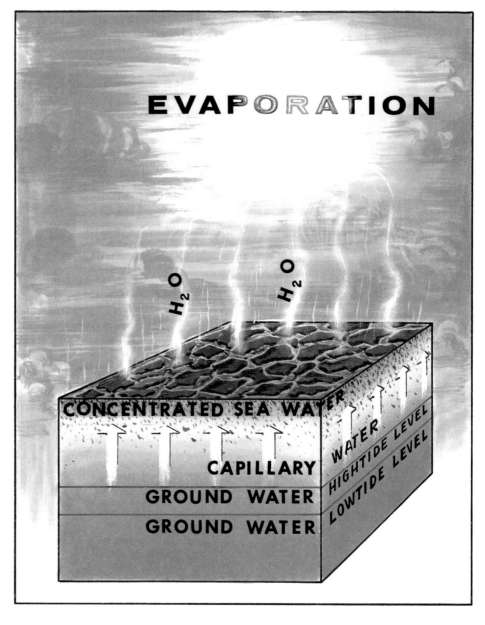

Figure 37—Detail of gypsum precipitated on the Trucial Coast tidal flats in the Persian Gulf. Such gypsum forms a mush at least 30 cm thick above a distinctive algal-laminated zone in the upper intertidal zone. See Figure 3 for relationships. This mush grades laterally landward into layered, nodular, often contorted anhydrite. See next figure.

carbonate tidal channels and those in siliciclastic fluvial deposits is the presence of burrowing. The classic fluvial "point bar" accumulation starts with a basal cobble or pebble layer (including sun-dried clay intraclasts), which grades upward into coarse-grained, festooned crossbedding, followed by finer-grained planar bedding containing "clay drapes" (thin layers of clay deposited during quiet conditions between floods). Similar accumulations might form in tidal carbonate channels were it not for the action of marine burrowing organisms. The observed sequence at Andros Island depicted in Figures 31 and 32 starts at the bottom with clasts of sun-dried supratidal sediment, fragments and slabs of supratidal dolomitic crust, and eroded bits and pieces of the underlying Pleistocene bedrock. Dolomitic and non-dolomitic crusts which form on channel levees, as shown in Figures 33 and 34, provide a local source of lithoclasts for these channel lags. These clasts occur in a matrix of cerithid gastropod sand which grades upward into finer-grained carbonate sand and packstone containing numerically abundant foraminiferal assemblages with limited species diversity (Fig. 35). Most primary bedding has been destroyed by burrowing organisms. It should be pointed out, however, that although bioturbated bedding may be difficult to see in an 80-cm-diameter core, it might be more obvious in larger outcrops. Migrating channel deposits, as shown in Figure 32, are eventually capped by burrowed intertidal sediments, and with continued lateral migration, non-burrowed laminated supratidal sediments will cap the sequence.

In the modern example at Andros Island, the underlying bedrock

Figure 38—A pit dug in the Trucial Coast sabkha showing white anhydrite resting on dark algal sediment at base of pit. Salt has precipitated on surface of algae after pit was dug. See next figure for detail of anhydrite. Giving scale to the photo is Godfrey Butler, to whom the author is indebted for acting as expert guide to the area.

Figure 39—Detail of sabkha trench shown in previous figure showing white "chicken wire" anhydrite at base and interbedding of quartz-rich dolomite and contorted anhydrite layers. Anhydrite also occurs as blebs and lenses.

prevents downward cutting beyond 3 m; along the Trucial Coast, however, some channels that have cut through submarine cemented crusts extend down to a depth of 18 m or more. The age and composition of the underlying pre-Holocene bedrock could not be determined, but in some channels the bottom was strewn with huge slabs of Holocene submarine crusts up to 30 cm thick and more

than 1 m across.

The important aspect of carbonate tidal channel sediments, whether burrowed or not, is that their basal parts are porous and permeable and surrounded by relatively less porous and less permeable tidal flat sediments. They extend more or less perpendicular to the supratidal and intertidal belt and, thus, may serve as pathways for the migration of interstitial fluids.

DIAGENESIS

Dolomite

Both transported and indigenous diagenetic minerals can occur on carbonate tidal flats. The most economically important mineral is dolomite because of its common association with enhanced porosity and permeability. Dolomite not only forms diagenetically on supratidal

Figure 40—Contorted diapiric anhydrite truncated by sabkha deflation surface. In this arid climate winds deflate sediment down to permanent capillary zone, where permanently wet sediment resists removal. Note that sediment has been added subsequent to erosion. Position of the capillary zone varies depending upon season and climatic variations. Note numerous nodules of anhydrite. Gypsum also occurs as large scattered rosettes.

flats but, as discussed earlier, detrital dolomite can find its way to these environments through eolian processes.

Much has been written concerning dolomitization on modern tidal flats (Wells, 1962; Illing et al, 1965; Deffeyes et al, 1965; Shinn et al, 1965; Shinn et al, 1969; Hardie, 1977; Kinsman, 1964; McKenzie et al, 1980, to name a few). Although the chemistry of dolomitization remains controversial, many of the controlling parameters are known, especially for supratidal dolomites which are by far the easiest to recognize in the geologic record. Dolomite tends to form at the surface on supratidal flats a few centimeters above normal high tide. In this setting, underlying sea water is brought to the surface by capillary action, or is supplied by surface flooding during storms, and evaporates to form a highly concentrated brine (Fig. 36). Through precipitation of gypsum ($CaCO_2$) and aragonite, calcium is selectively removed, thus elevating the Mg-to-Ca ratio to many times that in normal sea water. Changes in pH associated with respiration and photosynthesis of blue-green algae, as well as oxidation and the pumping action of fluctuating ground water, may also be contributing factors in supratidal dolomitization. Regardless of the exact chemistry, this kind of dolomite is readily identified in modern and ancient rocks because of its association with distinctive sedimentary structures and its small crystal size. Modern tidal flat dolomite, often called "protodolomite" because of poor ordering and high Ca-to-Mg ratio, is universally composed of small crystals ranging in size from 2 to 4 μm. In similar ancient tidal flats, dolomite crystals are also small, but generally in the 5 to 10 μm size range, and the crystal lattice is well ordered. Larger dolomite crystals do occur, however, especially in associated marine or channel deposits, where initial permeabilities were high. Whether the chemistry of ancient tidal flats was different, or whether poorly ordered high-calcium dolomite becomes ordered with age is not known.

Dolomitization probably also occurs by seepage of Mg-rich surficial brines into underlying supratidal, intertidal and subtidal sediments, as proposed by Adams and Rhodes (1960). Deffeyes and others (1965) believe that much of the dolomite in sediments underlying the salt pans on the island of Bonaire (Dutch Antilles) was formed by this mechanism. Steinen and Halley (1979) have reported protodolomite beneath a Florida Bay mud island that they believe may have formed by refluxation. An example of subsurface primary (non-replacement) dolomitization in brines beneath a porous quartz sand sabkha in the Persian Gulf was discussed by Shinn (1973b) and DeGroot (1973).

Detrital dolomite, quartz silt, and various clay minerals are present in most Persian Gulf sediments. The detrital dolomite contaminates not only areas of supratidal dolomitization but also practically all marine sediments of similar grain size. Pilkey (1966) discovered 10% detrital dolomite in fine-grained subtidal sediments from the central axis of the Persian Gulf. Such occurrences are not surprising. During offshore work in the Persian Gulf, the author's research vessel was often covered in a matter of hours with dolomite, clay, and silt dust resembling yellow-brown paint. Some of what had been reported as modern dolomite from Persian Gulf flats was most likely of detrital origin. It seems likely, therefore, that detrital dolomite should be abundant in many ancient tidal flat settings, especially those deposited during particularly arid

Figure 41—Abandoned and filled tidal channels (**A**) made visible by changes in grain size and temporary humidity conditions, Qatar Peninsula (see Shinn, 1973a, for details of abandonment and sedimentary sequence). Camel tracks in foreground. These subtle channels can be traced seaward to presently active tidal channels. Portion of a 15-m-long by 2-m-deep trench (**B**) dug across an abandoned channel like that shown in previous figure. During the latter stages of abandonment, the channels become filled with quartz and carbonate sand derived from nearby eolian dunes, giving a channel sequence with more porous and permeable sediment at the top. Note gray muddy subtidal sediment at base of trench below man's hand. Man is standing on water-covered cemented crust, which kept trench dry until crust was penetrated with a rock hammer.

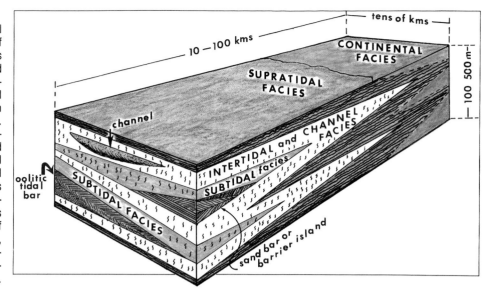

Figure 42—Idealized tidal flat model combining onlap and offlap features of Andros Island and Persian Gulf tidal flats to show stratigraphic traps and related porous accumulations. Trapping conditions occur where intertidal and channel facies pinch out beneath and within relatively impermeable supratidal facies. Supratidal facies in most ancient examples are sealed with anhydrite and other evaporitic minerals. Additional reservoir development may occur in tidal deltas, channel and barrier island sands that lie seaward of supratidal and continental facies. In most ancient examples supratidal facies are composed of relatively impermeable micro-dolomite, whereas subtidal facies tend to be composed of more permeable, coarser-grained dolomite.

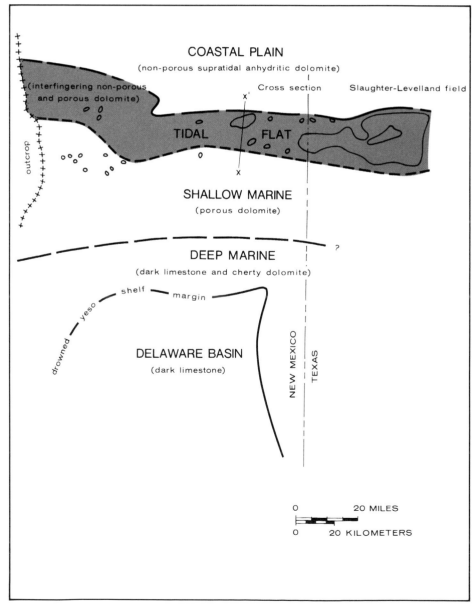

Figure 43A—An example of a tidal flat system developed parallel to the northern margin of the Delaware Basin (Middle Permian San Andres time). Section X-X' is shown in Figure 43B. Illustration courtesy of Fred Meissner (Bird Exploration Company).

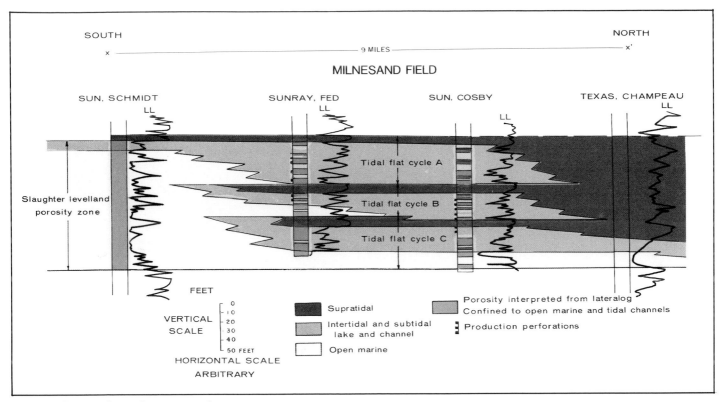

SOUTH

NORTH

9 MILES

MILNESAND FIELD

SUN, SCHMIDT
LL

SUNRAY, FED
LL

SUN, COSBY

TEXAS, CHAMPEAU
LL

LL

Tidal flat cycle A

Tidal flat cycle B

Tidal flat cycle C

Slaughter levelland
porosity zone

FEET

VERTICAL
SCALE

0
10
20
30
40
50 FEET

HORIZONTAL SCALE
ARBITRARY

Supratidal

Intertidal and subtidal
lake and channel

Open marine

Porosity interpreted from lateralog
Confined to open marine and tidal channels

Production perforations

Figure 43B—North-south cross section
showing three reservoirs sealed by impermeable anhydritic supratidal facies to
the north. Several of the fields along
the trend shown in Figure 43 owe
their existence to a combination of
hydrodynamic flow from the west and
the northward pinchout of porosity in the
supratidal anhydritic dolomites. The
largest of these fields, Slaughter-
Levelland, contains 3.5 billion barrels of
oil in place (Fred Meissner, personal
commun., 1981).

periods, such as the Permian.

Gypsum and Anhydrite
Some evaporitic minerals, such as
halite, anhydrite, and gypsum, are
present on all modern arid tidal flats.
Many ancient tidal flats contain
abundant anhydrite and salts,
especially those of Permian age. Gypsum may form temporarily on humid
tidal flats like those at Andros Island
in the Bahamas. It is often present
during the dry season but disappears
during the wet season and therefore is
not preserved for the geologic record.
Ancient tidal flats that do not contain
evaporitic minerals could have formed by this mechanism during humid
periods of the earth's history.

On the Persian Gulf tidal flats,
where annual rainfall is less than 2
cm, a gypsum mush, composed of
crystals resembling ice cream salt,
forms an extensive layer up to 30 cm
thick. The gypsum mush forms in the
transitional zone between intertidal
and supratidal conditions (Figs. 3, 37;
Kinsman, 1964; Butler, 1965; Kendall
and Skipwith, 1969). The gypsum
layer grades laterally to contorted,
fine-grained anhydrite (Figs. 38, 39,
40) farther landward, where it is
overlain by supratidal dolomite,
anhydrite, and quartz sand. The
lower part of the buried anhydrite
zone contains the familiar "chicken-
wire" texture so prevalent in many
ancient gypsum and anhydrite
deposits. The chicken-wire texture
grades upward into distorted layers of
fine-grained anhydrite (Figs. 38, 39,
40). How much of the anhydrite is
primary and how much has formed
by replacement of gypsum is not well
known. Farther landward, where the
sabkha is subject to flooding by rainwater outwash or influx of continental groundwater, anhydrite has locally
been hydrated back to gypsum
(Butler, 1965; Kendall and Skipwith,
1969).

Gypsum also occurs as large
isolated rosettes several centimeters
across. Such crystals are randomly
scattered throughout the intertidal
and algal mat zone underlying the
anhydrite. Gypsum rosettes increase
in abundance and size toward land
and appear to have a different origin
than the gypsum mush or anhydrite.
Celestite, a temporary evaporite
mineral, is sometimes associated with
anhydrite. Since this mineral is not
preserved, however, it is considered
unimportant for the purposes of this
article.

Cementation
Cementation of sediment occurs in
four main areas on tidal flats: (1) as
surficial crusts, often dolomitic, on
the lower parts of supratidal flats and
natural levees; (2) along the edges of
tidal channels in areas where channel
banks are composed primarily of carbonate sand (Fig. 34); (3) as
beachrock in the intertidal part of
beach ridges and spits; and (4) on
nearly horizontal, sandy flats in the
upper part of the subtidal and lower
part of the intertidal zones.
Of the four kinds of cementation,
the last three can be considered forms

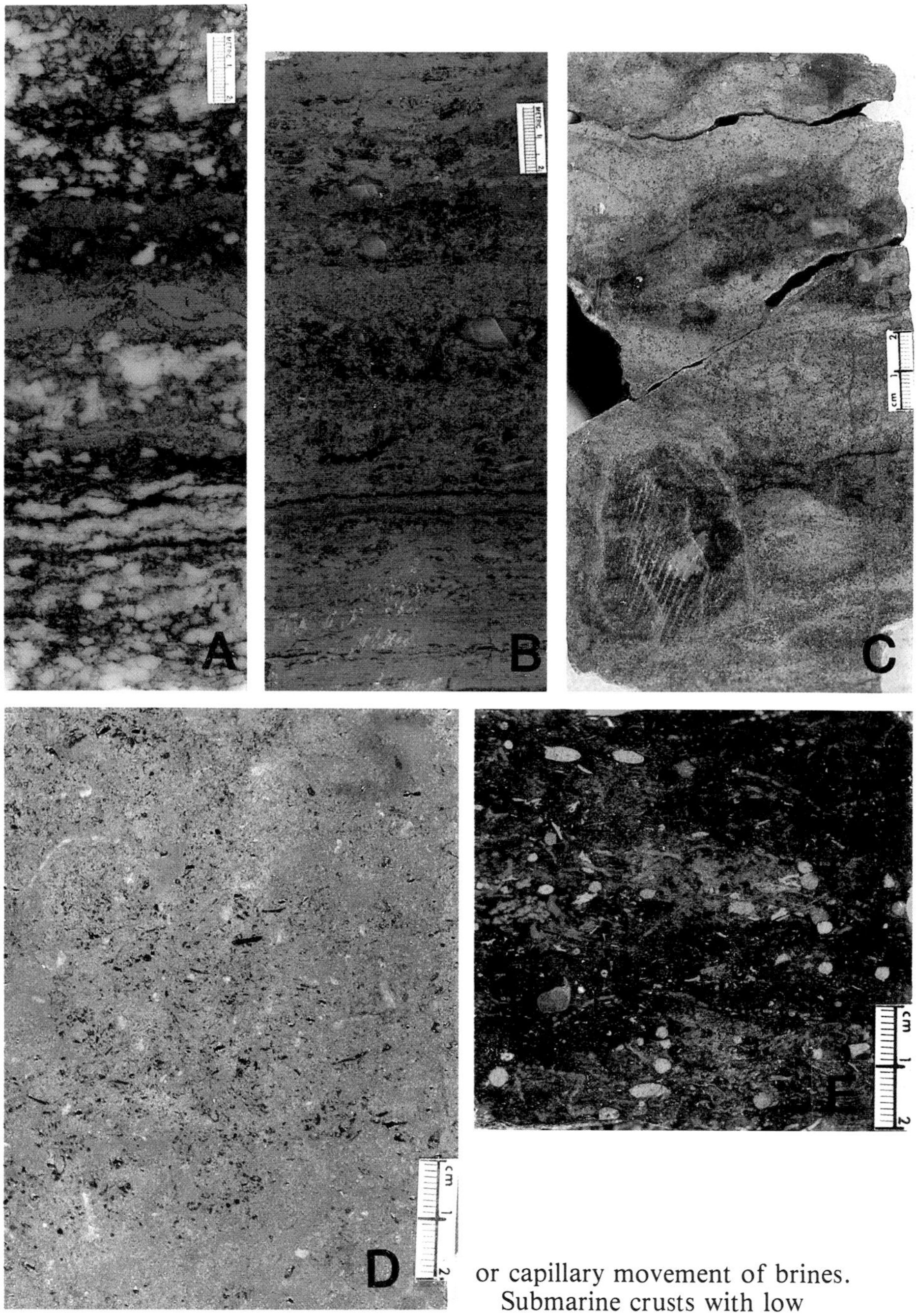

Figure 44—Selected facies from Permian Slaughter-Levelland-Millensand trend. (**A**) Supratidal nodular anhydrite facies. (**B**) Supratidal anhydrite and algal stromatolite facies. (**C**) Burrowed, high-energy intertidal or channel facies. (**D**) Leached skeletal shallow-marine dolomite wackestones. (**E**) Open-marine, porous brachiopod and echinoid spine lime wackestone.

of intertidal beachrock. These sand-sized sediments become cemented where the accumulations are stabilized and wetted and drained with each tide. Observations by the author in the Persian Gulf indicate that beachrock type of cementation by aragonite and Mg-calcite can take place in a few years. As described earlier, cementation is also taking place seaward of the Persian Gulf tidal flats under subtidal conditions (Shinn, 1969).

All the forms of cementation produce rocks that provide a source of lithoclasts. Beachrock is reworked and reincorporated into beach deposits and spits (Shinn, 1973a), while cemented material along the sides of channels is undermined by lateral migration and incorporated in basal channel deposits. Two forms of cementation occur on channel levees — beachrock formation, where the channel bank is coarse-grained and the resulting rock is non-dolomitic (Fig. 34), and cementation by dolomite and aragonite on the back sides of levees in muddy areas. The dolomite-and-aragonite-cemented levee crusts form where the surface is at approximate spring high tide level (Shinn et al, 1969). These crusts provide dolomitic flat pebble lithoclasts to channel lags (Fig. 33).

As noted earlier, low intertidal crusts (Fig. 3) can effectively limit the downward cutting of tidal channels. Figure 41A shows abandoned and filled tidal channels on a supratidal sabkha. Figure 41B shows part of a 15 m trench dug across one such channel. The trench in Figure 41B could not be dug deeper because of a cemented crust. The crust was so impermeable that the trench remained

dry even though the bottom was below sea level. When the crust was broken, however, water gushed up to fill the trench to a depth of approximately 30 cm above the crust. No leakage through the unbroken crust was observed, although it was exposed in the trench for a distance of 15 m. Such crusts, therefore, not only limit channel scour but undoubtedly have a controlling effect on such diagenetic phenomena as refluxation

or capillary movement of brines.

Submarine crusts with low permeabilities are also forming farther offshore in the Persian Gulf (Shinn, 1969). Given time, many portions of the various Persian Gulf tidal flats will accrete seaward over these submarine cemented crusts. Thus, the tidal flat system could incorporate and be affected by yet another permeability barrier. Similar crusts, recognizable mainly by fibrous calcite or dolomite cements, are undoubtedly a component of many ancient tidal flats. They can easily be misinterpreted as exposure unconformities because of their bored and eroded surfaces.

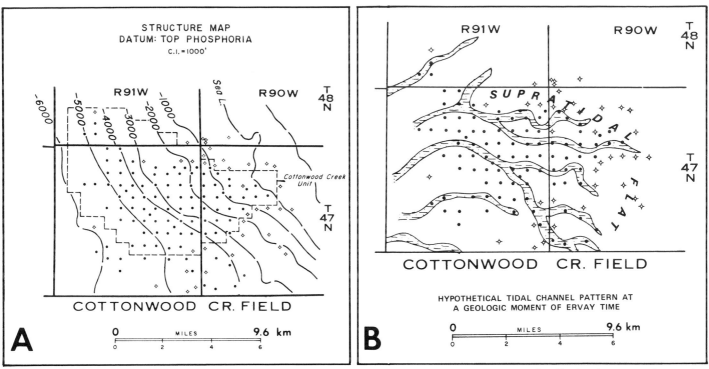

Figure 45A—Structure map on the top of the Phosphoria Formation in the Cottonwood Creek Field in the Bighorn Basin of Wyoming. Note dry holes updip to the northeast and producing wells downdip to the southwest. The field had produced 37 million barrels of oil as of 1971 (from Rogers, 1971).

Figure 45B—Facies interpretation of the Park City Formation in the Cottonwood Creek Field showing porous channel deposits extending from downdip permeable intertidal and marine rocks into updip impermeable evaporitic supratidal facies. Awareness of the possibility of channels in tidal flat reservoirs can aid hydrocarbon exploitation.

Hypersaline Caliche and Pisolites

Laminated crusts, similar to freshwater vadose caliche crusts of the American west and those described from Florida by Multer and Hoffmeister (1968) and Robbin and Stipp (1979), occur in several supratidal areas of the Persian Gulf. Pisolites are associated with the crusts (Purser and Loreau, 1973) or as individual pisoids floating in uncemented supratidal sediment (Shinn, 1973b). Scholle and Kinsman (1974) demonstrated that these features are, in essence, vadose caliche, but instead of owing their formation to fresh water, these formed by precipitation from hypersaline waters. Scholle and Kinsman (1974) also showed evidence that these features are modern analogues to the well known pisolites and pisolitic crusts associated with lagoonal and tidal flat facies connected with the Permian Capitan limestone, a reef complex of west Texas and New Mexico. Similar pisolitic facies associated with tidal flats are common in the subsurface platform areas of the Permian Basin of west Texas (Longacre, 1980).

In modern environments of the Persian Gulf, laminated crusts and pisolites occur in supratidal environments on Pleistocene bedrock (Scholle and Kinsman, 1974) and high intertidal-to-supratidal areas on beachrock (Purser and Loreau, 1973) or in unconsolidated siliciclastic sediment in the low supratidal zone (Shinn, 1973b). In all examples, the site of formation is in the seaward portion of the sabkhas exposed to sea spray and spring or storm tide flooding. If their origin in ancient tidal flats is similar, then one can assume that their presence may be a useful indicator for predicting facies relationships. Ancient pisolites and pisolitic crusts, usually associated with teepee structures in the Permian rocks, are suspected to have formed within 2 km of either a subtidal lagoon or the sea.

Evaporites and Silica Cement

Evaporites, either as rosettes of gypsum or banded anhydrite (Figs. 38, 39), can be considered a cement. These evaporites may cause some cementation of sediment by infilling of bird's-eye vugs or, in some cases, produce a hard rock by infilling intraparticulate void space. Whether or not these cements produce a hard rock, they are nevertheless important, because they can effectively block porosity and permeability, and they can flow rather than fracture during compaction or tectonic deformation. Supratidal sabkhas containing evaporites can therefore form effective seals. If leached by fresh water during later diagenesis, excellent porosity and permeability may be produced.

Opaline silica, or chert, has been repeatedly observed in ancient supratidal sediments, although true silica cement has not yet been reported from modern tidal flats. Silica cements in tidal flats are probably not important, as they are seldom extensive enough to form effective permeability barriers.

Figure 46—Core from a portion of the Permian San Andres dolomite in west Texas showing facies detail that can be extracted from tidal flat sequences. At least 18 supratidal zones (indicated by red color in the environmental interpretation column) were logged in this core. Intertidal and/or marine rocks are shown in blue, and root zones (either intertidal or high supratidal) are indicated by green. Minor unconformities are the rule. Because of exposure associated with tidal flat accumulation, significant portions of tidal flat sequences are probably eroded away before deposition of the next unit, thus perfect cycles, such as shown in Figure 50, are probably rare in the geologic record. Notice persistent color changes from oxidized brown supratidal dolomite to gray subtidal dolomites.

DISCUSSION AND ECONOMIC IMPLICATIONS

Porosity and permeability in tidal flat systems vary greatly from facies to facies. The various studies of modern and ancient examples combined suggest that porosity and permeability are best developed in subtidal-to-intertidal facies with greatest development in tidal channels. Supratidal sediments appear to offer the least potential for porosity and permeability development. The relative lack of porosity and permeability in this environment can be attributed to susceptibility to (1) early cementation, (2) formation of micro-crystalline dolomite (often associated with cementation by aragonite and Mg-calcite), (3) formation of evaporites, and (4) the development of algal mats.

Structural Traps
Knowledge of permeability variations and recognition of the facies that contain these variations can be useful to the subsurface explorationist, whether he or she is searching for structural or stratigraphic traps. The best documented ancient example involving structure is in the Stony Mountain Formation (Ordovician) and Interlake Formation (Ordovician and Silurian) in the western Williston

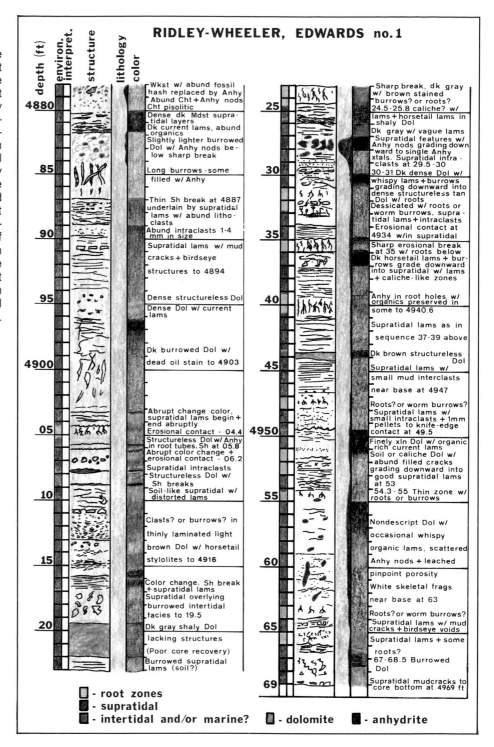

Basin, described by Roehl (1967). The Cabin Creek oil field is an anticlinal structure that uplifted rocks representing all common tidal flat facies, thereby making identification of those zones most economically exploitable an important criterion in the search for oil. Predictability of facies was

good in the Cabin Creek Field, but predictability of porosity associated with the facies was poor, probably because early subaerial exposure left a diagenetic porosity overprint on all facies. Roehl (1967) did note that, although all the facies are dolomitized, the larger crystal size and hence

Figure 47A—Photograph of the Cambrian Carrara Formation in Nevada demonstrating the persistence of color in supratidal and subtidal carbonates. The light-colored oxidized unit at the top of the cliff overlies gray reduced marine rocks. Such color changes are thought to be useful in interpreting tidal flat rocks of all ages. These color changes can be useful indicators to the subsurface geologist working only with cuttings. Photo courtesy of R. B. Halley (U.S. Geological Survey). For details, see Halley (1975).

the greatest porosity, up to 20%, and permeability of approximately 2.5 md were associated with interconnecting burrows in subtidal and intertidal facies.

Sediment contained in modern burrows is generally coarser-grained than the surrounding sediment (Shinn, 1968c). Filled burrows can therefore channel fluids and become dolomitized more quickly than the matrix. Such dolomite is likely to be of larger grain size than that in the surrounding matrix. Highly burrowed zones, as noted earlier, are restricted to subtidal ("infratidal" of Roehl, 1967) and upper intertidal facies. Coarse-grained channel deposits were thought by Roehl (1967) to occur within the intertidal zone. Even the supratidal zone, especially where it contained abundant intraclasts, also contained oil and sufficient porosity for production, but this was interpreted to be the result of early subaerial leaching. It should be remembered that tidal flats flanking a continually subsiding basin would be less likely to experience exposure and leaching.

Based on the Holocene models (Andros Island and the Persian Gulf) and excluding the effects of early exposure and leaching, the best chances of porosity development lie in subtidal marine, channel and intertidal deposits. Early cementation and

evaporite development leave supratidal sediments with the least potential for reservoir development. If tidal flat rocks have been structurally uplifted, one should expect maximum reservoir development where initially more porous zones have closure. The potential for multiple reservoirs is enhanced even more if several impermeable supratidal zones separate and seal structurally uplifted porous and permeable subtidal and intertidal facies. Such knowledge could be additionally important in tidal flat oil fields, where secondary recovery is necessary. When structurally deformed, intertidal and submarine-cemented crusts, such as those discussed earlier, probably would not provide effective seals because of the ease with which these brittle rocks fracture. Evaporitic supratidal zones, however, would be more plastic and less likely to fracture and leak.

The Stratigraphic Trap Model

Tidal flats are basically a special kind of shoreline deposit and, as such, are subject to repeated transgressions and regressions (onlap and offlap) and subsidence through time. On the flanks of subsiding epirogenic seas and basins, these special shoreline deposits have the potential to produce predictable belts of reser-

voirs and seals. The simplified conceptual model shown in Figure 42, based on studies of modern and ancient examples, shows the relationships between porous and non-porous facies, as well as trends in relation to the basin or adjacent land mass.

Meissner (1974) described the hydrologic and stratigraphic trap conditions controlling the Milnesand and Slaughter-Levelland fields in the Permian San Andres dolomite of the northern Delaware Basin in New Mexico and Texas. According to Meissner (personal commun., 1981), these originally discrete fields are now interconnected and controlled by an impermeable east-west supratidal and continental facies to the north. Oil migrated from the deeper Delaware Basin in the south and became trapped in a belt of more porous and permeable subtidal-to-intertidal sediments that pinch out to the north within and beneath the impermeable supratidal (sabkha-type) facies. The relationship of these facies is shown in Figure 43A and B. Core slices showing the major facies are shown in Figure 44A through E.

Rogers (1971) described tidal flat facies and its control on reservoir and trap strata in the Permian Phosphoria Formation Cottonwood Creek Field in the Bighorn Basin of Wyoming. Rogers' geologic interpretation is

Figure 47B—Closeup of the oxidized algal-laminated and mudcracked supratidal zone shown in previous figure (from Halley, 1975).

similar to that of Meissner (1974). At Cottonwood Creek, production (over 37 million barrels of sour crude have been produced from nearly 100 wells since the field's discovery in 1953) is from grain-supported dolomitized intertidal facies, including the deposits of tidal channels. The seal is an evaporitic supratidal zone to the northeast (Fig. 45A). Porous fingers penetrate into the non-porous supratidal zone, as interpreted in Figure 45B (Rogers, 1971). Rogers pointed out that tidal flat facies could be identified in cuttings but that cores were necessary to determine the vertical facies sequence. The author has logged cuttings and cores from several wells penetrating Permian San Andres tidal flat rocks in west Texas which support Rogers' conclusion. When the geologist is working with these kinds of rocks, recognition of facies relationships from a few cores, aided by electrical logs, should allow a rewarding use of cuttings if the wells are closely spaced. A large amount of facies detail can be extracted from cores, as shown in a 22 m (70 ft) core in Figure 46. More than 15 supratidal zones were recognized in the core shown in Figure 46. Subsequent study of cuttings from similar rocks showed that the characteristic oxidized brown color was especially useful in separating

supratidal and intertidal rocks from gray subtidal rocks. Figure 47, a photograph of the Carrara Formation (Cambrian) in Nevada described by Halley (1975), shows that even in these very old rocks, the supratidal zone can still be distinguished by its oxidized color.

Longacre (1980) described the combined structural and stratigraphic aspects of the Permian Grayburg and Queen Formations of the North McElroy Field on the eastern side of the Central Basin Platform of the Permian Basin in west Texas. At McElroy Field, platform margin faulting provided the pathway for oil migration from basinal source rocks into the numerous onlap and offlap tidal flat sequences on the platform. The seal is an anhydritic supratidal facies, whereas the reservoir rocks are "shallow shelf" and "exposed flats" (presumably synonymous with subtidal and intertidal flats) that pinch out to the west beneath supratidal facies (Longacre, 1980). Selected examples from the North McElroy Field are shown in Figure 48.

McElroy-type reservoirs are probably repeated in many locations along the eastern and western sides of the Central Basin Platform. Flannigan Field farther to the north is a good example of a reservoir developed in tidal flat facies (Jerry Lucia, personal

commun., 1981), as are numerous fields flanking the northern edge of the Delaware Basin (Allan Thompson, personal commun., 1981).

Many of these fields are near depletion and have been undergoing water flood and other secondary recovery programs for several years. Although seldom mentioned in geologic literature, exploration geologists often discuss the problems encountered by production engineers during water flooding. Basically, the problems have resulted from lack of recognition by petroleum engineers of the numerous horizontal permeability barriers. Understanding of tidal flat sedimentation and recognition of the facies variations, therefore, would be very useful to petroleum engineers.

Outcrop Examples

Similar transitions from rocks dominated by supratidal facies to those dominated by subtidal and intertidal facies have been reported from outcrop studies of rocks of Ordovician age in Kentucky (Cressman and Noger, 1976), (Fig. 49). Although environmental interpretations are slightly different, Howe (1968) provided a similar interpretation of lateral pinchout of tidal flat facies in Cambrian strata of the St. Francois Mountain area of Missouri (Fig. 50). The geological literature abounds

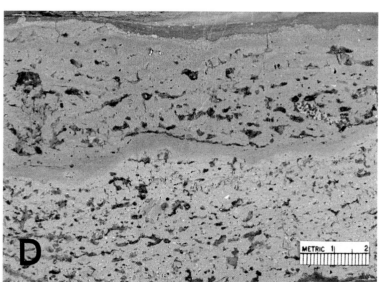

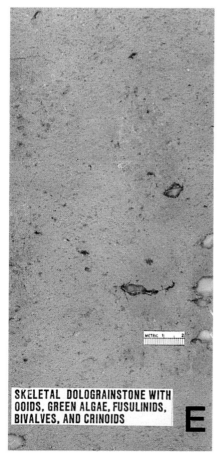

Figure 48—Examples of tidal flat facies from the Permian Grayburg and Queen Formations of the North McElroy Field on the east side of the Central Basin Platform of the Permian Basin, west Texas. (**A-D**) Examples of various supratidal facies with diagnostic features, such as algal mats, gypsum crystals, nodular anhydrite, and bird's-eyes. (**E**) Representative subtidal facies. Cores courtesy Susan Longacre (Getty Exploration).

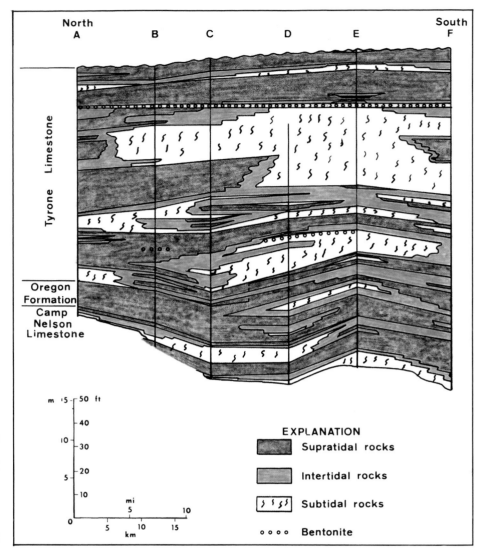

Figure 49—Idealized north-south stratigraphic cross section through the Ordovician Tyrone Limestone in Kentucky. Note numerous stratigraphic pinchouts, where intertidal and subtidal rocks pinch out against supratidal dolomite. Multiple reservoirs could be developed if the burrowed subtidal or intertidal rocks should be porous and permeable. Supratidal rocks would form seals (from Cressman and Noger, 1976).

EXPLANATION

Supratidal rocks

Intertidal rocks

Subtidal rocks

Bentonite

with other examples of tidal flat rocks not cited here, one of the more notable being the Lofer tidal flat cycles of the northern Italian Alps, described by Fischer (1964). Examples for this paper were chosen mainly on the basis of economic importance and regional stratigraphic extent.

Sedimentary and Diagenetic Features for Identifying Ancient Tidal Flats

No single sedimentary structure will serve as the key to unlock the secrets or even the presence of ancient tidal flats. Recognition of the sequence of sedimentary structures and facies commonly developed in tidal flats is probably more important than any single sedimentary feature. Figures 51 and 52 are attempts to combine what is known of both modern and ancient tidal flats into a usable sequence of sedimentary structures. Figure 51 shows the sequence and subtidal variations of the Persian Gulf-type of tidal flat, and Figure 52 shows important attributes of all known tidal flat cycles combined. It should be made clear that numerous and important variations can occur. Many of these variations are indicated in the onlap-offlap model in Figure 42 and in the sedimentary sequences shown in Figures 51 and 52. What the geologist encounters in any geologic basin is likely to differ, depending upon structural style, climate, age and latitude.

No attempt has been made to show all possible combinations of structural and diagenetic alterations that can occur. Such an effort is probably not possible and would be beyond the scope of this chapter. As in most geological problems, each area will have its own peculiarities and variations that can be understood only through thorough geological study. The models provided here are intended to serve as bases for geological departure. Modifications for any particular area will be necessary for better understanding, discovery, and exploitation of this economically important sedimentary system.

ACKNOWLEDGMENTS

The author gratefully acknowledges the constructive suggestions and discussions with Robert B. Halley and Peter A. Scholle of the U.S. Geological Survey. Special appreciation is extended to Barbara Lidz for editing, preparation of illustrations and continual review during all stages of manuscript preparation. The field work was done and many of the ideas expressed here were developed when the author was in the employ of Shell Development Company, USA, and the Royal Dutch Shell Laboratory in the Netherlands.

The following people provided either samples or slides that contributed greatly to the quality of the paper: Jack Moore of Shell Development Company (two figures); Earl Cressman of the U.S. Geological Survey (several color

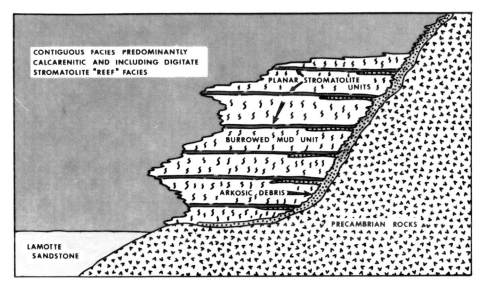

CONTIGUOUS FACIES PREDOMINANTLY CALCARENITIC AND INCLUDING DIGITATE STROMATOLITE "REEF" FACIES

PLANAR STROMATOLITE UNITS

BURROWED MUD UNIT

ARKOSIC DEBRIS

PRECAMBRIAN ROCKS

LAMOTTE SANDSTONE

Figure 50—The potential for several pay zones is apparent in this stratigraphic interpretation of Cambrian rocks in Missouri described by Howe (1968). What Howe called planar stromatolite units is interpreted by this author to represent offlap supratidal marsh or sabkha equivalents and thus might serve as seals. Reservoirs would likely be restricted to the burrowed mud unit interpreted to be undifferentiated intertidal and subtidal rocks, provided the entire section dips toward the basin.

photographs); Robert N. Ginsburg of the University of Miami Fisher Island Station (slides); Clyde Moore of Louisiana State University (samples of Cretaceous tidal flat rocks); Dennie Cody for preparation of color prints. Godfrey Butler of Exxon Company, USA, introduced the author to the Trucial Coast evaporites and served as guide to the area. Fred Meissner of Bird Exploration Company provided material and information concerning oil production from tidal flat rocks. Edward Evenson of Lehigh University provided the soilstone crust in Figure 16B. Susan Longacre of Getty Exploration provided samples from North McElroy Field.

Appreciation is also extended to Daniel M. Robbin and J. Harold Hudson of the U.S. Geological Survey Fisher Island Station for encouragement and discussions during manuscript preparation.

SELECTED REFERENCES

Adams, J. E., and M. L. Rhodes, 1960, Dolomitization by seepage refluxion: AAPG Bull., v. 44, p. 1912-1920.

Ball, M. M., E. A. Shinn, and K. W. Stockman, 1963, Geologic effects of Hurricane Donna: Abs., AAPG Bull., v. 47, p. 349.

Black, M., 1933, The algal sediments of Andros Island, Bahamas: Phil. Trans. Royal Soc. London, Series B, v. 222, p. 165-192.

Butler, G. P., 1965, Early diagenesis in the recent sediments of the Trucial Coast of the Persian Gulf: London Univ., Ph.D. dissert., 251 p.

Cressman, E. R., and M. C. Noger, 1976, Tidal-flat carbonate environments in the High Bridge Group (Middle Ordovi-

cian) of central Kentucky: Lexington, Univ. Kentucky, Kentucky Geol. Survey, Series X, Rept. of Invest. 18, p. 1-15.

Davies, G. R., 1970, Algal-laminated sediments, western Australia, *in* B. W. Logan et al, eds., Carbonate sediments and environments, Shark Bay, Western Australia: AAPG Mem. 13, p. 169-205.

Deffeyes, K. S., F. J. Lucia, and P. K. Weyl, 1965, Dolomitization of Recent and Plio-Pleistocene sediments by marine evaporite waters on Bonaire, Netherlands Antilles, *in* Dolomitization and limestone diagenesis: SEPM Spec. Pub. 13, p. 71-88.

DeGroot, K., 1973, Geochemistry of tidal flat brines at Umm Said, S.E. Qatar, Persian Gulf, *in* B. H. Purser, ed., The Persian Gulf–Holocene carbonate sedimentation and diagenesis in a shallow epicontinental sea: Heidelberg, Berlin, Springer-Verlag Pub., p. 377-394.

Evans, G., 1965, Intertidal flat sediments and their environments of deposition in the Wash: Quart. Jour., Geol. Soc. London, v. 121, p. 209-245.

_____ 1966, The recent sedimentary facies of the Persian Gulf region: Phil. Trans. Royal Soc. London, Series A, v. 259, p. 291-298.

_____ et al, 1969, Stratigraphy and geologic history of the sabkha, Abu Dhabi, Persian Gulf: Sedimentology, v. 12, p. 145-159.

Fischer, A. G., 1964, The Lofer cyclothems of the Alpine Triassic, *in* D. F. Merriam, ed., Symposium on cyclic sedimentation: State Geol. Survey Kansas Bull. 169, v. 1, p. 107-149.

Garrett, P., 1970, Phanerozoic stromatolites — non-competitive ecologic restric-

tion by grazing and burrowing animals: Science, v. 169, p. 171-173.

Ginsburg, R. N., 1975, Tidal deposits, a casebook of Recent examples and fossil counterparts: New York, Spring-Verlag Pub., 428 p.

Grover, G. Jr., and J. F. Read, 1978, Fenestral and associated vadose diagenetic fabrics of tidal flat carbonates, Middle Ordovician New Market Limestone, southwestern Virginia: Jour. Sed. Petrology, v. 48, no. 2, p. 453-473.

Hagan, G. M., and B. W. Logan, 1975, Prograding tidal-flat sequences — Hutchinson Embayment, Shark Bay, Western Australia, *in* R. N. Ginsburg, ed., Tidal deposits, a casebook of Recent examples and fossil counterparts: New York, Springer-Verlag Pub., p. 215-232.

Halley, R. B., 1975, Peritidal lithologies of Cambrian carbonate islands, Carrara Formation, southern Great Basin, *in* R. N. Ginsburg, ed., Tidal deposits, a casebook of recent examples and fossil counterparts: New York, Springer-Verlag Pub., p. 279-288.

Hardie, L. A., 1977, Sedimentation on the modern carbonate tidal flats of northwest Andros Island, Bahamas: Baltimore, The Johns Hopkins Univ. Press, The Johns Hopkins Univ. Stud. in Geology, no. 22, 202 p.

Hoffman, P. F., 1967, Algal stromatolites–use in stratigraphic correlation and paleocurrent determination: Science, v. 157, p. 1043-1045.

Hofman, H. J., 1969, Stromatolites from the Proterozoic Ahimikie and Sibley groups, Ontario: Geol. Survey of Canada Paper 68-69, 55 p.

Howe, W. B., 1968, Planar stromatolite

Diagnostic Sedimentary Structures

EOLIAN Dune Sands — Variable Thickness

SUPRATIDAL ⇕ ~1 Meter

INTERTIDAL ⇕ 1-2 Meters

SUBTIDAL ⇕ Variable Thickness

Crossbedded carbonate or quartz eolian sand with root marks.

↶ Unconformity.

↶ Anhydrite diapirs and contorted layers in dolomite with windblown quartz.

Anhydrite or gypsum (chicken wire texture).

↶ Gradational contact.

Highly organic algal laminations with birdseye voids and mudcracks. Laminations less algal downward grading into burrowed lime mud or dolomite.

Gray burrowed and pelletal lime mud.

↶ Bored grainstone crust may or may not be present.

(A) Crossbedded carbonate grainstone (tidal bar or beach). May be oolitic or siliciclastic sand possibly underlain by coral or other open water facies.

(B) Lagoonal facies. Gray burrow-mottled lime mud (may be dolomitic).

Figure 51—Ideal offlap sequence based on the Persian Gulf examples described in the literature and observed by the author. Such a complete sequence is unlikely in the geological record due to the potential for subaerial erosion during deposition. In addition, several variations are possible. First, notice that eolian sands above the supratidal zone are likely and where they occur, these sands probably blanket and preserve sabkha facies. Such sands commonly migrate over the supratidal zone during and soon after deposition in arid climates. This movement and protection by eolian sands is common on modern tidal flats in the Persian Gulf. Second, notice that subtidal facies can range (depending upon local conditions) from crossbedded carbonate sands, often oolitic, to coral reefs. In the geological record corals might be replaced by stromatoporids, rudistids or other reef builders, but oolitic facies are common. Subtidal facies may also be composed of pelleted and burrowed muds. Cemented crusts with bored and eroded upper surfaces may occur both within and at the top of subtidal facies. The most diagnostic facies, however, is the desiccated algal stromatolite zone and the nodular or chicken wire anhydrite zone.

and burrowed carbonate mud facies in Cambrian strata of the St. Francois Mountain area: Missouri Dept. Business Admin., Div. of Geol. Survey and Water Resources, Rept. of Invest. 41, 113 p.

Illing, L. V., 1954, Bahamian calcareous sands: AAPG Bull., v. 38, no. 1, p. 1-95.

_____ 1959, Deposition and diagenesis of some Upper Palaeozoic carbonate sediments in western Canada: Proc., Fifth World Petroleum Cong., Sec. 1, Paper 2.

_____ A. J. Wells, and J. C. M. Taylor, 1965, Penecontemporary dolomite in the Persian Gulf, in L. C. Pray and R. C. Murray, eds., Dolomitization and limestone diagenesis — a symposium: SEPM Spec. Pub. 13, p. 89-111.

James, N. P., 1977, Shallowing upward sequences in carbonates: Geoscience Canada, v. 4, no. 3, p. 126-136.

Kendall, G. St. C., and Sir P. A. D'E. Skipwith, 1969, Holocene shallow-water carbonate and evaporite sediments of Khor al Bazam, Abu Dhabi, southwest Persian Gulf: AAPG Bull., v. 53, p. 841-869.

Kinsman, D. J. J., 1964, The recent carbonate sediments near Halat el Bahrani, Trucial Coast, Persian Gulf, in Deltaic and shallow marine deposits–developments in sedimentology, v. 1: Amsterdam, Elsevier Sci. Pub., p. 185-192.

Klein, G. V., 1971, A sedimentary model for determining paleotidal range: Geol. Soc. America Bull., v. 82, p. 2585-2592.

_____, 1972, Determination of paleotidal range in clastic sedimentary rocks: Proc., 24th Internat. Geol. Cong., Sec. 6, p. 397-405.

Logan, B. W., R. Rezak, and R. N. Ginsburg, 1964, Classification and environmental significance of algal stromatolites: Jour. Geology, v. 72, p. 68-83.

Longacre, S. A., 1980, Dolomite reservoirs from Permian biomicrites, in R. B. Halley and R. G. Loucks, eds., Carbonate reservoir rocks: Denver, SEPM Notes for core workshop no. 1: p. 105-117.

Matter, A., 1967, Tidal-flat deposits in the Ordovician of western Maryland: Jour. Sed. Petrology, v. 37, p. 601-609.

McKenzie, J. A., K. J. Hsu, and J. F. Schneider, 1980, Movement of subsur-

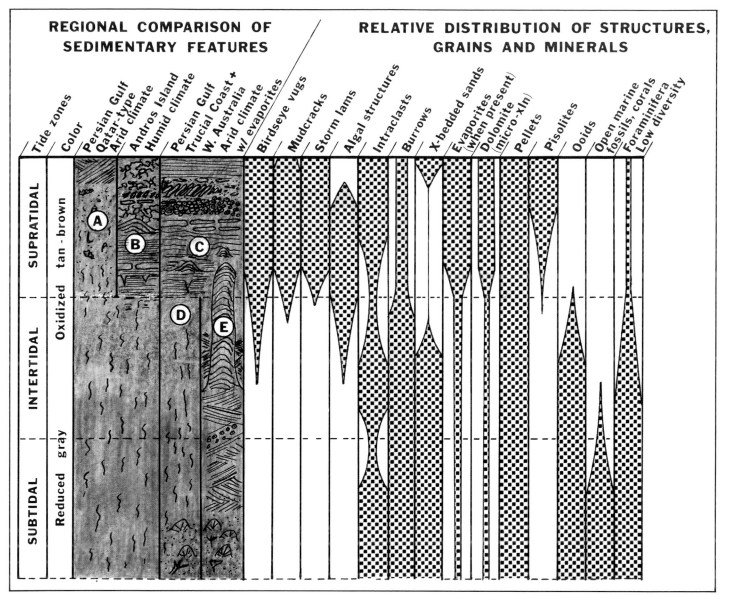

Figure 52—An attempt to compile all commonly accepted facets of facies variations associated with carbonate tidal flat accumulations. (**A**) shows the sequence on tidal flats around the Qatar Peninsula in the Persian Gulf. Note lack of a well developed algal mat or anhydrite zone due to slightly higher rainfall than on the Trucial Coast. (**B**) shows sedimentary structures, such as soil clasts, current-deposited intraclasts, minor algal heads and domes, mud polygons and mudcracks, developed in humid climates, such as at Andros Island in the Bahamas. (**C**) shows the sedimentary features associated with arid tidal flats, such as on the Trucial Coast in the Persian Gulf or in the Shark Bay region, western Australia. Note that the major difference in the last case is the presence of nodular or chicken wire anhydrite. The intertidal zones can range from oxidized muds to (**D**) coral reefs to (**E**) rippled crossbedded sands containing large club-shaped algal structures, such as those shown earlier in Figures 15A and B. The righthand section is an attempt to show relative abundance of various sedimentary structures, grains, minerals and fossils, based on the literature and personal observations of modern and ancient sequences. It should be emphasized that no geologic example is likely to fit exactly all the features shown here.

face waters under the sabkha, Abu Dhabi, United Arab Emirates and its relation to evaporative dolomite genesis: SEPM Spec. Pub. 28, p. 11-30.

Meissner, F. F., 1974, Hydrocarbon accumulation in San Andres Formation of Permian Basin, southeast New Mexico and west Texas: Abs., AAPG Bull., v. 58, no. 5, p. 909-910.

Multer, H. G., and J. E. Hoffmeister, 1968, Subaerial laminated crusts of the Florida Keys: Geol. Soc. America Bull., v. 79, p. 183-192.

Newell, N. D., and J. K. Rigby, 1957, Geologic studies in the Great Bahama Bank, *in* Regional aspects of carbonate deposition: SEPM Spec. Pub. 5, p. 15-79.

Pilkey, O. H., 1966, Carbonate and clay mineralogy of the Persian Gulf: Deep-Sea Research, v. 13, p. 1-16.

Purser, B. H., 1973, The Persian Gulf-Holocene carbonate sedimentation and diagenesis in a shallow epicontinental sea: Heidelberg, Berlin, Springer-Verlag Pub., 471 p.

_____ and G. Evans, 1973, Regional sedimentation along the Trucial Coast, SE Persian Gulf, *in* B. H. Purser, ed., The Persian Gulf-Holocene carbonate sedimentation and diagenesis in a shallow epicontinental sea: Heidelberg, Berlin, Springer-Verlag Pub., p. 211-213.

_____ and J. P. Loreau, 1973, Aragonitic, supratidal encrustations on the Trucial Coast of the Persian Gulf, *in* B. H. Purser, ed., The Persian Gulf-Holocene carbonate sedimentation and diagenesis in a shallow epicontinental sea: Heidelberg, Berlin, Springer-Verlag Pub., p. 343-376.

Read, J. F., 1975, Tidal-flat facies in carbonate cycles, Pillara Formation (Devonian), Canning Basin, Western Australia, *in* R. N. Ginsberg, ed., Tidal deposits, a casebook of Recent examples and fossil counterparts: New York, Springer-Verlag Pub., p. 251-256.

Reinick, H. E., 1967, Layered sediments of tidal flats, beaches, and shelf bottoms of the North Sea: Estuaries, Am. Assoc. Advan. Science, v. 83, p. 191-206.

Robbin, D. M., and J. J. Stipp, 1979, Depositional rate of laminated soilstone crusts, Florida Keys: Jour. Sed. Petrology, v. 49, no. 1, p. 175-180.

Roehl, P. O., 1967, Stony Mountain (Ordovician) and Interlake (Silurian) facies analogs of Recent low-energy marine and subaerial carbonates, Bahamas: AAPG Bull., v. 51, p. 1979-2032.

Rogers, J. P., 1971, Tidal sedimentation and its bearing on reservoir and trap in Permian Phosphoria strata, Cottonwood Creek Field, Big Horn Basin, Wyoming: Mtn. Geol., v. 8, no. 2, p. 71-80.

Schneider, J. F., 1975, Recent tidal deposits, Abu Dhabi, United Arab Emirates, Arabian Gulf, *in* R. N. Ginsburg, ed., Tidal deposits, a casebook of Recent examples and fossil counterparts: New York, Springer-Verlag Pub., p. 209-214.

Scholle, P. A., and D. J. J. Kinsman, 1974, Aragonitic and high-Mg calcite caliche from the Persian Gulf-a modern analog for the Permian of Texas and New Mexico: Jour. Sed. Petrology, v. 44, no. 3, p. 904-916.

Shinn, E. A., 1968a, Selective dolomitization of Recent sedimentary structures: Jour. Sed. Petrology, v. 38, no. 2, p. 612-616.

_____ 1968b, Practical significance of birdseye structures in carbonate rocks: Jour. Sed. Petrology, v. 38, no. 1, p. 215-223.

_____ 1968c, Burrowing in Recent lime sediments of Florida and the Bahamas: Jour. Paleontology, v. 42, no. 4, p. 879-894.

_____ 1969, Submarine lithification of Holocene carbonate sediments in the Persian Gulf: Sedimentology, v. 12, p. 109-144.

_____ 1972, Worm and algal-built columnar stromatolites in the Persian Gulf: Jour. Sed. Petrology, v. 42, no. 4, p. 837-840.

_____ 1973a, Carbonate coastal accretion in an area of longshore transport, NE Qatar, Persian Gulf, *in* B. H. Purser, ed., The Persian Gulf-Holocene carbonate sedimentation and diagenesis in a shallow epicontinental sea: Heidelberg, Berlin, Springer-Verlag Pub., p. 179-191.

_____ 1973b, Sedimentary accretion along the leeward, SE coast of Qatar Peninsula, Persian Gulf, *in* B. H. Purser, ed., The Persian Gulf-Holocene carbonate sedimentation and diagenesis in a shallow epicontinental sea: Heidelberg, Berlin, Springer-Verlag Pub., p. 199-209.

_____ D. M. Robbin, and R. P. Steinen, 1980, Experimental compaction of lime sediment: Denver, Abs., AAPG Ann. Mtg., p. 120.

_____ R. N. Ginsburg, and R. M. Lloyd, 1965, Recent supratidal dolomite from Andros Island, Bahamas, *in* L. C. Pray and R. C. Murray, eds., Dolomitization and limestone diagenesis-a symposium: SEPM Spec. Pub. 13, p. 112-123.

_____ R. M. Lloyd and R. N. Ginsburg, 1969, Anatomy of a modern carbonate tidal flat, Andros Island, Bahamas: Jour. Sed. Petrology, v. 39, p. 1202-1228.

Steinen, R. P., and R. B. Halley, 1979, Ground water observations on small carbonate islands of southern Florida, *in* R. B. Halley, ed., Guidebook to sedimentation for the Dry Tortugas: Southeastern Geol. Soc. Pub. No. 21, p. 82-89.

Tebbutt, G. E., C. D. Conley, and D. W. Boyd, 1965, Lithogenesis of a distinctive carbonate rock fabric, *in* R. B. Parker, ed., Contributions to geology: Laramie, Univ. Wyoming, p. 1-13.

Van Straaten, L. M. J. U., 1954, Sedimentology of Recent tidal flat deposits and the Psammites du Condroz (Devonian): Geol. Mijnbouw, v. 16, p. 25-47.

Wells, A., 1962, Primary dolomitization in Persian Gulf: Nature, v. 194, no. 4825, p. 274-275.

5

Beach Environment

Richard F. Inden
Clyde H. Moore

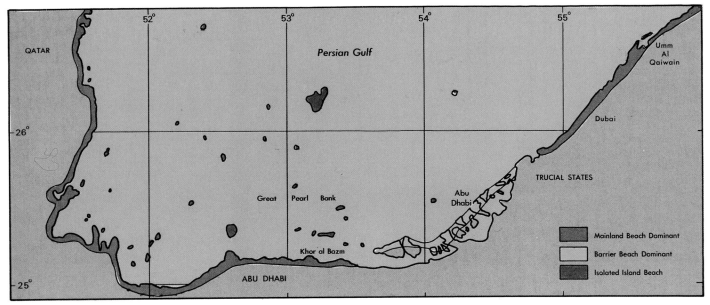

Figure 1—Diagram showing the variation in beach development along the Trucial Coast of the Persian Gulf (after Purser and Evans, 1973).

*M*ost ancient carbonate strata were deposited on extensive, warm, shallow, marine shelves, and record regressive rather than transgressive periods of sedimentation. These sequences commonly contain significant volumes of rocks deposited in low energy, shallow, subtidal-to-supratidal, strandline lithotopes. At the same time, however, higher energy wave-dominated, strandline deposits (for example, beaches) are surprisingly scarce in the geological record. We feel that this is probably due to lack of recognition, rather than an actual absence of beach deposits in carbonate strandline sequences.

Inasmuch as shallow marine siliciclastic sequences frequently contain both low energy tidal flat and higher energy beach units along their shorelines, we should expect the same to be true in many carbonate sequences, because energy flux varies along strandlines in response to basin hydrology and topography, irrespective of the mineralogy of the sea floor sediment. The only real difference between the two strandline types is that

those composed of carbonate grains are fed by a nearby offshore source of sediment that is being formed through physiochemical precipitation (oolites, grapestones), the breakdown of carbonate skeletal material, or the erosion of previously deposited lime mud or submarine cemented layers (intraclasts); siliciclastic beaches are, for the most part, fed directly by streams or by long shore currents carrying stream-derived sediment.

We suspect that many ancient bioclastic and oolitic calcarenite shoal deposits are comprised in part of unidentified beach lithofacies, and that many of the thin fossiliferous, oolitic, and intraclastic calcarenite zones found as interbeds between evaporites, laminated calcilutites and dololutites, and other limestones, are also beach deposits. Susceptibility of the grains in these beaches to the syn-diagenetic and early burial processes of cementation or exposure of carbonate beach sediment to weathering stabilization processes (recrystallization, cementation) commonly results in their lithification. The strandline sequence is thus rendered more resis-

tant to mechanical erosive processes during erosional transgressive episodes, or scour by tidal, alluvial, or distributary channels in laterally equivalent and overlying facies. Its chance of preservation, therefore, is greater than in equivalent siliciclastic systems.

DEFINITION

Beaches are wave-dominated coastal sedimentological systems composed of loose sediment, whose overall external morphology and internal character may contain modifications induced by tidal activity and/or longshore currents. Their upper and lower environmentally defined limits are thus set where wave current activity is no longer dominant. Laterally, they can grade into tidal flats because of a decrease in the availability of sand or coarser grain sizes, and/or a change from wave- to tidally-dominated processes along the coastline. Beach surfaces are inclined seaward at angles steeper than those characteristic of tidal flats because of the combined actions of intense tur-

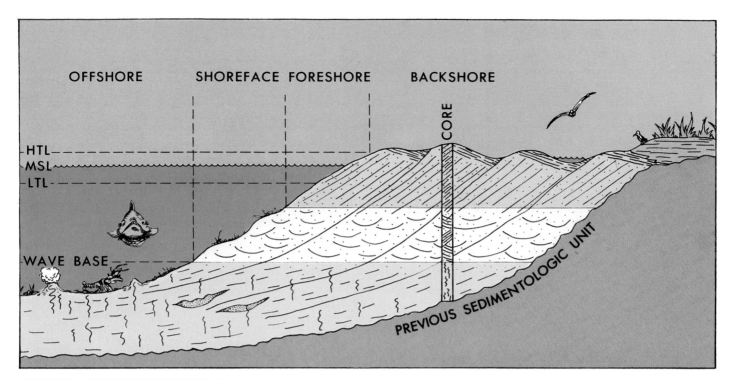

Figure 2—Generalized beach model illustrating the most simple and common beach setting. Conceptualized after Bernard and Majors (1962).

ZONES	TYPICAL E-LOG (Before Burial)		GRAIN SIZE		SORTING		LITHOLOGY	CORE	SEDIMENTARY STRUCTURES	PROCESSES
	SP	Res.	Coarse	Fine	Poor	Well				
FORESHORE							Grainstone		Parallel Laminations Small Scale Avalanche Cross Bedding Fine Graded Laminations Vertical Burrows	Wave Swash
SHOREFACE							Grainstone to Packstone		Small to Medium Tabular Festoon Crossbeds	Directed Tidal and Along Shore Currents
OFFSHORE							Packstone to Wackestone		Horizontal Branching Burrows	Biologic

bulence in the wave breaker zone at the toe of the beach, and the progressive loss of energy up the beach toward the upper limit of the swash zone. Finer grained sediment is winnowed away by incessant wave action, and, even if it were not, it could not stand at a steep enough angle of repose to be recognized as a beach if one were standing on it. The width and steepness of the active beach zone being affected by daily wave ac-

tion is dependent upon wave size, as well as tidal activity (tidal height and current strength) and the size of constituent grains. The general environment includes barrier islands, spits, and non-barrier mainland beaches, as well as isolated island beaches (Fig. 1). These isolated beaches may bear some relationship to the regional coastline, such as those associated with paralleling reef systems or horst blocks, or they may develop random-

ly in association with salt domes, mud mounds, patch reefs or drowned topographic features. Variations in beach settings and their internal and geometric characteristics will be presented in the following pages.

General Model

The basic process-response model for carbonate beaches generally parallels the clastic barrier island model proposed by Bernard and

Figure 3a—Pleistocene beach ridges showing seaward dipping carbonate sand rich accretion layers; South African coastline (photo by Dave Hobday).

Figure 3b—Seaward dipping beach lime grainstone accretion beds; Edwards Formation, Round Mountain, Texas.

Figure 3c—Arcuate beach ridges; Joulters Cay, Bahamas (courtesy Mitch Harris).

others (1962) for beaches that develop in a microtidal (0 to 2 m) setting. Since then, many sedimentologists working in numerous other (especially clastic) beach settings (Clifton et al, 1971; Hayes, 1976; Ball, 1967; Hubbard et al, 1979; Shinn, 1973; Kumar and Sanders, 1974) have added to our knowledge of how this basic model varies in response to different wave, longshore current, and tidal energy regimes. It is not our intention to describe all the sedimentary structures and structure sequences that result from varying grain size, and the types and amounts of energy input to strandlines. Instead, we will keynote those structures and sequences necessary for beach definition and recognition.

The beach depositional environment, exclusive of the backshore zone, represents, in its simplest form, a seaward-inclined depositional interface along which the high energy regimen found in the surf zone gradually decreases in a seaward (deeper water) direction. This interface may be divided into three gradational subzones (Fig. 2), identified by the dominant depositional processes operative in each zone. The foreshore is dominated by high energy wave swash and related processes. The shoreface is the zone in which longshore and tidal current activity is pronounced, and the offshore environment is a lower energy environment, where biological processes, such as bioturbation, predominate. The backshore environment includes

dunes, washovers, and supratidal complexes which form as a result of wind, storm, and tidal action. These grade laterally landward into lagoonal, sabkha, or continental settings. A soil profile may separate the main beach from the low energy backshore deposits, or it may develop across both beach and backshore sediments. In a vertical sequence representing one regressive cycle, backshore units lie directly above the main beach deposits, which in turn grade downward into finer-grained offshore units. The process-product relationships responsible for this characteristic sequence are outlined in Figure 2.

Vertical Textural Sequence

The main textural response to the energy flux across the beach zone is the removal of mud from the sediment, and the gradation from well-winnowed sands (lime grainstones) in the upper foreshore to progressively muddier sediments (lime packstones and wackestones) toward deeper, lower energy environments. Foreshore units most often have a small percentage of coarser grains mixed in with a dominant finer, well sorted grain population (negatively skewed), but they are also commonly very well or well sorted (Folk and Cotera, 1970). In general, size and sorting decrease vertically down the depositional interface, as illustrated in Figure 2. However, the coarsest sediment may occur in the middle of the interface (Folk and Robels, 1964; Folk and Cotera, 1970), which marks the transition zone between the lower foreshore and upper shoreface.

The grain sizes present, as well as their composition, are in part a function of wave intensity and current strength acting on the beach interface. To a large extent, however, these parameters are controlled by the variety and individual numbers of shell-bearing organisms living in the beach and in near offshore environments. This is true because these organisms are the primary sediment contributors to the beach, and each type has a shell which breaks down at

Figure 4—Uppermost portion of beach accretion beds showing gentle seaward dip (right) and uppermost bed dip reversal toward backshore (left). Cow Creek Limestone, Lower Cretaceous, Texas.

Figure 5a—Gently seaward dipping evenly laminated sequences in uppermost part of beach accretion unit; South Carolina coast (courtesy John Barwis).

a specific rate into preferred grain sizes. Grain roundness is also a function of energy input to the sediment as well as its composition, and in general, it will decrease downward into lower energy environments. This, of course, would not be the case if oolite shoals were the source of the beach sediment.

Figure 5b—Laminated foreshore beds, beach sequence in Mississippian Newman Limestone, Eastern Kentucky.

Figure 6a—Large vertical and subvertical burrows penetrating even, parallel laminations in upper part of beach accretion bed, Pleistocene, South Africa (photo courtesy of Dave Hobday).

Figure 6b—Worm burrows in coarse-grained mollusk lime grainstone of Lower Cretaceous beach foreshore units.

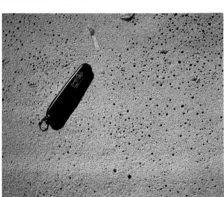

Figure 7a—Air-bubble holes in uppermost part of swash zone; holes are approximately 3 to 5 mm in diameter.

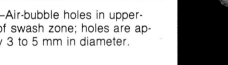

Figure 7b—Large irregular keystone vugs in slab of lime grainstone, Lower Cretaceous, Texas (scale in mm).

Sedimentary Features — Physical and Biological

The most obvious sedimentary structures in outcrops of ancient beaches are gentle, seaward-dipping (5 to 15°), large-scale planar accretion beds (Fig. 3a, b) which record the progradation of the beach over deeper, subtidal sediments. These units are associated with the beach ridges that mark prograding Holocene beaches (Fig. 3c). Each accretion unit contains all microfacies of the beach foreshore and shoreface and fade into typically lower energy offshore facies. At the top of the sequence, it is common to see berm deposits, with dip reversal toward the mainland (Fig. 4), that are associated with overlying and infilling carbonate muds.

Within each progradational accretion unit there is a distinct suite of smaller-scale sedimentary structures which reflect the changing energy regimes across the depositional interface of the beach. The uppermost part of the sequence represents deposition by wave swash processes in the beach foreshore zone (Fig. 2) and includes alternating fine- and coarse-grained, sometimes inversely graded, even laminations which have a gentle seaward dip (Fig. 5). A wide variety of vertical and subvertical mollusk, worm, and crustacean burrows, with morphologies characteristic of high

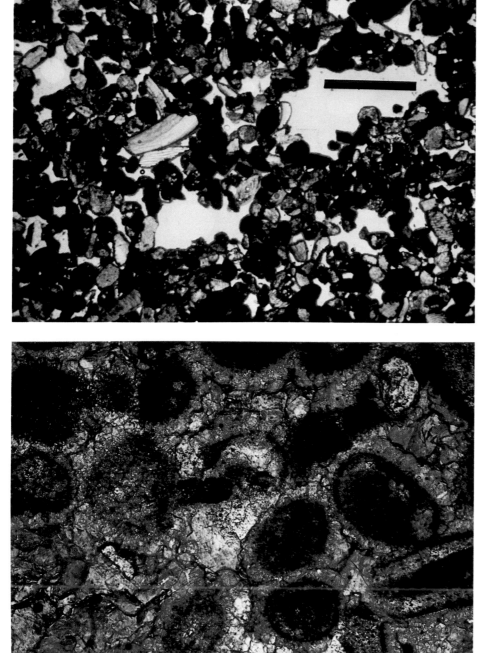

Figure 7c—Large ovoid keystone vugs in Recent carbonate beachrock; scale equals 1.0 mm.

Figure 7d—Keystone vug (lower left corner), in oolitic lime grainstone; Mississippian Newman Formation, Kentucky; scale equals .3 mm.

energy, high sedimentation environments, penetrate these laminations (Fig. 6a, b). The uppermost laminations of the beach accretion beds often contain tabular to somewhat spherical cavities referred to as keystone vugs (Dunham, 1970; Fig. 7). These represent voids left by air and gas bubbles, which form immediately above the wave swash zone as a result of air escaping from intergranular pores as they are flooded with marine waters during the flood tidal cycle. Oscillation and current ripples, as well as current lineations and antidunes, are present on accretion bedding planes in the laminated swash zone, but these are rarely observed in ancient sequences because of their modification into upper flow regime flat bedforms (Harms, 1979) before burial.

A zone in the center of the vertical sequence and just below the swash laminations (Fig. 2), is referred to as the shoreface, and is dominated by small- to large-scale tabular and festoon crossbeds (Fig. 8a, b). This zone represents deposition by currents flowing predominantly parallel to the beach (Fig. 8c), and thus most crossbed dip directions are oriented perpendicular to the beach accretion

Figure 8a—Large festoon (trough) cross-beds immediately below beach foreshore unit (upper left of photo). Orientation of crossbeds indicates flow parallel to depositional strike of beach, Lower Cretaceous Cow Creek Limestone, Texas.

Figure 8b—Cross-sectional view of large festoon (trough) crossbeds under laminated beach foreshore sequence; South African coastline (photo courtesy of Dave Hobday).

Figure 8c—Barrier beach system (lagoonal system in upper half of photo) showing large sand waves oriented perpendicular to beach trend. Sand waves are formed as a result of longshore currents flowing toward the left.

Figure 9a—Branching crustacean burrow fill from a low energy subtidal Florida Bay environment. Burrow fills may be finer or coarser grained than the surrounding matrix sediment, depending mostly on what the organism ingested, and what grain-size sediment the overlying unit is composed of.

Figure 9b—Crustacean burrows in shallow subtidal deposits beneath beach shoreface deposits. Burrow fills are slightly arenaceous pellet lime/packstones, whereas the matrix rock is a fossiliferous, dolomitic quartz sandstone, Lower Cretaceous Cow Creek Limestone, Texas.

beds (Stricklin and Smith, 1972). Landward oriented planar crossbeds may also be present, in which case the system probably represents the preserved ridge and runnel zone that commonly develops in mesotidal (2 to 4 m) beach settings (Hayes, 1976). Crossbeds may be composed of sediment coarser or finer grained than the overlying laminated zone. They have either a sharp scour contact with, or grade downward into, the offshore

deposits, and may be interbedded with the lowermost laminated foreshore deposits.

Underlying offshore deposits are typically composed of bioturbated and burrowed lime (or dolomite) packstones and wackestones. Distinct burrow traces within these units are either horizontal or branching, and reflect the prevailing lower energy conditions in offshore areas (Fig. 9a, b). Thin, rippled lime packstone and

grainstone interbeds and lenses (Fig. 9d), commonly with sharply scoured lower contacts and bioturbated or bored upper contacts, are often present immediately below the shoreface zone; they represent storm deposits formed through the winnowing of the underlying and laterally equivalent sediments. The biota contained within the offshore deposits may be highly varied and abundant or extremely limited, depending on environmental

Figure 9c—Subhorizontal *Thallasinoides* (crustacean) burrows, shallow offshore facies, Lower Cretaceous Cow Creek Limestone, Texas.

Figure 9d—Thin ripple crosslaminated lime packstone lenses with basal scour contacts into finer grained shallow offshore facies, Lower Cretaceous Cow Creek Limestone, Texas.

Figure 10a—Storm washovers behind a long straight barrier beach. Sediment from the beach zone and offshore environments is dispersed into the backshore zone via storm surge channels (arrow). Large washover in foreground is now stable and has a complex of subenvironments (supratidal marsh and pond, tidal creeks, eolian dunes and sand flats) developed on it (photo courtesy M. O. Hayes).

conditions. In either case, however, it will be similar to the biota found in the foreshore and shoreface sediments.

The backshore environment is made up of a myriad of sedimentary subenvironments because of the numerous processes that act in its formation. In the geologic record, however, only storm washovers and thin tidal flat and supratidal units are common. Dune deposits, although

present in the Holocene and Pleistocene, seem to be rarely preserved or recognized in older units.

Eolian dunes form as a result of strong landward directed winds eroding sediment from the foreshore and depositing it in the immediate backshore area. These dunes grade laterally into progressively finer grained sand flats toward the far backshore areas. Sediment in these sand flats is derived from both the dunes and the uppermost layers of washover fans.

Washover fans are deposited by wave induced currents carrying sediment through channels cut into low lying beach areas; they form during intense storm activity and can make up the bulk of the backshore sediment volume.

Fan deposits cover from a fraction to 4 sq km of backbeach supratidal marsh, intertidal flat, or lagoonal floor (Fig. 10a, b). They are primarily composed of evenly laminated grainstones and packstones. Normally, they are thin (.1 to 1 m), become finer-grained upward, and have scoured or sharp lower contacts. Basal fossil and intraclast-rich lags, as well as avalanche (tabular) style cross-bedding deposits, are common in the proximal and distal portions of fans, respectively (Fig. 10b; Schwartz,

Figure 10b—Washover fans prograding (from right to left) onto algal-matted, low energy tidal flat; steep frontal margin of washover indicates internal avalanche-style planar crossbedding.

Figure 11a—Seaward dipping slabs of Recent beachrock cemented foreshore sediment, Grand Cayman Island, British West Indies.

1975). Large branching crustacean burrows, smaller vertical "pencil" burrows, and roots typically penetrate their upper layers. Their distal portions, and areas between them, may be dominated by tidal flats or lagoonal deposits, or they may grade landward directly into sabkha or continental settings. Supratidal and/or intertidal lithologies are highly varied in character (Shinn, 1982), and supratidal sequences generally make up the bulk of the beach backshore sequence in ancient carbonate sequences.

Sedimentary Features — Diagenetic

Carbonate beach deposits are subject to certain diagenetic processes soon after their deposition. The resultant diagenetic products are somewhat unique to the beach setting, and are thus useful as criteria for beach recognition. The most common diagenetic process is penecontemporaneous beach cementation associated with the foreshore environment. The most obvious product of this early cementation is the development of large seaward dipping slabs of beachrock in the foreshore zone (Fig. 11a). During storms these slabs are undercut by wave and current action, causing collapse, breakage and reworking into tabular cobble to pebble size beachrock clasts (Fig. 11b, c). These clasts are recognized in ancient beach sequences by subtle textural, cementation, structural (disoriented laminations) and color contrasts with the surrounding beach deposits (Fig. 11c). On a smaller scale, the margins of the clasts display truncated grains and cements (Fig. 11d), and may be penetrated by the borings of various marine organisms (worms, clams, barnacles, sponges, etc.). Donaldson and Ricketts (1979) describe in detail the occurrence of beachrock clasts in a beach sequence from the Precambrian of Canada.

The waters from which beachrock cements are precipitated are generally marine, so that the mineralogy of the cements is dominantly aragonite, and occasionally high-magnesium calcite.

Figure 11b—Collapse zone, probably due to wave undermining, in laminated beach foreshore beds; Lower Cretaceous Edwards Limestone, Texas.

Figure 11c—Rounded cobble and boulder size slabs of beachrock contrast in color, texture, and structure with surrounding beach lime grainstones; Lower Cretaceous Edwards Limestone, Texas.

Figure 11d—Beachrock clast displaying truncated grains and isopachous, formerly aragonite(?) beachrock cement; Lower Cretaceous Cow Creek Limestone, Texas (scale equals .3 mm).

Aragonite beachrock cements usually occur as thick circumgranular acicular crusts which sometimes occlude all pore space (Fig. 12a, b). High-magnesium calcite cements are present as thin-bladed isopachous crusts (Fig. 12c) or, more commonly, as dark or golden brown, pelletal and micritic, irregular grain coatings and void fillings (Fig. 12d, e; Alexandersson, 1972). The latter often contains abundant microfossils and may form, at least in part, as a result of intergranular biological activity (Moore, 1973). These cements may exhibit microstalactitic and meniscus orientation fabrics (Fig. 12e, f, g), both of which are unique to the vadose zone if they form in the intertidal or supratidal portion of the beach. Below low tide level, acicular cements typically occur as somewhat regular or isopachous crusts (Fig. 12c). These cements, because of their mineralogical instability, may be dissolved or recrystallized during later diagenesis, and thus, are not always recognizable or present in ancient beach sequences. However, their "fossilized" occurrence in ancient beach sequences has been well documented (Petta, 1977a; Moore et al, 1972) and thus provides us with an additional set of criteria by which we may recognize ancient beaches.

Because of the slow rate of deposi-

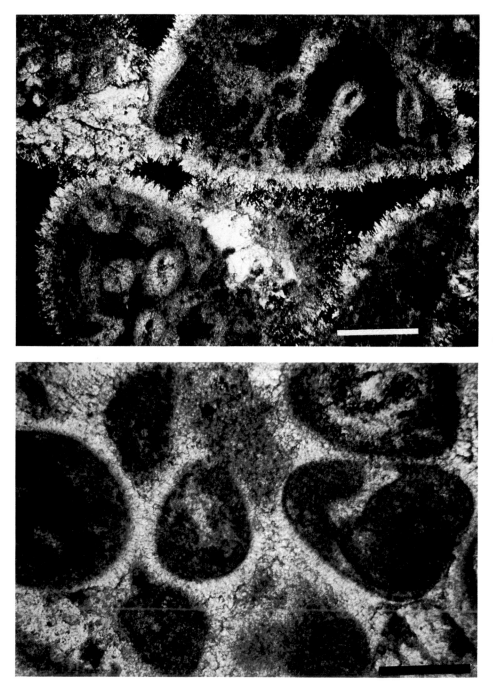

Figure 12a—Aragonite needle fringe cements rimming algal grains; note slight irregularity in thickness of cement; Grand Cayman Island, British West Indies (scale equals .25 mm).

Figure 12b—Fibrous, somewhat irregular fringe cements, originally thought to be aragonite, from Mississippian beach sequence in Kentucky (scale equals .25 mm).

tion on many beaches, a variety of soil features may develop on and in the foreshore units, finer grained backshore supratidal facies, and washover deposits. The soil forming processes normally lead to grain dissolution, microfracturing, micritization, and general destruction of the original texture and fabric of the deposits (Fig. 13a, b). Diagenetic fabrics associated with soil formation, such as vadose cements, pisolites, and coated grains occur in the upper 1 or 2 m of the foreshore grainstones and backshore units (Fig. 13c, d). Nodular caliches develop out of, and on top of, these beach deposits, and may be separated from them by a hummocky iron-stained dissolution surface (Fig. 13e; Inden, 1972). Laminated subaerial crusts (Fig. 13f, g; Multer and Hoffmeister, 1968) coat complexly brecciated black and brown micritized clasts and many extend downward along previously formed fractures that cut across and parallel foreshore bedding units (Ferm et al, 1971; Inden and Horne, 1973; Fig. 13h).

Beach Recognition

Identification of beach sequences rests on the recognition of the vertical mosaic of textures, fabrics, and sedimentary structures which together indicate an overall downward grada-

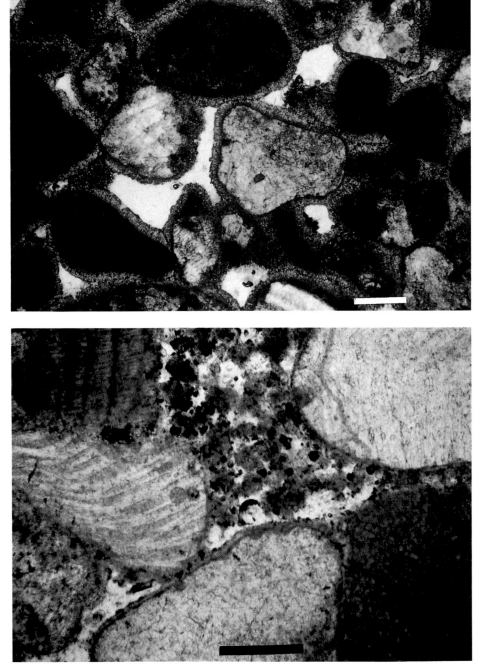

Figure 12c—High-magnesium calcite isopachous fringe cements; note how regular this cement development is compared to Figure 13a; Grand Cayman Island, British West Indies; scale equals .125 mm (stained with Clayton Yellow).

Figure 12d—High-magnesium calcite micritic and pelletal cements; if cementation of this type goes to completion the sediment will lose its original textural identity of being a lime grainstone and will be converted diagenetically into a lime packstone; Grand Cayman Island, British West Indies (scale equals .125 mm).

tion from high energy (upper flow regime) environmental conditions to normally lower energy environmental conditions. Evidence of exposure, such as fine-grained supratidal carbonates and/or evaporites, paleosols, or subaerial crusts, must be present above the characteristic high energy, low angle, evenly laminated foreshore lime/grainstone beds in regressive beach sequences. Unique sedimentary features, such as keystone vugs, reworked beachrock clasts, and marine vadose beachrock cements, are common and are generally useful in the recognition of ancient beach sequences.

GEOLOGIC SETTINGS, ASSOCIATED FACIES, AND BASIC BEACH MODEL VARIATIONS

Variations on the basic beach model (Figs. 2, 14a) are controlled by the geomorphic and sedimentologic setting of the beach. Major variations are formed when the beach: (1) is at-

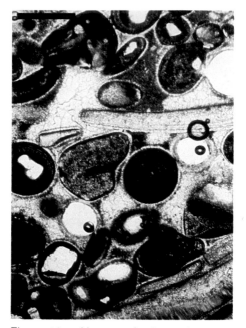

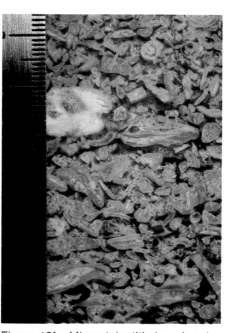

Figure 12e—Very regular isopachous crusts coated by micrite cements displaying meniscus and microstalactitic fabrics, both of which are characteristic of precipitation in the vadose zone; scale equals .25 mm; Jurassic (Dogger) of Great Britain.

Figure 12f—Microstalactitic beachrock cements (arrows) developed in a very coarse algal coated grain lime grainstone, Lower Cretaceous Stuart City Reef Trend, Texas (scale in mm).

Figure 12g—Microstalactitic cement (**MS**), (probably originally aragonite) displaying growth lines of (originally high-magnesium calcite) micrite (arrow); micrite below (**m**) may be infiltrated sediment or cement; scale equals .065 mm; Lower Cretaceous, Texas.

tached to a mainland, (2) forms a barrier with potential tidal inlets and a low energy lagoonal environment on its landward side, (3) has high energy (rather than low energy) offshore equivalent facies, such as a reef or bioclastic-oolitic sand sheet or shoal complex; or (4) develops in a low energy environment.

Mainland Beaches

Mainland beaches form regional strandlines along continents or large islands. Landward facies equivalents are carbonate, evaporite (sabkha), or clastic (coastal plain, alluvial fan) sequences depending on climate and the relative deliverability rate of offshore-derived marine carbonate sediments, and landward-derived siliciclastic detritus. A Holocene example is the regional beach system along the northeastern Trucial Coast area of the Persian Gulf (Fig. 1; Purser and Evans, 1973). The coastline in this region is exposed to the deep, open waters of the Persian Gulf, as well as being oriented at almost right angles to "shamal" winds which blow across the width of the gulf from a west-

northwesterly direction. Thus, without the baffling effect of shallow barriers, such as reefs or other structural and topographic controls, the unimpeded waves generated in the open gulf hit the coast with maximum fetch, and in the process generate strong longshore currents. Because of the continuous and rapid production of bioclastic debris in the offshore area, the mainland beach sequence has prograded from 5 to 10 km seaward during the Holocene (Purser and Evans, 1973). This has resulted in the development of a sheet of sand which is elongate parallel to the coast and up to 15 m thick and 10 km wide. It is underlain by bioturbated, finer grained, poorly sorted molluscan skeletal sands (lime packstones and wackestones). If progradation continues, it will eventually be overlain by sabkha deposits and distal alluvial fan muds and channels sands.

A mainland beach also exists on the western Trucial coast (Fig. 1; Purser and Evans, 1973), but here the composition of the beach sands varies tremendously from east (coral-algal),

where reefs exist offshore, to west (oolite), where the beach system is increasingly influenced by the wind shadow created by the Qatar Peninsula. Offshore deposits are rippled and crossbedded at shallow depths (1 to 2 m) and rippled, grading to bioturbated, at greater depths (Purser and Evans, 1973).

The Cow Creek Limestone and equivalents serve as an ancient example and will be treated in the last section of this chapter.

Barriers

Barrier beach sequences are the most common variant of the generalized beach model, and show the widest variety of associated facies. The major additional facies components are those deposits that form as a result of current activity through tidal inlets which pass through the barrier into the back barrier lagoon (Fig. 15a, b). These processes lead to the development of a downward coarsening sequence which is as thick or thicker (up to 12 m) than the laterally equivalent shoreface and foreshore units. The result is the for-

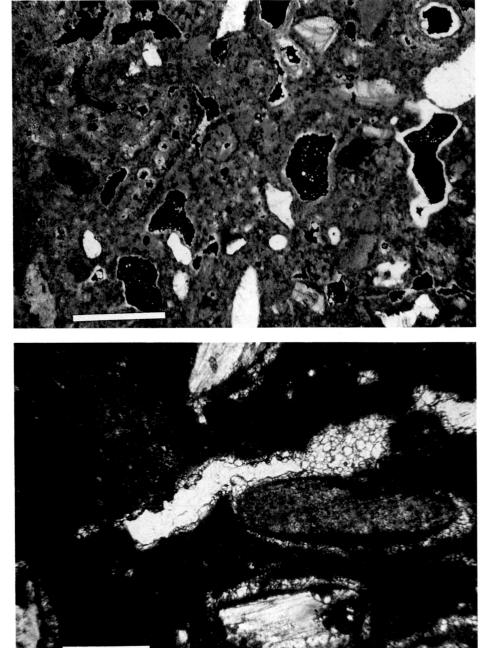

Figure 13a—Holocene beach lime grainstone being altered into soil (caliche) through processes of grain micritization and precipitation of micrite cement; small vugs are also forming as a result of dissolution; St. Croix, Virgin Islands (scale equals .5 mm).

Figure 13b—Paleocaliche from top of beach sequence. Original lithology was a mollusk lime grainstone, but precipitation of micrite cement, intense micritization, dissolution, and fracturing have destroyed its original texture and fabric and converted it into a lime wackestone; Lower Cretaceous Cow Creek Limestone, Texas (scale equals .25 mm).

mation of a couplet consisting of two high energy end components which may be separated by somewhat lower energy lime wackestones and packstones (Fig. 14b).

Because tidal inlets may migrate along the strike of the barrier, the following normal progression of sedimentary features is commonly observed in Holocene as well as ancient carbonate and clastic barrier complexes (Boutte, 1969; Hayes,

1976; Hubbard et al, 1979) deposited under a microtidal (0 to 2 m) or mesotidal (2 to 4 m) setting. The inlet sequence itself is bounded by a basal scour surface (Figs. 14b, 15c) which may have associated coarse intraclast or fossil lags immediately above it (Fig. 14b). These are overlain by large- and medium-scale tabular crossbedded lime grainstones (Fig. 16c, d), which in turn grade upward into small scale tabular and festoon

crossbedded, as well as ripple cross-laminated, deposits. Crossbed dip directions are typically unimodal in the upper and lower part of the sequence and trend perpendicular to the strike of the barrier. Ebb-oriented crossbeds typically occur in the lower part of the inlet fill. Wave-produced flood (landward) oriented crossbeds may well dominate its upper portions, but bimodal crossbeds are common in this zone as well. The sequence pro-

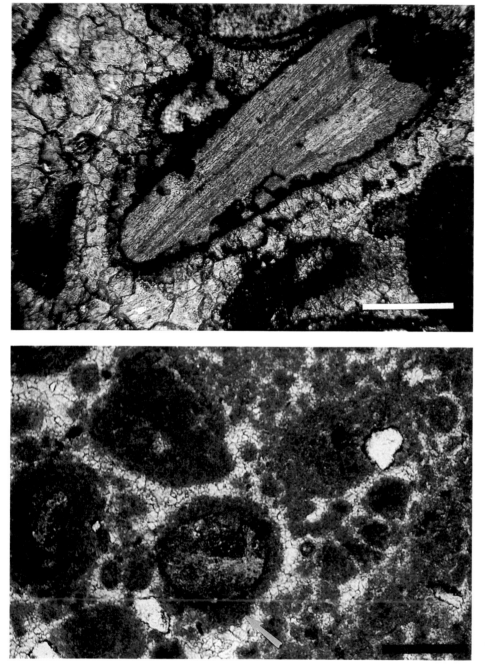

Figure 13c—Microstalactitic and dolomicrite cements in upper part of beach and immediately below lithology in photo 14b; Lower Cretaceous Cow Creek Limestone, Texas (scale equals .25 mm).

Figure 13d—Micrite coated grains, pelletal micrite cements and microstalactitic micrite cement (arrow) in paleosol capping Mississippian oolite shoal sequence, Kentucky (scale equals .125 mm).

gressively fines upward into offshore, lagoonal, or backshore tidal flat and supratidal facies, or is capped by coarsening upward shoreface and foreshore deposits, either abruptly, or with a thin intervening offshore unit (Fig. 16e). Tidal delta and tidal inlet fills are rare in microtidal barrier island deposits (Fig. 15a), but normally make up a large part of the mesotidal barrier sequence (Fig. 15b; Hayes, 1976).

Barriers are best developed in the Persian Gulf along the central Trucial Coast (Fig. 1; Purser and Evans, 1973). A structural high, referred to as the Great Pearl Bank, gave rise to an offshore topographic ridge prior to the onset of Holocene development; an intervening topographic low (now the Kor al Bazm lagoon) separated this ridge from the mainland. At the onset of the Holocene depositional cycle, reefs

began to grow on the crest of the ridge, and coral-algal skeletal sands began accumulating around them to form small subaerially exposed islands (Fig. 16a). These islands began to expand lengthwise along the trend of the ridge because of continued lateral accretion due to longshore drift (Fig. 16b). Eventually, tidal flat areas began to form in the protected areas behind the islands, while channels in the topographic low

Figure 13e—Hummocky iron-stained dissolution surface separating well-developed nodular caliche (N) from underlying partially calichified beach lime grainstones; Lower Cretaceous Cow Creek Limestone, Texas.

Figure 13f—Surface of Pleistocene iron-stained subaerial crust, Florida Keys.

behind the ridge progressively became more restricted (Fig. 16c). Tidal deltas and tidal inlet fills of bioclastic-oolite sands began to form, and the Kor al Bazm lagoon started to fill with low energy pelletal muds and tidal flat deposits (Purser and Evans, 1973; Fig. 16d). If the present system continues to operate, the lagoon will become a highly restricted environment and subaqueous

evaporites will begin to precipitate. The entire area will eventually be converted into, and capped by, a low energy, evaporitic, tidal flat-sabkha system. This portion of the Persian Gulf, and its depositional evolution, serves as an excellent Holocene analog for the Mission Canyon Formation, which is discussed in the latter part of this chapter.

Beaches Associated with Higher Energy Shelf Environments and Reefs

Often in carbonate provinces, the shelf is a high energy environment. A shoreline adjacent to such a shelf will not exhibit the normal coarse to fine-grained downward transition. Instead, beach foreshore and shoreface deposits will be underlain by oolitic or bioclastic marine sand sheets or

Figure 13g—Mississippian subaerial crust displaying blacken micrite clasts, multiple fracturing and *in situ* brecciation, pisolites, and micrite coatings (light colored) on black micrite clasts (scale in mm).

Figure 13h—Subaerial crust development along bedding plane and cross-cutting fractures in Mississippian (Newman Formation) of Kentucky.

shoal units (Fig. 17a, c) similar to those that form as marine sand belts and tidal bar belts near the Bahamian shelf margin (Ball, 1967). Hine (1977) and Harris (1979) describe in detail the development of these systems, their internal characteristics, and the morphology of the subtidal sand bodies which this beach type eventually caps.

Beach-dominated islands also form in reefal settings, such as around patch reefs on broad shallow shelves and depositional ramps (Folk and Robles, 1964; Purser, 1973; Flood, 1974; Petta, 1977a, b) or along shelf margin barrier systems (Bebout and Loucks, 1974). This depositional motif existed at the onset of Holocene deposition along the central Trucial Coast (see preceding barrier island discussion), and still exists farther to the west (Fig. 1) where islands are developing around reefs fringing salt domes or other structural highs (Purser and Evans, 1973). The sediments making up and surrounding these beach islands are inevitably composed of coral-algal rich sands, reflecting their derivation from the reefs upon which the beaches have developed. The beach sediments eventually grade into a variety of deeper water, finer grained sediments (for example, open marine mollusk lime packstones, lagoonal pellet-foram lime wackestones, or tidal bar oolite lime packstones-grainstones), depending on which environments flank the reefs (Purser, 1973). However, in areas where bottom currents were swift enough to remove most bottom sediment from the area while the reef grew to near sea level, foreshore deposits may rest directly on lithified reef rock or be separated from it by a thin, high-energy crossbedded sand or rubble zone (Figs. 14d, 17c, d, e).

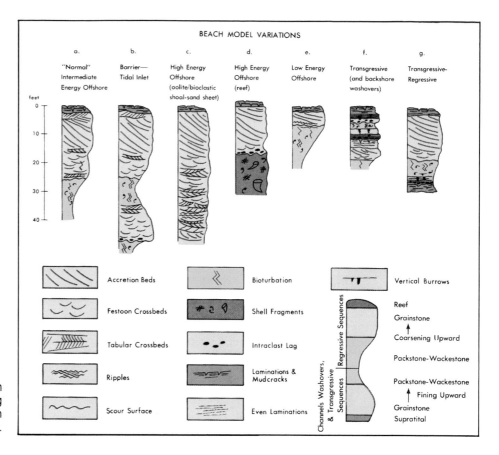

BEACH MODEL VARIATIONS

a. "Normal" Intermediate Energy Offshore
b. Barrier— Tidal Inlet
c. High Energy Offshore (oolite/bioclastic shoal-sand sheet)
d. High Energy Offshore (reef)
e. Low Energy Offshore
f. Transgressive (and backshore washovers)
g. Transgressive-Regressive

Accretion Beds
Festoon Crossbeds
Tabular Crossbeds
Ripples
Scour Surface

Bioturbation
Shell Fragments
Intraclast Lag
Laminations & Mudcracks
Even Laminations

Vertical Burrows

Regressive Sequences
Reef
Grainstone
Coarsening Upward
Packstone-Wackestone
Packstone-Wackestone
Fining Upward
Grainstone
Supratital

Channels Washovers, & Transgressive Sequences

Figure 14—Variations in general beach model that develop as a result of varying the overall depositional environment in which beaches occur.

Low Energy Beaches

Large waves, and thus effective longshore currents, commonly do not develop in lagoons or restricted marine shelves either because of limited fetch, or great loss of wave energy due to frictional drag along a shallow sea floor. The resulting beaches are formed with thin foreshore units (0.1 to 1.5 m thick; Fig. 17f) which grade directly into bioturbated fine-grained lime or dolomite wackestones-mudstones (Fig. 14e) that commonly contain beds or nodules of evaporites.

These carbonate units will carry a very restricted fauna, and in many cases the foreshore beds will be composed of intraclasts, irregularly shaped ooids and pisoids, or the slightly abraded shells of only one species (for example, the gastropod *Battillaria* almost exclusively makes up the beaches developed in Florida Bay). These beaches would be expected to evolve into, and be overlain by, low energy tidal flat and sabkha deposits. They would form thin lenses

which parallel depositional strike, or would be randomly distributed around the mud mounds that developed in the lagoon.

Preservability and Morphologies of Fossilized Beach Facies

Beach complexes may form significant hydrocarbon reservoirs. It is for this reason that it is important to understand which parts of a beach complex might be preserved under different burial conditions. Beach facies stand the best chance of being retained for burial during regressive phases of sedimentation, because deposits of dominantly low energy systems, such as those of lagoonal, supratidal, aeolian, or alluvial floodplain environments, prograde over each lithofacies. Deep channels associated with prograding alluvial or deltaic systems might cut through the underlying beach in places, but for the most part, the beach would be left intact.

Regressive barrier island sand bodies and mainland beach sands

would be predominantly elongate and aligned parallel to depositional strike; these units might have lengths of up to 5 to 50 km with widths of greater than 70 km (see Cretaceous example of strandline beach) in the case of long progradational histories over stable platforms. Most regressive beach sequences, however, tend to be relatively narrow and lenticular along depositional dip sections. In either case, they pinch out landward into lagoonal, sabkha, or continental units, and marineward into lithologies deposited in deeper, lower energy marine environments. The thickness of barrier sequences might reach 15 to 20 m in areas where tidal inlets and deltas formed, but would be less in areas not affected by tidal inlet migration. Tidal inlet sand bodies would also be somewhat pod-shaped perpendicular to the trend of the barrier.

Transgression occurs when the rate of sediment supply to the beach system is much lower than the rate of subsidence or rise in sea level. The

Figure 15a,b—Barrier island depositional motif showing offshore (**a**, upper photo), ebb tidal delta (**e**), flood tidal delta (**f**), tidal channel (**t**), and lagoonal (**b**, lower photo) depositional settings. Tides in offshore photo are microtidal, whereas in the lagoonal photo, tidal activity is intense.

Figure 15c—Basal scour surface of tidal channel in Mississippian (Newman Limestone) barrier sequence, Eastern Kentucky. Unit overlying scour surface is festoon crossbedded coarse-grained intraclast lime grainstone.

Figure 15d—Large festoon crossbeds at base of tidal channel; Lower Cretaceous Edwards Limestone, Texas.

Figure 15e—Depositional couplet displaying basal tidal channel fill unit (**c**) fining upward into dark colored offshore facies (**o**), which in turn is overlain by prograding beach accretion units (**b**).

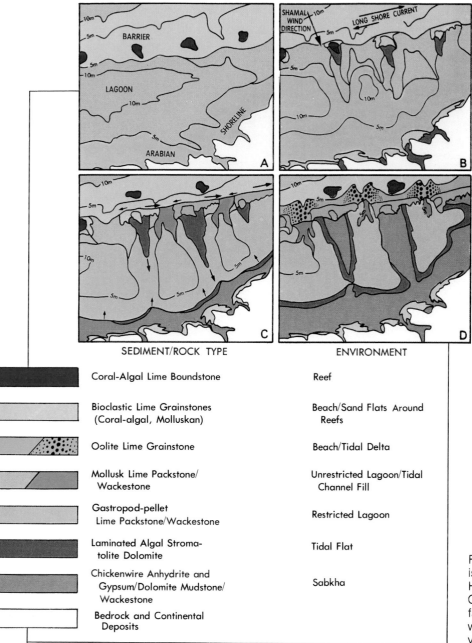

SEDIMENT/ROCK TYPE   ENVIRONMENT

	Sediment/Rock Type	Environment
	Coral-Algal Lime Boundstone	Reef
	Bioclastic Lime Grainstones (Coral-algal, Molluskan)	Beach/Sand Flats Around Reefs
	Oolite Lime Grainstone	Beach/Tidal Delta
	Mollusk Lime Packstone/ Wackestone	Unrestricted Lagoon/Tidal Channel Fill
	Gastropod-pellet Lime Packstone/Wackestone	Restricted Lagoon
	Laminated Algal Stroma- tolite Dolomite	Tidal Flat
	Chickenwire Anhydrite and Gypsum/Dolomite Mudstone/ Wackestone	Sabkha
	Bedrock and Continental Deposits	

Figure 16—Development of barrier island-lagoonal complex through the Holocene along the central Trucial Coast. Evolution of system shows how facies patterns and types change without seaward progradation of environments.

most commonly encountered effect of the transgressive processes is erosion of the sediment-starved beach foreshore and shoreface deposits by wave action, leaving behind a scour surface capped by a thin lag of coarse intraclasts (beachrock clasts) or fossil debris. This lag may in turn be overlain by a thin burrowed or laminated lime grainstone/packstone package which represents deposition in shallow offshore marine or washover fan environments, respec-

tively. If the transgression takes place regionally along the strandline, and landward marine inundation is extensive, a broad, regionally correlative crossbedded sand will develop. This sand would normally fine upwards into other low energy marine units. Although it would pinch out in the zone of maximum updip transgression, its thickness would vary depending on the topography upon which it was deposited, and the relative rates of local transgression. Laporte (1969)

studied a Lower Devonian (New York) transgressive carbonate sequence in which the upper Coeymans Formation (bioclastic lime grainstones and packstones) probably represents reworked strandline deposits which separated lagoonal (Manlius Formation) from deeper offshore marine (Kalkberg Formation) deposits.

In essence, preservability potential of beach complex facies increases with depth below mean low water. Essentially all facies deposited above

Figure 17a—Island development on top of elongate oolite shoal, Bahamas.

Figure 17b—Beach foreshore (**b**) capping tidal inlet (**t**) and oolite shoal (**s**) sequence. Overlying units, subaerial crust (dark layer), and supratidal dolomite (**d**), indicate shoal was subaerially exposed and formed an island.

Figure 17c—Elongate reef forming core of emergent beach-dominated island. Only a thin veneer of carbonate sand and rubble separate the beach foreshore from underlying consolidated reef rock; seaward facing side is toward bottom of photo.

Figure 17d—Coral rubble beach foreshore on island similar to Figure 17b. Rubble consists of coral and fragments of cemented reef rock which accumulated during high-energy storm activity.

effective wave base (beach foreshore and shoreface) are removed during transgression and marine inundation. The most preservable beach depositional facies, in descending order, are: (1) tidal inlet sequences, including both tidal channel and all but the uppermost units of flood-tidal delta deposits. The latter would occur as elongate or semi-circular lenticular sand bodies beneath a regionally transgressive sand sheet, if present; (2) lowermost parts of ebb tidal deltas; and (3) washover fans into the lagoon and backshore zone (these stand the best chance of being preserved if the beach is developed on a reefrock core, or behind a rigidly cemented beach). Small beaches that fringe a lagoon on the mainland and on the backshore side of a barrier might also be preserved because low-energy lagoonal or supratidal deposits will probably cap these sequences before effective wave scour takes place. However, their preservation, as is the case with washover fans, depends on the rate of subsidence and the rate at which transgression takes place. Beachrock formation, early fresh water cementation, and stabilization of beach sediment through soil formation all help preserve the beach foreshore and backshore units from erosion. They therefore stand a better chance of being preserved during marine transgression than do their siliciclastic counterparts.

Electric Log Response

The geophysical log response (SP-resistivity; gamma ray-neutron/density) for the general beach model, neglecting burial effects, is shown in Figure 2. The combined geophysical log shape can be described as an upright cone, and is identical in shape to its quartzose clastic counterpart (Shelton, 1973). However, because of the primary difference in mineralogic makeup of the sediment, early diagenetic alterations in carbonate beaches commonly create dramatic changes in their porosity and permeability characteristics, and thus give rise to geophysical log patterns that are drastically different from those predicted from the simple beach depositional model. It is for this reason that the subsurface geologist will, in most cases, have to initially describe a number of cores or well cuttings for the purpose of correlating log responses to specific lithologies or diagenetic modifications. If a spectrum of logs is available (sonic, neutron, density, etc.), then lithologic types, variations, and associations in specific beach settings can be identified and used to predict the trend and extent of beach associated porosity, and potential zones of stratigraphic and/or diagenetic trapping.

THE GENERAL BEACH ENVIRONMENT AS A POTENTIAL RESERVOIR

Single regressive beach sequences generally average 6 to 10 m thick, with sands comprising the upper 2 to 5 m (Shelton, 1973). Their length can usually be measured in kilometers parallel to depositional strike with the width of the beach measured from hundreds of meters to kilometers in the case of a long progradational history. Significant vertical carbonate beach reservoir development can be achieved through the depositional stacking of beach sequences. Stacking is commonly accomplished at a hinge

Figure 17e—Festoon crossbedded lime grainstones in abrupt contact with underlying lime boundstone facies of rudistid patch reef.

Figure 17f—Low-energy beach on lagoonal side of island; note small waves and very shallow offshore environment; backshore bedding features here would be destroyed by intense rooting.

line, with localized growth faulting, or with continuous salt structural movement in combination with continuous regional subsidence and appropriate sediment availability. In the case of beach sequence stacking, the potential reservoir will not be continuous, but will be broken into a series of isolated, stacked reservoirs separated by the low energy offshore and backshore facies of each beach package, as illustrated diagrammatically in Figure 18. The prime reservoir rock types are obviously the well-sorted grainstones of the foreshore and shoreface. As will be seen in the specific examples to follow, however, all beach associated lithologies are potential reservoir rocks because of many commonly occurring later diagenetic events.

Cretaceous Isolated Island Beach — Texas

Our first specific example is an isolated island beach associated with a Cretaceous rudist reef. This example was chosen because of the excellent quarry exposures and classic development of beach structures and facies. The exposure is located atop Round Mountain northwest of the village of Comanche, Comanche County, Texas (Fig. 19). The beach sequence is developed in the Lower Cretaceous Edwards Formation. It is

associated with an isolated rudist patch reef located on a shallow marine shelf on the eastern flank of the Concho Arch (see Figs. 19, 20). The beach itself is approximately 1 km long and built into normal marine shelf waters about 5 m deep (Fig. 20).

Three-dimensional quarry exposures allow a view of beach geomorphic landforms not commonly seen in the ancient record, such as the large scale beach accretion units (Fig. 21) which mark the progradational episodes of the foreshore. A berm position is indicated by subtle dip toward the backshore at the top of some accretion units (Fig. 22). The development of a thicker supratidal fill in the beach swales between major accretion bedsets is also displayed in Figure 22. A thin recessive clayey and chalky caliche dedolomite unit (Fig. 23) is exposed near the top of the quarry. This caliche dedolomite represents a Cretaceous soil profile (Moore et al, 1972).

Above this soil horizon, and capping the sequence, is finely crystalline, thinly laminated dolomite which is 3 to 7 ft thick and exhibits

mud cracks, rip-up clasts (Fig. 24), and bird's-eye fabrics. This dolomite is thought to represent the supratidal backshore depositional environment which prograded over the soil horizon shortly after it developed across the upper surface of the beach accretion units. The lithofacies developed in the sequence are shown diagrammatically in Figure 25, and can be recognized in the quarry wall photographs (Figs. 21, 22, 23). Figure 25 illustrates the accretion-bedded mollusk lime grainstones below the soil profile and dolomite. Accretion beds dip to the right (seaward) from 8 to 15°. The grainstones are coarse, laminated, and may be graded within individual accretion units (Fig. 26). Large collapse structures and disc-shaped, reworked cobbles (Fig. 27) represent Cretaceous beachrock collapsed and torn up by storms and reincorporated into the sediments. In thin section, these cobbles commonly exhibit originally aragonite microstalactitic beachrock cements (Fig. 28). The lime grainstones of the upper foreshore give way downward into more muddy rocks including mollusk lime packstones and wackestones (Fig. 29). The lower part of the sequence, which represents the marine offshore environment, is marked by small scale

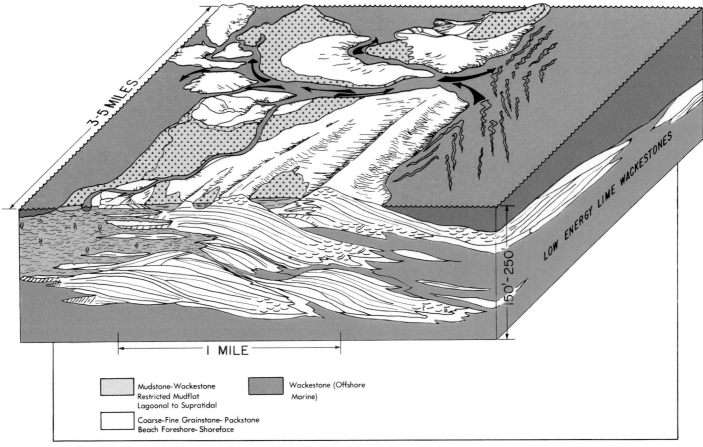

Figure 18—Diagram displaying the three-dimensional distribution of barrier island facies and their relationships to surrounding units in a vertically stacked sequence.

festoon crossbedding, and numerous subhorizontal feeding burrows (Figs. 30, 31). The beach associated depositional environments represented by the rocks of this quarry are summarized in Figure 32.

The porosity present in this beach at the time of deposition must have had a distribution similar to that illustrated in Figure 32, with relatively low effective porosity in the fine-grained supratidal muds, maximum effective porosity developed in the upper foreshore, and a progressive loss of this porosity downward into the offshore marine facies. A similar porosity distribution has been documented in Holocene quartzose clastic beach sequences. The actual measured porosity in this sequence (Fig. 32) is strikingly different from its original inferred distributional pat-

tern and illustrates dramatically the importance of diagenesis in porosity development in carbonate rock sequences. Dolomitization of the supratidal sequence and subsequent removal of remnant calcite resulted in the development of high porosities (up to 30%) and moderate permeabilities (30 to 40 md) (Fig. 33). The upper foreshore, the site of maximum effective porosity at the time of deposition, has undergone extensive early diagenesis, including beachrock cementation and early silicification associated with reflux of waters from the overlying supratidal environments. The result, as shown in Figure 34, is a tightly cemented silicified lime grainstone. The grainstones, packstones, and pelletal wackestones below the intertidal zone were relatively unaffected by very early diagenesis and hence retained maximum primary effective porosity into early burial. Lower Cretaceous unconformities along the margins of the East Texas Basin exposed the se-

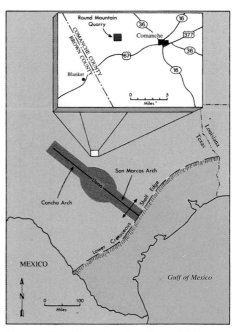

Figure 19—Map showing Round Mountain location and main Early Cretaceous structural elements.

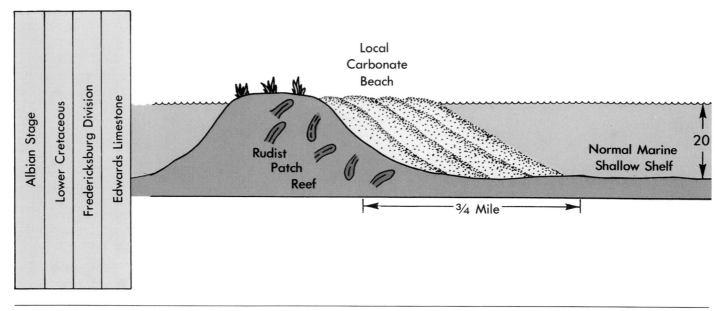

Figure 20—Diagrammatic sketch of beach setting at Round Mountain.

Figure 21—Exposure showing laminated seaward dipping (15°) beach foreshore accretion units. Note tabular foresets between dashed lines near center of photo.

Figure 22—Round Mountain quarry wall showing foreshore accretion beds (**f**), backshore dipping berm sequences (**b**), and thicker supratidal dolomites (**s**) in backshore topographic low.

Figure 23—Beach backshore caliche dedolomite (**c**) between underlying beach foreshore unit and overlying supratidal dolomite.

Figure 24—Laminated and mudcracked supratidal dolomite facies displaying disrupted and truncated laminations near its contact with the underlying caliche dedolomite.

Figure 25—Distribution of lithofacies types in the Round Mountain beach sequence, Comanche, Texas.

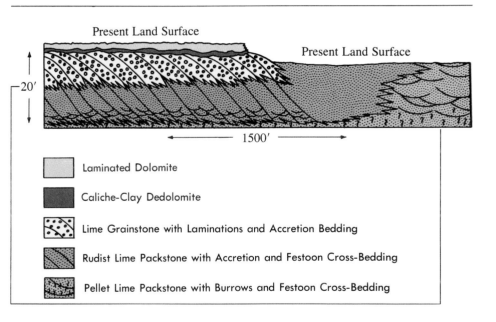

quence to fresh meteoric water before inversion of the aragonitic bioclastic grains of the lower foreshore to calcite, resulting in the development of extensive fabric-selective moldic porosity (Fig. 35). Molds were selectively developed in aragonitic caprinid (rudistid) grains. Calcite cement, originating from the dissolution of adjacent aragonitic grains, occluded primary intergranular porosity. Preserved porosity in the sequence is almost entirely secondary in origin, and all primary porosity has been destroyed. Figures 36 and 37 diagrammatically illustrate the diagenetic history and resulting porosity evolution of this sequence (published originally by Moore et al, 1971).

We will compare the above example with other ancient carbonate beach sequences, pointing out their contrasting depositional settings, sediment types, and diagenetic modifications, especially those which have modified the original primary porosity distribution.

Cretaceous Mainland Beach — Central Texas

The classic example of a mainland beach sequence is developed in the outcrop belt of the Cow Creek Limestone (Fig. 38). This unit was deposited during Middle Trinity (Early Cretaceous) time around the margin of the Llano Uplift in central Texas (Stricklin and Smith, 1972; In-

Figure 27—Cobbles and boulder-sized beachrock clasts at base of foreshore rubble zone.

Figure 26—Evenly laminated mollusk lime grainstone beds of beach foreshore sequence.

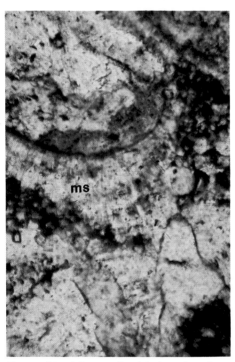

Figure 28—Microstalactitic beachrock cement (**ms**) thought to be originally aragonite (scale equals .125 mm).

den, 1972, 1974). The Middle Trinity is a carbonate–clastic package which represents deposits formed during one transgressive–regressive cycle (Fig. 39) on the broad Lower Cretaceous shelf that extended eastward toward the Gulf of Mexico. Carbonate strata in each Trinity cycle represent the marginal marine and offshore equivalents of alluvial-deltaic units deposited closer to the source areas (Figs. 39, 40).

The Cow Creek Limestone and Hammett Shale represent a high energy beach and its equivalent lower energy offshore deposits, respectively. The beach extends as a continuous regressive sheet of carbonate sand from its pinchout near Marble Falls, Texas, to over 5 m thick at its seaward-most (easternmost) extent in outcrop (Fig. 41); Loucks (1977) has published on its equivalent in the sub-surface to the south-southeast of this area. Rigorous waves and strong longshore currents are indicated by coarse-grained, steeply dipping (8 to 12°), evenly laminated lime grainstone beds (Fig. 42a, b, c) and underlying festoon crossbedded lime packstone-grainstone units (Fig. 43a,

b). Abundant siliciclastics in these deposits and finer grained quartz sands in the shallow offshore lime packstone equivalents (Figs. 40, 44a), indicate significant detrital influx from the nearby source area (Llano Uplift). Oysters and other mollusks constitute most of the bioclastic material in all facies, suggesting that the offshore environments were not conducive to either ooid formation or reef growth. A regionally extensive thin oyster biostrome, at the contact between the Cow Creek Limestone and Hammett Shale, and an overlying lime/packstone shoal deposit containing abundant disarticulated *Trigonia* valves and other large mollusks (Figs. 40, 44b), were the probable sources of most of the beach bioclastic debris. The Hammett Shale consists primarily of muddy lime and dolomite wackestone/packstone (Figs. 40, 44c) and represents units deposited in deeper offshore en-vironments which were below effec-tive wavebase.

The backshore facies of this beach consists of sporadically developed washover deposits (lime grainstones, Fig. 45a), nodular caliche paleosols

(Figs. 13e, 45b) and very thin, disrupted, supratidal marsh and pond deposits (dolomitic *Chara* lime wackestones, Fig. 45b). These grade vertically and laterally landward into alluvial-plain red mudstones and arkosic sandstone channel-fill deposits (Figs. 40, 41). The lack of a well-developed supratidal marsh facies probably reflects a paucity of car-

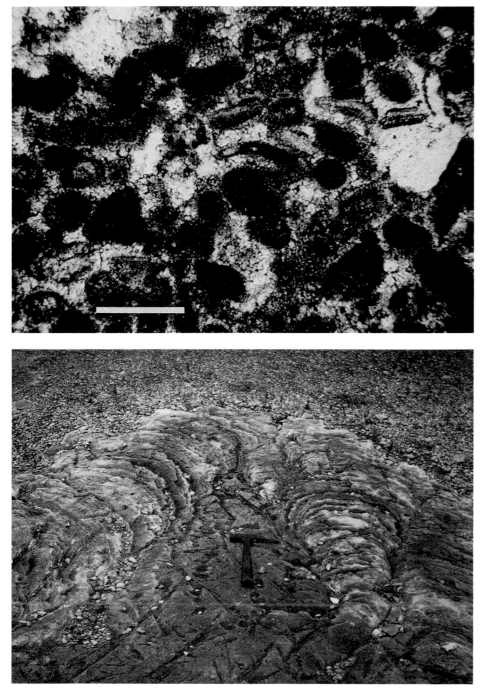

Figure 29—Fine grained, sorted fossil fragment-pellet lime packstone deposited in the offshore environment (scale equals .125 mm).

Figure 30—Small-scale festoon crossbeds developed in sub-beach units; these indicate that the offshore environment was periodically swept by wave-induced currents.

bonate mud in the offshore environment due to extremely high-energy conditions or lack of production.

Diagenesis

The following is an outline of the diagenetic modifications that were imposed on the beach sediment prior to any significant burial.

(1) Beachrock cementation occluded only a minor amount of original primary porosity in the foreshore and shoreface lime grainstones.

(2) Soil-forming processes resulted in the upper part of the foreshore units being cemented with, and replaced by, micrite and dolomicrite, and capped by a nodular caliche (Fig. 45b).

(3) The early development of a stable fresh ground water lens displaced sea water in the sediments to a depth of 10 to 40 ft and brought about the following modifications: (a) abundant secondary bimoldic porosity in the foreshore, decreasing downward into the shoreface sequence; (b) a downward increase, into shoreface and offshore facies, of recrystallization/inversion fabrics in grains that were originally aragonite; (c) a downward decrease in the frequency and thickness of bladed ce-

Figure 31—Branching burrows in lime packstones and wackestones below foreshore indicate deposition in low energy subtidal environment.

Figure 32—Diagram illustrating depositional environments and resultant lithofacies exposed in Round Mountain quarry, Comanche, Texas.

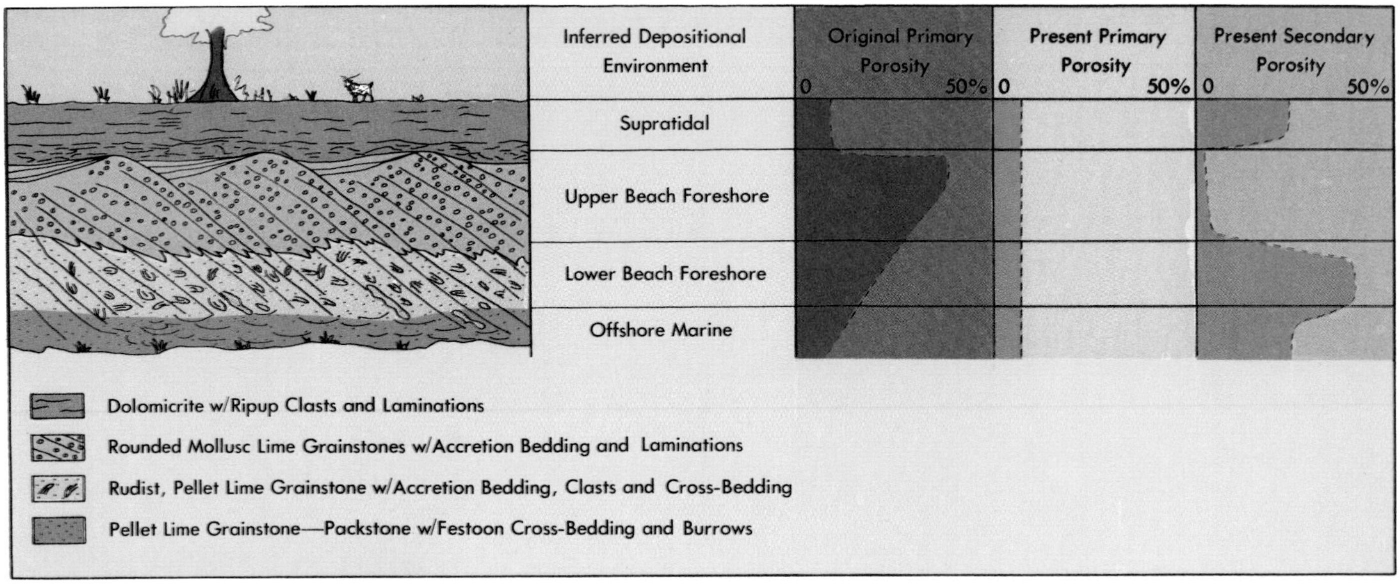

	Inferred Depositional Environment	Original Primary Porosity		Present Primary Porosity		Present Secondary Porosity	
		0	50%	0	50%	0	50%
	Supratidal						
	Upper Beach Foreshore						
	Lower Beach Foreshore						
	Offshore Marine						

Dolomicrite w/Ripup Clasts and Laminations

Rounded Mollusc Lime Grainstones w/Accretion Bedding and Laminations

Rudist, Pellet Lime Grainstone w/Accretion Bedding, Clasts and Cross-Bedding

Pellet Lime Grainstone—Packstone w/Festoon Cross-Bedding and Burrows

Figure 33—Laminated supratidal dolomite showing only rare moldic porosity (white circles); most porosity (30%) is microintercrystalline, however, and is not visible in this photograph (scale equals .065 mm).

ment grain coatings; and (d) plugging of some secondary porosity and small primary pores with iron-free equant calcite cements (Fig. 43b).

All of these diagenetic alterations and gradients are the result of undersaturated (with respect to aragonite and calcite) rain water becoming progressively more saturated with respect to calcite as it percolated downward through the sequence, and the mixing of this water with sea water near the fringes of the water lens, thereby increasing the magnesium content of the water and inhibiting the precipitation of calcite cements.

During progradation of the alluvial system over the beach, a regional ground water flow system replaced the pre-existing local one and caused: (1) dolomitization of the underlying Hammett Shale offshore sequence of clayed lime wackestones and mudstones, thus creating abundant secondary microintercrystalline

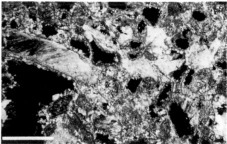

Figure 34—Partially silicified lime grainstone; silica cement fills primary and moldic voids after initial cementation with coarse bladed calcite cement, of presumably fresh water origin (scale equals .125 mm).

Figure 35—Extensive moldic porosity in the lowermost portion of the beach foreshore sequence; porosity is developed in leached rudistid (mollusk) grains. Note the lack of any remaining primary porosity between the molds (scale equals 25 mm).

Syndepositional Stage (Mixed Waters)

Diagenetic Processes	Mineralogy	Primary Porosity 0 % 50		Secondary Porosity 0 % 50	
Dolomitization	Dolo. & Arag.				
Heavy-Cementation Calcite Aragonite Silicification	Calcite Aragonite Silica				
Micrite Rim Light Cement	Aragonite Calcite				
	Aragonite Mag. Calcite				

Marine-Evaporative Brines | Fresh Meteoric and Marine

Figure 36—Diagenetic processes and products resulting from the syn-diagenetic modification of the Round Mountain beach sequence. Note especially the modification to porosity distribution.

porosity between the dolomite rhombs; and (2) precipitation of iron-free equant cement (in the foreshore) and iron-rich equant cement (lower foreshore and underlying units) occluding all remaining biomoldic and primary intergranular porosities.

Mississippian of Kentucky

Mississippian carbonates (Newman Formation) were deposited in a bar-rier complex of beaches and tidal bar belt deposits near the axis of the Waverly Arch in eastern Kentucky (Figs. 46, 47a, b) (Ferm et al, 1971). They separate pelletal lime mudstones and wackestones of lagoonal origin to the south from open marine, thinly bedded bioclastic carbonates and red and green shales to the north (Fig. 47b). This barrier complex illustrates the effect that local structural and erosional topography has on both beach development (Fig. 47b), and the early diagenesis within beaches and surrounding deposits (Canfield, 1974). Studies on similar Mississip-pian shoal deposits have been carried out in Indiana by Carr (1973) and in the Bridgeport oil field of eastern Il-linois by Choquette and Steinen (1980).

Most of the barrier complex is made up of lime grainstones contain-ing abundant, well-rounded and sorted intraclasts, oolites, bryozoans and mollusks (Fig. 48a, b). Individual depositional packages of these are up to 20 ft thick and 1 km long, but their width is unknown because of the lack of lateral outcrop control. They contain medium- to large-scale fes-toon crossbeds in their central and

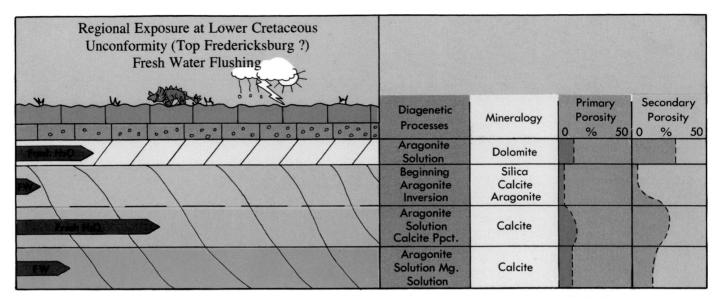

Figure 37—Late stage diagenetic model for the Round Mountain beach sequence, and final porosity distribution.

Diagenetic Processes	Mineralogy	Primary Porosity 0 % 50	Secondary Porosity 0 % 50
Aragonite Solution	Dolomite		
Beginning Aragonite Inversion	Silica Calcite Aragonite		
Aragonite Solution Calcite Ppct.	Calcite		
Aragonite Solution Mg. Solution	Calcite		

Regional Exposure at Lower Cretaceous Unconformity (Top Fredericksburg ?) Fresh Water Flushing

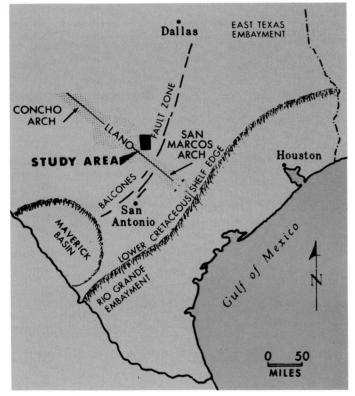

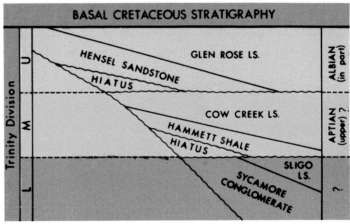

Figure 39—Stratigraphic relationships between Trinity age rocks on east flank of Llano Uplift; units on the right are carbonate-rich and were deposited farther away from the source areas than those on the left (after Lozo and Stricklin, 1956).

Figure 38—Diagram showing the Cow Creek Limestone study area on the eastern flank of the Llano Uplift, just north of the San Marcos Arch.

basal portions. These crossbeds are oriented perpendicular to, and grade upward and laterally into large scale, low angle (less than 10°) accretion beds (Figs. 47d, 48c) that are commonly laminated, or graded, and contain occasional keystone vugs (Fig. 7d). The basal and distal ends of these accretion beds are usually com-posed of poorly sorted fossil frag-ment lime packstone (Fig. 48e). Beaches and tidal channels developed on the immediate flanks of the topographic high, whereas tidal bar belt units formed in a topographic low to the east and eventually built up into islands defined by the beach facies and underlying units (Fig. 48d, f).

Beaches throughout this complex are capped by one or a combination of: (1) convex-up spillover crossbeds representing storm washovers; (2) low energy supratidal carbonates; (3) green clayey insoluble residue seams, and laminated subaerial calcrete crusts (Figs. 13g, 49) containing

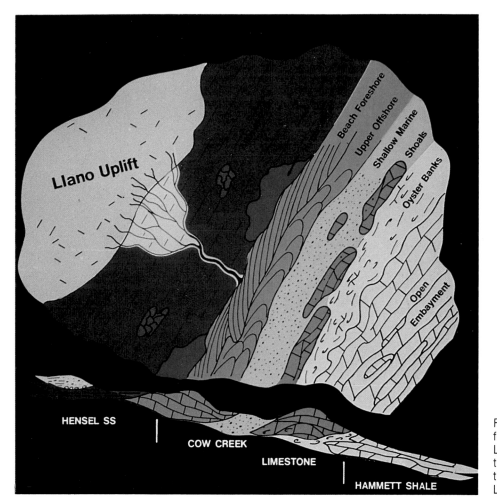

Figure 40—General depositional model for Hensel Sandstone-Cow Creek Limestone-Hammett Shale. This deposition pattern generally holds true for all the Trinity sequences deposited near the Llano Uplift.

coated grains, teepee structures (Fig. 50) and a variety of vadose features (Fig. 14d); and (4) low energy open marine (red and green shales; lime packstone) or lagoonal (lime wackestone) facies (Fig. 48d). The fact that low energy subtidal facies directly overlie these beaches without a basal lag indicates that beaches can be preserved during transgression if they form in a lagoonal setting or are protected in some manner (for example by spit accretion, early cementation) from erosion by direct wave approach.

Diagenesis

The diagenetic history of the sequence is outlined as follows.

(1) Beachrock cementation occluded minor intra- and intergranular porosities in the beach and shoal units (Fig. 12b).

(2) Intensive weathering took place in the upper shoal and beach deposits before burial (Fig. 49), resulting in: (a) formation of a thick brecciated, and partially silicified, laminated subaerial crust containing vadose (stalactitic, meniscus) cements, root tubules, and numerous other carbonate soil features (Fig. 13g); and (b) development of narrow bedding plane (desiccation?) fractures and small subvertical solution channels extending downward through the shoal and into the offshore bioclastic lime packstone facies (Fig. 13h).

(3) During the weathering period, a fresh water lens occupied the shoal and small portions of the underlying and surrounding facies, and caused (Fig. 49): (a) the precipitation of abundant iron-free bladed cements in only the upper part of the shoal (Fig. 48a); and (b) slightly later precipitation of iron-free equant cement which plugged much of the remaining primary porosity (Fig. 48a).

Little if any secondary porosity was created during this diagenetic episode except for the solution channels and fractures mentioned above.

During shallow burial, the lower part of the shoal and underlying packstone facies underwent compaction (tight packing fabric) and pressure solution (abundant grain interpenetrations and microstylolites; Fig. 48b) in response to the overburden pressure being exerted by the weight of the overlying prograding facies. Compaction effects were minimal in the upper shoal and beach grainstones because of cementation and formation of a rigid framework during earlier diagenesis. Cementation by iron-rich equant cements occluded all remaining porosity at a slightly later time (Fig. 48b).

Cretaceous Shelf Margin — Texas

The subsurface Stuart City Trend of the Texas Gulf Coast is a facies association that represents the

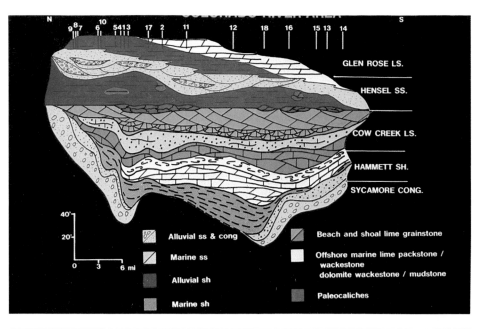

Figure 41—Stratigraphic cross-section of Middle Trinity rock sequence along eastern margin of Llano Uplift.

Figure 42a—Sequence of beach foreshore accretion beds dipping toward background of photo and indicating progradation to the east.

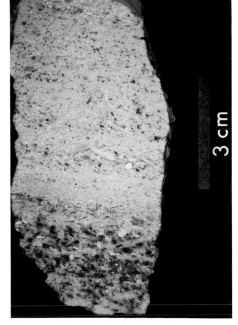

Figure 42b—Evenly laminated, well sorted, coarse- and medium-grained mollusk lime grainstone.

Aptian-Albian/Lower Cretaceous shelf margin (Figs. 19, 51). On the shelfward side of this complex (Bebout and Loucks, 1974), beaches are abundant in the uppermost part of the Edwards Formation (Figs. 51, 52). They occur in stacked, thin, narrow regressive or transgressive sequences parallel to depositional strike. These potential reservoir packages attain a thickness of 75 m and have a length of 10 to 15 km, and width of up to 4 km. Their distribution is largely controlled by the position of older structural highs and organic reef mounds, both of which provided the minor but significant topographic relief to the sea floor necessary for the intense winnowing action of waves and currents, and the concentration and buildup of carbonate sands in these areas.

The resultant sand bodies form significant gas reservoirs throughout a large part of the trend. The best production from these units occurs in beach foreshore and washover lime grainstones, in a relatively narrow zone paralleling depositional strike. These grainstones pinch out into tight finer-grained supratidal and lagoonal units in a shelfward direction.

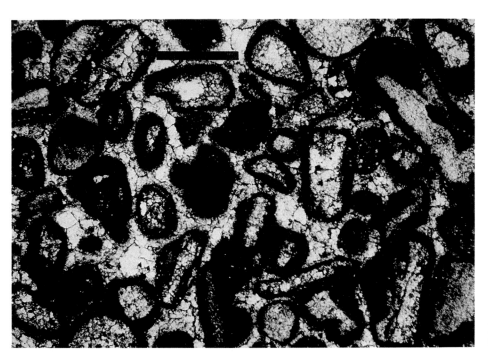

Figure 42c—Well sorted and rounded medium grained lime grainstone (scale equals .25 mm).

Figure 43a—Medium-large scale festoon crossbeds oriented perpendicular to trend of beach foreshore units.

Figure 43b—Medium-coarse, poorly sorted mollusk lime grainstone containing abundant quartz sand (scale equals .5 mm).

Toward the shelf edge they grade into offshore lime wackestones and packstones carrying a diverse marine fauna.

In cores, porous beach foreshore deposits .3 to 3 m thick, are typically tan, and have even horizontal laminations (Fig. 53b, c) or are homogeneous. They are composed of well-sorted or bimodal fossil fragment-intraclast-algal coated grain lime grainstones (Fig. 53b, c), many of which display large fenestral-like,

cement-filled, keystone vugs (Fig. 53a). In regressive sequences, these grainstones are overlain by thin supratidal algal-laminated lime mudstones and fractured soil horizons (Fig. 54a). They fine downward into offshore, bioturbated, algal-Toucasid and Chondradontid lime packstones (Fig. 54b, c) or lagoonal fine-grained, wispy bedded, calcisphere-algal-pellet lime wackestones (Fig. 54d, e). The lack of a crossbedded shoreface sequence indicates that longshore cur-

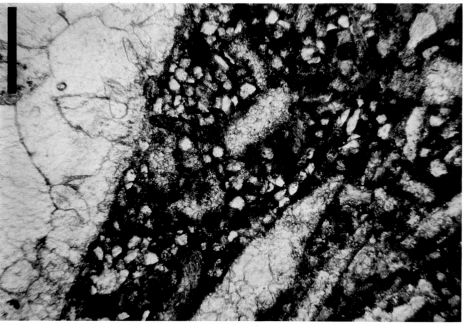

Figure 44a—Very arenaceous, bimodally sorted, mollusk lime packstone from shallow offshore marine environment; bimodality is probably due to transport of fine quartz sand into an area where coarse mollusk fragments were being produced; Middle Cow Creek Limestone (scale equals .5 mm).

Figure 44b—Upper surface of shoal lime packstone deposit displaying abundant large *Trigonia* values; Lower Cow Creek Limestone.

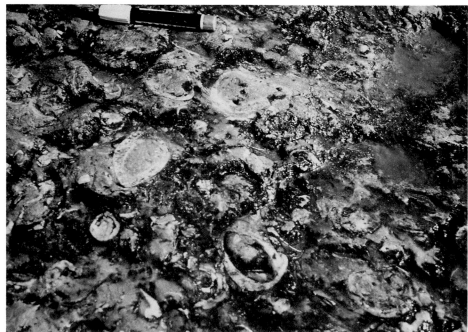

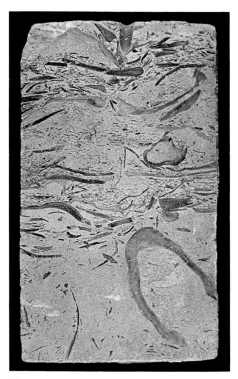

Figure 44c—Bioturbated, poorly sorted oyster fragment clayey dolomite wackestone-packstone; unit was deposited in very low energy environments; Hammett Shale; length of slab equals 10 cm.

rents were either nonexistent or so weak that deposits laid down by them were homogenized by later bioturbation. Tidal inlet fills (.3 to 7 m thick) are somewhat coarser overall, and contain abundant bimodal, oncolite lime grainstones (Fig. 55a, b). They typically become finer grained upward from a sharp basal scour contact with a wide variety of lithologies, and are overlain by beach foreshore,

supratidal, or shallow restricted lagoonal deposits.

Transgressive beach sequences are made up of multiple storm washovers that overlie scoured supratidal deposits. Individual washovers are thin (.3 to 1 m thick) and are represented by a lowermost thin, fossil-intraclast lime packstone lag and an overlying finer grained and better sorted foreshore-type lime

Figure 45a—Top of beach accretion units (dipping toward right) is partially truncated by channel deposit (left center) which in turn is overlain by landward (left) dipping washover unit (arrow).

Figure 45b—Nodular caliche paleosol, composed of diagenetically formed lime wackestone, overlain by distorted dolomite mudstone of supratidal marsh and pond origin.

grainstone. These units are capped by either supratidal or lagoonal (tan calcisphere and *Chondradonta* bearing black lime wackestones) sediments. Transgressive beach washovers cannot be differentiated from associated small tidal channel deposits in individual cores because the sediments and sequence for both are essentially the same.

Although these beaches occur in close association with rudist reefs, they contain only a limited on-reef fauna, thus suggesting that they formed in a somewhat restricted backreef environment. This might be the result of transport barriers, such as deeper water channels, across which sediment could not be moved shelfward from the reefs.

Diagenesis

The generalized early to late diagenetic modification of this sequence is outlined below.

(1) Mild weathering episodes created minor fracture porosity in rocks deposited in the supratidal zones of the stacked barriers (Fig. 54a).

(2) Fresh water lenses developed in these barrier beach complexes causing the leaching of aragonitic mollusk fragments and minor early cementation by bladed calcite cement. Most moldic porosity was eventually filled during this stage with bladed and equant cements (Fig. 53c).

(3) Shortly after deposition (Cenomanian time) and during shallow burial, sea level fell

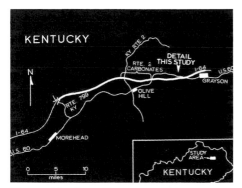

Figure 46—Location map of Newman Limestone barrier sequences in eastern Kentucky.

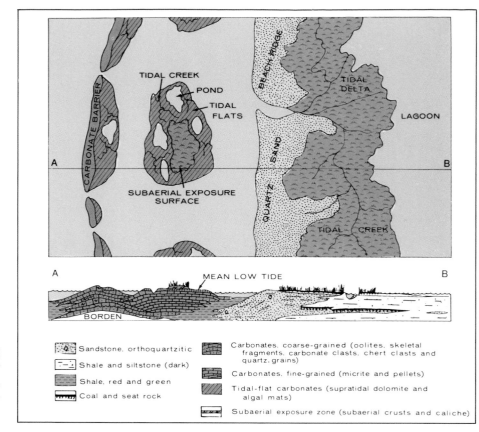

Figure 47a—General depositional model for offshore, barrier, and back barrier carbonate sequences (from Ferm et al, 1971).

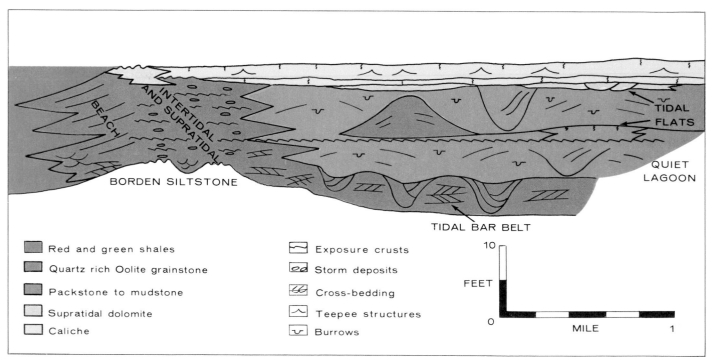

Figure 47b—Cross-section showing local facies variations in carbonates around erosional topographic high on Borden Siltstone.

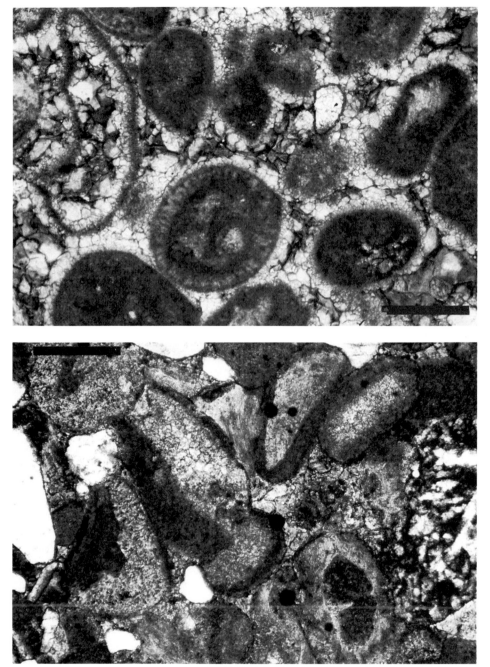

Figure 48a—Well sorted oolite-intraclast lime grainstone from barrier beach; early diagenetic precipitation of iron-free (pink) bladed and equant spar occluded most porosity; later precipitation of iron-rich (purple) equant calcite spar plugged remaining porosity (scale equals .25 mm).

Figure 48b—Sorted, oolite-fossil fragment lime grainstone from lower part of barrier sequence; because of the lack of precipitation of early iron-free cement a rigid framework was not created in the sediment; grain interpenetrations and close packing fabric resulted from compaction; late iron-rich (purple) equant calcite occluded remaining pore space (scale equals .25 mm).

significantly, resulting in the nearly total exposure of the Lower Cretaceous shelf. Fresh water displaced the previous pore waters in the rocks beneath this area and gave rise to a massive dissolution episode in the shelf margin zone. In the beach and tidal channel lime grainstone-packstone facies, earlier cements were partially dissolved and secondary porosity was created in the form of vugs (Fig. 55b), and more important-

ly, as microintercrystalline porosity in the algal coatings of individual grains. Solution channels and larger cavernous voids formed in the fine grained lagoonal and offshore marine facies surrounding the beaches.

(4) During deeper burial diagenesis, very coarse equant spar was precipitated in the large secondary voids, leaving only microintercrystalline porosity and minor small vugs as the effective porosity in the

beach sequence. Oil inclusions are present in the equant spar suggesting that hydrocarbon migration into the sequence occurred at approximately the same time as final cementation.

Mission Canyon Formation — Williston Basin

The Mission Canyon Formation is one of a series of regressive cycles that was deposited in the Williston Basin during Mississippian time (Fig.

Figure 48c—Low angle accretion beds of tidal bar and beach sequence. Dark unit in center of outcrop is lime packstone partially altered by weathering into subaerial crust.

Figure 48d—Lower half of diagram represents depositional characteristics noted in outcrop photos 48c and 48f.

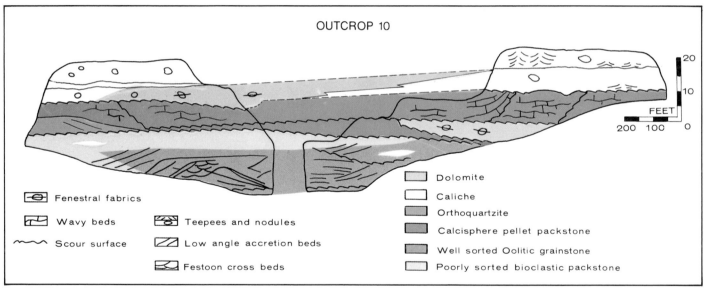

OUTCROP 10

Fenestral fabrics

Wavy beds

Scour surface

Teepees and nodules

Low angle accretion beds

Festoon cross beds

Dolomite

Caliche

Orthoquartzite

Calcisphere pellet packstone

Well sorted Oolitic grainstone

Poorly sorted bioclastic packstone

FEET
200 100 0

56a, b). Each cycle has a somewhat unique set of subfacies and porosity that characterize it, but along the Mississippian basin all margins contain a barrier system, a carbonate mud and evaporite filled lagoon, and low energy offshore carbonate muds (Fig. 56b, c). The cycles were formed in an environmental setting very similar to the Persian Gulf Coast (Purser and Evans, 1973) and presumably broad sabkhas extended outward from the basin margins.

The Mission Canyon is one of the most prolific hydrocarbon producers

of these cycles (Fig. 56c). It produces from many fields along the periphery of the northern half of the basin, and significant fields are presently being discovered in the southwestern part of the basin. Wittstrom and Hagemeier (1978) and Lindsay and Kendall (1980) have studied in detail the Mission Canyon sequence at Little Knife Field (Fig. 56a) and Gerhard and others (1978) have reported on the Glenburn Field in Bottineau County. Pinchouts of porous facies into the Bottineau evaporite (anhydrite), combined with anhydrite plugging, are the

most common reservoir seals in the Mission Canyon in both areas. The following discussion is a summary of the lithologic and reservoir characteristics that typify the sequence in the southwestern portion of the basin.

Either barrier sequences or lagoonal carbonates are found immediately beneath the Bottineau anhydrite (Fig. 56c). Barriers actually ranged from beaches to medium and low energy tidal flats depending on the amount of wave and tidal current energy the strandline was subjected

Figure 48e—Fine-medium grained, poorly sorted, pellet-fossil fragment lime packstone from basal and distal ends of accretion units in 48c (scale equals .25 mm).

Figure 48f—Tidal bar belt (**t**) sequence grading upward in middle of photo into beach (**b**) unit which is partially altered into and capped by a subaerial crust (**sc**) and supratidal dolomite (**s**).

to. Beaches are made up of buff, well-sorted and rounded pisolitic-oolitic-intraclastic lime grainstones (Fig. 57a, b, c), whereas tidal flat units are buff to brown, poorly sorted, lime packstones to wackestones with fenestrae and desiccation fractures (Fig. 57d). Both attain a maximum thickness of approximately 6 m. The ooids and pisoids in these deposits typically display highly

irregular micritic laminations and radial-fibrous structure, indicating that they formed in a hypersaline low energy environment (Loreau and Purser, 1973); however, many may have formed as a result of weathering and the formation of paleosols (Gerhardt et al, 1978). Units containing these grains have previously been referred to in the literature as algal reefs (Hansen, 1966).

Beaches and tidal flats grade downward on the basinward side into a restricted marine, tan, bioturbated and wispy laminated, mollusk-spicule-calcisphere dolomite wackestone/packstone (Fig. 58a, b), which was deposited in a restricted marine environment. Below this facies are brown and black, laminated and bioturbated, siliceous bryozoan-brachiopod-crinoid lime packstones

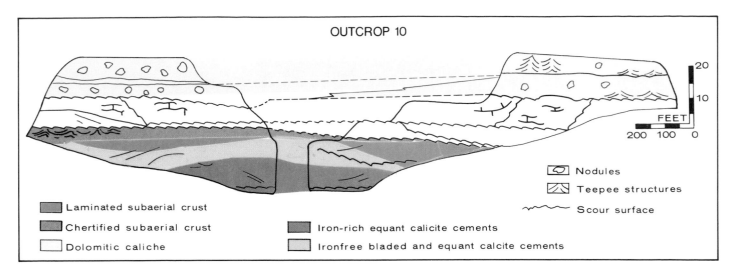

OUTCROP 10

20
10
FEET
200 100 0

☐ Nodules
⬚ Teepee structures
∿ Scour surface

■ Laminated subaerial crust
■ Chertified subaerial crust
□ Dolomitic caliche
■ Iron-rich equant calcite cements
□ Ironfree bladed and equant calcite cements

Figure 49—Diagenetic modifications to tidal bar belt and beach sequence. Position of permanent ground water table can be approximated by contact between iron-rich equant and iron-free bladed calcite cements.

Figure 50—Teepee structures developed in supratidal unit as a result of calichification.

(Fig. 58c, d), which presumably represent the open marine environment in the deeper part of the basin. In a landward direction, the barrier system grades into tan homogeneous and burrowed calcisphere-pellet dolomite mudstones and wackestones of restricted lagoonal origin (Fig. 59a, b, c). Both the offshore and lagoonal carbonate facies are similar to those that developed in association with Mississippian barriers in the Illinois Basin (Choquette and Steinen, 1980). Thin supratidal dolomites and pisolitic laminated subaerial crusts (paleosols) cap, and are intermixed with, the uppermost lagoonal and barrier facies.

The lagoonal dolomites commonly contain abundant microintercrystalline and moldic porosity (Fig. 59a, c) and form the main reservoirs at Rough Rider, Little Knife, Fryburg and many other fields. The high energy beach sequence proper, which is normally not dolomitized, contains little, if any, predictable porosity.

Diagenesis — Mission Canyon Formation, Little Knife Area, and Glenburn Field, Bottineau County

A comparison of the diagenesis in two widely separated reservoir sequences follows below.

(1) Minor beachrock cementation

occluded primary porosity in the barrier beach oolite-pisolite lime grainstones and packstones.

(2) Weathering took place mostly in the backshore and associated tidal delta areas, forming subaerial crusts with pisolites and other paleosol features.

(3) In some areas (Bottineau County oil fields), partial dolomitization took place in the backbarrier supratidal zone, thus creating excellent reservoirs with secondary porosity in paleosols, tidal delta, supratidal, and high subtidal facies.

(4) Cementation in poorly developed fresh?–brackish? water lenses filled most primary pores in the

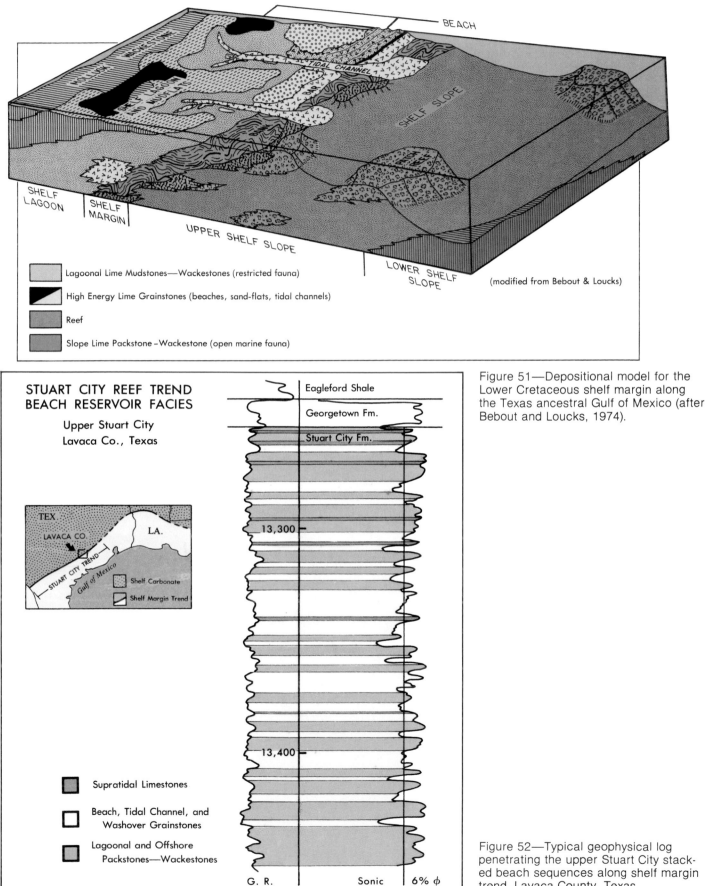

BEACH

HOLLOW WACKESTONE

DEEP LAGOON AND MUDFLATS

TIDAL CHANNEL

SHELF SLOPE

SHELF LAGOON   SHELF MARGIN

UPPER SHELF SLOPE

LOWER SHELF SLOPE

(modified from Bebout & Loucks)

Lagoonal Lime Mudstones—Wackestones (restricted fauna)

High Energy Lime Grainstones (beaches, sand-flats, tidal channels)

Reef

Slope Lime Packstone—Wackestone (open marine fauna)

Figure 51—Depositional model for the Lower Cretaceous shelf margin along the Texas ancestral Gulf of Mexico (after Bebout and Loucks, 1974).

STUART CITY REEF TREND BEACH RESERVOIR FACIES

Upper Stuart City
Lavaca Co., Texas

Eagleford Shale

Georgetown Fm.

Stuart City Fm.

13,300

13,400

TEX.
LAVACA CO.
LA.
STUART CITY TREND
Gulf of Mexico

Shelf Carbonate

Shelf Margin Trend

Supratidal Limestones

Beach, Tidal Channel, and Washover Grainstones

Lagoonal and Offshore Packstones—Wackestones

G. R.   Sonic   6% φ

Figure 52—Typical geophysical log penetrating the upper Stuart City stacked beach sequences along shelf margin trend, Lavaca County, Texas.

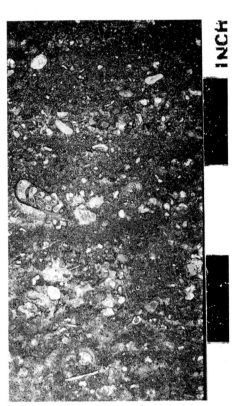

Figure 53a—Foreshore deposit of laminated fossil-pellet-algal coated grain lime grainstone. Coarser grained laminations display large, somewhat irregular keystone vugs now filled with sparry calcite cement; abundant microintercrystalline porosity in algal coatings on grains (scale in inches).

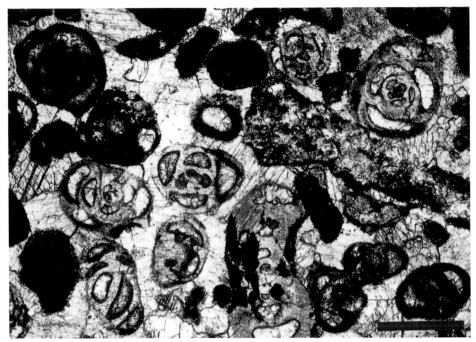

Figure 53b—Well-sorted pellet-miliolid lime grainstone from foreshore sequence; extensive recrystallization (especially miliolids) due to early fresh water diagenesis (scale equals .25 mm).

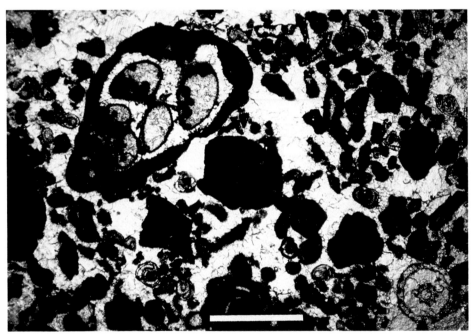

Figure 53c—Medium-very coarse fossil fragment-intraclast lime grainstone; gastropod has thick algal coating; abundant microintercrystalline porosity in algal coatings and intraclasts; loose packing is probably due to early diagenetic total grain dissolution, and partial collapse of loosely cemented sediment (scale equals 1.0 mm).

Figure 54a—Miliolid lime mudstone of supratidal origin displaying vertical fractures, small solution channels, and wispy, dark colored laminations formed through soil processes (scale in inches).

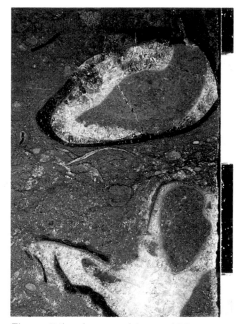

Figure 54b—Lagoonal toucasid lime packstone-wackestone with finer grained matrix of poorly sorted pellets, algal coated grains and fossil fragments (scale in inches).

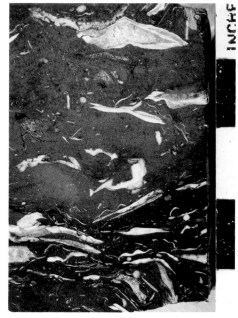

Figure 54c—Lagoonal chondradontid lime packstone-wackestone with fine grained fossil fragment and pellet-rich matrix (scale in inches).

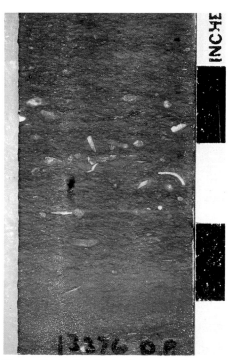

Figure 54d—Restricted lagoonal tan, wispy-laminated, fossil fragment-algal-pellet lime wackestone (scale in inches).

grainstone-packstone facies in fields such as Little Knife; minor excellent (20%) primary porosity is occasionally preserved (Fig. 57a), but generally cannot be predicted. In the Glenburn Field the beach sequence contains abundant primary porosity, thus making the beach lime grainstone facies one of the main reservoirs (Gerhard et al, 1978).

(5) During shallow burial beneath the Bottineau Anhydrite, dolomitization gave rise to biomoldic porosity (in calcispheres, spicules, and mollusks) and microintercrystalline porosity in the lagoonal and restricted marine fossil wackestones and mudstones (Figs. 58, 59). In the Little Knife area, these facies make up the main reservoir and have up to 15% porosity and good permeability. Dolomitization probably took place through seepage refluxion, a process in which lagoonal high-density, magnesium-rich brines percolate seaward through porous sediments, replacing them with dolomite along the way.

(6) During and shortly after dolomitization, anhydrite replaced grains and cements, and plugged much of the remaining primary porosity as well as some of the secondary porosity created through dolomitization.

(7) Deeper burial diagenesis resulted in minor dissolution of anhydrite, formation of stylolites, and fracturing.

CONCLUSIONS

Carbonate beach units are probably quite common in limestone sequences recording deposition on shallow marine platforms. They have only rarely been recognized, and have more than likely been grouped together with other calcarenite units of diverse origins, even though they carry a distinctive suite of sedimentary structures which readily allows them to be identified as beach deposits. Carbonate sand deposits of beach origin may be locally developed around structural or topographic highs, patch reefs, or shoals, or may extend for miles essentially along the depositional strike of a basin or a fault block system. Other depositional components of beach complexes (tidal deltas and channels, washover fans)

Figure 54e—Restricted lagoon calcisphere-foram-pellet lime wackstone; miliolid foram and calcispheres (upper right) have irregular algal coatings (scale equals .25 mm).

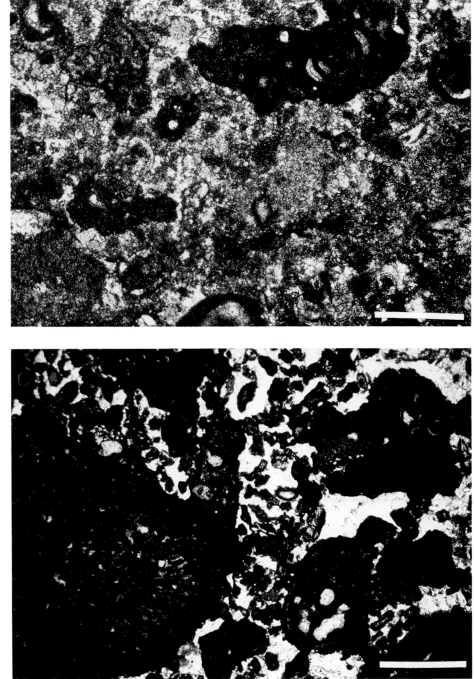

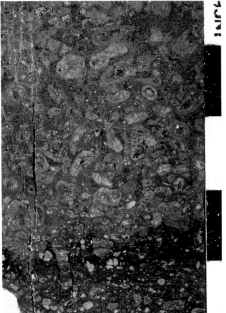

Figure 55a—Tidal channel deposit of bimodal oncolite lime grainstone (scale in inches).

Figure 55b—Bimodal, algal oncolite lime grainstone with fine-medium grained pellets and fossil fragments (scale equals 1.0 mm).

occur as lobes of carbonate sand which extend up depositional dip. Individual beaches may attain a thickness of up to 10 to 15 m, depending on energy conditions and the depth of water they prograde into.

In a regressive sequence, accretion beds, which are composed of laminated lime grainstone, mark the beach foreshore (wave swash) zone,

and are capped by any lithology carrying evidence of subaerial exposure (for example, paleosols, mudcracked supratidal dolomite). Commonly, the unit immediately beneath the laminated lime grainstones is a more poorly sorted lime grainstone to packstone which displays tabular and festoon crossbedding and represents deposition by longshore currents in the shoreface zone of the beach.

Whether or not this or other units are present in this stratigraphic position depends on the overall depositional setting (sea floor topography, wave and current energy flux) of the beach. In extremely low energy settings a crossbedded unit will not be present, and in reefal settings, beach foreshore units may directly overly a rubble zone which caps the lithified reef. In barrier island or high-energy shoal en-

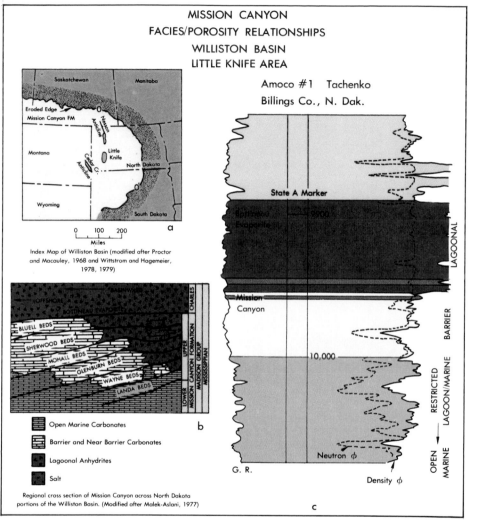

MISSION CANYON
FACIES/POROSITY RELATIONSHIPS
WILLISTON BASIN
LITTLE KNIFE AREA

Amoco #1 Tachenko
Billings Co., N. Dak.

Index Map of Williston Basin (modified after Proctor
and Macauley, 1968 and Wittstrom and Hagemeier,
1978, 1979)

Regional cross section of Mission Canyon across North Dakota
portions of the Williston Basin. (Modified after Malek-Aslani, 1977)

Open Marine Carbonates

Barrier and Near Barrier Carbonates

Lagoonal Anhydrites

Salt

Figure 56—Mission Canyon formation
location map and general stratigraphic
relationships.

Figure 57a—Well-sorted, fine-medium
grained oolite-intraclast lime grainstone.
Pores are partially filled with early
calcite cement, but abundant primary
porosity remains (scale equals 1.0 mm).

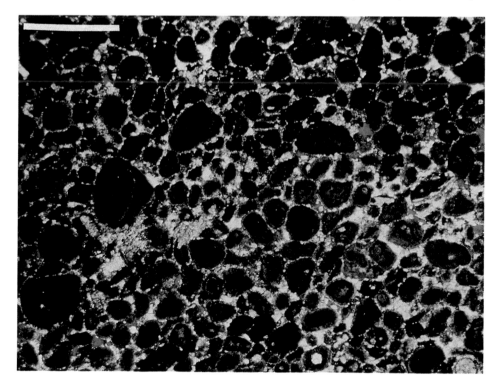

vironments, a thick (5 to 20 m) crossbedded sequence, which represents tidal channel-tidal delta deposition, or the carbonate shoal unit itself, will underlie the beach foreshore facies. Evidence of exposure must be present in order to validate the interpretation that any of the above sequences represents a beach. Early cementation, by either marine, mixed, or fresh pore waters helps stabilize the beach sands and prevent their removal by erosion.

Transgressive beach depositional units are rarely preserved intact because they are destroyed by wave scour and erosion as the strandline position migrates landward. Instead, the strandline facies is represented by a lag of fossil fragments or intraclasts which occurs directly above a surface scoured into the underlying (typically

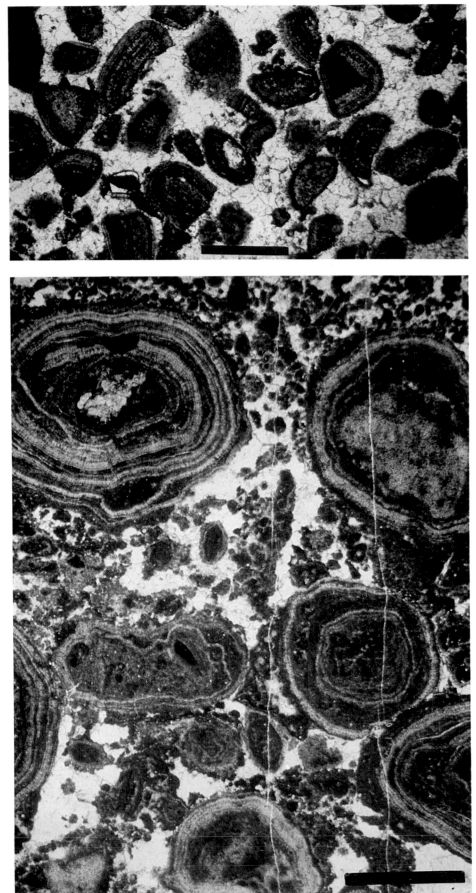

Figure 57b—Sorted oolite fragment lime grainstone. Most ooids have fibrous structure arranged radially about their nuclei, a factor which contributed to their fragmentary nature (scale equals .5 mm).

Figure 57c—Bimodal pellet-oolite-pisolite lime grainstone. Irregular fibrous laminae of pisoids indicates deposition in a hypersaline environment (scale equals 1.0 mm).

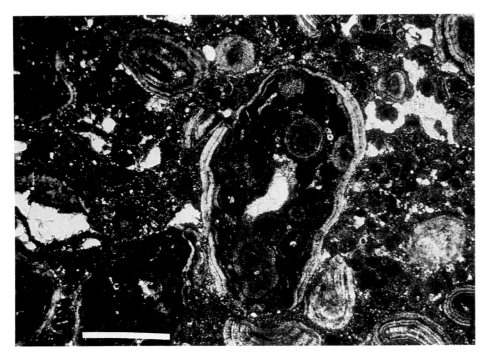

Figure 57d—Poorly sorted intraclast-pellet-pisolite lime packstone-wackestone of low energy tidal flat sequence (scale equals 1.0 mm).

Figure 58a—Tan, bioturbated, pellet, dolomite wackestone with wispy, organic-rich partings (scale in inches); restricted marine.

more landward deposited) facies. This lag is overlain by a thin laminated or bioturbated carbonate sand which fines upward into an offshore facies. In the end, the transgression is a fining upward, relatively continuous sheet of sand that extends over the area affected by the transgression.

Except for those beaches developed on reefs, mudmounds, salt diapirs, or other structurally controlled topographic highs, most beach dominated sequences are parallel to the basin margins, and typically increase in both frequency and gross thickness away from the center of the basin. Beaches may not be present continuously along depositional strike at any one time because of varying energy conditions and the availability of sand size (and coarser) carbonate in the offshore areas. If its progradational history across the shelf was interrupted frequently, it will be represented by carbonate sand lenses. Oil and gas production from these lenses is often from stratigraphically formed traps because the beach units are overlain and grade laterally updip into nonpermeable evaporite, lime and dolomite mudstones, or shales. If a beach dominated strandline has had a long uninterrupted progradational history, it will occur as a relatively

continuous sheet sand, separating landward from more open marine facies.

Cementation and other early diagenetic processes such as neomorphism and solution, give rise to dramatic changes in the distribution of porosity and alter the original permeability characteristics of the beach facies. In many cases, the units with the most original effective porosity (medium-coarse well sorted lime grainstones and packstones) are converted into rocks with good reservoir sealing potential whereas finer grained units (poorly sorted lime wackestones and mudstones) become the most desirable reservoirs, mainly as a result of dolomitization and solution. There is no set pattern of diagenetic history, porosity development, or porosity destruction, in beach sequences. Each stratigraphic unit is unique within itself and must be treated as such. However, some of the primary factors governing the types and trends of diagenesis, and in part the resultant porosity distribution of beach sequences are:

(1) Original mineralogy — Many early diagenetic cements which precipitate out of fresh water form from calcium carbonate which is derived from the contemporaneous

dissolution of aragonite contained within the sediment. Low- and high-magnesium calcite are not as susceptible to dissolution as is aragonite in fresh water environments. Therefore, a beach sand rich in aragonitic ooids, coral, or algal grains, which is subjected to early fresh water diagenesis, stands a greater chance of having its porosity and permeability characteristics changed than does, for example, a beach sand composed of calcitic mollusk, crinoid, or bryozoan grains.

(2) Beach setting — A regressive beach sequence will be bathed in widely varying water types during its early burial history, depending on whether it is attached to the

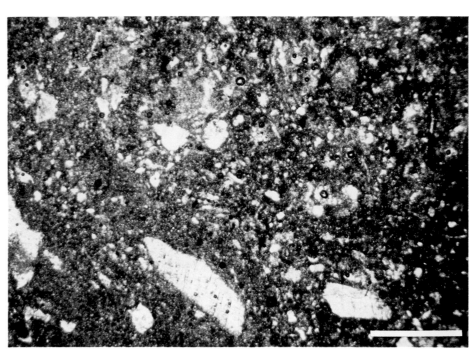

Figure 58b—Bioturbated fossil fragment-spicule dolomite wackestone (scale equals .065 mm); restricted marine.

Figure 58c—Black and brown, laminated, brachiopod-crinoid lime packstone-wackestone (scale in inches); restricted marine.

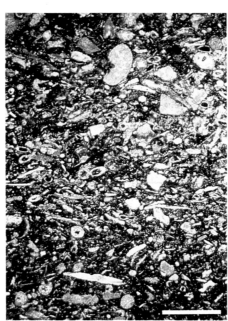

Figure 58d—Slightly sorted and bioturbated siliceous bryozoan-brachiopod-crinoid lime packstone from offshore open marine facies (scale equals .5 mm).

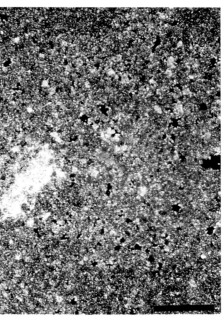

Figure 59a—Medium-coarsely crystalline calcisphere dolomite wackestone with abundant secondary and microinter-crystalline porosity (scale equals 1.0 mm).

mainland, backed by a lagoon, or developed on an atoll or offshore tectonic structure. For example, beaches attached to a mainland might be subjected to dissolution and cementation events that first take place in the local ground water systems that develop in the beach ridges, and secondly in the

fresh ground water flow systems that occupy continental coastal regions. Beaches developed on atolls would be affected by the diagenesis taking place in the local fresh water lens, but would then be submerged into a marine pore water system during its early burial.

(3) Climate — In wet climates, abundant fresh water would be expected to pass through barrier and mainland beaches during their early burial histories. In very arid climates, evaporites would form in the backshore tidal flats, lagoons, or sabkhas on the landward side of the

Figure 59b—Calcisphere dolomite wackestone with coarsely crystalline dolomite matrix (scale equals .125 mm).

Figure 59c—Calcisphere dolomite wackestone with coarsely crystalline dolomite matrix; moldic porosity developed out of leached calcispheres (scale equals .065 mm).

beach. Hypersaline waters derived from these environments would pass through the beach facies during their basinward flow, precipitate evaporites in primary and early secondary pores, as well as possibly dolomitize the finer grained near offshore facies. The resultant dolomites might form reservoirs or reservoir seals, depending on the degree of dolomitization, the size of the replacement dolomite crystals, and other factors.

(4) Resident time in each diagenetic environment — The degree of alteration the beach sequence suffers in each diagenetic environment (marine vadose and phreatic, mixed marine, local and regional fresh vadose and phreatic) is to a large extent dependent on both the sedimentation and subsidence rates in the areas of beach formation. For example, in areas of extended beach progradation (sedimentation rate much greater than subsidence rate), the beach system would be subjected to the shallow burial diagenetic environments and pore fluid types for a much longer period of time than in a zone where beaches would be stacked along a hingeline (sedimentation rate equals subsidence rate), in which case they would be subjected to the shallow diagenetic environments for a much shorter period of time, and would be more rapidly submerged into the regional pore fluid systems occurring at depth.

ACKNOWLEDGMENTS

The authors wish to express their appreciation to John Horne and Peter Scholle for critically reviewing the manuscript and making numerous suggestions which have led to its improvement. Thanks are also due to John Barwis, Mitch Harris, Miles Hayes, and numerous others for supplying the photographs which have enhanced the illustrative quality of this paper. Superior Oil permitted the release of data on the Stuart City and Mission Canyon formations. Kathy Brown, formerly of MRO and Associates, graciously typed numerous versions of this manuscript.

SELECTED REFERENCES

Alexandersson, T., 1972, Mediterranean beachrock cementation; marine precipitation of Mg-calcite, *in* The Mediterranean Sea; a natural sedimentation laboratory: Stroudsburg, Pa., Dowden, Hutchinson, and Ross Pub., p. 203-223.

Allen, S. H., 1970, Stratigraphy and diagenesis of carbonate beach complexes, central Texas: Baton Rouge, La. State Univ., unpub. Master's thesis.

Ball, M. M., 1967, Carbonate sand bodies of Florida and the Bahamas: Jour. Sed. Petrology, v. 37, p. 556-591.

Barwis, J. H., 1976, Internal geometry of Kiawah Island beach ridges, *in* M. O. Hayes and T. W. Kana, eds., Terrigenous clastic depositional environments: AAPG Field Course, Univ. South Carolina Tech. Rept. No. 11-CRD, p. II/115-II/125.

_____ and J. H. Makurath, 1978, Recognition of ancient tidal inlet sequences; an example from the Upper Silurian Keyser Limestone in Virginia: Sedimentology, v. 25, p. 61-82.

Bebout, D. G., and R. G. Loucks, 1974, Stuart City Trend, Lower Cretaceous, south Texas, a carbonate shelf-margin model for hydrocarbon exploration: Austin, Tex., Bureau Econ. Geol. Rept. No. 78, 80 p.

Bernard, H. A., R. J. Leblanc, and C. F. Major, 1962, Recent and Pleistocene geology of southeast Texas, *in* Geology of the Gulf Coast and central Texas and guidebook of excursions: Houston, Tex., Houston Geol. Soc.-Geol. Soc. America Ann. Mtg., p. 175-205.

Boutte, 1969, Callahan carbonate-sand complex, west-central Texas, *in* C. Moore, ed., Depositional environments and depositional history, Lower Cretaceous shallow shelf carbonate sequence, west central Texas: Dallas, Tex., Dallas Geol. Soc., p. 40-74.

Campbell, C. V., 1971, Depositional model–Upper Cretaceous Gallup beach shoreline, Ship Rock area, northwestern New Mexico: Jour. Sed. Petrology, v. 41, p. 395-409.

Canfield, J. C., 1975, A depositional

6

Shelf Environment

Paul Enos

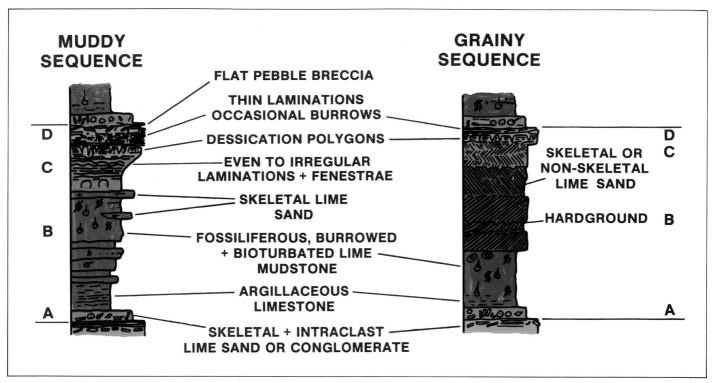

Figure 1—Hypothetical shoaling-upward carbonate sequences of relatively low energy (left) and somewhat higher energy (right) (from James, 1977). Letters indicate: **A**. Basal transgressive lag; **B**. Subtidal, restricted to open marine; **C**. Intertidal; **D**. Supratidal.

Restricted shelf, bay, or lagoon environments can be reasonably well defined geographically or hydrographically in modern settings. In the geologic record, these terms are typically used somewhat more loosely to form a "waste-basket" category for low-energy, shallow-water carbonates, especially if a demonstrable barrier occurs seaward, or if lateral facies equivalents reflect severe restriction, as in the case of evaporitic or euxinic (organic carbon-rich) sequences.

Geographically, bays and lagoons are partly enclosed bodies of water. A bay is a shoreline recess or inlet between headlands. Headlands in carbonate settings may be formed by depositional topography from the current depositional cycle as well as by erosional or depositional topography from a previous cycle. Lagoons are more completely enclosed (for example by barrier islands or reefs) so that connection to the sea is distinctly restricted. Lagoons are normally shallow, although lagoons

of some larger atolls reach 70 m deep; bays have no particular depth connotation.

Restricted shelves may be defined as any part of a continental or island shelf with slow water circulation resulting in abnormal salinity, depleted nutrients, or temperature extremes. Restriction that reduces normal wave or current energy may result from any physical barrier such as reefs, islands, skeletal or oolitic sand shoals, or from the damping effect of vast expanses of shallow water. Abnormal salinities resulting from restriction may be elevated, diluted, or seasonally fluctuating depending on climate and access of terrestrial runoff. Restricted waters are commonly subject to temperature extremes, both high and low, especially if shallow, and to oxygen and nutrient depletion.

In ancient settings, restricted shelf, bay, or lagoonal depositional environments may be inferred from stratigraphic evidence of paleogeography similar to that described for modern settings, for example by

transition to reefal or sand-shoal (ooid, skeletal, terrigenous) deposits on the one hand and to shoreline, tidal flat, or evaporitic facies on the other. With less stratigraphic resolution, restriction may be inferred from impoverishment of fauna or from lithofacies, such as muddy sediment, especially if it contains abundant organic matter, pyrite, or evaporites.

Summary of Diagnostic Criteria

No single criterion can be diagnostic for environments as diverse as restricted shelves, bays, and lagoons. Sequence and transitions are the most reliable guides. Lateral transition to a barrier facies in one or more directions — for example to a reef, barrier bar/island, or oolite shoal — coupled with transitions to low-energy shoreline facies in other directions is typical. Some possible transitions are outlined in Table 1.

Vertical sequence can be used to in-

TABLE 1

Seaward or Windward Environment		Restricted Shelf, Bay, or Lagoon	Landward or Leeward Environment	
Environment	Example (Age)/Reference		Environment	Example (Age)/Reference
Reef	Belize Shelf (Holocene); Wantland & Pusey, 1975 El Abra Fm. Carrillo, 1971		Tidal Flats	Great Bahama Bank, Andros Lobe (Holocene); Shinn & others, 1969 Black River Grp. (Ordovician); Walker, 1972
Oolite Shoal	Little Bahama Bank (Holocene); Hine, 1977 Smackover Fm. (Jurassic); Wilson, 1975		Salt Flats (Sabkha)	Persian Gulf (Holocene); Purser, 1973 Edwards Grp.; Rose, 1972
Skeletal Sand Shoal	Florida Shelf Margin (Holocene); Enos, 1977		Evaporites	El Abra Fm. (middle Cretaceous); Carrillo, 1971
Barrier Island	No. Biscayne Bay, Fla. (Holocene); Wanless, 1969		Euxinic (Shelf Basin)	McKnight Fm., Edwards Grp. (Lower Cretaceous); Rose, 1972
Mud Bank	So. Biscayne Bay, Fla. (Holocene); Wanless, 1969 West Florida Bay (Holocene); Enos & Perkins, 1979		Swamp (Peat, Coal)	Florida Bay, Everglades (Holocene); Enos & Perkins, 1979
Island (Prior topography)	Florida Bay (Holocene); Enos & Perkins, 1979 Great Bahama Bank (Holocene); Enos, 1974b Shark Bay, W. Austr. (Holocene); Logan & others, 1970		Beach	Cow Creek Mbr., Pearsall Fm. (Lower Cretaceous); Loucks, 1977
			Erosional Coast	Great Bahama Bank, Eleuthera Lobe (Holocene)
			Karst Plain	Campeche Bank & Yucatan (Holocene)

Table 1. Typical lateral transitions for restricted shelf, bay, and lagoon environments.

fer lateral relations. A typical vertical sequence is a low-energy shoaling-upward cycle (Fig. 1) consisting of a basal transgressive unit, muddy carbonate with impoverished fauna, and capped by intertidal and/or supratidal deposits. Although they have not received equivalent notice in recent compilations of depositional models, many cycles containing restricted intervals are transgressive or include transgressive intervals. In these, disconformity surfaces, tidal-flat deposits, or shoreline facies are overlain by low-energy, semi-restricted deposits and in turn by higher energy, more open-marine deposits.

In the absence of well-defined sequences or cycles, rocks from restricted environments may be characterized by a "facies mosaic" (Laporte, 1967) of more or less random transitions among muddy, low-energy lithofacies and/or impoverished organic communities. Such transitions are generated readily in very shallow water where minor changes in sea level or depositional topography can cause significant changes in environment. In contrast, some expansive settings with little relief, such as the interior of the modern Great Bahama Bank, develop monotonous muddy facies that may extend for hundreds of kilometers laterally or for many meters vertically.

Depauperate biota is a common, important clue to restricted settings. Restricted settings typically have low faunal diversity, but relatively great numbers of those few organisms adapted for survival. Reduced size of individuals (dwarf faunas) and aberrant growth forms are also clues to adverse conditions. Extreme adversity may result in few organic remains of any kind.

An abundance of burrows is characteristic, although certainly not diagnostic, of restricted subtidal deposits. Burrows may be of one or a few distinctive types, indicative of a restricted fauna, or the deposits may have an indistinct mottled or swirled appearance indicating extensive bioturbation. The end member is complete homogenization of the deposits with loss of definition of individual burrows, so that, ironically, direct evidence of burrowing may be lost. Not all structures in shallow restricted environments are biogenic, however. In settings where macrobenthos are severely reduced in number, or lacking, lamination may be

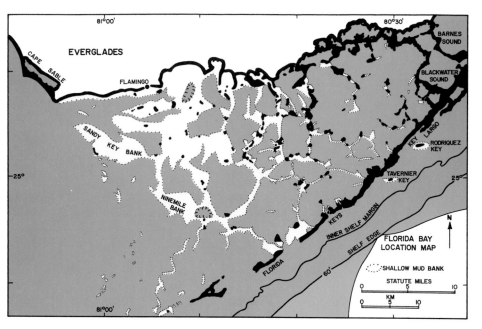

Figure 2a—Modern semi-restricted settings: Florida Bay, restricted by Pleistocene coralline limestone forming the Florida Keys (after Enos and Perkins, 1979).

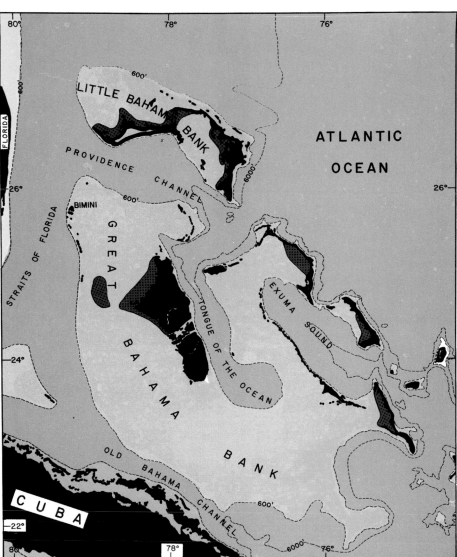

Figure 2b—Modern semi-restricted settings: The Bahama Banks showing areas of maximum restriction (stippled).

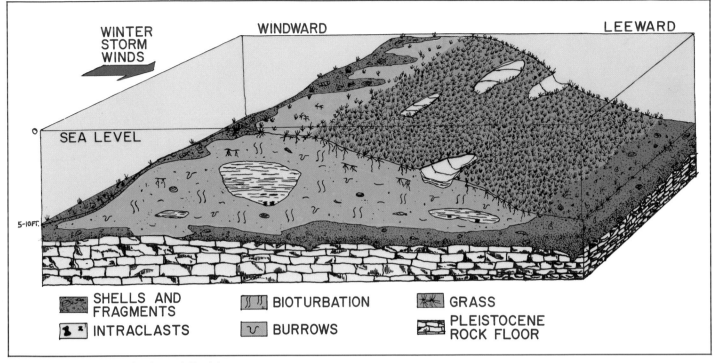

Figure 3—Diagrammatic view of Florida Bay mud bank. Vertical exaggeration is about 100x.

preserved. Fine millimeter-scale lamination characterizes deposits of anaerobic settings (Byers, 1977). In hypersaline settings, lamination or bedding is commonly distorted by interstitial crystal growth, dehydration shrinkage, or solution of evaporites.

Non-skeletal grain composition is not very diagnostic for recognition of semi-restricted shallow-water environments, but fecal pellets and peloids of diverse origins are characteristic sedimentary particles; locally they may comprise virtually the entire sediment. "Grapestone" aggregates and small intraclasts are also common constituents.

Mineralogy is scarcely indicative of depositional environment, not even if the metastable sedimentary mineralogic assemblages were preserved. However, evaporites and penecontemporaneous dolomites are common in more restricted portions of shelves and lagoons or in laterally equivalent deposits.

SETTING

Lateral Facies Relationships

Lateral facies relations may be the most reliable clue in recognition of restricted shelf, lagoon, or bay deposits in the rock record. Seaward or windward equivalent deposits in many instances produce the restrictions, but landward or leeward equivalents may record the restricted conditions more drammatically. Both produce more distinctive facies than the commonly nondescript rocks representing semi-restricted subtidal environments.

Laterally equivalent facies representing more open water conditions (Table 1) include reef complexes and sand shoals consisting of ooids, skeletal grains or even quartz. For example, the inner part of the south Florida shelf margin (Fig. 2a) is a semi-restricted, low-energy shelf within 5 km of a major slope break. The immediate cause of restriction is skeletal sand shoals, although these are probably directly dependent on the skeletal productivity and wave transport of skeletal grains from reefs

nearer the slope break (Enos, 1977). The transition from cross-laminated skeletal sand to burrowed muddy sediments is abrupt (Swinchatt, 1965), especially where there is a marked change of relief from the sand shoals (1 to 4 m water depth) to level-bottom muds (5 to 15 m deep).

Islands formed by depositional or erosional topography of contemporaneous or earlier depositional episodes may also contribute to restriction. Islands formed by sediment accretion such as the sand and rubble cays associated with barrier bars or reefs are actually only a minor part of the restricting mass; submarine shoals or reef growth produce the major bulk. Islands may form from erosional remnants of older rocks or from eolian dunes, sand shoals, or reefs of a previous depositional episode. Examples of circulation-restricting islands formed by pre-existing depositional topography are the upper Florida Keys, consisting of slightly eroded Pleistocene coral reefs that seal off one flank of Florida Bay (Fig. 2a; Enos and Perkins, 1979); the lower

Figure 4—Plastic cast of burrow excavated by the ghost shrimp *Callianassa*. Photograph courtesy E. A. Shinn.

Florida Keys, consisting of Pleistocene oolite bars (Perkins, 1977) with leeward carbonate banks (Jindrich, 1969; Basan, 1973); the major Bahaman islands, consisting of Pleistocene eolianite ridges leeward of which are the most restricted portions of the Bahama Banks (Fig. 2b, Enos, 1974b); and Bermuda, also consisting of Pleistocene eolianite dunes that enclose Harrington Sound (Neumann, 1965) and Great Sound — semi-restricted environments in an open oceanic setting. Because the margins of carbonate platforms are typically elevated, whether from karstic erosional processes (Purdy, 1974a) or

from differential accumulation of carbonates, rocky islands may have caused restriction in many ancient settings. Recognition of the existence of a paleo-island, which involves lateral transition to a non-depositional setting, is often more difficult than recognizing a facies transition, especially given the facies complexities of a carbonate shelf margin. The criteria for recognizing disconformity surfaces are now receiving attention (Perkins, 1977; Read and Grover, 1977; Esteban and Klappa, 1982).

Restricted settings may also originate without significant physiographic barriers. Broad ex-

panses of shallow water, such as existed in epeiric seas, can damp out tidal and wave energy. Modification of the water mass may follow in the form of hypersalinity from prolonged evaporation, dilution from rainfall and runoff, or salinity fluctuations from strongly seasonal or variable climates. Shoal waters have large temperature fluctuations in response to air temperature changes. Long residence time of water on shallow shelves, because of poor circulation, may lead to severe nutrient depletion. Facies patterns in idealized epeiric seas have been considered by Shaw (1964) and Irwin (1965). Basically,

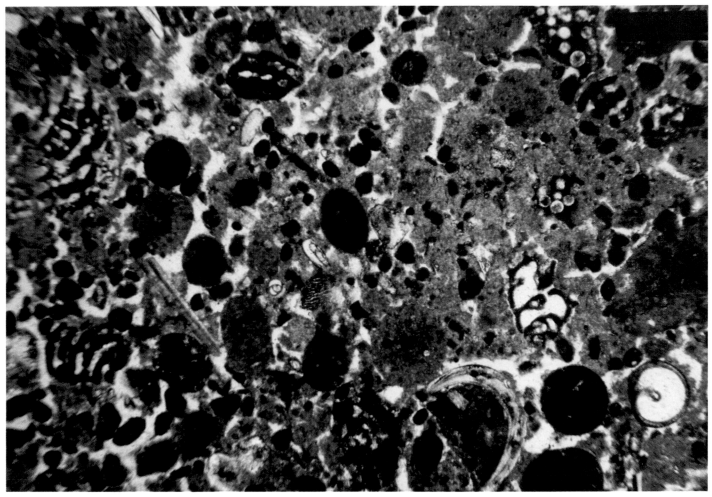

Figure 5—Peloidal sediment from the Great Bahama Bank, including both hardened (dark) and soft peloids (light brown with indistinct outlines). Definition of soft peloids was enhanced by drying during sample preparation. Scale bar is 0.3 mm. Photograph courtesy of R. D. Perkins.

low-energy, muddy carbonates (or mixed terrigenous muds) are deposited in more open, deeper water. Higher energy, winnowed sediments accumulate as the bottom slope intersects wave base, a geometry termed the carbonate ramp by Ahr (1973). Very shallow-water, restricted environments develop further leeward or landward and may cover vast expanses with little facies differentiation. Lateral transitions from such low-relief restricted settings are likely to be very gradual and to fluctuate in position with time.

Landward facies changes from restricted shelves and lagoons are from subtidal settings to shoreline environments. Shoreline facies may be narrow belts of beach-complex deposits which pass in turn into terrestrial deposits (deltaic or fluvial) or non-depositional surfaces (a karst plane). Higher-energy facies, typically shelly sands, may be developed at the shoreline by focusing of local waves that develop even within restricted lagoons.

More typical shoreline transitions are from restricted subtidal settings to tidal-flat complexes. These very distinctive deposits (Shinn, 1982) may include sediments ranging from humid salt marshes to sabkhas with hypersaline interstitial waters. Typically subtidal and tidal-flat facies are extensively interbedded through many vertical and lateral transitions. Well-documented examples occur in the Manlius Formation (Lower Devonian) of New York (Laporte, 1967) and in the Lofer cycles (Upper Triassic of the Alps, see below; Fischer, 1964).

Although there are apparently no documented examples of lateral transitions from limestone to coal, several modern sub-tropical carbonate lagoons are bordered by mangrove swamps where peat is actively accumulating. A notable example is Florida Bay and the adjacent Everglades (Spackman et al, 1964; Enos and Perkins, 1979).

Depositional Geometry

Depositional geometry of restricted carbonate settings may range from simple level-bottom expanses exemplified in the Holocene by the Great Bahama Bank, to complex, compartmentalized environments where depositional relief has developed, as in Florida Bay (Ginsburg, 1956; Enos and Perkins,

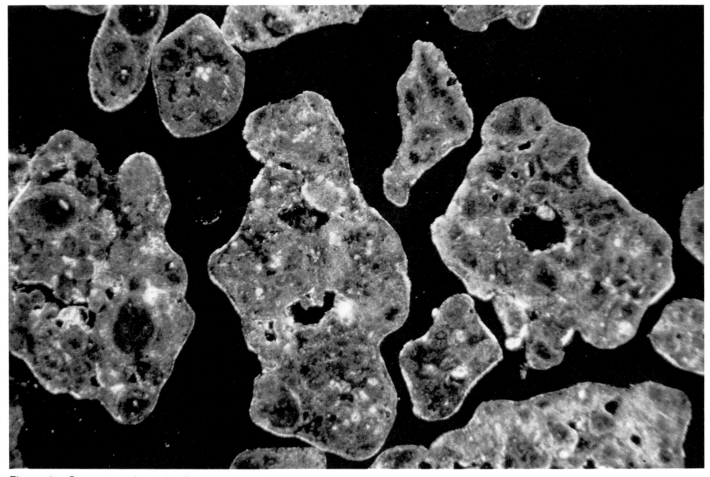

Figure 6—Grapestone from the Great Bahama Bank. Cross-polarized light. Largest grapestones are about 3 mm in maximum dimension. Photograph courtesy of R. D. Perkins.

Figure 7a—Requienid rudists: a Cretaceous bivalve roughly analogous to oysters in life style. Cluster of requienids in lime packstone with peloids and miliolid foraminifers (white specks). El Abra Formation (mid-Cretaceous). Sierra de El Abra, San Luis Potosi, Mexico. Slab is 12 cm wide. Sample courtesy of C. J. Minero.

1979), the coastal lagoons of the north coast of Cuba (Price, 1967), and the Belize lagoon (Wantland and Pusey, 1975). Compartmentalized settings may produce rapid lateral and vertical facies transitions which appear random in detail, aptly termed "facies mosaics" by Laporte (1967). Large-scale progradational or transgressive patterns may be recognizable despite the small-scale complexity (Laporte, 1969).

In contrast, the monotonous level-bottom facies so typical of epeiric-sea deposits might be called "facies prosaic." The Ellenburger Group (Lower Ordovician) of Texas is an example (Cloud and Barnes, 1956). Similar sheet-like geometries develop in the interior of large isolated banks or platforms that may have precipitous sides with sharp facies gradients, but broad flat interiors. Examples are the modern Great Bahama Bank, the mid-Cretaceous (Albian–Cenomanian) Valles-San Luis Potosi and Golden

Lane platforms of central Mexico (Carrillo, 1971; Enos, 1974a), and the Middle Permian of the northwest shelf of the Permian basin in Texas and New Mexico (Meissner, 1972). These units are tens to hundreds of kilometers wide and cover thousands of square kilometers with very little interior facies differentiation. Typical lithofacies are lime mudstones, muddy peloidal packstone and wackestones, and foraminiferal (miliolid, fusulinid) wackestone. On isolated platforms surrounded by deep water, islands and tidal-flat complexes typically develop at the raised platform margins, leaving slightly deeper water in the interiors without much facies differentiation. If relief is developed all around the platform, evaporites may develop in the interior, as in the mid-Cretaceous platforms of Mexico (Viniegra, 1971). In the modern, the raised platform margin of the Bahama Banks is much more pronounced on the windward

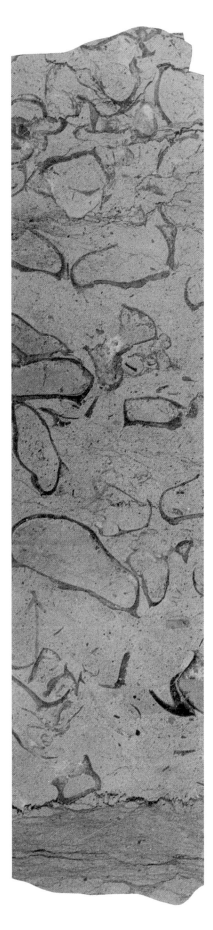

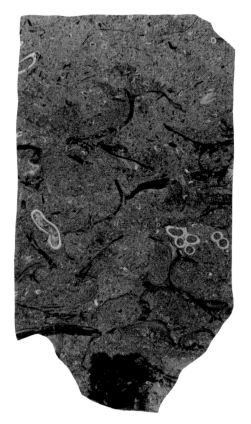

Figure 8—Poorly sorted packstone that contains large fragments of oysters (dark shells) and serpulid worm tubes (white). Lower part of Bexar Shale Member, Pearsall Formation (Lower Cretaceous), Medina County, south Texas. Slab of core, 8 cm wide, from Tenneco No. 1 Ney. From Loucks (1977); photograph courtesy of R. G. Loucks.

Figure 7b—Requienid rudists: a Cretaceous bivalve roughly analogous to oysters in life style. Requienid lime wackestone. Whole and broken shells in a matrix of lime mud cut by horsetail stylolites (prominent at top and bottom). Lower part of Bexar Shale Member, Pearsall Formation (Lower Cretaceous), Medina County, south Texas. Slab of core, 8 cm wide, from Tenneco No. 1 Ney. From Loucks (1977); photograph courtesy of R. G. Loucks.

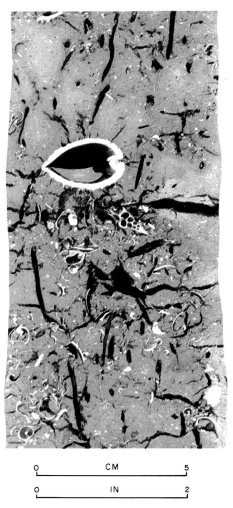

| 0 | CM | 5 |
| 0 | IN | 2 |

Figure 9a—Burrowed and bioturbated lithologies. Bioturbated Holocene sediment from a Florida Bay mud bank. Depositional texture corresponds to a pelletal-molluskan lime wackestone. Vertical tubes were formed by grass roots. Core slab, impregnated by blue plastic (from Enos and Perkins, 1979).

margins (Fig. 2b), giving rise to asymmetric facies distribution in the lee, with maximum restrictions and muddiest sediments adjacent to the large windward islands (Smith, 1940). However, in the present semi-humid, subtropical climate of the Bahamas, salinities reach only about 42 ppt,

20% above normal seawater, far short of the 4-fold concentration necessary for the onset of the evaporite precipitation. The facies pattern is a simple gradation from soft pellet mud in the most restricted areas to hard pellet sand toward the leeward bank margins. In the areas of slow deposition the pellets are bound together in "grapestone" aggregates.

With the development of embayed shorelines and/or compartmentalization through depositional relief, the depositional geometry becomes much more complex. Florida Bay is a familiar modern example (Fig. 2a).

Figure 9b—Burrowed and bioturbated lithologies. Burrow-mottled ostracod foraminifer lime wackestone. Burrows are filled with clean peloid miliolid lime packstone. Structure at lower right is a soil pipe from an immediately overlying subaerial erosion surface. El Abra Formation (mid-Cretaceous), Sierra de El Abra, San Luis Potosi, Mexico. Slab is 9 cm wide. Sample courtesy C. J. Minero.

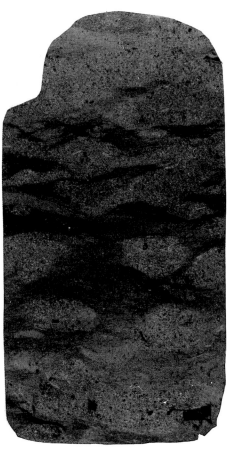

Figure 9c—Burrowed and bioturbated lithologies. Burrowed packstone containing miliolid foraminifers, echinoids, and mollusks. Pearsall Formation (Lower Cretaceous), south Texas. Slab of core is 8 cm wide. Photograph courtesy of R. G. Loucks.

The low-relief Florida mainland forms the northern side of the triangular bay. The southeastern side is enclosed by a ridge of Pleistocene coralline limestone and the southwestern limit is formed by depositional relief of broad mudbanks, developed in the Holocene, rising to sea level. The interior of the bay is compartmentalized into shallow basins by anastomosing linear mudbanks. Mud is winnowed from the basins by periodic storm waves and trapped on the shallow grass-covered banks. The result is a basal shelly lag in the basins and muddy banks which contain virtually all of the sediment. The relief of the banks is very small scale (about 2 m) with gentle slopes (< 1°), but nevertheless produces more facies differentiation (Fig. 3). The narrow windward margins of the banks comprise shelly packstone; however, the bulk of the sediment is bioturbated pelletal molluscan wackestone with local pockets of laminated pellet mudstone or very fine-grained packstone. These pockets result from the filling of negative micro-relief, probably erosional scours, as well as tractional deposition of soft, low-density pellets in low-amplitude dunes (Enos and Perkins, 1979). Islands with supratidal algal flats and mangrove forests dot the mudbanks. The resulting sediments are lenses of stromatolitic, mud-cracked pellet wackestone and organic-rich lime mud or peat. Local higher energy shelly lenses develop as island beaches, as lag deposits and sand waves in channels cutting the mudbanks, and as low-relief spillovers spread from the windward margins onto the bank tops by storm currents. In more open settings at the seaward margins of the bay, bank edges act as miniature shelf breaks populated by a shoal-fringe community characterized by finger-corals. The resultant sediment is a muddy skeletal gravel.

Tidal deltas, developed on the channels into Florida Bay and Biscayne Bay, have most of the facies characteristic of Florida Bay mud banks (Ebanks and Bubb, 1975). The overall geometry is, of course, much different; the triangular plan and wedge-shaped cross section of classical deltas are developed. Other facies packages within the generally low-energy delta complex are associated with the tidal channels, active and abandoned, and the natural levees which may have mangrove forests and islands. The seaward margins of deltas also form mini-shelf breaks with the development of the shoal-fringe finger-coral community and coarser sediment (Enos, 1977).

Other Modern Settings

Other modern semi-restricted carbonate environments apparently fall between the extremes of the level-bottomed Bahama Banks and compartmentalized Florida Bay, although 3-dimensional data are generally sparse. Patterns of depositional relief, however, vary considerably.

In the somewhat less restricted setting of the inner part of the south Florida shelf margin, depositional relief is more subdued, but still exerts important influence on facies patterns. Linear anastomosing mudbanks are lacking, but a few broad, low-relief banks have developed. Rodriquez Key bank is the best known (Turmel and Swanson, 1976). It is a flat-topped mudbank elongate transverse to depositional strike with

Figure 10a—Carbonate cycles in the Cambrian of the Canadian Rocky Mountains. Transition zone from Arctomys Formation (below, Middle Cambrian) to Waterfowl Formation (Upper Cambrian); Chaba River section on Alberta-British Columbia boundary. Cycles delineated by color contrasts consist of dark, thin-bedded lime mudstone grading upward into yellow-weathering dolomite cryptalgal laminite. Cross-laminated dolomitic siltstones at the top of the dolomite interval contain calcite blebs which may have replaced evaporites.

a well-developed seaward shoal-fringe community producing a linear bank of skeletal gravel. The finger corals are joined in this less restricted setting by branching red algae and plate-producing green algae. The muddy, grass-covered bank interior is a favorite haunt of burrowing ghost shrimp who have converted it into a monument to burrow architecture

(Shinn, 1968b). A small mangrove island near the landward edge of the bank has produced a peat lens. A storm ridge of skeletal sand has developed on an analogous mangrove island on neighboring Tavernier bank. Other slightly deeper muddy banks which dot the Florida inner shelf margin have developed patch reef caps; one has developed into a shoal of rippled skeletal sand surrounded by level-bottom muds (Enos, 1977). A much larger mass of muddy sediment with little facies differentiation is contained in low-relief wedges which have accreted seaward from the Florida Keys. These range from wedges with relief of a few decimeters along embayments in the shoreline to a broad wedge extending several kilometers seaward with relief of 5 m or more (Enos, 1977).

The Gulf of Batabano, a large em-

bayment on the south coast of Cuba, is partially blocked by reef barriers, but salinity is near normal throughout. Surface sediment in nearly half the bay contains more than 50% mud ($< 62\mu$m, Daetwyler and Kidwell, 1959). Several muddy shoals as much as 50 km long and several kilometers wide snake across the embayment. There are, unfortunately, no data on sediment thickness and it appears possible that the shoals may reflect structure in the underlying rock.

The Belize lagoon is enclosed between the mainland of the Yucatan Peninsula and the longest barrier reef in the Western Hemisphere, but salinities do not rise above normal marine values (Wantland and Pusey, 1975). Abundant runoff from adjacent highlands results in brackish rather than elevated salinities in the

Figure 10b—Carbonate cycles in the Cambrian of the Canadian Rocky Mountains. Lyell Formation (Upper Cambrian), Mount Ogden, Yoho National Park, British Columbia. Cycle (delineated by ticks) begins with erosive base, overlain by laminite clasts and by dark, nodular lime mudstone with microcrystalline dolomite partings. The top of the cycle is dolomite cryptalgal laminite (Aitkin, 1978). Photographs courtesy of J. D. Aitken.

nearshore parts of the lagoon and a lagoon floor blanketed with mont-morillonitic mud. Terrigenous muds grade seaward into calcareous mud derived from the reef tract. The seaward and southern portions of the lagoon are studded with steeped sided "pinnacle" reefs and micro-atolls or "faroes," (Purdy, 1974b) which grow on pinnacles in Pleistocene rock that in turn reflect deeper structure. Thus Holocene depositional relief here merely amplifies previous relief of

possible karstic origin (Purdy, 1974b).

Shark Bay, Western Australia, is a modern example of a hypersaline lagoon with salinities reaching 70 ppt, twice normal sea water (Logan et al, 1970). The bay is cut off from the open Indian Ocean (except at the north end) by ridges of Pleistocene calcareous eolianite. Although the bay lies in the belt of southeast tradewinds and has a considerable fetch over its area of 12,000 sq km, the northward opening keeps waves subordinate to tides in moving water masses. Sediments on the flat floor of the bay are sand-to-silt-sized skeletal fragments (foramineral-molluscan microcoquinas). Around the shallow margins of the bay, the skeletal sediments are diluted by quartz sand and ooids. The bulk of the sediment in the bay has accumulated in a grass-covered bank that extends along the mainland at the northeast margin of

the bay (Davies, 1970). The bank is a wedge-shaped deposit 130 km long, about 8 km wide, and 9 m thick at its seaward edge. The seaward edge is less than 2 m deep and the bank gradually shoals into an extensive tidal flat. Numerous tidal channels with well-developed levees cut the bank. Bank and levee sediment contains about 30% mud ($< 62\mu$m), consisting of foraminifers and red-algae that encrusted the sea grass. Sand-sized particles are foraminifers (about 50%), mollusks (about 25%), coralline algae, peloids, and quartz sand (Davies, 1970). Channel and intertidal sediments contain few fines and relatively more quartz. Tidal-flat sediments have prograded over the subtidal bank sediments at its landward margin to produce a shoaling and coarsening-upward sequence.

The Persian Gulf is also hyper-saline (40 to 45 ppt, in the open gulf)

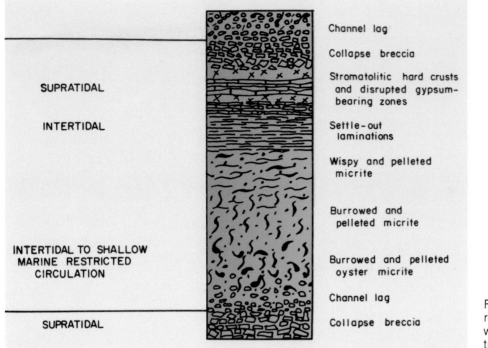

SUPRATIDAL

INTERTIDAL

INTERTIDAL TO SHALLOW
MARINE RESTRICTED
CIRCULATION

SUPRATIDAL

Channel lag

Collapse breccia

Stromatolitic hard crusts
and disrupted gypsum-
bearing zones

Settle-out
laminations

Wispy and pelleted
micrite

Burrowed and
pelleted micrite

Burrowed and pelleted
oyster micrite

Channel lag

Collapse breccia

Figure 11—Typical cycle representing restricted environments from the Edwards Group (Lower Cretaceous) of central Texas (from Rose, 1972).

with a moderately high-energy carbonate province along the north-facing portions of the Arabian coast (Purser, 1973). The gulf floor slopes gently northeastward to form a trough more than 80 m deep along the Iranian coast that receives terrigenous sediment from the Zagros Mountains in Iran and the Tigris-Euphrates delta at the north end of the gulf. The coastal zone along the shallow Arabian side of the Gulf is a complex of sabkhas, small tidal lagoons, channels, ooid bars and beaches, and reefs, that are discussed elsewhere in this volume. The offshore sediments are a more orderly blanket of skeletal sand with a gradual seaward increase in the content of mud, both carbonate and terrigenous (Houbolt, 1957; Purser, 1973). Virtually the only relief on this broad shelf results from actively rising salt domes which produce small shoals of winnowed skeletal sand and/or living reefs.

Depositional relief characterizes and profoundly influences so many modern muddy carbonate settings that it appears likely to have been a factor in many restricted ancient settings. The gentle relief associated with mudbanks is difficult to document in ancient rocks, however. Most

recognized banks and mounds from the rock record are steeper sided organic buildups with well-developed flank facies. In view of the abundance of mudbanks and comparable low-relief features in modern environments, it seems that many ancient carbonate sequences may have been influenced by penecontemporaneous depositional relief, particularly those in which facies mosaics are well developed.

Microfacies
Facies of modern and ancient restricted subtidal settings are characterized by a limited number of grain types: carbonate or terrigenous mud, fecal pellets, peloids, grapestones, intraclasts, and a limited range of skeletal components. Peloids (Fig. 5) are ovoid micritic grains of various origins such as micritization of skeletal and oolitic grains; biological, including fecal, aggregation of lime mud; and mechanical erosion of carbonate mud or lithified mudstone. In areas of slow deposition, grains of all these origins converge in appearance through rounding, micritization, and hardening (intra-particle cementation), so that the mode of origin is commonly indeterminate. Grapestones (Fig. 6) are clusters of

carbonate sand grains bound together by encrusting foraminifers, filamentous algae, and/or micritic cement (Windland and Matthews, 1974). The resultant clusters are generally flakes as much as a centimeter in diameter with spongy, porous interiors. "Encrusted lumps," "constituent grains," "organic aggregates," "aggregates," and "lumps" of various authors are virtually synonymous. Such grains are typical of shallow-water settings with slow sedimentation rates. Characteristic skeletal components (Table 2) are benthic foraminifers (for example, miliolids, Fig. 7a and 9a; fusulinids); ostracods; gastropods, particularly high-spired types; bivalves such as oysters and requienid rudists (Fig. 7); algal oncoids; serpulid worms (Fig. 8); and lingulid brachiopods. Mechanical abrasion, reworking, rounding, and sorting are at a minimum, reflecting low-energy depositional environments. In contrast, micritization, boring, metaloxide staining, encrustation, and other indicators of slow accumulation rates are generally extensive.

Wilson (1975) listed the following standard microfacies (SMF) types as typical of restricted carbonate environments: peloidal grainstones (Fig. 5), locally with ostracods or

TABLE 2

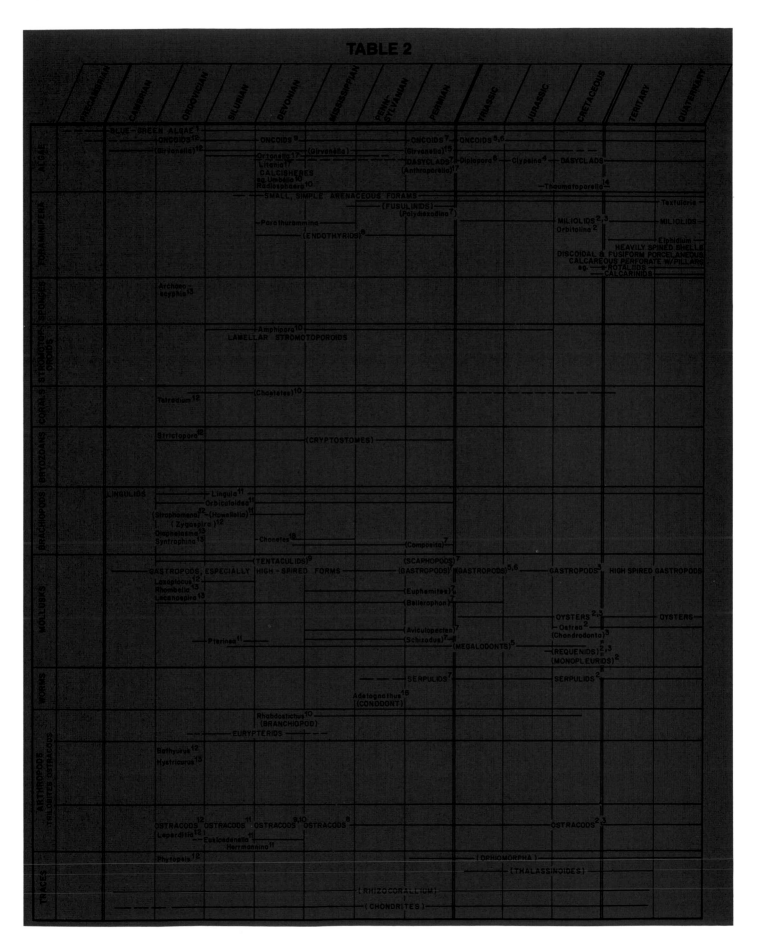

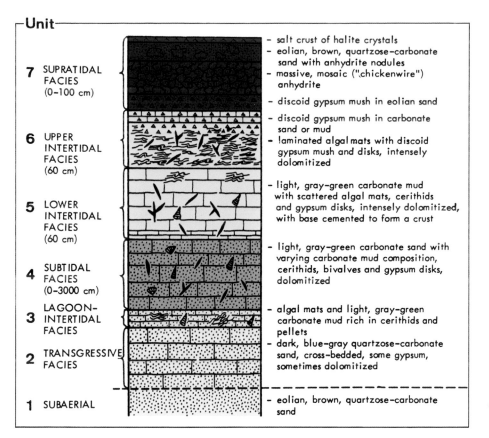

Unit		
7	SUPRATIDAL FACIES (0–100 cm)	– salt crust of halite crystals – eolian, brown, quartzose–carbonate sand with anhydrite nodules – massive, mosaic ("chickenwire") anhydrite – discoid gypsum mush in eolian sand
6	UPPER INTERTIDAL FACIES (60 cm)	– discoid gypsum mush in carbonate sand or mud – laminated algal mats with discoid gypsum mush and disks, intensely dolomitized
5	LOWER INTERTIDAL FACIES (60 cm)	– light, gray–green carbonate mud with scattered algal mats, cerithids and gypsum disks, intensely dolomitized, with base cemented to form a crust
4	SUBTIDAL FACIES (0–3000 cm)	– light, gray–green carbonate sand with varying carbonate mud composition, cerithids, bivalves and gypsum disks, dolomitized
3	LAGOON–INTERTIDAL FACIES	– algal mats and light, gray–green carbonate mud rich in cerithids and pellets
2	TRANSGRESSIVE FACIES	– dark, blue–gray quartzose–carbonate sand, cross–bedded, some gypsum, sometimes dolomitized
1	SUBAERIAL	– eolian, brown, quartzose–carbonate sand

Figure 12—Idealized cycle of Holocene sediments beneath the Sabkha at Abu Dhabi, adjacent to the Persian Gulf. Units 2, 3, and 4 were deposited primarily in restricted subtidal environments. Compiled from various sources by McKenzie and others (1980).

Table 2. Summary of some taxa common in restricted shelf, bay, lagoon environments. (**A**) Group or genus which has broader ecologic range but is common in restricted environments is enclosed in parentheses. (**B**) Long-ranging groups are shown by vertical lines; the name is entered and footnoted under the geologic period where specific references have been found. Ranges indicated are from Treatise on Invertebrate Paleontology, R. C. Moore, ed.

[1]Environmental range has become restricted by competition in Phanerozoic time so that "stromatolites: have become virtually synonymous with tidal-flat disposition. Blue-green algae persist in a variety of marine environments although micritization of grains may be their only recognizable geologic record.
[2]Perkins, 1974.
[3]Loucks, 1977.
[4]Wood and Wolfe, 1959.
[5]Fischer, 1964.
[6]Bosellini and Rossi, 1974.
[7]Newell and others, 1953.
[8]Edie, 1958.
[9]Laporte, 1967.
[10]Wilson, 1967.
[11]Zenger, 1971.
[12]Walker, 1972.
[13]Cloud and Barnes, 1956.
[14]Conrad, 1977.
[15]Flugel, 1977.
[16]Heckel and Baesemann, 1975.
[17]Tsien and Dricot, 1977.
[18]Heckel, 1973.

foraminifers (SMF 16); grapestone–peloid–intraclast grainstone (SMF 17, Fig. 6); foraminiferal or dasyclad algal grainstone with peloids (SMF 18); laminated to bioturbated pellet mudstone or wackestone (Fig. 9), locally with fenestral fabric (SMF 19); oncoid wackestone or floatstone (SMF 22), and homogeneous unfossiliferous mudstone which may have gypsum crystals or casts (SMF 23). These are only a sampling of the many combinations of composition and textures from the components mentioned above. The variation in characteristic skeletal components with geologic age (Table 2) greatly expands the possibilities. This is true despite the fact that the relatively primitive, hardy forms able to cope with restricted environments are among the most persistent facies fossils known.

One suite of facies not commonly ascribed to restricted lagoons, shelves, and bays are euxinic or organic-rich carbonates. These restricted environments can develop as brackish-water, anoxic swamps, such as modern mangrove swamps, or as hypersaline shelf basins (Rose, 1972). Their apparent scarcity in the geologic record may be in part due to oversight, although abundant fresh water and organic-rich, acid-producing sediments are certainly not the most compatible settings for carbonate accumulation. A typical lithofacies is black, laminated or slightly burrowed lime mudstone, rich in organic matter, containing few or specialized fossils such as ostracods, lingulid brachiopods, or thin-shelled bivalves.

Sequences

Far more diagnostic than individual lithofacies are associations of lithofacies, specifically the vertical succession in which they are ordered. As Johannes Walther noted long age (Walther, 1894), the vertical sequence of facies in the rock record reflects lateral sequence in the depositional setting, a principle that has become known as Walther's Law. This principle is especially important in subsur-

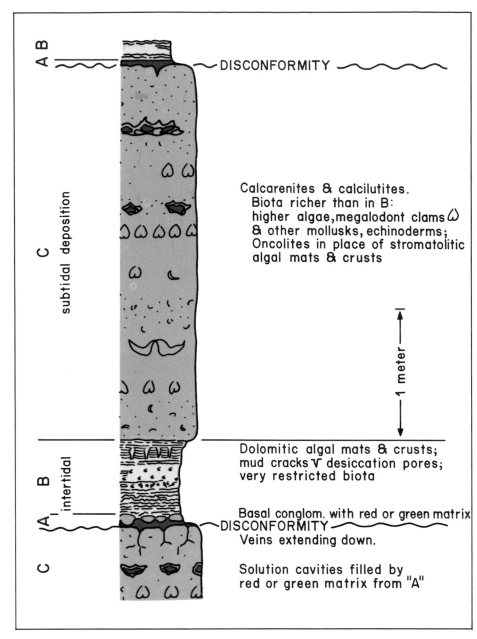

Figure 13—Diagrammatic representation of the "Lofer cyclothem," (Upper Triassic) Calcareous Alps. Units are described in text (from Fischer, 1964).

face exploration with limited well data where it has found many applications. The principle is also extremely useful in carbonate rocks from restricted settings in that the lateral equivalents tend to be more distinctive than the restricted facies themselves.

Considerable attention has been devoted to shoaling-upward sequences deposited on carbonate shelves (James, 1977; Wilson, 1975). A typical sequence (Fig. 1) begins with a "basal transgressive lag" representing a relative rise in sea level that is more rapid than accumulation rates. The overlying unit commonly represents the deepest water and lowest energy in the sequence. Vertical accretion or lateral progradation produces shoaling which may eventually reach the intertidal zone. Further accretion leads to supratidal sedimentation with possible subaerial alteration of the sequence. The actual elements present in a given sequence vary considerably, depending in part on whether shoaling leads to dampening or focusing of energy. Thus the shallower portions of a cycle may be beach deposits, reflecting relatively high energy, or tidal-flat deposits where energy is generally very low. James (1977) termed these examples "grainy sequences" and "muddy sequences," respectively. This terminology is apt, as it recognizes that designations such as "fining upward" and "coarsening upward" are not generally applicable in carbonate rocks where in situ particle production results in highly variable size, shape, and hydraulic density. Grain-sized parameters are thus unreliable if used indiscriminately.

Basal-lag units may be comparable

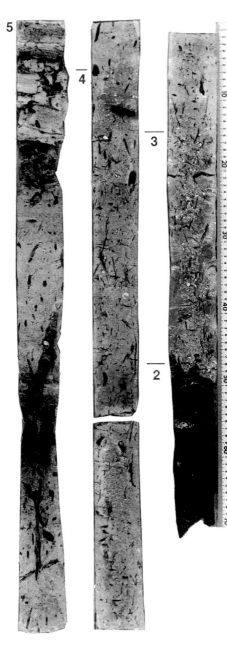

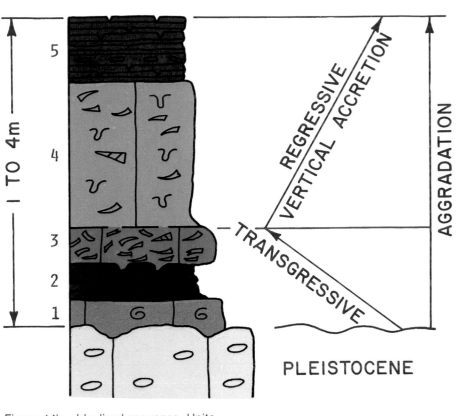

Figure 14b—Idealized sequence. Units are: (1) Basal calcitic lime mud of fresh-water ponds (not represented in core); (2) Mangrove swamps that produce peat; (3) "Lake" sediments are shelly packstone lag with a few lithoclasts; (4) Mud-bank sediments are pelleted mollusk wackestone; and (5) Island sediments are stromatolitic pellet wackestone to mudstone, with peat lenses. From Enos and Perkins (1977).

Figure 14a—Asymmetric transgressive-regressive sequence of Florida Bay. Continuous plastic-impregnated core; top is left.

from both grainy and muddy sequences. Specific examples range from conglomerates through floatstones to mud with a thin layer of coarser particles. Typical particles are intraclasts, lithoclasts from underlying sequences, and skeletal debris. Boring, breakage, encrustation, and metal-oxide staining of particles are common because accumulation is slow. In fact, slow accumulation because of low in situ production and/or periodic winnowing is probably more characteristic than surf-zone reworking mentioned by James (1977).

Subtidal deposits may include any of the spectrum of lithofacies discussed above. Care must be taken to distinguish restricted deposits from deeper water, open-shelf deposits in this part of the sequence. Either may eventually be succeeded by intertidal sequences; the energy level of the shoaling succession may provide a clue to seaward barriers or restrictions in the lower part of the sequence. High-energy sands grading up into beach deposits or developing as an offshore shoal (oolite bar) suggest the lack of a significant barrier. A narrow shelly beach zone may, however, develop in a restricted setting such as the landward side of an enclosed lagoon. Beaches are developed locally in low energy settings such as Biscayne Bay in south Florida, on islands in Florida Bay, along the west coast of Florida, and bordering the tidal flats of Andros Islands, Bahamas. Beach deposits may be overlain by supratidal deposits, ranging from muddy algal laminites or carbonaceous swamp deposits to eolian dunes, depending upon climate as well as energy level. A non-depositional weathering zone may cap the sequence. Characteristic features of such disconformities are root casts, soil structures, soil breccias, laminated calcrete, caliche, etc. (Perkins, 1977; Esteban and Klappa, 1982).

Figure 15—Shoaling-upward sequence or "punctuated aggradational cycle" within the Lowville Limestone of the Black River Group (Middle Ordovician), Inghams Mills, New York. The base of the cycle is at the color change just below the head of the hammer. The dark gray lime wackestone is bioturbated and horizontally burrowed, contains gastropods, corals, and other marine fossils, and is apparently of shallow subtidal origin, in contrast to the underlying, unfossiliferous, vertically burrowed lime mudstone. The prominent tidal channel in the center of the photo is overlain by vertically and burrowed mudcracked lime mudstone of tidal-flat origin. Dark gray subtidal wackestones at the top of the picture mark the base of the next cycle. Photograph courtesy of Peter Goodwin and E. A. Anderson.

Figure 16—Shoaling-upward sequence within Manlius limestone, Helderberg Group (Lower Devonian), Perryville Quarry, New York. The cycle begins with an erosive base, 10 cm above the lower end of the hammer handle, overlain by reworked stromatoporoid heads. This grades up into fossil-bearing lime wackestone and faintly laminated lime mudstone. The brown dolomitic algal laminites form the top of the sequence. Photograph courtesy E. A. Anderson and Peter Goodwin.

patch-reef or biohermal sequences of back-reef or atoll lagoons.

Examples of shoaling-upward sequences within restricted environments have been described from the Cambro-Ordovician of the Canadian Rocky Mountains (Aitken, 1978), the Ordovician Stony Mountain Formation of the western Williston Basin (Roehl, 1967), the Devonian Duperow Formation of the Williston Basin (Wilson, 1967), the Middle Carboniferous Windsor carbonates of the Maritime Provinces (Schenk, 1969), the Permian San Andres Formation of the Permian Basin (Chuber and Pusey, 1967; Meissner, 1972), the Upper Permian

Beach deposits are probably less common in restricted intertidal settings than are muddy sediments characterized by lamination, fenestrae, oncolites, mud cracks and/or vertical burrows. The muddy supratidal zone is characterized by stromatolites, mud cracks, flat-pebble conglomerate, interstratal dolomite, and evaporites. Other variants of shoaling-up sequences from a restricted subtidal setting include

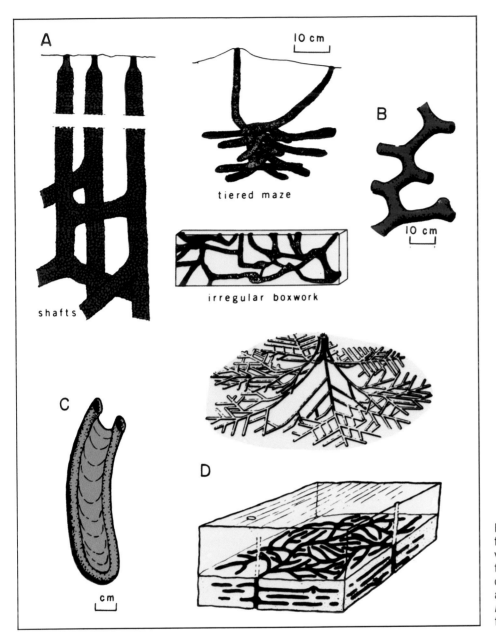

Figure 17—Trace fossils commonly found in shallow-water carbonate environments, including semi-restricted settings. (**A**) *Ophiomorpha,* showing some of its morphologic range (from Frey et al, 1978). (**B**) *Thalassinoides.* (**C**) *Rizocorallium.* (**D**) *Chondrites.* B to D from Hantzschel, 1975.

Bellerophon Formation of the southern Alps (Bosellini and Hardie, 1973), the Alpine Triassic (Assereto and Kendall, 1971; Bosellini and Rossi, 1974), the Jurassic (?) Arab/Darb Formation, Persian Gulf (Wood and Wolfe, 1969), the Lower Cretaceous Sligo-Hosston Formations, south Texas (Bebout, 1977), the Lower Cretaceous Edwards Group, central Texas (Rose, 1972), and the Holocene sediments of Abu Dhabi adjacent to the Persian Gulf (McKenzie et al, 1980).

Deepening upward or transgressive cycles characterize some restricted set-tings. A well-known example is the Lofer cyclothems, lagoonal deposits formed behind the Dachstein barrier reef, Upper Triassic of the northern Alps (Fischer, 1964). In these cycles, red or green pelleted weathering residue (A, in Fig. 13) overlies and fills cavities within the deepest-water part of the cycle, the megalodont (bivalve) limestone (C). The weathering residue variously forms an argillaceous parting, fills solution and shrinkage cavities, or forms the matrix of a basal conglomerate. The succeeding "intertidal" deposits (B) contain crinkled and laminated stromatolites, flat-pebble breccia, fenestrae, prismatic mud cracks, sheet cracks and a very restricted fauna of foraminifers, gastropods, and ostracods. The thickest member of the sequence is commonly an overlying muddy to calcarenitic limestone (C) with a relatively diverse biota dominated by megalodont clams, red and green algae, oncoids, and encrusting foraminifera, interpreted as restricted subtidal deposits in a few meters of water (Fischer, 1964). The cycle is abruptly terminated with the appearance of a weathering residue.

A sequence showing progressively

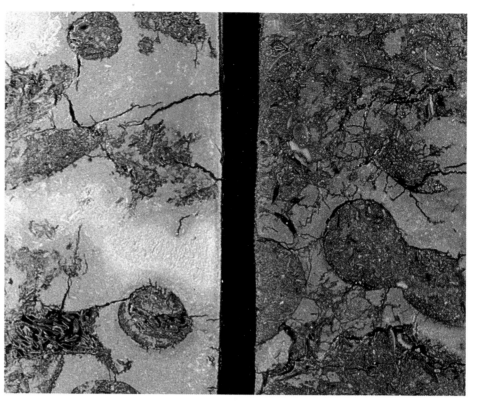

Figure 18—Burrows produced by burrowing shrimp, probably *Callianassa* (see Fig. 4). (**A**, right) 7-cm impregnated core, south Florida inner shelf margin (Fig. 2a). (**B**, below) Underwater excavation, Great Bahama Bank, southwest of Andros Island (Fig. 2b). Water depth about 6 m.

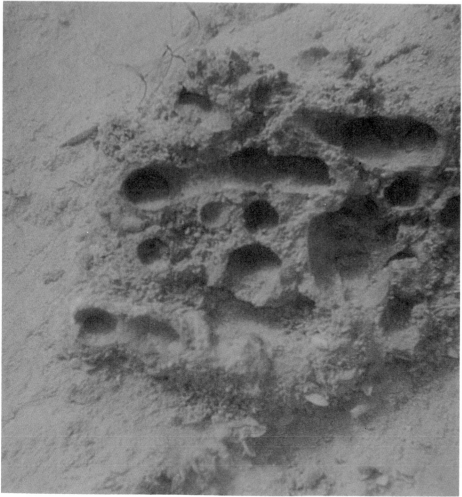

less restriction and probably a slight deepening upward has formed during the Holocene transgression of the south Florida inner shelf margin just seaward of the Florida Keys (Enos, 1977). A mollusk-fragment, lithoclastic lag deposit overlies a microkarstic weathering surface on Pleistocene coralline limestone. The lag deposit is overlain by muddy fine-peloidal sediment (wackestone texture) containing a sparse foraminiferal fauna, few megafossils, and few burrows. This grades up into pelletal-skeletal "wackestone" (using the lithified equivalent) and ultimately packstone. Both are thoroughly bioturbated and contain diverse molluscan and foraminiferal faunas, codiacian green algae, and sea-grass roots. The grain-supported sediment represents the present depositional regime in the semi-restricted inner shelf margin.

An asymmetrical sequence comprising a thin transgressive interval and regressive (progradational) interval records the Holocene transgression on the landward side of the Florida Keys in Florida Bay (Fig. 14). Where mud banks have been colonized by mangroves to form islands, the

Figure 19—Plastic cast of burrow excavated by the shrimp *Alpheus.* Photograph courtesy of E. A. Shinn.

underlying sequence consists of (1) a basal calcitic mudstone of freshwater ponds, (2) peat from coastal mangrove swamps, (3) aragonitic skeletal packstone lag from the open bay and shallow basins between mud banks, (4) skeletal-pelletal wackestone of the mud banks with minor lenses of laminated pellet mudstone and skeletal packstone, and (5) stomatolitic pelletal wackestone with local peat lenses from supratidal islands (Enos and Perkins, 1979). This idealized sequence is transgressive through elements 1 to 3 and regressive through mud-bank and supratidal island phases, which are generally much thicker.

The Black River (Middle Ordovician) and Helderberg (Lower Devonian) Groups in the Appalachian Basin are well-documented transgressive, upward-deepening sequences (Walker, 1972; Laporte, 1969; Walker and Laporte, 1970). Both sequences measure a few tens of meters thick. Each begins with an intertidal-supratidal facies mosaic that grades upward into shallow-water, subtidal muddy limestones with relatively diverse biota. The basal Manlius Formation of the Helderberg Group consists of complexly interbedded pellet mudstones, wackestones, and dolomites with stromatolites, mud cracks, intraclasts, vertical burrows, and a restricted fauna including stromatoporoids, tentaculids, and gastropods (Laporte, 1967). Low-energy subtidal, intertidal, and supratidal depositional environments are represented. The succeeding Coeymans Formation is higher energy deposits including many encrinites that formed skeletal sand shoals and beaches (Anderson, 1972). The overlying Kalkberg Formation is a cherty lime packstone and wackestone with a diverse fauna including brachiopods, bryozoans, and trilobites, deposited on a low-energy, shallow-water "open shelf." The Kalkberg grades up into the more argillaceous, sponge-bearing New Scotland Formation (Laporte, 1969).

Recent, more detailed work in progress suggests that these overall transgressive sequences consist largely or entirely of small (a few meters) shoaling-upward intervals (Figs. 15, 16; Anderson et al, 1978). These intervals or "punctuated aggradational cycles" represent progradational episodes superposed on progressive deepening of water and development of less restricted environments. A similar progression of small shoaling-upward cycles was noted by Bebout (1977) within the generally transgressive Hosston-Sligo sequence (Lower Cretaceous, south Texas). This re-emphasizes the importance of the

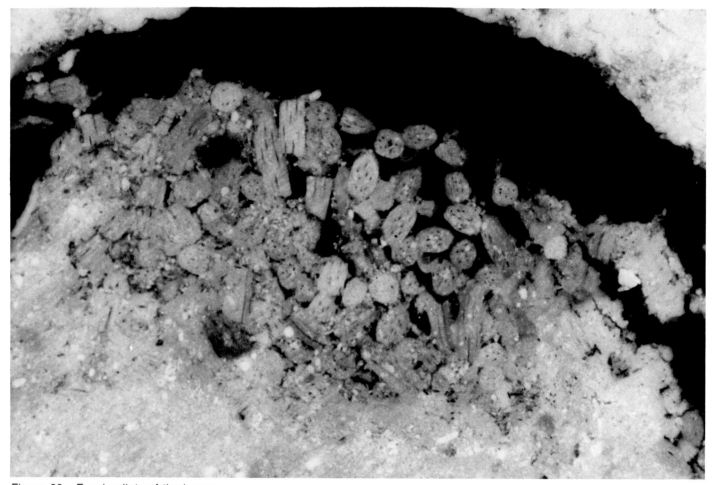

Figure 20—Fecal pellets of the burrowing shrimp *Callianassa,* filling a burrow in a plastic-impregnated core. Holocene, south Florida inner shelf margin, 4 km east of Rodriquez Key (Fig. 2a). Pellets are about 1 mm in diameter.

shoaling-upward cycles beginning in shallow subtidal settings, but does not alter the more regional significance of overall transgressive units or negate the possible development of small-scale transgressive cycles.

The importance of sequence and cyclicity of the various types described in placing shallow subtidal limestones in their proper paleogeographic settings can scarcely be overemphasized.

Sedimentary Structures

Burrows are the most common structures of restricted subtidal environments. Incidence of burrowing in restricted subtidal carbonates ranges from rare or lacking to thorough bioturbation. In fact, the presence of well-preserved discrete

burrows may indicate a lower population of burrowers, or slower rates of burrowing relative to sedimentation, than do mottled or homogeneous sediments that can result from thorough bioturbation. A burrowing infauna persists in oxygen-deficient environments even where oxygen restriction has become too severe to support a shelly epifauna (dysaerobic facies, Byers, 1977). Corresponding observations have not been made for other environmental restrictions such as salinity.

Restricted shelf, bay, and lagoonal environments span parts of the shallow-water to intertidal *Skolithus* ichnofacies and shallow-shelf *Cruziana* ichnofacies of Seilacher (1967). Shallow-water carbonates in general are characterized by *Ophiomorpha* (Permian to Recent; *Skolithus* facies) and by *Thalassinoids* (Triassic to Tertiary), *Rhizocorallium* (Cambrian to Tertiary), and *Condrites* (Cambrian?,

Ordovician to Tertiary), all of the *Cruziana* facies (Fig. 17). However, suites of ichnofossils specifically characteristic of restricted carbonate environments within this bathometric range have not been identified (Kennedy, 1975). Rhoads (1967) has observed that under environmental stress imposed by very shallow water and intermittant exposure, vertical burrows predominate over horizontal burrows and feeding traces. Vertical burrows are indeed more common in many of the environments of interest (Laporte, 1969). Small shoaling-upward cycles in the Black River Group show repeated alternations from horizontal feeding traces in subtidal rocks to long vertical burrows in presumed intertidal rocks (Fig. 15).

Shinn (1968b) described burrow architecture in modern shallow-water carbonate environments. Shrimp, notably *Callianassa* and *Alpheus,* construct distinctive burrows (Figs.

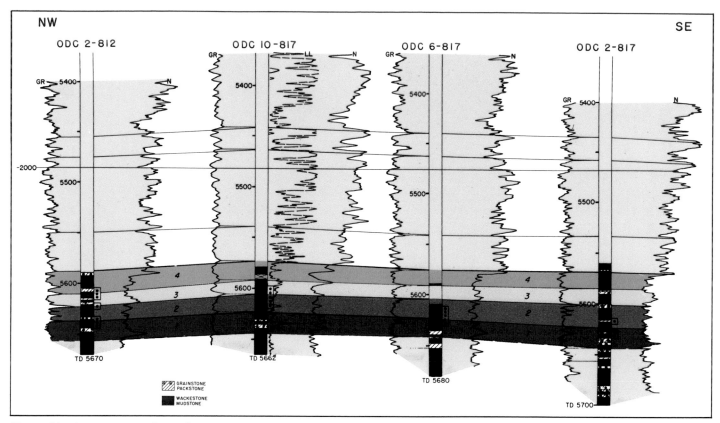

Figure 21—Log response in cyclic carbonates from San Andres Formation (Guadalupian, Permian), Reeves field, west Texas. Four numbered intervals near the base of each well are "backshelf cycles" from the landward part of the Reeves field. The top of cycle 4 coincides with the top of the main porosity zone. Logs are gamma ray (**GR**), neutron (**N**), and lateral log resistivity (**LL**) (from Chuber and Pusey, 1969).

18, 19) that are most numerous in semi-restricted environments, although they persist to the shelf edge at depths less than 10 m (Shinn, 1968b). *Callianassa* burrows are of particular interest in that the complex spiral burrows with galleries and radiating dead-end tunnels (Fig. 4) are restricted to muddy environments. In sandy sediment, *Callianassa* constructs simple straight or predominantly vertical tube networks with thick pelleted mud lining the walls. Identification is enhanced by *Callianassa's* habit of stuffing dead-end tunnels with one type of particle, for example coarse skeletal grains, grass blades, or fecal pellets. *Callianassa's* perforated cylindrical

pellets (Fig. 20) further aid burrow recognition. Similar pellets are known at least as early as Triassic under the form name *Favreina*. The ichnofossil, *Ophiomorpha,* includes callianassid burrows but spans a wider range of environments and probably was produced by other organisms as well (Frey et al, 1978). The basis for definition of *Ophiomorpha* is a knobby burrow lining, not burrow pattern (Fig. 17a).

Lamination is more typical of intertidal and supratidal sediments than shallow subtidal sediments, but they are occasionally encountered in modern subtidal sediments and in ancient sediments interpreted as subtidal. Laminites are common in organic-rich limestones where anoxic conditions have excluded burrowing organisms (Byers, 1977), but most known examples are interpreted as basinal rather than restricted shelf in origin. The McKnight Formation (Cretaceous) of south Texas (Rose, 1972) and the Holzmaden beds of the Lower Jurassic of southern Germany may represent restricted shelf environments. Macrobenthos are also

lacking or severely reduced in numbers of hypersaline lagoons. Here lamination or bedding is commonly distorted by dehydration shrinkage of gypsum or by solution of evaporites. Lenses of laminated carbonate mud occur locally in less restricted settings such as Florida Bay. These may result from sedimentation from suspension in local depressions and from current reworking of low-density, aggregated sediment (pellets) as a traction carpet in slightly more exposed settings (Fig. 3; Enos and Perkins, 1979). The resulting deposits of either of these features, one of negative relief and the other of positive relief, will appear as a small lens of pelleted or clotted, laminated mudstone within coarser, bioturbated, pelletal-skeletal wackestone.

Fenestral or "bird's-eye" structures — generally elongate synsedimentary pores larger than the surrounding grains and lacking obvious framework support — are likewise commonly associated with supratidal deposits but are also reported from restricted subtidal rocks. One example is from "unwinnowed shelf car-

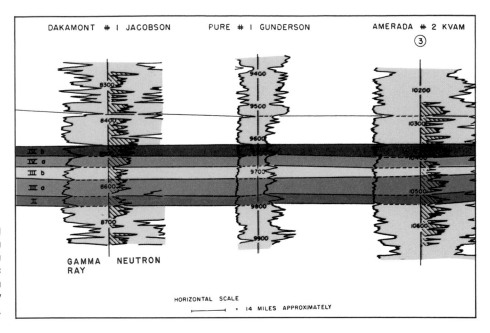

Figure 22—Gamma ray-neutron log response in cyclic Duperow Formation (Upper Devonian), southwestern North Dakota. Roman numerals indicate cyclic units defined on gamma ray-neutron logs, bounded by evaporites and silty dolomites (see text) (from Wilson, 1967).

bonates" of the Upper Permian Tansill Formation, Guadalupe Mountains, New Mexico (Tyrrell, 1969). The Tansill rocks are pellet, grapestone, and intraclast dolopackstones interpreted as comparable to the grass-covered inner shelf-margin sediments of south Florida (Tyrrell, 1969), although grains and textures are more analogous to those of the interior of the Great Bahama Banks. Shinn's statement (1968a) in reference to distribution of fenestral pores in modern sediments that they ". . . are preserved in supratidal sediments. . . sometimes in intertidal sediments. . . and never in subtidal sediments. . ." is valid as a generalization with only a few exceptions.

In summary, sedimentary structures are not the best guides in restricted subtidal environments, with the possible exception of a few distinctive burrow types. In fact, lack of hydrodynamic structures and high incidence of burrowing are typical.

Log Response

The common occurrence of restricted shelf deposits in cycles that include adjacent facies suggests that characteristic mechanical log responses should be recognizable. However, in the few published examples in which good log definition is noted, it is non-carbonate material punctuating the cycle which produces the response. The San Andres Formation (Guadalupian, Permian Basin, Texas-New Mexico) in the Reeves field, West Texas, exhibits marked cyclicity in both shelf-edge, high-energy carbonates and equivalent lower energy "backshelf deposits" (Chuber and Pusey, 1969). The characteristic log response is a positive gamma-ray peak coupled with a negative "dense" deflection on the neutron log (Fig. 21). This response is produced by a dark organic-rich shale interval at the base of the back-shelf cycle.

In correlation of shelf carbonates within the Pearsall Formation (Lower Cretaceous, Texas; Loucks, 1977), R. G. Loucks (personal commun., 1979) noted that cycles or packages could be distinguished on mechanical logs only in the updip portions where tidal-flat deposits contain terrigenous clastic influxes.

Well-developed cycles of normal marine carbonate, through restricted marine to evaporitic unfossilferous dolomite and anhydrite, characterize the lower Duperow Formation (Frasnian, Upper Devonian) of the Williston Basin (Wilson, 1967). The dolomite-anhydrite lithology produces a sharp gamma-ray peak normally coupled with a dense response on the neutron log (Fig. 22), although locally the dolomite has more than 10% porosity (Wilson, 1967). The anhydrite is very pure, but the dolomite commonly contains a few percent of detrital clay and silt which apparently produces the gamma-ray peak. Some cycles in each of these units exhibit tantalizing shapes suggesting characteristic bell-shaped or funnel-shaped curves, but the shapes vary from cycle to cycle (Fig. 22).

Economic Considerations

Porosity Potential

In general, the economic potential of the restricted facies in question is low. Carbonate mud has high initial porosity (60 to 75%; Enos and Sawatsky, 1981), but it is not preservable through the neomorphic transition from the original elongate or irregular, predominantly aragonite grains to the equant, loaf-shaped, calcite crystals typical of ancient lime mudstones (Folk, 1965). Moreover, initial sediment permeabilities are relatively low, (0.6 to 100 md) in Florida Bay subtidal sediments (Enos and Sawatsky, 1981). Pelleted sediments do not have much higher permeabilities nor potential for porosity retention unless a significant number of the peloids are hard and therefore act as sand grains. Peloidal sands of the Great Bahama Bank have frame-supported porosities of 40

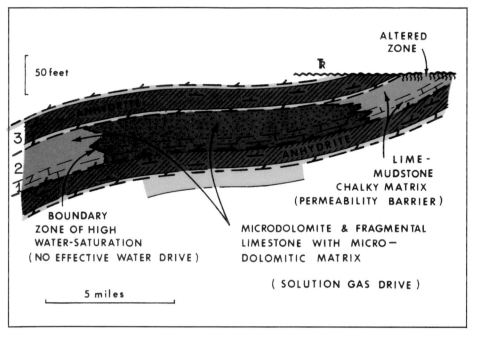

Figure 23—Diagrammatic representation of stratigraphic trap in Midale trend (Mississippian) Saskatchewan. Numbers 1 to 3 are intervals in shoaling-upward sequences described in text (from Illing et al, 1967).

to 50% and initial permeabilities of 1 to 10 darcys, the upper limit approaching that of oolitic sand. Common association of restricted carbonates with terrigenous clays or evaporites further reduces reservoir potential by introducing barriers to fluid migration. Evaporites, moreover, commonly cause secondary plugging of initial pore space beginning with penecontemporaneous diagenesis. Thus most economically important restricted carbonates owe their productivity to diagenetic or tectonic processes.

Common Diagenetic Patterns

Sea level forms the most important datum for early diagenesis, which in many cases is the major determinant of ultimate economic potential. Of submarine diagenetic processes, cementation is the most relevant to reservoir potential and, in general, is more intense in shallow water, culminating in beachrock settings. The effects can be extensive plugging of pore space (Shinn, 1969) or formation of a framework that can aid in retention of primary porosity (Purser, 1969). Subaerial processes can also be either beneficial (leaching) or detrimental (cementation) to porosity and permeability development.

Carbonates from restricted environments considered here are sub-

tidal and are thus less readily effected by exposure than are shoreline facies and various kinds of buildups such as reefs and sand shoals. The cyclic nature of many shallow-water carbonates, however, commonly leads to subaerial exposure and consequent diagenesis at the end of each cycle. Therefore, the lower parts of the cycle, the normal domain of restricted subtidal deposits, may share in the action through the development of a fresh water lens even if they are not actually exposed. Of course, the initially low permeability would continue to retard diagenetic alteration.

Wilson (1975) characterized one of his three types of shelf cycles as "platform cycles with intense diagenesis." The results of exposure or fluctuation in sea level are most pronounced in the top portions of the cycle, typically supratidal or beach deposits, but may extend down to lower portions of the cycle. The effects include large scale shrinkage features such as sheet cracks (Fischer, 1964) and desiccation polygons (Assereto and Kendall, 1971), soil and caliche horizons (Esteban and Klappa, 1982), internal sediment (Dunham, 1969a), soil pipes, "vadose pisoids" (Dunham, 1969b), teepee structures, Neptunian dikes (Fischer, 1964), solution-collapse breccia, cave networks, and dolomitization. The

extent to which the net result shows enhanced or reduced porosity varies considerably from case to case.

Dolomitization is particularly important to the reservoir properties of many carbonates from restricted environments. By most accounts, productive dolomites are coarsely crystalline with connected intercrystalline porosity ("sucrosic" dolomites; Murray, 1960). Dolomitization is typically on a regional scale and is probably attributable to seepage-refluxion (Adams and Rhodes, 1960) or mixing-zone (Badiozamani, 1973) processes. In such cases, favorable porosity development in restricted carbonates stems from their location downdip from extremely restricted environments where evaporites are produced, or their location adjacent to areas of subaerial exposure where a fresh-water lens is developed. Of course, both processes may also be superimposed from a later depositional regime whose facies distribution may be very different.

Fracturing is another method by which reservoir potential of otherwise unfavorable facies may be enhanced. Restricted subtidal facies are as likely to be affected by fracturing as the next facies, especially if they have been made more brittle through dolomitization.

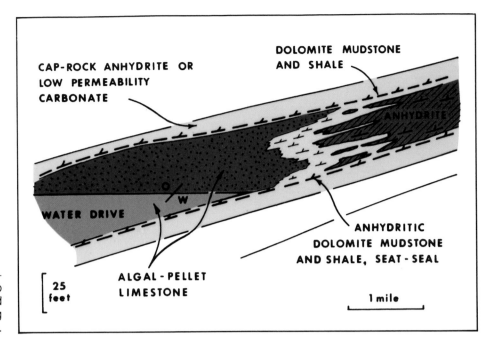

Figure 24—Seal formed by up-dip pinch-out of algal-peloidal limestone into dolomitic-evaporative rocks, Midale trend (Mississippian) Saskatchewan (from Illing et al, 1967).

Petroleum Trap Potential

Trap potential in the simplest cases is controlled by relation either to depositional or structural highs. Restricted subtidal facies are neither depositional buildups nor the updip limits of carbonate units, thus their inherent potential is further reduced. Restricted facies are most likely to be favorable for trapping where adjacent, normally more favored facies are rendered unfavorable by secondary processes. For example, landward facies may become sealed so that hydrocarbons migrating updip will be arrested in the restricted facies. This is the case in some major fields of the Midale trend (Mississippian) of the northeastern Williston basin, Saskatchewan and North Dakota.

Production from Restricted Shelf, Bay, and Lagoon Facies

(1) Steelman-Lampman field, Midale trend (Mississippian) Williston Basin, Saskatchewan (Edie, 1958; Illing et al, 1967) — Carbonate rocks about 500 m thick were deposited during Early Mississippian in the Williston Basin. Maximum accumulation was in the area near the present center of the basin and depositional strike parallels the present outline of the basin in many areas. Mississippian strata are bevelled by a pre-Mesozoic unconformity and overlain by generally impermeable Triassic redbeds.

In the Mission Canyon Formation, open-shelf limestones of the center of the basin grade into restricted carbonates and eventually evaporitic rocks at the northeast margin of the basin. The sequence is generally regressive or progradational. At the northeastern margin, progradation is accomplished by rhythmic shoaling-upward sequences (Fig. 23) consisting of (1) peloidal and bioclastic packstones and wackestones, dolomitized toward the top of the interval, overlain by (2) microdolomite with remnants of ostracods, oncoids, and gastropods, and (3) massive anhydrite and microdolomite with anhydrite stringers. Reservoir rock is developed in the dolomitized limestones of unit 1 with porosities of 10 to 20% and permeabilities of 0.5 to 10 md, and in the dolomite of unit 2 with porosities of 20 to 35% and permeabilities of 5 to 50 md. Seals are formed by the overlying anhydrite (unit 3) and by updip transition into porous but impermeable "chalky" lime mudstone (15 to 25% porosity, 0.3 to 3 md permeability) which is further sealed by redbeds above the unconformity (Fig. 23). Rocks of this configuration contain an estimated 600 million barrels of oil in place at Steelman-Lampman. Comparable fields are Weyburn (1.1 billion barrels) and Midale (500 million barrels). A different type of reservoir is formed where the updip evaporite rhythms are entirely tight and form a seal above algal-peloidal limestones with porosities of 5 to 17% and permeabilities of 5 to 50 md (Fig. 24). This type of trap forms the Riva, Wiley, and Glenburn fields in North Dakota and subsidiary reservoirs at Steelman and Weyburn.

(2) Reeves field (Permian), Yoakum County, Texas (Chuber and Pusey, 1969) — Many moderate-sized oil reservoirs have been located in the cyclic San Andres facies (Guadalupian, Permian) in West Texas and New Mexico (Chuber and Pusey, 1969; Meissner, 1972). One example in which reservoir facies have been shown to be the more restricted part of the cycle is the Reeves field in Yoakum County, Texas (Chuber and Pusey, 1969). The field is located near the shelf edge at the north end of the Midland basin. Shelf-edge cycles are characterized by dolostone sequences 4 to 10 m thick with (1) a basal burrowed organic shale overlain by (2) dark-gray sparsely fossiliferous dolomudstone, (3) gray increasingly fossiliferous dolowackestone and (4) dolopackstone, and capped by (5) oolitic and fossiliferous

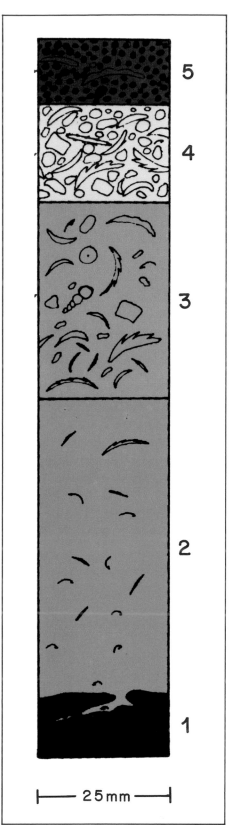

Figure 25—Shelf-edge depositional cycle from San Andres Formation (Guadalupian, Permian), Reeves field, west Texas. Numbers correspond to intervals described in text (from Chuber and Pusey, 1969).

dolograinstone (Figs. 21, 25). Further onto the shelf, in part within the Reeves field, this cycle passes into one in which the shale, dolomudstone, and dolowackestone grade up into stromatolites with shrinkage cracks and birdseyes. Oil is produced primarily from moldic, fracture, and intercrystalline porosity in the dolowackestone; it is occluded from the dolograinstones by secondary plugging of the pores by anhydrite, apparently derived from adjacent sabkha cycles. Draping over a gentle pre-San Andres shelf edge has produced about 13 m of structural closure on the field. An oil column in excess of 30 m indicates that porosity pinchouts updip into the shelf cycles play an important role in reservoir development in this 20 million barrel field.

Other examples in which major production is obtained from rocks probably deposited under restricted marine conditions include the Neocomian limestones of the Lacq gas field, Aquitaine basin, France (Winnock and Pentalier, 1970) and parts of the Asmari Limestone in southeastern Iran (Hull and Warman, 1970). Many of the studies of shoaling-upward cycles mentioned above stem from interest in petroleum potential of these rocks (Duperow Formation, Wilson, 1967; San Andres, Meissner, 1972; Arab-Darb, Wood and Wolfe, 1969), although detailed field studies are not available.

SELECTED REFERENCES

Adams, J. E., and M. L. Rhodes, 1960, Dolomitization by seepage refluxion: AAPG Bull., v. 44, p. 1912-1920.
Ahr, W. M., 1973, The carbonate ramp, an alternative to the shelf model: Trans., Gulf Coast Assoc. Geol. Soc., v. 23, p. 221-225.
Aitken, J. D., 1978, Revised models for depositional grand cycles, Cambrian of the southern Rocky Mountains, Canada: Bull. Canadian Petroleum Geology, v. 26, p. 515-542.
Anderson, E. J., 1972, Sedimentary structure assemblages in transgressive and regressive calcarenites: Montreal, 24th Int. Geol. Cong., Sec. 6, p. 369-378.
Anderson, E. A., P. W. Goodwin, and B. Cameron, 1978, Punctuated aggredational cycles (PACS) in Middle Ordovician and Lower Devonian sequences, in D. F. Merriam, ed., 50th Ann. Mtg. Guidebook, New York State Geol. Assoc., p. 204-224.
Assereto, R. L., and C. G. St. C. Kendall, 1971, Megapolygons in Ladinian limestones of Triassic of southern Alps; evidence of deformation by penecontemporaneous desiccation and cementation: Jour. Sed. Petrology, v. 41, no. 3, p. 715-723.
Badiozamani, K., 1973, The dorag dolomitization model–application to the Middle Ordovician of Wisconsin: Jour. Sed. Petrology, v. 43, p. 965-984.
Basan, P. B., 1973, Aspects of sedimentation and development of a carbonate bank in the Barracuda Key, south Florida: Jour. Sed. Petrology, v. 43, no. 1, p. 42-53.
Bebout, D. G., 1977, Sligo and Hosston depositional patterns, subsurface of south Texas, in D. G. Bebout and R. G. Loucks, eds., Cretaceous carbonates of Texas and Mexico; applications to subsurface exploration: Austin, Univ. Texas, Bureau Econ. Geol., Rept. Invest. 89, p. 79-96.
Bosellini, A., and L. A. Hardie, 1973, Depositional theme of a marginal marine evaporite: Sedimentology, v. 20, p. 5-27.
_____ and D. Rossi, 1974, Triassic carbonate buildups of the dolomites, northern Italy, in L. F. Laporte, ed., Reefs in time and space: SEPM Spec. Pub. 18, p. 209-233.
Byers, C. W., 1977, Biofacies patterns in euxinic basins; a general model, in H. E. Cook and Paul Enos, eds., Deepwater carbonate environments: SEPM Spec. Pub. 25, p. 5-18.
Carrillo, B. J., 1971, La plataforma Valles-San Luis Potosi: Asoc. Mexicana Geologos Petroleros Bol., v. 23, p. 1-101.
Chuber, S., and W. C. Pusey, 1967, Cyclic San Andres facies and their relationship to diagenesis, porosity, and permeability in the Reeves field, Yoakum County, Texas, in J. C. Elam and S. Chuber, eds., Cyclic sedimentation in the Permian Basin: Midland, West Texas Geol. Soc., p. 136-151.
Cloud, P. E., Jr., and V. E. Barnes, 1956, Early Ordovician sea in central Texas, in H. S. Ladd, ed., Treatise on marine ecology and paleoecology: Geol. Soc. America Mem. 67, v. 2, p. 163-214.
Conrad, M. A., 1977, The Lower Cretaceous calcareous algae in the area surrounding Geneva (Switzerland); biostratigraphy and depositional environments, in E. Fluegel, ed., Fossil algae: Berlin, Springer-Verlag Pub., p. 295-300.
Daetwyler, C. C., and A. L. Kidwell, 1959, The Gulf of Batabano, a modern carbonate basin: Proc., 5th World

Petroleum Cong., Geology and Geophysics, Sec. 1, p. 1-22.

Davies, G. R., 1970, Carbonate bank sedimentation, eastern Shark Bay, Western Australia, *in* B. W. Logan et al, eds., Carbonate sedimentation and environments, Shark Bay, Western Australia: AAPG Mem. 13, p. 85-168.

Dunham, R. J., 1969a, Early vadose silt in the Townsend mound (reef), New Mexico, *in* G. M. Friedman, ed., Depositional environments in carbonate rocks: SEPM Spec. Pub. 14, p. 139-181.

_____ 1969b, Vadose pisolite in the Capitan Reef (Permian), New Mexico and Texas, *in* G. M. Friedman, ed., Depositional environments in carbonate rocks: SEPM Spec. Pub. 14, p. 182-191.

Ebanks, W. J., Jr., and J. N. Bubb, 1975 Holocene carbonate sedimentation, Matecumbe Keys tidal bank, south Florida: Jour. Sed. Petrology, v. 45, no. 2, p. 422-439.

Edie, R. W., 1958, Mississippian sedimentation and oil fields in southeastern Saskatchewan: AAPG Bull., v. 42, p. 94-126.

Enos, Paul, 1974a, Reefs, platforms, and basins in middle Cretaceous in northeast Mexico: AAPG Bull., v. 58, p. 800-809.

_____ 1974b, Surface sediment facies map of the Florida-Bahamas Plateau: Geol. Soc. America, Map Series, MC-5, 4 p.

_____ 1977, Holocene sediment accumulations of the south Florida shelf margin, *in* P. Enos and R. D. Perkins, eds., Quaternary sedimentation in south Florida: Geol. Soc. America Mem. 147, pt. I, p. 1-130.

_____ and R. D. Perkins, 1979, Evolution of Florida Bay from island stratigraphy: Geol. Soc. America Bull., Part I, v. 90, p. 59-83.

_____ and L. H. Sawatsky, 1981, Pore space in Holocene carbonate sediments: Jour. Sed. Petrology, v. 51, no. 3, p. 961-985.

Esteban, M., and C. Klappa, 1982, Subaerial exposure surfaces, *in* P. A. Scholle, D. Bebout, and C. Moore, eds., Carbonate depositional environments: AAPG Mem. 33, this volume.

Fischer, A. G., 1964, The Lofer cyclothems of the Alpine Triassic, *in* D. F. Merriam, ed., Symposium on cyclic sedimentation: Kansas Geol. Survey Bull. 169, v. 1, p. 107-149.

Fluegel, E., 1977, Environmental models for Upper Paleozoic benthic calcareous algal communities, *in* E. Flugel, ed., Fossil algae: Berlin, Springer-Verlag Pub., p. 314-343.

Folk, R. L., 1965, Some aspects of

recrystallization in ancient limestones, *in* L. C. Pray and R. C. Murray, eds., Dolomitization and limestone diagenesis: SEPM Spec. Pub. 13, p. 14-48.

Frey, R. W., J. D. Howard, and W. A. Pryor, 1978, *Ophiomorpha*; its morphologic, taxonomic, and environmental significance: Palaeogeography, Palaeoclimatology, Palaeoecology, v. 23, p. 199-229.

Ginsburg, R. N., 1956, Environmental relationships of grain size and constituent particles in some south Florida carbonate sediments: AAPG Bull., v. 40, p. 2384-2427.

Hantzschel, W., 1975, Trace fossils and problematica, *in* C. Teichert, ed., Treatise on invertebrate paleontology: Geol. Soc. America, Univ. Kansas Press, Part W, suppl. 1, 2nd ed., 269 p.

Heckel, P. H., 1973, Nature, origin, and significance of the Tully Limestone; an anomalous unit in the Catskill delta, Devonian of New York: Geol. Soc. America Spec. Paper 138, 244 p.

_____ and J. F. Baesemann, 1975, Environmental interpretation of conodont distribution in Upper Pennsylvanian (Missourian) megacyclothems in eastern Kansas: AAPG Bull., v. 59, p. 486-509.

Hine, A. C., 1977, Lily Bank, Bahamas; history of an active oolite sand shoal: Jour. Sed. Petrology, v. 47, no. 4, p. 1554-1581.

Houbolt, J. J. H. C., 1957, Surface sediments of the Persian Gulf near the Qatar Peninsula: The Hague, Mouton, 113 p.

Hull, C. E., and H. R. Warman, 1970, Asmari oil fields in Iran, *in* M. T. Halbouty, ed., Geology of giant petroleum fields: AAPG Mem. 14, p. 428-437.

Illing, L. V., G. V. Wood, and J. G. C. M. Fuller, 1967, Reservoir rocks and stratigraphic traps in nonreef carbonates: 7th World Petroleum Cong., v. 2, p. 487-499.

Irwin, M. L., 1965, General theory of epeiric clear water sedimentation: AAPG Bull., v. 49, p. 445-459.

James, N. P., 1977, Facies models 8; shallowing-upward sequences in carbonates: Geoscience Canada, v. 4, no. 3, p. 126-136.

Jindrich, V., 1969, Recent carbonate-sedimentation by tidal channels in the Lower Florida Keys: Jour. Sed. Petrology, v. 39, p. 531-553.

Kennedy, W. J., 1975, Trace fossils in carbonate rocks, *in* R. W. Frey, ed., The study of trace fossils: New York, Springer-Verlag Pub., p. 377-398.

Laporte, L. F., 1967, Carbonate deposition near mean sea-level and resultant facies mosaic; Manlius Formation

(Lower Devonian) of New York State: AAPG Bull., v. 51, p. 73-101.

_____ 1969, Recognition of a transgressive carbonate sequence within an epeiric sea; Helderberg Group (Lower Devonian) of New York State, *in* G. M. Friedman, ed., Depositional environments in carbonate rocks: SEPM Spec. Pub. 14, p. 98-119.

Logan, B. W., et al, 1970, Carbonate sedimentation and environments, Shark Bay, Western Australia: AAPG Mem. 13, 223 p.

Loucks, R. G., 1977, Porosity development and distribution in shoal-water carbonate complexes–subsurface Pearsall Formation (Lower Cretaceous) south Texas, *in* D. G. Bebout and R. G. Loucks, eds., Cretaceous carbonates of Texas and Mexico: Austin, Univ. Texas, Bureau Econ. Geol., Rept. Invest. no. 89, p. 97-126.

McKenzie, J. A., K. J. Hsu, and J. F. Schneider, 1980, Movement of subsurface waters under the sabkha, Abu Dhabi, UAE, and its relation to evaporative dolomite genesis, *in* D. H. Zenger, J. B. Dunham, and R. L. Ethington, eds., Concepts and models of dolomitization: SEPM Spec. Pub. 28, p. 11-30.

Meissner, F. F., 1972, Cyclic sedimentation in Middle Permian strata of the Permian Basin, West Texas and New Mexico, *in* J. C. Elam and S. Chuber, eds., Cyclic sedimentation in the Permian Basin: Midland, West Texas Geol. Soc., 2nd ed., p. 203-232.

Moore, R. C., ed., Treatise on invertebrate paleontology: Geol. Soc. America, Univ. Kansas Press, 24 parts.

Murray, R. C., 1960, Origin of porosity in carbonate rocks: Jour. Sed. Petrology, v. 30, p. 59-84.

Neumann, A. C., 1965, Processes of Recent carbonate sedimentation in Harrington Sound, Bermuda: Bull. Marine Sci. Gulf Caribbean, v. 15, p. 987-1035.

Newell, N. D., et al, 1953, The Permian reef complex of the Guadalupe Mountains region, Texas and New Mexico: San Francisco, Freeman and Co. Pub., 236 p.

Perkins, B. F., 1974, Paleoecology of a rudist reef complex in the Comanche Cretaceous Glen Rose Limestone of central Texas: Geoscience and Man, v. 8, p. 131-173.

Perkins, R. D., 1977, Pleistocene depositional framework of south Florida, *in* Paul Enos and R. D. Perkins, eds., Quaternary sedimentation in south Florida: Geol. Soc. America Mem. 147, p. 131-198.

Price, W. A., 1967, Development of the basin-in-basin honeycomb of Florida Bay and the northeastern Cuba lagoon: Trans., Gulf Coast Assoc. Geol. Soc.,

v. 17, p. 368-399.

Purdy, E. G., 1974a, Reef configuration; cause and effect, in L. F. Laporte, ed., Reefs in time and space: SEPM Spec. Pub. 18, p. 9-76.

_____ 1974b, Karst determined facies patterns in British Honduras; Holocene carbonate sedimentation model: AAPG Bull., v. 58, p. 825-855.

Purser, B. H., 1969, Syn-sedimentary marine lithification of Middle Jurassic limestones in the Paris Basin: Sedimentology, v. 12, p. 205-230.

_____ ed., 1973, The Persian Gulf; Holocene carbonate sedimentation and diagenesis in a shallow epicontinental sea: Berlin, Springer-Verlag Pub., 471 p.

Read, J. F., and G. A. Grover, Jr., 1977, Scalloped and planar erosion surfaces, Middle Ordovician limestones, Virginia; analogues of Holocene exposed karst or tidal rock platforms: Jour. Sed. Petrology, v. 47, p. 956-972.

Rhoads, D. C., 1967, Biogenic reworking of intertidal and subtidal sediments in Barnstable Harbor and Buzzards Bay, Massachusetts: Jour. Geology, v. 75, p. 461-476.

Roehl, P. O., 1967, Stony Mountain (Ordovician) and Interlake (Silurian) facies analogs of Recent low-energy marine and subaerial carbonates, Bahamas: AAPG Bull., v. 51, p. 1979-2032.

Rose, P. R., 1972, Edwards Group, surface and subsurface, central Texas: Austin, Univ. Texas, Bureau Econ. Geol., Rept. Invest. 74, 198 p.

Schenk, P. E., 1969, Carbonate-sulfate-redbed facies and cyclic sedimentation of the Windsorian Stage (Middle Carboniferous) Maritime Provinces: Can. Jour. Earth Sci., v. 6, p. 1037-1066.

Seilacher, A., 1967, Bathymetry of trace fossils: Marine Geol., v. 5, p. 413-428.

Shaw, A. B., 1964, Time in stratigraphy: New York, McGraw-Hill Pub., 353 p.

Shinn, E. A., 1968a, Practical significance of birdseye structures in carbonate rocks: Jour. Sed. Petrology, v. 38, no. 1, p. 215-223.

_____, 1968b, Burrowing in recent lime sediment of Florida and the Bahamas: Jour. Paleontology, v. 42, no. 4, p. 879-894.

_____, 1969, Submarine lithification of Holocene carbonate sediments in the Persian Gulf: Sedimentology, v. 12, p. 109-144.

_____, 1982, Ancient carbonate tidal flats, in P. A. Scholle, D. Bebout, and C. H. Moore, eds., Carbonate depositional environments: AAPG Mem. 33, this volume.

_____ R. M. Lloyd, and R. N. Ginsburg, 1969, Anatomy of a modern carbonate tidal-flat, Andros Island, Bahamas: Jour. Sed. Petrology, v. 39, no. 3, p. 1202-1228.

Smith, C. L., 1940, The Great Bahama Bank: Jour. Marine Research, v. 3, p. 1-31, 147-189.

Spackman, W., D. W. Scholl, and W. H. Taft, 1964, Environments of coal formation in south Florida: Miami Beach, Geol. Soc. America Guidebook Field Trip 5, 67 p.

Swinchatt, J. P., 1965, Significance of constituent composition, texture, and skeletal breakdown in some Recent carbonate sediments: Jour. Sed. Petrology, v. 35, no. 1, p. 71-90.

Tsien, H. H., and E. Dricot, 1977, Devonian calcareous algae from the Dinant and Namur basins, Belgium, in E. Fluegel, ed., Fossil algae: Berlin, Springer-Verlag Pub., p. 344-350.

Turmel, R. J., and R. G. Swanson, 1976, The development of Rodriguez bank, a Holocene mudbank in the Florida reef tract: Jour. Sed. Petrology, v. 46, no. 3, p. 497-518.

Tyrrell, W. W., Jr., 1969, Criteria useful in interpreting environments of unlike but time-equivalent carbonate units (Tansill-Capitan-Lamar), Capitan Reef Complex, West Texas and New Mexico, in G. M. Friedman, ed., Depositional environments in carbonate rocks: SEPM Spec. Pub. 14, p. 80-97.

Viniegra O., F., 1971, Age and evolution of salt basins of southeastern Mexico:

AAPG Bull., v. 55, p. 478-494.

Walker, K. R., 1972, Community ecology of the Middle Ordovician Black River Group of the New York State: Geol. Soc. America Bull., v. 83, no. 8, p. 2499-2524.

_____ and L. F. Laporte, 1970, Congruent fossil communities from Ordovician and Devonian fossil communities of New York: Jour. Paleontology, v. 44, p. 928-944.

Walther, J., 1894, Einleitung in die Geologie als historische Wissenschaft: Jena, Fischer Verlag, v. 3, 1055 p.

Wanless, H. R., 1969, Sediments of Biscayne Bay-distribution and depositional history: Univ. Miami (Fla.), Inst. Marine Atmos. Sci., Tech. Rept. 69-2, 260 p.

Wantland, K. F., and W. C. Pusey, III, eds., 1975, Belize shelf-carbonate sediments, clastic sediments, and ecology: AAPG Stud. in Geology No. 2, 599 p.

Wilson, J. L., 1967, Carbonate-evaporite cycles in lower Duperow Formation of Williston basin: Canadian Petroleum Geol. Bull., v. 15, p. 230-312.

_____ 1975, Carbonate facies in geologic history: New York, Springer-Verlag Pub., 471 p.

Winland, H. D., and R. K. Matthews, 1974, Origin and significance of grapestone, Bahama Islands: Jour. Sed. Petrology, v. 44, no. 3, p. 921-927.

Winnock, E., and Y. Pentalier, 1970, Lacq gas field, France, in M. T. Halbouty, ed., Geology of giant petroleum fields: AAPG Mem. 14, p. 370-387.

Wood, G. V., and M. G. Wolfe, 1969, Sabkha cycles in the Arab-Darb Formation off the Trucial Coast of Arabia: Sedimentology, v. 12, p. 165-191.

Zenger, D. H., 1971, Uppermost Clinton (Middle Silurian) stratigraphy and petrology east-central New York: N.Y. State Mus. and Sci. Service Bull. 417, 58 p.

7

Middle Shelf Environment

James L. Wilson
Clif Jordan

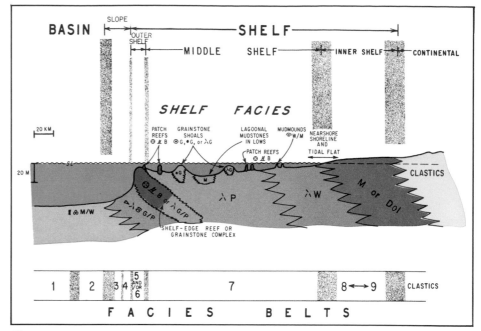

Figure 1a—Profile of shelf facies showing major subdivisions of shelf and facies subdivision of middle shelf such as patch reefs (coral and red algal boundstones), mud mounds (algal plate wackestones to mudstones), grainstone shoals (oolite, pellet, or bioclastic grainstones) and depressions filled with carbonate mud.

*I*n the constant struggle of tectonic uplift versus subsidence and sedimentation, a number of recognizable sedimentary patterns may become established during the various stages of basin-filling. Carbonate shelf facies form one of these patterns. They develop in shallow subtropical to tropical seas where little or no terrigenous material is introduced. The rate of sedimentation is rapid and, though sporadic, can keep pace with tectonic subsidence. For this reason the bulk of carbonate sediment which builds out from cratonic positive areas and fills adjacent basins is deposited in the shallow shelf environment.

The term shelf has always been used somewhat loosely by geologists; here it refers to areas of extensive shallow-water environments bordered on their shoreward side by nearshore or continental sediments and offshore by slope and basinal sediments. Subdivision of the idealized shelf spectrum of carbonate facies, first outlined by Irwin (1965), was expanded by Wilson (1970, 1974) who defined nine basic facies belts. Shelf facies may include six of these belts (Nos. 4 through 9) as shown in Figure 1. In general, the term shelf facies is used here for middle or open shelf sediments. The deposits are continuous, widespread sheets formed predominantly from carbonate sediments produced in shallow-water environments.

The distinctive sediments of the inner shelf (restricted marine), outer shelf, and slope are discussed in other chapters in this volume. Deposits of the middle shelf, mostly of normal marine salinity, are the subject of this chapter. They have the following general characteristics: (1) sea water mostly of normal salinity; (2) depths varying from a few tens of meters

(actually local developments of emergent bars or islands are only 0 to 2 m above sea level) to one or two hundred meters; (3) temperatures ranging from 10 to 30°C (subtropical to tropical); (4) generally well oxygenated water; and (5) bottom conditions commonly below normal wave base, but above storm wave base.

There are several diagnostic criteria for the recognition of shelf facies:

(1) The biota include various stenohaline forms (that is, those with a low tolerance to salinity variations from normal sea water), commonly referred to as "normal marine faunas."

(2) Carbonate textures are generally muddy, with packstones and wackestones dominating shelf environments; but patch reefs and sand shoals also occur, producing local accumulations of boundstones and skeletal grainstones.

(3) Bedding is of variable thickness with lens- or wedge-shapes common. Thin shale beds may interrupt shelf

sequences of limestone or dolomite.

(4) Sedimentary structures such as extensive bioturbation, burrows, nodular bedding, and flaser bedding are typical.

SETTING

Tectonically, carbonate shelves are best developed on the margins of cratonic blocks, in intracratonic basins, across tops of major offshore banks, and on localized positive elements on wide shelves. The thickness of shelf deposits is a function of the stability of the shelf relative to sedimentation rates. Carbonate depositional rates are high, and subsidence is rarely rapid enough to inhibit the system's ability to produce sediment. In most cases, sedimentation far exceeds subsidence. Some of the thickest carbonate sections in the world consist of prograding shelf facies deposited on continental margins. In the Triassic Dolomites of northern Italy and in the Cretaceous of central Mexico, 2,000 to 3,000 m of carbonate shelf deposits were produced over large offshore banks during the time of one or two stages.

The particular depositional setting of shelf facies is influenced by tectonic controls, the outer shelf profile,

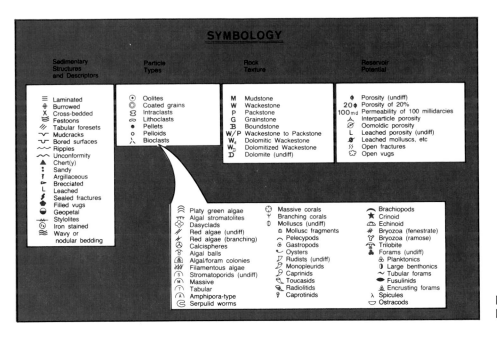

Figure 1b—List of symbols used in Figure 1a.

and to some extent, by the age of geologic time represented. Thus, the type of carbonate facies formed at the shelf edge depends on the character of reef-building organisms, and this in turn controls circulation over the entire shelf. The width of the shelf and the effectiveness of barriers at the shelf edge directly influence the composition of shelf facies. On especially wide, shallow shelves or on shelves with a very well-developed barrier at the shelf edge, restricted shelf facies are deposited (Enos, 1982).

"Restricted" Conditions

Geographic restriction — This invokes the existence of barriers to form a lagoon; the barriers may be shelf-edge reefs or oolite bars, coastal irregularities of the inner shelf, dense reef clusters of the middle shelf, or simply an excessive shelf width not allowing adequate water circulation. This last type of barrier results when frictional forces completely consume marine energies trying to move large volumes of water across broad, shallow shelves. These energy losses across shelf lagoons most strongly affect sedimentation in inner shelf environments. Shaw (1964) discussed this phenomenon in depth and described examples mainly from intracratonic basins.

Salinity restriction — The existence

of shelf edge barriers effective enough to restrict the exchange of water across middle and inner shelf environments is common throughout geologic history. Most examples of severe salinity restriction are from inner shelf environments, where typical barriers include mudbanks, islands, mangroves, swamps, and broad, sluggish shelf lagoons. Depending on climate, the type of salinity restriction may be hypersaline (36 ppm NaCl — usually associated with arid regions) or hyposaline (36 ppm — usually associated with tropical areas with considerable fresh water runoff).

Faunal restriction — Geographic and salinity restrictions result in little or no biotic content in the sediment or in simple fossil assemblages with low diversity, high-number populations. This is caused by the ability of one or two well-adapted organisms to multiply in a stress environment with essentially no competition from other organisms. With regard to complete faunal restriction, there are several examples in the geologic record of basically non-bioclastic shelf sedimentation. The modern-day Great Bahama Bank is a good example, with its pelletal grainstone/packstone facies broadly extending across most of the shelf environment. Hypersaline conditions, lack of nutrients, slight stagnation, and elevated temperatures

of the vast expanse of shallow shelf waters result in faunal restriction.

STRATIGRAPHY AND FACIES

Shelves and their margins follow two main patterns which we term ramp and drop-off models.

Shelves with profiles of the ramp model have broad, gradational facies belts. These sediments are widespread and were deposited in layercake fashion as broad sheets or wedges. Isopach maps of these sediments should reflect the relatively uniform sedimentation rates. The width of these shelves is highly variable and ranges from ten to several hundred kilometers. The width of ramp-model facies belts depends on the angle of inclination of the ramp (the sloping seafloor); gently inclined ramps generally occur on wide shelves and yield irregular facies belts.

Shelves with profiles of the drop-off model have well-developed linear facies belts, trending parallel to the shelf edge. Isopach maps indicate that carbonate facies of the drop-off model are thickest at the outer shelf, especially at the prograding shelf margin. Thinner (and more compactable) beds of the basin occur on one side of the thick outer shelf facies, and thinner beds of the inner and middle shelf facies occur on the other side. The internal geometry of drop-

Figure 2—Upward shoaling sedimentary cycles in middle Pennsylvanian strata in Sierra de las Palomas, northernmost Chihuahua. Recessive bands consist of slightly argillaceous dark limestones with some marine fossils, lower parts of cliffs consist of fusulinid limestones and massive top of each ledge has a grainstone to packstone texture. Very top of ledge commonly displays a lithoclastic sedimentary breccia overlain directly by argillaceous strata of next cycle (see also Fig. 27).

Figure 3—Variable and irregular bed thickness typical of shelf limestones. Alternation lime-wackestone and dark shale beds of the shallow Koessen basin below the Late Triassic Steinplatte Reef, northern Limestone Alps, Waidring, Austria. Deposited in 100 m of water.

off model carbonates usually reflects deposition under classic regressive conditions — somewhat as a large scale foreset/topset type of basin-filling.

Several identifiable stratigraphic patterns occur across shelf areas. The first is a widespread sheet limestone with more or less uniform normal marine faunas and considerable consistency in sedimentary facies over vast areas. The Wolfcampian Hueco Limestone of west Texas and southern New Mexico and Trenton-Galena-Red River-Big Horn-Viola-Montoya unit of Middle to Late Ordovician age are good examples.

A second stratigraphic pattern is typified by upward shoaling sedimentary cycles (Wilson, 1975) wherein normal marine units comprise the early part of each cycle and are replaced upward by restricted marine facies, and, in some examples, ultimately by evaporite beds (Figs. 2, 27). Both uniform sheets and upward shoaling sequences may change facies to more restricted marine facies over slight paleo-highs on the shelf and in shallow mud-filled depressions on the shelf where dark laminated, argillaceous limestones accumulate. Usually narrow linear facies belts occur close to barriers at the shelf margin, and wide, irregular patterns occur on shallow shelves.

Stratigraphic patterns of carbonate sedimentation in middle shelf environments may be strikingly interrupted by shale units representing periodic incursions of clay. These shales are commonly gray-green or brown across the high parts of the shelves; they are noticeably darker in the basins.

Shallow intracratonic basins within broad shelves may be filled in two ways (Wilson, 1975). Slight, gradual, but continuous subsidence results in stratigraphic units passing into the basin without much facies change, but thickening two or three times basinward. Cyclic-reciprocal sedimentation forms another type of basin-filling pattern. During high sea level stands, carbonate fringes or shallow platforms develop around the basin, its central part being generally starved of sediment. The very thin basinal section equivalent to the carbonate fringe is submerged in somewhat deeper water. At lower sea level stands, this basin becomes filled with terrigenous clastics which by-pass the shelves, or thick basinal evaporites may form if proper climate and restriction prevail.

DESCRIPTION OF SHELF FACIES

Regional Facies

Wackestone textures dominate shelf facies; mudstones and packstones are also very common. Shelf carbonates range in color from light to dark gray with lighter shades being typical since most shelves are well oxygenated. The periodic incursions of clay result in discrete stratified shales and are probably due to mild distant tectonic uplift and erosion or climate change on a distant land mass. These shales are commonly calcareous (marls) and may be extremely fossiliferous. The basic association of tabular shelf limestones with gray-green or brown

Figure 4—Even rhythmic bedding with argillaceous siliceous partings in pelagic limestones of Middle Cretaceous Cuesta del Cura Formation. Canyon de los Chorros, 30 km south of Saltillo, Coahuila, Mexico. Contrast this bedding with irregular bed thickness of open-marine shallow shelf facies.

Figure 5—Close up of Portlandian (Late Jurassic) strata in southeast Paris Basin at Ville Arvonal near Chaumont. Wavy nodular bedding caused by differential compaction of varying carbonate and argillaceous content of partings between strata.

shales is most typical of the shelf environment.

The biota of middle shelf carbonates is usually diverse and indicates widespread normal-marine salinities. A lack of fossils or severely limited fossil diversity is caused by a stress environment resulting from an effective shelf barrier of some sort restricting water circulation and salinity across the shelf.

Microfacies

In the vast blanket of skeletal/pelletal wackestones that occurs across most carbonate shelves, three special microfacies occur. These textures are illustrated in photomicrographs of ancient examples of shelf limestones.

Mudstones and skeletal wackestones — These fine-grained carbonates are deposited in localized low places on the shelf to form mud-filled depressions. Hawk Channel behind the Florida Reef Tract is a good Holocene example of a linear depression being filled with muddy skeletal carbonates. Similar low areas with muddy carbonates are known across the Great Pearl Bank of the Persian Gulf.

Grainstones and packstones — Shallow shoals are common in shelf settings. The main particle types are ooids, peloids, or bioclasts. Cross-bedding and shallow-water fossils reflect the relatively high energy conditions. Emergent shoals may develop local beach and supratidal facies. Thicknesses of these middle shelf facies are much less than their facies counterparts on the outer shelf.

Boundstones and flanking skeletal grainstones/packstones — Several types of localized organic buildups occur across carbonate shelf environments, becoming sites of deposition for everything from a rigid reef framework, to piles of reef detritus, to mud mounds with algal plates or mollusks. These buildups may be categorized by their geometry and their position on the shelf. Patch reefs have an irregular, but basically circular pattern in plan view and are found only in shelf environments. Details of their facies have been studied in Bermuda, the Florida Reef Tract, and on the shelf behind the Belize barrier reef (Rhomboid Shoals). Patch reefs are generally very steep-sided and rise sharply from a flat lagoon floor. Their height or vertical relief is naturally limited by shallow water depths.

In contrast, small reefs in deeper water on the outer shelf and slope are referred to as knoll reefs. They develop best on gently dipping slope sediments. They generally are low-relief features and contrast with pinnacle reefs which in basinal environments form very steep-sided structures with considerable relief of up to hundreds of meters.

Patch reefs and lime mud mounds within shelves tend to be more irregular both in shape and trend than such structures at shelf margins. At the shelf edge, they are commonly elongate or loaf-shaped, and directed either parallel to the shelf margin or perpendicular to it. In middle shelf environments, buildups may be controlled by underlying topography or previous reef positions; they may also occur as piles of sediment built out from fault scarps or growing positive areas. In such cases, thicker sediments commonly accumulate on the downthrown sides of faults or on the flanks of paleohighs.

Paleoecology

Fossils found in deposits on shelves with open circulation and normal marine salinities are stenohaline. This

Figure 6—Argillaceous thin-bedded limestone and marl of open shelf fossiliferous wackestone at base of Callovian (Jurassic) upward shoaling cycle, Etrochey Formation, Cure Valley, southeast Paris Basin. Note nodular bedding and irregular thickness of strata.

is true for most carbonate shelves throughout geologic time, except for restricted environments of some inner shelves where a very limited biota exists (Enos, 1982).

Normal marine salinity is indicated by the following fossil groups: brachiopods, echinoids, crinoids, nautiloids, and ammonites, as well as by many individual genera of other fossil groups. The presence of marine calcareous algae indicates water depths within the photic zone which extends down to about 100 m. Common algae found in shelf environments include phylloid (platy) algae of the Late Paleozoic and codiacean algae in general. Red algae also generally thrive in water of normal marine salinity but can flourish in deeper water than can codiaceans. Filamentous blue-green and dasycladacean algae are also present in normal marine environments but are more tolerant of salinity variation and are commonly found in water of less than 5 m. Ginsburg and others (1972) and Wilson (1975) show the distribution of various algal types across carbonate shelves.

Similar schemes exist for foraminifera and mollusks. Planktonic foraminifera such as *Globorotalia* and *Globigerina* occur in basinal deposits from Cretaceous time onward, whereas benthonic foraminifera occur in shelf environments and are larger, more

varied, and commonly thicker-walled. These include fusulinids and endothyrids in the Paleozoic, and miliolids and alveolinids in younger strata. Encrusting forms such as *Homotrema* and cornuspirids/calcitornellids and the dubious foraminifer *Tubiphytes* require hard substrates on which to attach. This limits their distribution on the shelf to patch reefs, rocky crags, or possibly submarine cemented crusts.

Mollusks are also sensitive to conditions of salinity and marine energy across the shelf. Basically, the thick-shelled, robust, and commonly larger mollusks live in high-energy shoal environments. On the other hand, smaller thinner-shelled forms live in muddy parts of the shelf, burrowing through sediment of mudstone to wackestone texture. One of the best examples of mollusk zonation in the geologic past is that of the rudists in Cretaceous carbonates throughout the Tethyan Belt. Caprinids and radiolitids lived on the outer to middle shelf, monopleurids thrived in argillaceous sediment of the middle shelf, and requienids and oysters lived mainly on the inner shelf.

Sedimentary Structures

A wide variety of sedimentary structures occurs in carbonate shelf environments. The few that are specifically diagnostic of the environment in which they formed include

Figure 7—Nodular bedding in *Orbitolina* wackestone from *Salenia* zone of the Cretaceous Glen Rose Formation, Seco Creek, Medina County, Texas. The irregular bedding results from solution-compaction along numerous tiny solution seams concentrating clay layers between purer calcium carbonate lumps.

structures related to surfaces of exposure or to types of cross-bedding.

Bedding Types

Irregular bedding (Fig. 3) is typical of shelf sedimentation. Bedding thicknesses vary from thick or massive homogeneous strata to thinner layers less than 30 cm. Shale interbeds are common. Beds of shelf sediment may be tabular and retain a constant thickness over large areas, or they may gradually thicken or thin

Figure 8—Same as Figure 7. Closer view, showing abundant *Orbitolina*.

Figure 9—Planar sedimentary lamination in shallow subtidal to intertidal Upper Cambrian strata in Canadian Rockies above Ice Field Chalet, Banff-Jasper Highway.

Figure 10—Sedimentary lamination caused by algal stromatolite in shallow subtidal to intertidal strata of uppermost Lower Cretaceous Cupido Formation, Canyon San Lorenzo near Saltillo, Coahuila, Mexico.

over distances of several tens of meters. Local variations in rates of sediment productivity and/or transport result in irregular, lens-shaped beds. Thin clay partings cause bedding planes and separate units of presumably uninterrupted sedimentation. Minor hiatuses then are confined to periods of fine-grained terrigenous sedimentation which seemingly occur between spurts of more rapid carbonate deposition. Vertical sequences of irregular bed thicknesses then must result from irregular episodic introduction of fine-grained argillaceous material onto the shelf. Contrast this irregularlity of bedding with the well-known rhythmites of certain basinal areas (Fig. 4).

Wavy bedding is very common in shelf environments and consists of simple undulose planes at the top and bottom of beds (Fig. 5). This type of bedding grades into nodular bedding which displays undulose patterns throughout the bed, giving an internal nodular or lumpy appearance (Fig. 6). Nodules and matrix material may be all carbonate or of different composition (for example, limestone nodules in shale).

Flaser bedding, also called sedimentary boundinage or ball-and-flow structure, is the result of differential compaction and solution in inhomogeneous mixtures of shale and limestone (Figs. 7, 8). The shales compact readily, but carbonates which lithify early in their history resist compaction. The resulting texture is an irregular, closely spaced, almost nodular structure caused by the disruption of layers by solution and compaction giving the impression of stretching and flowage. Microstylolites are present at contacts between clay seams and carbonate masses indicating that the solution of

CaCO3 may be an attendant process. Inhomogeneity in the original sediment may be attributed to burrowing, the presence of shell beds, or local accumulations of pebbles or intraclasts on the sea floor.

Laminations form in quiet water as sedimentary particles fall from suspension. They develop best in clay and silt-size sediment, but are found in sediments of all size ranges. No inferences can be made as to water depth, for quiet-water deposition can occur in very shallow as well as very deep water. However, in most shelf environments, laminations are destroyed by burrowing. Figures 9 to 12 show a variety of laminations associated with shallow subtidal to intertidal environments.

Algal stromatolite refers to a type of plant-induced curvy lamination that occurs mainly in inner shelf to supratidal sediments. Modern species of blue-green filamentous algae (*Schizophytes*) thrive at the top of the intertidal zone; they also extend downward into near-shore subtidal environments. Actually, no calcified skeletal material exists in stromatolite-forming algae. Studies of Holocene species indicate that these algae form intertidal mats and, at ebb tide, trap fine-grained sediment between their soft extended filaments. In order to photosynthesize, every day the filaments grow through the

Figure 11—Sedimentary lamination including a graded encrinite bed, in shallow subtidal strata, lower part of an upward shoaling cycle in Lodgepole (Mississippian) Formation of Timber Creek Canyon, Big Snowy Mountains, Montana.

Figure 12—Algal stromatolitic lamination overlain by shallow subtidal burrowed unit in El Abra shelf facies of middle Cretaceous exposed at Cuesta de El Abra, 20 km east of Valles, San Luis Potosi, Mexico, Shadow at base is hammer.

fine layer of trapped sediment and form new layers.

Patch reefs, bioherms and mud mounds affect bedding by building structures that interrupt tabular or wavy bedding. For patch reefs, the massive unbedded reef framework provides the source of sediment that is eroded to form an apron of reef flank material (Fig. 13). These beds are inclined as they onlap the sides of the reefs; dips average about 25° but may be as steep as 45°. Beds that cap the reef thin over the reef crest and

characteristically consist of a specialized microfacies adapted to conditions of periodic exposure. An example is the common cap of cornuspirids which forms a dense tangled network of tubular foraminiferal boundstone on top of Late Pennsylvanian and Early Permian bioherms in southern New Mexico and west Texas. Mud mound accumulations in Paleozoic and Mesozoic strata interrupting shallow subtidal bedded lime wackestones are illustrated in Figures 14 and 15.

Cross-bedding refers to inclined sets of sedimentary laminations caused by wind, wave, and current energies transporting sediment. In carbonate shelf environments, the common cross-bedded rock texture is a grainstone commonly forming in sand bars (Fig. 16). The dominant particle types are ooids, peloids, crinoidal particles or other bioclasts (Fig. 17).

Festoon cross-bedding forms as medium-scale cross-beds in sets up to one-half meter thick (Figs. 18, 19).

Figure 13—Patch reef core of caprinid boundstone overlain by gently dipping flank beds of rudist debris and other bioclasts. Rudist mound along Red Bluff Creek, Bandera County, Texas.

They are caused by the scouring of a channel and its subsequent fill by megaripples in moderately strong currents. Festoon cross-bedding is characteristic of channel deposits and occurs in outer shelf environments in channels leading down the slope and also in inner shelf environments in tidal channels or in nearshore sand sheets or belts formed by longshore currents (Ball, 1967).

Tabular, foreset, or accretion cross-bedding forms by the addition of sediment on the downstream side of megaripples. This type of cross-bedding is typical of advancing sand bars (Fig. 20).

Ripple marks are the expression of cross-bedding at the top surface of a bed (Fig. 21). They are of two types: (1) Symmetrical ripples have symmetric profiles and parallel, linear crests. The pattern is like a sheet of corrugated metal. Oscillation ripples are symmetric and form in shallow water by the in-and-out motion of bidirectional waves on sandy bottoms. (2) Asymmetric ripples have asymmetric profiles with short steep slopes on the down-current side and long gentle slopes on the up-current side. Current ripple marks (Fig. 22) are asymmetric and form by unidirectional currents in either shallow or deep waters.

Exposure Surfaces

Sedimentary structures associated with disconformities yield a variety of features reflecting weathering processes, early diagenesis by fresh water, or the availability of hard substrates for nearshore marine organisms — that is, subtidal hardgrounds which are surfaces of non-deposition over long periods of time (Figs. 23 to 26). At the time of formation, these structures occur mainly in inner shelf environments and also on emergent shoals and reef flats across the shelf. However, as regressive cycles develop, they are formed on exposed shelf deposits.

Collapse breccias form by the dissolution of more soluble lithified beds and the ensuing collapse of less soluble strata. Many collapse breccias are formed by evaporite dissolution, but the dissolution of limestone may be equally important. Collapse breccias are commonly developed on the tops of exposed mud mounds although some brecciation may have formed by subtidal slumping. Figure 27 shows a bedding-plane view of limestone breccia persistently seen at tops of massive open marine middle shelf limestones in Pennsylvanian strata of northern Mexico and southern New Mexico (see also Fig. 2).

Bored surfaces are created mostly by mollusks and worms that chemically dissolve lithified carbonate rock (Figs. 23 through 25). Usually bored surfaces are best developed along rock shorelines, but they may occur anywhere across the shelf where hard surfaces are available.

Similarly, encrusting organisms require a hard substrate on which to at-

Figure 14—Lime and mound in Late Jurassic strata in railway cut. Le Roudel near St. Sulpice in Jura Mountains of western Switzerland. Transitional strata between Effinger marls and *Natica* marls.

Figure 15—Small lime mud mound about 5 m thick imbedded in thin bedded encrinites flanking a much larger mound. Upper and east branch of Deadman Canyon. From Lake Valley Group (Mississippian) of Sacramento Mountains, near Alamogordo, New Mexico.

tach themselves. Certain species of bivalves, corals, and algae behave in this manner, oysters providing the best example (Figs. 25, 26). Clear desiccation features such as mud cracks and curled mud chips, and hailstone and raindrop imprints, are obvious indicators of subaerial exposure and are rare to absent in open shelf limestone.

Planar corrosion zones form by the physical abrasion and truncation of surfaces by marine currents and bioerosion, or by meteoric water dissolution. Such surfaces form in a variety of ways and may occur in exposed parts of the inner shelf or as a consequence of submarine erosion in shelf environments (Figs. 23, 24).

Burrows and bioturbation are the most common sedimentary structures observed in limestone of ancient shelves (Fig. 28). In fact, the normal shelf wackestone has a thoroughly bioturbated structure. Burrow mottling of anastomosing trails is marked by color or textural differences, or by dolomitization. Burrow tubes filled with coarser and more permeable sediment are preferentially dolomitiz-ed (Fig. 29). Isolated burrows and burrows with nested pellets are readily recognizable, and the mere presence of discrete burrows probably indicates less burrowing activity than normal. Anastomosing burrow trails are more commonly observed in subtidal sediment (Fig. 30), and vertical burrows occur more in hardened intertidal areas. The net effect of burrowing is to homogenize sediment, destroy primary sedimentary structures, oxidize buried layers, and remove organic material from the sediment.

Figure 16—Lime sand bars in upper Ain Tobi Formation (Middle Cretaceous) in the section at Tarhuna, Jebel Nefusa, near Tripoli, Libya.

Figure 17—Small sand bar of fusulinid-echinodermal grainstone, Leavenworth limestone, upper Pennsylvanian of Oread megacyclothem, Osage County, Oklahoma.

Figure 18—Festoon cross-bedded oolite of Lower Stanton Limestone (upper Pennsylvanian), Tyro quarry, Kansas. Sets are about 8 in thick.

A number of secondary or late diagenetic structures (such as stylolites, fractures, and certain calcite fillings or geopetal cavities) occur commonly in shelf limestones. These are not discussed here, because they are not depositional features, and are not specific to normal marine, middle shelf environments.

HOLOCENE EXAMPLES OF SHELF FACIES

Modern carbonate shelves do not offer as much similarity to deposits on ancient shelves as might be liked; basically what is missing in Holocene shelf deposits is the vast expanse of fossiliferous lime mud seen in the geologic record. The sporadic and periodic Late Pleistocene glaciations and non-equilibrium conditions ex-

isting at present (after the drastic and sudden post-Wisconsin sea level rise of more than 100 m in 18,000 years, or about 5 m per 1,000 years) indicate that carbonate facies patterns on most modern shelves are atypical. Indeed, certain broad shelves are merely veneered with a thin coating of relict calcarenitic sediment. This has been demonstrated for the Campeche Bank, the west Florida Shelf, and the Persian Gulf. The short period of

Figure 19—Cross-bedded (medium scale) festoons in oolite-encrinite strata in sand bar at top of upward shoaling cycle (TC 290) in Lodgepole Formation, Timber Creek Canyon, Big Snowy Mountains, Montana.

Figure 20—Foreset bedding off front of large sand bar, Oolite Blanche Formation in Massangis quarry. Lighter colored, horizontal beds above foresets comprise Comblanchien Formation, restricted marine strata, at top of Callovian Middle Jurassic cycle in Paris Basin.

5,000 to 7,000 years since sea level stabilization has been insufficient for much infilling and covering of the recently transgressed shelves. Yet parts of modern carbonate shelves are very useful in the recognition of ancient sedimentary environments, as well as in studying process/response models. This is true despite the fact that modern shelves may contain special features caused by the paroxysms of Pleistocene sea level change. South Florida, for example, conforms with the standard facies pattern (Wilson, 1974, 1975) seen on carbonate shelf margins (a drop-off model), inasmuch as semi-restricted to normal marine muds occur in Florida Bay (inner and middle shelf), normal marine skeletal muds occur outside the Keys in Hawk Channel (inner part of the outer shelf), marginal lime sands occur in White Bank (outer part of the outer shelf), and coralgal boundstones occur along patches at the shelf edge drop-off into the basin, the Straits of Florida (Fig. 31). But certain features, such

as the protective Pleistocene Keys, the localization of mud banks, and the special biota of *Acropora* reefs, may limit analogies with most ancient shelves.

Another problem in making comparisons is that most well-known modern carbonate environments (exceptions being the Persian Gulf and Shark Bay off Western Australia) are strongly influenced by open ocean conditions with periodic seasonal storms, definite windward/leeward directions, and more open marine circulation.

The Great Bahama Bank: Drop-Off Model, Offshore Bank

Little carbonate mud is deposited at present on banks open to oceanic influences. This is illustrated by the site of several classic studies of carbonate bank facies, the Great Bahama Bank (Fig. 32). A fringe of high energy, normal marine bioclastic grainstone (coralgal facies) or oolite grainstone surrounds the bank. Across the broad expanse of middle shelf environments on the bank, the

Figure 21—Extensive rippled surface developed on top of a mollusk fragment grainstone. Highway 36 quarry, Edwards Limestone (Middle Cretaceous), Bell County, Texas.

Figure 22—Small-scale ripple cross lamination (TC-240) in fine grained peloidal carbonate sand in lower part of upward shoaling cycle, Lodgepole Formation, Timber Creek Canyon, Big Snowy Mountains, Montana.

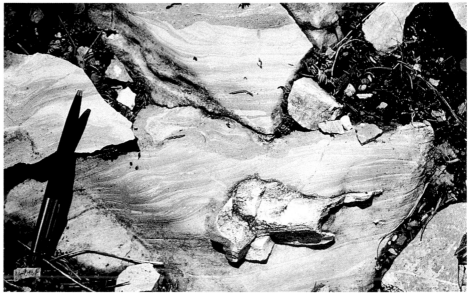

pelletal grainstone/packstone facies occurs (grapestone facies of Purdy, 1961). This exists even though the water is generally of normal marine salinity (except where the pelletal packstone/wackestone facies is being deposited in the lee of Andros Island). Circulation is sufficient to transport interstitial carbonate mud to the tidal flats of western Andros or westward in the opposite direction, off the bank.

Sediments of the inner and middle shelf environments contain mostly non-skeletal material. The bulk of this facies consists of pellets, peloids, lumps, aggregates, indeterminate grains, and carbonate mud. A very sparse fauna and flora (accounting for no more than 30% of the sediment) includes mollusks, *Halimeda,* peneroplid foraminifera, and other foraminifera. In contrast, during the Pliocene, this offshore bank was covered by skeletal packstones and wackestones containing mollusks and foraminifera. Beach and Ginsburg (1980) attribute the Holocene decrease in organic productivity across the

bank to inhibition of growth caused by cooler water and inhibited circulation due to lower sea level during Pleistocene glaciation.

The combined effects of Pleistocene topography and Holocene wind directions are important controls on modern carbonate facies patterns. A prime example is the rim of built-up Pleistocene eolianites on the windward side of the bank, as seen in the Exuma Islands, a discontinuous chain of Pleistocene dune ridges formed at the shelf edge of Exuma Sound. With the Holocene rise in sea

level, saddles between the crests of the dune ridges have been flooded, and now tidal currents pass through these breaks. The bottleneck effect here causes swift currents and agitation sufficient for oolite deposition in the form of tidal spits and bars.

The controlling factor of paleotopography on modern sands is clearly demonstrated here. The distribution of potential reservoir facies in the Exumas is parallel to the shelf edge, but in a discontinuous lobe pattern that occurs only in passes between the Pleistocene islands.

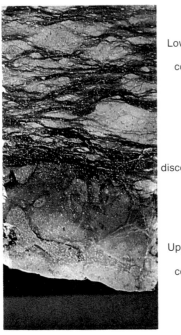

Lower unit
of
couplet

disconformity

Upper unit
of
couplet

Figure 23—Hardground and brecciated surface below ball-and-flow, nodular, or flaser structure in Lower Cretaceous strata of Thamama Group in a Shell well of Id el Shargi Field, offshore Qatar.

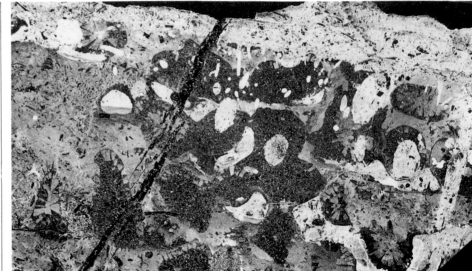

Figure 24—Polished slab of subtidal hardground on coralline limestone bed in Late Triassic Koessen strata at Gaissau section near Adnet, Austria. Notice multiple borings in lithified rock and geopetal filling of cavity below surface (courtesy H. Zankl, Marburg Univ.). Corallites are about 1 cm in diameter.

Figure 25—Top view of extensively bored Cretaceous hard ground surface on oolite with encrusting oysters. Holes are *Lithophaga* borings. Surface is limonitized. Whitestone Quarry, Cedar Park, Travis County, Texas. Hammer handle is 12 cm long.

Belize: Drop-Off Model

The western Caribbean coast from Guatemala to northeastern Yucatan is fringed by a narrow and generally open marine shelf, at most 30 to 40 km wide and as little as 5 km across. Off Belize, the shelf is divisible into several belts (Fig. 33). The outermost is a well developed, thriving barrier reef whose reef flat sediments of coarse sand and rubble are nearly built up to sea level. Shoreward of this, in water depths of a few tens of meters, is a platform up to 35 km wide with numerous patch reefs. The interreef areas are being filled with aragonitic muddy sands.

Farther shoreward, in deeper-water lagoonal environments, foram-molluscan wackestones and mudstones are being deposited. The finer fraction of this sediment contains abundant coccoliths washed across the outer shelf from the open Caribbean. Water over the entire shelf is normal marine, except close to shore where effluent from the Belize River moves southward, and in Chetumal Bay behind Ambergris Cay,

an effective barrier north of Belize City. Here brackish waters with restricted circulation contain very limited biota.

The steep drop-off profile along downfaulted blocks (Glovers Bank, Turneffe Keys, and Cinchorra Bank) and the high energy, open marine environment at the shelf edge are basically similar to the south Florida model. The present carbonate pattern is superimposed on a Pleistocene karst surface (Purdy, 1974). The model demonstrates that with open sea circulation and major storm patterns, most sedimentation remains normal marine and strongly calcarenitic. The only carbonate mud accumulation occurs in the very protected, restricted marine lee areas of the shelf. These limited shelves behind coastal reef-sand barriers and high energy carbonate fringes are hardly analogous to widespread sheet deposits of normal marine wackestones found through much of the geologic record. However, they are very similar morphologically to the curvilinear fringe of Smackover grainstone and associated facies around the Gulf of Mexico.

Persian Gulf: Ramp Model

The Persian Gulf is the only existing shallow inland sea in low enough latitudes and with clear enough water to permit carbonate deposition (Fig. 32). This area is a large (1,000 km by 350 km) shallow sea (20 to 80 m deep) almost completely land locked in a totally desert climate, largely protected from oceanic influence. A hydrographic barrier caused by upwelling at the Straits of Hormuz largely confines connection with the Indian Ocean to surface water. The Gulf has its own

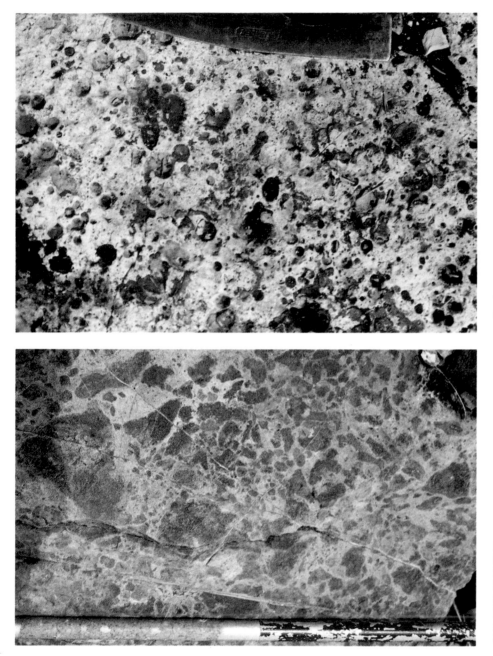

Figure 26—Bored and limonitized surface at hardground on Jurassic Callovian strata, southeast Paris Basin, France. Note blackened nodules on surface.

Figure 27—Top view of irregular blackened clasts (?) believed to result from exposure and partial desiccation. Such surfaces cap grainstone-packstone at top of major upward shoaling cycles in middle Pennsylvanian shelf strata of southern New Mexico and northern Chihuahua. Staff is about 30 cm long.

moderately strong tides with a range of several meters. Minimal ingress of fresh water occurs at the northern end. The Tigris, Euphrates, and Karun Rivers flow only part of the year and do not bring much fine-grained terrigenous material into the basin. The sea bottom is markedly asymmetric with a deep (less than 100 m) axis close to and paralleling the Iranian coast. The vast expanse of the southeastern Gulf (about 200 km across) is the shallow Great Pearl Bank which geographically ap-

proaches very closely some ancient inland epeiric seas. Its depth ranges from 10 to 50 m. Surface salinity here is about 39 to 40%. Salinity increases slightly with depth across this shelf. Tectonically, the Persian Gulf is a foredeep area fronting the Zagros mobile belt; its carbonate shelf is built out from the Arabian Shield. This shallow shelf was apparently laid bare by Late Pleistocene sea level lowering. In the southeast portion of the Gulf, the bottom topography is controlled by westward-facing cuestas

of Eocene and Miocene strata and by Pleistocene karst drainage and terraces. However, regional dip is less than 0.5 m per kilometer, so the surface is essentially flat.

The embayment southeast of the Qatar Peninsula (offshore Abu Dhabi-Dubai-Sharja) has a veneer of various carbonate sediments. The slight swales contain carbonate mud accumulations; the higher areas are shoals of bioclastic grainstones (mainly bivalve fragments), muddy packstones, and coralgal grainstones.

Figure 28—Burrows in argillaceous limestone of Jurassic shelf strata (Lias 5 beds) in Valley of Siz River, High Atlas of south-central Morocco.

Figure 29—Burrow-controlled dolomite and ferruginous mottling on top of Khumi (Lower Cretaceous) limestone in Tang-i-Gurguda, Zagros Mountains, Iran.

The grainstones are presently being cemented on the sea floor. Crudely defined facies belts extend out from the Qatar Peninsula eastward across the northern part of the Great Pearl Bank toward the axis of the Gulf some 200 km offshore (Fig. 32). Water depths increase from a few meters near shore to about 50 m out into the Gulf.

In sequence from shallow to deep, these facies are: rounded mollusk-fragment and gastropod grainstones, angular molluscan grainstones, molluscan wackestones and packstones, and dark argillaceous carbonate mudstones and wackestones.

Depth of water on the open shelf, in which wave action can winnow mud and can round sand-size bioclasts, depends on the degree of protection by the Qatar Peninsula, which shields the area from the northwest shamal wind and inhibits wave action in its lee. This depth generally varies from 10 to 20 m. Holocene sediments are generally thin across this area, averaging about 2 m thick, with 10 to 15 m thickness present in a few low areas as determined from sparker profiles. The fauna is essentially normal marine along this traverse (Houbolt, 1957), but lacks globigerinid foraminifera and codia-cean algae.

In a general way, the facies pattern of the southeast Persian Gulf is spread out along a very gently sloping ramp—one that is almost flat but with local irregularities. Some finer-grained sediments are trapped in coastal lagoons, but for the most part, coarse-grained sands are deposited close to shore, and bioclastic wackestones occur farther offshore. Fringing reefs are beginning to be established in some places close to shore and along windward edges of some shoals, but time has been too short and subsidence too slight since the end of the last transgression

Figure 30—Cemented *Calianassa*-type burrows in the *Globigerina* Limestone (Miocene) at Fomm-ir-Rih Bay, Malta.

Figure 31—Landsat image of southern Florida, Florida Bay, and the Florida Keys. The main landmass shown is the southern tip of Florida, with the Everglades vegetation appearing red. Field checks indicate that light blue in image corresponds to carbonate sediments in very shallow nearshore (less than 10 ft deep) to supratidal environments; medium blue corresponds to subtidal carbonate environments (less than 20 ft deep); very dark blue to black corresponds to deep ocean waters in the Straits of Florida which occupy about 2/3 of the area shown. Clouds are white and their shadows black in the image. The distance represented across the photo is approximately 115 mi.

(5,000 to 7,000 years ago) for a carbonate platform margin (drop-off model) to be established at the gentle break in slope of the bank about 200 km offshore.

SHELF FACIES THROUGH GEOLOGIC TIME

The control of biological evolution on carbonate shelf facies through geologic time is indicated by the fact that certain shelf facies are very characteristic and occur in strata of the same age around the world. For example, there are the famous Jurassic oolite grainstones from the U.S. Gulf coast, northwestern Europe, and the Middle East; Cretaceous rudist boundstones from the Tethyan seaway and the Gulf of Mexico; Mississippian crinoidal grainstones from the midcontinental U.S.A., Alaska, and western Europe; and glauconitic sandstones and trilobite hash beds from the Cambrian.

Early Paleozoic — During this time, carbonate shelf facies consisted mainly of bioclastic grainstones, packstones, and wackestones (Figs. 34, 35); oolites were common, especially in the Cambrian of North America. Normal marine conditions were widespread in the Early Paleozoic, and shelf bioclasts were dominated by brachiopods, bryozoans, pelmatozoans, and trilobites, the latter being particularly abundant in the Cambrian and Lower Ordovi-

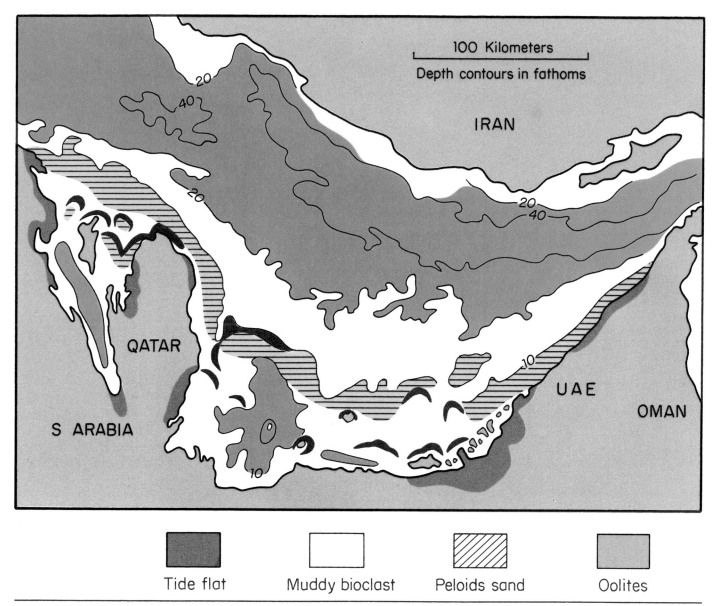

Tide flat Muddy bioclast Peloids sand Oolites

Figure 32—Comparison of modern carbonate facies of Persian Gulf Great Pearl Bank (ramp model) and Great Bahama Banks (drop-off model) on opposite page.

cian. Patch reefs and micritic mounds before Middle Ordovician time contained an assemblage of sponges, calcareous algae, stromatolites, and gastropods.

Middle Paleozoic — Shelf facies at this time were similar to those of the Early Paleozoic, except that in Later Ordovician, Silurian, and Devonian times, rugose corals were added to the brachiopod-bryozoan-pelmatozoan-trilobite faunas (Figs. 36, 37). All of these indicate normal marine

salinities across the shelf. It is interesting that the original mineralogy of these dominantly Paleozoic faunas was all low-magnesian calcite, except for echinoderms which are high-magnesian calcite. Aragonite-secreting organisms are hardly represented as fossils except for nautiloids, gastropods, and quite possibly stromatoporoids and receptaculitids. In the Middle Paleozoic, these presumed aragonitic forms were most common in and around carbonate

build-ups such as mud mounds and patch reefs. Across the North American craton, the textures of these shelf deposits are mostly micritic, occurring commonly as skeletal wackestones or even as calcareous marls. Both generally contain whole fossils. The patch reef facies consist of boundstones with bryozoans, tabulate corals, rugose corals, receptaculitids, and stromatoporoids (Fig. 38). The diverse biota in these reefs, their im-

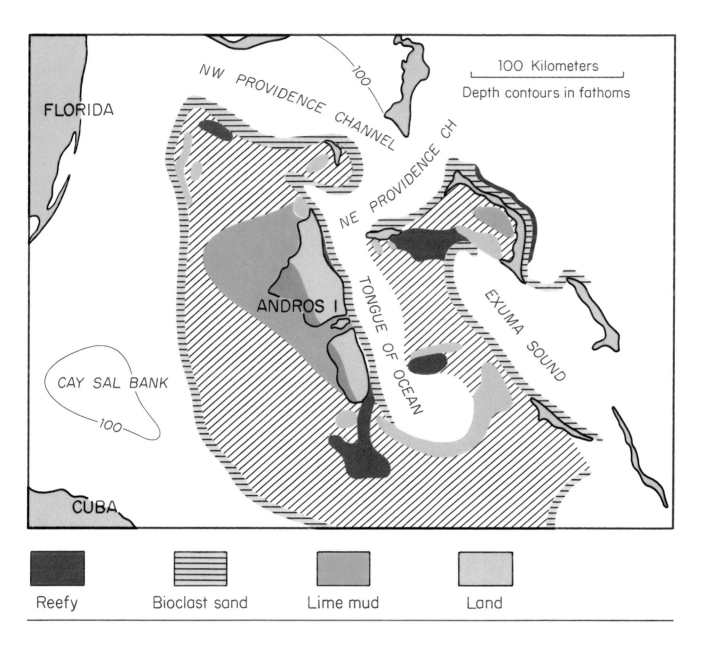

Reefy   Bioclast sand   Lime mud   Land

pressive dimensions, and growth potential contrast strikingly with those of the earlier Paleozoic.

Late Paleozoic — In Mississippian and Early Pennsylvanian time, outer shelf facies were dominated by stalked pelmatozoans (crinoids and blastoids) and bryozoans to the exclusion of most other invertebrates (Figs. 39, 40). Platy green algae, fusulinids, and other diverse and abundant foraminifera were the main additions to Early and Middle Paleozoic shelf

facies from Pennsylvanian to Permian time; they are common faunal elements in micritic carbonates and marls spread across the shelf (Fig. 42 a through f). Patch reef faunas contain numerous encrusting organisms such as certain tubular foraminifera, algae, hydrozoans, and several prominent forms of uncertain biological affinity (*Tubiphytes,* Fig. 41; *Colinella; Archeolithoporella*). Corals and stromatoporoids declined noticeably in the Late Paleozoic. Mississippian

and Early Pennsylvanian patch reefs contained bryozoans and the corals *Chaetetes, Lithostrotionella,* and *Syringopora,* but these reefs were generally small. All of these corals apparently thrived in muddy environments. Pennsylvanian and Early Permian build-ups were formed by phylloid (platy) algae in wackestone textures (Figs. 42f, 43, 44); beds flanking the mounds were typically crinoidal or consisted of other bioclastic packstone or grainstone

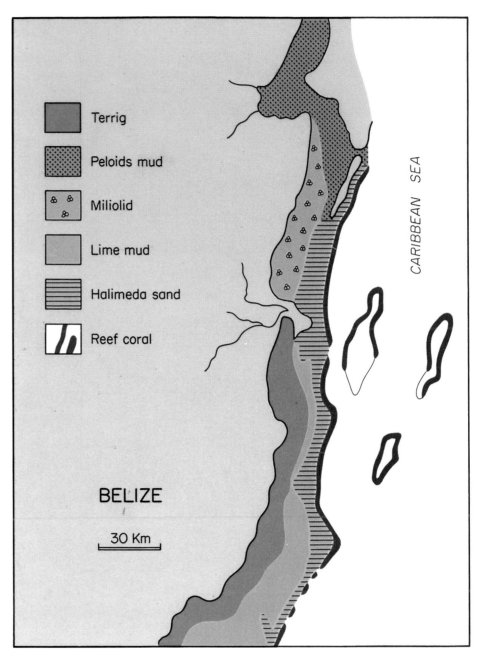

Figure 33—Belize shelf margin drop-off model showing facies belts.

debris derived from communities on top of the mounds. Later Permian reef biota are very distinctive, containing many encrusting algal and algal-like forms, hydrozoans, and a large array of sponges. These faunas continued well into the Triassic.

Mesozoic — Characteristic Mesozoic shelf facies are molluscan wackestones with abundant echinoids and codiacean and/or blue-green algae (*Cayeuxia, Lithocodium,* or *Bacinella*). In the Early Mesozoic,

crinoids and brachiopods were also still abundant. Large shelf foraminifera became prominent, with large valvulinids and pseudocyclaminids in the Jurassic and older Cretaceous, and *Orbitolina* and alveolinids (Figs. 45, 46, 47) later in the Mesozoic. Molluscan and coral biostromes are common in Mesozoic strata and are illustrated in Figures 48 to 51. The Cretaceous also has patch reefs which, early in the period, consist of corals and stromatoporoids and later

of various rudist boundstones with flanking strata of reef-derived debris. From Cenomanian to Danian time, another characteristic outer shelf facies is pelagic coccolith chalk. The same facies with much more diagenetic alteration is known from the Tethyan seaway at the Jurassic-Cretaceous boundary; the Solnhofen, Oberalm, Biancone, and Maiolica limestones.

Tertiary — Skeletal wackestones dominate the shelf facies of the Ter-

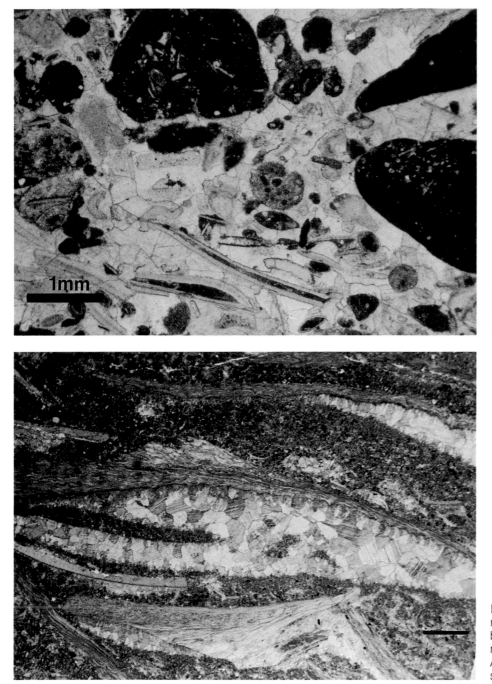

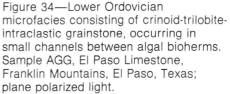

Figure 34—Lower Ordovician microfacies consisting of crinoid-trilobite-intraclastic grainstone, occurring in small channels between algal bioherms. Sample AGG, El Paso Limestone, Franklin Mountains, El Paso, Texas; plane polarized light.

Figure 35—Middle Ordovician microfacies. Brachiopod packstone, bioclastic hash partly infilled with lime mudstone. Salona Limestone in central Appalachians, Pennsylvania, 6X. (bar scale = 2 mm).

tiary with abundant mollusks, echinoids, and large foraminifera such as *Nummulites* and lepidocyclinids (Figs. 52 to 55). These large forms constitute a special outer shelf shoal facies of grainstones and packstones. Patch reefs and biogenic banks of corals, red algae, and bryozoans represent a cooler, more temperate climate in parts of the Mediterranean, more during the Ter-

tiary than in the Cretaceous. The universality of Eocene and Oligocene coral faunas change in the Miocene to more provinciality.

Pleistocene — Following the Holocene rise in sea level, Pleistocene shelves were generally exposed only at outer shelf edges where heaped up eolianites or rapidly growing marginal reefs formed. This is the case in Florida, the Yucatan, Bermuda, and

the Bahamas. However, recent drilling across the Great Bahama Bank (Beach and Ginsburg, 1980) indicates that Pleistocene facies are similar to overlying Holocene pelletal grainstones. Pre-Pleistocene facies, on the other hand, consist of more typical shelf facies of skeletal packstones and wackestones.

Figure 36—Strata of Late Devonian (Frasnian) age between patch reefs near Erfoud, South Morocco, with ammonites and orthocerid nautiloids. View is about 1 m across.

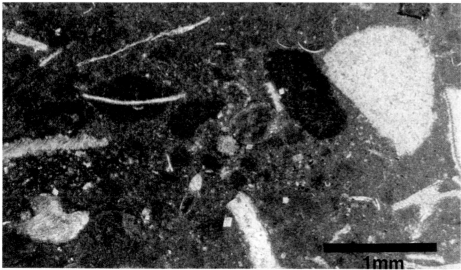

Figure 37—Middle Paleozoic shelf-shallow basin microfacies. Skeletal wackestone with crinoids, brachiopods, *Umbellina* fragments from the Tenneco, Tange 1 in northeastern Montana is the open marine phase of typical Duperow (Frasnian) cycles.

ANCIENT EXAMPLES OF SHELF FACIES

A few regional facies studies of shelf sedimentation with sufficient detail for satisfactory interpretation are available. Several of these are discussed below.

Leavenworth Limestone — In Late Pennsylvanian (Virgilian) time, widespread shelf carbonates were deposited as thin sheet limestones across the mid-continent of the United States (Fig. 56). These limestones are constituents of typical Carboniferous cyclothems, representing chiefly the inundative phases when maximum spread of the seas occurred. The complex meandering shoreline trends roughly north-to-south, and gradual facies changes in the sheets of limestone are east-to-west (Wilson, 1975). Because the strike of the outcrops is roughly NNW-to-SSE and only slightly oblique to the depositional strike, very uniform facies sequences may be traced in Virgilian strata from Iowa and southern Nebraska to northern Oklahoma.

Detailed studies by Toomey (1969a, b, 1972) on the 1 m-thick Leavenworth Limestone of the Oread megacyclothem describe three shelf facies and the distribution of algae and foraminifera. Sample control is from 29 stations along the outcrop belt and all locations are in middle shelf facies. The eastern shore strata are eroded, and the offshore subsurface extent has not been well studied lithologically. Skeletal wackestones (Toomey's skeletal mud facies) comprise the dominant shelf sediments and occur throughout the outcrop belt. Two additional facies were encountered but only at the extreme northern and southern outcrops: (1) skeletal wackestones associated with mudstone and (2) onkoidal packstone to wackestone and fusulinid bioclastic packstone — termed the aggregate grain facies (Fig. 17). From the narrow strip of outcrop, one assumes the broad irregular facies pattern typical of wide shelves. Toomey's distribution charts demonstrate that most algal and foraminiferal species are uniformly spread across the shelf and show irregular patterns of distribution. The presence of *Epimastopora* (a dasycladacean alga) indicates

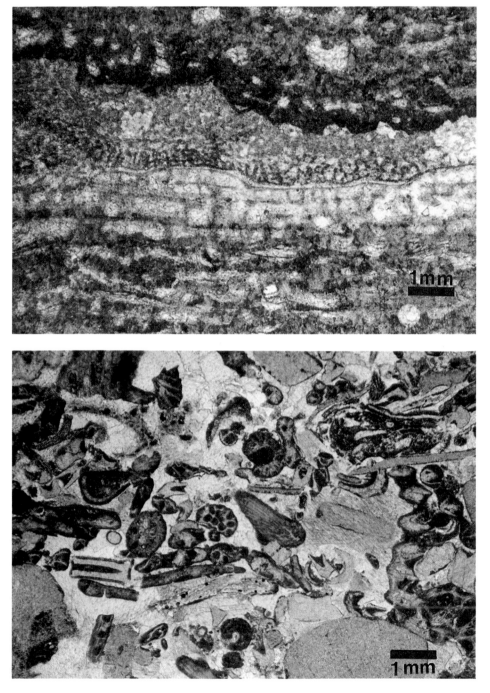

Figure 38—Middle Paleozoic stromatoporoid boundstone from a patch reef in the Duperow Formation (Frasnian) of northeastern Montana.

Figure 39—Crinoid-bryozoan grainstone from a bioclastic bank or shoal in the Lodgepole Formation (Mississippian) shows micrite rims around crinoid and blastoid particles and shows rounded, abraided bryozoan fragments. Timber Creek section, Jenks 1-1, Big Snowy Mountains, central Montana.

shallow water depths of hardly more than 5 to 10 m along the whole 500-km width of the shelf.

In summary, the Leavenworth Limestone is a uniform sheet, 500 km long and at least many tens of kilometers broad, trending into the subsurface of central and western Kansas. It shows second and third order variations that occur in middle shelf environments, but such facies patterns are irregular with a non-

systematic trend. Associated limestones in the Oread megacyclothem, like the Toronto and Plattsmouth, show similarly uniform patterns.

Hueco Limestone — The Hueco Limestone of Lower Permian (Wolfcampian) age and its equivalents throughout southern New Mexico, west Texas, and northern Mexico illustrate a gently inclined ramp profile across the Orograne Basin (Jordan

and Wilson, 1971; Jordan, 1975). Figure 57 shows the paleoenvironmental setting during the Upper, Middle, and Lower Wolfcampian. In contrast, carbonates deposited on the shelves surrounding the Pedregosa Basin change facies rapidly, forming a steeply inclined shelf margin (drop-off model) in southwestern New Mexico and in northern Chihuahua.

Normal marine skeletal wackestones and packstones form the bulk

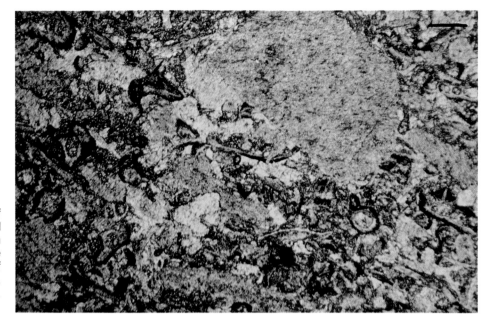

Figure 40a—Encrinite-packstone of crinoidal fragments and triturated brachiopod-bryozoan debris. 6X, from beds flanking lime mud mounds of Lake Valley Group (Mississippian) of Sacramento Mountains, New Mexico. (scale bar = 2 mm).

Figure 40b—Crinoid-bryozoan grainstone from Mississippian strata of southeastern Missouri.

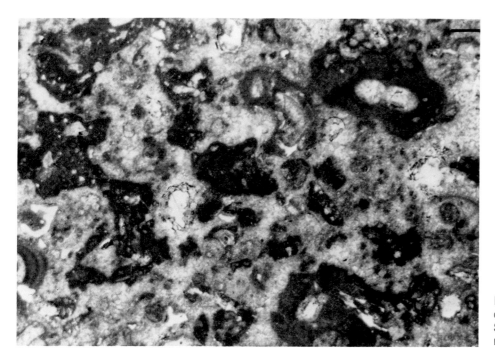

Figure 41—*Tubiphytes* encrusting organism, 8X, from Townsend reef trend, Shell Lusk No. 1, 10,626 ft, Lower Permian (scale bar = 1 mm).

of middle shelf sediment (Fig. 42 a to f). Fossil material includes whole and fragmented crinoids, brachiopods, algal plates, large foraminifera (especially fusulinids and endothyrids), bryozoa, pelecypods, gastropods, ostracods, and trilobites. Peloids are also common constituents; they were most likely soft pellets formed in place by sediment-injesting burrowers and detrital feeders. Typically, these middle shelf facies exhibit some degree of burrowing.

Shoal environments are best developed in Upper Wolfcampian time in west Texas (Hueco Mountains and Sierra Diablo). Bioclastic grainstones and packstones, especially rich in large benthonic foraminifera, were deposited on a broad, submerged southwestward extension of the Pedernal Landmass. Obvious rounding of grains indicates high energy, shallow water conditions; very shallow waters (less than 20 ft) are further substantiated by the occurrence of dasycladacean algae fragments.

A second type of shoal deposit is the cross-bedded fusulinid grainstone or packstone facies. The elongate, football-shape of the fusulinid tests, along with their low bulk density, made them ideal current indicators.

With the proper conditions of current velocity and grain mobility, the fusulinid could pivot and be oriented by currents with its long diameter parallel to the direction of current flow.

Patch reef or biohermal facies in the Hueco consist of brecciated algal plate and *Tubiphytes* wackestones. Algal plate mud mounds formed best in the Middle and Upper Wolfcampian in the central part of the Orogrande Basin. Influx of clastics from the north (Abo) and from the east (Powwow Conglomerate) off the Pedernal Landmass probably muddied inner shelf waters, pushing algal mound development out into the shallow water Orogrande Basin. Algal plates acted as baffles for suspended carbonate mud and aided in the trapping and accumulation of fines. *Tubiphytes,* an encrusting foraminifer(?)[1] (Figs. 41, 42f) attach themselves to algal plates and can form a considerable part of the sediment. Brecciation occurs early in the rock's history as a subaerial breccia, formed during exposure and desiccation.

As a whole, Hueco shelf facies are best described as a facies mosaic involving various mixtures and combinations of skeletal and pelletal particle types in muddy-matrix carbonates. Included here are crinoidal

packstones, brachiopod-crinoid wackestones, and pelletal-skeletal packstones, to name only a few.

Edwards and Glen Rose Limestones — Middle Cretaceous (Albian) rocks in central Texas were deposited on the wide, extensive Comanche Shelf (Fig. 58) which covered most of the state. During Aptian through Cenomanian time, the seaward margin of the Comanche Shelf developed as a shelf edge rudist complex extending in a remarkably even curvilinear pattern along the northern boundary of the Gulf of Mexico from Mexico to Louisiana. This belt is referred to as the Stuart City Reef and consists largely of rudist, coral, and algal debris. The presence of a long, linear thin area in the overlying carbonates and shales, a "thin" which tracks the trend of the reef belt, indicates that the Stuart City Reef formed a ridge with some relief. Immediately seaward of the shelf edge position of the reef trend, the sea floor dropped off abruptly and deeper water carbonate mudstones and wackestones were deposited on the adjacent forereef slope. This well-developed reef and

---

[1]Its origins are undetermined; it has been classed by various workers as a hydrozoan, foraminifer, and a type of algae.

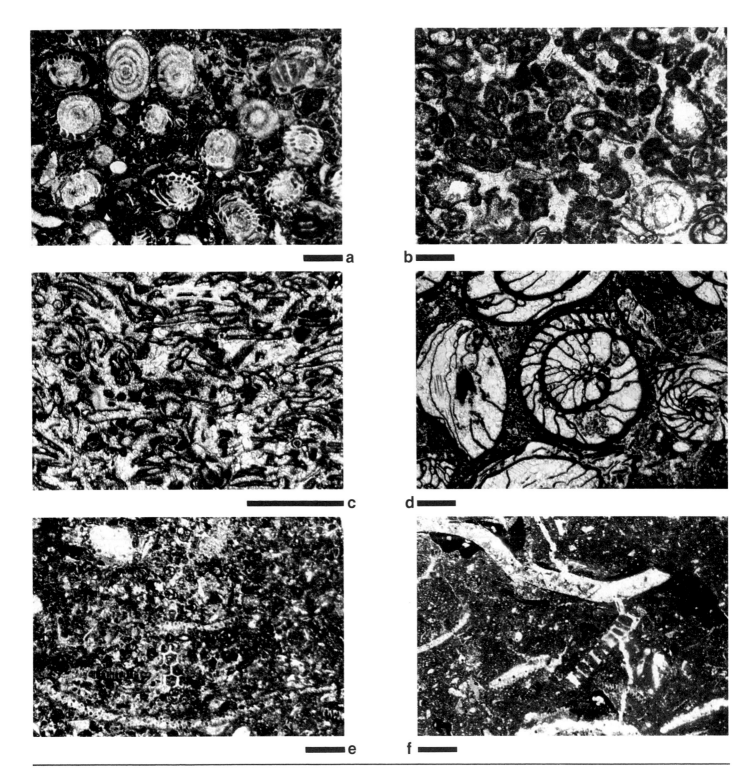

Figure 42—Bioclastic microfacies of the Wolfcampian Hueco Limestone, Permian of southern New Mexico and West Texas (all scale bars = 1 mm). (**a**) Foraminiferal packstone, Hueco Mountains (HM 25), Texas. Large benthonic forams are staffellids, deposited in a shoal sequence; they are commonly recrystallized. (**b**) Foraminiferal grainstone, Hueco Mountains (HM 63), Texas. Forams include cornuspirids (tubular forams), staffellids, and other benthonic genera. (**c**) Tubular foram grainstone, San Andres Mountains (SA 13), New Mexico, commonly forms a capping bed on bioherms. (**d**) Fusulinid packstone, Jarilla Mountains, New Mexico; note compaction effects — deformed and crushed fusulinids. (**e**) Algal plate-foraminiferal packstone/grainstone. Hueco Mountains (HM 73), Texas. Algae include *Epimastopora,* a dasycladacean indicative of very shallow water. (**f**) Algal plate-*Tubiphytes* wackestone, Franklin Mountains (FM 89), Texas. *Tubiphytes* (dark brown) are shown in the upper left encrusting a recrystallized plate of green algae; generally, *Tubiphytes* appear as a detrital grain (lower left near the scale bar), separated from a formerly encrusted surface.

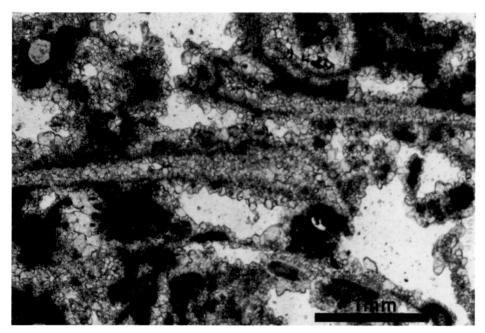

Figure 43—Leached algal plate grainstone showing codiacean(?) algae in Middle Pennsylvanian (Desmoinesian) reefs along the San Juan River. Note porosity development between as well as within algal plates. Sample SJ-128-8.

Figure 44—Silicified codiacean (?) platy algae on outcrop of Late Pennsylvanian strata, Horquilla Formation of Big Hatchet Mountains showing abundance of these accumulations.

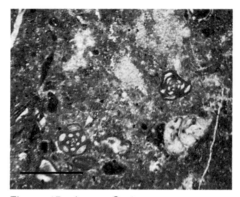

Figure 45—Lower Cretaceous wackestone microfacies of miliolid foraminifera, echinoderm, and molluscan debris, 25X. Middle Cupido Formation, Potrero de Garcia north of Monterrey, Mexico, PG-IV-32 (scale bar = 1 mm).

the broad shelf to the northwest behind it represent a typical drop-off model of shelf-to-basin transitions. This shelf edge pattern commonly is regressive and, with several variations (Wilson, 1974; Hine and Neumann, 1977), occurs throughout geologic time in many stratigraphic sequences.

In contrast, the Comanche Shelf represents a ramp model of sedimentation. This is indicated by a steadily deepening profile of water depths and carbonate facies on the shelf. In the middle shelf environments, broad depressions and swells exerted some influence on thickness and type of carbonate facies deposited. The major

depressions were the south Texas–Maverick Basin and north Texas–Tyler Basin, both of which were filled with evaporites at times in the Albian. Separating these two depressions was a broad elongate swell, the central Texas–San Marcos Platform, bearing southeasterly and seaward from the vicinity of San Angelo. The San Marcos Platform was drowned by the Albian transgression.

The Albian strata across this shelf contain many thin, discrete, widely correlatable units, part of them extensively dolomitized; patch reefs occur both toward the north and southwest perimeter of the dome-shaped Llano Uplift which lies across the axis of the platform (Rose, 1972; Fisher and Rodda, 1969).

The lower half of the Glen Rose, as exposed on the San Marcos Platform (Figs. 59, 60) provides a good example of shelf margin facies. Its inner shelf facies fringing the Llano Uplift is as little as 10 m thick although it thickens to the south; it contains argillaceous beds, algal stromatolites, and ripple-marked pelletal-molluscan grainstones. Its middle shelf facies is as thick as 100 m and is dominated by skeletal wackestones, marly mudstones, calcareous shales, and localized reefs and biostromes; there occurs one very fossiliferous unit

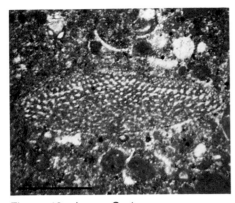

Figure 46—Lower Cretaceous wackestone microfacies with peloids and Orbitolinid, 32X. Upper Cupido Formation, Potrero de Garcia, north of Monterrey, Mexico. PG-V-6 (scale bar = 1 mm).

Figure 48—Biostrome of megalodont bivalves uppermost Triassic strata (Rhaetian at Brauneck, Bavaria, Northern Limestone Alps).

Figure 47—Lower Cretaceous wackestone microfacies, oyster biostrome, 9X, from uppermost Cupido Formation, Potrero de Garcia, north of Monterrey, Mexico. PG-VI-5 (scale bar = 1 mm).

Figure 49—Biostrome of lithiotid bivalves in Jurassic shelf limestone (Lias 3) of High Atlas, Siz River Canyon, Morocco.

toward the top, the Salenia Zone. Glen Rose patch reefs and biostromes are zoned across the inner and middle shelf according to reef framework with oyster, monopleurid, and caprinid reef facies occurring in northeast-to-southwest trending bands paralleling the early Glen Rose shoreline (Figs. 61, 62).

Figure 63 is a cross section somewhat higher in the Albian (Fredericksburg Group) which trends obliquely across the shelf of the San Marcos Platform into the north Texas–Tyler Basin. Over the central Texas High (Llano Uplift), the Edwards facies contains tidal flat mudstones, dolomites, miliolid grainstones, and requenid biostromes.

The inner shelf facies prograde over the middle shelf, where a series of high energy grainstone bars developed not far offshore. Figure 64 shows the distribution of carbonate facies and reconstructed environments of deposition at Moffat Mound near Belton, Texas (Kerr, 1977). The bank measures 40 m thick, several km wide, and about 70 km long. It is basically an oolitic grainstone with skeletal wackestones to the north in front of the bank. Farther offshore, circular caprinid rudist patch reefs occur, surrounded by chalky wackestone. Behind the bank, lenses of scattered patch reefs, reef-derived sands, and skeletal packstones occur. These facies show regressive development of the mound from normal marine shelf through grainstone shoals and rudist patches to exposed beach calcarenites and restricted marine lagoonal sediments, to intertidal mudflats and supratidal dolomites (Fig. 65). These same strata are developed southward across the San Marcos Platform to the Stuart City Reef Trend which is another example of the drop-off model presented originally by Griffith and others (1969) (Fig. 66).

ECONOMIC CONSIDERATIONS

Original shelf carbonate sediments consist dominantly of the minerals aragonite and Mg-rich calcite as carbonate mud with a certain percentage of grains. Generally, initial porosity is high (40 to 70%), and permeability is very low. The mud matrix alters to micrite, which consists of a mosaic of tiny (4 to 10 micrometers) calcite rhombs with much reduced porosity (less than 5 to 10%) and low permeability (less than 10 md). Many ancient limestones of the shelf facies are completely dense, owing to a welding process involving grain solution and reprecipitation of calcite on the grains of the originally porous sediment. Only certain depositional environments and special diagenetic

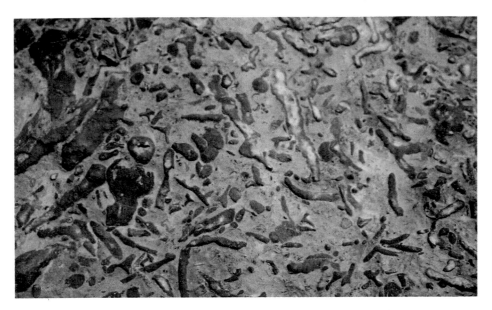

Figure 50—Biostrome of dendroid finger corals, base of Middle Jurassic below pinnacle reefs near Rich, above Siz River Valley, High Atlas, Morocco.

Figure 51—Biostrome of rudist bivalves (*Ichthyosarcolites* bed) in middle of Ain Tobi Formation (Middle Cretaceous), Tarhuna section of Jebel Nefusa, near Tripoli, Libya.

conditions result in preserved porosity and permeability sufficient to form hydrocarbon reservoir rock in middle shelf environments. In view of this, it is interesting that the world's largest single reservoir (the Jurassic Arab D Zone of Saudi Arabia) is in coarse carbonate grainstones of middle shelf facies (Fig. 67).

In the progression of facies across carbonate shelves, potential middle shelf reservoir facies for oil, gas, water, and ore-forming fluids occur in shoal sands, patch reefs, dolomites, and leached horizons beneath unconformities. Widespread pelagic chalks form on deeper shelves and are also middle shelf reservoir targets. They are very porous (up to 45% porosity), but due to their very fine grain size, have low permeabilities (less than 2 md).

Many shelf carbonates have ineffective porosity attributed to either finely disseminated void space in the matrix or to isolated moldic porosity from dissolved bioclasts or oolites. In such tight facies, well developed fractures function as primary conduits for fluid flow with the main enhancement of reservoir quality being improved permeability, accompanied by a

modest increase in porosity. Even extensively fractured reservoirs show only about a 6% maximum increase in porosity due to the storage capability of fractures.

Log response — Traditionally in the petroleum industry, the use of electrophysical logs has been more extensive and more quantitative in production/exploitation situations than in exploration. Generally, interpretations derived from log analysis in clastic sequences are more straightforward than those in carbonates. Several factors account for this: (1) Complex mineralogies occur in car-

bonates — for example, limestone-to-dolomite gradations and a wide range of rock types that includes interbedded or interspersed evaporites, shales, and siltstones; (2) Even pure limestone sequences contain basically inhomogeneous textures — for example, wackestones grading into packstones; (3) Diagenetic effects in carbonate sections are usually more severe or dramatic than in clastics, since metastable carbonate minerals are susceptible to alteration by surface or subsurface fluids. Thus, the distribution of diagenetically induced porosity can be very complex; and (4)

Figure 52a—Typical Tertiary bioclastic wackestone with large benthonic foraminifera, Conoco No. 1 Rubah, 4135 ft, Indonesia; natural size.

With the recognized variability in carbonate systems, the effects of averaging the response of thin beds with mechanical logs can be problematical.

With these limitations in mind, it is evident that running a suite of logs is superior to depending on any single log response (Figs. 68, 69). Having a suite of logs permits the use of Schlumberger cross-plots, as described by Asquith (1979), to determine lithologies. The most cost-effective suite is the ISF Sonic. This suite produces a combination of the following traces:

(1) Sonic log — Used for porosity evaluation (slow zones correspond to porosity development). Sonic log is used to make synthetic seismic traces at the well which are then used to tie in with seismic lines near the well site.

Figure 52b—Foraminiferal packstone of middle shelf facies of the Mokattam Formation (Middle Eocene); sampled in Cairo from quarries that once supplied building stone for the pyramids. The larger foraminifera are microspheric forms (called *Nummulites gizehensis*), and the smaller ones are megalospheric forms (called *Nummulites curvispira*).

(2) Gamma ray log — Useful in indicating shale beds by their naturally high radioactivity.

(3) Induction log — Gathers resistivity data for evaluating fluid content and saturations.

(4) Spherically focused resistivity log — Very shallow penetration is useful for formation evaluation. Aids in the interpretation of resistivity logs with deeper penetration. More quantitative interpretations can be made if the density log is also run. In carbonate sequences, the main density differences are between limestone (2.71 gm/cu cm) and dolomite (2.87 gm/cu cm). Density logs measure bulk porosities with a precision of 1 to 2%.

(5) Neutron log — Neutron logs are also commonly used. Combined with sonic and density logs, lithologies and porosities can be determined by cross-plot techniques.

Figure 53—Red algal colonies 4 to 6 cm across with bryozoans in shallow sub-tidal shelf sediments of Oligocene. Lower Coralline Limestone, Blue Grotto, Malta.

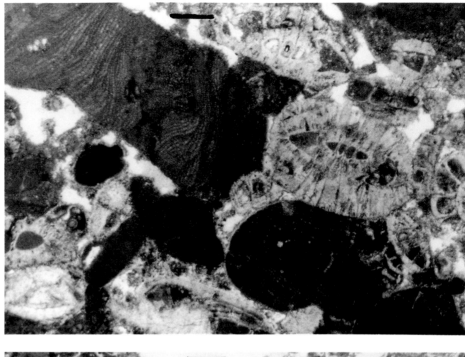

Figure 54—Red algal and large foraminiferal fragments in grainstone shoals of shelf facies, 6X, Lower Coralline Limestone of Malta (scale bar = 2 mm).

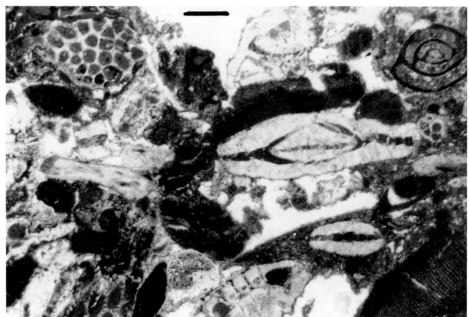

Figure 55—Large foraminiferal, bryo-zoans, miliolid packstone of shelf facies, 6X, Lower Coralline Limestone of Malta (scale bar = 2 mm).

Log responses across carbonate shelves tend to be most uniform in middle shelf environments where broad sheet deposits predominate. Reservoir targets on the shelf include patch reefs, grainstone shoals, and tabular dolomites. Mechanical logs indicate porosity and basic lithologies; however, detailed facies information is available only through sample examination. For example, recognizing Dunham textures or major carbonate particle types cannot be accomplished with log analysis; the same holds true for distinguishing porous boundstone textures of patch reefs from porous oolite grainstones. Obviously, it is necessary to combine both types of information to identify and map carbonate facies with reservoir potential.

Reservoir Facies

Linear shoals of porous grainstone commonly form in outer shelf environments, at or behind the shelf margin and in trends parallel to the margin (Ball, 1967). In middle shelf environments, shoals are less common and form either as irregular, elongate bars or as broad subcircular or ring-like shoals deposited over or around positive areas on the shelf. The Great Pearl Bank of the Persian Gulf contains such bodies of sediment encir-cling weak positive elements across

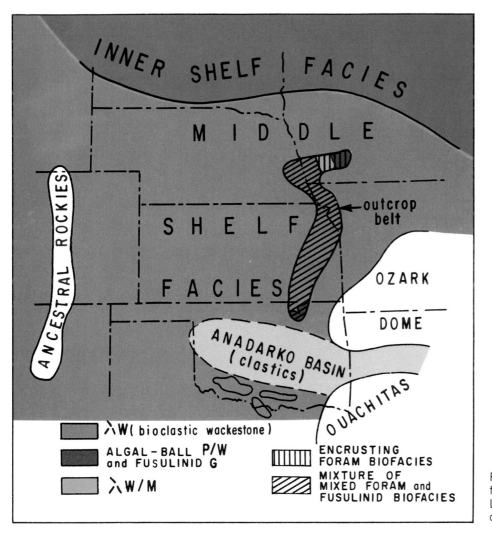

Figure 56—Regional interpretation of facies of the Upper Pennsylvanian Leavenworth Limestone of Kansas outcrops after D. F. Toomey (1969).

the bank, some of which are growing salt domes. Shoal facies in the inner shelf environment include tidal deltas, nearshore bars, beaches, and dunes. Porosities in Holocene shoal facies range from 40 to 50% with an estimated 5 to 10% of the total due to intragranular void space.

Grainstone bodies of the inner and outer shelf are deposited in trends parallel to the shoreline and shelf margin respectively, and such orientation may be of great use in predicting trends of reservoir rock. Shoals of the middle shelf, however, are less abundant and more complexly distributed. An understanding of former circulation patterns in a basin may be necessary to predict the distribution of shoal facies in middle shelf environments. Such an understanding may be possible only in densely drilled areas. Prolific carbonates of the Arab Zones in Saudi Arabia produce

from primary porosity in shoal sands of the middle shelf environment, but wide spacing of drill holes and drilling exclusively along structure in these highly permeable sands has precluded any delineation of trends of component sand bodies.

Patch reefs form as steep-sided boundstone build-ups that rise abruptly from a lagoon floor. They may be common in middle shelf areas. Erosional debris from the reef framework forms sands which are trapped as internal sediment pockets and channels within the reef itself. An apron of reef-derived sediment, formed as flanking beds, grades down-dip into bioclastic wackestones of the lagoon. Several types of porosity (see Choquette and Pray, 1970) combine in these facies to make almost any part of the patch reef system a good reservoir target: cavities constructed and sheltered by

boundstone growth (SH), intraparticle void space in internally porous organic framework (WP), interparticle voids between grains (BP), and true cavities formed by pillared reef growth (GF).

Exploration in patch reef facies encounters several problems, the first of which is that the origin of patch reef development (what makes a patch reef grow where it does?) is simply not well understood. Purdy (1974) proposed that with a rising sea level irregular paleotopography is commonly encrusted with corals and red algae to form patch reefs. Water conditions (circulation patterns and salinities), as well as substrate type, affect patch reef development. In some areas, the vagaries of distribution of coral spat and larvae may control the ultimate shape of reefs, their spacing, and the general reef trend — making prediction of reef

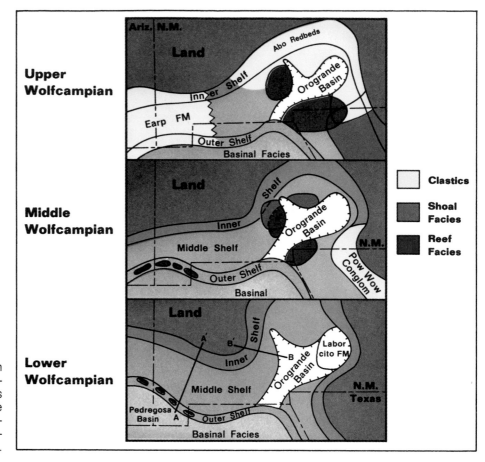

Figure 57—Facies of Wolfcampian Hueco Limestone of southern New Mexico and West Texas. Line A-A′ indicates a "drop-off" profile from the shelf to the Pedregosa basin, whereas line B-B′ indicates a ramp profile into the intracratonic Orogrande Basin.

location essentially a paleoecological problem.

A second problem in patch reef facies exploration is that the term reef is confused in the literature. Following Dunham (1970), ecologic reefs (having true boundstone textures and a typical reef biota) may be distinguished from stratigraphic reefs whose textures and fossils are largely unknown. Ginsburg (in Wilson, 1975) has used the general term carbonate build-up to include both types of reef bodies. The Miocene of Salawati, Irian Jaya is a case in point. Several "reefs" recognized on seismic lines as build-ups were also indicated in bore holes by SP deflection and other log responses. Detailed sample examination reveals that few of these are true coralgal reefs and that many are muddy biogenic bank deposits, mainly packstones with abundant fragments of coral, red algae, and foraminifera. The texture is definitely detrital rather than the growth framework texture of true bound-

stone. The red algae here is the thin branching type, not the typical massive encrusting *Lithothamnion* of the Pacific, and the foraminifera are large benthonic forms such as *Lepidocyclina* and *Gypsina*.

A third problem in patch reef exploration is that a good paleontological understanding of the particular reef fauna and flora is essential to the recognition of reef proximity indicators. Halos of dispersed reef-derived fragments occur around patch reefs. For example, around Holocene patch reefs on the Bermuda Platform, the concentration of coral and *Homotrema* (a large encrusting foraminifera) in the sediment increases near the reef and decreases lagoonward (Jordan, 1973). But the volumetrically dominant reef flanking material is the marine alga *Halimeda*.

Dolomites form predominantly in inner shelf environments, in trends parallel to the general shoreline. Such dolomites are typical of tidal flat environments, with complex facies

variations involving deposits formed in tidal channels, levees, channel bars, beach ridges, ponds, and intertidal flats. Across some shelves, the inner shelf processes of dolomitization apparently acted at a much larger scale and continued to the extent of dolomitizing the entire shelf out to the shelf margin (the Upper Cambrian–Lower Ordovician Knox Dolomite of Tennessee, described by Harris, 1973).

Generally, when fossiliferous subtidal normal marine facies are dolomitized, dissolution of bioclasts begins with fossils originally composed of aragonite and high magnesium calcite; in some cases, all calcite has been dissolved. The resultant moldic porosity is commonly enlarged by further dissolution of the rock matrix to form vuggy porosity. Both intercrystalline and vuggy types of porosity are commonly associated with dolomitized shelf facies. Thus, relatively early syngenetic dolomitization can result in concordant sheets

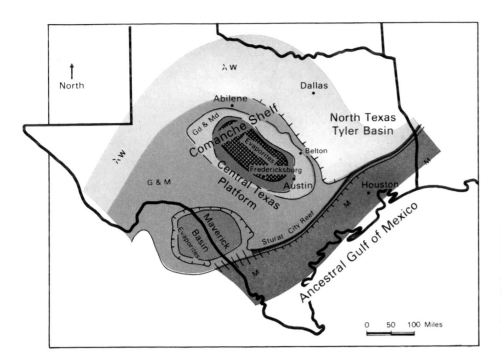

Figure 58—Paleogeography of the Middle Cretaceous (Albian) Edwards Formation of Texas. Major structural and regional depositional features and distribution of lithological types are indicated (from Kerr, 1977). Broad middle shelf environments are bounded on the southeast by the shelf-edge Stuart City Reef Trend (after Perkins, 1974).

Figure 59—Section of middle Glen Rose Limestone, Cretaceous (Albian) along Seco Creek, Medina County, Texas. Transgressive-regressive sequence consisting of four lithologic subunits: (**1, 2**) algal stromatolite on rippled grainstone of inner shelf, (**3**) nodular-bedded, very fossiliferous, normal marine bioclastic wackestone and calcareous shale of middle shelf, (**4**) *Corbula* beds and clay of inner shelf, and solution and collapse zone of shelf interior (see also Fig. 60).

of porosity following intertidal facies. This dolomite may be followed by later discordant patterns of somewhat coarser dolomite which cuts across bedding and represents former water tables. Apparently, these formed through migrating lenses of meteoric water moving interstitial brines above or below them. Such dolomite may form completely uncontrolled by original sediment or grain permeability and may overgrow dolomite formed earlier. Its coarse grain-size may enhance permeability, but its tendency toward overgrowth and vein-filling (in addition to replacement of precursor calcite) may destroy earlier porosity.

The presence of fractures may completely control the economics of prospects in dolomites with minor amounts of vuggy porosity. For ex-

ample, large Anadarko Basin gas fields in the Arbuckle Limestone and in the Hunton Group produce from fractured dolomites deposited originally as shelf facies with vuggy porosities of only 5 to 10% and intercrystalline porosities of but 3 to 6%.

Chalky-textured limestones of shallow water origin may also be reservoirs, especially if fractured. Because the original carbonate sedi-

ment is highly porous, consideration to what processes effectively prevent cementation in buried shelf sediments should be taken. These include lack of burial below 1,000 m, absence of solution welding, and lack of a hydraulic head so that little water flows through the system. Furthermore, where early slight cementation has evenly rimmed calcarenite grains, the fringe of cement can prevent

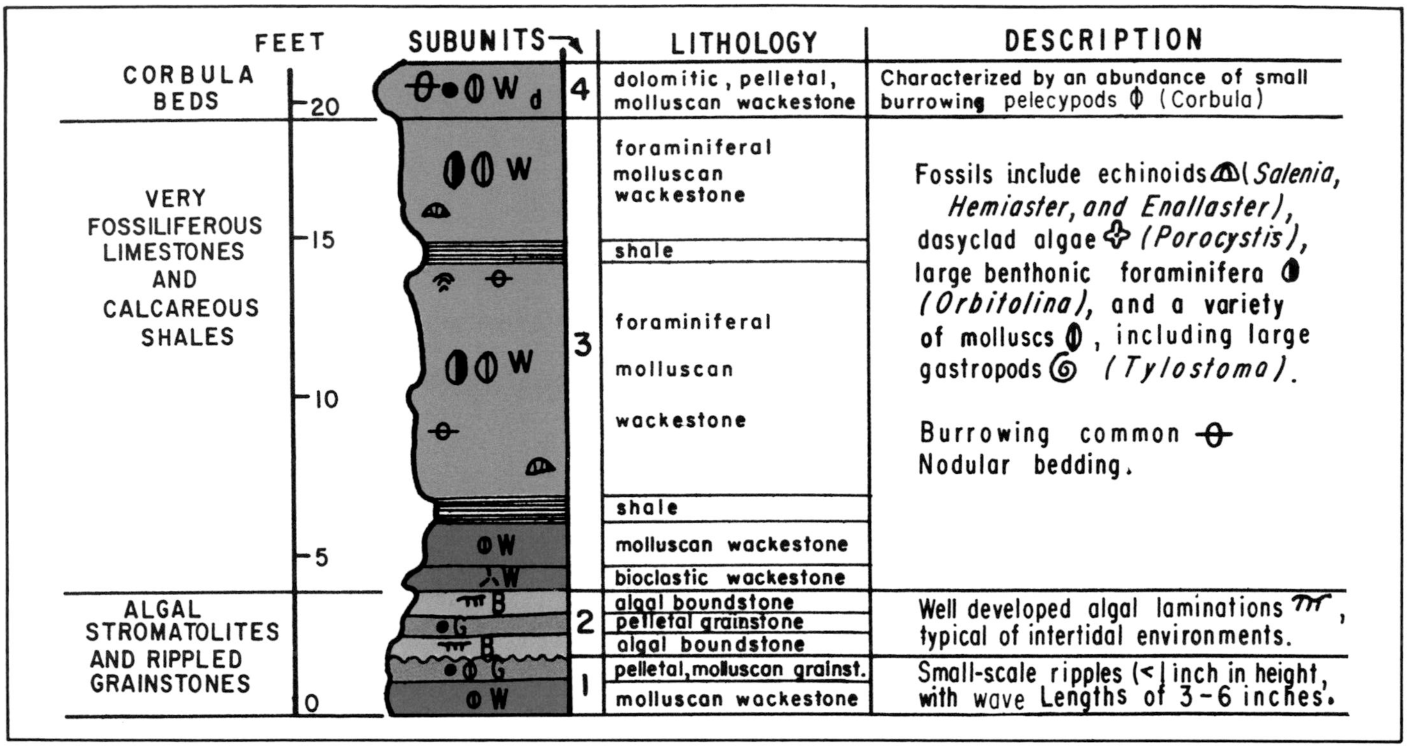

	FEET	SUBUNITS	LITHOLOGY	DESCRIPTION
CORBULA BEDS	20	4	dolomitic, pelletal, molluscan wackestone	Characterized by an abundance of small burrowing pelecypods Φ (Corbula)
VERY FOSSILIFEROUS LIMESTONES AND CALCAREOUS SHALES	15		foraminiferal molluscan wackestone	Fossils include echinoids (Salenia, Hemiaster, and Enallaster), dasyclad algae (Porocystis), large benthonic foraminifera (Orbitolina), and a variety of molluscs, including large gastropods (Tylostoma).
			shale	
	10	3	foraminiferal molluscan wackestone	Burrowing common. Nodular bedding.
	5		shale	
			molluscan wackestone	
			bioclastic wackestone	
ALGAL STROMATOLITES AND RIPPLED GRAINSTONES		2	algal boundstone	Well developed algal laminations, typical of intertidal environments.
			pelletal grainstone	
			algal boundstone	
		1	pelletal, molluscan grainst.	Small-scale ripples (<1 inch in height, with wave Lengths of 3-6 inches.
	0		molluscan wackestone	

Figure 60—Columnar section of lower part of cliff section on Seco Creek, Glen Rose Formation, Medina County, Texas (see also Fig. 59).

solution-compaction welding after burial and the "locked-in fabric" can remain highly porous. This porosity, however, is commonly destroyed by a second, later generation of clear, blocky calcite. If brine rich in magnesium and low in calcium permeates carbonate muds and stagnates within the porous but not very permeable layers, it may prevent solution-compaction welding and the sediment retains the original high chalky porosity under some minimum of overburden.

Shelf Carbonates as Source Rocks

Carbonate source rocks occur only in muddy carbonate facies. This excludes grainstones, packstones with a high grain content and low mud content, and most boundstones — a notable exception being the typically organic-rich algal stromatolites. Apparently, this relationship is a function of preservation of organic material rather than of the original organic content. In modern lagoonal environments where carbonate mud is accumulating today (Florida Bay and northeastern Yucatan), the content of organic carbon, much of it humic as well as algal, may range from 0.3 to 3%. Sediment cores taken in Florida Bay show a thin, light colored oxidation zone that extends down only a few inches from the sea floor. Beneath this, there exists a reducing zone of gray carbonate mud that yields a noticeable odor of hydrogen sulfide. In contrast, modern skeletal grainstones (for example, the Florida Reef Tract) do not display this obvious reducing zone. This results from the higher grainstone permeabilities allowing greater exchange of surface waters through sediment. The degree of bioturbation may also affect the ability to preserve organic material in carbonates.

Basinal carbonate facies are most likely the richest carbonate source rocks (the Permian Bone Springs Limestone of West Texas; the Cogollo of the Maracaibo Basin; and the Jurassic Diyab-Darb of Saudi Arabia), although muddy limestones of the inner and middle shelf may also be adequate source beds to charge nearby reservoir facies. In southern Florida's Sunniland Trend, inner to middle shelf mudstone facies of Cretaceous age are apparently source rocks for overlying reservoir facies. Data from Palacas (1978) and Pontigo and others (1979) indicate an average of 0.55% organic carbon, 200 to 800 ppm of C_{15+} hydrocarbons, and an algal origin for kerogen that is moderately mature. Total production from the Sunniland Trend amounts to approximately 55 million barrels of oil through 1977, the largest fields being Sunniland Field with an estimated ultimate recovery of 18.8 million barrels and West Sunoco-Felda Field with an estimated ultimate recovery of 50 million barrels of oil. In addition, Palacas (personal commun.) observed an even carbon number preference in the C_{20}-C_{30} n-paraffins. This trend of even number preference may in time prove to be characteristic of carbonate source rocks; in contrast, clastic

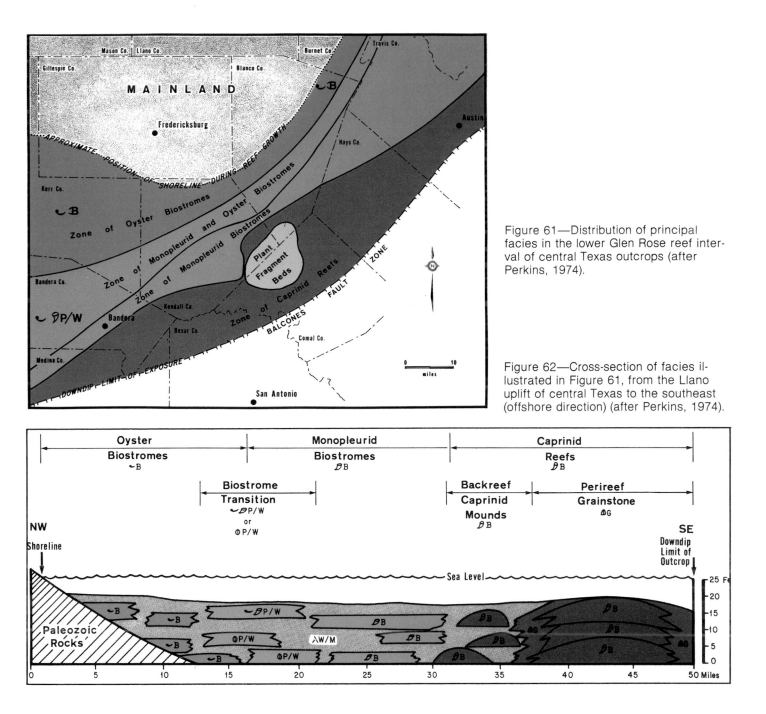

Figure 61—Distribution of principal facies in the lower Glen Rose reef interval of central Texas outcrops (after Perkins, 1974).

Figure 62—Cross-section of facies illustrated in Figure 61, from the Llano uplift of central Texas to the southeast (offshore direction) (after Perkins, 1974).

source rocks tend generally to display an odd-number preference.

In general, carbonates have been discounted or underestimated as source rocks. However, it appears that carbonates, with their high content of non-calcified algal material, are more efficient hydrocarbon producers than are clastics. Data from Gehman (1962) and Claypool and Reed (1976) indicate that the hydrocarbon content per unit of organic carbon is 3 to 4 times higher in carbonates than in shales. This suggests that the lower limit cutoff of 1% organic carbon in traditional clastic source rocks would be about 0.25 to 0.30% for carbonates. The increasing acceptance of fine-grained carbonates as source rocks was demonstrated by a recent symposium on this topic entitled "Petroleum Geochemistry and Source Rock Potential of Carbonate Rocks" (Geological Society of America 1980 Annual Meeting, Atlanta).

The problem often arises as to the ability of carbonate shelf facies to act as seals and to trap fluids or gas. The sheet-like geometry of shelf deposits and their basic "layer cake stratigraphy" certainly encourage the formation of seals for reservoirs. The progradational pattern of the typical upward-shoaling cyclic sediments across shelves aids in sealing off middle shelf reservoirs with both up-dip and overlying fine-grained carbonates and shoreward sabkha evaporites.

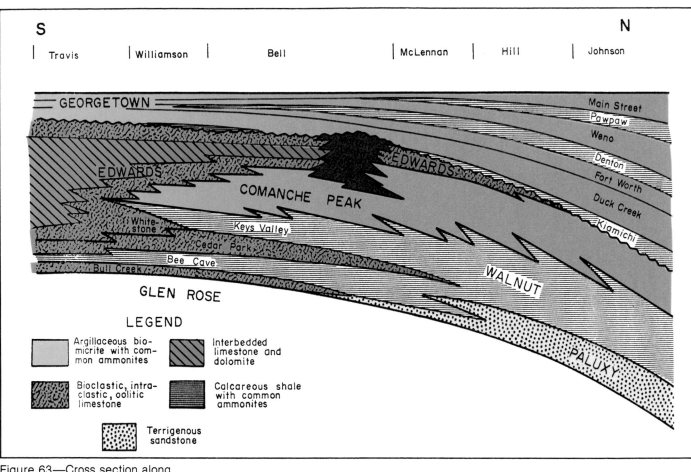

S   N

| Travis | Williamson | Bell | McLennan | Hill | Johnson |

Figure 63—Cross section along Fredericksburg outcrops from central to north Texas indicating progradation of upward shoaling facies in a northerly direction (from Rose, 1972).

However, the brittleness of dolomite and limestone exceeds that of other sediment types so that carbonates tend to fracture more readily than terrigenous clastics. There is, then, no straightforward answer concerning the quality of sealing carbonates. Many carbonate rocks probably are of intermediate quality and leak fluids continuously. If the leaks are slow enough to permit the temporary accumulation of fluids, the beds are effective seals.

The lack of strata with sealing capacity is the main weakness seen in evaluating some 3,000 m of Upper Cretaceous and Tertiary carbonates of south Florida and the Bahama Banks. Seawater invasion has occurred throughout this part of the section. In a well on Long Island in the Bahamas, salt water extends down

past 5,500 m, as determined by a complete lack of any temperature gradient to the total depth of the well (Meyerhoff and Hatten, 1974). However, dead oil stains in these beds indicate the presence of migrated hydrocarbons high in the section.

Predictive Stratigraphic Models of Shelf Sedimentation

Several models exist designed to predict the distribution of carbonate reservoir facies. In most carbonate sections, primary facies patterns control the distribution of porosity and permeability, and the description of carbonate shelves is carried out through extensive sample examination programs. Where diagenetic facies control the porosity and permeability, the overprint of diagenesis may be concordant with primary facies pat-

terns or discordant. An example of a concordant relationship occurs in the Jurassic Smackover Formation of southwestern Arkansas. At Walker Creek Field, Becher and Moore (1976) found porous oolite grainstones only at high parts of oolite bars, parts subaerially exposed to meteoric diagenesis. The remainder of the bars were non-porous owing to calcite cement. In contrast, discordant relationships exist with regard to the distribution of dolomitic facies in the Hunton Group of the Anadarko basin where irregular facies boundaries cut across what must have been primary trends parallel to the strike of the basin (Amsden and Rowland, 1967). Similar discordant trends of dolomitization are observable in the Madison of central and western Montana.

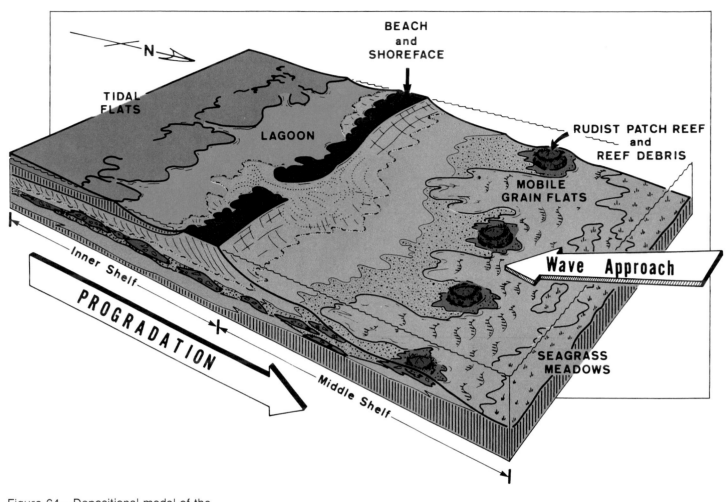

Figure 64—Depositional model of the Edwards Limestone (upper Fredericksburg and equivalent strata) in the area of Belton, Texas, showing progradational inner and middle shelf facies (Kerr, 1977).

The steep shelf margin or drop-off model is the most reliable and useful in predicting facies. In such situations, the most favorable reservoir facies are commonly found at or near the shelf edge, but inner shelf dolomites are also prime reservoir facies. Patch reefs and shoals across the middle shelf are commonly productive, but their distribution is usually non-systematic. On the other hand, the less regular facies developed on the ramp model are determined by the low ramp gradient. Very gentle sea floor inclinations of far less than 1° occurred in broad epicontinental seas and intracratonic basins. The resultant facies pattern reflects the broad, subtle facies varia-

tions found in widespread shallow water environments (Irwin, 1965; Wilson, 1975). In contrast, steeper dips observed in the Smackover of the Gulf Coast demonstrate that narrow more distinct facies belts form with steeper ramp gradients.

The shelf margin or drop-off model is well established and discussed in many papers. The basic facies belts of inner, middle, and outer shelf are generally recognized, and outer shelf reservoir facies are common exploration targets.

A real problem, however, is the lack of predictive depositional models within the middle shelf. Most of the shelf consists of irregular belts of facies and local patch reefs with no

trends. The only method for identifying proximity to isolated organic buildups is through rather sophisticated ecological study of contiguous biota, and this usually furnishes only very local control.

There are two regional stratigraphic situations across broad middle shelves where prediction of reservoir rocks is possible:

(1) In stratigraphic traps in progradational sedimentary cycles (inner to middle shelf). The normal spectrum of carbonate facies, seen in upward-shoaling cycles across shelves, places (following Walther's Law) muddy carbonate source rocks of middle shelf environments in lateral sequence next to high energy

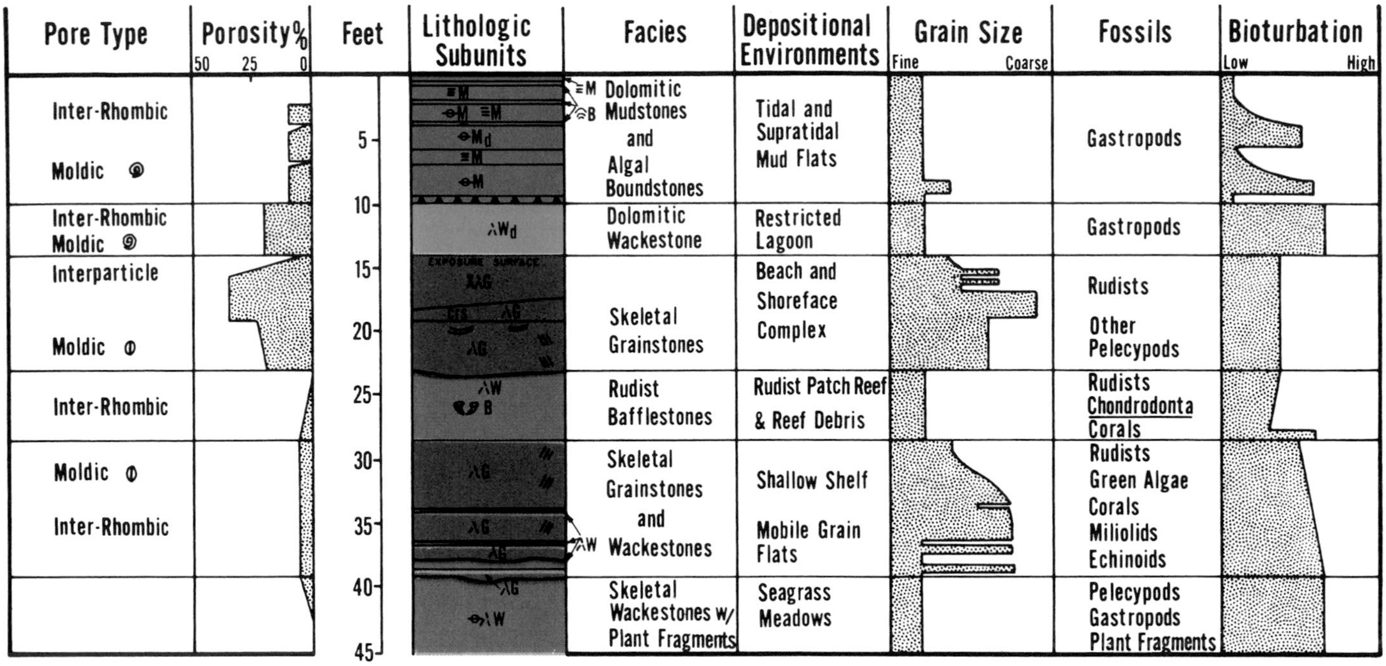

Pore Type	Porosity % (50 25 0)	Feet	Lithologic Subunits	Facies	Depositional Environments (Fine — Coarse)	Grain Size	Fossils	Bioturbation (Low — High)
Inter-Rhombic; Moldic		5	Dolomitic Mudstones and Algal Boundstones		Tidal and Supratidal Mud Flats		Gastropods	
Inter-Rhombic; Moldic		10	Dolomitic Wackestone		Restricted Lagoon		Gastropods	
Interparticle; Moldic		15 / 20	EXPOSURE SURFACE	Skeletal Grainstones	Beach and Shoreface Complex		Rudists; Other Pelecypods	
Inter-Rhombic		25		Rudist Bafflestones	Rudist Patch Reef & Reef Debris		Rudists; Chondrodonta; Corals	
Moldic; Inter-Rhombic		30 / 35		Skeletal Grainstones and Wackestones	Shallow Shelf; Mobile Grain Flats		Rudists; Green Algae; Corals; Miliolids; Echinoids	
		40 / 45		Skeletal Wackestones w/ Plant Fragments	Seagrass Meadows		Pelecypods; Gastropods; Plant Fragments	

Figure 65—Composite section of the Edwards Limestone showing lithofacies, depositional environments, and porosity distribution. Note shoaling upward, regressive sequence from middle to inner shelf environments (Kerr, 1977).

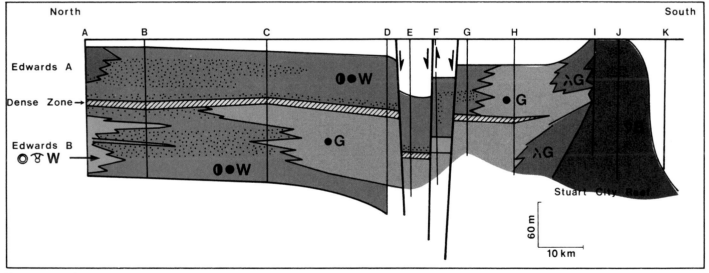

Figure 66—Shelf to basin profile of the Edwards Limestone (total Fredericksburg Group) northwest to southeast from the San Marcos Arch to the Stuart City Reef Trend, after Griffiths and others (1969), from Wilson (1975). Dotted pattern indicates porosity distribution.

grainstones of the inner shelf, these in turn being adjacent to dolomitized intertidal mudstones. The most shoreward facies is commonly impermeable sabkha anhydrite-gypsum which acts as an effective up-dip seal. The progradational sequence of the cycles generally results in basinward outstepping of the sabkha facies, placing the seal above, as well as up-dip, to excellent reservoirs. The Mississippian of the Williston Basin and Arab zones of the Jurassic across the Arabian Shield are classic examples.

(2) Where regional disconformities across shelves produce important reservoirs. Leaching and dolomitization of subjacent horizons can transform originally impermeable

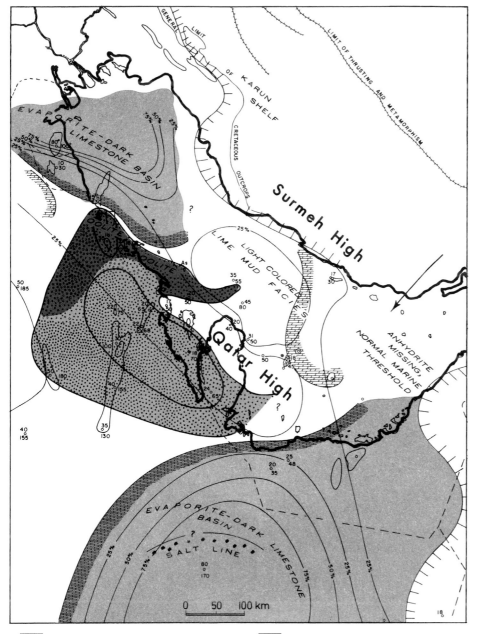

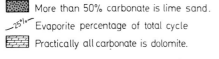

More than 50% carbonate is lime sand.

_.25%._ Evaporite percentage of total cycle

Practically all carbonate is dolomite.

More than 25% carbonate is lime sand.

Well─$\begin{array}{c}20\\ \circ \\ 45\end{array}$─Percentage anhydrite of total cycle
─Thickness of anhydrite

Edge line of dense brown lime mudstone facies

Thickness of cycle (including anhydrite) varies from 300-200 feet (91.5-61 m) thinning away from west central area.

Figure 67—Facies of Late Jurassic Arab D zone in Persian Gulf area (from Wilson, 1975).

micritic sediments into acceptable reservoirs. The top of the Edwards over the San Marcos Arch is an example of such a situation. The widespread Arbuckle-Ellenburger surface at the top of the Lower Ordovician of the mid-continent is another instance where complex phases of dolomitization plus fracturing have resulted in sub-unconformity reservoirs.

Examples of Oil and Gas Production in Middle Shelf Carbonates

As indicated above, the world's greatest single reservoir formation, the cyclic succession of Jurassic Arab Zones of Saudi Arabia, exists in coarse shoal grainstones and associated dolomitized wackestones of middle shelf facies. These were described in detail by Powers (1962), followed by a regional facies description by Wilson (1975). A vast shoal area lying between the interior of the

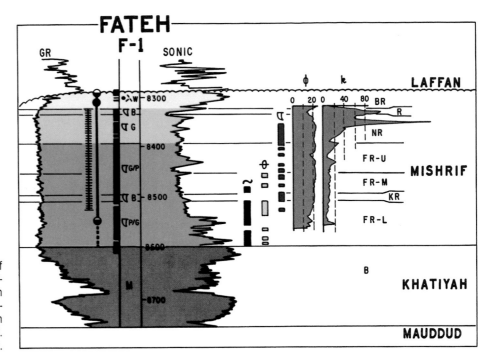

Figure 68—Core analysis of Mishrif Limestone in Fateh Field of Dubai offshore illustrating use of sonic log with gamma ray log for distinguishing porosity. Note permeability connected with coarse flanking beds of the near reef. After Jordan and others (1981).

Arab craton and the Qatar-Dirang-Namak High to the east (beneath the Qatar Peninsula and the Iranian coast) contains, in each of the six Arab and Jubaila cycles, middle shelf bioclastic grainstones, some oolite grainstones, and a minimum of mudstones and bioclastic wackestones. This area also separates the northern Basrah Trough and the southeastern Rub al Khali Basin (Fig. 67). Giant anticlines such as Ghawar, Abqaiq, Manifa, Khurais, and the Dukhan cross this area. Wells are unusually productive in these Jurassic strata, being capable of initial oil production of 8,000 to 12,000 b/d from coarse carbonate grainstones. Dukhan area on Qatar has less well-developed grainstones, with wells capable of half this production; farther east, into the Persian Gulf offshore, fields like Id el Shargi and Maydan Mazdam produce from Arab Zone dolomite partly of intertidal origin. These fields lie in an area of more restricted circulation over the Qatar-Dirang-Namak High and away from middle shelf environments. It is noteworthy that although the great reservoir is produced by lack of cementation of sands produced in open shelf environment, the seal of each cycle is created by extensively prograded, restricted

marine and sabkha evaporites.

Another great carbonate shelf exists along the Trucial coast bordering the Rub al Khali Basin on the north. The thick and relatively pure Thamama Limestone of early Cretaceous age stretches over the eastern Persian Gulf; the shelf area was tentatively outlined by Wilson (1975). The sequence begins with a mostly pelagic lime mudstone (Habshan or Sulaiy Formations) both to the east and west of the Trucial Coast area, and develops upward through 300 m into a dolomitic shoal facies which is partly intertidal. The overlying Lekhwair (Fig. 70) and Kharaib units represent a total of almost 500 m of alternating chalky, somewhat argillaceous, locally dense, stylolitic lime mudstones and wackestones with porous packstones and grainstones (Fig. 71). Porosity is both inter- and intragranular. Particle types include (1) algal and peloid types with numerous lumps and (2) worn bioclastic debris with certain large foraminifera. The dominance of codiacean green algae and some red algae indicates normal marine circulation and the development of widespread shoals.

There are about 20 such cycles across the shelf trending regularly from east to west within the Lekhwair

and Kharaib Formations (Hassan et al, 1975). The complex is capped by another cycle, one with more strongly differentiated facies, the Hawar Shale-Shuaiba Limestone. The Shuaiba forms a distinct east-to-west algal platform at the top of the Thamama with very well-developed, porous rudist facies. The major disconformity at the top of the Shuaiba and the localized influence of rising deep-seated salt domes resulted in meteoric alteration of the upper Thamama sediments. Oil and gas are present in great quantity across the middle of this platform in a broad north-to-south belt transecting the middle Trucial Coast. Hydrocarbons occur throughout the Thamama but are most abundant at the top of the Shuaiba and Kharaib Zone B. The source for this production is most probably the bituminous Hawar Shale (Murris, 1979). Some of these Thamama fields, such as Bab, Bu Hasa, Id el Shargi, and Zakum are in the billion barrel range of ultimate production. Discoveries were made from 1954 to 1970.

The geological examples of well-studied middle shelf carbonates discussed above, the Late Paleozoic Leavenworth and Hueco Limestones, are typical shallow shelf deposits sur-

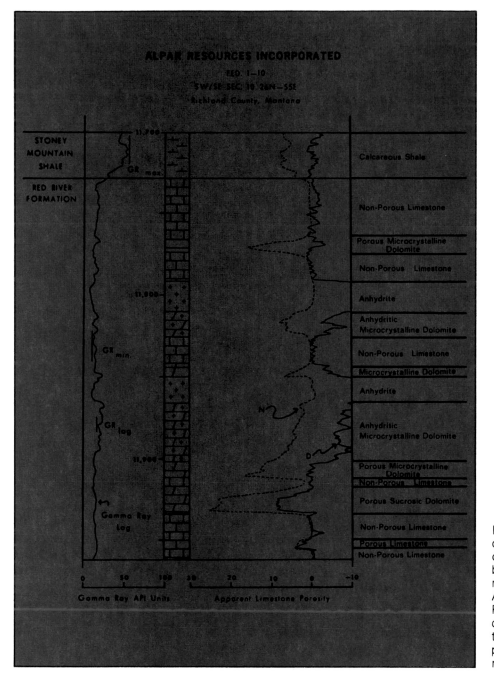

Figure 69—Use of neutron and electron density logs in interpreting porosity in carbonate strata. Schlumberger combination gamma ray, compensated neutron, and formation density log from Alpar Resources Fed. 1-10 well, Richland County, Montana; from the Ordovician Red River Formation. Illustrates the relationship of lithology (checked by petrographic analysis of cuttings) to log responses (from Asquith, 1979).

rounding Early Pennsylvanian block-faulted uplifts throughout the southwestern and mid-continent of the United States. Many fields produce from organic buildups in such carbonate strata. These porous and leached mounds include the Scurry County complex of north-central Texas, one of the largest fields in the continental United States, and hundreds of smaller isolated fields along shelf margins, within the shelf, and as

pinnacles offshore in shallower Pennsylvanian basins. Pennsylvanian and Wolfcampian shelves were mostly subject to open circulation as indicated by the common brachiopods, crinoids, corals, and bryozoans found in benthonic limestones and shales. The reservoir rock is of a more specialized biota, mainly of phylloid algae, encrusting tubular foraminifera, *Tubiphytes,* and *Chaetetes.*

Many fields have been studied in detail and several are reviewed by Wilson (1975) and Greenwood (in press). Greenwood estimates that Pennsylvanian strata in the southwestern United States should ultimately produce 8 billion barrels of oil.

An excellent review of the Pennsylvanian shelf around the Paradox Basin is found in Bass and Sharps (1963); see Figure 72 for the distribu-

Figure 70—Thickness of Lekhwair Formation, Thamama Group, early Cretaceous shelf facies across the Trucial Coast of eastern Arabia, United Arab Emirates. Note.that the unit doubles in thickness into the Rub al Kali basin (from Hassan, Mudd, and Twombley, 1975).

Figure 71—Lekhwair Formation of Thamama Group; cross section west to east across United Arab Emirate states showing multiple productive horizons in early Cretaceous open shelf grainstones and packstones (from Hassan, Mudd, and Twombley, 1975).

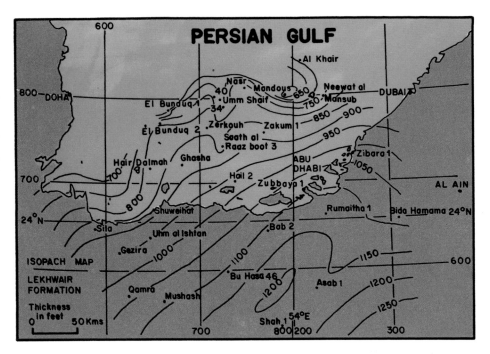

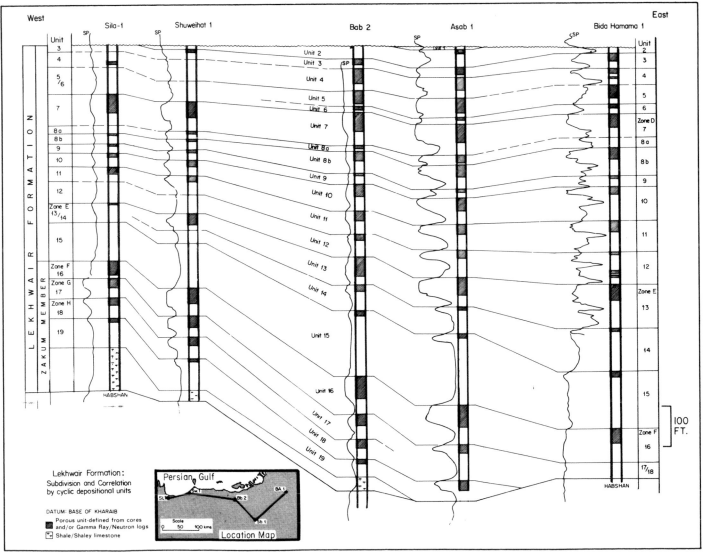

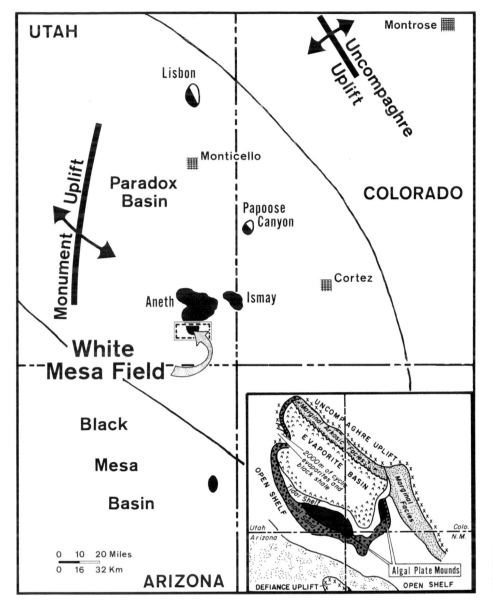

Figure 72—Location map and facies distribution in the Paradox basin in Pennsylvanian time showing Aneth and White Mesa fields on its southwestern shelf.

tion of facies here. The cyclic Middle Pennsylvanian strata of this shelf are well exposed in the Monument Uplift of southern Utah. East and north of this Laramide uplift, the buried portion of the shelf surrounds the deeper Paradox evaporite basin and contains numerous small oil fields (White Mesa, Desert Creek and Ismay) and one large field, Aneth (Fig. 73). Reservoirs are formed of piles of phylloid algae with caps of grainstone (see also Fig. 43). Meteoric water has leached through the sediment, dissolving algal plates of probable aragonite composition and increasing the porosity and permeability.

The Viburnum Trend

Extensive lead-zinc deposits in Upper Cambrian carbonates of southeastern Missouri formed in oolite grainstones on the flanks of the Ozark Uplift. The regional setting consists of a broad stable sandy area south of the Canadian Shield (Fig. 74), the classical Upper Cambrian (Croixan) area. To the south, these nearly littoral sands of the Canadian Shield grade into limestones and dolomites deposited on the flanks of the Ozark Uplift which rose episodically from early Cambrian through Pennsylvanian time (Thacker and Anderson, 1977).

The Upper Cambrian section in Missouri is about 650 m thick and consists of a basal sandstone, the Lamotte, which is overlain by a thick sequence of mostly carbonate rocks. The oldest of these is the ore-bearing Bonneterre Formation. In the Viburnum Trend, ore distribution is related to the geometry of an Upper Cambrian oolite belt about 30 mi long and 2 mi wide (Fig. 74). The Bonneterre facies here is a brecciated, dolomitized oolite grainstone. Cross-bedding, exposure surfaces, and abundant intraclasts indicate frequent emergence of the oolite bar. To the east, in middle shelf environments behind the

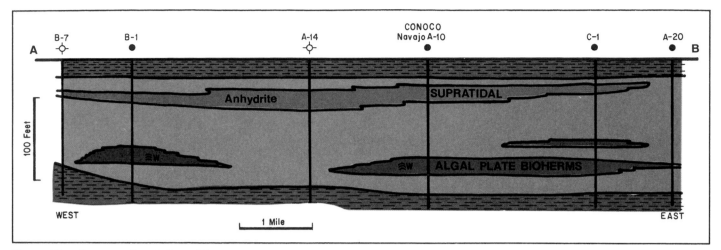

Figure 73—Cross section A-B west to east across White Mesa Field, southern Aneth Field, Paradox basin, New Mexico.

bar, burrowed, dolomitized lime mudstones and planar algal stromatolites were deposited in a very shallow, backbar lagoon (Howe, 1968). Inner shelf sediments lapped up onto the Ozark Uplift which had considerable relief at the time. In front of the oolite bar, lime mudstones, wackestones, and green shales were deposited.

The whole system can be considered a "shelf within a shelf." Thus, the basic elements of inner, middle, and outer shelf environments are present at different scales. At the larger scale, there is a progression from inner shelf sandstone facies near the Canadian Shield to mixed sandstone-limestone facies of the middle shelf in the mid-continent area. Around the Ozark Uplift, similar facies developed on a shelf built outward from the Precambrian granite core of the uplift. Thus, the Bonneterre oolite bar is a local outer shelf deposit, but in the larger sense is a middle shelf grainstone shoal, formed on the flanks of a local high.

SELECTED REFERENCES

Amsden, T. W., and T. L. Rowland, 1967, Silurian-Devonian relationship in Oklahoma, in International symposium on the Devonian system, Vol. 1: Calgary, Alberta Soc. Petroleum Geols.

Asquith, G. B., 1979, Subsurface carbonate depositional models; a concise review: Tulsa, Petroleum Pub. Co., 121 p.

Ball, M. M., 1967, Carbonate sand bodies of Florida and the Bahamas: Jour. Sed. Petrology, v. 37, p. 556-591.

Bass, R. O., and S. L. Sharps, eds., 1963, Shelf carbonates of the Paradox Basin: 4th Field Conf., Four Corners Geol. Soc., 273 p.

Beach, D. K., and R. N. Ginsburg, 1980, Facies succession, Plio-Pleistocene carbonates, northwestern Great Bahama Bank: AAPG Bull., v. 64, p. 1634-1642.

Becher, J. W., and C. H. Moore, 1976, The Walker Creek Field, a Smackover diagenetic trap: Trans., Gulf Coast Assoc. Geol. Socs., v. 26, p. 34-56.

Claypool, G. E., and P. R. Reed, 1976, Thermal-analysis technique for source rock evaluation; quantitative estimate of organic richness and effects of lithologic variation: AAPG Bull., v. 60, p. 608-612.

Dunham, R. J., 1970, Stratigraphic reefs versus ecologic reefs: AAPG Bull., v. 54, p. 1931-1932.

Enos, Paul, 1982, Restricted shelves, bays, lagoons, in P. Scholle, D. Bebout, and C. Moore, eds., Carbonate depositional environments: AAPG Mem. 33, this volume.

Fisher, W. L., and P. U. Rodda, 1969, Edwards Formation (Lower Cretaceous), Texas; dolomitization in a carbonate platform system: AAPG Bull., v. 53, p. 55-72.

Gehman, H. M., Jr., 1962, Organic matter in limestones: Geochim. et Cosmochim. Acta, v. 26, p. 885-897.

Ginsburg, R. N., R. Rezak, and J. L. Wray, 1972, Geology of calcareous algae: Univ. of Miami, Miami, Fla., Sedimenta 1, Comp. Sed. Lab., 40 p.

Greenwood, E., in press, Pennsylvanian production in southwest U.S.A.: Urbana, Ill., Proc., 9th Intern. Carboniferous Cong.

Griffith, L. S., M. G. Pitcher, and G. W. Rice, 1969, Quantitative environmental analysis of a Lower Cretaceous reef complex, in G. M. Friedman, ed., Depositional environments in carbonate rocks: SEPM Spec. Pub. 14, p. 120-138.

Harris, L. D., 1973, Dolomitization model for Upper Cambrian and Lower Ordovician carbonate rocks in the eastern United States: Jour. Research, U.S. Geol. Survey, v. 1, p. 63-78.

Hassan, T. H., G. C. Mudd, and B. N. Twombley, 1975, The stratigraphy and sedimentation of the Thamama Group (Lower Cretaceous) of Abu Dhabi: Dubai, 9th Arab Petroleum Cong., v. 107 (B-3) p. 1-11.

Hine, A. C., and A. C. Neumann, 1977, Shallow carbonate bank margin growth and structure, Little Bahama Bank, Bahamas: AAPG Bull., v. 61, p. 376-406.

Houbolt, J. J. H. C., 1957, Surface sediments of the Persian Gulf near the Qatar Peninsula: Utrecht, Netherlands, Univ. Utrecht, Master's thesis, 113 p.

Howe, W. B., 1968, Planar stromatolite and burrowed carbonate mud facies in Cambrian strata of the St. Francois Mountain area: Mo. Geol. Survey and Water Resources, Rept. Inv. 41, 113 p.

Irwin, M. L., 1965, General theory of epeiric clear water sedimentation: AAPG Bull., v. 49, p. 445-459.

Jordan, C. F., Jr., 1973, Carbonate facies

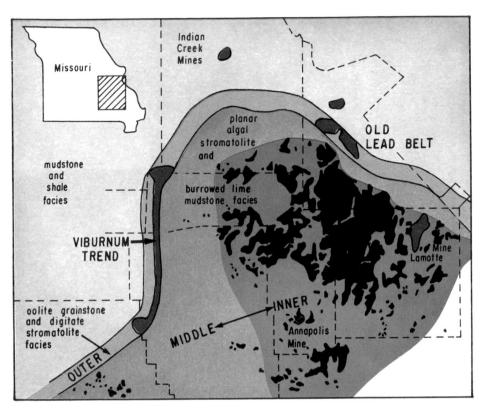

Figure 74—Facies distribution of the Bonneterre Formation (Upper Cambrian), southeast Missouri Lead Mining District; note the position of the Old Lead Belt and the newer Viburnum Trend in the "outer shelf" environments which formed on the flank of the Ozark uplift as a small shelf system within a larger shelf (the stable mid-continent area). Black areas mark Precambrian outcrops.

and sedimentation of patch reefs off Bermuda: AAPG Bull., v. 57, p. 42-54.

———— 1975, Lower Permian (Wolfcampian) sedimentation in the Orogrande Basin, New Mexico: 26th Field Conf. Guidebook, New Mexico Geol. Soc., p. 109-117.

———— and J. L. Wilson, 1971, The Late Paleozoic section of the Franklin Mountains: Field Conf. Guidebook, Permian Basin Sec., SEPM, p. 77-86.

———— T. C. Connally, Jr., and H. A. Vest, in press, Upper Cretaceous carbonates of the Mishrif Formation, Fateh Field, Dubai, U.A.E., in P. O. Roehl and P. W. Choquette, eds., Carbonate petroleum reservoirs: New York, Springer-Verlag Pub.

Kerr, R. S., 1977, Development and diagenesis of a Lower Cretaceous bank complex, Edwards Limestone, north central Texas, in D. Bebout and L. Loucks, eds., Cretaceous carbonates of Texas and Mexico: Austin, Tex., Bureau Econ. Geology, p. 216-233.

Meyerhoff, A. A., and C. W. Hatten, 1974, Bahamas salient of North America; tectonic framework, stratigraphy, and petroleum potential: AAPG Bull., v. 58, p. 1201-1239.

Murris, R. J., 1979, Hydrocarbon habitat of the Middle East, in A. D. Miall, ed., Facts and principles of world petroleum occurrence: Can. Soc. Petroleum,

Geols., Mem. 6, p. 765-800.

Palacas, J. G., 1978, Preliminary assessment of organic carbon content and petroleum source rock potential of Cretaceous and Lower Tertiary carbonates, South Florida Basin: Trans., Gulf Coast Assoc. Geol. Socs., v. 28, p. 357-381.

Perkins, B. F., 1974, Paleoecology of a rudist reef complex in the Comanche Cretaceous Glen Rose Limestone of central Texas, in Geoscience and man, v. 8, Aspects of Trinity Division Geology: Baton Rouge, La. State Univ. School Science, p. 131-173.

Pontigo, F. A., Jr., et al, 1979, South Florida's Sunniland oil potential: Oil and Gas Jour., v. 77, p. 226-232.

Powers, R. W., 1962, Arabian Upper Jurassic carbonate reservoir rocks, in W. E. Ham, ed., Classification of carbonate rocks, a symposium: AAPG Mem. 1, p. 122-192.

Purdy, E. G., 1961, Bahamian oolite shoals: AAPG Bull., v. 45, p. 53-62.

———— 1974, Karst-determined facies patterns in British Honduras; Holocene carbonate sedimentation model: AAPG Bull., v. 58, p. 825-855.

Rose, P. R., 1972, Edwards Group, surface and subsurface, central Texas: Austin, Univ. of Texas, Bur. Econ. Geol., Rept. Invest. 74, 198 p.

Shaw, A. B., 1964, Time in stratigraphy:

New York, McGraw-Hill Pub., 353 p.

Thacker, J. L., and K. H. Anderson, 1977, The geologic setting of the southeast Missouri lead district–regional geologic history, structure, and stratigraphy: Econ. Geology, v. 72, p. 339-348.

Toomey, D. F., 1969a, The biota of the Pennsylvanian (Virgilian) Leavenworth Limestone, mid-continent region, part I; stratigraphy, paleogeography, and sediment facies relationships: Jour. Paleontology, v. 43, p. 1001-1018.

———— 1969b, The biota of the Pennsylvanian (Virgilian) Leavenworth Limestone, mid-continent region, part II; distribution of algae: Jour. Paleontology, v. 43, p. 1313-1330.

———— 1972, The biota of the Pennsylvanian (Virgilian) Leavenworth Limestone, mid-continent region, part III; distribution of calcareous foraminifera: Jour. Paleontology, v. 46, p. 276-298.

Wilson, J. L., 1970, Depositional facies across carbonate shelf margins: Trans., Gulf Coast Assoc. Geol. Socs., v. 20, p. 229-233.

———— 1974, Characteristics of carbonate platform margins: AAPG Bull., v. 58, p. 810-824.

———— 1975, Carbonate facies in geologic history: New York, Springer-Verlag Pub. 439 p.

Reef Environment

Noel P. James

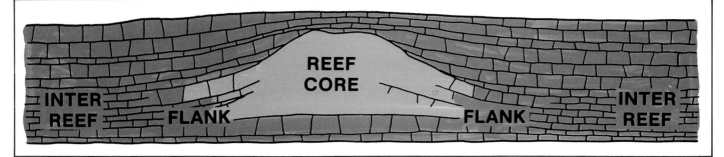

Figure 1—A sketch illustrating the major facies of a fossil reef (after James, 1979, with permission of Geol. Assoc. Canada).

Figure 2—An interpretative sketch of the different types of reef limestone recognized by Embry and Klovan (1971) (after James, 1979, with permission of Geol. Assoc. Canada).

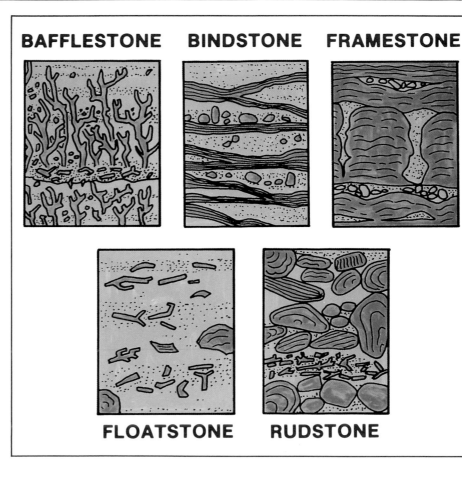

A reef, rising above the sea floor, is an entity of its own making — a sedimentary system within itself. Numerous, large calcium carbonate secreting organisms stand on the remains of their ancestors and are surrounded and often buried by the skeletal remains of the many small organisms that once lived on, beneath and between them.

Because they are built by organisms, fossil reefs are storehouses of paleontological information and modern reefs are natural laboratories for the study of benthic marine ecology. Furthermore, fossil reefs buried in the subsurface contain a disproportionately large amount of our oil and gas reserves compared to other types of sedimentary deposits. These two facts alone have resulted in reefs being studied in more detail by paleontologists and sedimentologists, than perhaps any other type of sedimentary deposit. Consequently, there is a large and diverse source of data for the student of reefs to draw upon when studying these deposits.

Unlike other sedimentary deposits, a reef is not entirely the product of physical sedimentation but is almost entirely the physical expression of a community of organisms, growing in one place for an extended period of time. These communities have, however, changed dramatically through the course of geologic time, so that a reef at any one time may be quite different from a reef only a few million years younger or older.

Because of this, any synthesis of reefs must be largely comparative; not only a comparison of modern and ancient reefs, but a comparison of different fossil reefs as well. With this guiding tenant I have, in the following pages, outlined the attributes of different types of modern reefs. These examples are from the Atlantic-Caribbean area, the region with which I am most familiar. This section is followed by a discussion of the terminology used in describing ancient reefs, and then by a synthesis of the characteristics of these fossil

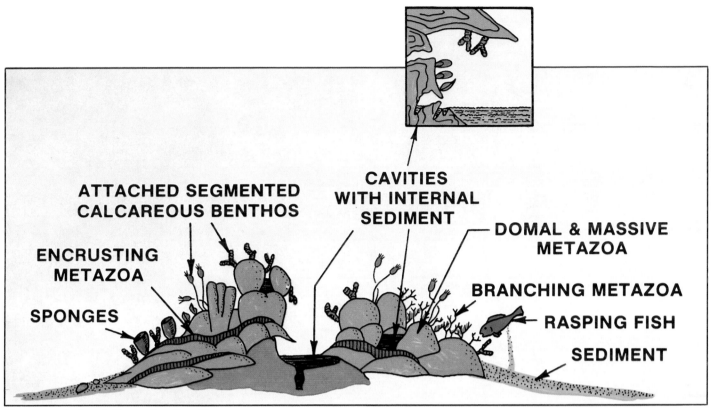

Figure 3—A sketch illustrating the different aspects of the organism/sediment mosaic that is a reef (after James, 1979, with permission of Geol. Assoc. Canada).

structures. To illustrate the pertinent textures and fabrics of those carbonates, the text is supplemented by examples of modern and ancient reef carbonates.

CLASSIFICATION OF REEF LIMESTONES

Before discussing the deposits themselves there should be some consensus as to the classification of different types of reefal carbonates. Because reef rocks are often composed of large components and are commonly formed in place by the growth of organisms and therefore not physically deposited, they are not amenable to the type of classification schemes outlined by Folk (1962) and Dunham (1962). Folk recognized reef limestones as different and called them biolithites, recognizing that this was a general term and that these rocks needed to be subdivided. Dunham, on the other hand, sug-

gested the term boundstone for those reef rocks showing signs of being bound during deposition.

The most descriptive and widely accepted scheme, however, is a modification of Dunham's (1962) classification of carbonate sand and mudrocks proposed by Embry and Klovan (1971) at the University of Calgary (Fig. 2). They recognize two kinds of reef limestone, allochthonous and autochthonous. The allochthonous limestone is the same as the finer grained sediments, but with two categories added to encompass large particles. If more than 10% of the particles in the rock are larger than 2 mm and they are matrix supported, it is a floatstone; if the rock is clast supported it is a rudstone. The autochthonous limestone is more interpretative; framestones contain in-place, large fossils that formed the supporting framework; bindstones contain in-place, tabular or lamellar fossils that encrusted or bound the sediment together during deposition; and bafflestones contain in-place, stalked fossils that trapped sediment by baffling.

MODERN REEFS

Reefs in the modern ocean are commonly envisaged in terms of the prolific, shallow water, coral reefs of the Caribbean and Indo-Pacific. In the geologic record, there is a much wider concept of reef or carbonate build-up, and, under the heading of Modern Reefs, we must include not only the familiar coral reefs, but also most other in-place accumulations of skeletal calcium carbonate that rise above the sea floor. This wider definition includes other shallow reefs composed almost entirely of algae, banks of branching coral and calcareous algae, banks of mostly lime mud and build-ups of coral, and skeletal sediment in deep water.

Dynamics of Reef Sedimentation

The present state of any reef is a delicate balance between the upward growth of large skeletal metazoans, the continuing destruction of these same organisms by a host of rasping, boring and grazing organisms, and the prolific sediment production by rapidly growing, short-lived, attached

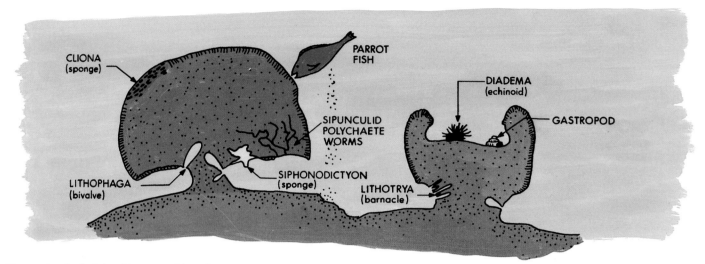

Figure 4—A sketch of two coral heads and some of the organisms responsible for bioerosion (after Ginsburg and James, 1974, with permission of the University of Miami).

calcareous benthos (Fig. 3).

The large skeletal metazoans (corals) generally remain in place after death, except when they are so weakened by bioeroders that they are toppled by storms. The irregular shape and growth habits of these reef-builders results in the formation of roofed-over cavities inside the reef that may be inhabited by smaller, attached calcareous benthos, and may be partly to completely filled with fine-grained "internal" sediment. Encrusting organisms grow over dead surfaces and aid in stabilizing the structure. Branching reef-builders frequently remain in place, but just as commonly are fragmented into skeletal conglomerates around the reef.

Most reef sediment is produced by the post-mortem disintegration of organisms that are segmented (for example, crinoids, in the Paleozoic and Mesozoic, calcareous green algae in the Cenozoic) or non-segmented (bivalves, brachiopods, foraminifers, etc), that grow in the many nooks and crannies between the larger skeletal metazoa. The remainder of the sediment is produced by the various taxa that erode the reef — boring organisms (worms, sponges, bivalves) producing lime mud; or rasping organisms that graze the surface of the reef (echinoids, fish) producing copious quantities of carbonate sand and silt (Fig. 4). These sediments are deposited around the reefs as an apron and also filter into the growth cavities to form internal sediment, which is characteristically geopetal.

Coral-Algal Reefs

The most widespread and volumetrically important reefs in modern oceans are constructed by hermatypic corals and calcareous algae. They are best developed and most successful on the windward sides of shelves, islands and platforms where wind and swell are consistent and onshore. The asymmetry of many ancient reefs and distribution of sediment facies suggests that this was so in the past as well. The reason for the preferential development of reefs on the windward side is by no means established, but sedimentation is likely the most important control. Shallow water reef-building species characteristically produce abundant fine sediment, yet the major reef-builders, because they are filter feeders and micropredators are intolerant of fine sediment. The open-ocean, windward locations are the only ones in which fine sediment is continuously swept away.

Growth of reefs into this zone of onshore waves and swell forms a natural breakwater of linear barrier reefs. If the shelf is narrow, this structure is often hard up against the shoreline and is therefore called a fringing reef. Because the open-ocean swells and storm waves are absorbed by this barrier, the shelf and lagoon behind is commonly a relatively quiet environment, with the water stirred up only by wind waves, tidal currents, and occasional cyclonic storms. Here reefs are generally isolated structures, called patch reefs or lagoon reefs.

Patch Reefs

These reefs (Figs. 6 to 13) have been particularly well documented in Bermuda, and the following description is summarized from Garrett and others (1971).

The smallest elements of the Bermuda patch reefs are coral knobs (Fig. 5); they are intergrowths of corals, algae, and associated organisms that may be up to 5 m across and rise up 1 to 3 m above the surrounding sand-floored interareas. Larger reefs are for the most part aggregates of coral knobs. These reefs are generally unzoned, rising from depths as much as 20 m to within a few meters of mean low water. The patch reefs range in shape from pinnacles, to walls, to micro-atolls, which enclose a small sandfloored lagoon.

Corals cover from 10 to 45% of the hard substrate and comprise 40 to 80% of the solid reef mass. The bulk of most reefs is made up of domal or massive corals (*Montastraea* and *Diploria*) as revealed by dynamited sections. Coral knobs on the reef tops

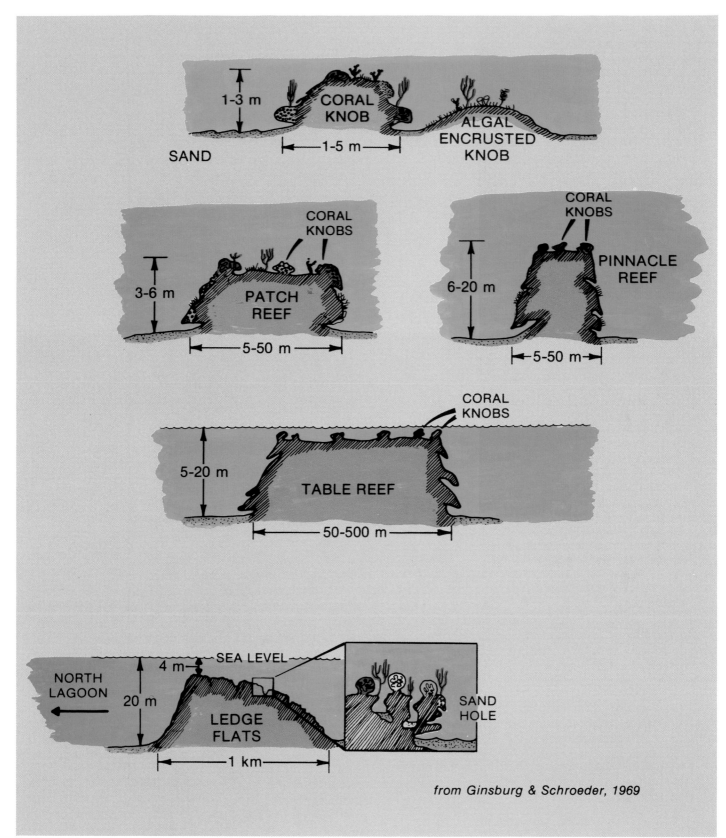

Figure 5—A series of sketches illustrating the different types of coral-algal reefs on the Bermuda Platform (from Ginsburg and Schroeder, 1969).

Figure 6—An areal view of numerous coral patch reefs on the Belize shelf rising from depths of about 5 m. The dark brown areas are living coral and the buff colored areas are skeletal sand. The off-reef areas are muddy carbonate sands colonized by meadows of *Thalassia* grass. The two reefs at the lower left are about 30 m in length.

Figure 7—An areal view of a complex of rhomboid-shaped platform reefs on the Belize shelf. The complex in the foreground is about 6 km in length. These structures rise from water more than 30 m deep and their steep (more than 30°) sides support a wide variety of corals. Their shallow margins are either zoned like other reefs or a community of delicate branching corals and algae while the centers are mostly muddy skeletal sands.

are dominantly the corals *Montastraea, Diploria* and *Porites astreoides. Millepora,* an encrusting and bladed calcareous hydrozoan, is to be found on the reef face and reef top where it encrusts flat surfaces or the branched stubs of dead sea fans and sea ships. Branching corals such as *Oculina* and *Madracis decactis* grow near the base of the reef face. The thin saucer-shaped coral *Agaricia fragilis* occurs only beneath or behind overhanging projections. Bun-shaped *Siderastraea* and

Isophyllia grow near the bases of coral knobs.

Most of the rock surface is covered by coralline algae which also encrusts dead coral and skeletal rubble on the reef top. Other encrusting organisms such as a few ectoproct bryozoa and bivalves (*Chama, Pseudochama,* and *Spondylus*) are also common.

Flabelleform algae and soft corals bush out over much of the reef rock surface. The two most common algae, *Sargassum* and *Dictyota,* are geologically unimportant as they have

no calcareous hard parts. However, the shrub-like calcareous alga *Stypopodium, Galaxaura, Padina, Udotea,* and *Neomeris* produce fine sediment, while the branching forms *Goniolithon, Amphiroa,* and especially *Halimeda* produce abundant sand-sized sediment. The alcyonarian or "soft" corals *Gorgonia* and *Plexaurella* also produce fine sand-size calcareous spicules.

Boring organisms can be found in every piece of rock broken from the reef. Predominant among these are

Figure 8—A truck tire in 5 m of water in Eniwetok lagoon on which corals (*Acropora*) have settled and begun to grow. This tire has been in place less than 20 years (photo J. Warme).

Figure 9—A coral knob on top of a small patch reef near Goulding Cay, Bahamas, illustrating a diverse coral community composed of *Diploria* (large dome about 1 m across at center), *Porites astreoides* (yellow coral at right and left), *Porites porites* (white branching coral at center), *Agaricia* (leaf-like coral at lower left) and *Acropora palmata* (top left). The dark branching organism at the top center left is an octocoral which has no skeleton, but only small calcareous spicules embedded in the soft tissue.

the rock-boring clams, (*Lithophaga, Nigra,* and *Spenglaria rostrada*), several species of polychaete worms, endolithic algae, and boring sponges (*Cliona* and *Siphonodictyon*). The spiny echinoderm *Diadema antillarum* grazes the reef surface while the parrot fishes (*Sparisoma virde* and *S. abillgardi*) graze the surface flora, ingest sand from the valleys, and occasionally take bites of hard coral and coralline algae.

The interiors of these reefs contain many cavities, inside corals, from the boring by organisms and from the iregular overgrowths of encrusting organisms. The larger cavities (growth cavities) are due to biological roofing-over, chiefly by corals. The catacomb of chimneys, tunnels and cavities inside the reef contains a distinctive community of organisms. Near the outside, cavity walls are encrusted with the right valve of the bivalve *Spondylus americanus,* coralline algae, ectoprocts, serpulid tubes, and the red foraminifer *Homotrema rubrum.* Deep within the dark recesses of the reef, the proportion of the wall covered by living organisms decreases to zero. Garrett and others (1971) have estimated that from 30 to 50% of the reef volume is occupied by open cavities and sediment fill.

Sediments on top of the reef are mainly coarse to very coarse gravels and sands composed of whole to fragmented corals, coralline algae, the attached foraminifer *Homotrema*, the segmented calcareous alga *Halimeda* and bivalves. Every piece of the larger particles is bored by endolithic

Figure 10—The top of a patch reef near Andros Island, Bahamas, composed almost entirely of small, irregular *Porites astreoides* colonies. The white areas are dead coral encrusted with coralline algae while the dark areas in shadow are entrances to an extensive network of growth cavities inside the reef.

Figure 11—The crest of a small coral knob on top of a patch reef in 1 m of water on the Belize shelf. The large multilobate colonies are *Montastraea annularis* (compare the shape of this colony to the growth form of one in deeper water illustrated in Figs. 22 and 23) while the smaller finger-like colonies are *Porites porites*. The areas of green between corals are lush growths of the segmented, platey, calcareous algae *Halimeda* (photo P. Scholle).

algae and encrusted by coralline algae, ectoproct bryozoa, and *Homotrema*. Some of the reef sands are washed down the reef face to build a steep sediment slope which grades outward into lagoon sediments; at a distance of 10 m or more from the reef, reef-derived constituents become quantitatively unimportant. The grain-size of the sediment decreases rapidly down the sediment slope and out onto the lagoon

floor. The fine grained sediments either filter down into the reef cavities where the amount of fine sediment is considerable (as evidenced when blasting open the structures) or it is winnowed, settling out of suspension as a halo of fines around the reef.

While the distribution of different corals and calcareous algae is more or less uniform over the patch reefs of the Bermuda platform, in other areas

the structures are distinctly zoned. Patch reefs in the lagoons and shelves of the Belize barrier and atoll reef complex illustrate this well (Fig. 13). In some areas, the zonation is simply one of more abundant coral growth on one side (windward) than the other (Shinn et al, 1977). In others, there is a pronounced assymetry to the structures, and the corals are distinctly zoned.

In most cases the limestone formed

Figure 12—The wall-like margin of a low-relief patch reef on the Belize shelf. The 2 m deep channel, floored by skeletal sand, is bounded by numerous multilobate *Montastraea annularis* colonies (photo P. Scholle).

Figure 13—A series of sketches illustrating the different profiles of and principal corals on coral-algal patch reefs in Glovers Atoll lagoon, Belize (data from Wallace and Schafersman, 1977, and personal observation).

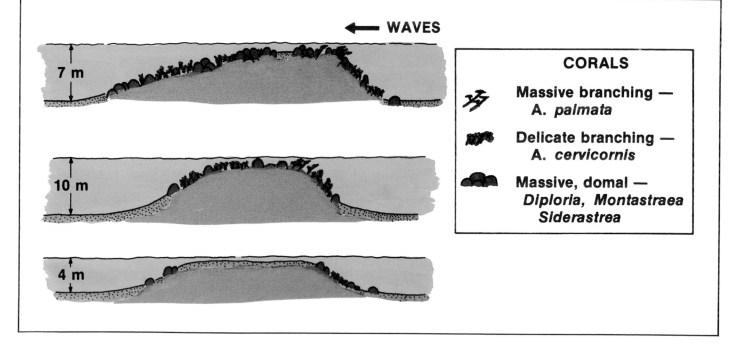

⬅ WAVES

7 m

10 m

4 m

CORALS

Massive branching —
   A. *palmata*

Delicate branching —
   A. *cervicornis*

Massive, domal —
   Diploria, Montastraea
   Siderastrea

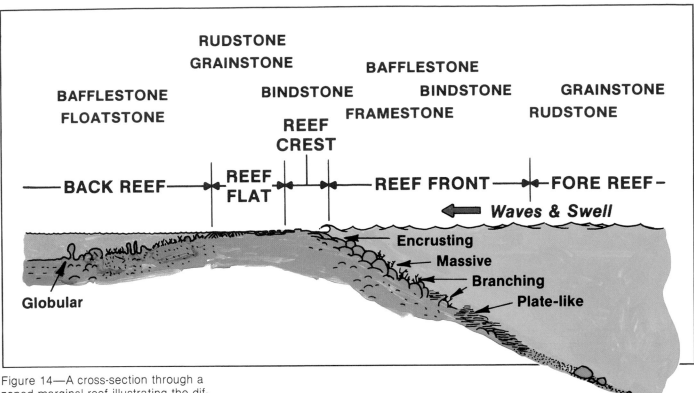

Figure 14—A cross-section through a zoned marginal reef illustrating the different reef zones, spectrum of different limestones produced in each zone, and environment of different reef-building organisms. In many modern reefs, the reef crest is occupied by the massive branching coral *Acropora palmata,* a situation that is rare in the fossil record (after James, 1979, with permission of Geol. Assoc. Canada).

Figure 15—An areal view looking north along the Belize barrier reef on a calm day. The reef crest is the sharp dark line curving toward the upper right with deep water to the lower right. The wide, light-green area shelfward of the very narrow reef crest is a 1 to 2 km wide blanket of skeletal sand (mostly *Halimeda*) with zones of coral rubble in the lee of the reef crest. Passes through the barrier (upper right) are as much as 20 m deep. Numerous patch reefs can be seen on the shelf behind (upper left). (Photo R. N. Ginsburg).

in the patch reef is largely coral framestone to coral bindstone with obvious cavities filled with geopetal skeletal wackestone to sometimes packstone. The reefs are surrounded by skeletal, mainly green and red calcareous algae, foraminifer, and coral rudstones and grainstones. In the rock record, crinoid skeletal grains commonly take the place of the calcareous algae. These grade outward into progressively more muddy and well-bedded sediments.

Platform Margin Reefs
Reefs along the platform edge (Fig. 14) range from almost continuous barriers to isolated patches of coral, irregularly distributed along the margin. In some areas, such as Bermuda, the platform margin is in water 5 m or more deep and the sea floor is covered by wide areas of coral growth (the ledge flats in Fig. 5),

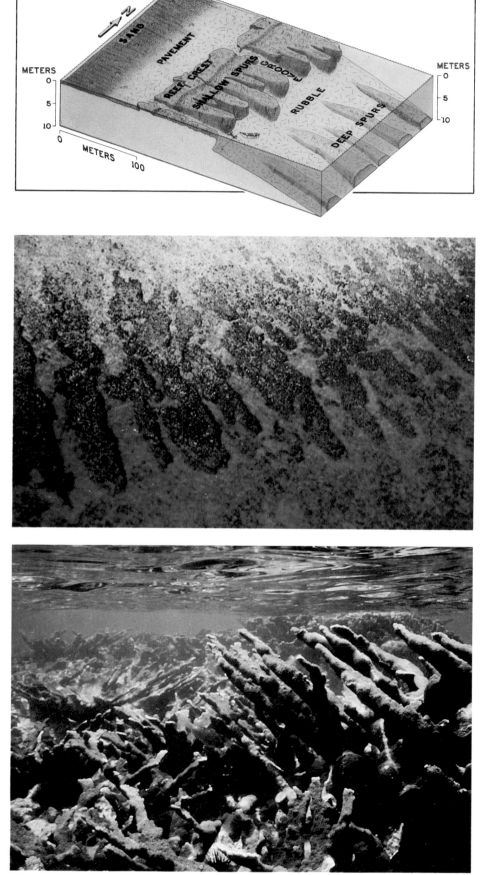

Figure 16—An isometric diagram illustrating the main morphological elements of the shallow Belize barrier reef margin (modified from James et al, 1976).

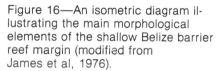

Figure 17—An areal view of the spur and groove zone off Sand Key, Florida. The reef flat is at the top and deep water at the bottom. Individual spurs are 5 to 10 m across (photo E. A. Shinn).

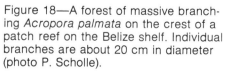

Figure 18—A forest of massive branching *Acropora palmata* on the crest of a patch reef on the Belize shelf. Individual branches are about 20 cm in diameter (photo P. Scholle).

Figure 19—The transition between two zones of reef growth on the reef front off Discovery Bay, Jamaica; the shallow zone to the right is dominated by *Acropora palmata* growth and the deeper zone in water depths of about 7 m to the left exhibits a wide variety of corals. The rounded multilobate coral colonies in the foreground are *Montastraea annularis.*

Figure 20—A diverse coral community in the lee of the reef crest behind Carysfort Reef, Florida, in water 2 to 3 m deep. The three most obvious corals are *Acropora palmata* (large branching form), *Acropora cervicornis* (delicate branching form), and *Montastraea annularis* (massive multilobate form).

mainly massive and hemispherical colonies with only algal cup reefs breaking the surface. More commonly, the platform margin reefs, as seen from the air, are true barrier reefs, well-zoned complexes of living reefs and the sediments derived from them (Fig. 15). The shallowest part of the complex is the reef crest, growing in the zone of breaking waves. Most of the coral-algal reef is found seaward of the reef crest, growing on an irregularly sloping ramp called the reef front. Seaward and basinward of the zone of coral and algal growth is the fore reef, an area of reef debris, carbonate sand, blocks of limestone and coral fragments, as well as a variety of different benthic organisms. In the lee of the reef crest is a very shallow zone that is often awash at low tide, called the reef flat. This style of platform margin complex is often segmented by channels bisecting the reef. Skeletal sands are sometimes funneled basinward through these channels.

Figure 21—The margin of a deep-water (45 m) coral spur covered almost entirely by an intergrowth of *Acropora cervicornis,* off Discovery Bay, Jamaica. The coral sticks here are longer and much more delicate than in shallow water. The blue color is caused by the low light intensity and absorbtion of all but the blue end of the light spectrum by the overlying water column.

Figure 22—Looking down on a large colony of *Montastraea annularis* composed of numerous overlapping plates of coral at a depth of 55 m off Discovery Bay, Jamaica (hammer on top of the colony at right center for scale).

Figure 23—The margin of a large colony of plate-like *Montastraea annularis* at a depth of 55 m off Discovery Bay, Jamaica. The whispy growths in the left foreground are octocorals (diver for scale).

Figure 24—A small stand of stubby *Acropora cervicornis* about 1 m high in the lagoon of Eniwetok Atoll at a depth of 4 m (photo J. Warme).

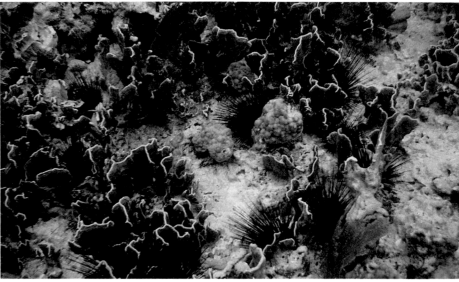

Figure 25—A close view of the surface of a seaward-facing spur at a depth of 7 m on the Florida reef tract. The bladed colonies are the hydrozoan *Millepora* and between these colonies are scattered *Porites astreoides* colonies. The black, spiney organism is the echinoid *Diadema antillarum* which continuously scrapes the surface of the reef rock grazing on algae and eroding the coral. The white areas are reef rock partially coated with coralline algae.

Figure 26—A huge single colony of *Montastraea annularis* at a depth of 20 m on the Flower Gardens Bank in the Gulf of Mexico off Texas (photo E. A. Shinn).

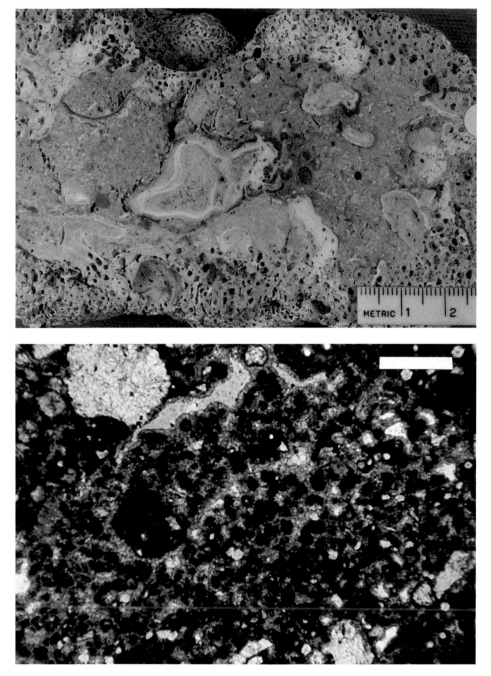

Figure 27—A slabbed and polished piece of reef limestone from a shallow coral spur in 2 m of water on the eastern ocean-facing reef front of Glovers Atoll, Belize. The rock is composed of branches of the hydrozoan *Millepora* (center) which are encrusted by the red foraminifer *Homotrema rubrum*. The sediment between the branches is a submarine cemented skeletal grainstone (see thin section photograph in Fig. 29). The holes at the edge of the sample are excavated by boring sponges, with holes cutting both skeletal elements and cemented sediment.

Figure 28—A thin section under plane polarized light of cemented reef limestone from shallow Belize reefs. The section is stained with Clayton Yellow so all aragonite components are white and all Mg-calcite components are red or pink. The rock is a packstone with sand-size grains of coral (aragonite, upper left) in a matrix of fine sand and silt-size peloids. The peloids are Mg-calcite micrite and the cement around the peloids is Mg-calcite micrite and bladed spar. Scale bar 200 microns.

Reef Crest Zone

This is the highest part of the reef at any stage in its growth, and if in shallow water, it is that part of the reef top that receives most of the wind and wave energy. The composition of the reef crest depends on the degree of wind strength and swell. In areas where wind and swell are intense, only those organisms that can encrust, generally in sheet-like forms, are able to survive. When wave and swell intensity are only moderate to strong, encrusting forms still dominate but are commonly also bladed or possess short, stubby branches. In localities where wave energy is moderate, hemispherical to massive forms occur with scattered clumps of branching reef-builders, although the community is still of low diversity. The lithologies formed in these three cases would range from bindstones to framestones.

Reef Front Zone

This zone extends from the surf zone to an indeterminate depth, generally no deeper than 100 m. On modern reefs there is commonly a steep cliff at this point whereas in fossil reefs, the zone of abundant skeletal growth grades into sediments of the forereef zone. Direct comparison between modern reefs and ancient reefs is difficult because today the sea floor from the surf zone to a depth of 12 m or so is often

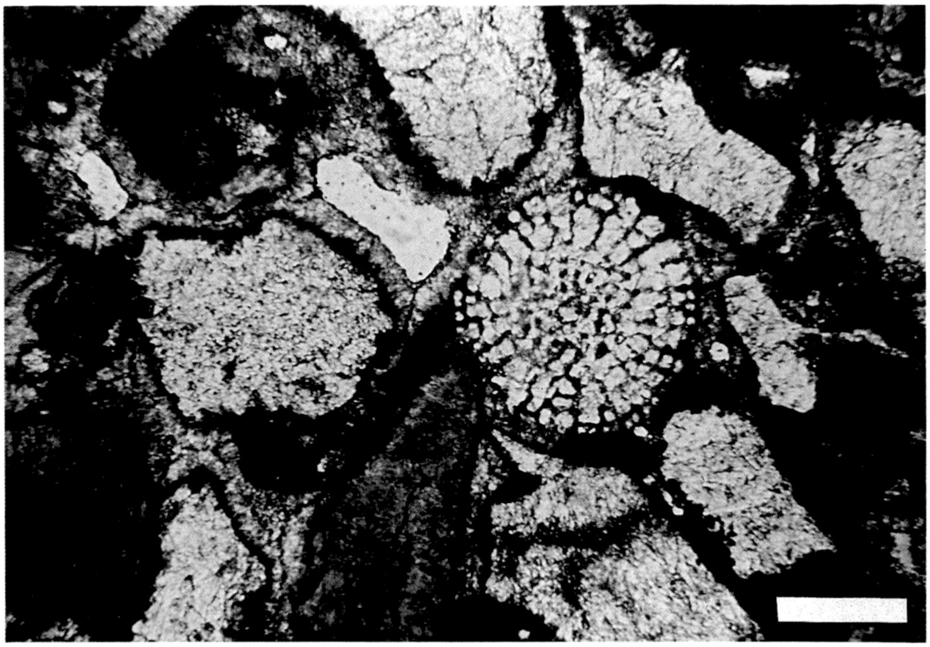

Figure 29—A thin section under plane polarized light and stained with Clayton yellow of the grainstone illustrated in Figure 27. The angular aragonite grains are coral while the Mg-calcite grains are branching coralline algae, benthic foraminifers and echinoid particles. The cement is isopachous rinds of Mg-calcite bladed spar. Scale bar 100 microns.

Figure 30—A sample of reef wall limestone from the Belize barrier reef (circa 13,000 years old). In this example a cavity (the walls of which can just be seen at upper right and upper left) is almost completely filled with spherulitic growths of botryoidal aragonite. The open space remaining in the center of the void is floored with a thin layer of marine sediment which partially buries some of the aragonite (scale in cm).

dominated by the robust branching form *Acropora palmata,* a species which developed only recently, in the late Pleistocene. Such branching forms are rarely found in ancient reefs. Instead, the most abundant forms are massive, laminar to hemispherical skeletons, forming framestones and sometimes bindstones.

This zone commonly has a distinct topography called the "spur and grove" consisting of a series of linear reef ridges running seaward and separated from one another by sediment floored channels. This is best developed in shallow water or at breaks in slope on the ramp.

The main part of this zone supports a diverse fauna with reef-builders ranging in shape from hemispherical to branching to colum-nar to dendroid to sheet-like. Accessory organisms and various niche dwellers such as brachiopods, bivalves, coralline algae, and green segmented calcareous algae (*Halimeda*), are common. On modern reefs where the reef-builders are corals, this zone commonly extends to a depth of 30 m or so. The most common rock type formed in this zone would still be framestone but the

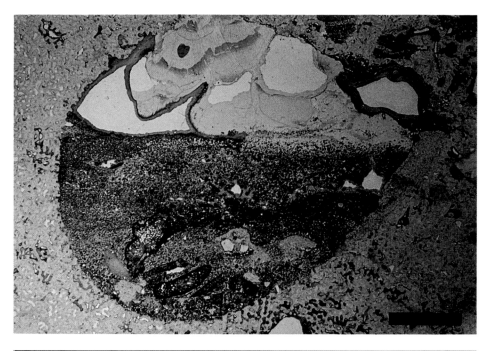

Figure 31—A thin section of Belize reef wall limestone (circa 9,000 years old) under plane polarized light in which coral has been bored by a bivalve and the boring has been filled by a complex suite of marine sediments and cements. The section is stained with Clayton Yellow so that all aragonite components are white or light gray and all Mg-calcite components are red or pink. The cavity is partially filled with fine-grained geopetal sediment some of which is cemented by aragonite and some of which is cemented by Mg-calcite. The void at the top of the cavity contains a first generation of botryoidal aragonite (the holes in which are due to thin section preparation) and a second generation of fibrous Mg-calcite cement (scale bar 0.5 cm).

Figure 32—A slab of coral rudstone from the Belize barrier reef, cemented together to form a pavement of coral conglomerate in the lee of the reef crest. The corals in this limestone are less than 500 years old (scale in cm).

variety of growth forms also leads to the formation of many bindstones and bafflestones.

Below 30 m or so, wave surge is almost non-existent and light is very attenuated. The response of many reef-building metazoans is to increase their surface area by having only a small basal attachment and a large, but delicate, plate-like shape. Rock types from this zone look like bindstones, but binding plays no role in the formation of these rocks and perhaps another term is needed.

The deepest zone of growth of coral and green calcareous algae on modern coral reefs is around 70 m. The lower limit may depend on many factors, perhaps one of the most important being sedimentation, so that this lower limit should be used with caution in the interpretation of fossil reefs.

Sediments on the reef front are of two types: (1) internal sediments within the reef structure, generally lime mud, giving the rocks a lime mudstone to wackestone matrix; and (2) coarse sands and gravels in channels running seaward between the reefs. These latter deposits have rarely been recognized in ancient reefs.

As a result of numerous observations on modern reefs, it appears that most of the sediment generated on the upper part of the reef front and

Figure 33—A sample of reef limestone (circa 9,000 years old) from the Belize barrier reef wall. The coral is a *Montastraea cavernosa* colony while the sediment below is a platey algae (*Halimeda*)-rich lime wackestone to packstone. The numerous cavities are mostly shelter voids (scale in cm).

Figure 34—A polished slab of reef limestone (circa 13,000 years old) from the Belize barrier reef wall. This algal-plate floatstone to rudstone occurs between large coral colonies and is composed of numerous *Halimeda* plates (white) often roofing shelter pores and skeletal packstone to wackestone.

on the reef crest is transported episodically by storms up and over the top and accumulates in the lee of the reef crest. Sediments on the intermediate and lower regions of the reef front, however, are transported down to the fore-reef zone. Shallow-water material is contributed to the fore-reef zone only when it is channelled by way of passes through the reef.

Reef Flat Zone
The reef flat varies from a pave-ment of cemented, large skeletal debris with scattered rubble and coralline algae nodules to shoals of well-washed lime sand in areas of moderate wave energy. The sand is mainly plates of the calcareous alga *Halimeda*, which grows on the reef

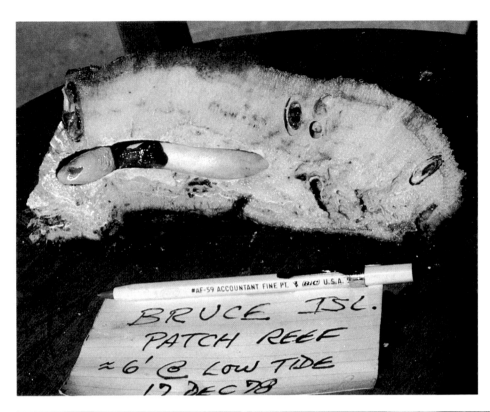

Figure 35—A small coral colony which has been sawed in half revealing several bivalve borings, the most prominent of which, running horizontally across the specimen, still contains the living mollusk *Gastrochaena hians.* This sample is from a patch reef on Eniwetok Atoll (photo J. Warme).

Figure 36—Looking down on the center of a small patch reef in 2 m of water near Goulding Cay, Bahamas. At the center is a hemispherical coral (*Diploria*) the base of which has been intensively bored and was easily broken off with a light push of the hand.

Figure 37—The shore of a small island on the Belize barrier reef complex that is composed entirely of coral rubble, eroded from the reef just offshore and swept together by currents and waves.

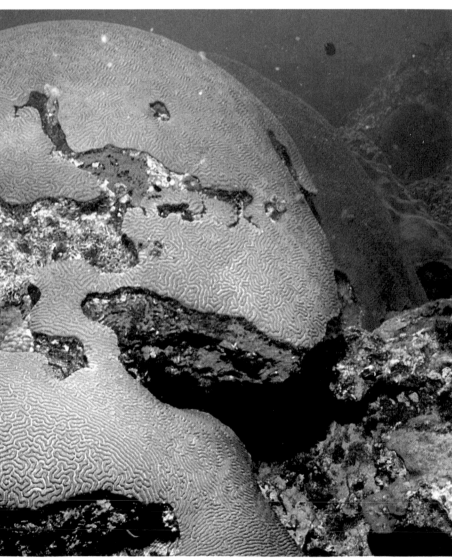

Figure 38—A large colony (circa 1.5 m across) of *Diploria* that has been injured and died in several places. Each of these areas has been infested by boring sponges (the bright orange tissue). The shadows are the entrances to small cavities roofed over by the coral. This colony is on the Flower Garden Bank in 20 m of water off the Gulf of Mexico coast (photo E. A. Shinn).

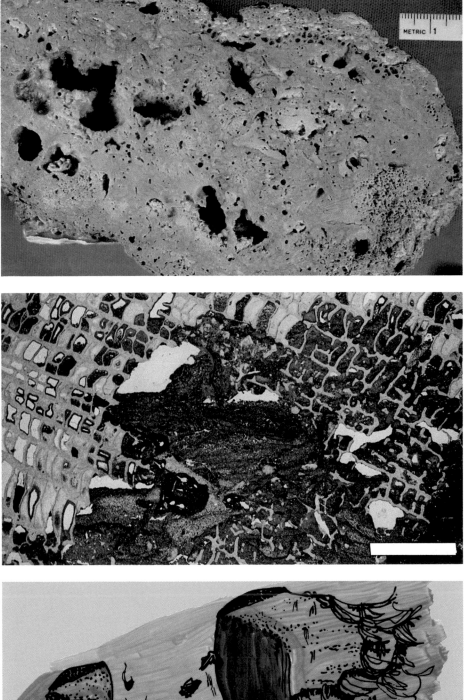

Figure 39—A polished slab of reef wall limestone (circa 9,000 years old) from the Belize barrier reef. This was originally a coral (*Montastraea annularis*) which has been subject to several generations of sediment infill, lithification and boring by sponges. The original coral skeleton can just be seen at the center. In this sample there are several different types of fine-grained sediment infill (light brown and dark brown) as well as different types of sponge borings (tiny holes at the lower right, medium-size galleries at the top, and large irregular voids in the center). The process of sponge boring sediment infill and early lithification repeated many times has almost completely altered this coral to a skeletal wackestone.

Figure 40—A thin section in plane polarized light of a coral partially altered in a manner similar to that illustrated in Figure 39. The section is stained with clayton yellow so that all aragonite components are white and all Mg-calcite components are red or pink. The coral pores are partially to completely filled with either aragonite cement, Mg-calcite cement, sediment, or all three. The irregular cavity at the center was excavated by a sponge (*Siphonodictyon*) and partially filled with fine-grained peloidal lime wackestone, mostly cemented by Mg-calcite (scale bar 1 cm).

Figure 41—An idealized sketch of a circular algal cup reef from Bermuda. The elevated rim is exposed at low tide and the depth of the surrounding coral floor is 3 to 8 m. Individual cup reefs vary from 10 to 30 m in their long dimension and the central depression is 2 to 3 m deep at high water (modified from Ginsburg, Schroeder and Shinn, 1971, by permission of the Johns Hopkins Press).

Figure 42—An areal view of the south coast of Bermuda with numerous algal cup offshore reefs in the lower foreground. Each cup reef has a raised rim and slightly deeper interior. The platform extending from shore is also rimmed by a raised lip of similar composition (photo R. N. Ginsburg).

Figure 43—The margin of an isolated cup reef near the northern margin of the Bermuda Platform exposed at spring low tide. This cup reef rises from water 8 to 10 m deep and the rim is composed of an intergrowth of coralline algae and vermetid gastropods (photo R. N. Ginsburg).

seaward. Sand shoals may also be present in the lee of the reef pavement. Vagaries of wave refraction may sweep the sands into cays and islands. These obstructions in turn create small protected environments very near the reef crest. Water over this zone is shallow (only a few meters deep at most) and scattered clumps of reef-building metazoans are common. The resulting rock types

range from clean skeletal lime grainstones to rudstones.

Back Reef Zone

In the lee of the reef flat, conditions are relatively tranquil and much of the mud formed on the reef front comes out of suspension. This, coupled with the prolific growth of mud and sand-producing bottom fauna such as calcareous green algae,

brachiopods, and ostracodes, among others, commonly results in mud-rich lithologies. The two most common growth habits of reef-builders in these environments are stubby, dendroid forms, often bushy (Fig. 24) and knobby, and/or large globular forms that extend above the substrate to withstand both frequent agitation and quiet muddy periods.

The rock types characteristic of this

Figure 44—An underwater view of the margin of a cup reef looking up toward the breaking surf. The reef is generally devoid of coral growth and has the appearance of bare rock (photo R. N. Ginsburg).

Figure 45—A cut and polished slab of cup-reef limestone composed of an intergrowth of coralline algae and *Millepora* encrusted by the foraminifer *Homotrema* (red). The pores and cavities are filled with cemented skeletal sand (brown) which is, in turn, coated with a rind of manganese (black lines). Remaining cavities are filled with yet a final stage of fine-grained wackestone which is also cemented (scale in cm) (photo R. N. Ginsburg).

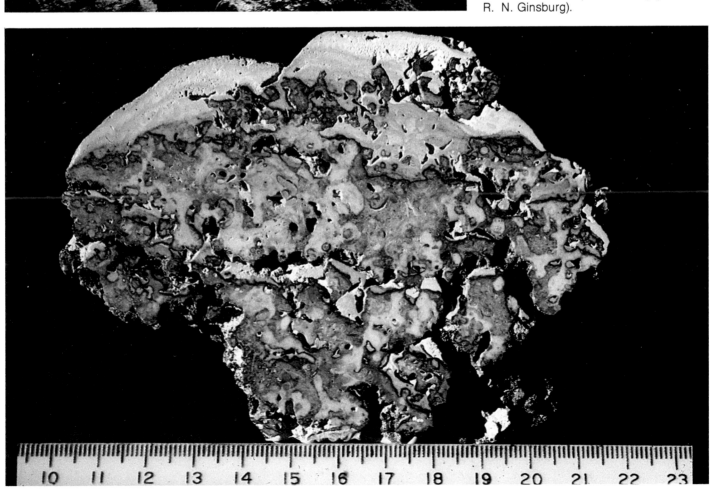

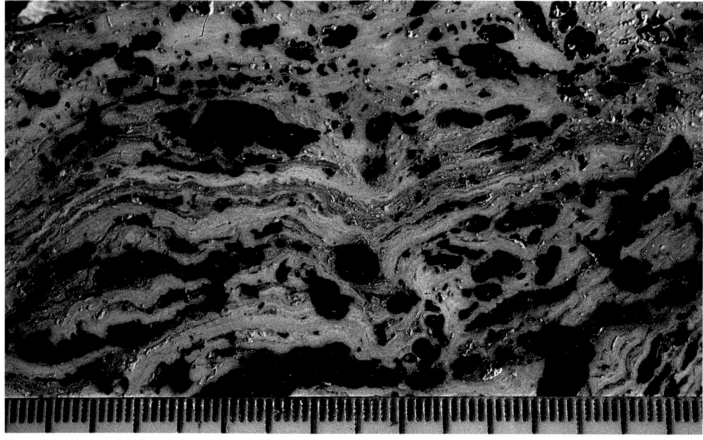

Figure 46—A cut and polished slab of limestone from an algal cup reef composed of coralline algae (white layers) in which the voids between layers (black areas) are not filled with cemented sediment (scale in cm) (photo R. N. Ginsburg).

Figure 47—An areal view of Rodriguez Key, Florida, showing a shallow bank composed of muddy skeletal sediments, fringed by a rim of branching corals and calcareous algae and topped by a forest of mangrove trees. The bank is 2.5 km long and located on the inner part of the Florida reef tract in shallow water; just seaward of the Florida Keys (top), a ridge of Pleistocene limestone (photo E. A. Shinn).

environment are bafflestones or floatstones to occasional framestones with a skeletal wackestone to packstone matrix.

Early Diagenesis

Two processes that tend to affect reef sediments along the platform margin in particular and that can be recognized in the rock record are early cementation and bioerosion. Early cement, in the form of Mg-calcite (as isopachous rinds of spar and as micrite) (Figs. 28, 29) and/or aragonite (as needles and large spherulites) (Figs. 30, 31), has been found to be particularly common in

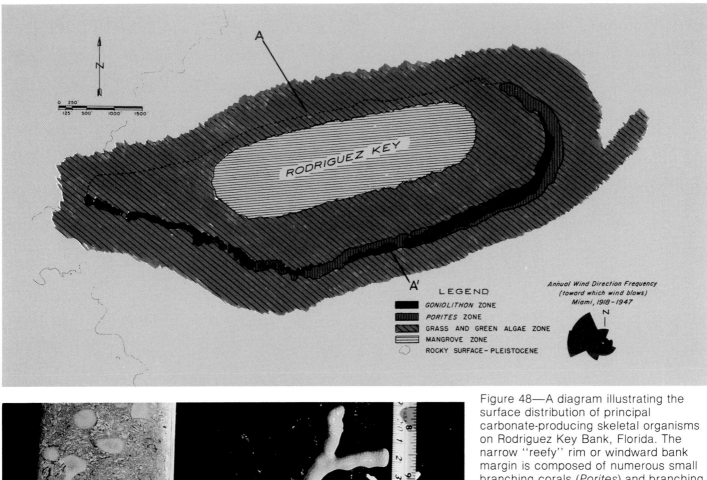

Figure 48—A diagram illustrating the surface distribution of principal carbonate-producing skeletal organisms on Rodriguez Key Bank, Florida. The narrow "reefy" rim or windward bank margin is composed of numerous small branching corals (*Porites*) and branching coralline algae (*Goniolithon*). The line of section through A-A' is illustrated in Figure 50 (after Turmel and Swanson, 1976, with permission of Soc. Econ. Paleontologists and Mineralogists).

Figure 49—A plastic impregnated and slabbed core through the branching coral (*Porites porites*) and calcareous algae (*Halimeda* and *Goniolithon*) seaward rim of Rodriguez Key Bank. To the right of the core are fragments of the coral *Porites porites*, the white portions being dead and the upper yellow portions alive; scale in inches (right) and centimeters (left) (photo R. N. Ginsburg).

ocean-facing reefs. These cements occur both in the reef itself, in the wall below the reef, sometimes in the fore-reef zone, and sometimes in the conglomerates of the reef flat (Figs. 32, 33, 34).

Bioerosion (Figs. 35, 36), while just as common in the lagoon reefs, can, with the occurrence of early lithifica-tion to create ever-new substrates for infestation, with time often alter the depositional fabric of the reef limestone (Figs. 39, 40). This process involves repeated boring by sponges, bivalves, and other endoliths, death of the organisms, infill of the cavity by sediment, lithification of that sediment, and then reboring by another generation of endoliths.

Algal Cup Reefs

Cup-shaped reefs (Figs. 41 to 46, 50) up to 10 m high and a few tens of meters in diameter are known from Bermuda, Yucatan and Brazil and occur on the seaward margins of pronounced breaks in slope off islands,

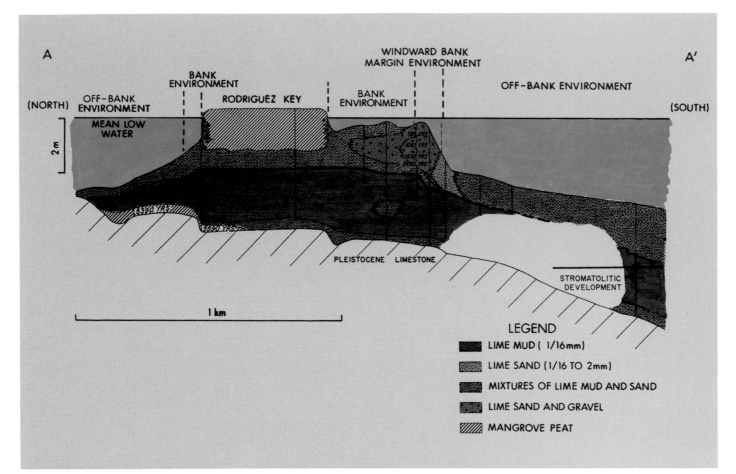

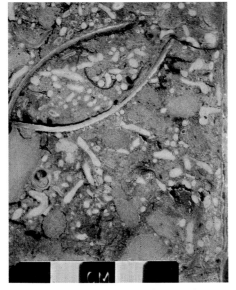

Figure 51—A close view of a plastic-impregnated and slabbed core through the seaward rim of Rodriguez Key Bank. The branching corals are *Porites porites* and almost all of the platey algal skeletons are from the green alga *Halimeda*. This sediment would be classified as a coral-platey algal floatstone to rudstone with a skeletal packstone matrix.

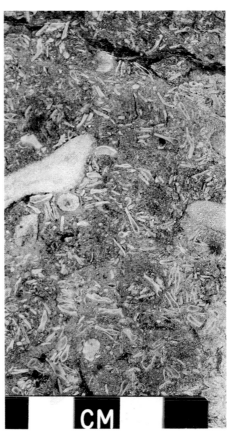

Figure 50—A diagrammatic cross-section through Rodriguez Key Bank, Florida (after Turmel and Swanson, 1976, with permission of Soc. Econ. Paleontologists and Mineralogists).

Figure 52—A close view of the plastic-impregnated and slabbed core through the *Porites-Goniolithon* zone just in the lee of the seaward rim of Rodriguez Key Bank. The corals are *Porites porites*, the white rod-like skeletons are the branching coralline alga *Goniolithon* and between these major components are bivalve shells, small gastropods and the platey alga *Halimeda*. This sediment would be classified as a coral-coralline alga rudstone with a skeletal packstone to grainstone matrix.

Figure 53—An areal view looking across Florida Bay. The sinuous features are linear mud banks topped occasionally by islands which are colonized by mangrove trees. The water in the areas between banks is 3 to 4 m deep and the long spit in the foreground is about 2 km in length. The white areas off the banks are regions in which fine-grained bottom sediment has been stirred up.

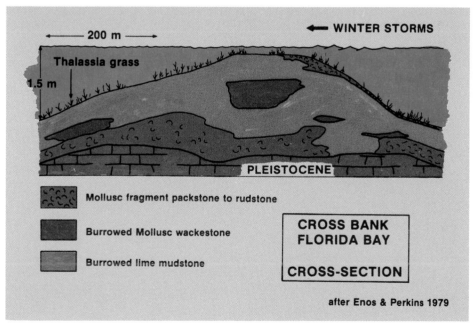

Figure 54—A diagrammatic cross-section through Cross Bank, Florida Bay (modified from Enos and Perkins, 1979, with permission of Geol. Soc. America).

Figure 55—A plastic-impregnated and slabbed core of sediment from one of the mud banks (the horizontal cracks are an artifact of sample preparation). The sediment is mostly gray lime mudstone to wackestone with occasional small concentrations of shells or individual gastropods. The linear brown tubes are rhyzomes of the marine grass *Thalassia* which commonly covers large portions of the bank tops (scale in cm).

shelves, and platform margins.

The best described examples occur in Bermuda (Ginsburg and Schroeder, 1973), and in plan view these cup reefs are circular, elipsoidal, oblate or crescentic. All the examples have an elevated rim that is awash at low tide and surrounds a central depression, a micro-lagoon, usually only a few meters deep but occasionally up to 5 m deep (Iams, 1970). In three dimensions the reefs are cup- or vase-shaped, with rough inward sloping sides. Near the bases of many cup reefs there is a sand-floored moat-like depression and numerous cavities 1 m or more in diameter extending from the moat into the base of the reefs.

A first impression on inspection of the underwater surfaces of the cup reefs is the sparseness of growth on them. There are sheet-like growths of yellow-brown *Millepora* (a calcareous hydrozoan) often with protruding knobs a few centimeters in diameter. There are occasional smaller en-

Figure 56—An idealized sketch of a deep-water carbonate mound or lithoherm from a depth of 600 to 700 m in the northeast Straits of Florida. The sketch is constructed from observations accumulated during dives on the submersible ALVIN. An exposed hardground (foreground) is partially covered with a veneer of muddy carbonate sand. Attached organisms cluster on the rock mound composed of a sequence of concentric crusts of submarine lithified sediment (after Neumann et al, 1977, with permission of Geol. Soc. America).

Figure 57—A sketch illustrating the difference in geometry between a biostrome, which is a bedded unit composed of a concentration of generally inplace carbonate skeletons and a bioherm which is a lens-like body of organic origin within rocks of different lithology.

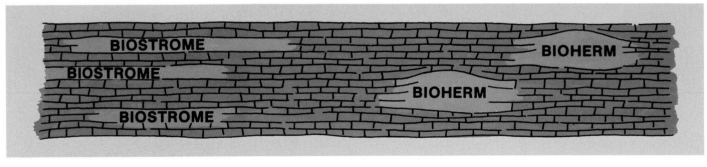

crusting and dome-shaped corals, (*Diploria, Porites astreoides*), and there are shrublike growths of the brown algae *Stypopodium* and *Sargassum.* But the prevailing impression is that the reef is a massive limestone with a sparse cover of organisms. Closer inspection of the surface shows that it is an almost continuous living skin of crustose coralline algae and *Millepora* in which numerous Vermetid gastropods are imbedded. Examination of numerous specimens from the surfaces of cup reefs shows that the principle carbonate secreting constituents are crustose coralline algae, *Millepora* and attached *Vermetid* gastropods

(*Dendropoma irregulare*). There are large variations in the relative proportions of these encrusting organisms, even within large hand specimens, but lamellar growths of crustose algae and/or *Millepora* with imbedded gastropods generally form well over two-thirds of the surface.

By sectioning several of these algal cup reefs, Ginsburg and Schroeder (1973) have found that they are built by an intergrowth of these encrusting organisms, principally crustose coralline algae, the encrusting hydrozoan *Millepora,* and the attached gastropod *Dentropoma irregularae.*

The growth framework of these algal cup reefs has extensive voids —

large and intermediate size growth framework and shelter pores, borings of bivalves and sponges, and both intra- and interparticle pores. A variety of vagile and sessile organisms (coelobites) inhabit these pores, with an encrusting foraminifer, *Homotrema rubrum,* the most abundant attached coelobite, and the tests of a variety of benthic foraminifera and ostracods are common; branched coralline algae, barnacles, bivalves, ahermatypic corals, bryozoans and burrowing crustaceans occur in varying abundance.

Beginning millimeters below the living surface, internal sediments accumulate in these extensive voids.

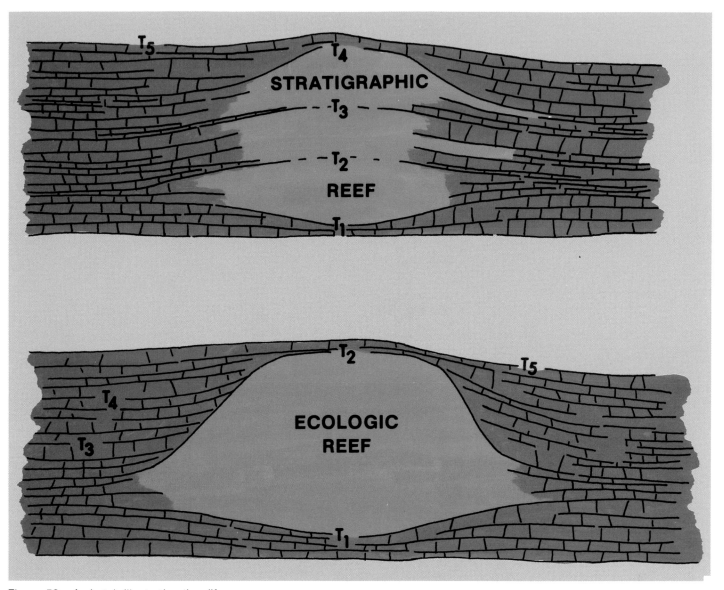

Figure 58—A sketch illustrating the difference between a stratigraphic reef, which is a thick, laterally restricted mass of carbonate rock, often composed of several superimposed bioherms which individually had little relief above the surrounding sea floor and an ecologic reef which is a rigid, wave-resistant topographic structure generally formed during one specific period of time.

Coarse-grained skeletal sand derived from the surface of the reefs is characteristic of the larger voids; lime mud with the tests of planktic foraminifera and planktic algae generally occurs in the smaller voids. Most specimens from the interior of the reefs show multiple generations of internal sediment that vary in grain size, composition and color. The sand size sediments are pumped into the voids by the frequent and intense wave action; the lime mud settles out in the smaller, less agitated pores.

Cementation of internal sediments and surrounding growth frame begins centimeters below the living surface. It is so pervasive that marble-hard reef rock is developed within 1/2 m or less of the surface. The cement is principally magnesium-calcite.

The rock types in these cup reefs are primarily coralline alga-coral bindstones with numerous cavities partly to completely filled with multi-generation, geopetal sediment.

Inshore Banks of Branching Coral and Algae

These reefs are typified by a series of banks in the nearshore zone along the Florida Keys (Figs. 47 to 52) which are covered by a community of calcareous algae and branching corals. The banks, up to 3 km long and 1 km wide, and oriented more or less parallel to the Florida Reef tract, rise as much as 4 m above the surrounding sea floor and are exposed during spring low tide. They are protected from ocean swells by numerous reefs and shoals, but they are washed by wind waves from the northeast.

All of the banks show a marked zonation (Fig. 57). The windward margin (east facing) is a zone of

GROWTH FORM AND ENVIRONMENT OF REEF BUILDING SKELETAL METAZOA			
GROWTH FORM		**ENVIRONMENT**	
		Wave Energy	Sedimentation
	Delicate, branching	low	high
	Thin, delicate, plate-like	low	low
	Globular, bulbous, columnar	moderate	high
	Robust, dendroid, branching	mod-high	moderate
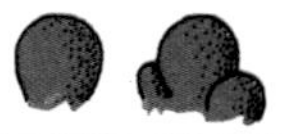	Hemispherical, domal irregular, massive	mod-high	low
	Encrusting	intense	low
	Tabular	moderate	low

Figure 59—A sketch illustrating the growth form of reef-building metazoans and the types of environments in which they most commonly occur.

STAGE	TYPE OF LIMESTONE	SPECIES DIVERSITY	SHAPE OF REEF BUILDERS
DOMINATION	bindstone to framestone	low to moderate	Laminate encrusting
DIVERSIFICATION	framestone (bindstone) mudstone to wackestone matrix	high	domal massive lamellar branching encrusting
COLONIZATION	bafflestone to floatstone (bindstone) with a mud stone to wackestone matrix	low	branching lamellar encrusting
STABILIZATION	grainstone to rudstone (packstone to wackestone)	low	skeletal debris

Figure 60—A sketch of the four divisions of the reef core facies with a tabulation of the most common types of limestone, relative species diversity and shape of reef-builders found in each stage (after James, 1979, with permission of Geol. Assoc. Canada).

branching finger corals (*Porites porites var. divericata*) and branched twig-like coralline algae (*Goniolithon strictum*). The top of the banks are gardens of marine grass (principally *Thalassia testudinium*), calcareous green algae (*Halimeda, Acetabularia*), infaunal bivalves, and burrowing crustaceans. Off the bank in the surrounding water is a similar community, but with numerous echinoids and large sponges as well.

Many banks have a composite internal structure, early Holocene mud bank cores veneered with 2 to 3 m of coral/algae deposits (Fig. 59), while others are all corals and algae.

Because most of the organisms that contribute to these banks are branched and segmented, they commonly break down on deposition, leaving no evidence of a coral and algal "frame." Cores through these banks show that most of the windward fringe is recorded as a branching coral floatstone to rudstone with a *Halimeda* (platy green algae) and *Goniolithon* (branching red algae), grainstone to packstone matrix. Subsurface sediments on the bank top are burrowed *Halimeda* and/or *Goniolithon* and bivalve packstones to wackestones.

Linear Mud Banks

Accumulations of lime mud (Figs. 53 to 55) that stand some 3 to 4 m above the sea floor have been documented in Florida Bay and on the northern shelf of Belize.

Florida Bay is compartmentalized into a series of "lakes" by shallow, sub-linear mudbanks. The sea floor in the lakes consists of a winnowed lag of molluscan shell fragments and/or a thin veneer of bioturbated mud over Pleistocene bedrock.

The tops of the banks are awash at low tide and covered by a thick growth of the sea grass *Thalassia testudinium*. They are occasionally

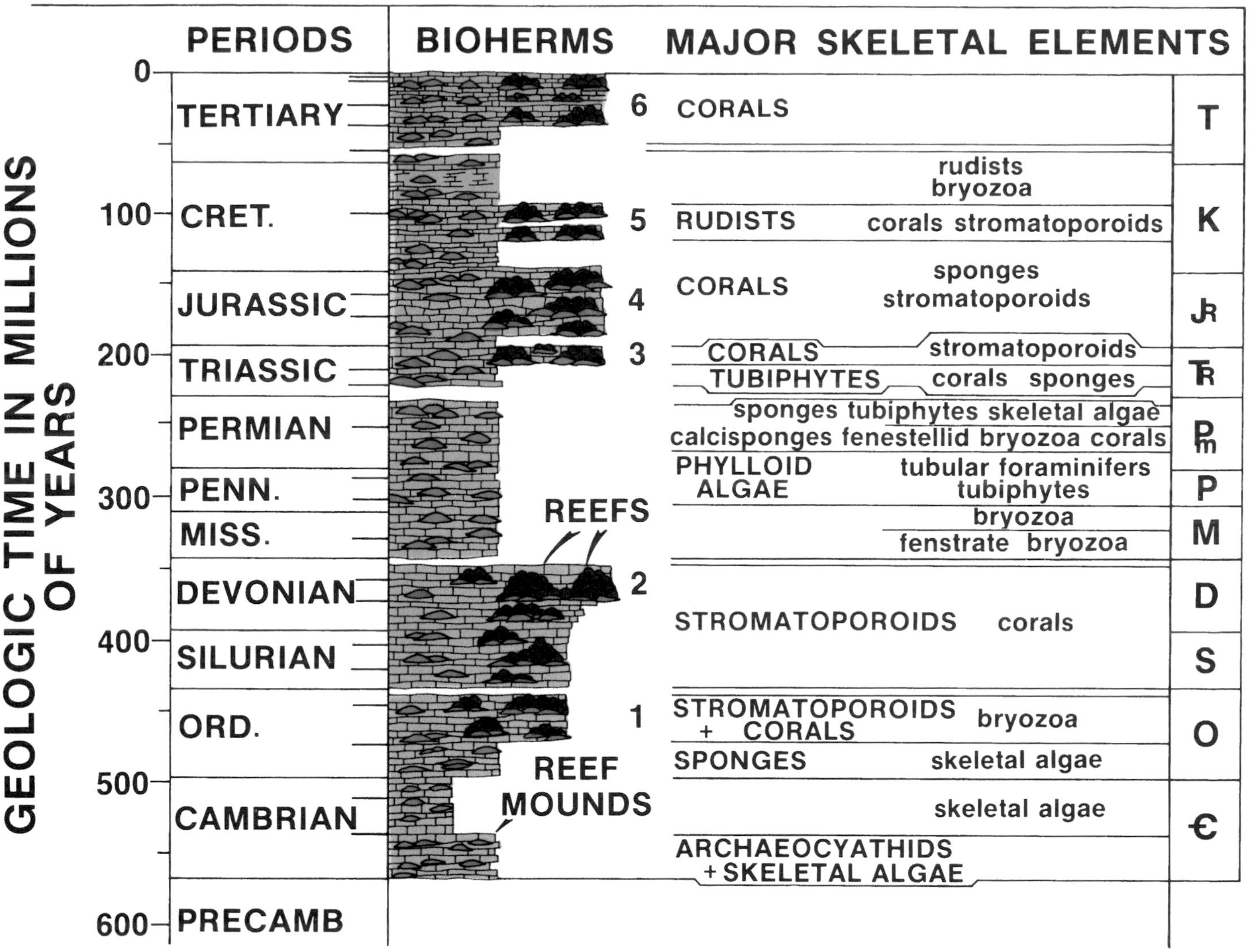

PERIODS	BIOHERMS	MAJOR SKELETAL ELEMENTS	
TERTIARY	6	CORALS	T
CRET.	5	rudists bryozoa	K
		RUDISTS corals stromatoporoids	
JURASSIC	4	CORALS sponges stromatoporoids	Jr
TRIASSIC	3	CORALS stromatoporoids	Tr
		TUBIPHYTES corals sponges	
PERMIAN		sponges tubiphytes skeletal algae	Pm
		calcisponges fenestellid bryozoa corals	
PENN.		PHYLLOID tubular foraminifers ALGAE tubiphytes	P
MISS.		bryozoa	M
		fenstrate bryozoa	
DEVONIAN	2	STROMATOPOROIDS corals	D
SILURIAN			S
ORD.	1	STROMATOPOROIDS + CORALS bryozoa	O
		SPONGES skeletal algae	
CAMBRIAN		skeletal algae	€
		ARCHAEOCYATHIDS + SKELETAL ALGAE	
PRECAMB			

GEOLOGIC TIME IN MILLIONS OF YEARS

0 — 100 — 200 — 300 — 400 — 500 — 600

REEFS

REEF MOUNDS

Figure 61—An idealized stratigraphic column representing the Phanerozoic and illustrating times when there appear to be no reefs or bioherms (gaps), times when there were only reef mounds, and times when there were both reefs and reef mounds; the numbers indicate different associations of reef-building taxa discussed in the text.

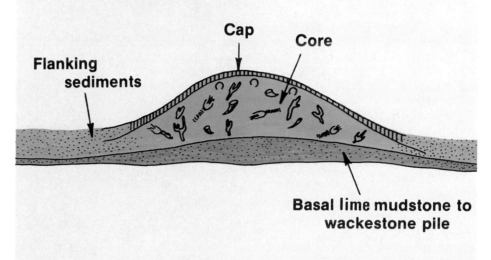

REEF MOUND

Flanking sediments

Cap

Core

Basal lime mudstone to wackestone pile

Figure 62—A cross section through a hypothetical reef-mound illustrating the geometry of the different facies and stages, based on Wilson (1975) (after James, 1979, with permission of Geol. Assoc. Canada).

Figure 63—A dense region of club-shaped stromatolites in the intertidal zone of Hamelin Pool, Shark Bay, Western Australia (photo P. Playford).

Figure 64—A cross section through a columnar algal stromatolite from the intertidal zone of Hamelin Pool, Shark Bay, Western Australia (photo J. L. Wray).

topped by islands and often cut by tidal channels. Banks that intercept effective winter storms have different windward and leeward sides (Fig. 54); the windward sides have steep slopes mantled by a skeletal lag, while the leeward have a gentle slope and a thick carpet of grass associated with muddy sediment. Bare areas with positive relief are traction deposits of pelleted mud with a distinct surface asymmetry, ornamented by low relief ripples. These areas are most common in the form of spits that project from the leeward sides of mud banks.

Most of the sediments are skeletal wackestones thoroughly burrowed and penetrated by roots and rhizomes of turtle grass (*Thalassia*).

Deep Water Carbonate Buildups

At the other end of the spectrum, a combination of seismic, dredging and direct observations from submersibles indicates the presence of elongate mounds up to 100 m long and 50 m high in depths of 600 to 700 m in the

Figure 65—A cluster of stromatolites in 3 m of water on the floor of Hamelin Pool, Shark Bay, Western Australia. The "beard" around the lower parts of the columns is mostly the calcareous alga *Acetabularia* and the rippled sand is mainly oolitic (photo P. Playford).

straits of Florida (Neumann, Kofoed, and Keller, 1977) (Fig. 56). These mounds are composed of surface-hardened concentric crusts of submarine lithified muddy to sandy sediment and have therefore been called "lithoherms."

The mounds rise from a flat sea floor composed of hardgrounds veneered with rippled, muddy skeletal sand (pelagic forams and pteropods). The most common biota are stalked crinoids and sea pens. The crinoids are oriented into the bottom currents that flow at rates of from 2 to 7 cm/sec.

The mounds themselves have steep sides, a pronounced northward elongation, smooth margins, and ir-

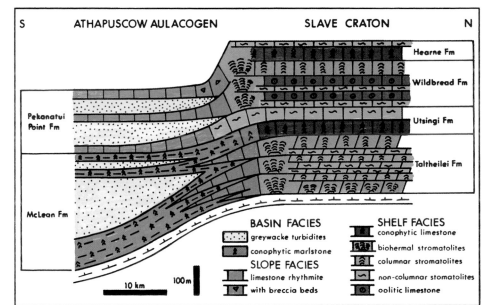

Figure 66—A diagrammatic cross section illustrating facies transitions in the Pethei Group of early Proterozoic age from the Slave craton into the Athapuscow aulacogen (modified from Hoffman, 1974).

regular tops. The crest is commonly populated by a dense community of deep-water organisms such as unstalked crinoids, branching ahermatypic corals, sponges and alcyonarians.

Figure 67—An outcrop photograph of part of a 20 m thick biostrome of low-relief (laterally linked) columnar stromatolites in a back-reef, littoral environment. The section here is perpendicular to the elongation of the columns in plan. Wildbread Formation, early Proterozoic, East Arm Great Slave Lake, North West Territories, Canada (photo P. Hoffman).

Individual layers or crusts are 10 to 30 cm thick and conform to the shape of the mound. The top of each crust is smooth. The sediments are more cemented at the exposed upper surface and become progressively less cemented and more irregular toward the base. Voids between crusts produced by winnowing out of soft sediment and by boring, if later filled with spar, may resemble some stromatactoid-like structures in the rock record.

The rock ranges from a mud-supported to grain-supported coral biomicrudite to pelagic foraminifer pteropod biomicrite to biopelmicrite. Micropelleting is common in the sediments and the rock is characterized by manganese stains, geopetal fills, and multi-generation boring and sediment infill. The sediments are cemented by Mg-calcite micrite and intensively bored by sponges.

FOSSIL REEFS

Nomenclature

While it is easy to recognize a modern reef or bank while swimming or flying over it, in the rock record it is an entirely different matter. There are several reasons for this, but the main complicating factor is time. First, most of our observations are in the vertical dimension, in quarry walls, in mountain exposures, in roadcuts, in drill core, or in cross sections, all of which show the reef as a limestone body constructed of different components generally formed at different times. Second, organisms have changed and evolved with time, so that the reef-building metazoans for one period are different from those in another period. Another complicating factor is diagenesis, the main component of the reef that allows us to differentiate stages of reef growth and different facies may be dolomitized at best or completely

Figure 68—An outcrop photograph of a biostrome composed of isolated high-relief columnar stromatoids. The inter-column areas are filled with edge-wise intraclasts. Wildbread Formation, Early Proterozoic, East Arm Great Slave Lake, North West Territories, Canada (photo P. Hoffman).

dissolved away at worst. The final compromise is exposure — cover, tectonics and partial well control often result in only a partial picture of the whole structure, and so the true geometry of the body cannot be determined with certainty. The upshot of all these is that the description of a reef or a reef-like structure in the rock record should be approached with caution. Perhaps the best and most rigorous approach is to first describe the shape and size of the limestone body and then, if possible, on the basis of detailed examination of the internal structure, postulate as to its origin, evolution and structure.

Another factor that has led to confusion in nomenclalture results from the misconception of scale. A complete reef in the Lower Paleozoic or Cretaceous which is formed by a variety of organisms and displays several stages of growth may be only as large as a single coral colony in a Jurassic or Pleistocene reef.

Because of these complications, the problem of fossil reef nomenclature is one that has pervaded the literature on carbonate rocks for over 80 years. For detailed reviews of this problem, the interested reader is referred to scholarly treatments by Nelson and others (1962), and Heckel (1974).

The word bioherm has long been a useful non-genetic term for lens-like bodies of organic origin that are embedded in rocks of different lithologies (Fig. 57) (Cummings, 1932). These structures should be differentiated from biostromes, which are purely bedded structures, such as shell beds, coral-rich beds, and the like, consisting of and built by sedentary organisms that do not swell into mound-like or lens-like forms (Fig.

57), Wilson (1975) uses the term "carbonate build-up" for a body of locally formed (laterally restricted) carbonate sediment which can be demonstrated to possess topographic relief. This term likewise carries no inference as to the internal composition or size of the structure; for example, a carbonate build-up may be composed of both bioherms and biostromes.

Dunham (1970) suggested separating the different types of carbonate build-ups in the rock record into two types. The first of these he called stratigraphic reefs or thick

Figure 69—The flank of a small (3 m wide) bioherm of columnar stromatoids from a tidal channel between much larger bioherms. Sediment adjacent to the bioherms is cross-bedded intraclast grainstone (scale in tenths of feet). Taltheilei Formation, Early Proterozoic, East Arm Great Slave Lake, North West Territories, Canada (photo P. Hoffman).

laterally restricted masses of pure or largely pure carbonate rock (Fig. 58). Implicit in this concept is its objectivity and reference only to the geometry of the build-up. On the other hand, an ecologic reef is a rigid, wave-resistant topographic structure produced by actively building and sediment binding organisms (Fig. 58). The binding in this structure must be organic and not inorganic. A stratigraphic reef may

include ecologic stages.

When viewed in the experience of the rock record, there is in fact a large class of structures which could be called stratigraphic reefs and commonly comprise a succession of localized biostromes and/or bioherms which had little or no relief above the sea floor at their time of growth (Fig. 58).

While the term ecologic reef is often used, the prerequisite that it be an organically-bound, wave-resistant topographic structure is unsettling. This concept comes from a discussion by Lowenstam (1950) in his classic paper on the Silurian reefs of the Niagara region, where he felt that the biologic potential of frame building and sediment binding organisms to

actually build wave-resistant structures defined a reef. If this concept were rigidly applied, many modern reefs would not qualify because most are not organically bound, especially those in the platform interior and deep water (where there are no waves). In addition, it is quite clear that the bulk of many reefs is a pile of in-place coral rubble and sediment which are certainly not wave resistant. On the other hand, fossil buildups composed of delicate branching organisms which would appear on the surface to have no wave-resistant potential may have been cemented very early and so have been hard structures of limestone on the sea floor at their time of growth. With these facts in mind, the term

Figure 70—An outcrop photograph of the lateral margin of a large (2 m wide by 15 m high, but less than 2 m synoptic relief) bioherm of columnar stromatoids. The bioherms are elongate, perpendicular to depositional strike as are the constituent columns. Taltheilei Formation, Early Proterozoic, East Arm Great Slave Lake, North West Territories, Canada (photo P. Hoffman).

ecologic reef should be used with caution and rigorously defined.

Perhaps a useful alternative is use of the term stratigraphic reef as defined, but use of the terms reef and reef mound only as discussed in the next section, once the detailed composition of the structure is known.

Analysis of Reef Limestones

At this point, it should be clear that to interpret any fossil reef or reef complex, the key is to extract as much information as possible from the limestones that make up the reef core. A concise description of the type of reef limestone is a necessary first step. Several other observations, however, such as (1) the relative diversity of reef-building organisms, (2) the growth form of the reef builders, and (3) the nature of the in-

ternal sediments and cements are also needed to give a complete picture.

Diversity Among Reef-Building Organisms

Very diverse faunas in terms of both growth form and taxa occur when a community is well-established and conditions for growth are optimum, that is, nutrients are in good supply, while chemical and physical stresses are low. In such optimum environments, the division of biomass among various species is due mainly to complex biological controls.

In contrast, low diversity environments commonly fall into three general categories — unpredictable environments; new environments (fauna moving into a new environment), and severe environments (high chemical and physical stress). Among

the factors most likely to stress modern and fossil reef-building communities are: (1) temperature and salinity fluctuations — most modern and likely most ancient reef-builders grow or grew best in tropical sea water of normal salinity; (2) intense waves and swell — the skeletons of most reef-builders will be broken or toppled by strong wave surge; (3) low-light penetration — in modern reef-building organisms, rapid calcification takes place because symbionts, which are light dependent,

Figure 71—A bedding surface of elongate columnar stromatolites from a 20 m thick undulose biostrome which grades seaward a few kilometers into discrete or linked bioherms. Taltheilei Formation, Early Proterozoic, East Arm Great Slave Lake, North West Territories, Canada (photo P. Hoffman).

take over some of the bodily functions of the host; and (4) heavy sedimentation — all reef-builders are sedentary filter-feeders or micropredators and water filled with fine-grained sediments would clog the feeding apparatus.

The Growth Form of Reef-Building Metozoans

The relationship between organism shape and environment is one of the oldest and most controversial topics

in biology and paleobiology. In terms of reef-building metazoans, however, many observations of the interrelationship between organisms and surrounding sediments from the rock record combined with studies of modern coral distribution on tropical reefs allow us to make some generalizations about form and environment that are very useful in reef facies analysis (Fig. 59).

Internal Sediment and Cement

The identification of internal sediment, or that sediment that has filtered into, and partially to completely filled cavities in the reef, is often a key to identification of reefal carbonates. The most common expression of this sediment is its

laminated geopetal nature with the overlying void now occluded with cement. This may range from an obvious filling between corals or stromatoporoids to the much more subtle partial filling of a cavity in a mudstone mound with a mud of a slightly different composition. Documentation of internal sediment of clearly marine origin is good evidence that there were holes or cavities in the structure. This in turn implies that the structure probably stood above the sea floor and that there was a structure formed by an irregular accumulation of large skeletal metazoa and/or some combination of smaller metazoans, early lithification, and possibly dewatering.

Early lithification is common in

Figure 72—An outcrop photograph of a biohermal cluster of deep-water stromatoids in fore-slope rhythmites composed of brown terrigenous marlstone, yellow allodapic dolomite and gray carbonate. Pen 10 cm in length at top right for scale. McLean Formation, Early Proterozoic, East Arm Great Slave Lake, North West Territories, Canada (photo P. Hoffman).

modern reefs and also appears to have been widespread in ancient reefs as well. This cement is often in the form of tiny micrite-size crystallites and is therefore almost impossible to differentiate from mud in the rock record. More easily recognizable is radiaxial (Bathurst, 1959), fibrous or botryoidal calcite associated with the voids in reef rock. Because these cements (or their Mg-calcite and aragonite precursors) may also be precipitated in environments other than on the sea floor, great care should be exercised before identifying them as syn-sedimentary. Among the criteria that can be useful in this regard are: (1) Is this cement localized to the reef and not present in the surrounding strata?; (2) Is this cement interlayered with internal marine sediment in cavities? (3) Are organisms seen growing on the cement as a substrate?; (4) Is the cement cut by borings; (5) Are clasts with this cement or clasts of this cement seen in surrounding off-reef sediments?; and (6) If this cement is finely crystalline, are the cemented sediments multi-generation and cut by numerous borings.?

Among common structures in many Paleozoic reefs is *stromatactis*. Once thought to be a recrystallized reef-building organism, these structures are now generally regarded as narrow sedimentary cavities with flat bases and irregular roofs that are sometimes floored with internal sediment but filled mostly with submarine cement now neomorphosed to radiaxial or fibrous calcite (Bathurst, 1959).

Internal Structure of Fossil Reefs

It has long been recognized that there is an ecological succession in many Paleozoic reefs (Lowenstam, 1950), that is, the replacement of one community of reef-building organisms by another as the reef grew. A recent synthesis by Walker and Alberstadt

Figure 73—An areal view, looking down on a Lower Cambrian bioherm exposed on the low tide platform at Point Amour, Labrador, Canada. The mound-shaped bioherm (center), composed mostly of archaeocyathans, is about 8 m in diameter. The surrounding joints are in inter-reef limestones and shales.

Figure 74—A Lower Cambrian bioherm exposed in cliffs along the Strait of Belle Isle, Labrador, Canada. The white nodular limestone of the bioherm is rich in archaeocyathans.

Figure 75—The margin of a Lower Cambrian bioherm complex exposed in cliff section along the shores of the Strait of Belle Isle, Labrador. The irregular nodular limestone to the right is the reef complex and the massive well-bedded sediments to the left are off-reef rudstones composed of skeletal debris. These beds grade rapidly into inter-reef shales and nodular limestones at the far left.

Figure 76—The surface of an archaeocyathan biostrome in southern Labrador. This plan view illustrates that almost all of the cup-shaped archaeocyathans are in upright position. The sediment between skeletons in this archaeocyathan bafflestone is a skeletal lime wackestone.

Figure 77—A close view of sheet-like, prone archaeocyathans at the basal, pioneer stage of an archaeocyathan biostrome in western Newfoundland, Canada.

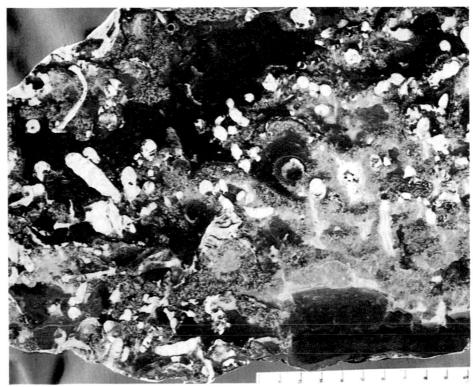

Figure 78—A slabbed and polished portion of the archaeocyathan floatstone/bafflestone that makes up the Lower Cambrian bioherms from southern Labrador. The white skeletons are archaeocyathans, the white speckles *Renalcis* and the dark red sediment skeletal wackestone to packstone. At the lower right is a cavity filled with geopetal sediment (scale in cm).

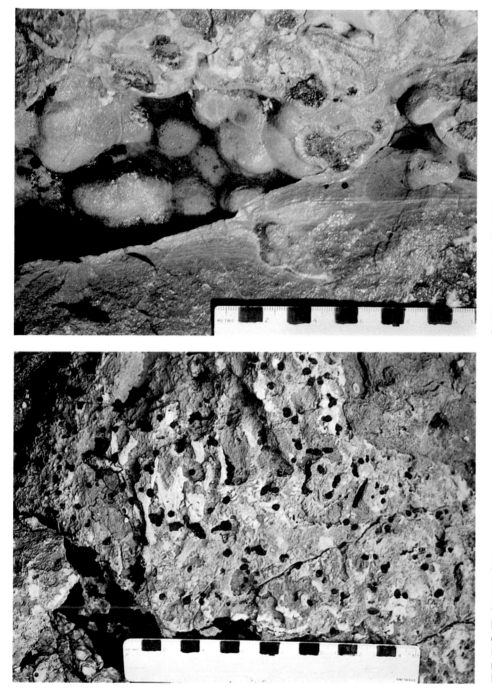

Figure 79—A close view of a partially filled cavity from one of the Lower Cambrian bioherms in southern Labrador. Clusters of *Renalcis* hang from the roof and are in turn encased in fibrous calcite cement that has developed a botryoidal appearance. The geopetal sediment at the base is interlayered with cement and has buried some of the botryoids that grew down from the roof (right center) (scale in cm).

Figure 80—An outcrop photograph of the top of a small mound in one of the Lower Cambrian bioherms from southern Labrador. The dark spots are the tops of tubular borings (*Trypanites*) that have been filled with terrigenous silt and dolomite (scale in cm); note that the borings penetrate both skeletons and matrix, indicating that the sediment was lithified.

(1975) of reefs ranging in age from Early Ordovician to Late Cretaceous suggests that a similar community succession is present in reefs throughout the Paleozoic and Mesozoic. Application of this concept to Oligocene reefs (Frost, 1977) which are dominated by scleractinian corals (the reef-builders in today's oceans) allows us now to equate ancient reef community succession with observa-

tions on modern reef communities with some measure of confidence.

In most cases, four separate stages of reef growth can be recognized, and these stages, along with the types of limestone, relative diversity of organisms and growth form of reef-builders in each, are summarized in Figure 60.

Pioneer (Stabilization) Stage

This first stage is commonly a series of shoals or other accumulations of skeletal lime sand composed of pelmatozoan or echinoderm debris in the Paleozoic and Mesozoic, and plates of calcareous green algae in the Cenozoic. The surfaces of these sediment piles are colonized by algae (calcareous green), plants (sea grasses), and/or animals (pelmato-

Figure 81—The surface of one of the flanking skeletal calcarenite beds that surrounds a small Lower Cambrian bioherm in southern Labrador. The rock is a pelmatozoan, trilobite grainstone to rudstone. The pelmatozoan grains are the white fragments in the sediment (scale in cm).

Figure 82—An algal bioherm, composed of stromatolitic algae in the Middle Cambrian March Point Formation, western Newfoundland, Canada.

Figure 83—The upper surface of a small algal bioherm in the Wheeler Shale, Middle Cambrian, western Utah. The structure is composed of stromatolitic algae and *Renalcis*.

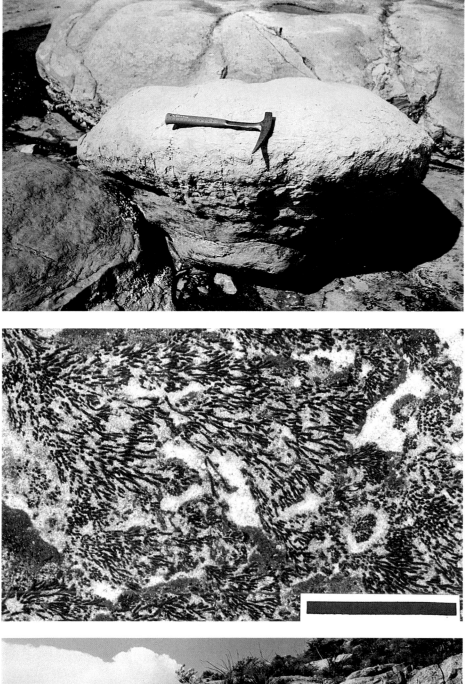

Figure 84—A single algal head from a series of algal bioherms (background) of Upper Cambrian age, west Newfoundland. This head is a thrombolite (Aitken, 1967) because it is not laminated and has a clotted and burrowed internal structure.

Figure 85—A thin section photomicrograph under plane polarized light of *Epiphyton,* geopetal sediment, and calcite cement from an Upper Cambrian bindstone, western Newfoundland. The branching *Epiphyton* grows in tiny clusters and is often associated with the laminar form *Girvanella* (scale 1 cm).

Figure 86—A bioherm (called Lechuguilla Mound) exposed in the lower part of the McKelligan Canyon Formation (Lower Ordovician) southern Franklin Mountains at El Paso, west Texas. This structure is approximately 15 m wide and composed of lithistid sponges (*Archaeoscyphia*), receptaculitid algae (*Calathium*), and the enigmatic encrusting coelenterate *Pulchrilamina* (photo D. Toomey).

Figure 87—An outcrop photograph of *Pulchrilamina spinosa*, one of the principal biotic components of Lower Ordovician mounds in the southern United States (photo D. Toomey).

Figure 88—An outcrop photograph of numerous silicified specimens of the receptaculitid alga *Calathium* from Lower Ordovician reef mounds in west Texas; coin is 2.4 cm in diameter (photo D. Toomey).

zoans) that send down roots or holdfasts to bind and stabilize the substrate. Once stabilized, scattered branching algae, bryozoans, corals, soft sponges and other metazoans begin to grow between the stabilizers.

Colonization Stage

This unit is relatively thin when compared to the reef structure as a whole, and reflects the initial colonization by reef-building metazoans. The rock is generally characterized by few species, sometimes massive or lamellar forms but more frequently thickets of branching forms, often monospecific. In Cenozoic reefs, the one characteristic common to all corals in this stage of reef growth is that they are able to get rid of sediment and clean their polyps, and so are able to grow in areas of high sedimentation. The branching growth form creates many smaller subenvironments or niches in which numerous other attached and encrusting organisms can live — forming the first stage of the reef ecosystem. *Stromatactis* is common in rocks representing this stage.

Diversification Stage

This stage usually provides the bulk of the reef mass and is the point at which most pronounced upward-building towards sea level occurs and easily definable, lateral facies develop. The number of major reef-building taxa is usually more than doubled, and the greatest variety in growth habit is encountered. With this increase in form and diversity of

Figure 89—An outcrop photograph of the lithistid sponge *Archaeoscyphia* (above quarter) and *Pulchrilamina spinosa* (right) from one of the Lower Ordovician reef mounds in west Texas; coin is 2.4 cm in diameter (photo D. Toomey).

Figure 90—A large bioherm (80 m high, 300 m long) in the Middle Ordovician (Whiterock Stage), Antelope Valley Limestone at Meiklejohn Peak, southern Nevada. This structure contains few skeletal organisms and is composed of mottled lime mudstone and *Stromatactis*.

framework and binding taxa, comes increased nestling space (surfaces, cavities, nooks and crannies) leading to an increase in diversity of debris-producing organisms.

Domination Stage

The change to this stage of reef growth is often abrupt. The most common lithology is a limestone dominated by only a few taxa with only one growth habit, generally en-crusting to laminated. Most reefs show the effects of surf at this stage, in the form of beds of rudstone.

The reason for this ecologic succession is at present a topic of much debate. Some workers feel that the

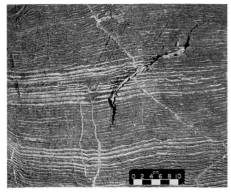

Figure 91—An outcrop photograph of part of the lower third of the Meiklejohn bioherm composed of "Zebra Limestone" made up of layers of lime mudstone and calcite spar.

Figure 92—Two polished slabs of "Zebra Limestone" illustrating dark gray-brown lime mudstone, cavity-filling gray-green geopetal sediment, and two stages of radiaxial calcite cement, one dark brown and the other white (photo R. J. Ross).

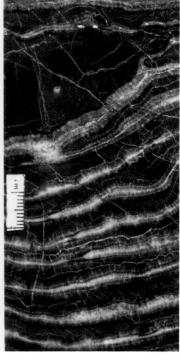

Figure 93—A small bioherm (recessively weathering massive limestone at center) in the Holston Formation, Middle Ordovician (Chazy Stage), Deane Quarry, Tennessee. The bioherm is composed almost entirely of bryozoans while the surrounding massive calcarenites (cut by the row of drill holes) are predominantly pelmatozoan grainstones (hammer at the center is scale).

Figure 94—The surface of a small bioherm like the one from the Deane Quarry (Fig. 192) exposure illustrating abundant bryozoan skeletons (white) in red lime mudstone.

control is extrinsic and reflects a progressive replacement of deep-water communities by shallower water ones as the reef grows to sea level and into more turbulent water, yet there is often abundant evidence that the first two stages are developed in shallow water. Other workers feel that the control is intrinsic and reflects a natural succession as the organisms gradually alter the substratum and change the energy flow pathways as the commumity develops, yet there is abundant evidence of increasing water turbulence as the structure grows.

Superimposed Reefs

Reef structures in the rock record are often impressive because of their size, not only laterally but vertically. Careful examination of stratigraphically thick reefs, however, often reveals that they are not a

Figure 95—An outcrop photograph of the numerous superimposed sheets of bryozoans (*Batostoma*) comprising a small bioherm in the Laval Formation (Chazy Stage) near Montreal, Quebec.

Figure 96—A small coral bioherm in the Lourdes Limestone, Middle Ordovician (Black River Stage), western Newfoundland.

single structure, but a series of superimposed or stacked reefs that grew on top of one another in more or less the same place. Individual episodes of reef growth are commonly separated by periods of exposure, reflected in the rock by intensive diagenesis, calcrete horizons, or shales (paleosols). When the ocean floods one of these surfaces that has been exposed, reef growth begins at the diversification stage because there is already a hard, often elevated, substrate present.

Reef Mounds

This model of reef development is predicated on the assumption that a full spectrum of reef-building organisms are present, as we see in the tropical oceans today — but such was not the case for much of the Phanerozoic. The critical element that is often missing, and without which the four stages of development in the reef core cannot occur, is the presence of skeletal metazoans that secrete large robust, branching, hemispherical or tabular skeletons.

Without them the reef cannot exist in the zone of constant turbulence, usually wave induced, because smaller and more delicate forms would be broken and swept away (unless submarine cementation is very rapid, pervasive and near-surface). This zone of turbulence is the optimum area for growth and diversity because sediment is constantly removed, water is clear, and nutrients are constantly swept past the sessile organisms. Such large skeletal metazoa were, however,

present only at certain times during the Phanerozoic, and each period has its own specialized group of framebuilders: (1) Middle and Upper Ordovician — bryozoa, stromatoporoids, tabulate corals; (2) Silurian and Devonian — stromatoporoids, tabulate corals; (3) Late Triassic — corals, stromatoporoids; (4) Jurassic — corals, stromatoporoids; (5) Middle Cretaceous — rudist bivalves; and (6) Middle and Late Tertiary — scleractinian corals (Fig. 61).

Figure 97—An outcrop photograph of the interior of a bioherm in the Middle Ordovician Lourdes Limestone from western Newfoundland. The framestone is composed of large corals (*Labyrinthites*) sometimes encrusted by stromatoporoids (*Labechia*) as at the center. The sediment between the corals and stromatoporoids is skeletal packstone to grainstone (scale in cm).

Figure 98—A small Silurian (early Wenlock) bioherm about 2 m high composed of corals and stromatoporoids in the upper Visby shales, Gotland, Sweden (photo P. Copper).

What then of the rest of the Phanerozoic record — were there no reefs? While there were certainly periods when no reefs at all formed, these periods were generally short and represent either climatic/tectonic crises or the complete lack of any reef builders, even small ones (as during Middle and Upper Cambrian). During most of the Phanerozoic there were structures that some workers call reefs, some call mounds, some call banks; they lack many of the characteristics we ascribe to reefs, yet they were clearly rich in skeletal organisms and had relief above the sea floor. The origin of these structures, which I have called reef mounds, has probably caused more discussion than any other topic in the literature on reefs (Heckel, 1974). When viewed against the backdrop of the general reef model, however, I think of them as half-reefs or incomplete reefs because they represent only stages one and two of the model. These structures did not develop the other upper two stages either because the environment was not conducive to the growth of large skeletal metazoa, or because these larger metazoans simply did not exist at the time the structure formed.

Reef mounds are, as the name suggests, flat lenses to steep conical piles with slopes up to 40° consisting of poorly sorted bioclastic lime mud with minor amounts of organic boundstone. With this composition they clearly formed in quiet water en-

Figure 99—A view of the Flathead Range in the Canadian Rocky Mountains illustrating dolomitized bioherms (**R**) in the Upper Devonian Peechee member of the Southesk Formation (photo B. Pratt).

Figure 100—A view of the margin of the Upper Devonian Miette Reef complex at Marmot Cirque in the Canadian Rocky Mountains (view to the northwest). The complex comprises (**1**) a basal platform carbonate unit, the Flume Formation, overlain by (**2**) the massive light-colored dolomites of the reef margin, the Cairn Formation, that are in turn buried by (**3**) the basinal shales of the Perdrix Formation (photo E. W. Mountjoy).

vironments and from the rock record appear to occur in three preferred locations; (1) arranged just downslope on gently-dipping platform margins; (2) in deep basins; and (3) spread widely in tranquil reef lagoons or wide shelf areas. When viewed in section, reef mounds display a similar facies sequence in each case (Wilson, 1975) (Fig. 62).

Reef Mound Core Facies
Stage 1 — Basal bioclastic lime mudstone to wackestone pile; very muddy sediment with much bioclastic debris but no baffling or binding organisms.

Stage 2 — Lime mudstone or bafflestone core; the thickest part of the mound, consisting of delicate to dendroid forms with upright growth habits in a lime mudstone matrix. The limestone is frequently brecciated, suggesting partial early lithification, dewatering and slumping, and contains *stromatactis*. Each

geologic age has its own special fauna that forms this stage: Lower Cambrian — archaeocyathans; Middle to Lower Ordovician — sponges and algae; Middle Ordovician, Late Ordovician, Silurian, Early Carboniferous (Mississippi) — bryozoa; Late Carboniferous (Pennsylvanian) and Early Permian-platey-algae; Late Triassic — large fasciculate dendroid corals; Late Jurassic — lithistid sponges; Cretaceous — rudist bivalves (Fig. 61).

Figure 101—An areal view of Winjana Gorge and the Napier Range reef complex, Canning Basin, Western Australia. The soft shales that filled the basin (upper left) and covered the reef complex have been removed by erosion so that the carbonates are exhumed and are exposed now much as they were in the Upper Devonian (photo P. Playford).

Figure 102—One wall of Winjana Gorge illustrating from right to left: horizontally bedded back-reef and reef-flat strata; massive reef margin; and steeply dipping fore-reef sediments. Upper Devonian Napier Range reef complex, Canning Basin, Western Australia (photo P. Playford).

Stage 3 — Mound cap; a thin layer of encrusting or lamellar forms, occasional domal or hemispherical forms, or winnowed lime sands.

Reef Mound Flank Facies

These massive, commonly well-bedded carbonates comprise extensive accumulations of archaeocyathan, pelmatozoan, fenestrate bryozoan, small rudist, dendroid coral, stromatoporoid, branching red algae or tabular foraminifer debris and chunks of wholly to partly lithified lime mudstone. Volumetrically these flank beds may be greater than the core itself and almost bury it.

While in many reef mounds the core is massive carbonate, in others, particularly stratigraphic reef mounds, the core is a heterogeneous assemblage of mound-shaped or pillow-shaped masses, commonly 0.5 to 1.5 m across and 0.3 to 1.0 m thick. Smith (1981b) has proposed the term "saccolith" as an epithet to describe these irregular, sack-shaped, pillow-shaped or bun-shaped, discrete, boulder-size masses of autochthonous reef rock, of whatever composition within a reef or reef complex. These elements are most easily seen in natural outcrops where weathering has accentuated mutual contacts.

In summary, there are times when the model is applicable because there are no reefs at all, there are times then only reef mounds form, and

Figure 103—An upper Devonian reef margin at Najee Cave, Bugle Gap, Canning Basin, Western Australia. The massive reef front limestones (left) interfinger with the well-bedded inclined fore-reef strata (right) (photo P. Playford).

Figure 104—The abrupt but interfingering transition in the lower 150 m of a dolomitized upper Devonian reefal carbonate platform margin (Fairholme Group), and basinal shales and limestones (Mount Hawk Formation) at Burnt Timber in the Front Ranges of the southern Canadian Rocky Mountains (photo I. A. McIlreath and G. E. Tebbutt).

there are times when both reef mounds and reefs occur, but in different environments.

EXAMPLES OF FOSSIL REEFS

Stromatolite Buildups

During the Precambrian and earliest Paleozoic, prior to the appearance of herbivorous metazoa, stromatolites formed impressive buildups. These stromatolite complexes clearly had relief above the sea floor and in terms of morphology were surprisingly similar to later skeletal reefs.

There are no stromatolite bioherms documented from modern oceans, but in Shark Bay, Western Australia, where all environments are hypersaline and stromatolites abound, the interrelationship between stromatolite morphology and environment has recently been documented (Hoffman, 1976). In the intertidal zone columnar to club-shaped forms, up to 1 m high are found rimming headlands (Figs. 63, 64). In relatively high energy, exposed environments the relief of the columns is proportional to the intensity of wave action. These grade laterally away from the headlands to the lower energy bights where the

Figure 105—A disphyllid coral colony (*Syringopora*) in growth position and encased in predominantly nodular *Alveolites* and lamellar stromatoporoids from the core of a bioherm in the Mercy Bay Member of the Weatherall Formation (upper Devonian), Banks Island, North West Territories.

Figure 106—Massive domal stromatoporoids and stromatoporoid rubble in the upper few meters of a massive biostromal unit in the Upper Devonian Cairn Formation, Slide Creek, Jasper National Park, Canadian Rocky Mountains (photo E. W. Mountjoy).

Figure 107—A close view of a coral (white) being encrusted by a stromatoporoid (dark gray) in a small bioherm in the Upper Devonian Blue Fiord Formation, Ellesmere Island, North West Territories, Canada.

Figure 108—The Upper Devonian reef flat facies at Winjana Gorge, Canning Basin, Western Australia, composed of the stromatoporoid *Actinostroma,* the alga? *Renalcis* (speckled) and laminated internal sediment (photo P. Playford).

Figure 109—An outcrop photograph of thin platy stromatoporoid framestone of the reef facies from the Silurian West Point Formation, Gaspe, Quebec; hammer for scale (photo P. Bourque).

stromatolites are more prolate and elongate, oriented normal to the shoreline. In tidal pools, digitate columnar structures abound.

These growth forms are the result of active sediment movement; algal mats only grow on stabilized substrate, thus columns are nucleated upon pieces of rock; growth is localized there and does not occur on the surrounding shifting sands. Early lithification of the numerous superimposed layers of mat and sediment turns the structures into resistant limestone. Moving sand continuously scours the bases of the stromatolites. The mounds or pillars are largest in subtidal or lower intertidal en-

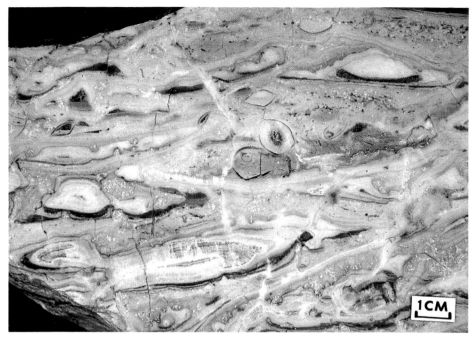

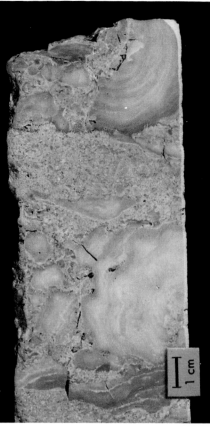

Figure 110—A polished slab of reef limestone from the Upper Devonian Lloyd Hill platform atoll-pinnacle reef composed of laminar stromatoporoids, the alga? *Renalcis* (speckled) and a large system of cavities filled with fibrous calcite (photo P. Playford).

Figure 111—A polished core slab from the reef facies of the Upper Devonian Redwater Reef complex, Alberta, Canada. The rock is a framestone of massive stromatoporoids, and skeletal packstone. S.W.D.W. no. 5, depth 3689 ft (photo J. L. Wray).

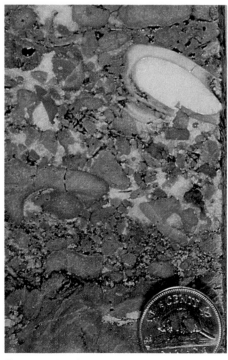

Figure 112—A polished slab of laminar stromatoporoids in skeletal rudstone to grainstone of the reef flat facies; Upper Devonian Leduc Formation, Golden Spike reef complex, Alberta, 10-27-51-27W4 (photo R. A. Walls and E. W. Mountjoy).

Figure 113—A stromatoporoid rudstone from the reef facies in which interparticle spaces are filled with cement; Swan Hills Reef Complex, Beaverhill Lake Formation, Alberta (10-15-67-10W5); depth 7959 ft. The interparticle spaces are filled with cement (coin 2.1 cm diameter) (photo R.A. Walls).

Figure 114—A porous, dolomitized skeletal sand from the reef facies; Upper Devonian Nisku Formation in West Pembina Field, Alberta (6-25-50-10W5), depth of 8488 ft (coin 2.5 cm diameter) (photo R. A. Walls).

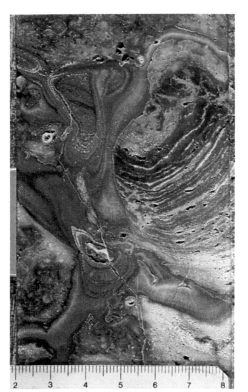

Figure 115—A stromatoporoid rudstone from the reef facies in which the pores are filled with banded radial cement and subsequent laminated internal sediment; Middle Devonian Keg River Formation, Rainbow Field, Alberta (2-27-108-9W6), depth 6529 ft (photo R. A. Walls).

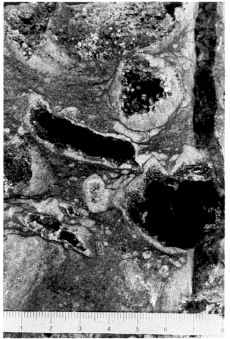

Figure 116—A stromatoporoid rudstone from the reef facies in which the whole rock is dolomite and the stromatoporoids have been leached resulting in excellent moldic porosity; Upper Devonian Beaverhill Lake Formation, Kabob South Field, Alberta (1-11-59-18W5), depth 10,069 ft (photo R. A. Walls).

Figure 117—A polished core slab of massive stromatoporoids and a dolomitized mudstone matrix from the reef facies; Upper Devonian Leduc Formation, Strachan Field, Alberta (10-31-37-9W5); depth 13,572 ft (photo R. A. Walls).

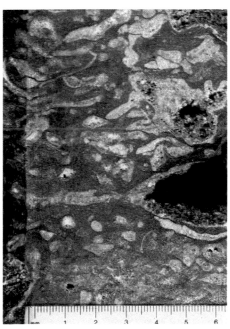

Figure 118—A dolomitized and porous stick-like stromatoporoid (*Stachyoides*) floatstone with a mudstone matrix and excellent moldic porosity; Upper Devonian Beaverhill Lake Formation, South Kabob Field (1-11-59-18W5); depth 10,945 ft (photo R. A. Walls).

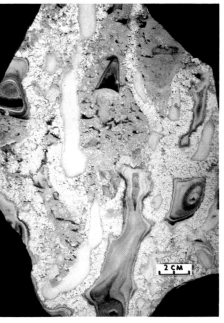

Figure 119—A polished slab of reef limestone (reef flat facies) composed of the stromatoporoid *Stachyoides* (white), the alga? *Renalcis* (speckled), skeletal packstone and a large cavity system filled with interlaminated fibrous calcite spar and pelleted lime mud; Upper Devonian of the Canning Basin, Western Australia (photo P. Playford).

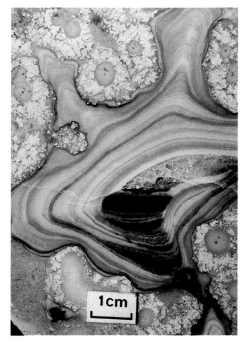

Figure 120—A polished slab of *Stachyoids* (white)-*Renalcis* (speckled) reef framestone with a large cavity infilled with banded fibrous cement and red pelleted mud, which is in turn covered with geopetal sediment and finally equant blocky spar; Upper Devonian of the Canning Basin, Western Australia (photo P. Playford).

Figure 121—A coral (*Disphyllum*) bafflestone, dolomitized and with excellent moldic porosity; Upper Devonian Nisku Formation, West Pembina, Alberta (6-25-50-10W5); depth 8520 ft (coin 2.5 cm diameter) (photo R. A. Walls).

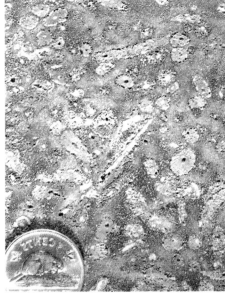

Figure 122—A polished core slab of *Amphipora* (stromatoporoid) wackestone from the back-reef facies; Upper Devonian Beaverhill Lake Formation, Swan Hills, Alberta (10-16-67-10W5); depth 7980 ft (coin 2.1 cm diameter) (photo R. A. Walls).

Figure 123 (above)—An outcrop photograph of a bioherm containing huge branching colonies of corals; Silurian West Point Formation, Gaspe, Quebec; hammer for scale (photo P. Bourque).

Figure 124 (below)—An outcrop photograph of a biostrome composed of rugose corals; Silurian West Point Formation, Gaspe, Quebec; pen knife 10 cm in length (photo P. Borque).

vironments and decrease in synoptic relief upwards, finally merging with stratiform mats in upper intertidal zones above the zone of active sediment movement.

These stromatolites continue seaward into the subtidal environment (Playford and Cockbain, 1976) where they grow to depths of at least 3.5 m and possibly more. They occur in patches for several hundred meters offshore. These stromatolites range from tiny, bulbous bodies developed on otherwise flat algal mats to large complex mounds. Columnar forms almost 1 m high are most common and range from conical to club-shaped (Fig. 65). Some structures are

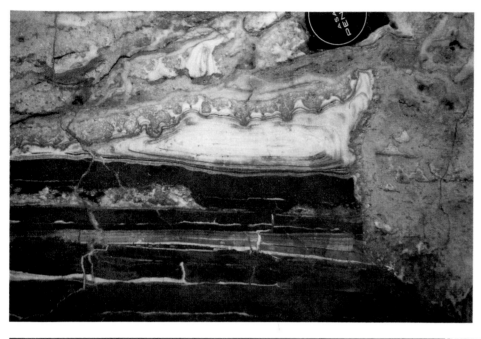

Figure 125—A cavity in reef-flat facies limestones in which the base contains layered geopetal sediment and the ceiling is encrusted with pendant *Renalcis;* Upper Devonian of the Canning Basin, Western Australia; lens cap 5 cm diameter (photo P. Copper).

Figure 126—A cut and polished slab of a toppled *Stachyoides* (stromatoporoid) colony encrusted with the alga? *Renalcis* around which a cavity system became filled with laminated pelletal lime mud and synsedimentary cement; Upper Devonian of the Canning Basin, Western Australia (photo P. Playford).

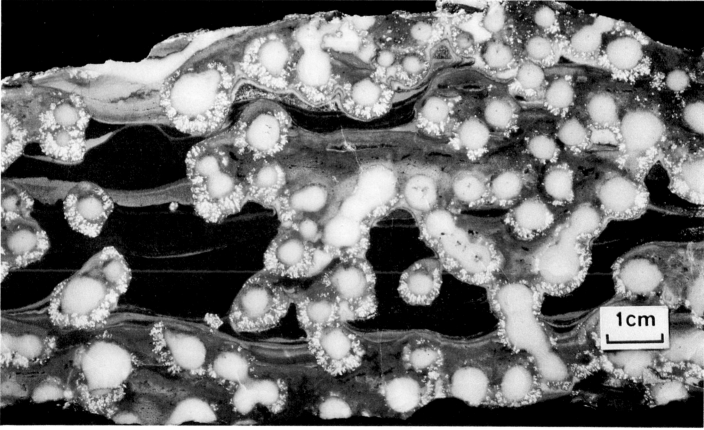

elongate, either parallel or perpendicular to the shoreline. Although the living surface of the stromatolite is soft, lithification begins a few millimeters to centimeters below the surface and the interior and lower parts of the structures are hard.

Although there has been much study of Precambrian stromatolites, much less attention has been paid to their role in the formation of reefs and reef mounds. Stromatolites are rare but do occur in Archean (more than 2,600 m.y. old) carbonates. By Aphebian time (2600 to 1700 m.y. ago), stromatolites are present in a wide variety of environments and bioherms of different types can be

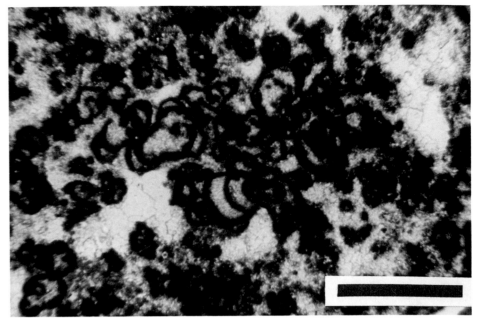

Figure 127—A thin section photograph in plane polarized light of *Renalcis*, characterized by inflated chambers with microcrystalline calcite walls. Scale 1 mm (photo J. L. Wray).

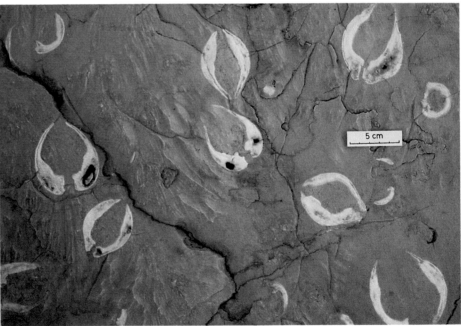

Figure 128—An outcrop photograph of in place bivalves (*Megalodon*) in back reef deposits immediately behind the reef facies; Upper Devonian reef complex near Brooking Springs, Western Australia (photo P. Playford).

discerned. One impressive variety of stromatolites in a platform-to-basin transition has been documented by Hoffman (1974) from the Pethei Group (circa 1800 Ma) at Great Slave Lake, North West Territories, Canada (Fig. 66).

In this sequence, the edge of the platform is generally occupied by a narrow zone, the "mound and channel belt" similar to barrier reefs that would form later. Large elongate stromatolites with up to 3 m relief are separated by channels filled with cross-bedded and mega-rippled coarse-grained sands and conglomerates composed of stromatolite fragments and clasts of oolitic grainstone. This belt separates platform facies (a complex array of laminar to columnar stromatolites and intervening ooid, intraclast and oncolitic limestones) from a slope-to-basin facies (lime mudstone-shale rhythmites with bedded slump breccias and siliclastic mudstones contain-

ing poorly laminated, small, calcareous columnar stromatolites).

The vertical succession of stromatolites in a shelf edge buildup from contemporaneous strata of the Goulburn Group in the Kilohigok Basin has recently been documented by Cecile and Campbell (1978). Here a subtidal unit of clastic carbonates with evidence of deposition in a high energy environment is overlain by isolated stromatolite mounds generally in the form of elongate

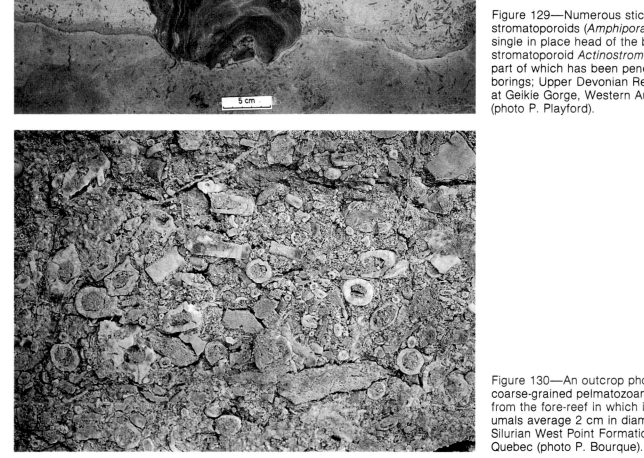

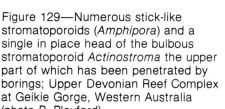

Figure 129—Numerous stick-like stromatoporoids (*Amphipora*) and a single in place head of the bulbous stromatoporoid *Actinostroma* the upper part of which has been penetrated by borings; Upper Devonian Reef Complex at Geikie Gorge, Western Australia (photo P. Playford).

Figure 130—An outcrop photograph of a coarse-grained pelmatozoan rudstone from the fore-reef in which individual columals average 2 cm in diameter; Silurian West Point Formation, Gaspe, Quebec (photo P. Bourque).

hemispheroids with synoptic relief of from 30 to 40 cm. The main part of the buildup is composed of large mounds greater than 1 m wide and more than 1 m high that coalesce into thick laterally extensive sheets separated by intraclast rich carbonates, calcareous siltstones and sandstones. Elongate sets of mounds attain heights of greater than 10 m and have synoptic relief of up to 2 m. The upper part of the buildup is composed of extensive sheets of laterally linked hemispheroidal columnar and branching stromatolites. The sheets and columns are cut by narrow channels filled with clastic carbonate. The buildup can be traced for 100 m along strike and is associated with thick deposits of clastic carbonate.

Early Paleozoic Reef Mounds

The first metazoan reefs developed at a time of plate fragmentation, wide cratonic seas and open continental margins. During this period of remarkable continental stability sediments were deposited in response to a gradual inundation of extensive, relatively flat cratonic areas. These shallow, Cambro-Ordovician epeiric seas were bounded by narrow continental margin facies of ooid/skeletal lime sand shoals; the small pioneer buildups appear to have developed in the lee of, or slightly downslope, from these shoals.

Figure 131—An areal view of the Lloyd Hill platform atoll-pinnacle reef complex about 0.5 km in diameter, Upper Devonian, Western Australia, which has weathered in relief in contrast to the surrounding soft basinal shales (photo P. Playford).

Figure 132—An outcrop photograph of a typical stromatactis calcilutite; Silurian West Point Formation, Gaspe, Quebec; pencil for scale (photo P. Bourque).

Early Cambrian

Reefs, or more precisely reef mounds, are found in the earliest Cambrian strata, even prior to the appearance of trilobites, on the Siberian platform. These small structures, meters in diameter, are composed primarily of lime mud and calcified algae (*Epiphyton, Renalcis* and *Girvanella*) while the relatively large skeletal metazoans, goblet-shaped archaeocyathans, are found only as accessory elements scattered in and around the structures. Very

quickly, however, a wide variety of sessile and vagrant calcareous benthos developed on and near these sea-floor reliefs so that by the end of Early Cambrian time small but complex reef mound ecosystems exhibiting all of the sedimentological features of modern reefs, such as prolific skeletal growth, internal sedimetnation and early cementation and bioerosion, had developed.

Individual bioherms and biostromes in the latest phases of Early Cambrian time are usually composed of

numerous, small stacked mounds (saccoliths), and are surrounded by skeletal sands composed of pelmatozoan debris together with brachiopod and hyolithid skeletons. Archaeocyathans are the most conspicuous skeletal elements of these mounds. Between these fossils are skeletons of trilobites, hyolithids, brachiopods and sponge spicules set in a matrix of lime mud. *Renalcis* and *Epiphyton* encrust archaeocyathans and line the walls of growth cavities. These cavities, which are often par-

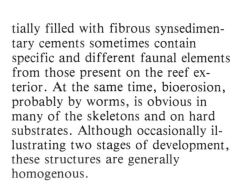

Figure 133—An outcrop photograph of "Zebra-type" *stromatactis* which occurs sporadically in the Gros Morbe facies of the Silurian West Point Formation (Silurian), Gaspe, Quebec (photo P. Bourque).

Figure 134—A peloidal lime mudstone to wackestone in which cavities are in places filled with a second generation of sediment (brown) and/or geopetal sediment and cement ("stromatactis"); a deep-water bioherm of Upper Devonian age, Canning Basin, Western Australia (photo P. Playford).

tially filled with fibrous synsedimentary cements sometimes contain specific and different faunal elements from those present on the reef exterior. At the same time, bioerosion, probably by worms, is obvious in many of the skeletons and on hard substrates. Although occasionally illustrating two stages of development, these structures are generally homogenous.

Middle Cambrian to Lower Ordovician

With the extinction of archaeocyathans at the end of early Cambrian time, reef mounds lost their major skeletal elements. This seems to have had a profound affect on reef growth because most of the buildups in succeeding Cambrian strata are algal, with scattered invertebrate skeletal elements, but much reduced in diversity compared to the early Cambrian.

Bioherms in the wide epeiric seas were mostly stromatolitic, and while some illustrate well-laminated internal structure, others show the effects of associated invertebrates whose burrowing has destroyed the laminations. Those stromatolite-shaped structures with a non-laminated clotted, and fenestral internal fabric have been called thrombolites. The stromatolites and thrombolites are often an intergrowth of non-calcified (blue-green) algae and calcified algae, par-

Figure 135—A thin section of a "stromatactis" cavity. Each cavity is filled with bladed fibrous calcite spar (submarine cement?) interlaminated at the base of some cavities with pelleted lime mud; Upper Devonian, Canning Basin, Western Australia (photo P. Playford).

Figure 136—A "neptunian dike" in a reef facies filled with red laminated peloidal marine sediment that was cemented early; Upper Devonian, Winjana Gorge, Western Australia (photo P. Playford).

ticularly *Girvanella, Renalcis* and *Epiphyton.* There is a suggestion that those algal buildups at or near the continental margin were composed primarily of calcified algae and abundant fibrous synsedimentary cement while those on-shelf buildups were dominated by blue-green algae.

As more and more sessile skeletal invertebrates appear in Late Cambrian and Early Ordovician time, these predominantly algal buildups are gradually populated by a more diverse skeletal biota. Among the more important of these organisms are siliceous (lithistid) sponges, the stromatoporoid-like metazoan *Pulchrilamina,* the large receptaculitid alga *Calathium* and sometimes the primitive coral *Lichenaria.* Thus, bioherms in Lower Ordovician strata near the continental margin are once again largely skeletal in composition.

Middle Paleozoic Structures

The onset of Middle Ordovician time coincided with a dramatic change in carbonate environments. Many continental margins were destroyed by plate collision and orogenesis while on the continents major intracratonic downwarps began to develop. Bioherm and carbonate bank development, while still present along some passive margins became widespread in the new intracratonic basins. Buildups of this age are best known from North America, Western Europe, North Africa, and Australia.

Figure 137—The Muleshoe Canyon bioherm, a reef mound composed primarily of lime mudstone (Waulsortian Type) in the Lake Valley Formation (Mississippian), Sacramento Mountains, New Mexico. The structure is approximately 60 m thick.

Figure 138—A view of the right side of Muleshoe bioherm pictured in Figure 137, illustrating two smaller satellite bioherms each about 15 m high that grew on the flanking beds which dip to the right from the main buildup.

The acme of Paleozoic reef growth, in terms of variety of structures, geographic and paleotectonic settings, and diversity of reef forming taxa, occurred during the Siluro-Devonian. This is largely due to the appearance, at the beginning of Middle Ordovician time, of large sessile benthic organisms such as corals and stromatoporoids which grew in many different shapes. The reefs were built by these metazoans together with a diverse fauna of sponges, bryozoans, calcareous red and green algae, brachiopods and pelmatozoans which increase in number and variety throughout the Silurian until latest Devonian when, coincident with extinction of many of the important groups, the reef ecosystem collapsed.

When viewed as a whole these buildups illustrate the complete spectrum of reef types, from small reef mounds on the platform interior to extensive reef-rimmed platform atolls and pinnacles to platform-margin barrier reef complexes to downslope mounds to basin-center pinnacles.

Middle Ordovician to Late Devonian

With the varied and complex groups of organisms present, these Middle Paleozoic buildups commonly exhibit well-developed vertical and lateral ecological zonation.

The reef mound, whether as an isolated structure or as the base upon which more complex reefs grew, is characteristically well developed. In

Figure 139—A close view of the limestone that makes up the core of Muleshoe bioherm. In this instance, the fenestrate bryozoan fragments have been silicified and so stand out as brown skeletons and particles.

Figure 140—A close view of the coarse crinoid-rich limestones which make up the flanking beds of Muleshoe bioherm.

some cases, such as the Middle Ordovician of the western United States, Silurian of Scandinavia, eastern United States and Canada, and the Devonian of western Europe. These structures are carbonate mud mounds with few if any skeletal components, but abundant *Stromatactis* and evidence of early cementation. In other situations they contain small skeletal elements. In Ordovician and Silurian mounds these skeletons are primarily ramose and encrusting bryozoans, but appear to be replaced

by corals in the Devonian.

The more complex buildups are composed mostly of stromatoporoids, and to a lesser extent, corals. The stromatoporoids are most numerous and largest and exhibit the most diverse forms in the Devonian but show a wide variety of growth forms throughout. They appear to be most prolific in the upper parts of reefs. Tabulate corals occur with the stromatoporoids and dominate the downslope environments in more turbid water. They are relatively small in

size, but have a wide variety of growth habits, in particular such forms as the Devonian genus *Alveolites* which is much like the modern *Montastraea*. While solitary rugose corals were best adapted to life on soft substrates, colonial rugose corals attained a wide variety of morphologies, for example the spectacular dendroid to fasciculate Devonian forms such as *Disphyllum* (up to 2 m) which resembles the later Triassic and Jurassic coral *Thecosmilia*.

Figure 141—Massive reef limestone (right) of the Nansen Formation (Permo-Pennsylvanian) extending downward and basinward into dark argillaceous limestones of the Hare Fiord Formation (left); arrows point out small reef-mounds developed on the seaward slopes of the reef front; western side of Blind Fiord, Ellesmere Island, North West Territories.

Figure 142—A small reef mound, approximately 30 m high in the Nansen Formation (Permo-Pennsylvanian) on the eastern side of Blind Fiord, North West Territories, Canada. The massive reef mound limestones are mainly phylloid algae floatstones to bafflestones.

Although dominated by skeletal invertebrates, the calcareous algae which occur in the earliest Cambrian reefs are still present in these structures. In particular, the tiny, chambered, encrusting form *Renalcis* occurs as a major and diagnostic component of many shelf margin reefs. The calcareous algae *Solenopora* and *Parachaetetes* are also common, but as nodules.

The flanking beds in both reefs and reef mounds are predominantly composed of debris from pelmatozoans (crinoids, blastoids and cystoids) and brachiopods. Crinoid flank beds are not so abundant in Devonian as Silurian buildups.

Late Paleozoic–Early Mesozoic Buildups

Near the end of late Devonian time, the complex reef ecosystem that gave rise to the extensive Siluro-Devonian buildups collapsed. By the close of Frasnian time many of the marine invertebrates that populated the seafloor had vanished. The bulk of the coralline skeletal elements and reef-builders were hardest hit; stromatoporoids were reduced to a few genera, rugose corals were dramatically altered, and tabulate corals became extinct. Brachiopods were decimated, and although they recovered in the later Paleozoic they were never again as diverse.

The following early Mississippian had few if any "reef-building" taxa, with the place of the corals and stromatoporoids being taken by pelmatozoans and bryozoans.

Figure 143—The margin of a reef mound in the Late Pennsylvanian Holder Formation, Dry Canyon, Sacramento Mountains, New Mexico. The structure, some 8 m in height, is composed of phylloid algae and lime mud.

Figure 144—The flanking beds, in an outcrop some 12 m thick, of a large reef mound exposed in the Upper Pennsylvanian Holder Formation in the walls of Dry Canyon, Sacramento Mountains, New Mexico. These beds, with a steep westward dip (to the left) drape off a mound which is out of the plane of the photograph and is composed of algal plates and pelmatozoan debris (photo B. Pratt).

organisms, scleractinian corals (the reef builders of modern seas) appear in the Middle Triassic and together with calcareous sponges and stromatoporoids alter the make up of buildups dramatically from reef mounds to true reefs.

Most Late Paleozoic buildups occur in and around the margins of intracratonic basins. In Permian time, with the breakup of Pangea, a large circumglobal seaway, the Tethys, gradually developed and the majority of subsequent Mesozoic reef growth took place along the margin of or in basins adjacent to this ocean.

Mississippian

Although the reef ecosystem collapsed during Devonian time, bioherms of Mississippian age are often substantial structures, rising as much as 150 m above the sea floor and having slope deposits dipping as much as 50°. They are found chiefly near shelf margins and in deep water

Gradually though, during the ensuing Mississippian and early Pennsylvanian, a variety of new carbonate-producing organisms appeared. These generally small organisms evolved continuously through early Permian time into the late Triassic. Among the most important of these new taxa are phylloid calcareous algae (*Archaeolithophyllum, Eugoniophyllum* and *Ivanovia*), and *Tubiphytes,* an enigmatic calcareous organism which grew as tiny, laminated epiphytic encrustations in reefs from Early Mississippian to Late Jurassic, but were principal components only in Permian, Triassic and some Jurassic buildups. Other important forms are Opthalmidid–Calcitornellid tubular foraminifrs, small dendroid stromatoporoids (for example, *Komia*) and the encrusting coralline alga? *Archeolithoporella.* Calcareous sponges also become a conspicuous part of the reef biota in Permian time as do spongiomorph hydrozoans (stromatoporoids) in the Triassic. The first large, massive frame-building

Figure 145—A detailed outcrop photograph of phylloid algae bafflestone in the Gobbler Formation, Upper Pennsylvanian, Sacramento Mountains, New Mexico. The alga is almost all the genus *Archaeolithophyllum* and the matrix between the plates is skeletal lime wackestone.

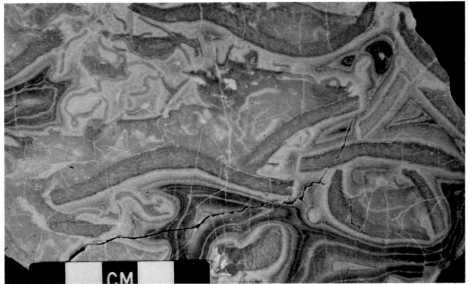

Figure 146—A cut and polished slab of phylloid alga-hydrozoan bafflestone from the core of a Lower Permian reef mound in the Nansen Formation, Ellesmere Island, North West Territories, Canada. Most of the large skeletal plates here are of the hydrozoan *Palaeoaplysina,* and each is surrounded by thick rinds of fibrous calcite cement (photo G. Davies).

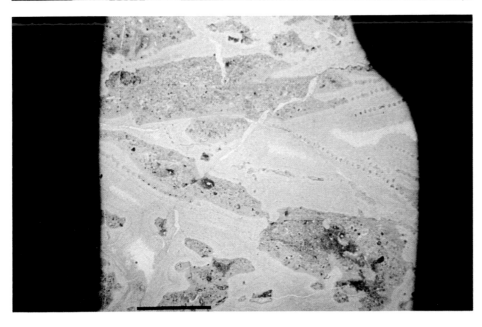

Figure 147—A thin section photograph of Fenestellid bryozoan (lines of small dots) floatstone in which the bryozoa are surrounded by thick rinds of cement and the voids are filled with skeletal packstone, from the Nansen Formation, Pennsylvanian, Ellesmere Island, North West Territories, Canada (scale 1 cm) (photo G. Davies).

Figure 148—An outcrop photograph of byrozoan framestone with boytryoidal calcite cements from reef mounds in the Nansen Formation, Lower Permian, Ellesmere Island, North West Territories, Canada.

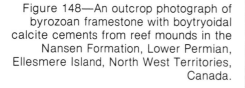

Figure 149—A cavity filled with laminated, dolomitized sediment in a Pennsylvanian bryozoan reef mound in the Nansen Formation, Ellesmere Island, North West Territories, Canada.

structure itself.

In late Mississippian time, isolated corals became locally important constituents in these bioherms.

Pennsylvanian

By mid-Pennsylvanian time, the newly evolving calcareous benthos had begun to populate the mud mounds, and structures of this age contain phylloid algae as a major component. These buildups are somewhat smaller, on the average, than their Mississippian counterparts, generally up to 30 m high with skeletal flanking beds dipping off the cores at about 25°. The phylloid algae with their carbonate-encrusted leaves probably grew upright in thickets and gardens on and around the reliefs on the sea floor. In some instances, encrusting bryozoans and foraminifers grew on the mounds as well and these were often as abundant as the algae.

Although most of the buildups are small reef-mounds, in the subsurface these buildups form a major component of what were large offshore

settings.

The scarcity of large skeletal biota is dramatically illustrated in the style of these buildups. They are primarily lens-like mounds composed 50 to 80% massive, vaguely clotted, peloid lime mudstone with only scattered fragments of crinoids and bryozoans. These structures have come to be called "Waulsortian mud mounds" after a locality in Belgium. The mud is often multigeneration with later phases filling cracks as geopetal sediment in what must have been partial-ly lithified reef rock. Original cavities are also filled with laminated sediment and fibrous cement forming multiple, regular and parallel *Stromatactis* structures. Extensive early lithification is also illustrated by occasional fragments of mound rock containing *Stromatactis* in flank beds. In European examples, flanking beds are present but not extensive, while in North American examples, the flanking beds made up of coarse encrinite and rare lithoclasts are thick, at times comprising more than half of the

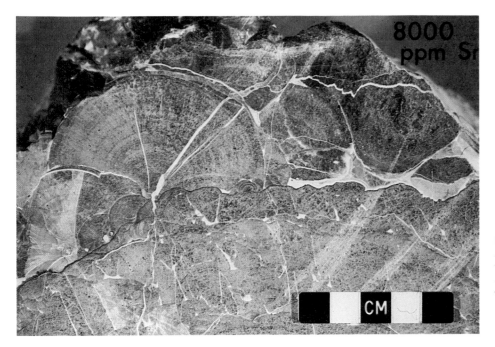

Figure 150—Botryoidal calcite in reef mounds of Pennsylvanian age from the Nansen Formation, Ellesmere Island, North West Territories, Canada. On the basis of similar microfabrics, geometry and high strontium content (8000 ppm) these are interpreted to be neomorphosed botryoidal aragonite, precipitated within the reef mound as it grew on the sea floor (photo G. Davies).

Figure 151—Looking northeast toward El Capitan, a massive carbonate that forms the western margin of the Permian reef complex in the Guadalupe Mountains of west Texas and New Mexico. In this photograph, the massive cliff-forming limestones at the center and left are the older fore-reef facies of the underlying Goat Seep reef. The massive limestone at the right is the platform margin, Capitan reef to fore-reef facies of this complex.

banks growing in deep basins. Here they are characteristically zoned with red algae, fusilinids, algal calcispheres and crinoids most common on the oceanward margins, and green algae, rugose corals and brachiopods usual on the leeward sides.

Permian

Buildups in earliest Permian strata are still predominantly phylloid algal mounds but these plants are soon augmented by a variety of low-lying, delicate, branching to encrusting organisms such as sponges, *Tubiphytes,* and the encrusting form *Archaeolithoporella* (which may be an alga or an inorganic precipitate), together with bryozoans and other calcareous algae so that most Permian bioherms are dominated by this new biota.

Perhaps the most famous of these buildups is the "Permian Reef Complex" in the Guadalupe Mountains of west Texas and new Mexico, southwest United States. This uplifted western margin of the Delaware Basin exposes an extensive carbonate platform complex which in the subsurface to the east is one of the world's great oil-producing structures. These rocks have been described in several major works (see Wilson, 1975, for an excellent summary). The marginal facies of this complex has long been an enigma because the massive unit which forms such an impressive

"rim" at the edge of the shelf is a muddy carbonate with abundant fine debris and small encrusting organisms and in many places an almost complete absence of relatively large frame-building taxa. It is, therefore, difficult to imagine it as a barrier reef similar to modern structures even though the location is right. Rather, it now appears that the original high point on the marginal profile was a lime sand barrier island complex and the "reefy" facies rich in sponges, *Tubiphytes* and cement-lined cavities is a down-slope deposit of coalesced reef mounds.

Figure 152—Looking south parallel to the margin of the Permian Reef Complex in McKittrick Canyon that runs normal to the trend of the complex. In the upper part of the wall on the skyline is the transition from bedded back-reef sediments (left) to the massive platform margin or "reef" carbonates (center) to the fore-reef strata (right) that dip basinward. The inclination of beds from upper left to lower right in the canyon wall is the dip of older fore-reef strata.

Figure 153—An outcrop photograph of an inclined surface in the massive "reefal" portion of the Permian Reef Complex near the mouth of Walnut Canyon, New Mexico. The limestone here is a framestone composed of sponges and *Tubiphytes* in a skeletal wackestone matrix (hammer blade for scale).

The only skeletons larger than a few centimeters in these reef mounds are the conspicuous calcareous sponges, especially the sphinctozoan or chambered forms. The matrix of lime mud makes up over half of the rocks while the other biota present are encrusting algae and/or foraminifers, *Tubiphytes* and locally, phylloid algae. Most of the growth cavities are filled with spectacular growths of fibrous cement which was probably marine. Some earlier Permian mounds in this area appear to be over one-half marine cement.

In other regions, such as Tunisia, Sicily and eastern Asia, calcareous sponges are also the primary components of Permian reef mounds. In eastern England, however, the Magnesian Limestone, which also forms an extensive shelf-margin complex, appears to contain mostly bryozoans and abundant encrustations identical to *Archaeolithoporella.* Likewise, lens-shaped reef mounds in the southern Alps, which reach thicknesses of over 300 m in

downslope faces are dominated by *Tubiphytes, Archaeolithoporella* and bryozoans and contain abundant synsedimentary cement, but lack sponges.

Triassic

Reefs of Triassic age are developed mostly along the northern and southern margins of the widening Tethys and along the western coast of North America. These reefs are best studied in the Alps where they occur on and marginal to huge carbonate

Figure 154—An outcrop photograph of an inclined surface in the massive "reefal" portion of the Permian Reef complex near the mouth of Walnut Canyon, New Mexico, in which sponges and other skeletons are bound by numerous encrustations (irregular dark lines) of the calcareous alga *Archaeolithoporella.*

Figure 155—An outcrop photograph of the massive reef facies of the Permian Reef Complex exposed in Dark Canyon. Here a pocket of skeletons rich in sponges and large gastropods occurs in an otherwise massive wackestone to mudstone (hammer blade for scale at top).

platforms which are similar to those of the preceding Permian.

There appears to be no record of early Triassic reefs and a scarcity of reef-building biota world-wide. The first Triassic reefs, of mid-Triassic age, are small deep-water buildups with sparse corals and echinoderms. These are followed by extensive reef complexes in late Middle Triassic strata in which the most important reef-building biota are calcisponges, *Tubiphytes,* some stromatoporoids and many colonial corals; essentially the same faunal elements as in the preceding Permian, but with the important addition of colonial corals. Although most reefs contain this varied biota some are dominated by *Tubiphytes.* The most extensive carbonate platforms developed during the late Triassic (Norian-Rhaetian) with marginal facies composed of sand shoals and reef complexes. Many of these platforms, however, were destroyed during the latest

Triassic and transformed into basins with terrigenous sediment and areas of small reefs.

Those structures studied to date along the west coast of North America appear to be small patch reefs without ecological zonation and may be deep-water, ahermatypic buildups.

The main biotic constituents of Upper Triassic reefs are corals, stromatoporoids (hydrozoans), calcisponges and calcareous algae.

The corals and calcisponges are the most important components but are restricted to different parts of the reefs and different zones of water energy; corals in shallow, high energy environments (especially the branching form *Thecosmilia*); calcisponges in the more protected and sheltered central parts of the reef. Colonial corals in these structures exhibit the same wide range in growth form as seen in modern reefs. Stromatoporoids are also abundant and oc-

Figure 156—The back-reef facies of the Permian Reef Complex (Tansill Formation) in Dark Canyon illustrating a large overturned colony of *Collenella* (hammer blade for scale).

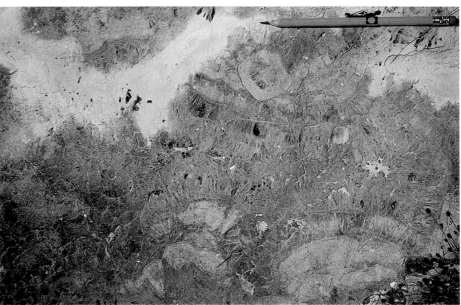

Figure 157—An outcrop photograph of an inclined surface in the massive "reefal" portion of the Permian Reef Complex near the mouth of Walnut Canyon, New Mexico. This part of the outcrop contains many spherulites or fan-shaped growths of calcite which were probably precipitated as aragonite into open spaces of the sea floor while the limestone was forming (pencil for scale).

Figure 158—The 200 m thick Steinplatte Reef of Upper Triassic age near Waidring in Tirol in which the massive reef limestone (right) passes into the fore-reef slope (left) dipping at about 30° into the Kossen basin (photo H. Zankl).

Figure 159—The massive limestone which comprises the core of the Dachsteinkalk Reef, Upper Triassic (Norian) in the Gosaukamm Range, Austria (photo E. Flugel).

Figure 160—The mature or diversification stage of Upper Rhaetian patch reefs at Tropfbruch near Halleim, Salzburg, Austria, here illustrating abundant corals; the large branching coral in the center is 0.5 m high (photo E. Flugel).

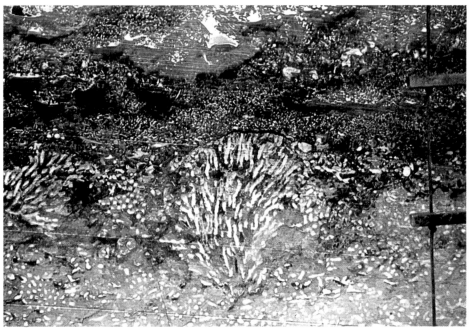

cupy a variety of different environments, mostly in protected parts of the reef. Together with these, bryozoans and calcareous algae contribute to the bulk of the reef mass. The major sediment producing organisms are brachiopods, bivalves, gastropods, cephalopods, serpulid worms, crustaceans, ostracodes, crinoids, ophiroids and holothurians. These buildups exhibit, for the first time, extensive bioerosion by bivalves and algae. Although boring organisms are known from the earliest reefs, these are the first structures in which they play a significant role as a diagenetic agent in reef development. As in the Permian many of the growth cavities are occluded by extensive fibrous spar interpreted to be synsedimentary cement.

These late Triassic structures, like those in the middle Paleozoic, exhibit the complete spectrum of reef growth from on-shelf reef mounds and patch reefs, to marginal reefs with excellent lateral zonation, to deep-water reef mounds. Many of the reefs have the characteristic four stage development but are commonly "drowned" after reaching the domination or climax phase.

Jurassic

Reefs, although known from the Lias of Morocco, only become a common element of Tethyan carbonate facies in Middle Jurassic time, reaching a maximum in Late Jurassic. They occur both as patch reefs in shallow marine basins in western Europe and as marginal facies around the northern edge of the Tethys from eastern Canada to the Middle East and along the rims of large but isolated platforms within the Tethys. The structures themselves appear to be mostly isolated deep water reef mounds or numerous small patch reefs and reef knolls.

Deep water reef mounds, which may have relief of up to 100 m, are composed of siliceous sponges, algae and tubular foraminifera set in a lime mud matrix. The most conspicuous associated fauna in these structures are bryozoans and brachiopods.

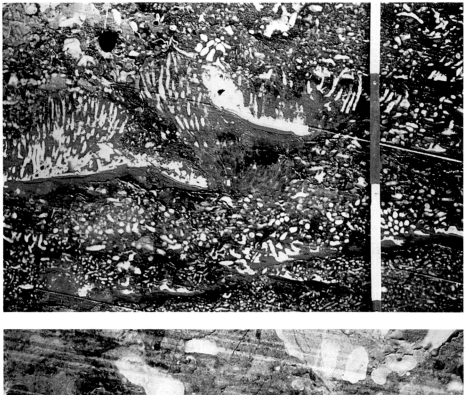

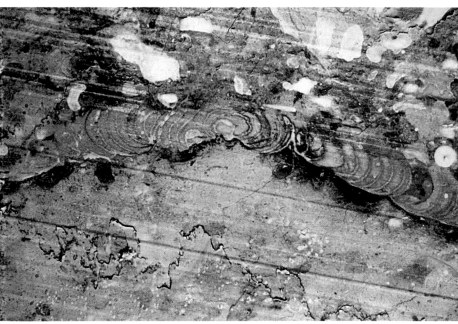

Figure 161—The same locality as illustrated in Figure 160 but here the limestone has suffered partial dissolution and the holes are filled with geopetal sediment and white calcite spar; divisions on the meter stick are 20 cm (photo E. Flugel).

Figure 162—The same locality as above illustrating a hardground surface encrusted by a sphinctozoan sponge which is about 2 cm thick. The subhorizontal lines across the photo are saw-marks on the quarry wall (photo E. Flugel).

Coral/stromatoporoid reefs grew as patches on wide shallow platforms, as belts of patch reefs near shelf margins, and as caps on pre-existing sponge reef mounds. In some areas these reefs are coral-dominated and exhibit excellent vertical zonation, while in other regions they are thinner with the principal elements being stromatoporoids and the encrusting coral *Microsolena.* The red calcareous algae *Solenopora* as well as the green Dasycladacean algae reach the height of their development in Late Jurassic time. In addition, the modern types of Codiacean algae and articulated Corallinacea first appear associated with these buildups. Bioerosion is conspicuous in these structures, with sponges now being one of the principal endolithic taxa together with algae and bivalves. The marginal patch reefs are often up to 80% skeletal debris from the remains of crinoids, snails, solitary corals and large bivalves.

Cretaceous

The late Jurassic reef-building community persisted into early Cretaceous time with shelf-margin and ramp buildups composed primarily of the coral-algal-stromatoporoid community. At the same time one group of mollusks, the rudist bivalves, evolved rapidly and by mid-Cretaceous time were important biotic constituents of most buildups. These bizzare bivalves usually had one large valve attached to the substrate while the other

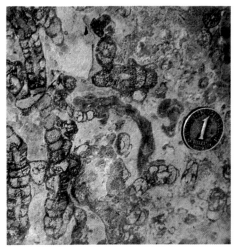

Figure 163—Numerous sphinctozoan sponges from the Dachstein Reef limestone (Fig. 159), Gosaukamm, Austria; penny scale 1.5 cm diameter (photo E. Flugel).

Figure 164—A thin section photomicrograph of typical Upper Rhaetian reef limestone illustrating calcisponges (circular cross-sections) and encrusting foraminifers (dark irregular mosaic between sponges); individual sponge cross sections average 75 mm in diameter (photo E. Flugel).

smaller valve was a lid or cap. They adapted to a wide variety of habitats from muddy lagoons to platform margins to foreslopes and in these different environments developed shapes that varied from prolate and planispiral (Requinids) to clumps of tall, twisted and slender (Caprinids) to massive and keg-shaped (Radiolotids) individuals with some shells as large as 1.5 m in length. In shape they closely resemble archaeocyathans, rugose corals and richtofenid brachiopods.

During middle Cretaceous time (Aptian through Cenomanian) the rudists diversified into several reef and back-reef habitats and locally in the surf zone. Mid-Cretaceous reefs grew at the margin of broad shelves and on the inner shelf, often in areas of restricted circulation. The high-diversity shelf-margin reefs consisted of a core community of corals, algae and a few rudists. The rudist communities tended to occupy the shallow, high-energy reef flat zone. Inner-shelf, low-diversity reefs were dominated by one or a few species of rudists that tended to form biostromes, patch reefs or high-energy shoreface beds marginal to high salinity lagoons or sabkhas.

Upper Cretaceous buildups are dominated by rudists, particularly radiolitids, with corals and encrusting algae only minor constituents. On-shelf buildups were generally biostromal while patch reefs developed near the shelf margin. Corals apparently diminished as frame builders because of secular environmental changes, perhaps in salinity, temperature or oxygen content. Competition by rudists does not seem likely because in Albian time corals and rudists occupied different environments.

Cenozoic Reefs

The catastrophic mass extinctions at the end of the Cretaceous, while affecting benthic calcareous taxa dramatically, did not wipe them out completely. Rudist bivalves disappeared completely while colonial corals survived but were greatly reduced in diversity from about 90 to 30 genera. Calcisponges and stromatoporoids (spongiomorph hydrozoans) also survived, but the enduring forms were few in number and never again important as reef-building taxa.

Cenozoic reefs are essentially the same as modern reefs, dominated by scleractinian corals. Their record of development, however, is not well known, largely because (1) in tectonically stable areas they are still buried while in tectonically active areas the facies relationships are obscured by faulting, and (2) they occur in the tropics where poor outcrops are altered by deep weathering.

The distribution of Cenozoic reefs in time and space is a direct result of changes in oceanic circulation patterns brought about by the progressive blockage of the Tethyan seaway by plate interaction. Although there are reefs of Paleocene age, most of the corals in them are survivors from the Cretaceous. *Solenopora,* an important coralline alga in reef growth since the Ordovician, disappears at the end of the Paleocene. Beginning in the Eocene there was a new radiation of hermatypic coral genera. It seems that these were spread throughout the diminishing Tethys because the Northwest Indian

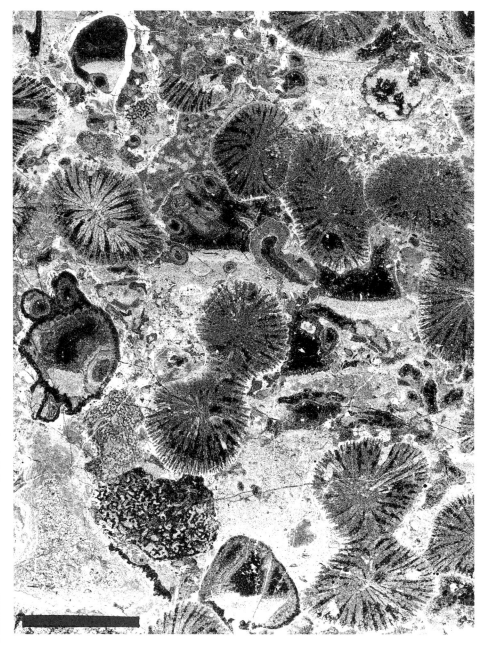

Figure 165—A negative peel of an Upper Triassic Dachstein Reef limestone. The branching coral *Thecosmilia* together with other corals and calcareous sponges forms the framework. Encrusting foraminifera (light rims) and algal crusts "Spongiostromata" (lower left) surround the corals and line the internal cavities. Fine geopetal sediment is found within and between the corals and sponges (scale 1 cm) (photo H. Zankl).

Figure 166—The ridge on the skyline is composed of a series of small bioherms with each pinnacle being the massive resistant core of a reef. The section of Middle Jurassic strata exposed here at Jebel Assameur n'Ait Fergane, Morroco, is about 200 m thick. The recessive slopes are fine-grained marls and limestones (photo J. Warme).

Ocean was still connected to the Mediterranean across a shallow seaway in the Middle East. This evolution continued through into the Oligocene at which time Cenozoic reefs reached the height of their development, in terms of range of reef communities, luxuriance of growth and numbers of buildups reported. The major change in reef distribution took place during the Miocene with the formation of the Mediterranean as a separate sea and the appearance of polar ice caps. By the end of Miocene time there were no coral reefs left in the Mediterranean. During ensuing Pliocene time the compression of climatic belts together with the rise of the Isthmus of Panama restricted reef growth to two regions, the Indo-Pacific and the Caribbean, the sites of modern reef growth.

Oligocene-Miocene Reefs

During the acme of Cenozoic reef growth, the Oligocene, reefs of all types can be recognized and in most

Figure 167—A partially eroded bioherm core (center) and inclined off-reef strata in the Middle Jurassic of the High Atlas, Morocco. The core is composed of scleractinian corals and skeletal wackestone while the flanking beds are mainly coral debris (photo J. Warme).

Figure 168—A large bioherm in Middle Jurassic strata at Jebel Taferdout, Morocco. The reef core is at center right 12 m high, and inclined off-reef strata are at center left (photo J. Warme).

cases a vertical as well as horizontal zonation discerned. The early stages of reef growth are characterized by low diversity assemblages of corals with light, porous, quickly-growing skeletons that could reject sediment and colonize muddy substrates; for example, *Goniopora,* a branching to encrusting form and *Actinacis* with superposed branches and plates like the modern *Acropora palmata.* The diversification stage of growth is

characterized by great numbers of species, but as in modern Caribbean reefs where seven of the 42 species dominate the reef framework, of the 30 Oligocene genera massive *Goniopora* and *Favia* as well as *Montastraea, Diploria, Pavona, Colpophyllia* and *Antiguastrea* form most of the reef frame. Other abundant forms such as the rapidly growing branching forms *Acropora* (or *Dendracis), Actinacis, Goniopora,*

Dictyaraea and *Stylocoenia* and the nodular forms *Alveopora* and *Astreopora* are also found in great luxuriance. Other reef-dwelling fauna such as coralline algae, foraminifers (especially encrusting forms) and bryozoans are similar to those in modern reefs. A major difference, however, is the relative scarcity of skeletal plates from the Codiacean alga *Halimeda.* These particles, which comprise most of the skeletal sand

Figure 169—An outcrop photograph of the core of one of the bioherms illustrated in Figure 166 composed of a massive scleractinian coral with an eroded upper portion and several in-place boring bivalves (center left); pen cap is 6 cm in length.

Figure 170—A massive branching coral from the core of one of the bioherms illustrated in Figure 166 (photo J. Warme).

produced on modern reefs, appear to be relatively rare in pre- mid-Miocene reefs.

The Miocene reefs of Indonesia illustrate similar complexity but those in the late Miocene of the Mediterranean reflect the gradual onset of conditions inimical to growth. These Messinian structures are generally narrow fringes around the steep margins of islands. Most Miocene reefs of the region contain a wide variety of species (5 to 15) with such forms as *Tabellastraea, Porites,* and *Montastraea* as well as other reef-dwelling biota including echinoids, bivalves, gastropods, barnacles, foraminifers, bryozoans and coralline algae. Those of late Miocene age are built almost exclusively by *Porites,* in long vertical sticks up to 4 m high and 2 to 3 cm in diameter, together with some bryozoans, coralline algae and serpulid worms entombed in marine cement. The extensive fore-reef deposits are mostly coralline algae and *Halimeda.* This dominance of one genera in the late Miocene likely reflects the first incursions of now cold Atlantic waters into the western Mediterranean and foretells the disappearance of reefs from this region.

One of the most spectacular attributes of modern reefs is the reef crest zone, which is dominated by rapidly-growing species of *Acroporidae, Poritidae* and *Seriatoporidae* and which is accreting at rates of up to 10 to 15 m every 1000 years. These species are not present in pre-Pleistocene structures,

Figure 171—A close view of a massive branching scleractinian coral from the flanks of one of the bioherms illustrated in Figure 166 (photo J. Warme).

Figure 172—The basal portion of a middle Jurassic scleractinian coral bioherm in the High Atlas Mountains of Morocco. The rock is a bafflestone to floatstone of branching and stick-like corals; cap of pen is 5 cm in length (photo J. Warme).

and *Acropora,* for example, although found in Eocene-Pliocene coral communities is rarely abundant, not thickly branched as in modern structures, and localized to quiet environments. As a result, Cenozoic reef accretion rates are much lower, more in the order of 1 to 3 m every 1000 years.

Cycles of Reef Growth

When the major trends in reef growth together with the variation in reef-building taxa are set against geologic time, and the short-term variations ignored, two major cycles of reef development seem to emerge, an early short CYCLE I of about 240 m.y., and a later longer CYCLE II of about 340 m.y. (Fig. 204). The general aspects of each of these cycles are the same, an early period in which only a few reef-building taxa are present, mainly small branching or encrusting organisms which form reef mounds, and a late period in which large skeletal metazoans become important and construct com-

plex reefs of all types. Superimposed on these general trends are two aberrant episodes of reef growth, one in the early Cambrian when archaeocyathans were the principal elements of reef mounds and one in the middle to late Cretaceous when rudist bivalves were among the main biotic constituents of reefs and reef mounds.

In the early part of Cycle I, calcified algae as tiny, branching, shrub-like clusters and anastomosing crustose masses (*Epiphyton, Gir-*

Figure 173—The core of a small bioherm from the Middle Jurassic of Morocco, composed of numerous small stick-like corals, often bored, in a matrix of lime mud (photo J. Warme).

Figure 174—A negative thin section print of an intensively bioeroded coral from the Middle Jurassic of the High Atlas Mountains, Morocco. The white areas are lime mudstone and the curved shells within the white mud are the valves of the boring pelecypods that excavated the holes (photo J. Warme).

complex reefs. In Cycle II, bryozoans are present throughout, alone as the principal elements of reef mounds at the beginning and then as reef dwellers, but rarely as reef builders.

The major period of reef growth in each cycle occurs when the consortium of corals and stromatoporoids appears. In the wholly Paleozoic Cycle I stromatoporoids tend to dominate reefs in the later phases while tabulate and colonial rugose corals are important, but not primary elements. In the Mesozoic part of Cycle II, scleractinian corals are most often the main reef elements together with spongiomorph hydrozoans or stromatoporoids and calcareous sponges as important but rarely dominant elements. Following the late Cretaceous mass extinction event, reefs are mostly built by scleractinian corals.

vanella, Renalcis) are the most essential elements of reef mounds. They continue to be important until the end of Devonian time. Reefs in the early parts of Cycle II are often dominated by the small encrusting organism *Tubiphytes* as well as encrusting foraminifers which remain fundamental components of the reef ecosystem through to the present day, even though the forms have changed. The buildups in the Paleozoic part of Cycle II also contain phylloid algae as conspicuous elements.

In both cycles the major change in reef biota is heralded by the ap-

pearance of sponges in abundance, siliceous sponges in the Early Ordovician of Cycle I and calcareous sponges in the Permian of Cycle II. These elements at first form buildups with other small, diminutive taxa and later are either accessory to dominant elements in more complex reefs or form bioherms on their own when the larger reef biota are excluded by environmental constraints.

Bryozoans do not appear until midway through Cycle I when they dominate bioherms for a short time and from then on are accessory elements except in the early stages of

REEFS AND CARBONATE PLATFORM GEOMETRY

In their synthesis of modern carbonate sediments on continental shelves, Ginsburg and James (1974) recognized two intergrading categories of shelves or platforms: (1) open shelves, those in which the outer shelf to slope break is in tens of meters of water; and (2) rimmed shelves, those which have a continuous or semicontinuous barrier along the shelf margin. At times when a full spectrum of reef-building organisms is present, both types of shelves may

Figure 175—A reef mound of Upper Cretaceous age (Campanian) exposed at Djebel Mokta, Tunisia. The mound is rich in the giant rudist *Hippurites lapourizi* (photo S. Frost).

Figure 176—The external molds of *H. lapourizi* in a reef mound similar to that in Figure 175 at Jebel Merieg, Tunisia.

Figure 177—Radiolitid rudist bivalves from the reef margin at the eastern edge of the Valles, San Luis Potosi platform in the Taninul facies of the El Abra Limestone (Albian-Cenomanian), Laguna Colorada, Queretaro, Mexico; small black divisions on scale in cm (photo P. Enos).

form and the margin on rimmed shelves may be either a barrier reef or a series of carbonate sand shoals or both.

The most complex type of platform at such times will be a reef-rimmed platform (Fig. 203). The barrier itself is well-zoned if the front is steep and wave action intense, but zonation is weak if the front slopes gradually seaward and the seas are relatively quiet. Patch reefs on the platform in the lee of the barrier reef range from circular to elliptical to irregular in plan and are sometimes large enough to enclose a lagoon themselves. Reef mounds occur on the inner, shallow parts of the platform, in areas of normal salinity but turbid water. Reef mounds also occur at depth, in front of the barrier reef, down on the reef front or forereef.

Patch reefs or reef mounds commonly form a very widespread lithofacies compared to the barrier

Figure 178—A cluster of radiolitid rudists from the same reef complex as illustrated in Figure 177 forming a reef framestone (photo P. Enos).

Figure 179—A "bouquet" of radiolitid rudists from the bioherm illustrated in Figure 175. These skeletons have toppled from their growth position and are now upside down (photo S. Frost).

Figure 180—Radiolitid rudist bivalves in an arborescent cluster from the Taninul facies of the El Abra Limestone (Albian-Cenomanian), Mexico (photo P. Enos).

reef. The stratigraphic thickness of these reefs is dependent on the rate of subsidence — if subsidence rate is low, reefs are thin; if subsidence rate is high, reefs may be spectacular in their thickness.

At times when large skeletal metazoa are absent, the small and delicate organisms cannot live in the turbulent shallow water at the shelf edge. As a result, the margin of the shelf or platform is normally a com-

plex of oolitic or skeletal (generally crinoidal) sand shoals and islands (Fig. 203). The only reef structures are reef mounds which occur below the zone of active waves down on the seaward slopes of the shelf or platform, and if conditions are relatively tranquil, behind the barrier, on the shelf itself.

SUMMARY

When documenting bioherms and biostromes one must not lose sight of the fact that they are primarily biological in origin. These structures are the fossil record of a community of organisms which, by continued growth in one spot, formed a sedimentary unit. Because of this, one reef is likely to be quite different from another, even in the same for-

Figure 181—Caprinid rudist bivalves filled with coarse calcite spar from the Cerro Angel reef zone, El Doctor Platform, Queretaro, Mexico (Albian-Cenomanian); scale 10 cm in length (photo P. Enos).

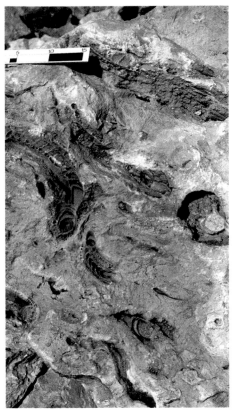

Figure 182—Large caprinid rudist bivalves in the Taninul facies of the El Abra Limestone (Albian-Cenomanian), Mexico (photo P. Enos).

mation. Consequently, the themes outlined in this synthesis should be viewed not as a rigid classification scheme into which all reefs must somehow fit, but rather as a guide to the study of these fascinating structures.

ACKNOWLEDGMENTS

Many of the concepts on the deposition and diagenesis of modern reefs and reef-like structures presented in this article were developed while at the Comparative Sedimentology Laboratory, University of Miami, with Robert N. Ginsburg. Much of the data that has allowed me to find the threads that bind reefs of different ages together is documented in the scholarly text on "Carbonate Facies in Geologic History," by James L. Wilson. This paper could not have been written without the generous co-operation of the following colleagues who gave me photographs to use and/or spent time discussing the vagaries of reef growth through geologic time: P. A. Bourque, P. Copper, G. Davies, P. Enos, M. Esteban, E. Flugel, S. Frost, P. Hoffman, D. R. Kobluk, C. F. Klappa, C. St. G. C. Kendall, R. Louckes, R. Lighty, M. Longman, I. G. McIlreath, E. W. Mountjoy, A. C. Neumann, P. Opalinski, B. R. Pratt, P. E. Playford, L. C. Pray, M. Rees, J. K. Rigby, A. J. Rowell, S. Rowland, R. J. Ross, P. A. Scholle,

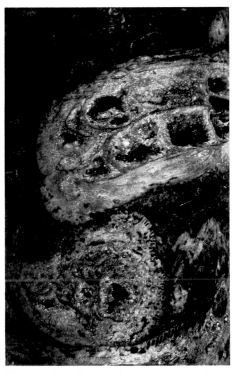

Figure 183—A core from a deep well in southern Florida composed of dolomitized Aptian caprinid rudists and wackestone matrix. Dolomitization has substantially increased matrix porosity and permeability (photo S. Frost).

Figure 184—A bioherm of Upper Miocene age (Messinian) near Almeria, Spain. The top of the hill is a reef composed predominantly of *Porites.* This structure developed during a gradual sea level fall and so lies upon basinal muds and calcisiltites (photo M. Esteban).

Figure 185—A large reef of Upper Miocene age (Messinian) near Almeria, Spain. The present day topography reflects the original reef topography with the massive limestone wall being the original reef wall composed mainly of *Porites* (photo M. Esteban).

R. W. Scott, C. W. Stearn, R. K. Stevens, D. F. Toomy, R. A. Walls, J. Warme, J. L. Wray and H. Zankl. R. W. Scott kindly provided me with data on the nature of Cretaceous buildups through time, which I used in the preceding text. The manuscript was written during the tenure of an Industrial Fellowship at the Denver Research Center of Marathon Oil Company. The diagrams were kindly drafted at the Research Center by Richard Nervig, Barbara Steele and Arzell Thompson. Many of the photographs were skillfully transformed from slides into prints by Wilf Marsh of Memorial University. My own recent work on fossil reefs has been generously funded by the National Science and Engineering Research Council of Canada. Parts of this article and several of the figures have already been published by the Geological Association of Canada and I am very grateful for permission to reproduce them here. Finally, this section is much enhanced by the three examples of reef reservoirs described by Dick Walls, Phil Choquette and Don Bebout, and I thank them for undertaking this endeavor.

SELECTED REFERENCES

Classification and Nomenclature

Cummings, 1932, Reefs or bioherms?: Geol. Soc. America Bull., v. 43, p. 331-352.

Dunham, R. J., 1962, Classification of carbonate rocks according to depositional texture, *in* W. E. Ham, ed., Classification of carbonate rocks: AAPG Mem. 1, p. 108-122.

Embry, A. F., and J. E. Klovan, 1971, A Late Devonian reef tract on northeastern Banks island, N.W.T.: Bull.

Figure 186—The spur and groove structure of the reef front in an Upper Miocene reef at Cap Blanc, Llucmajor, Mallorca. The spurs (arrows) are constructed by sticks and dishes of *Porites*; the man is standing in groove (photo M. Esteban).

Figure 187--A close view of the *Porites* reef-wall framework (plan view) in which the coral sticks are leached (diameter 1 to 2 cm) and filled with red volcanic sand; Cabo de Gata, southeastern Spain (photo M. Esteban).

Canadian Petroleum Geology, v. 19, p. 730-781.

Folk, R. L., 1962, Spectral subdivision of limestone types, *in* W. E. Ham, ed., Classification of carbonate rocks: AAPG Mem. 1, p. 62-85.

Nelson, H. F., C. W. Brown, and J. H. Brineman, 1962, Skeletal limestone classification, *in* W. E. Ham, ed., Classification of carbonate rocks: AAPG Mem. 1, p. 224-253.

General Articles Reviewing Modern Reefs
Bathurst, R. G. C., 1975, Carbonate sediments and their diagenesis: Amsterdam, Elsevier Sci. Pub., 658 p.

Darwin, C., 1842, Structure and distribution of coral reefs: repr. by Univ. of Calif. Press, from 1851 ed., 214 p.

Frost, S. H., M. P. Weiss, and J. B. Saunders, 1977, Reefs and related carbonates–ecology and sedimentology: AAPG Stud. in Geology, No. 4, 421 p.

Ginsburg, R. N., and N. P. James, 1974, Spectrum of Holocene reef-building communities in the western Atlantic, *in* A. M. Ziegler et al, eds., Principles of benthic community analysis (notes for a short course): Univ. of Miami, Fisher Island Station, p. 7.1-7.22.

Stoddart, D. R., 1969, Ecology and morphology of recent coral reefs: Biol. Rev., v. 44, p. 433-498.

Taylor, D. E., ed., 1977, Proceedings of Third International Coral Reef Symposium, Miami Florida; 2. Geology: Miami Beach, Fla., Fisher Island Station, 628 p.

Different Types of Modern Reefs
Pacific Atolls
Emery, K. O., J. I. Tracey, and H. S.

Figure 188—Upward-growing masses of *Porites* from the reef-wall facies, Cabezo Maria Reef, Fortuna, southeastern Spain (photo M. Esteban).

Figure 189—Numerous, plate-like colonies of *Tarbellastraea* from the deep reef-front facies in a Serravallian-Langhian coral reef from Olerdola (Barcelona), Spain (photo M. Esteban).

Ladd, 1954, Geology of Bikini and nearby atolls: U.S. Geol. Survey, Prof. Paper 260-A, 265 p.

Marginal Reefs

Goreau, T. F., 1959, The ecology of Jamaican coral reefs; I. Species, composition and zonation: Ecology, v. 40, p. 67-90.

———— and N. I. Goreau, 1973, The ecology of Jamaican coral reefs; II. Geomorphology, zonation and sedimentary phases: Bull. Marine Sci., v. 23, p. 399-464.

James, N. P., et al, 1976, Facies and fabric specificity of early subsea cementation in shallow Belize (British Honduras) reefs: Jour. Sed. Petrology, v. 46, p. 523-544.

———— and R. N. Ginsburg, 1979, The seaward margin of Belize Barrier and Atoll Reefs: Internat. Assoc. Sedimentologists Spec. Pub. 3, 196 p.

Maxwell, W. G. H., 1968, Atlas of the Great Barrier Reef: Amsterdam, Elsevier Sci. Pub. Co., 258 p.

Pinnacle Reefs

Korniker, L. A., and D. W. Boyd, 1962, Shallow water geology and environment of Alacran reef complex, Campeche Bank, Mexico: AAPG Bull., v. 46, p. 640-673.

Logan, B. W., et al, 1969, Carbonate sediments and reefs, Yucatan shelf, Mexico: AAPG Mem. 11, p. 1-196.

Patch Reefs

Garrett, P., et al, 1971, Physiography, ecology and sediments of two Bermuda patch reefs: Jour. Geology, v. 79, p. 647-668.

Ginsburg, R. N., and J. H. Schroeder, 1969, Notes for NSF seminar on carbonate cements, Bermuda Biological

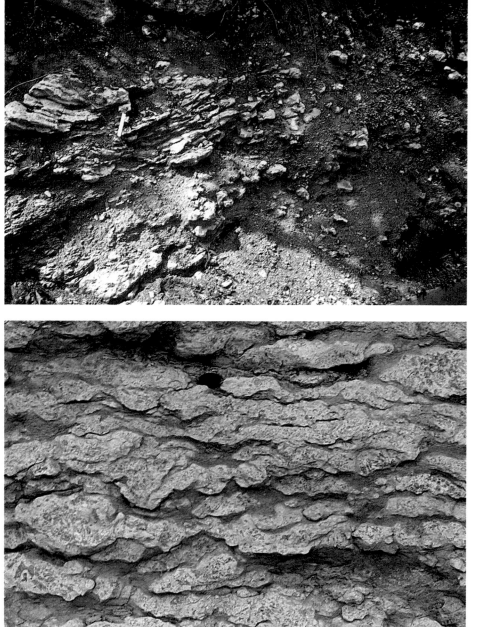

Figure 190—Plate-like colonies of *Actinacis rollei* are oriented pointing into and away from probable directions of wave surge in earliest Oligocene coral banks exposed in the lower slopes of Monti Grumi, near Castelgomberto, Venetian Alps region, Italy. The coralline marl matrix contains prolific growths of branching and nodular *Goniopora* (photo S. Frost).

Figure 191—Thick branches of the poritid coral *Goniopora* in patch reefs from the late Miocene reef complex exposed at Gunnung, Sappe, Salwesi. Most of the species found in these reefs in the southwest Pacific are still living today on modern reefs of the region (photo S. Frost).

Station: unpub.

Maiklem, W. R., 1970, The Capricorn Reef complex, Great Barrier Reef, Australia: Jour. Sed. Petrology, v. 38, p. 785-798.

Reef Mounds

Turmel, R., and R. Swanson, 1976, The development of Rodriguez Bank, a Holocene mudbank in the Florida Reef Tract: Jour. Sed. Petrology, v. 46, p. 497-519.

Algal Reefs

Adey, W., and R. Burke, 1976, Holocene bioherms (algal ridges and bank-barrier reefs) of the eastern Caribbean: Geol. Soc. America Bull., v. 87, p. 497-519.

Ginsburg, R. N., and J. H. Schroeder, 1973, Growth and submarine fossilization of algal cup reefs, Bermuda: Sedimentology, v. 20, p. 575-614.

Stromatolites

Hoffman, P., 1976, Stromatolite morphogenesis in Shark Bay, Western Australia, *in* M. R. Walter, ed., Stromatolites: Amsterdam, Elsevier Sci. Pub., p. 261-273.

Playford, P. E., and A. E. Cockbain, 1976, Modern algal stromatolites at Hamelin Pool, a hypersaline barred basin in Shark Bay, Western Australia, *in* M. R. Walter, ed., Stromatolites: Amsterdam, Elsevier Sci Pub., p. 389-413.

Deepwater Mounds

Mullins, H. T., et al, 1981, Modern deepwater coral mounds north of Little Bahama Bank; criteria for recognition of deep-water coral bioherms in the rock record: Jour. Sed. Petrology, v. 51, p. 999-1013.

Neumann, A. C., J. W. Kofoed, and G. H. Keller, 1977, Lithoherms in the

Figure 192—A view of the core of an Upper Eocene (Priabonian) reef mound near Barcelona, Spain. These mud-rich mounds are composed of floatstones comprising *Porites* (photo M. Esteban).

Figure 193—A bioherm rich in branching coralline algae (white shrubs) and having a matrix of lime mud from the Upper Miocene near Llucmajor, Mallorca (photo M. Esteban).

Heckel, P. H., 1974, Carbonate buildups in the geologic record; a review, *in* L. F. Laporte, ed., Reefs in time and space: SEPM Spec. Pub. No. 18, p. 90-155.

James, N. P., 1979, Reefs, *in* R. G. Walker, ed., Facies models: Geosci. Canada Repr. Ser. 1, p. 121-133.

Laporte, L. F., ed., 1974, Reefs in time and space: SEPM Spec. Pub. No. 18, 256 p.

Toomey, D. F., 1981, European fossil reef models: SEPM Spec. Pub. No. 30, 545 p.

Walker, K. R., and L. P. Alberstadt, 1975, Ecological succession as an aspect of structure in fossil communities: Paleobiol., v. 1, p. 238-257.

Wilson, J. L., 1975, Carbonate facies in geologic history: Heidelberg, Springer-Verlag Pub., 471 p.

Reefs at Different Times in Geologic History
Pleistocene

Chappell, J., and H. A. Polach, 1976, Holocene sea-level change and coral-reef growth at Huon Peninsula, Papua New Guinea: Geol. Soc. America Bull., v. 87, p. 235-239.

James, N. P., C. S. Stearn, and R. S. Harrison, 1977, Field guidebook to modern and Pleistocene reef carbonates, Barbados, West Indies: Univ. of Miami, 3rd Internat. Coral Reef Symp., Fisher Island, 30 p.

Mesolella, K. J., H. A. Sealy, and R. K. Matthews, 1970, Facies geometries within Pleistocene reefs of Barbados,

Straits of Florida: Geology, v. 5, p. 4-11.

Linear Mud Banks

Enos, P., and R. Perkins, 1979, Evolution of Florida Bay from island stratigraphy: Geol. Soc. America Bull., pt. 1, v. 90, p. 59-83.

Pusey, W. C., 1975, Holocene carbonate sedimentation on northern Belize Shelf, *in* K. F. Wantland and W. C. Pusey, eds., Belize Shelf–carbonate sediments, clastic sediments, and ecology: AAPG Stud. in Geology No. 2, p. 131-234.

General Articles Reviewing Fossil Reefs

Copper, P., 1974, Structure and development of early Paleozoic reefs: Proc., 2nd Internat. Coral Reef Symp., v. 6, p. 365-386.

Dunham, R. J., 1970, Stratigraphic reefs versus ecologic reefs: AAPG Bull., v. 54, p. 1931-1932.

Hartman, W. D., J. W. Wendt, and F. Widenmayer, 1980, Living and fossil sponges: Univ. Miami, Sedimenta VIII, Comparative Sedimentology Lab., 274 p.

Figure 194—The core of a small bioherm rich in serpulid worm tubes from the Upper Miocene of Mallorca; the diameter of the tubes is about 1.5 cm (photo M. Esteban).

Figure 195—The core of a late Pleistocene barrier reef exposed along the northeast coast of Barbados, West Indies. The large coral fronds and logs are all *Acropora palmata* (packsack for scale).

West Indies: AAPG Bull., v. 54, p. 1899-1917.

Stanley, S. M., 1966, Paleoecology and diagenesis of Key Largo limestone, Florida: AAPG Bull., v. 50, p. 1927-1947.

Cenozoic

Dabrio, C. J., M. Esteban, and J. M. Martin, 1981, The Coral Reef Model of Nijar, Messinian (uppermost Miocene) Almeria Province, southeast Spain: Jour. Sed. Petrology, v. 51, p. 521-541.

Esteban, M., 1979, Significance of the Upper Miocene coral reefs of the western Mediterranean: Paleogeography, Paleoclimatology, Paleoecology, v. 29, p. 169-189.

Forman, M. J., and S. O. Schlanger, 1957, Tertiary reefs and associated limestone facies from Louisiana and Guam: Jour. Geology, v. 65, p. 611-627.

Frost, S. H., 1977, Cenozoic reef systems of Caribbean-prospects for paleoecological synthesis: AAPG Stud. in Geology No. 4, p. 93-110.

———— 1981, Oligocene reef coral biofacies of the Vicentin, northeast Italy, *in* D. F. Toomey, ed., European fossil reef models: SEPM Spec. Pub. No. 30, p. 483-541.

Pedley, H. M., 1979, Miocene bioherms and associated structures in the Upper Coralline Limestone of the Maltese Islands; their lithification and paleoenvironment: Sedimentology, v. 26, p. 577-593.

Riding, R., 1979, Origin and diagenesis of lacustrine algal bioherms at the margin of the Reis Crater, Upper Miocene, southern Germany: Sedimentology, v. 26, p. 645-681.

Cretaceous

Bebout, D. G., and R. G. Loucks, 1974, Stuart City trend, Lower Cretaceous, south Texas: Austin, Tex., Bureau Econ. Geol. Rept. No. 78, 80 p.

Enos, P., 1974, Reefs, platforms, and basins of Middle Cretaceous in northeast Mexico: AAPG Bull., v. 58, p. 800-809.

Kauffman, E. G., and N. F. Sohl, 1974, Structure and evolution of Antillean Cretaceous rudist frameworks: Verhandl. Naturi. Ges. Basel., v. 84, p. 399-467.

Perkins, B. F., 1974, Paleoecology of a rudist reef complex in the Comanche Cretaceous Glen Rose Limestone of central Texas, *in* B. F. Perkins, ed., Aspects of Trinity Division Geology, Geoscience and Man VIII: La. State Univ., p. 131-173.

Philip, J., 1972, Paleoecologie des formations a rudistes du Cretace Superior-l'example du sud-est de la France: Paleogeography,

Figure 196—A succession of three coral communities in the back-reef zone of an 83,000-year-old reef complex on the northeast coast of Barbados. The lower facies is composed of the delicate branching coral *Porites porites* (see hammer, lower right) overlain by *Acropora palmata* fronds (center) which in turn serve as a base of growth of large *Montastraea annularis* colonies. This sequence suggests coral growth in a leeward protected environment (*Porites porites*), a reef crest setting (*Acropora palmata*) and a reef front environment (*Montestraea annularis*).

Figure 197—A multilobate colony of Pleistocene *Montastraea annularis* exposed in the sawed quarry wall at Windley Key, Florida. The lens cap is 5 cm in diameter.

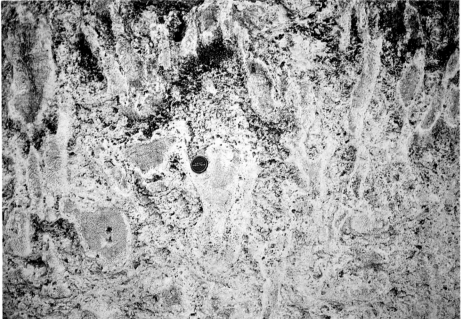

Paleoclimatology, Paleoecology, v. 12, p. 205-222.

Scott, R. W., 1979, Depositional model of Early Cretaceous coral-algal-rudist reefs, Arizona: AAPG Bull., v. 63, p. 1108-1128.

Jurassic

Eliuk, L. S., 1979, Abenaki Formation, Nova Scotia Shelf, Canada; a depositional and diagenetic model for Mesozoic carbonate platforms: Canadian Petroleum Geology Bull., v. 24, p. 424-514.

Gwinner, M. P., 1968, Carbonate rocks of the Upper Jurassic in southwest Germany, *in* G. Muller, ed., Sedimentology of parts of central Europe: Heidelberg, 8th Internat. Sedimentol. Cong., p. 193-207.

––––– 1976, Origin of the Upper Jurassic limestones of the Swabian Alb (southwest Germany): Contrib. Sedimentology, v. 5, 75 p.

Palmer, T. J., and F. T. Fursich, 1981, Ecology of sponge reefs from the Upper Bathonian (Middle Jurassic) of Normandy: Palaeontology, v. 24, p. 1-25.

Rutten, M. D., 1956, The Jurassic reefs of the Yonne (southeastern Paris Basin): Amer. Jour. Sci., v. 254, p. 363-371.

Warme, J. E., R. G. Stanley, and J. L. Wilson, 1975, Middle Jurassic reef tract, central High Atlas, Morocco: Proc. Internat. Assoc. Sedimentol., Theme VIII, 11 p.

Triassic

Bosellini, A., and D. Rossi, 1974, Triassic carbonate buildups of the dolomites, northern Italy: SEPM Spec. Pub. 18, p. 209-233.

Flugel, E., 1981, Paleoecology and facies of Upper Triassic reefs in the northern Calcareous Alps, *in* D. F. Toomey, ed., European fossil reef models: SEPM Spec. Pub. No. 30, p. 291-361.

Leonardi, P., 1967, Le Dolomiti; geologic dei monti tra Isarco e Piave: Rome, Nat. Research Council, v. 1 and 2, 1010 p.

Stanley, G. D., 1979, Paleoecology, structure and distribution of Triassic coral buildups in western North America: Article 65, Univ. Kansas Paleontol. Contrib., 58 p.

Zankl, H., 1971, Upper Triassic carbonate facies in the northern Limestone Alps, *in* G. Muller, ed., Sedimentology of parts of central Europe: Heidelberg, Guidebook 8th Internat. Sedimentol. Cong., p. 147-185.

Permian

Davies, G. R., 1970, A Permian hydrozoan mound, Yukon Territory: Canadian Jour. Earth Sci., v. 8, p. 973-988.

Dunham, R. J., 1972, Guide for study

Figure 198—A single large massive colony of *Montastraea annularis* in the back-reef facies of a late Pleistocene reef complex on the northeast coast of Barbados. The sediments surrounding the coral are bioturbated lime mudstones (hammer for scale).

Figure 199—A hemispherical colony of *Diploria* from the late Pleistocene Key Largo limestone exposed in Windley Key Quarry, Florida. The colony has been bored by bivalves yielding elongate holes. The boring next to the finger is due to boring by the date mussel *Lithophaga* while the others are due to the bivalve *Gastrochaena*.

and discussion for individual reinterpretation of the sedimentation and diagenesis of the Permian Capitan Geologic Reef and associated rocks, New Mexico and Texas: Permian Basin Sec., SEPM Pub. 72-14, 235 p.

Flugel, E., 1981, Lower Permian *Tubiphytes/Archaeolithoporella* buildups in the southern Alps (Austria and Italy), *in* D. F. Toomey, ed., European fossil reef models: SEPM Spec. Pub. No. 30, p. 143-161.

Hileman, M. E., and S. J. Mazzulo, 1977, Upper Guadalupian Facies, Permian Reef Complex, Guadalupe Mountains New Mexico and Texas: Permian Basin Sec., SEPM Pub. 77-16, 508 p.

Malek-Aslani, M., 1970, Lower Wolfcamp reef in Kemnitz Field, Lea County, New Mexico: AAPG Bull., v. 54, p. 2317-2335.

Mazzullo, S. J., and J. M. Cys, 1979, Marine aragonite sea floor growths and cements in Permian phylloid algae mounds, Sacramento Mountains, New Mexico: Jour. Sed. Petrology, v. 49, p. 917-937.

Newell, N. E., et al, 1953, The Permian Reef complex of the Guadalupe Mountains region, Texas and New Mexico: San Francisco, Freeman and Co., 236 p.

———— et al, 1976, Permian Reef Complex, Tunisia: Brigham Young Univ. Geol. Stud., v. 23, p. 75-112.

Smith, D. B., 1981a, The Magnesian Limestone (Upper Permian) Reef Complex of northern England, *in* D. F. Toomey, ed., European fossil reef

models: SEPM Spec. Pub. No. 30, p. 161-187.

———— 1981b, Bryozoan-algal patch reefs in the Upper Permian Magnesian Limestone of Yorkshire, northeast England, *in* D. F. Toomey, ed., European fossil reef models: SEPM Spec. Pub. No. 30, p. 187-203.

Pennsylvanian

Heckel, P. H., and J. M. Cocke, 1969, Phylloid algal mound complexes in outcropping Upper Pennsylvanian rocks of mid-continent: AAPG Bull., v. 53, p. 1084-1085.

Pray, L. C., J. L. Wilson, and D. F. Toomey, 1977, Geology of the Sacramento Mountains, Otero County, New Mexico: West Texas Geol. Soc. Guidebook, 216 p.

Toomey, D. F., and H. D. Winland, 1973, Rock and biotic facies associated with a Middle Pennsylvanian (Desmoinesian) algal buildup, Neca Lucia Field, Nolan County, Texas: AAPG Bull., v. 57, p. 1053-1074.

Mississippian

Bathurst, R. G. C., 1959, The cavernous structure of some Mississippian

Figure 200—A huge, partly eroded *Montastraea annularis* colony forming part of the Key Largo limestone in Windley Key Quarry, Florida. The colony stretches from left of the lens cap across the photograph to beyond the right-hand side. The concave upper surface was likely eroded by browsing echinoids (see Fig. 4) so that the center of the colony was dead while the periphery was still alive. The holes in the rock are diamond drill holes (lens cap is 5 cm in diameter).

Figure 201—A community of branching and small domal coral colonies forming part of a late Pleistocene patch reef on the southern part of Barbados. The branches or sticks are *Acropora cervicornis* while the two head corals in the center are *Porites porites* and *Diploria.* The upper part of the outcrop is occupied by a calcrete (caliche) crust, an excellent criteria for subaerial exposure.

Stromatactis reefs in Lancashire, England: Jour. Geology, v. 67, p. 506-521.

Cotter, E., 1965, Waulsortian-type carbonate banks in the Mississippian Lodgepole Formation of central Montana: Jour. Geology, v. 73, p. 881-888.

Lees, A., 1964, The structure and origin of the Waulsortian (Lower Carboniferous) "reefs" of west-central Eire: Phil. Trans. Royal Soc. London, Ser. B., No. 740, p. 485-531.

Devonian

Burchette, T. P., 1981, European Devonian Reefs; a review of current concepts and models, *in* D. F. Toomey, ed., European fossil reef models: SEPM Spec. Pub. No. 30, p. 85-143.

Davies, G. R., ed., 1975, Devonian reef complexes of Canada, I and II: Canadian Soc. Petroleum Geols. Repr. Ser. No. 1, 229 and 246 p., respectively.

Elloy, R., 1972, Reflexions sur quelques environments recafaux du paleozoique: Bull. Centre Rech. Pau-SNPA, v. 6, p. 1-105.

Jansa, L. F., and N. R. Fischbuch, 1974, Evolution of a Middle and Upper Devonian sequence from a clastic coastal plain-deltaic complex into overlying carbonate reef complex and banks, Sturgeon-Mitsue area, Alberta: Geol. Survey Canada Bull. 234, 105 p.

Klovan, J. E., 1974, Development of western Canadian Devonian reefs and comparison with Holocene analogues:

AAPG Bull., v. 58, p. 787-799.

Krebs, W., 1971, Devonian reef limestones in the eastern Rhenish Shiefergebirge, *in* G. Muller, ed., Sedimentology of parts of central Europe: Guidebook, Heidelberg, 8th Internat. Sedimentol. Cong., p. 45-81.

_____ and E. W. Mountjoy, 1972, Comparison of central European and western Canadian Devonian reef complexes: 24th Internat. Geol. Cong., sec. 6, p. 294-309.

Mountjoy, E. W., and R. Riding, 1981, Foreslope *Renalcis*-stromatoporoid bioherm with evidence of early cementation, Devonian Ancient Wall reef complex: Sedimentology, v. 28, p. 299-321.

Playford, P. E., 1980, Devonian "Great Barrier Reef" of the Canning Basin, Western Australia: AAPG Bull., v. 64, p. 814-840.

_____ and D. C. Lowry, 1966, Devonian reef complexes of the Canning Basin, Western Australia: Geol. Survey Western Australia, Bull. 118, 50 p.

Silurian

Bourque, P. A., 1979, Facies of the Silurian West Point reef complex, Baie des Chaleurs, Gaspesie, Quebec: Geol. Assoc. Canada Field Guidebook B-2, 29 p.

Fisher, J. H., ed., 1977, Reefs and evaporites-concepts and depositional models: AAPG Stud. in Geology No. 5, 196 p.

Heckel, P. H., and D. O'Brien, eds., 1975, Silurian reefs of Great Lakes Region of North America: AAPG Repr. Ser. No. 14, 243 p.

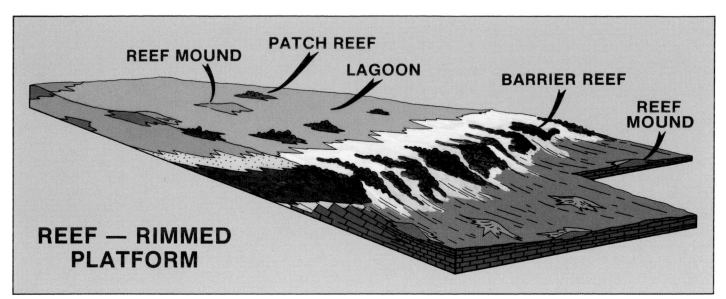

REEF MOUND PATCH REEF LAGOON BARRIER REEF REEF MOUND

REEF — RIMMED PLATFORM

Figure 202—An idealized isometric diagam illustrating the facies on a carbonate shelf or platform at times in geologic history when a full spectrum of reef-building skeletal metazoans are present (see Figure 61). The platform margin is a zoned barrier reef and isolated patch reefs dot the shallow platform behind. Reef mounds are found on the inner part of the platform and on the fore-reef slope.

Lowenstam, H. A., 1950, Niagaran reefs in the Great Lakes area: Jour. Geology, v. 58, p. 430-487.

Manten, A. A., 1971, Silurian reefs of Gotland, in Developments in Sedimentology, No. 13: Amsterdam, Elsevier Sci. Pub., 539 p.

Riding, R., 1981, Composition, structure and environmental setting of Silurian bioherms and biostromes in northern Europe, in D. F. Toomey, ed., European fossil reef models: SEPM Spec. Pub. No. 30, p. 41-85.

Scoffin, T. P., 1971, The conditions of growth of the Wenlock reefs of Shropshire, England: Sedimentology, v. 17, p. 173-219.

Shaver, R. H., et al, 1978, The search for a Silurian reef model; Great Lakes Area: Spec. Rept. No. 15, Indiana Geol. Survey, 36 p.

Ordovician

Klappa, C. F., and N. P. James, 1980, Small Lithistid sponge bioherms, Early Middle Ordovician Table Head Group, western Newfoundland: Bull. Canadian Petroleum Geology, v. 28, p. 425-451.

Pitcher, M., 1961, Evolution of Chazyan (Ordovician) reefs of eastern United States and Canada: Canadian Petroleum Geology Bull., v. 12, p. 632-691.

Ross, R. J., V. Jaanson, and I. Fried-

man, 1975, Lithology and origin of Middle Ordovician Calcareous mudmound at Meiklejohn Peak, southern Nevada: U.S. Geol. Survey Prof. Paper No. 871, 45 p.

Toomey, D. F., 1970, An unhurried look at a Lower Ordovician mound horizon, southern Franklin Mountains, west Texas: Jour. Sed. Petrology, v. 40, p. 1318-1335.

_____ and R. M. Finks, 1969, The paleoecology of Chazyan (lower Middle Ordovician) "reefs" or "mounds" and Middle Ordovician (Chazyan) mounds, southern Quebec, Canada: Plattsburg, N.Y., New York Assoc. Guidebook to Field Excursions, College Arts, Scies. (a summary rept.), p. 121-134.

_____ and M. H. Nitecki, 1979, Organic buildups in the Lower Ordovician (Canadian) of Texas and Oklahoma: Fieldiana N. Ser. 2, 181 p.

Walker, K. R., and K. F. Ferrigno, 1973, Major Middle Ordovician reef tract in eastern Tennessee: Am. Jour. Sci., v. 273-A, p. 294-325.

Cambrian

Ahr, W. M., 1971, Paleoenvironment, algal structures and fossil algae in the Upper Cambrian of central Texas: Jour. Sed. Petrology, v. 41, p. 205-216.

James, N. P., and D. R. Kobluk, 1978, Lower Cambrian patch reefs and associated sediments, southern Labrador, Canada: Sedimentology, v. 25, p. 1-32.

_____ and F. Debrenne, 1980, Lower Cambrian bioherms; pioneer reefs of the Phanerozoic: Acta. Palaeontologica Polonica, v. 25, p. 655-668.

Precambrian

Cecile, M. P., and F. H. A. Campbell, 1978, Regressive stromatolite reefs and

associated facies, middle Goulburn Group (Lower Proterozoic) in Kilohigok Basin, N.W.T.; an example of environmental control of stromatolite form: Bull. Canadian Petroleum Geols., v. 26, p. 237-267.

Hoffman, P., 1974, Shallow and deepwater stromatolites in lower Proterozoic platform-to-basin facies change, Great Slave Lake, Canada: AAPG Bull., v. 58, p. 856-867.

Semikhatov, M. A., et al, 1979, Stromatolite morphogenesis-progress and problems: Canadian Jour. Earth Sci., v. 16, p. 992-1016.

Other References Cited

Aitken, J. D., 1967, Classification and environmental significance of cryptalgal limestones and dolomites with illustrations from the Cambrian and Ordovician of southwestern Alberta: Jour. Sed. Petrology, v. 37, p. 1163-1187.

Frost, S. H., 1977, Ecologic controls of Caribbean and Mediterranean Oligocene reef coral communities, in D. L. Taylor, ed., Miami, Fla., Proc. 3rd Internat. Coral Reef Symp., p. 367-375.

Ginsburg, R. N., and N. P. James, 1974, Holocene carbonate sediments of continental shelves, in C. A. Burk and C. L. Drake, eds., The geology of continental margins: New York, Springer-Verlag Pub., p. 137-157.

_____ J. H. Schroeder, and E. A. Shinn, 1971, Recent synsedimentary cementation in subtidal Bermuda reefs, in O. P. Bricker, ed., Carbonate cements: The Johns Hopkins Press, p. 54-56.

Iams, W. J., 1970, Boilers on Bermuda's South Shore, in R. N. Ginsburg and S. M. Stanley, eds., Reports of research, 1969, seminar on organism-sediment in-

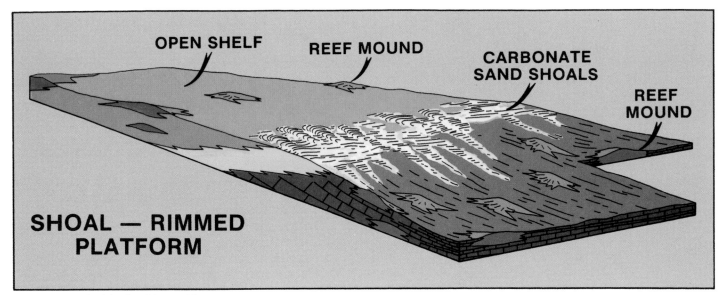

Figure 203—An idealized isometric diagram illustrating the facies on a carbonate shelf or platform at times in geologic history when the only skeletal invertebrates with reef-building potential are delicate ramose or encrusting forms and so reef mounds are the only types of buildups. The platform margin is a series of carbonate sand shoals and the reef mounds are restricted to the tranquil inner parts of the platform and down slope in deeper water in front of the platform.

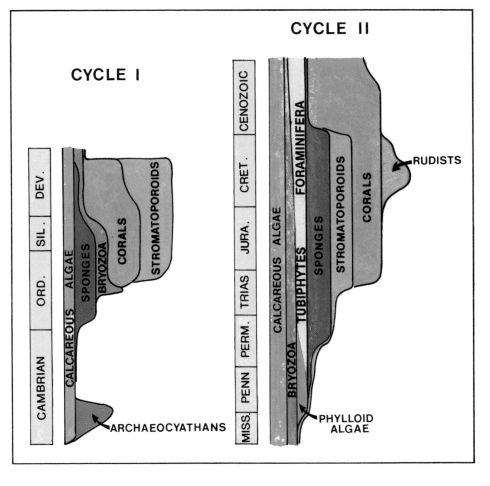

Figure 204—A generalized plot of the main biotic constituents of carbonate buildups against geologic time for the Phanerozoic. The balloons represent the relative abundance and importance of the different taxa. The plot is separated into two apparent cycles, each of which illustrates a similar evolution in reef-building taxa with time.

terrelationships: Bda. Bio. Stn. Spec. Pub. No. 6, p. 91-99.

Shinn, E. A., et al, 1977, Topographic control and accumulation rate of some Holocene coral reefs; south Florida and Dry Tortugas, in D. L. Taylor, ed.,

Proceedings of third international coral reef symposium, Miami: Geology, p. 1-9.

Wallace, J., and S. D. Schafersman, 1977, Patch-reef ecology and sedimentology of Glovers Reef Atoll, Belize, in

S. H. Frost, M. P. Weiss, and J. B. Sanders, eds., Reefs and related carbonates-ecology and sedimentology: AAPG Stud. in Geology No. 4, p. 37-53.

Lower Cretaceous Reefs, South Texas

Don G. Bebout
Robert G. Loucks

*L*ower Cretaceous shelf carbonates of Aptian, Albian, and Cenomanian age accumulated in a broad band which completely circled the Gulf of Mexico (Figs. 1, 2). These carbonates are well known on the outcrop in eastern Mexico and central Texas and in the subsurface of the Yucatan Peninsula in eastern Mexico, south and east Texas, central and south Louisiana, south Florida, and the Bahamas.

Throughout this Lower Cretaceous shelf area, water depths ranged from a few feet to one or two hundred feet and warm-water marine organisms flourished. Along a narrow band on the basinward edge of this broad shelf, biogenic growth climaxed and a complex of reefs, banks, bars, and islands developed. These Lower Cretaceous shelf-margin carbonates attained a thickness of 2,000 to 2,500 ft and have been identified as an almost continuous belt around the entire Gulf of Mexico.

The limestone making up this shelf-margin complex, the Stuart City Formation, is composed of reef and bank carbonates constructed primarily of rudist pelecypods and various types of carbonate sand bodies. Rudists are a group of sessile pelecypods which attached to each other on hard substrate to form boundstones and framestones. Other organisms which aided the rudist in forming boundstones and framestones of the shelf edge are encrusting algae, encrusting bryozoans, stromatoporoids, and corals.

Five major environmental units are recognized: open marine, lower shelf slope, upper shelf slope, shelf margin,

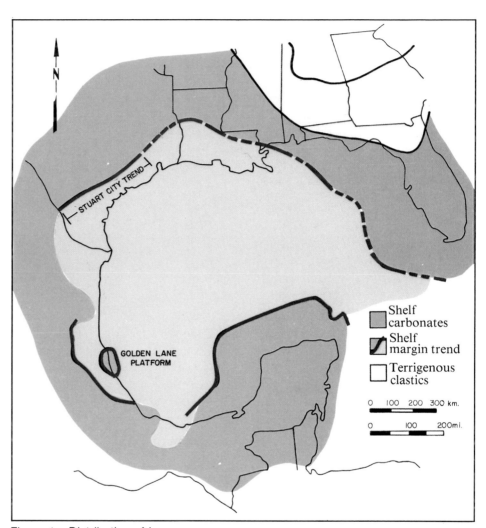

Figure 1—Distribution of Lower Cretaceous shelf-margin facies around the Gulf of Mexico (after Meyerhoff, 1967).

STUART CITY TREND

GOLDEN LANE PLATFORM

Shelf carbonates

Shelf margin trend

Terrigenous clastics

0   100   200   300 km.

0   100   200 mi.

and shelf lagoon (Fig. 3). The sea floor in the open-marine environment was probably under more than 60 ft of water, and reducing conditions existed within the sediment. The amount of influx of siliceous material varied greatly from time to time and from place to place. Tests of planktonic foraminifers rained down to accumulate on the bottom along with very small plates and spines of echinoids and small mollusks, animals which lived there.

Landward, the sea shallowed very gradually, resulting in the slightly shallower conditions on the shelf slope. On the lower shelf slope, echinoids and mollusks became more abundant and the individuals larger. Water depths probably ranged be-

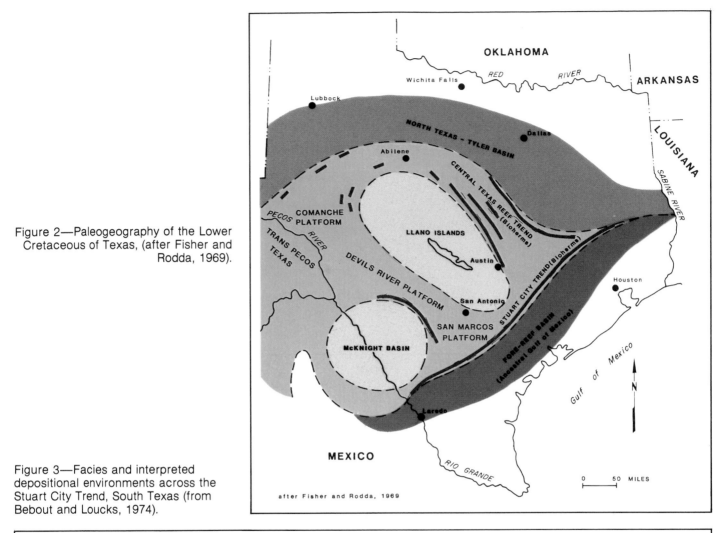

Figure 2—Paleogeography of the Lower Cretaceous of Texas, (after Fisher and Rodda, 1969).

Figure 3—Facies and interpreted depositional environments across the Stuart City Trend, South Texas (from Bebout and Loucks, 1974).

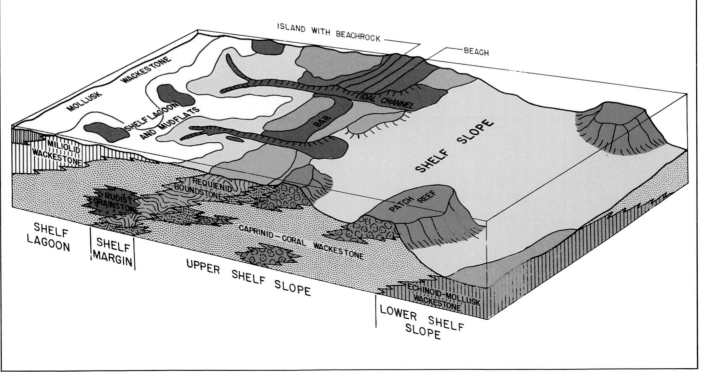

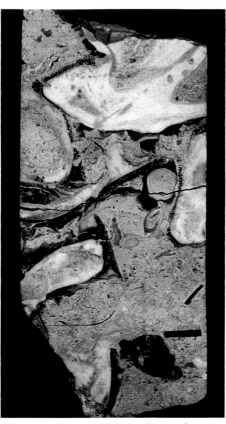

Figure 4—Requienid boundstone (bar scale = 1 cm).

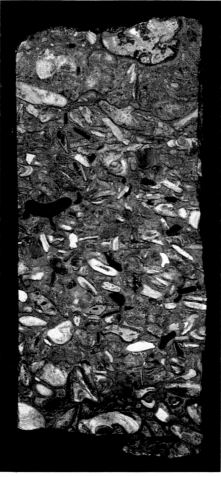

Figure 6—Rudist grainstone (bar scale = 1 cm).

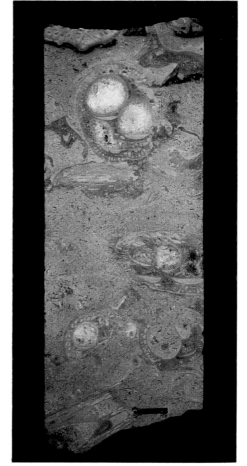

Figure 7—Caprinid-coral wackestone (bar scale = 1 cm).

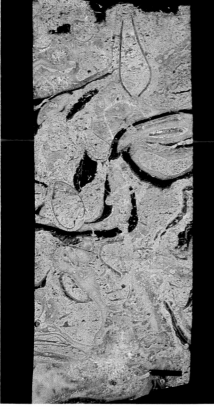

Figure 5—Requienid boundstone (bar scale = 1 cm).

tween 30 and 60 ft; local submarine currents concentrated these skeletal elements into channels although most of the bottom was mud. The tendency toward shallower water and more oxidizing conditions continued landward onto the upper shelf slope where water depths ranged between 10 and 50 ft. The bottom continued to be predominantly mud, but a larger number and greater variety of animals were able to live in this area. The most significant fauna added to those of the lower shelf slope are caprinid rudists and branching and massive corals. Locally, patch reefs or banks composed of corals and stromatoporoids grew. These patch reefs were isolated from the shelf-margin banks and reefs by wide areas of mud bottom with caprinids and corals.

The shelf-margin facies include coral-caprinid boundstone, requienid boundstone (Figs. 4, 5), rudist grainstone (Fig. 6), algae-encrusted miliolid-coral-caprinid packstone, and caprinid-coral wackestone (Fig. 7). The shelf edge is characterized as having been composed of a series of requienid banks and patch reefs with associated rudist sand bodies. Structures within the sand units indicate that a complex of environments is represented, including bars, tidal channels, islands, and beaches. On the seaward side of this band, coral-caprinid banks and reefs occurred more sporadically. None of the shelf-edge units formed a continuous band; instead, they were elongate, narrow bodies aligned parallel to the shelf edge, except for the channels which were perpendicular to the edge. Between the banks, reefs, and bars, caprinid rudists and branching corals lived on a mud bottom, thus resulting in the accumulation of the caprinid-coral wackestone facies. In some areas, a thick sequence of algae-coated grains accumulated on a broad flat covered by very shallow water.

The shallow-water shelf-lagoon facies include miliolid wackestone, mollusk wackestone, toucasid wackestone, and mollusk-miliolid grainstone. The facies accumulated in shallow water under low-energy conditions. Local variations in currents or bottom topography resulted in the formation of small islands typified by well-developed bird's-eye structures; these islands were commonly fringed by mollusk grainstone bodies. Grapestone and crusts developed on broad, very shallow-water stable flats in areas protected from normal surf and tidal currents. Small channels served as passageways for the slow movement of sea water on and off the flats during tidal changes but significant sediment transport took place on these flats only during major storms.

Cementation between grains, within voids in boundstone areas, and within body cavities of rudist and other mollusks occurred at several places in this environmental setting, but primarily within the facies of the shelf lagoon and shelf margin. Micrite rims, the result of both grain-edge alteration and algal coating, are present on all the skeletal fragments and intraclasts in the grainstone facies. These rims formed as the grains lay on the sea floor. Thin layers of marine fibrous to bladed isopachous cement were deposited within voids and between grains in facies which were accumulating in the very shallow, warm water. Dripstone and meniscus cements were deposited within the sediments exposed to vadose or intertidal conditions on small, local islands. The micrite rims, isopachous cement, and dripstone and meniscus cement, however, make up a very small part of the final calcite cementation and merely serve to partially lithify the sediment and protect it from later compaction.

After additional growth of the shelf margin, accompanied by slow subsidence of the shelf, these partially cemented sediments were buried to shallow depths. Leaching of aragonite and high-magnesium calcite shells took place. All voids were lined with thick layers of impure radiaxial cement deposited from phreatic-meteoric water. Later, after additional subsidence, the remaining voids were filled with equant calcite.

SELECTED REFERENCES

Bebout, D. G., and R. G. Loucks, 1974, Stuart City Trend, Lower Cretaceous, South Texas—a carbonate shelf-margin model for hydrocarbon exploration: Austin, Univ. Texas, Bur. Econ. Geology Rept. Inv. 78, 80 p.

Fisher, W. L., and P. U. Rodda, 1969, Edwards Formation (Lower Cretaceous), Texas—dolomitization in a carbonate platform system: AAPG Bull., v. 53, p. 55-72.

Meyerhoff, A. A., 1967, Future hydrocarbon provinces of Gulf of Mexico-Caribbean region: Trans., Gulf Coast Assoc. Geol. Socs., v. 17, p. 217-260.

Golden Spike Reef Complex, Alberta

Richard A. Walls

Upper Devonian (Frasnian) carbonate buildups in Alberta have been the subject of numerous studies since oil was discovered in them over 30 years ago. Of particular importance are Leduc carbonates which are known to contain 4.5 billion barrels of oil and 17 trillion cubic feet of gas in place. Leduc carbonates comprise a series of platformal (biostromal), reef and biohermal, and skeletal bank deposits that occur in the subsurface of the Western Canada basin and in various thrust sheets of the Rocky Mountains.

This paper discusses the depositional and diagenetic history as well as the reservoir properties of Golden

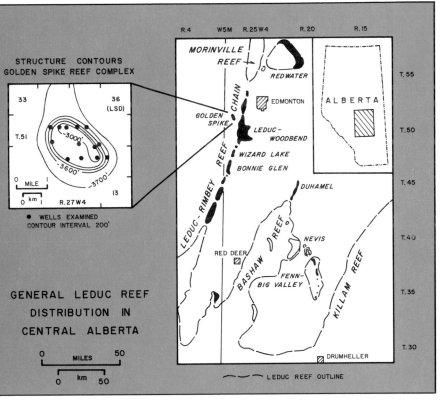

Figure 1—Location of the Golden Spike reef complex in the Alberta Basin, with structure contours on the Leduc Formation.

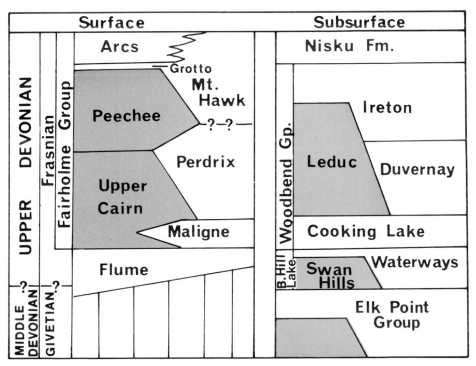

Figure 2—Subsurface and surface stratigraphic terminology for the Frasnian (Upper Devonian) in Alberta. Reef complexes are colored blue.

TABLE 1

Zones	Major Facies	Subfacies	General Environment
Reef Margin	Reef flank	*Detrital coral	Seaward of active reef growth; shallow-water low-relief fore-reef slope
		Tabular stromatoporoid	Fore-reef, similar to detrital coral but more in situ growth forms
		Argillaceous, skeletal rubble	"Deeper" fore-reef with mixture of fore-reef detritus and basin mud
	Reefs	Massive stromatoporoid	Shallow-water, well-washed zone of reef growth
		Stromatoporoid rubble	Accumulations of "eroded" reef framework
Reef Interior	Subtidal skeletal sand flats	Skeletal sand	Shallow-water, back-reef flat; "apron" of skeletal detritus
		Skeletal, peloidal sand	Well-washed, subtidal, back-reef shoals and lagoons
	Peritidal algal laminites		Intertidal to supratidal algal mud flats
	Patch reefs		Localized reef growth in subtidal, back-reef flats and lagoons

*Detrital refers to reef-building skeletal components which are not in situ.

Table 1. Summary of the major facies, subfacies, and general environments of deposition for Golden Spike.

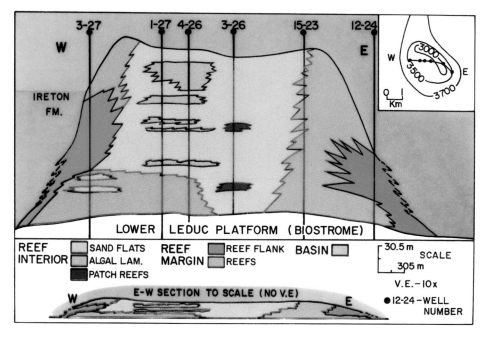

Figure 3—East-to-west cross section through Golden Spike illustrating the major facies. Datum is the top of the Beaverhill Lake Group.

Spike, one of these Leduc buildups. The Golden Spike reef complex (limestone) is located 34 km southwest of Edmonton, and just 4 km west of the dolomitized Leduc-Rimbey reef trend of central Alberta (Fig. 1). It is one of the smallest and most prolific oil producing (213 million barrels, recoverable reserves), Upper Devonian carbonate complexes in Western Canada, being only 10 sq km in area of closure but ranging up

to 197 m thick in the interior and thinning to less than 30 m on the flanks. It was discovered in 1949 by Imperial Oil Ltd.

STRATIGRAPHY: MAJOR FACIES

Rocks of the Golden Spike reef complex are part of the Upper Devonian Leduc Formation (Fig. 2) and have been subdivided by McGillivray and Mountjoy (1975) and McGillivray

(1977) into a lower Leduc platform, or biostromal unit, a middle Leduc bioherm unit, and an upper Leduc reef-fringed bank. The lower Leduc platform sequence is comprised of nonreservoir rocks that cover a 31 sq km area and are correlative with the Cooking Lake Formation to the east. The main reef complex (middle and upper Leduc) is considerably smaller (10 sq km), and is subdivided morphologically into a reef margin and

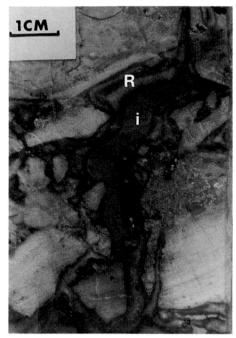

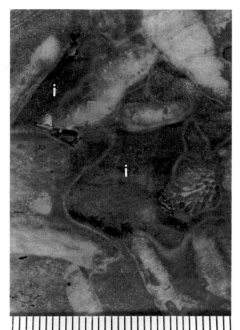

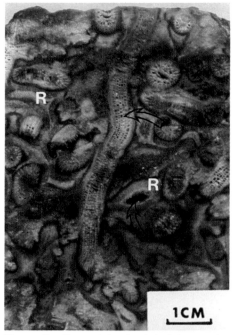

Figure 4—Core slab of a tabular stromatoporoid lime packstone containing a submarine fracture that is filled with an ostracod, peloidal grainstone internal sediment (i). The fracture truncates a banded radiaxial calcite cement (R). Well 12-24, 5829 ft.

Figure 5—Core slab of a tabular stromatoporoid lime packstone with several generations of marine internal sediments (i) filling large cavity systems. Bar scale is in millimeters. Well 9-27, 5558 ft.

Figure 6—Core slab of a tabulate coral lime packstone from the detrital coral facies. Radiaxial calcite cement (R) is common in this facies, and porosity (arrows) generally is low. Well 12-26, 5733 ft.

reef interior (Table 1, Fig. 3). Basin or off-reef sediments are referred to as the Duvernay and Ireton formations (Fig. 2).

Reef Margin

Golden Spike's reef margin (Figs. 4 to 8) is characterized by a reef flank (fore-reef) and reef facies.

Reef Flank

Rocks of the reef flank are composed of a mixture of poorly sorted skeletal debris, minor in situ stromatoporoids, encrusting stromatoporoids and algae, and abundant fine grained carbonate sediments (fine skeletal grainstones, lime mudstones, and calcareous shales). Banded radiaxial calcite is a common "binding" agent (cement) for the sediments of the reef flank.

A tabular stromatoporoid facies (Fig. 4), observed in most wells of the reef margin, is characterized by both detrital and in situ tabular stromatoporoids. Renalcids (problematic algae or foraminifera) often appear as encrustations on stromatoporoids and as discrete grains. Other fossil constituents include abundant crinoid plates and less common corals, brachiopods, pelecypods, and gastropods generally set in poorly sorted lime packstones and wackestones. Marine geopetal sediments (Fig. 5) are also common in this facies, generally occurring as complete to partial fillings in small (5 mm to 5 cm) shelter and growth framework voids. The tabular stromatoporoid facies occurs basinward of the reef in shallow water of a low relief, fore-reef slope.

A detrital coral facies (Fig. 6) is observed locally on the reef margin and is less common than the tabular stromatoporoid facies. It is composed of corals (*Thamnopora, Coenites, Phillipsastrea, Alveolites*) as well as stromatoporoids, crinoids, renalcids, and rare brachiopods, pelecypods, gastropods and ostracods. The detrital coral facies represents localized fore-reef deposition laterally equivalent to the tabular stroma-

toporoid facies.

An argillaceous skeletal rubble facies contains massive and tabular stromatoporoids, rugose and tabulate corals, and crinoid fragments in an argillaceous lime packstone matrix. Green to tan calcareous shales that occur throughout this facies as matrix, geopetal sediments, and as thin partings were probably deposited as a suspended load from the nearby basinal calcareous shales.

Reef

Rocks of the reef facies are composed of massive and tabular stromatoporoids occurring in poorly sorted lime grainstones (rudstones) or packstones (Fig. 7). A massive stromatoporoid boundstone and stromatoporoid rubble facies comprise the reef zone. Boundstones are found locally in all wells of the reef margin, but are probably best developed in wells on the northwest margin. They are characterized by an abundance of massive, and less commonly tabular stromatoporoids. Well

Figure 7—Core slab of bulbous and massive stromatoporoids from the reef facies. Well 12-24, 5918 ft.

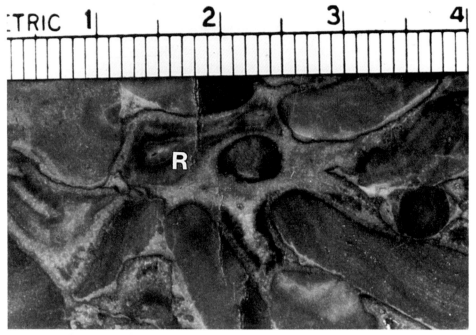

Figure 8—Core slab of a stromatoporoid rubble that is cemented by banded radiaxial calcite (**R**). Well 9-27, 5798 ft.

developed growth framework cavity systems are present in some cases, commonly filled with internal sediments and/or cement.

Boundstones are intercalated with coarse skeletal grainstones (stromatoporoid rubble facies) composed of poorly sorted massive and branching stromatoporoids, corals, and crinoids (Fig. 8).

Reef Interior

In Golden Spike, the reef interior (Figs. 9 to 11) is designated as a zone of subtidal to supratidal sedimentation that lies behind and is, in part, surrounded by an accretionary reef margin. Major facies of the reef interior include subtidal skeletal sand flats, peritidal algal laminites, and minor stromatoporoid patch reefs.

The subtidal skeletal sand flats are thick accumulations of both coarse and fine grained skeletal, peloidal grainstones and packstones (Fig. 9). Skeletal components include strongly abraded stromatoporoids and corals as well as crinoids, brachiopods, gastropods, pelecypods, ostracods, and calcispheres. Coarse grained skeletal grainstones are often interbedded with fine grained, well sorted peloidal grainstones (Fig. 10). These deposits represent deposition in the shallow subtidal portions of well-washed reef flats or shoals with skeletal material derived from the reef margin.

Peritidal algal laminites are most common in the upper Leduc (Fig. 11) where they are characterized by light gray to tan laminated, peloidal lime mudstones and fine grained peloidal lime grainstone (Fig. 11). Algal lamination is discontinuous and undulatory micritic bands that parallel bedding contain "trapped" or bound peloidal and skeletal grains. Laminoid, tubular, and irregular fenestrae (bird's-eyes) occur throughout this facies. These laminites represent high subtidal to supratidal semi-restricted deposition. Several laterally discontinuous paleoexposure surfaces have been documented in this facies.

Patch reefs are identified by the isolated occurrence of massive and branching (*Stachyodes*) stromatoporoids within the skeletal sand flat facies. Patch reefs range up to 10 m in thickness and represent localized biohermal growth within back reef areas of skeletal sand deposition.

DEPOSITIONAL HISTORY

The middle and upper Leduc at Golden Spike developed as a small reef complex isolated in a shale basin. Its history is similar in many aspects to other Upper Devonian reefs in Alberta (Andrichuck, 1958; Klovan, 1964, 1974; Mountjoy, 1967; Noble, 1970; Mountjoy and MacKenzie, 1973). The distribution of the aforementioned facies through Leduc time indicates (Fig. 12): (1) shallower water conditions occurring over the "muddy" lower Leduc platform (reef foundation); (2) widespread reef facies at the base of the middle Leduc; (3) extensive shallow water deposition of reef interior facies throughout the middle Leduc with

Figure 9—Core slab of a skeletal lime grainstone from the reef interior that contains moldic and interparticle porosity. Well 10-27, 5560 ft.

Figure 10—Core slab of a burrow mottled peloidal lime grainstone from the reef interior. Well 11-23, 5695 ft.

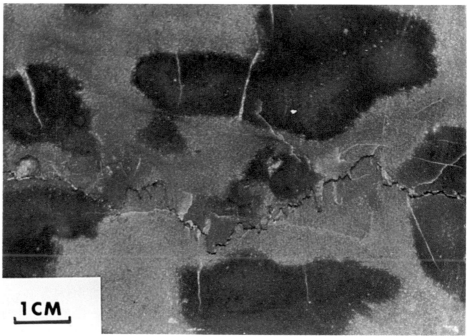

seaward accretion of the subtidal skeletal sand flats resulting in only localized reef development on the margins of the complex by upper Leduc time; (4) relatively continuous sedimentation on the margins with possible periodic exposure in the interior (peritidal laminite facies) throughout the upper Leduc; and (5) termination of reef development and eventual basin filling and burial by Ireton shales.

RESERVOIR QUALITY

Reservoir development in Golden Spike involved the combination of several sedimentological and diagenetic processes that influenced (enlarged or reduced) primary porosity. In the reef margin, shelter, growth framework, interparticle, and intraparticle pores are often filled by a combination of cements and internal sediments which leads to an occlusion of primary porosity (Table 2). On the other hand, skeletal, peloidal lime grainstones and laminated fenestral mudstones of the reef interior often show an enlargement of interparticle and fenestral pores by solution as well as some reduction by carbonate cements (Table 2).

Porosity distribution shows several trends throughout the reef complex

(Table 2):

(1) Porosity in the margin, particularly in reef flank facies is, with few exceptions, consistently low. Tabular stromatoporoids and detrital coral facies illustrate some of the lowest porosity in the reef due to pore occlusion by cementation and deposition of marine internal sediments. Argillaceous skeletal rubble facies generally contain low

primary porosity due to a matrix (mud) supported fabric.

(2) Reef facies illustrate both solution enlarged and cement reduced primary pores, as well as unaltered primary porosity.

(3) Porosity in the reef interior is variable. Subtidal skeletal sand flat deposits contain some of the highest porosity in the reef complex, due to preservation of original interparticle

TABLE 2

		Porosity (ϕ)		Permeability (K)		Major Diagenetic Effects on ϕ, Kh	
		Mean (%)	Mode (%)	Mean (md)	Mode (md)	ϕ, Kh Increase	ϕ, Kh Decrease
Reef Margin	Detrital Coral	4	2	1	1	Rare Solution	Extensive Radiaxial & Fibrous Calcite Cementation.
	Tabular Stromatoporoid	5.5	4	10	5	Rare Solution	Radiaxial & Fibrous Calcite Cementation Deposition Of Internal Sediment.
	Massive Stromatoporoid & Stromatoporoid Rubble	10	8.5	100	10; 100	Moldic Solution	Clear, Non-Ferroan & Ferroan Calcite Cements.
Reef Interior	Skeletal, Peloidal Sands	12.5	10	50	20	Moldic Solution	Fibrous, Non-Ferroan & Ferroan Calcite Cements.
	Algal Laminites	7	6	60	10	Solution Enlargement Of Fenestral Pores	Microstalactitic Cement, Non Ferroan & Ferroan Calcite Cementation

Table 2. Reservoir properties and major diagenetic influences on porosity and permeability for the different facies in Golden Spike.

and fenestral porosity, solution of early cements, and solution enlargement of primary pores.

DIAGENETIC CONTROLS ON RESERVOIR QUALITY

The diagenetic history of Golden Spike is complex but may be subdivided into at least four general stages (Fig. 13): (1) submarine and early burial (marine); (2) subaerial (during exposure periods, particularly in the upper Leduc); (3) middle or intermediate burial; and (4) late or deeper burial.

These stages are interpreted on the basis of certain diagnostic textural characteristics of different cements, their distribution, superposition, and relationships with associated sediments. For detailed discussions on the diagenetic history of Golden Spike see Walls (1977) and Walls,

Mountjoy and Fritz (1979).

The distribution of porosity in Golden Spike is affected by varying degrees of cementation and solution that occur during the four general diagenetic stages. The relative importance of these processes in reducing or increasing depositional porosity in the reef interior and reef margin are as follows:

(1) Reef Interior: (a) minor precipitation of marine cements (fibrous calcite cements, and early burial, non-ferron clear calcite cements); (b) partial solution of these early cements and associated sediments during exposure producing vug and moldic porosity; (c) vadose and/or phreatic cementation by fibrous, micro-stalactitic and non-ferroan calcites, followed by further subaerial solution, and cementation by clear calcites during middle burial, and ferroan calcites during late

burial; and (d) minor late solution associated with stylolite formation and fracturing.

(2) Reef Margin: (a) precipitation of submarine fibrous cements, now represented by radiaxial calcite, in large interconnected shelter, growth framework, and "stromatactis" type cavity systems; (b) precipitation of submarine and early burial (marine) non-ferroan syntaxial overgrowths (on crinoids) and minor clear calcites; (c) minor early solution (subaerial) which affects the reef facies but generally does not affect reef flank facies; (d) cementation by non-ferroan and minor ferroan calcites during middle and/or late burial diagenetic stages; and (e) minor late solution associated with stylolite formation and fracturing.

Submarine and subaerial cementation stages are important in early (precompaction) occlusion of primary

Figure 11—Core slab of an algal laminated lime mudstone with fenestral porosity. Well 11-23, 5556 ft.

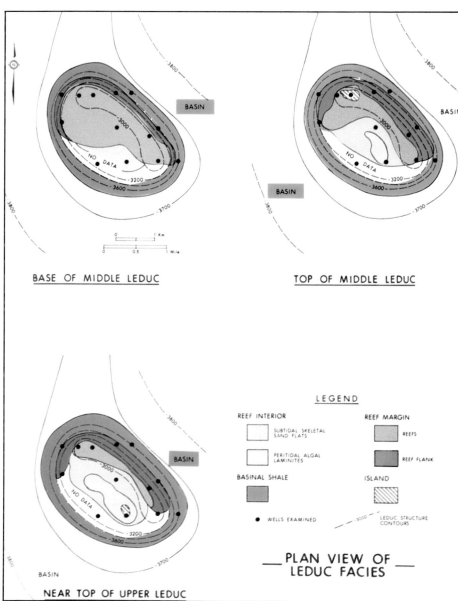

Figure 12—Plan view of the depositional history of the Golden Spike reef complex.

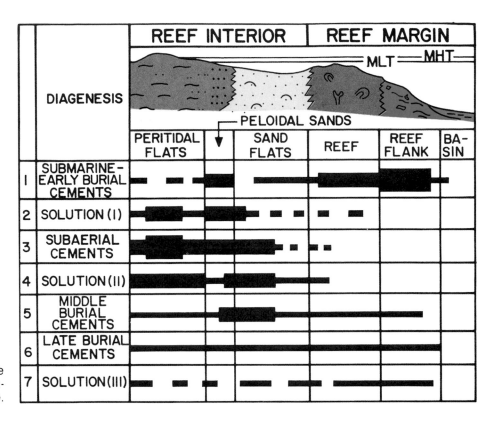

Figure 13—Schematic summary of the different stages of cementation and solution in Golden Spike.

porosity. However, high porosity zones in the reef interior are predominantly the result of solution which partially or completely removed some of the early cements as well as enlarged other primary pores. Zones of porous rocks intercalated with zones of cemented rocks occur throughout the reef interior, and these cemented zones form vertical permeability barriers. Permeability barriers of a cementation origin are obstacles to hydrocarbon recovery; therefore a complete knowledge of the diagenetic history of carbonate reservoirs is necessary for efficient reservoir management.

SUMMARY

Golden Spike is an Upper Devonian limestone reservoir that developed as a small reef complex in a shallow open marine basin only 4 km from a major linear reef barrier (Leduc-Rimbey reef trend). Although only 10 sq km in area of closure, it contains 213 million barrels of recoverable oil.

The major facies comprising Golden Spike are the reef margin and reef interior. The reef margin is composed of: (a) the reef flank—tabular stromatoporoid, detrital coral, and argillaceous skeletal rubble; and (b) the reef—massive stromatoporoid boundstone, and stromatoporoid rubble. The reef interior consists of subtidal skeletal sand flats, peritidal algal laminites, and patch reefs.

Porosity in reef margin facies, although extremely variable, is consistently lower than in the reef interiro. Large shelter and growth framework pores in the reef flank are commonly filled with "early" cements and geopetal sediments. On the other hand, subtidal skeletal sand flat deposits and peritidal laminites of the reef interior contain some of the highest porosity due to preservation of original interparticle and fenestral porosity, solution enlargement of primary pores, and solution of early cements.

The distribution of porosity in Golden Spike is affected by varying degrees of cementation and solution that occurred during four general diagenetic stages: (1) submarine and early burial (marine); (2) subaerial; (3) middle or intermediate burial; and (4) late or deeper burial.

Submarine and early burial cementation reduced porosity considerably in the reef margin. In the reef interior, periodic exposure and accompanying subaerial diagenesis produced alternating zones of tight (cemented), and porous rocks.

REFERENCES CITED

Andrichuck, J. M., 1958, Stratigraphy and facies analysis of Upper Devonian reefs in Leduc, Stettler, and Redwater areas, Alberta: AAPG Bull., v. 24, p. 1-93.

Klovan, J. E., 1964, Facies analysis of the Redwater reef complex, Alberta, Canada: Canadian Soc. Petroleum Geols. Bull., v. 12, p. 1-100.

_____ 1974, Development of western Canadian Devonian reefs and comparison with Holocene analogues: AAPG Bull., v. 58, p. 787-799.

McGillivray, J. G., 1977, Golden Spike D3A pool, in I. A. McIlreath and R. D. Harrison, eds., The geology of selected carbonate oil, gas, and lead-zinc reservoirs in western Canada: Canadian Soc. Petroleum Geols., p. 67-88.

_____ and E. W. Mountjoy, 1975, Facies and related reservoir characteristics, Golden Spike reef complex, Alberta: Canadian Soc. Petroleum Geols. Bull., v. 23, p. 753-809.

Mountjoy, E. W., 1967, Factors governing the development of Frasnian Miette and Ancient Wall reef complexes (banks and biostromes), Alberta, *in* D. H. Oswald, ed., International symposium on the Devonian system: Calgary, Alberta Soc. Petroleum Geols., v. 2, p. 387-408.

_____ and W. S. MacKenzie, 1973, Stratigraphy of the southern part of the Devonian Ancient Wall carbonate complex, Jasper National Park, Alberta: Geol. Survey Canada, paper 72-70, 121 p.

Noble, J. P. A., 1970, Biofacies analyses, Cairn Formation of Miette reef complex (Upper Devonian), Jasper National Park, Alberta: Canadian Soc. Petroleum Geols. Bull., v. 18, p. 493-543.

Walls, R. A., 1977, Cementation history and porosity development, Golden Spike reef complex (Devonian), Alberta: Montreal, McGill Univ., Ph.D. thesis, 307 p.

_____ E. W. Mountjoy, and P. Fritz, 1979, Isotopic composition and diagenetic history of carbonate cements in Devonian Golden Spike reef, Alberta, Canada: Geol. Soc. America Bull., v. 90, p. 963-982.

Platy Algal Reef Mounds, Paradox Basin

Philip W. Choquette

Reef mounds in which platy codiacean algae were important in the construction of carbonate-sediment build-ups are common in Late Paleozoic (Middle Pennsylvanian to Wolfcampian) shelf carbonates of the southwestern United States. According to Wilson (1975), they generally occur along shelf margins of that region, rimming oil and gas producing basins in the subsurface of New Mexico and Utah, Texas and Oklahoma; they are exposed in outcrops at basin margins in areas of Colorado (Eagle basin) and New Mexico (Orogrande basin).

In the Paradox basin, southwestern United States (Fig. 1), reef-mound buildups of platy-algal carbonate packstone and grainstone have received particular attention because of their importance as oil and gas reservoir rocks in some two dozen fields. The largest of these, Aneth field, had produced about 292 million barrels of oil and gas condensate and 281 billion cubic feet of gas by the end of 1978 (Petroleum Data Systems, Dept. of Energy, 1981). Other fields such as Desert Creek, Barker Dome, Cahone Mesa, White Mesa, Ratherford, and Ismay are smaller with cumulative productions on the order of 1 to 10 million barrels of oil and 1 billion cu ft of gas. The economic importance specifically of the reef-mound reservoirs in these fields is difficult to estimate because in most cases other carbonate facies are producing as well and production is comingled. It seems probable, however, that production of oil and gas condensate from reef-mound reservoirs in the Paradox basin to date is in the hundreds of millions of barrels.

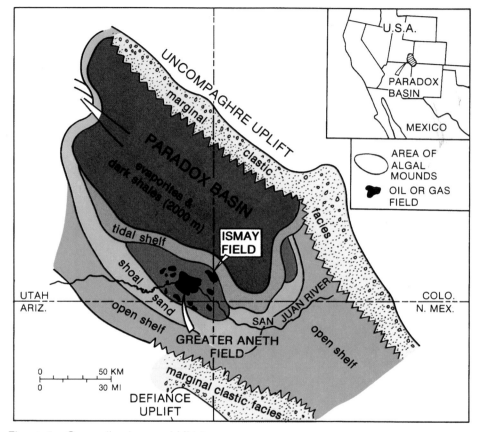

Figure 1—Generalized map of Middle Pennsylvanian (Desmoinesian) facies in the Paradox basin, showing areas of platy-algal reef mounds. Modified from Wilson (1975) and Choquette and Traut (1963).

GENERAL SETTING AND OCCURRENCE

The reef mounds occur along the southwest margin of the Paradox basin, in an area about 180 km in length by as much as 50 or 60 km in width (Fig. 1). The Paradox basin is a broad, asymmetrical sag of Late Paleozoic age. In Desmoinesian time, the central part of the basin received upwards of 2000 m of evaporites (both salt and anhydrite) with rhythmically interstratified black shales; this sequence is the Paradox Formation. The broad tract of reef-mound and associated carbonates and cyclically repeated thin dark shales equivalent to this sequence along the southwest margin of the basin is about 300 to 700 m thick and is known there as the Hermosa Formation. Siliciclastic sediments were shed into the basin from at least one bordering land mass, the Uncompahgre uplift (Fig. 1), but in the algal reef-mound belt along the opposing margin of the basin, siliciclastic

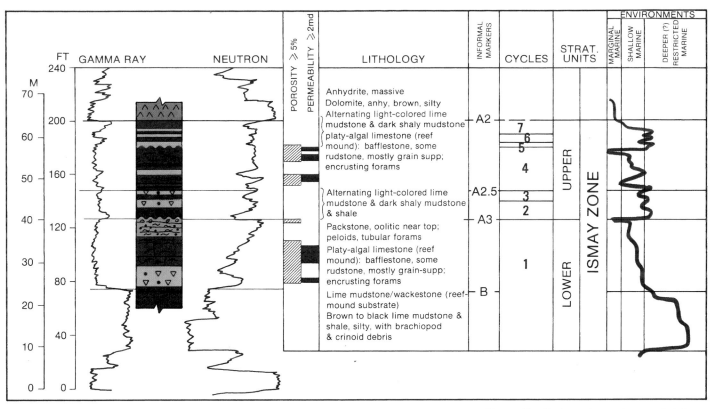

Figure 2—Carbonate cycles containing two algal reef-mound buildups in the Ismay Zone, Paradox Formation, in an oil-producing well, Ismay field. Well is the Pure Oil Company (Union Oil) East Aneth No. 28 D-1 in NW¼ NW¼ section 28, T35N, R26E. Cycles 1 and 4 contain reef mounds and terminate in subaerial exposure here and in most other parts of the area. A third reef-mound cycle, in the interval A2.5-A3 represented here by two inter-mound cycles, is developed in many other parts of the area (see Fig. 6). Modified from Choquette and Traut (1963).

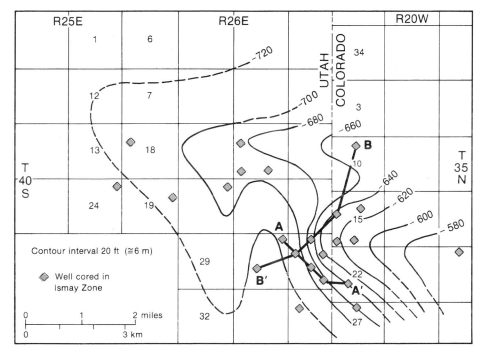

Figure 3—Map showing wells cored in Ismay Zone and structure drawn on top of lower Ismay shale (B horizon in Fig. 2). Carbonate buildups are localized along axis of structure (from Choquette and Traut, 1963).

detritus is restricted to the dark shales and to scattered grains in the carbonate rocks.

GENERAL FEATURES OF THE REEF MOUNDS

Synthesis of the regional relationships and general features of the mid-Pennsylvanian carbonate build-ups has been made by Wilson (1975), who described some of the variations in geologic setting, distribution, shape and geometry, and lithology that have been reported by many workers on carbonates in the Paradox basin

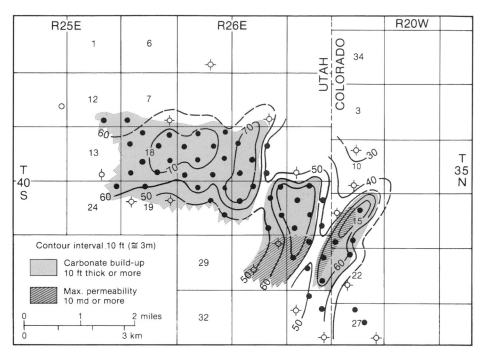

Figure 4—Map showing thickness of the A3-B interval. Thick areas are areas of thick algal reef mounds. Thin channel-like areas contain spicule mudstone and fine-grained calcareous quartz sandstone. A3-B interval is mantled by dark shales and is overlapped by evaporites to the north and east (see Fig. 6) (from Choquette and Traut, 1963).

Figure 5—Map showing thickness of the A2-A2.5 interval. Thick areas are areas of thick algal reef mounds. Reef mounds are mantled by dark shales and evaporitic dolomites grading upward to, and overlapped by, evaporites (from Choquette and Traut, 1963).

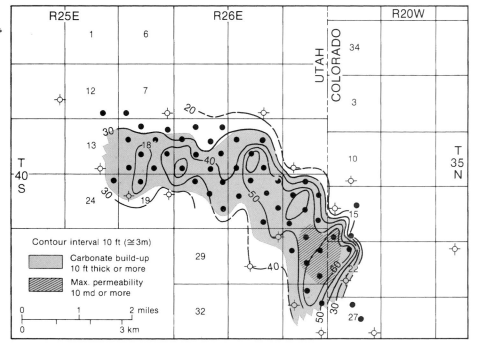

(Wengerd, 1951, 1955, 1963; Peterson, 1959; Peterson and Ohlen, 1963; Pray and Wray, 1963; Choquette and Traut, 1963; Elias, 1962, 1963; Gray, 1967; Hite and Buckner, 1981) and the Orogrande basin (Wilson, 1967, 1972, 1975). In general, the reef mounds appear to be flat-bottomed, convex-upward lenses with as much as 30 m of relief and regular to un- dulatory tops. Pray and Wray (1963) reported broad-scale cross bedding or accretionary bedding in biostromal banks of platy-algal grainstone which are exposed in the "Goosenecks" of the San Juan River, a series of large entrenched meanders about 110 km west of Aneth field. Although many of the build-ups are biostromal in the sense of having great lateral extent compared with their thickness, slopes of 25 to 30° exist locally on the tops and flank edges of some reef mounds. The mounds may be capped by ooid-rich grainstone which is also an important reservoir lithology in many of the Paradox basin fields, and/or by grainstone and packstone containing abundant fusulinids and foraminifers or encrusting life habit (ophthalmidids); this latter facies also occurs at the margins of some mounds and may represent in part the more mobile debris derived from the mounds. Some mounds terminate upward in platy-algal carbonates, at what are interpreted as erosional disconformities. Inter-mound facies are commonly fine-grained mudstone and wackestone rich in sponge spicules. Some reef mounds have been dolomitized but many remain limestone. A characteristic aspect of the reef mounds is the occurrence within them of breccias and conglomerates of algal-plate material and entrained fine carbonate sediment, suggesting partial induration followed by episodes of slumping or spalling of build-up sediments. Another characteristic feature is an association of reef-mound lithologies with multiple, shoaling-upward cycles which begin at their bases with transgressive

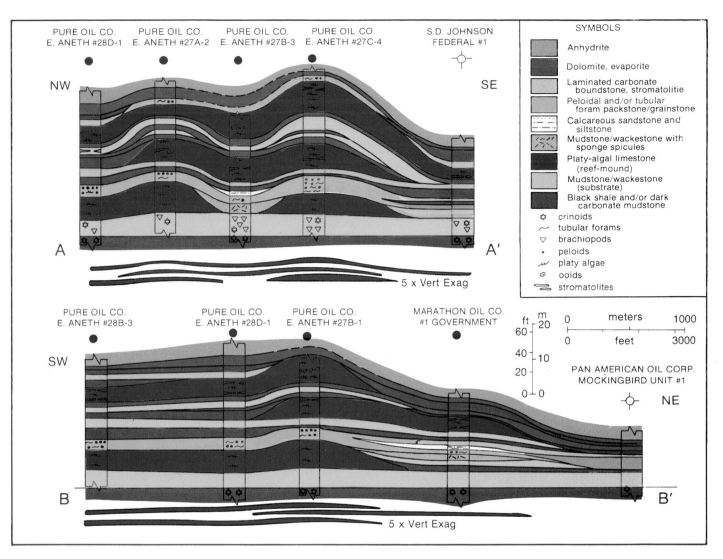

SYMBOLS

- Anhydrite
- Dolomite, evaporite
- Laminated carbonate boundstone, stromatolitie
- Peloidal and/or tubular foram packstone/grainstone
- Calcareous sandstone and siltstone
- Mudstone/wackestone with sponge spicules
- Platy-algal limestone (reef-mound)
- Mudstone/wackestone (substrate)
- Black shale and/or dark carbonate mudstone

☼ crinoids
∼ tubular forams
▽ brachiopods
• peloids
⤳ platy algae
⊘ ooids
⬂ stromatolites

Figure 6—Stratigraphic cross sections of Ismay Zone approximately parallel (A-A') and normal (B-B') to regional depositional strike (Figs. 4, 5). Well intervals here were cored in their entirety. Note that the buildups are subtle, low-relief features when seen even at 5X vertical exaggeration. Sag of upper reef mounds in A-A' probably reflects differential compaction of inter-mound sediments, cored in the Pure East Aneth No. 27B-3, 27-40S-26E (modified from Choquette and Traut, 1963).

shales and carbonate mudstones and culminate in peritidal or high-energy shoal carbonates or in exposure surfaces.

Wilson (1975) believes that the Desmoinesian algal-plate reef mounds accumulated somewhat downslope (basinward) from "tectonically induced shoals." Paleoslopes seem to have been gentle, rarely more than 2 to 3°, and successively younger reef mounds tend to overstep each other basinward. Subtle relief on the sea floor may have been enough to cause the initiation and colonization of individual stands of leaf-like platy algae. Careful study of excellent exposures along the San Juan River led Pray and Wray (1973) to conclude that algal-plate biostromal bank sequences there developed by the lateral coalescing of more abrupt build-ups.

REEF MOUNDS IN THE ISMAY FIELD: A CASE STUDY

A series of stacked algal reef-mounds has been cored extensively in the Ismay field situated about 5 km northeast of Aneth near the edge of the algal-mound fairway outlined earlier (Fig. 1). These build-ups have been described by Elias (1962, 1963), and in detail by Choquette and Traut (1963) from which much of the following sumary has been taken. Ismay is a relatively small field by comparison with Aneth and many other fields in the Paradox basin, having produced about 9.7 million barrels of high-gravity (41° API) oil and condensate, and 16.5 million cubic ft of low-sulfur gas through 1978. Nevertheless, the reef mounds there have features in common with other mid-Pennsylvanian buildups in the southwestern United States as outlined earlier.

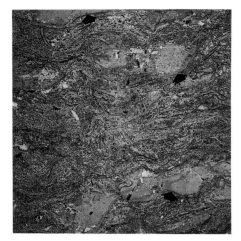

Figure 7A—Platy-algal limestone, bafflestone of dark phylloid algae (*Ivanovia*), some of which are shown by arrows, that have trapped light-colored micrite. Black specks are pore spaces of primary and solution origin. Pure East Aneth No. 28D-1, 28-40S-26E, 5806 to 6807 ft, about 8 ft above base of A3-B (cycle 1) reef mound.

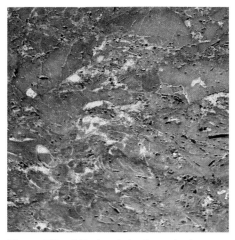

Figure 7B—Micrite-rich platy-algal limestone with algal-mold porosity (black) and patches of later pore-filling anhydrite (white). Pure East Aneth No. 28D-1, 5793 ft near top of A3-B (cycle 1) reef mound. Exposure to meteoric water is suggested by the moldic porosity; platy algae are inferred to have been aragonitic at the time.

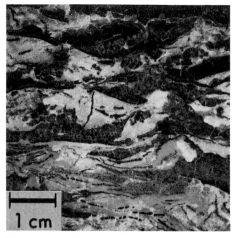

Figure 7C—Dolomitized platy-algal bafflestone. Phylloid algae (dark) entrapped lime mud (now micrite, pale tan) and peloids, now peloid grainstone and packstone. Diagenetic infill and partial replacement by anhydrite destroyed primary porosity. Pure East Aneth No. 28D-1, 5742 ft, from the upper (A2-2.5) reef mound in this well.

Setting and Large-Scale Features

The Ismay field reef-mounds occur in shoaling-upward, marine to peritidal cycles in three informally designated intervals of the Ismay Zone, Paradox Formation. The intervals are limited by the so-called A2, A2.5, A3, and B horizons (Fig. 2). In the section summarized in Figure 2 only two reef-mounds occur, in cycles designated informally as 1 and 4. The rest of the cycles shown there are interpreted as inter-mound sequences. We will return to the subject of these cycles after examining other aspects of the reef-mounds.

The field itself lies along the axis of a northwest-plunging anticline as defined on the B horizon (Fig. 3). Available well control indicates that the reef-mounds are flat-bottomed, thinly lensoid, biostromal accumulations up to about 15 m in thickness. Some of the buildups in the A3-B interval are .75 to 1.5 km across and at least 3 km long, elongated northeast-to-southwest almost at right angles to the anticline axis (Fig. 4). Build-ups in the two younger intervals of the Ismay Zone, illustrated by the isopach map of the A2-A2.5 interval in Figure 5, are more extensive features some 2 to 3 km across and at

Figure 7D—Complex breccia grainstone (mudstone) of gray micrite fragments and brown peloid-rich fragments in a groundmass dominated by anhydrite (white). Both replacement and pore-filling anhydrite occur. Most clasts are bordered by phylloid algae (arrows), suggesting that they were derived by breakup (off-mound slumping?) of partly-indurated sediment during reef-mound construction. Pure Aneth No. 28D-1, 5742.4 ft.

least 9 km long northwest-to-southeast. These mounds also display a "cross grain" nearly at right angles to the dominant trend. The cross grain may result from some combination of tidal currents and prevailing winds. Given the regional paleogeography (Fig. 1), it seems likely that tidal currents swept back and forth across the margin of the basin. Recently published plate-tectonic reconstructions (Scotese et al, 1979) suggest that in mid-Pennsylvanian time the basin was situated near the paleoequator. Comparison with present global wind patterns suggests that there may not have been strong prevailing winds in the region at that time.

Reef-Mound Facies

In cross sections (Fig. 6), the flat-bottomed form of the reef mounds is apparent for the oldest (A3-B) build-ups. The undulose appearance of the upper two build-ups in section A-A' may in part be a result of compaction draping over the lower reef-mounds.

The dominant lithology of the reef mounds is algal bafflestone (Embry and Klovan, 1971) in which loosely to closely packed fragments of the leaf-like or phylloid codiacean alga *Ivanovia* were the "constructing" elements. Pray and Wray (1963) provided an interpretation of the probable life form and the sediment-baffling, sediment-trapping effects of

Figure 8A—Dark lime mudstone with pelmatozoan debris and signs of bioturbation, from the basal "shale" unit of cycle 1 (Fig. 2), Pure East Aneth No. 28D-1, 5831 ft.

Figure 8B—Shelly lime mudstone, the usual substrate on which reef-mound deposition began. Some brachiopods were broken by compaction. Dolomitization in places of this substrate facies created good reservoir rocks with intercrystal and biomoldic porosity. Core is from cycle 1 (A3-B interval) in Pure East Aneth No. 28D-1, 5820.5 ft.

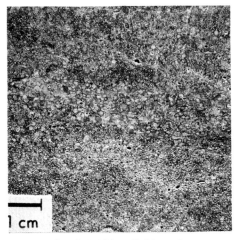

Figure 8C—Peloid-fusulinid packstone capping the A3-B (cycle 1) algal reef-mound in Pure East Aneth No. 28D-1, 5782 ft. The smaller grains are peloids and peloid molds (black); the larger grains are fusulinids and lithoclasts. Subaerial exposure of the reef-mound "cap" at the close of cycle 1 (Fig. 2) probably caused the solution porosity.

this alga. *Ivanovia* was probably an upright leaf-like marine plant which became calcified and upon death broke into somewhat rubbery, stiff but not rigid, platy fragments. Because of their broad, curved shapes, these fragments served as sediment traps in which lime mud, peloids and other particles collected (see Fig. 7A and 7C). The platy fragments also served as sheltering "umbrellas" beneath which shelter voids stayed free of sediment. Some of these original pores survived compaction and cement infill; others were filled by precipitated cements composed mainly of blocky calcite spar and/or coarse anhydrite (see Fig. 7B, 7C, 8A, 8B). Variations of baffle-stone fabrics with partly filled or occluded interparticle and shelter porosity are well illustrated by Choquette and Traut (1963).

Other biotic constituents of the reef mounds in addition to *Ivanovia* include rare fenestrate bryozoans which also serve as sediment baffles, pelmatozoan debris, an occasional fusulinid, ostracodes, and ophthalmidid foraminifers. The ophthalmidids rarely occur in life positions encrusting *Ivanovia* fragments and more commonly occur as loose skeletal grains. Peloids and micrite are the only common non-skeletal constituents.

Reef-mound breccias of synsed-

Figure 8D—Burrowed, organic-rich, dolomitized lime mudstone. This lithology commonly occurs at or near the tops of more evaporitic, off-mound cycles. It is overlain gradationally or abruptly by euxinic carbonate mudstone. Pure East Aneth No. 28D-1, 5743.4 ft.

imentary origin are common as noted earlier. Typically they consist of angular fragments of lime mudstone and/or peloid packstone or grainstone bounded on one side by a phylloid algal plate (Fig. 7D). The nature of these clasts suggests only partial induration when the breccias accumulated. The nature of the breccias suggests that some combination

of break-up by storms and off-mound slumping may have been responsible for them.

Reef-Mound and Associated Cycles

The thickest and most complete sequence in which *Ivanovia* reef mounds are developed in the Ismay area is in the lower part of the Ismay Zone referred to as cycle 1 in Figure 2. This sequence begins with a widespread unit of black shale and dark grayish-brown lime mudstone containing scattered crinoid columnals and phosphatic brachiopods, the so-called "B Shale" (Fig. 8A). This lithology suggests anoxic marine conditions across a broad but not necessarily deep basin of rather low relief. It grades upward to light-colored lime mudstone (or wackestone in places) of less anoxic origin containing a more normal marine but still sparse fauna, or thin-shelled calcitic brachiopods, scattered crinoid debris, and scattered ramose fenestrate bryozoans. This light-colored muddy lithology which Choquette and Traut (1963) termed "shelly calcilutite," is the substrate upon which the algal reef mounds of cycle 1 became established or colonized. An example of the lithology is shown

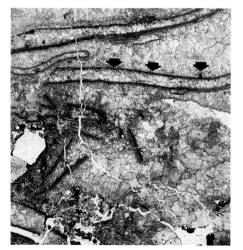

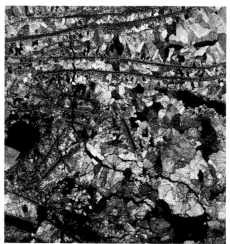

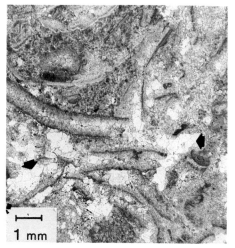

Figure 9A—Photomicrograph of calcite cements in *Ivanovia* grainstone (plane light). Finely crystalline radiaxial cement (arrows) of marine origin borders exterior of phylloid algae (dark). Coarser blocky calcite cement fills the remaining pore spaces between algal fragments as well as interiors of algae. Original porosity was probably more than 70%. Large white areas at left and lower right are artifacts, as are the cracks. Texaco Navajo No. J-1, 20-40S-26E, 5623-24 ft.

Figure 9B—Same photomicrograph as A, in cross-polarized light. Fringes of radiaxial cement are more clearly apparent here.

Figure 9C—Photomicrograph of *Ivanovia* grainstone (plane light). The phylloid algae are better preserved than in A or B. Darker material in upper left quadrant is peloidal silt. Most cement is fine blocky calcite, but the white areas (arrows) are occupied by anhydrite cement. Pure East Aneth No. 27A-2, 27-40S-26E, A3-B build-up, 5679 ft.

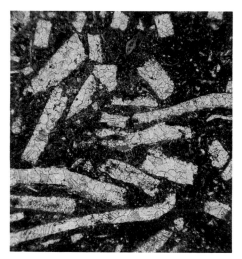

Figure 9D—Photomicrograph of algal packstone (plane light). Coarse blocky calcite fills former molds of calcareous (possibly aragonitic) phylloid algae (genus unknown but possibly *Ivanovia*). Pure East Aneth No. 27C-4, 27-40S-26E, near base of A2-A2.5 build-up, 2445 ft.

in Figure 8B and 8C. The upward change from this lithology to algal carbonates of the reef-mound facies is variable but generally rather abrupt over a few millimeters or centimeters, and in fact is often obscured by stylolites (Fig. 8C).

Carbonate "sand," either packstone or grainstone, generally caps the cycle 1 reef mounds as well as younger reef mounds (A-A' in Fig. 6). This lithology typically contains fusulinids, ophthalmidids, peloids, and occasional fragments of phylloid algae, bryozoans, and brachiopods. Rarely, small heads of chaetetid corals occur. This lithology is similar to the "cap facies" of Pray and Wray (1963). It also resembles in most respects the "oolite" capping facies of Peterson and Ohlen (1963) at Aneth, a lithology later redescribed by Wilson (1975) as rich in "tubular" (ophthalmidid) forams and leached "oolites." Leached-grain molds are indeed common in the capping facies of cycle 1 (Fig. 8C) and were probably aragonitic when they were dissolved; they may have been ooids.

Capping carbonate sands spill downslope for short distances of 1 km or less from the reef-mound flanks, and give way to muddy spiculite and siliclastic siltstone and fine-grained sandstones. These lithologies also fill shallow "channels" between mounds (A-A' in Fig. 6) and extend short distances into the basin (B-B' in Fig. 6). The capping carbonate sand is overlain abruptly, at an apparent disconformi-

ty, by dark crinoidal lime mudstone interpreted as the basal anoxic unit of the next younger cycle (cycle 2 in Fig. 2).

Younger reef mounds commonly terminate upward in, and may also contain intervals of, laminated peloidal limestone (or dolomite) containing mudcracks, mud-chip conglomerates, fenestral fabrics, and other products of deposition on high-intertidal flats that were periodically exposed. These tidal-flat lithologies are in some areas overlain by dolomitized mudstones, some of which are organic-rich and burrowed (Fig. 8D), and others having abundant nodular anhydrite.

The cycle in which lithologies like those just described occur are more evaporitic than the "ideal" reef-mound cycle outlined earlier (Fig. 2, A3-B buildup). Similar evaporitic cycles can be seen in areas immediately basinward from the reef mounds. There the cycles contain thicker units of dark shale and carbonate mudstone, evaporitic dolomite, and massive to bedded anhydrite. Units of anhydrite in these cycles thicken into the basin, in a general northeasterly direction (Fig. 6). Non-evaporitic sequences in areas between the reef mounds but along the reef-mound trend itself generally are thin. They begin with the lower shale and substrate mudstone units found in reef-mound cycles, but these are overlain by rather dark-colored mudstones made up of sponge spicules and other fine-grained skeletal debris together with micrite. Thin units of siliciclastic siltstone and fine to very fine grained sandstone also occur in some inter-mound cycles, particularly in the lower Ismay Zone (Fig. 6, section A-A′). The siliciclastic beds may be products of lowered sea level during which fine detritus reached the basin from land to the south.

Diagenesis and Porosity of Reef-Mounds

As noted earlier, some cycles in which algal reef mounds were developed apparently ended in subaerial exposure. Disconformities occur at the tops of these cycles (see Fig. 2). The limestones in these cycles show signs of dissolution which was selective only to some of the original grains, presumably those that were aragonitic. Moldic porosity is quite common, particularly in lime-mud-rich facies of reef mounds where phylloid algae were selectively dissolved (Fig. 7B), and also in capping limestones where ooids and possibly aragonitic peloids(?) were dissolved (Fig. 8C) leaving micrite or early cement intact. Molds like these in the reef mounds are commonly filled, however, by blocky calcite of phreatic origin (Fig. 9D). Cement-filled moldic porosity is more common in the lower parts of reef mounds than in their upper parts or in capping limestones, suggesting that the cementation by blocky calcite may have begun early, perhaps before deposition of the next cycle.

Phylloid-algal grainstones have at least two generations of carbonate cement — an earlier one of fine radiaxial calcite in fringes up to 0.2 to 0.3 mm thick along the exteriors of algal plates (Fig. 9A, 9B), and a later phase of coarse, blocky calcite which filled or partly filled remaining pore space (Fig. 9A to 9D). The radiaxial calcite is previsionally judged to be of marine origin. Examination in

cathodoluminescence of a few thin sections of grainstones cemented by blocky calcite has shown the calcite to have multiple zones of varying luminescence. This suggests that detailed study in cathodoluminescence would reveal a more complex cementation history perhaps relatable to repeated exposure as successive cycles accumulated.

Coarse anhydrite cement is also common in the reef-mound carbonates, especially the grainstones and open-work breccias (Figs. 7C, 7D, and 9C), in places completely occluding both original porosity and solution porosity. This evaporite cement post-dates the blocky calcite. It may be a result of sulfate reprecipitation from waters circulated through evaporite cycles in areas basinward from the Ismay area, or in evaporites overlying the Ismay field reef mounds. More focused study of diagenesis in the reef mound and associated rocks is needed to test this possibility and alternative models. Also poorly understood is the distribution and origin of dolomitization which has affected parts of reef mounds and underlying substrate lime mudstones in places enhancing the reservoir-rock quality of both (Choquette and Traut, 1963).

CONCLUSIONS

Reef mounds in which platy or phylloid algae were important sediment-constructing organisms are significant features of mid-Pennsylvanian cyclic carbonate sequences in the southwestern United States. In the Paradox basin, algal reef mounds occur mainly in a northwest-trending "fairway" (Wilson, 1975) up to 40 km in width and 85 to 90 km in length along the southwest margin of the basin (Fig. 1). Many of these reef mounds are reservoir rocks for oil and gas, with reserves ranging from small, as in Ismay field, to giant, as in Aneth.

The reef mounds are biostromal in dimensions, with maximum thicknesses of about 10 m and areal extents ranging up to a few kilometers in any one direction. They consist predominantly of algal baf-flestone (grainstone-packstone) in which loosely packed semi-rigid plates

of *Ivanovia* baffled and entrapped carbonate sediment, creating high original shelter and interparticle porosity, and synsedimentary breccia composed of algal-limestone intraclasts. Phylloid-algal wackestone and packstone with abundant micrite is also abundant, particularly in the lower parts of reef mounds. The breccias appear to be the only significant "reef-flank" material and are so closely associated with in-place baf-flestone that they have not been mapped separately. Grainstone and packstone carbonate sand composed mainly of fusulinids, ophthalmidid foraminifers, and peloids is a common capping lithology on the reef mounds.

Reef mounds occur in asymmetrical, shoaling-upward cycles 30 to 40 m thick which began with widespread basal dark shales and lime mudstones of anoxic marine origin, and terminated with foram-peloid-ooid carbonate sands of shallow-marine to beach or possibly dune origin, or peritidal carbonates. Subaerial exposure was common at the ends of cycles. Much of the diagenetic modification of reef mounds involved early selective leaching of aragonitic grains (vadose?) and later partial to complete cementation by block calcite (phreatic?) and/or coarse anhydrite. Parts of many reef mounds were also dolomitized.

ACKNOWLEDGMENTS

The author acknowledges the encouragement of Noel P. James, Memorial University of Newfoundland, at whose suggestion this summary of earlier work by John D. Traut and myself was written. Susan C. Hartline and other members of the Illustrations Department, Marathon Oil Company Research Center, prepared the illustrations. Published by permission of Marathon Oil Company.

SELECTED REFERENCES

Choquette, P. W., and J. D. Traut, 1963, Pennsylvanian carbonate reservoirs, Ismay field, Utah and Colorado, *in* R. O. Bass and S. L. Sharps, eds., Shelf carbonates of the Paradox basin, a symposium: 4th Field Conf., Four

Corners Geol. Soc., p. 157-184.

Elias, G. K., 1962, Paleoecology of lower Pennsylvanian bioherms, Paradox basin, Four Corners area: Guidebook, 27th Ann. Field Conf., Kansas Geol. Soc., p. 124-128.

_____, 1963, Habitat of Pennsylvanian algal bioherms, Four Corners area, *in* R. O. Bass and S. L. Sharps, eds., Shelf carbonates of the Paradox basin, a symposium: 4th Field Conf., Four Corners Geol. Soc., p. 185-203.

Embry, A. F., and J. E. Klovan, 1971, A Late Devonian reef tract on northeastern Banks Island, Northwest Territories: Canadian Petroleum Geols. Bull., v. 19, p. 750-781.

Gray, R. S., 1967, Cache Field—a Pennsylvanian algal reservoir in southwestern Colorado: AAPG Bull., v. 51, p. 1959-1978.

Hite, R. J., and D. H. Buckner, 1981, Stratigraphic correlations, facies concepts and cyclicity in Pennsylvanian rocks of the Paradox basin, *in* D. L. Wiegand, ed., Geology of the Paradox basin: Rocky Mtn. Assoc. Geols., p. 147-160.

Krivanek, C. M., 1981, New fields and exploration drilling, Paradox basin, Utah and Colorado, *in* D. L. Wiegand, ed., Geology of the Paradox basin: Rocky Mountain Assoc. Geols., p. 77-81.

Peterson, J. A., 1959, Petroleum geology of the Four Corners area: Proc., Fifth World Petroleum Cong., Sec. 1, Paper 27, p. 499-523.

_____ and H. R. Ohlen, 1963, Pennsylvanian shelf carbonates, Paradox basin, *in* R. O. Bass and S. L. Sharps, eds., Shelf carbonates of the Paradox basin, a symposium: 4th Field Conf., Four Corners Geol. Soc., p. 65-79.

Pray, L. C., and J. L. Wray, 1963, Porous algal facies (Pennsylvanian), Honaker Trail, San Juan Canyon, Utah, *in* R. O. Bass and S. L. Sharps, eds., Shelf carbonates of the Paradox basin, a symposium: 4th Field Conf., Four Corners Geol. Soc., p. 204-234.

Scotese, C. R., et al, 1979, Paleozoic base maps: Jour. Geology, v. 87, p. 217-277.

Wengerd, S. A., 1951, Reef limestones of Hermosa Formation, San Juan Canyon, Utah: AAPG Bull., v. 35, p. 1038-1051.

_____, 1955, Biohermal trends in Pennsylvanian strata of San Juan canyon, Utah, *in* Geology of parts of the Paradox, Black Mesa, and San Juan basins: Field Conf. Guidebook, Four Corners Geol. Soc., p. 70-77.

_____, 1963, Stratigraphic section at Honeker Trail, San Juan Canyon, San Juan County, Utah, *in* R. O. Bass and S. L. Sharps, eds., Shelf carbonates of Paradox basin, a symposium: 4th Field Conf., Four Corners Geol. Soc., p. 235-243.

Wilson, J. L., 1967, Cyclic and reciprocal sedimentation in Virgilian strata of southern New Mexico: Geol. Soc. America Bull., v. 78, p. 805-818.

_____, 1972, Influence of coral structure in sedimentary cycles of Beeman and Holder formations, Sacramento Mountains, Otero County, New Mexico, *in* J. C. Elam and S. Chuber, eds., Cyclic sedimentation in the Permian Basin, 2nd ed.: West Texas Geol. Soc., p. 41-54.

_____, 1975, Carbonate facies in geologic history: New York, Springer-Verlag Pub., 471 p.

9

Bank Margin Environment

Robert B. Halley
Paul M. Harris
Albert C. Hine

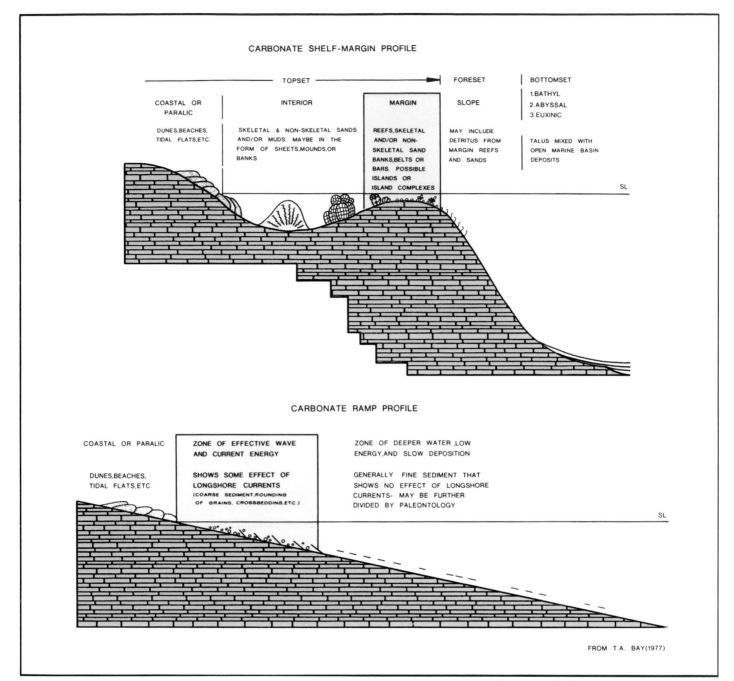

CARBONATE SHELF-MARGIN PROFILE

TOPSET →			FORESET	BOTTOMSET

COASTAL OR PARALIC | INTERIOR | MARGIN | SLOPE

1.BATHYL
2.ABYSSAL
3.EUXINIC

DUNES,BEACHES, TIDAL FLATS,ETC.

SKELETAL & NON-SKELETAL SANDS AND/OR MUDS. MAYBE IN THE FORM OF SHEETS,MOUNDS,OR BANKS

REEFS,SKELETAL AND/OR NON-SKELETAL SAND BANKS,BELTS OR BARS. POSSIBLE ISLANDS OR ISLAND COMPLEXES

MAY INCLUDE DETRITUS FROM MARGIN REEFS AND SANDS

TALUS MIXED WITH OPEN MARINE BASIN DEPOSITS

SL

CARBONATE RAMP PROFILE

COASTAL OR PARALIC

ZONE OF EFFECTIVE WAVE AND CURRENT ENERGY

ZONE OF DEEPER WATER ,LOW ENERGY,AND SLOW DEPOSITION

DUNES,BEACHES, TIDAL FLATS,ETC.

SHOWS SOME EFFECT OF LONGSHORE CURRENTS
(COARSE SEDIMENT,ROUNDING OF GRAINS, CROSSBEDDING,ETC.)

GENERALLY FINE SEDIMENT THAT SHOWS NO EFFECT OF LONGSHORE CURRENTS- MAY BE FURTHER DIVIDED BY PALEONTOLOGY

SL

FROM T.A. BAY(1977)

Figure 1—Bank-margin sands (highlighted in yellow) play an important part in geologic models, whether deposition occurred over a shelf-margin or a ramp profile.

Carbonate sand accumulations of reservoir size commonly occur on or near the seaward edge of banks, platforms and shelves. They may also form within the platform interiors or on topographically high areas in regionally deep water, but these occurrences are not as common as those along the margins. Bank-margin sand accumulations may grade landward or seaward within a fraction of a kilometer of other environments and, thus, do not have wide lateral extent in a dip direction. Such accumulations are sufficiently distinct and economically important as carbonate reservoirs to warrant their treatment in some detail. This paper provides an overview of modern and ancient carbonate sand bodies, and refers the reader to detailed work covering many aspects of carbonate sands.

Bank-margin carbonate sands occur repeatedly throughout the geologic record and are a prominent component of carbonate facies models (Fig. 1). Shaw (1964) and Irwin (1965) recognized the persistence of this

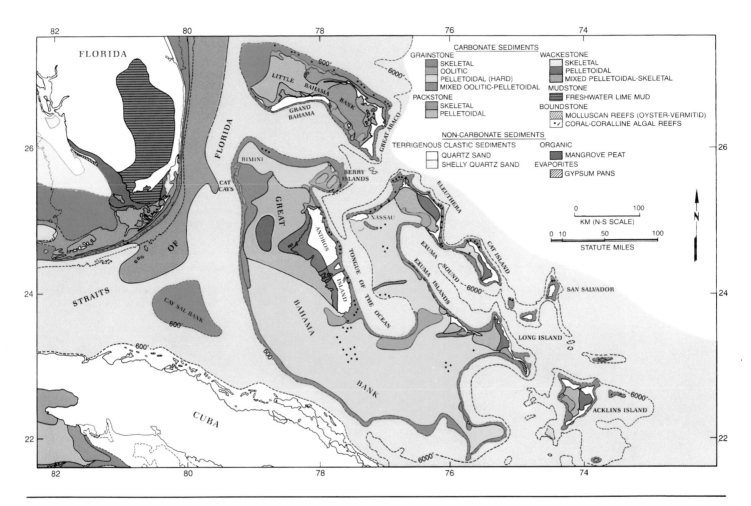

Figure 2—Location map and sediment distribution map for south Florida, and the Bahamas by Enos (1974). Note that skeletal grainstones, oolitic grainstones, and mixed oolitic-pelletoidal grainstones (shown in shades of green) rim most of the bodies.

facies in epeiric sea models as a seaward high-energy zone separating low-energy, deeper-water sediments from low-energy, shallow-water lagoonal deposits. More recent models by Heckel (1972), Lees (1973), and Wilson (1975) emphasize the importance of the zone in which carbonate sands accumulate. In nature, sand accumulations are not distinct and isolated from other facies (as are the chapters of this volume from each other, for example), and there will necessarily be some overlap between this chapter and those concerning reefs, beaches and islands, and lagoons.

Our understanding of carbonate sand deposition is biased toward bank-margin deposits because most modern studies have focused on these. The distribution of bank-margin sands in the Bahamas, where most studies have taken place, is well illustrated on a map showing sediment distribution (Fig. 2). It is tempting to interpret ancient deposits to fit a modern example, but we must remind ourselves that the Bahamian examples are not representative of all accumulations in the geologic record. In this review, most modern observations come from the Bahamas, but readers are encouraged to review those references to carbonate sand studies from other areas. These studies broaden our concepts of the variability of carbonate sand deposits. For example, the Persian Gulf (Arabian Gulf) provides an excellent modern example of carbonate sands

deposited on a gently sloping shelf or ramp (Ahr, 1973), but these deposits have received much less study than Bahamian sands.

PREVIOUS WORK

Before the turn of the century, most research on modern carbonate sediments centered on reefs. After 1900, other types of carbonate accumulations were examined in increasing detail. These studies are summarized in part by Illing (1954). By 1950, the ground work had been laid for more detailed investigations of carbonate sands, with studies supported by strong interests from both academia and the petroleum industry.

During the 1950s and early 1960s, carbonate sand studies emphasized particle analyses and morphology of sand accumulations. The composition, origin, size, shape, and distribu-

Table 1. Major types of carbonate sand grains (modified from Illing, 1955).

Figure 3—Gemini photograph of the south end of Tongue of the Ocean (TOTO) and Exuma Sound in the Bahamas. A tidal bar belt rims southern TOTO and the northeastern rim of Exuma Sound. The field of view across the base of the photograph is about 120 km.

tion of carbonate sands were documented over wide areas of the Bahamas and southern Florida. Much of this work was the result of projects undertaken by Norman D. Newell, Leslie V. Illing, Robert N. Ginsburg, and their colleagues and students. By the late 1960s and early 1970s, the emphasis had moved to internal structure of the sand bodies and early diagenesis of sand grains. Leaders in these aspects of carbonate sand studies were John Imbrie, Edward G. Purdy, Mahlon M. Ball, Hugh Buchanan, and Paul Enos.

Studies entered another stage in the late 1970s, concentrating on the hydrodynamics of carbonate sand movements (Hine, 1977; Harms et al, 1978). If these inquiries continue, they should provide a firm basis for comparisons between carbonate and siliciclastic sand deposits.

Carbonate Sands

Illing (1954) proposed a subdivision for calcareous sands into categories of "skeletal" and "non-skeletal" (Table 1). The reader will be familiar with most of the subdivisions, with the possible exceptions of the non-skeletal grain types — grains, lumps, and aggregates. Grains are aragonitic, micritic grains of indeterminate origin. Aggregates are weakly cemented clumps of sand grains that can easily be separated. Lumps are well-cemented grain-aggregates that include grape-like clusters (grapestone), botryoidal lumps, encrusted lumps, and irregular forms. Purdy (1963) and Winland and Matthews (1974) have shown that no genetic distinction exists between aggregates and lumps.

Numerous photomicrographs of these and other carbonate grain types have been illustrated by Milliman

(1974), Bathurst (1975), and Scholle (1978). Unfortunately, some of the criteria by which modern non-skeletal grains are recognized are not applicable to ancient carbonate rocks. McKee and Gutschick (1969) have suggested the term "peloids" to embrace all cryptocrystalline grains of uncertain origin. Although this term is admittedly a label of ignorance, it has proven an extremely useful term in the description of ancient limestones. The grain types of Illing (1954) are now recognized to comprise over 90% of the grains in modern shallow-water carbonate sands. Besides the four skeletal grain types listed by Illing, other organisms contributing to carbonate sands are echinoderms, brachiopods, arthropods, annelids, byrozoans, and sponges. Of these, echinoderms, brachiopods, and arthropods were conspicuously more important sand

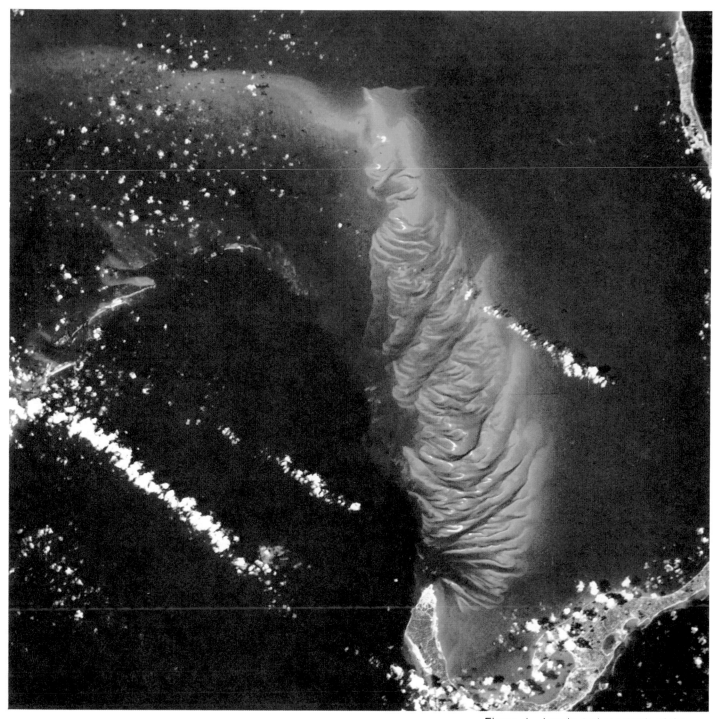

Figure 4—Landsat photograph of the Schooner Cay tidal-bar belt, rimming northeastern Exuma Sound in the Bahamas. The abrupt change to deeper water is clearly shown by a color change. The field of view across the base of the photograph is approximately 65 km.

producers during the Paleozoic.

MODERN SAND BODIES

Recent studies have shown that modern carbonate sand bodies associated with bank, platform, or shelf margins reflect the orientation, exposure, and topographic complexity of those margins. Winds, waves, currents, tides, biogenic barriers, and rock ridges or terraces along specific margins control sediment type, sediment production, sediment transport, and sand-body geometry, orientation, and size. Because knowledge of these last three characteristics is paramount to subsurface paleo-environmental

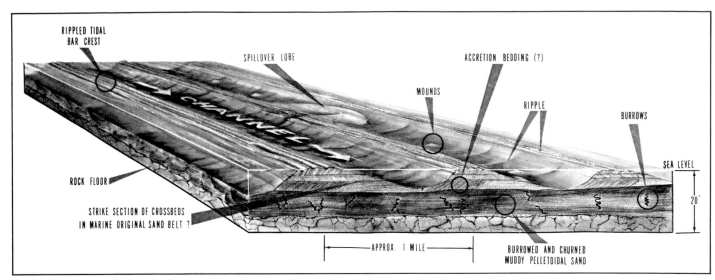

Figure 5—Schematic block diagram from Ball (1967) showing sedimentary structures that typify tidal-bar belts like those rimming southern Tongue of the Ocean and northeastern Exuma Sound.

Figure 6—Aerial photograph of tidal bars at southern rim of the Tongue of the Ocean. The abrupt change in water color toward the left side of the photograph reflects rapidly increasing water depth. The individual tidal bars are long (up to 20 km), narrow (1 to 1.5 km wide) and covered with smaller sand waves.

analysis, our presentation of modern carbonate sand bodies will follow Ball's (1967) classification stressing geometry and orientation rather than adopting a scheme based on some other factor, such as dominant physical process or bank-margin exposure.

Ball (1967) classified sand accumulations of the Bahamas and south Florida into four groups: (1) tidal-bar belts; (2) marine sand belts; (3) aeolian ridges (dunes); and (4) platform interior blankets. The dunes and platform interior sand bodies are discussed elsewhere in this volume, so

they will not be stressed here. The emphasis here will principally be on the wide variety of tidal bar and marine sand belts that develop along carbonate bank margins.

The primary difference between tidal-bar and marine sand types of linear sand belts is their orientation with respect to the trend of the bank-margin. Tidal-bar belts develop perpendicular to the margin, whereas marine sand belts form parallel to the bank edge. Both types of belts respond primarily to daily tidal flows, along with wave and storm-generated currents; but significant differences

exist between them, as well as between individual sand accumulations within each sand belt. Such differences result from antecedent topographic control, response to sea-level change, response to storms, bedform distribution, role of diagenesis, development of benthic communities, sediment type, sediment thickness, and lateral facies changes.

The critical factor determining whether a sand body develops into a tidal-bar belt or a marine sand belt is the dominance of daily tidal flows. When peak tidal currents are high (>100 cm/sec), parallel-to-flow sand

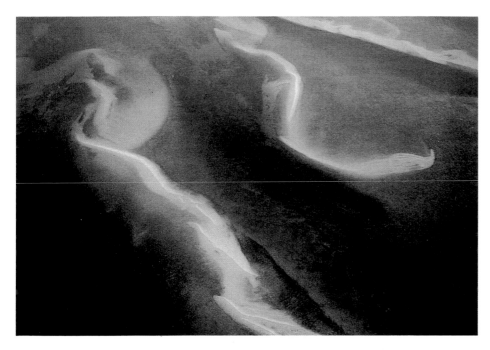

Figure 7—Aerial photograph showing detail of tidal bars from Figure 4. Light-colored ooid sands mark the active portions of the tidal bars. The bars are separated by deeper channels, also current swept but commonly covered with grass or cemented-sand bottoms.

Figure 8—Aerial photograph showing detail of tidal bars from Figure 6. Transverse sand waves, oriented approximately 45° to the length of the tidal bar, adorn the nearly intertidal, highly-agitated sea bottom of clean ooid sands.

bodies develop. When peak tidal currents are not as great (50 cm/sec) but still well above critical threshold levels, a sand belt normal to the tidal flow and parallel to the bank edge can develop. Along margins where high tidal currents form tidal-bar belts at or near the bank edge, these same currents are reduced further on the bank top by frictional effects. As a result, a bank-normal tidal-bar belt may merge into a bank-parallel

marine sand belt.

Some other modern sand accumulations along bank, platform, or shelf margins do not occur as belts. Tidal deltas, lobate in form, are generated by tidal currents flowing through gaps between islands. Also, back-reef sands may migrate into lagoons as lobate forms. Finally, there are composite sand bodies, which exhibit attributes of several end member types.

TIDAL-BAR BELTS

The classic examples of carbonate tidal-bar belts are located along the southern end of the Tongue of the Ocean (Fig. 3), and the north end of Exuma Sound (Schooner Cay area; Fig. 4), on Great Bahama Bank, Bahamas. The tidal-bar belt along Tongue of the Ocean dominates the bank-margin for nearly 100 km; the tide-dominated zone along Exuma

Figure 9—Underwater photograph of the lee slope of a mobile sand wave located in a tidal channel near Joulters Cays on Great Bahama Bank. The slip face is about 1 m high and the water depth is 4 m.

Figure 10—Underwater photograph showing detail of the lee slope of a sand wave like that of Figure 9. Note the sorting produced down the avalanche slope (Imbrie and Buchanan, 1965) as larger grains cascade to the base of the bed form and bury rippled ooid sands. The contact along the base of the slope is not straight.

Sound is nearly 45 km wide. These sand bodies have been studied and illustrated by Newell (1955), Newell and Rigby (1957), Ball (1967), Dravis (1977), and Palmer (1979). The sedimentary structures and bar morphology of tidal-bar belts are summarized schematically on a block diagram (Fig. 5).

Specific tidal-current data have not been published for these two areas, but daily tidal currents surely dominate these bank-margin settings. Charts indicate that velocities of up to 200 cm/sec occur. Also, these sand bodies are similar in geometry and size to siliciclastic sand bodies found on known tide-dominated shelves such as the embayments surrounding the North Sea (Off, 1963; Houboldt, 1968; Caston, 1972; Hayes, 1979). The north end of Exuma Sound and south end of Tongue of the Ocean are open (no island rims) embayments. The setting augments the tidal range, thus further increasing bank-normal tidal currents at the margin (Southard and Stanley, 1976).

The individual tidal bars within the tidal-bar belts range from 0.5 to 1.5 km in width, 12 to 20 km in length, and 3 to 9 m in thickness (Figs. 6, 7). Tidal bars are separated by 1 to 3 km wide channels dominated by tidal currents. The channels are 2 to 7 m deep and are generally floored by sea

Figure 11—Aerial photograph of a small portion of the Cat Cay marine sand belt along the western margin of Great Bahama Bank. From bottom to top, the color changes reflect water depth and bottom vegetation: grass-covered, dark, shallow-subtidal sands; white, highly-agitated, nearly intertidal ooid sands; and rapidly increasing water depth toward the bank-margin. The broad shoal to the left of the prominent tidal channel is about 1 km wide.

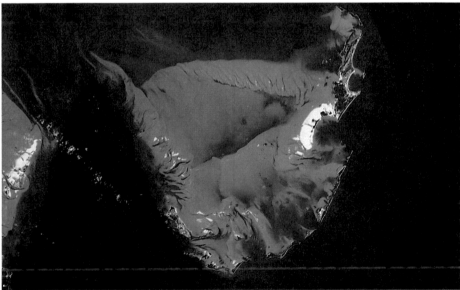

Figure 12—Landsat image of the Berry Islands, northern Great Bahama Bank. The open bank-margin to the north and west permits vigorous tidal exchange which is sufficient to maintain the sand belt. Field of view is 90 km.

grasses and algae. Despite the high tidal-current velocities, there are only a few places where sand waves (terminology of Boothroyd and Hubbard, 1975) are active on the floors of these channels.

Sand waves are the dominant bed forms covering the tidal bars themselves (Fig. 8). The axes of the bars support large symmetrical sand waves whose crests are oriented oblique to or roughly parallel to the bar axis. Smaller-scale mega-ripples and ripples dominate the flanks of these

bed forms. These sand waves indicate little net transport and probably support slipfaces which reorient themselves as the tidal flow changes direction. This bed form orientation also indicates that flow on the tops of the tidal bars is oblique or normal to their trend. Even though the great mass of tide-driven seawater flows parallel to the bars, the currents near the bars are refracted toward their crests, thus developing the observed bed form orientation and distribution. In places, smaller channels have

actually cut through these sand bodies, resulting in the development of lateral bulges or spillover lobes (Ball, 1967).

Along the flanks of the tidal bars in slightly deeper waters, upslope-oriented, asymmetric sand waves provide sand to the bar crest (Figs. 9, 10). This bed form distribution indicates that the channels on either side of the sand bodies may be flood or ebb dominant. Tidal-current segregation and resulting time-velocity asymmetry is important to the

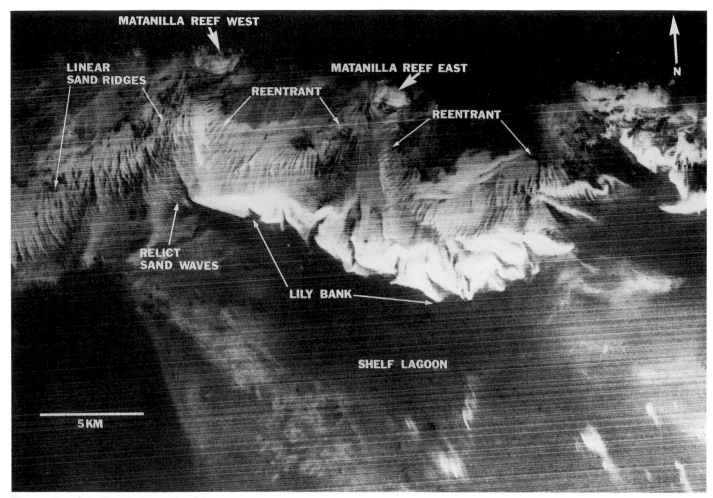

Figure 13—Satellite photograph of the northern rim of Little Bahama Bank from Hine (1977) highlighting the Lily Bank marine sand belt and other sedimentologic features. Note the linear sand ridges adjacent to the marine sand belt.

development and maintenance of these sedimentary units. Storm effects on tide-dominated sand bodies such as tidal-bar belts are largely unknown. The small channels that cut across individual bars and the development of spillover lobes may be storm-dominated events.

Variations in surface sediments and biota have been well documented for the bank-margin surrounding Bahamian tidal-bar belts. Seaward of a tidal-bar belt, the sediment changes abruptly from ooid sands to a mixture of skeletal, peloidal, and aggregate sands. In this area, the sea bottom is not highly agitated due to greater water depth and supports a variable covering of sea grasses and algae. A similar flora is also present bankward of the ooid bars, but the sediments are pellet sands or mud and are intensely burrowed. Sediment changes in the bankward direction are generally more gradual than those in a seaward direction.

Other Examples

Other tidal-bar belts in the Bahamas include portions of the Joulters ooid shoal and the Frazers Hog Cay area, north of Andros Island on Great Bahama Bank, and linear sand ridges adjacent to Lily Bank on the northern margin of Little Bahama Bank. These sand ridges, up to 5 km long, 2 to 5 m in relief, and 200 to 600 m apart, are much smaller than those along Tongue of the Ocean and Exuma Sound. They are now partially covered with sea grasses and are thought to be relict, perhaps being active only during large hurricanes. Surface sands are mostly lumps and micritized ooids with some peloids and skeletal fragments. The orientation of the bars, roughly normal to the bank edge (actually radiating away from a reentrant), indicates that they were at one time responding to tidal and storm-generated flows.

Another tide-dominated zone occurs between Grand Bahama and Great Abaco Islands on Little Bahama Bank. Although smaller in size, these sand ridges are similar to those along Tongue of the Ocean and Exuma Sound. The ridges respond to both ebb and flood currents as shown by the sharply defined symmetrical and asymmetrical sand waves on their crests and flanks.

An example of a carbonate tidal-bar belt not located in the Bahamas is the linear sand-shoals formed along

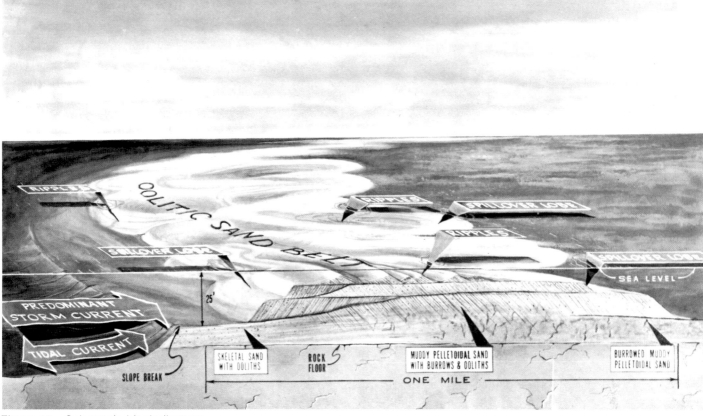

Figure 14—Schematic block diagram from Ball (1967) indicating sedimentary structures and grain types that typify marine sand belts like those along western Great Bahama Bank and northern Little Bahama Bank.

the Trucial Coast, offshore of Abu Dhabi in the eastern Persian Gulf. Ooid shoals are located between offshore islands and the mainland where tidal flows are strong (Loreau and Purser, 1973). There are numerous other modern examples of small, linear, parallel-to-flow carbonate tidal-bar belts, particularly in areas where the tidal range is high. However, the Bahamian examples remain the largest and best documented of modern analogues.

MARINE SAND BELTS

Marine sand belts, those linear sand bodies oriented roughly parallel to the bank, platform, or shelf edge, demonstrate a much greater degree of variability in geometry, size, sediment type, and origin than do tidal-bar belts. They occur both along wind-ward as well as leeward bank-margins and may consist of skeletal or non-skeletal sands.

Three well-documented marine sand belts occur in the Bahamas: (1) the Cat Cay ooid shoal lying south of Bimini along the western margin of Great Bahama Bank (Purdy, 1961; Ball, 1967; Fig. 11); (2) the platform behind the Berry Islands (Buchanan, 1970; Hine et al, 1981; Fig. 12); and (3) Lily Bank on northern Little Bahama Bank (Hine, 1977; Fig. 13). Figure 14 is a schematic block diagram that summarizes the sedimentary structures and bar morphology of sand belts. These marine sand belts range from 1 to 4 km in width and from 25 to 75 km in length. They are maintained by daily tidal flows, but each is also significantly influenced by storms (Perkins and Enos, 1968; Hine, 1977). Storm-generated flows may be responsible for developing channels normal to the axis of the sand body. These short channels are either floor or ebb dominated and commonly terminate in spillover lobes (Fig. 13).

The dominant bed forms are sand waves (both symmetrical and asymmetrical) with spacings ranging from 10 to 100 m and heights of 0.5 to 4 m. Smaller-scale ripples and mega-ripples are ubiquitous on the stoss slopes of the sand waves. The crests of some of the sand waves reach within 0.5 m of low tide.

As in the case of tidal-bar belts, important sea bottom and sediment changes have been documented for the bank-margin adjacent to marine sand belts. In a seaward direction, the sediments are a mixture of pellet, aggregate, and superficially coated ooid sands with an admixture of skeletal grains. The bottom is stabilized by sea grasses and green algae. On the bankward side of the sand belt, grasses and algae cover a burrowed bottom of lime mud or muddy peloid and superficially coated ooid sand. Both the seaward and bankward sediment changes occur rather abruptly at the margins of the sand belt.

Other Examples
Along open, leeward margins in the

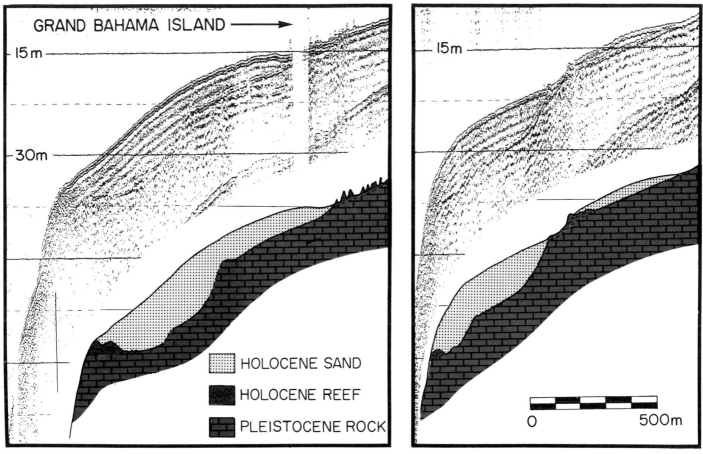

HOLOCENE SAND

HOLOCENE REEF

PLEISTOCENE ROCK

Figure 15—Seismic profiles located seaward of Grand Bahama Island on Little Bahama Bank. Holocene skeletal sands have been transported off the shallow bank and trapped in deeper water behind reefs. Accumulations up to 7 m thick have formed.

Figure 16—Aerial photograph of Escollo de Arenas sand belt off Viegras Island east of Puerto Rico. Note the numerous sand waves on top of the shoal. Puerto Rico is in the background.

Bahamas, carbonate sands originating in the bank interior may be transported to the margin, where they accumulate to form bank-parallel sand belts. Lumps, peloids, and micritized ooids, along with algae and mollusks, are predominant grain types. As indicated by large sand waves, sediments are carried to the bank edge by dominant offbank-oriented currents that are generated by storm winds. At the bank edge, a wedge of sand, up to 21 m thick, may form (Hine and Neumann, 1977; Palmer, 1979). Much of the shallow-

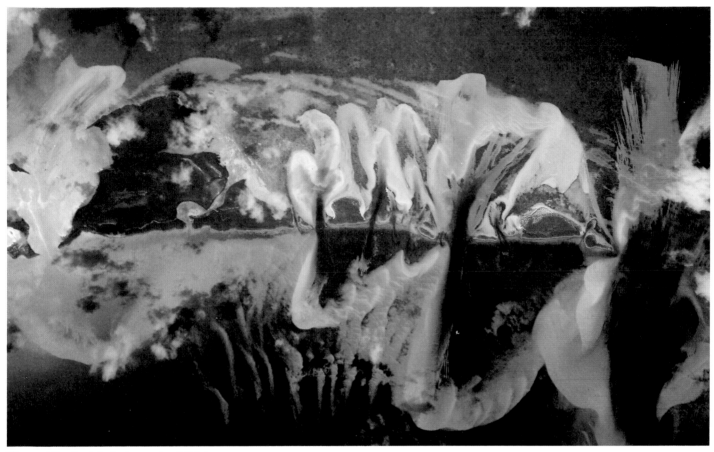

Figure 17—Aerial photograph of tidal deltas of carbonate sand formed in channels between islands near Curlew Cay, Little Bahama Bank.

water sand is transported off the marginal escarpment, ultimately to be deposited on the deep flanks of the carbonate platforms.

Along windward margins, particularly those fronting large islands such as Grand Bahama Island on Little Bahama Bank, marine sand belts can form at the bank margin in water depths of 30 to 50 m. There, skeletal sands (mostly algal) are transported over the bank edge by currents initiated by storms. Deeper water reefs along the margin may act as dams, behind which the sands accumulate as thick as 7 m (Fig. 15). Presumably during storms, water piled up against the large islands sets up seaward return flows at the bottom. The combination of a sloping sea bottom and the seaward-directed currents can transport skeletal sands to the bank edge and over the escarpment.

Marine sand belts can form in association with longshore transport in the nearshore zones of large islands. The mixed carbonate-terrigenous sand body Escollo de Arenas off Viegras Island east of Puerto Rico is a good example (Rodriguez, 1979). This sand body extends 6 km off the west end of the island; maximum width is 1 km and maximum thickness is 15 m (Fig. 16). The top of the shoal lies in 4 m of water and is dominated by a series of widely-spaced, symmetrical sand waves that respond to nearly equal ebb and flood tidal currents. During a former low stand of sea level, siliciclastic sands were brought to the west end of the island by longshore currents. As sea level flooded the platform, these sands were left on the shelf in a manner similar to the cape-associated shoals or cape-retreat massifs of the United States east coast shelf (Swift et al, 1972). Tidal currents reworked these sands forming a large sand-wave field, and carbonates (primarily algae and mollusks) were introduced from the shelf lagoons.

A carbonate bank, which occurs along the northeast shore of Shark Bay, Australia (Davies, 1970), might also be considered a sand belt. This sediment body, 7 to 12 m wide and about 100 km long, consists of a continuous intertidal sand sheet paralleling the shoreline and cut by numerous tidal channels with associated levees and fans leading offshore. The bank, composed largely of quartz, algal, foraminiferal, ooid, and mollusk sands, contains lesser amounts of muddier sediments.

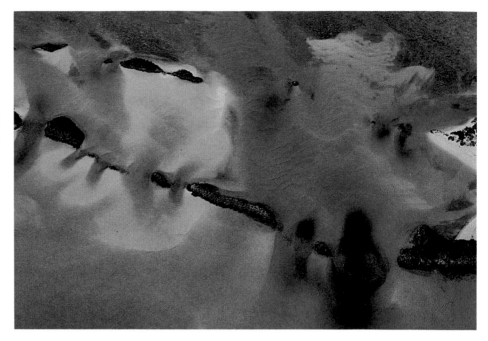

Figure 18—Aerial photograph of tidal deltas in the Exuma Islands. Ooid sands are forming on many of the deltas. Deep waters of Exuma Sound are to the bottom of the photograph.

Figure 19—Geologic map of Florida shelf by Enos (1977). The area mapped as bare sand (white and blue diagonal stripes) corresponds with White Bank. Note that active reefs and dead "rubble" reefs are shown in red, and patch reefs are shown in black.

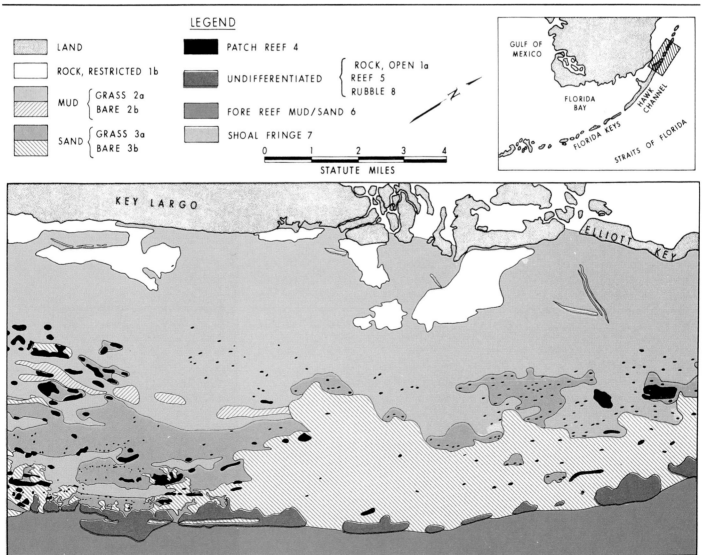

LEGEND

LAND

ROCK, RESTRICTED 1b

MUD { GRASS 2a / BARE 2b }

SAND { GRASS 3a / BARE 3b }

PATCH REEF 4

UNDIFFERENTIATED { ROCK, OPEN 1a / REEF 5 / RUBBLE 8 }

FORE REEF MUD/SAND 6

SHOAL FRINGE 7

0 1 2 3 4
STATUTE MILES

GULF OF MEXICO

FLORIDA BAY

FLORIDA KEYS

HAWK CHANNEL

STRAITS OF FLORIDA

KEY LARGO

ELLIOTT KEY

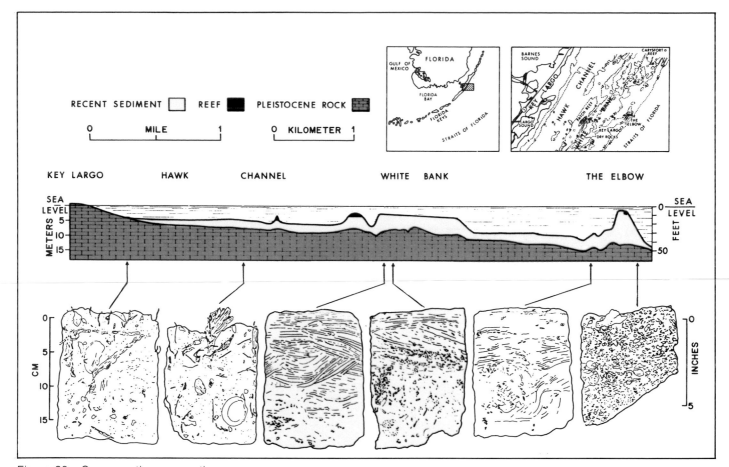

Figure 20—Cross section across the Florida shelf by Enos (1977) showing sediment thickness (in yellow) and sedimentary structures as revealed by box coring. The box core from White Bank (also highlighted in yellow) indicates that the clean skeletal sands are cross-bedded. The structures have been partially obliterated by burrowing and by grass roots toward the base of the core on the right (from Paul Enos, 1977).

TIDAL DELTAS

Lobate or fan-shaped flood and ebb tidal deltas formed of ooid or skeletal sand may occur opposite inter-island gaps or rocky tidal inlets. Small islands along carbonate bank margins are commonly lithified. As a result, the passes or gaps between these islands do not change position substantially with time. The associated sand bodies therefore, are not exact analogues of the tidal inlets found along siliciclastic barrier island coasts.

The shallowest portions of tidal deltas respond to daily tidal flows. Bed forms (ripples, mega-ripples, and sand waves) consisting of ooid sands are common in these areas. Deeper portions may consist of skeletal sands, partially stabilized by sea grasses and algae. The bedform distribution, including symmetrical sand waves, indicates that significant migration of the tidal delta does not occur during normal tidal events. Migration probably occurs during storm events, when water piled up against the islands forms a steep hydraulic gradient. As a result, strong currents may be generated in the channel between the islands. Between storms the tidal delta is essentially stabilized, except for the shallowest portions over which water flow is accelerated.

Large ebb-dominated tidal deltas, illustrated by Purser and Evans (1973) from the Trucial Coast area of the Persian Gulf, extend 1 to 2 km off- shore and are 5 to 6 km wide. They are wave-modified as shown by the numerous subtidal bars superimposed on them. Ooid sands forming on the deltas are transported to the adjacent islands. Many tidal deltas are quite small, sometimes only a fraction of a square kilometer, such as those illustrated by Jindrich (1969) and Basan (1973) from the southern Florida Keys.

Considering the numerous small islands and island chains on the Bahama Banks, the number of tidal deltas is certainly large. The best examples are found in the Carter and Strangers Cays area of northeastern Little Bahama Bank (Fig. 17) and in the Exuma Islands of Great Bahama Bank (Fig. 18).

Figure 21—Aerial photograph of the skeletal sand belt that lies immediately landward of the Belize barrier reef (darker band along left margin of sand belt). The reef and back-reef sands, about 100 m wide, form the major topographic feature between the Atlantic Ocean and the Belize shelf-lagoon.

Figure 22—Aerial photograph of a carbonate sand shoal and island, located on a topographic high and surrounded by deep water, in the Persian Gulf east of Qatar peninsula. The island is about 1 km in length. The topographic high is probably the surface expression of a salt dome.

BACK-REEF SANDS

Sand bodies frequently occur landward of reefs, because reefs often produce an abundance of sand. White Bank, a shoal area about 5 km east of Key Largo, Florida, is such a sand body (Enos, 1977). Algal, mollusk, and coral sands form a belt that is 1 to 2 km wide and 40 km long (Fig. 19). Although large portions of the belt are stabilized by sea grass, white rippled sands are common (Fig. 20). Both active and stable bottoms occur at similar water depths, 1 to 3 m, suggesting that scattered large sand waves must be due to localized hydrologic conditions. The sand body is asymmetrical in profile; the steeper lagoon side suggests bankward transport of sand.

Large elongate algal sand lobes have also built into the back-reef lagoon of Bermuda (Garrett and Scoffin, 1977; Garrett and Hine, 1979) and project as much as 4 km into the lagoon. They are up to 1 km wide and 15 m thick. Small reefs, which contribute sand grains other than algae, are common on top of the lobes. Seismic data indicate that the positions of the lobes are controlled in part by underlying topography of Pleistocene bedrock highs.

A much more extensive sand body occurs landward of the Belize barrier reef (Fig. 21). Although only 100 to 200 m wide, it is continuous and uniform with a total length of about 160 km and unbroken lengths of about 30 km. This sand belt probably owes its exceptional uniformity to the well-developed barrier reef that lies immediately seaward of it. The sediment surface is very flat. The sands abut the reef in 1.5 m of water and drop off into a shelf lagoon toward the west. The slope into the lagoon may be as steep as 35°, and the sands appear to be burying lagoonal patch reefs in some areas. Islands occur within the sand belt, usually near breaks in the reef.

Figure 23—Aerial photograph of large ooid and skeletal sand waves in Mexico located between Isla Mujeres on the right and the Yucatan mainland on the left. Isla Mujeres is approximately 8.5 km long. Individual sand waves are as high as 5 m.

Figure 24—Landsat image of the Joulters ooid shoal, lying north of Andros Island (shown in red) on Great Bahama Bank. Deep-water Tongue of the Ocean is to the upper right. The ooid shoal is a vast sand flat, 400 sq km, that is penetrated partially by tidal channels and fringed on the eastern border by mobile sands. Holocene islands are scattered along the mobile fringe and throughout the sand flat. Contrast the size and morphology of the Joulters shoal with shoals pictured in Figures 6 and 13.

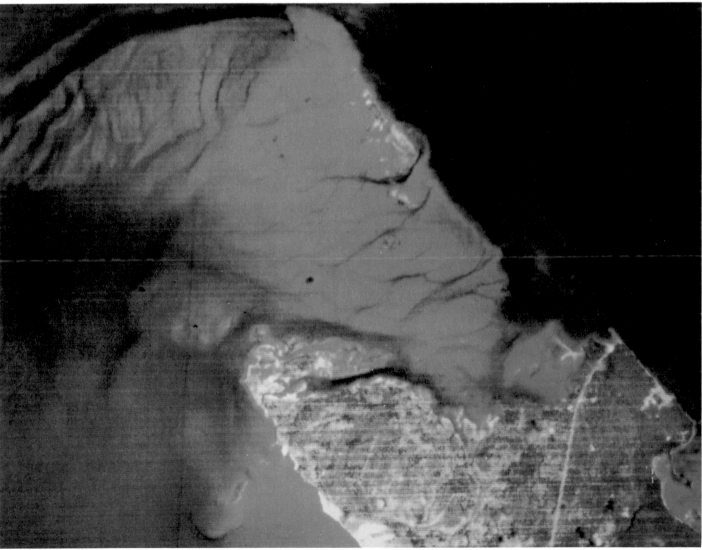

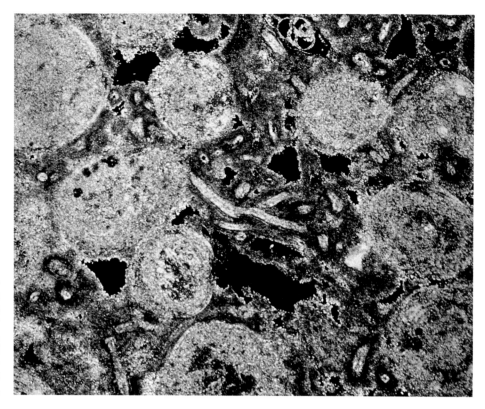

Figure 25—Photomicrograph of filamentous, calcifying algae that are acting as the primary binding and cementing agent in ooid sands from the Schooner Cays area, Great Bahama Bank. Magnesium-calcite is artificially stained reddish-brown; aragonite remains unstained. Width of the photograph represents 3 mm.

MISCELLANEOUS SAND BODIES

Within the interior of carbonate banks and platforms, active sand shoals, commonly oolite, may be situated on or near topographic highs. These sand bodies respond primarily to wind-generated currents; tidal flow is substantially diminished by the time it travels 20 to 50 km across the platform. Mackie shoal on northern Great Bahama Bank in the Bahamas is an excellent example. Seismic data indicate that a rock ridge may have localized this 1.5 km wide, 7 m thick, 30 km long linear sand shoal (Hine et al, 1981).

Reefs and sand shoals form on structural rises in the Gulf of Mexico and the Persian Gulf. Many of these accumulations overlie structures produced by salt diapirs, an example being the Flower Gardens reef about 200 km south-southeast of Galveston, Texas (Edwards, 1971). Purser (1973) has classified similar bathymetric highs in the Persian Gulf as being outer (in the basin center), intermediate (Fig. 22), or inner (coastal). The sediments on these highs vary depending on their location. In the basin center, they are characterized by open-marine, foraminiferal or coral/algal sands. On highs located near the coast, sediments are composed of pelletoidal sands, carbonate muds with restricted faunas, and algal stromatolites and evaporites. Where the highs exceed 5 km in width, double sediment spits or "bull horns" form curving elongate barrier sand ridges. These tails produce sheltered lagoonal conditions over the center of the structure. Islands develop on many of these shoals, and the exposed sediments are subjected to diagenesis while in contact with fresh waters. Unexposed sands which occur on highs in more than 10 m of water are extensively lithified by marine cements.

The geometries (lobate or linear) described in the preceding sections are relatively simple and represent end members in a continuum of shapes of possible carbonate sand deposits. Undoubtedly, many carbonate sand accumulations do not fall into one of these described categories because they are unique, their geometry controlled by the peculiarities of geographic location. An example of such a shelf-edge sand accumulation is one located off the Yucatan Peninsula of Mexico (Harms et al, 1978; Fig. 23). Other sand bodies may be combinations of two or more of the end-member forms, or they may include large areas of inactive, stabilized sand.

SUMMARY OF FACTORS INFLUENCING SAND BODIES

Because bank-margins are the primary zones of energy absorption and because the duration and magnitude of physical energy can fluctuate as a result of changes in bank orientation and topography, important variations in sand-body size, geometry, and sediment type should be expected. The previous discussion has pointed out these variations.

The development of sand bodies requires, in the simplest view, a source of sand-size sediment and a mechanism to exclude or remove sediment smaller than sand-sized material. On shallow carbonate

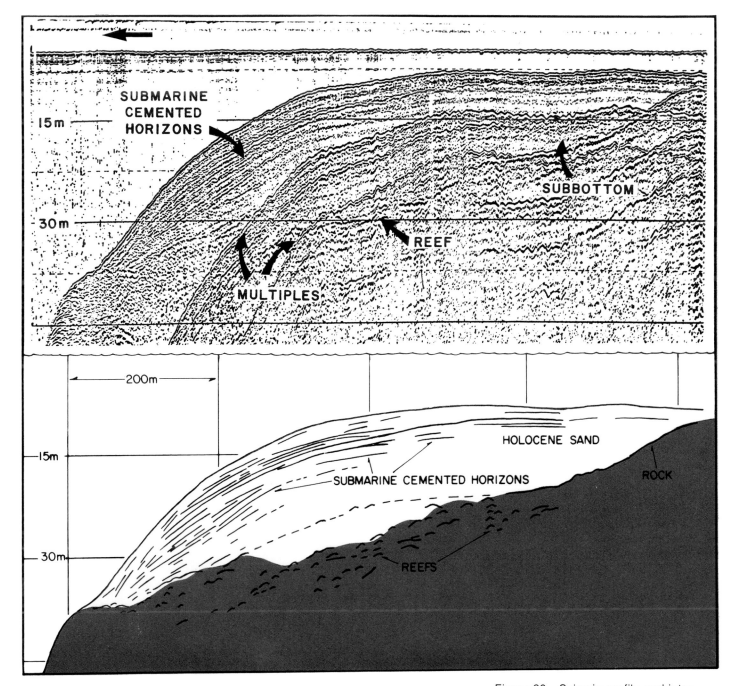

Figure 26—Seismic profile and interpretation along a leeward bank margin in the Bahamas. Numerous reflectors within the thick sand accumulation are the result of hardgrounds, thin laterally continuous lithified layers in the otherwise uncemented sands. The hardgrounds are formed by submarine cementation of carbonate sands.

banks, platforms, or shelves, the sand source is almost always nearby (autochthonous). The sand may be of biological or physicochemical origin. A change in the shelf slope may provide a focus for factors that promote carbonate sand production including wave impingement on the bottom, upwelling, strong tidal currents, and enhanced biotic activity. If the slope break faces open water that is more than 10 m deep, there exists enough wave energy commonly to remove particles finer than sand size. Given these two generalities — a sediment source and a mechanism for removing the fines — then sand bodies will develop near the shelf edge. Several other factors contribute to form sand bodies with differing attributes.

The predominant influences on sand production and the development of carbonate sand bodies are: (1) antecedent topography; (2) physical

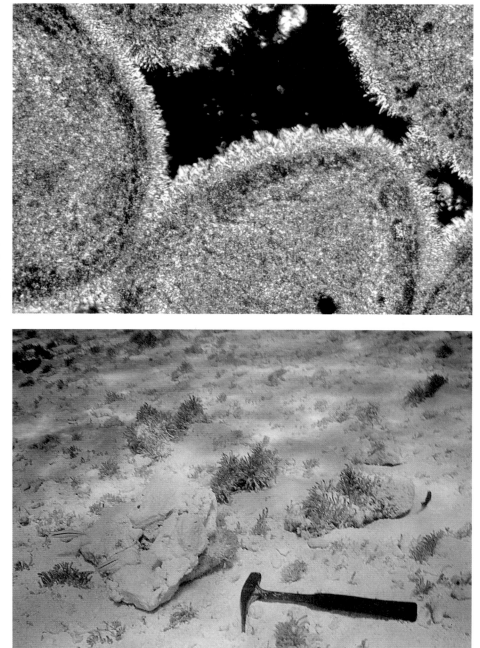

Figure 27—Photomicrograph of a thin fringe of aragonite cement forming a cemented ooid crust, from the Joulters ooid shoal on Great Bahama Bank. This habit of aragonite cement is termed fibrous or needle. Width of the photograph represents 1 mm.

Figure 28—Underwater photograph showing friable clasts and boulders from the Schooner Cays area of Great Bahama Bank. The clasts, cemented with fibrous aragonite marine cement, are common on the bottom tidal channels that lie between active sand shoals (see Fig. 7).

processes — tidal, wind, and wave-generated currents; (3) sea-level history; and (4) diagenesis. All the factors provide varying feedback effects on one another, resulting in a strongly integrated system of environmental controls. No one factor can be examined without considering the influence of the others.

Antecedent Topography

The positioning of a sand body along a bank, platform, or shelf margin may be the result of underlying topography that localized agitated bottom conditions. The antecedent topography may be in the form of submarine rock ridges, gaps between islands, reentrants within the bank-margin, gradient and elevation of the bank-margin zone, and sheltering effects of other sand bodies. For example, the Cat Cay ooid shoal on Great Bahama Bank (Fig. 11) has formed on submarine rock ridges that connect small Pleistocene islands along that margin. The rock ridges, after being flooded during sea-level rise, locally focused tidal flows on and off the bank into small areas. This in turn stimulated ooid production and the sand body began to develop. Another example is the tidal deltas produced at the mouths of channels between Pleistocene islands in the lower Florida Keys (Jindrich, 1969). The gaps or passes between the islands are obvious examples of antecedent topography controlling the

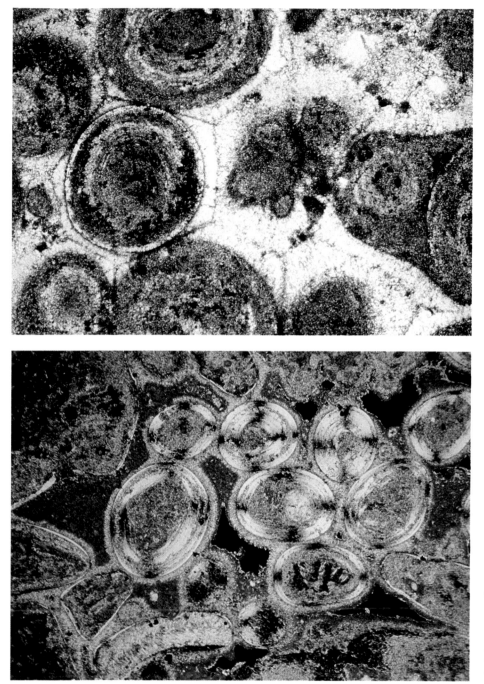

Figure 29—Photomicrograph of cemented ooid crust with all of the primary interparticle porosity filled with fibrous aragonite cement. Note the polygonal cement boundaries surrounding each ooid. Sample from the Yellow Bank area of Great Bahama Bank. Width of photograph represents 2.5 mm.

Figure 30—Photomicrograph of cemented oolite from the Schooner Cays, Great Bahama Bank. Micritic magnesium-calcite cement, artificially stained reddish-brown, fills most of the porosity. Note that it postdates an earlier generation of fibrous aragonite cement that initially cemented the rock. Width of the photograph represents 3 mm.

placement and development of the tidal deltas.

The location, size, or orientation of some sand shoals shows no obvious correlation with underlying preexisting topography, but important genetic relationships may still exist. Seismic profiling indicates that Lily Bank on Little Bahama Bank was not controlled by underlying rock structure. However, antecedent topography in the form of two broad topographic lows or reentrants within the outer bank-margin has played a key role in shoal development (Fig. 13). These high-energy windows along the windward margin provided the currents necessary to create an agitated bottom at the shoal location.

Whether the relation between bedrock topography and initial deposition is obvious or difficult to document, the effects may only be fleeting as sediment accumulates and buries the bedrock inrregularities. Apparently with the accumulation of sufficient sand, the sand body assumes its own characteristics, so the resultant size and orientation of Holocene examples are to a large degree products of the hydrodynamic regime and related depositional buildup.

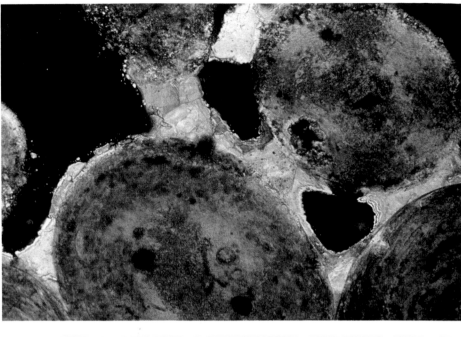

Figure 31—Photomicrograph of oolite from Joulters Cay, Great Bahama Bank, cemented with calcite which has precipitated from fresh-waters held by capillarity above the water table (vadose zone). Note the tendency for the cement to occur at grain contacts and to have curved surfaces that are concave toward the pore interiors. Width of photograph represents 1.5 mm.

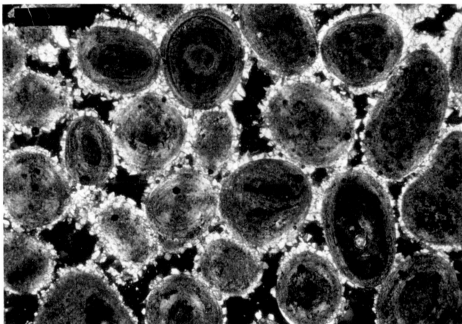

Figure 32—Photomicrograph of oolite from Joulters Cay, Great Bahama Bank, cemented below the fresh-water table (phreatic zone) with calcite. Cement occurs as rims of crystals that are uniformly distributed around each ooid. Note that the rims are not of even thickness (non-isopachous). Width of photograph represents 3 mm.

Physical Processes

The orientation of a bank-margin with respect to the dominant winds, waves, tides, and storms imparts a significant effect on sand body type. The duration, magnitude, and direction of water movement shape the sediment body and produce physical characteristics which may be preserved in the rock record.

Bank margins can be classified as: (1) windward, (2) leeward, or (3) tide-dominated. Back-reef sheets, belts, and lobes of skeletal sand form along open windward margins, where sand transport is generally toward the bank. Where small islands exist, oolitic tidal deltas occur opposite the inter-island gaps. Commonly the flood tidal delta is much larger as a result of storm-generated currents. If reentrants occur along the shallow margin, tidal and storm-generated flows can generate wide belts of tidal bars or marine sands. Along windward margins dominated by large islands, skeletal sands generated within the fore-reef environment may be carried seaward to the marginal escarpment. There they accumulate behind rocky barriers or are carried further seaward into deep water. Leeward, open margins are dominated by offbank sand transport and wide belts or sheets of non-skeletal sands form at the bank edge.

Figure 33—Surface photograph at low tide of extensively rippled bottom of Joulters ooid shoal on Great Bahama Bank.

Figure 34—Surface photograph showing detail of ripples. A pattern of nearly parallel ripples has been modified by currents during the falling tide. The fecal mound in the center is nearly 3 cm high.

At the ends of embayments, where tidal currents are accelerated, large tidal-bar belts characteristically form.

Sea-Level Changes

Sea-level changes can significantly control the location and development of carbonate sand bodies because the water depth over the bank controls the level of physical energy imported to an area. Rapid changes in sea level may isolate or even drown sand bodies on different parts of the shelf, while slow changes may produce widespread sand sheets. For example, Pleistocene oolites deposited about 20 thousand years ago occur at the shelf edge of the eastern Florida peninsula, now about 63 m below sea level. During a sea level high stand, about 130 thousand years ago, oolitic deposits of the Miami Oolite were formed 16 to 24 km bankward from the shelf edge. These distinctly separate sand bodies, deposited during a very brief span of geologic time, owe their origins to a rapidly fluctuating sea level. In contrast, slow migration of a Devonian crinoidal sand shoal during a relative sea level change produced the widespread encrinites of the Becraft Limestone in New York State (Laporte, 1969).

The geologic histories of both Lily Bank on Little Bahamas Bank (Hine, 1977) and the Joulters ooid shoal on Great Bahama Bank (Harris, 1979) are tied very closely to the Holocene rise in sea level. In both cases, a wide

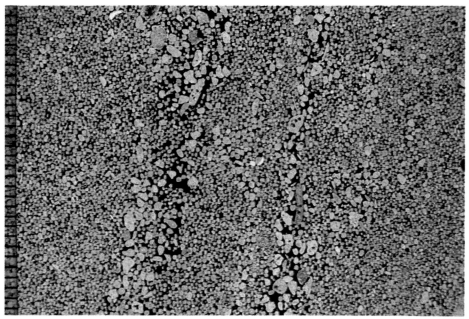

Figure 35—Plastic-impregnated surface of a box core, approximately 15 cm across, illustrating the small-scale, ripple cross-bedding that characterizes active portions of sand shoals.

Figure 36—Photomicrograph of clean sands from a core taken on the active sand shoal shown in Figure 33. The extremely well-sorted, medium sand-size ooids are in laminations that dip slightly to the right. Thin beds of loosely-packed, coarse grains (algae, foraminifera, and aggregate grains) also dip to the right. The inclined laminae are defined clearly where there are variations in grain size; they are more subtle in a well-sorted ooid sand (for instance in the bottom one-third of the photograph). The scale at the bottom is in millimeters.

active sand sheet consisting both of parallel-to-flow bed forms and large sand waves was developed soon after sea level had flooded the bank top with a few meters of water. Parts of these sand bodies became relict and stabilized by vegetation as they were not able to build vertically as quickly as the rising sea. Portions did remain active, though. Lily Bank today is a mobile sand belt that has been reduced considerably in width; in the Joulters example, the active shoal has shed considerable sand bankward to form a large sand flat and has also built above sea level along much of its length (Fig. 24).

Diagenesis

Several diagenetic processes may greatly alter the character of carbonate sands while they are still essentially within the depositional environment. These processes, occurring before or shortly after final deposition, while the sands are at or within a meter of the original depositional surface, fall into categories of grain alteration and cementation.

Grain alteration begins essentially as the grains are produced and is a

continuing but sporadic process while sands accumulate. Alteration of grains by repeated microboring and infilling of microbores can completely destroy original grain fabrics (Purdy, 1968; Harris et al, 1979). Often such alteration is confined to the grain edge and produces micrite envelopes (Bathurst, 1966). Micrite rims can also form by the addition of fine-grained material to grain surfaces (Kobluk, 1977). These material additions are usually aragonite or Mg-calcite. The added material may bridge the space between grains, resulting in grain-to-grain cementation (Dravis, 1979; Fig. 25).

Unlike their clastic counterparts, carbonate bank-margin sand bodies may be subjected to syndepositional submarine cementation. These cements result in the formation of clasts, hardgrounds, and "tepee"

structures in shallow marine water (Shinn, 1969). Although the details of inorganic marine cementation in carbonates are unclear, the general processes which enhance cementation are: (1) good water circulation (shallow-marine waters overlying carbonate sediments are typically supersaturated with respect to the major carbonate minerals); (2) high sediment permeability; (3) slow sedimentation rates; and (4) hypersalinity.

Marine cementation may exert a significant influence on the subsequent development of the sand body. Submarine hardgrounds that commonly develop on the flanks of sand bars and lithified layers that form on beaches combat erosion and play a role in maintenance of steep slopes (Fig. 26). These hardened bottoms also aid in the establishment of a benthic community that could not survive on an unstable bottom.

Bliefnick (1980) has documented the growth of sponge and bryozoan buildups on marine-cemented sea bottoms of the Joulters ooid shoal on Great Bahama Bank. To the north of the shoal, the open bank margin has changed through time from an environment of active tidal bars to one of an essentially stabilized bottom with the widespread occurrence of lithified layers and patch reefs. A similar scenario may explain the

Figure 37—Surface photograph at extremely low tide exposing spill-over lobes on the backside of an active ooid shoal at Joulter Cays on Great Bahama Bank. Clean, crossbedded ooid sands are covering bioturbated, muddy ooid sands that have been partially stabilized by marine grasses and algal mat.

Figure 38—Aerial photograph of sand flat lying west of the Joulters Cays on Great Bahama Bank. The mounds, produced primarily by burrowing shrimp and crabs, are spaced less than 1 m apart. Photograph taken from about 70 m altitude.

association of ooid shoals and certain organisms reported by various authors such as: stromatolitic bioherms associated with late Precambrian oolite of the Canadian Archipelago (Young and Long, 1977); archaeocyathid-*Renalcis* bioherms interbedded with Lower Cambrian oolites of the Poleta Formation of western Nevada (Rowland, 1978); and *Epiphyton* boundstones with Cambrian and Ordovician oolite shoals of the Central Appalachians (Pfeil and Read, 1980).

Physicochemical marine cements commonly take the form of fibrous aragonite or micritic Mg-calcite. Narrow fringes of aragonite-needle cements are sufficient to lithify carbonate sand into friable clasts and hardgrounds (Figs. 27, 28). Fibrous, isopachous cements may form polygonal cement sutures around grains (Shinn, 1975; Fig. 29) resulting from competitive cement growth into pore spaces. Micritic Mg-calcite cement can occur as the sole cement in a rock or it can be associated with aragonite cement, often as a second stage cement after the aragonite (Fig. 30). Mg-calcite cement may also occur in a fibrous or bladed habit, but this form does not appear to be common in Holocene sands. Continued cement growth can greatly reduce porosity and permeability and destroy

Figure 39—Surface photograph at low tide showing details of burrow mounds. The white mounds of pelleted sand are surrounded by grass- and algal-covered bottom. The mounds have been reworked by the tidal currents, giving them a flattened appearance.

Figure 40—Photomicrograph of muddy sands from a core taken on the sand flat shown in Figure 39. Ooid sands are bioturbated and mixed with mud and other carbonate sand grains. The mud, appearing a darker gray, has an irregular distribution within the sediment due to the burrowing. Width represents about 30 mm.

the reservoir property of the sands very early in the history of the rocks.

Fresh-water cements may also significantly affect the development of sand shoals. Islands and dune ridges are rapidly cemented in Bahamian ooid shoals to form erosion-resistant features. The islands initiate growth of beaches on their seaward sides. Also, they may significantly change current patterns and depositional environments by forming rigid barriers and creating energy shadows

in their lee. The islands also serve as catchment areas for fresh waters, and mixing zones form around the islands where the fresh water and marine water intermix.

Although investigations are continuing, it is believed that calcite cementation, dissolution, and possibly dolomitization characterize the diagenesis associated with island formation. Islands act as traps for ground water that differs significantly in chemical composition from the

marine waters surrounding the island. Often, as in the case of Bermuda (Plummer et al, 1976), clearly recognizable hydrologic zones develop beneath an island. These include: (1) the vadose zone (that portion of the island above the standing water table): (2) the phreatic zone (those sediments saturated with ground water below the water table); and (3) a mixing zone (the zone of mixed ground water and sea water). The waters in these zones and the

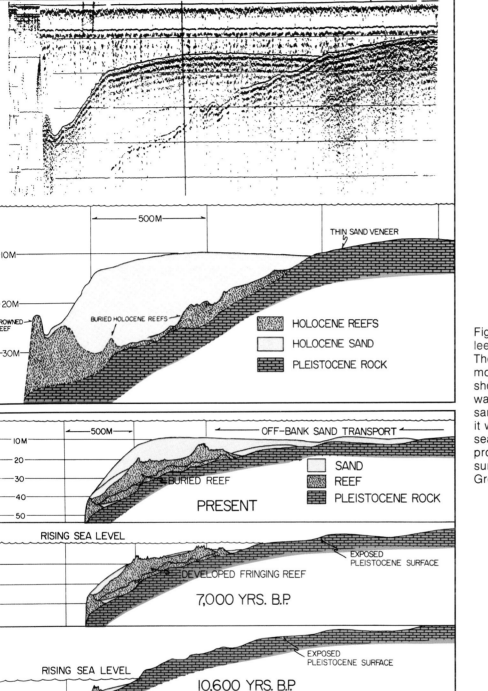

Figure 41—Seismic profile from the leeward margin of Little Bahama Bank. The interpretation of the profile and the model for its development through time show that an early Holocene coral reef was buried by later Holocene sands. The sands formed on the platform only after it was substantially flooded during rising sea level. A similar scenario has been proposed for portions of the platform surrounding Tongue of the Ocean on Great Bahama Bank (Palmer, 1979).

geometry of the zones are controlled by sediment properties and hydrologic conditions which are beyond the scope of the present paper. However, illustration of some of the early fresh-water diagenetic products will serve as analogues to some carbonate reservoir rocks.

Carbonate sands within fresh ground water commonly develop meniscus cements above the water table (Dunham, 1971; Fig. 31). These cements form where water is held by capillary forces between sand grains.

Below the water table, cement crystals grow freely in pore waters between sand grains (Fig. 32). These cements result from the dissolution of aragonite grains and the reprecipitation of calcite as cement. The process results, in part, from the greater

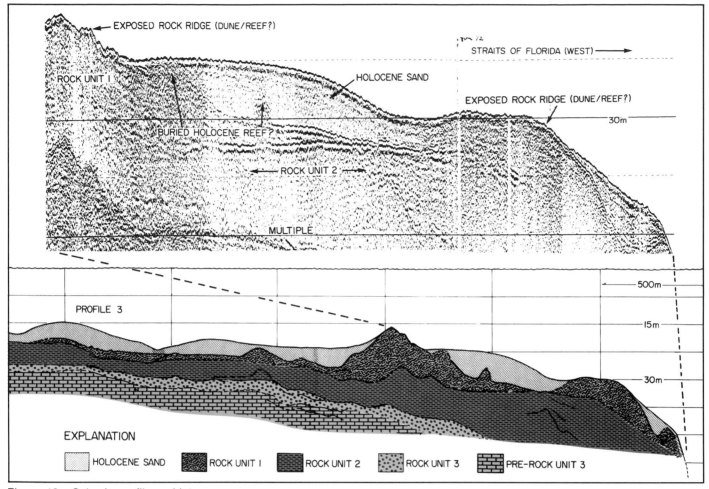

Figure 42—Seismic profile and interpretation from leeward bank margin of Little Bahama Bank. The Holocene accretion overlying Pleistocene limestone "basement" is a complex package of bank-margin sands and reefs, with both unlithified and lithified carbonate sands.

solubility of aragonite than calcite in fresh water (Land, 1967). Because secondary porosity is created as aragonite dissolves and as cement precipitates in pores, little overall porosity change takes place in the sediment. However, a significant permeability decrease results from this early cementation (Halley and Harris, 1979).

Mixing zones may localize dolomitization (Land, 1973) or calcite dissolution and precipitation (Runnels, 1969). The mixing may take place early or late in the history of the sand body (Hanshaw et al, 1971). Of course, the mixing zone is not the only possible mechanism for dolomitization. It is also possible that hypersaline ground waters dolomitize sediment while sinking through a sandy body (Adams and Rhodes, 1960), but this process has not been well documented in the Holocene. Hsu and Schneider (1973) have shown that the dolomitization found along the southern shores of the Persian Gulf occurs when sea water is drawn landward due to extremely high evaporation rates on the tidal flats.

PRESERVATION OF MODERN SAND BODIES

The geologic record of modern sand bodies provides us with sedimentary criteria that form the basis for the environmental reconstruction of outcrop and subsurface core data. Recent studies in the Bahamas by Buchanan (1970), Hine (1977), Harris (1979), and Palmer (1979) have added greatly to our understanding of the facies, sedimentary structures, and geometry of oolitic deposits and associated sediments.

Surface Sediments

Early studies of surface sediments documented the variations of organisms, surface structures, grain types, grain size, and amount of mud that typify bank-margins dominated by sand shoals. The variations are complex and somewhat variable between different localities, but general trends can be recognized. For example, the crest of a shoal itself is usually devoid of organisms and biologically produced structures, as physical sedimentary structures predominate in the clean, well-sorted ooid sands (Figs. 33 to 36). Bioturbation and burrows are more common seaward and bankward of the shoal,

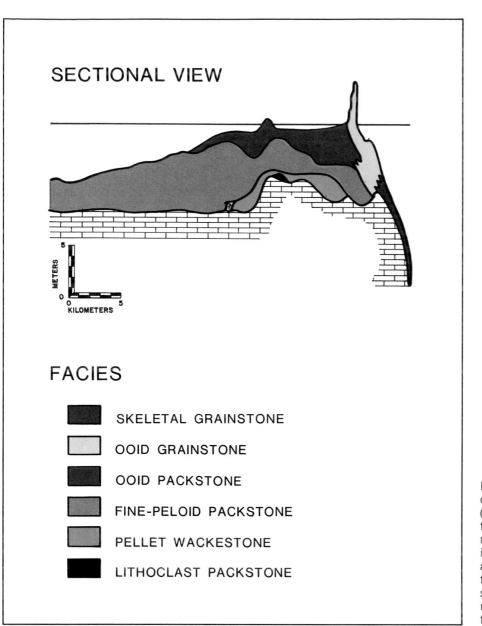

SECTIONAL VIEW

FACIES

■ SKELETAL GRAINSTONE

□ OOID GRAINSTONE

■ OOID PACKSTONE

■ FINE-PELOID PACKSTONE

■ PELLET WACKESTONE

■ LITHOCLAST PACKSTONE

Figure 43—Cross section through the ooid shoal that is pictured in Figure 24 (Harris, 1979). Beginning on the right, the section extends from the bank-margin for about 35 km into the bank-interior. The actual buildup of the shoal above the surrounding sea bottom is due to the accumulation of ooid sands in several facies. The facies relationships reflect environmental changes through time (from P.M. Harris, 1979).

where ooid sands are mixed with other grain types and mud (Figs. 37 to 40).

Subsurface Facies Relations

Seismic profiling, along with sediment coring and probing, have shown that modern sand bodies in the Bahamas are less than 20 m thick and commonly less than 10 m. The sand bodies have built either landward or seaward through time, depending on the predominant direction of sediment transport. The complexity of platform margins has been graphically shown by Hine and Neumann

(1977), and Palmer (1979), who presented seismic and core data along a leeward margin showing early Holocene coral reefs that have been subsequently buried by thick accumulations of carbonate sands during subsequent offbank transport later in the Holocene (Fig. 41). This change through time is largely a result of the flooding of a greater portion of the carbonate platform during a rising sea level. Continued fluctuation of sea level might result in several accretions to the margin (both vertical and lateral growth; Fig. 42).

The mosaic of facies that can be

recognized in studies of surface sediments forms an even more complicated pattern when viewed in three dimensions; the complexities arise from sediment variation that is the result of topographic irregularity, localized depocenters, and, most importantly, changing depositional patterns through time. Bahamian ooid sands originate in several settings — primarily in lobe-shaped or elongate submarine bars, but also in tidal deltas localized in passes between islands. The sands may also accumulate in a variety of subenvironments associated with these sites

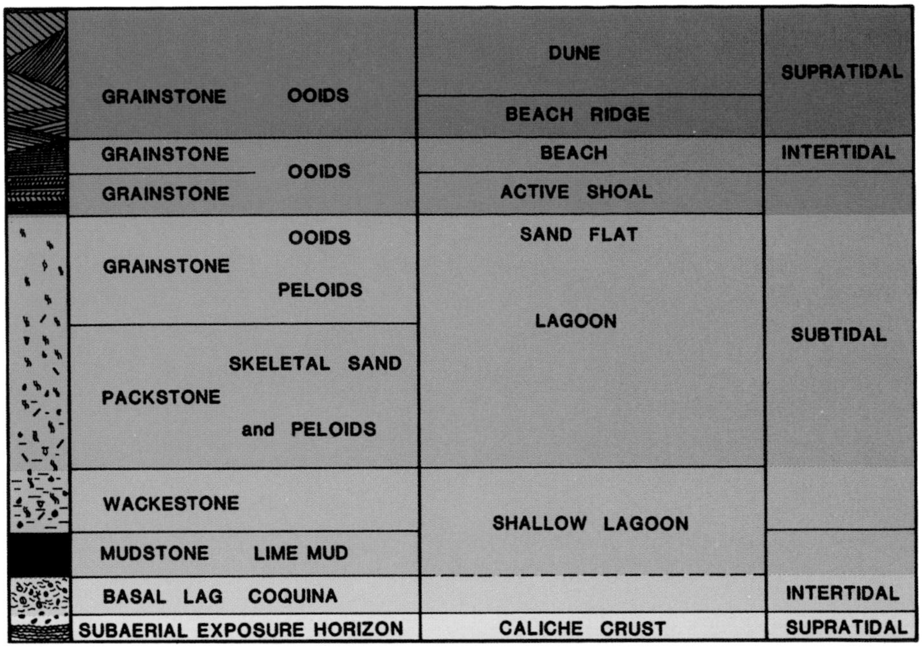

	GRAINSTONE	OOIDS	DUNE	SUPRATIDAL
			BEACH RIDGE	
	GRAINSTONE	OOIDS	BEACH	INTERTIDAL
	GRAINSTONE		ACTIVE SHOAL	
	GRAINSTONE	OOIDS	SAND FLAT	SUBTIDAL
		PELOIDS		
	PACKSTONE	SKELETAL SAND	LAGOON	
		and PELOIDS		
	WACKESTONE		SHALLOW LAGOON	
	MUDSTONE	LIME MUD		
	BASAL LAG	COQUINA	- - - - - -	INTERTIDAL
	SUBAERIAL EXPOSURE HORIZON		CALICHE CRUST	SUPRATIDAL

Figure 44—Composite stratigraphic column showing the lithologies that compose the shallowing-upward depositional sequence that has been recognized in Pleistocene limestones of the Bahamas (Pierson, 1980). The environmental interpretations of each lithology are based on studies of their Holocene counterparts.

Figure 45—Outcrop photograph of Pleistocene Miami Oolite exposed in Florida. The cross-bedded ooid sand was deposited as a marine bar like that shown in Figure 33. Note the excellent preservation of the cross-bedding as a result of selective cementation and subsequent weathering.

of formation. An ooid shoal is comprised of a succession of these environments produced through time during changing sea level.

Changes in sedimentation through time have best been documented by coring on Great Bahama Bank north of Andros Islands. Buchanan (1970) found that deposition in the Frazers Hog Cay area was initially influenced by an elevated rim of Pleistocene limestone along the seaward edge of the platform which effectively restricted circulation during early flooding of the platform. Lime muds accumulated in the lagoon behind the

ridge. As the ridge was flooded by the rising sea, lime sands formed. Ooid shoals and skeletal sands occur at the platform margin while pellet, organic aggregate, and grapestone sands are found bankward. The scenario is similar for the Joulters ooid shoal located nearby.

The Joulters shoal today is not a tidal bar or marine sand belt, but rather a vast sand flat fringed on the ocean-facing borders by mobile sands where ooid formation occurs. Harris (1979) has documented the geologic history of the shoal, showing one possible evolutionary path of the bar

and channel pattern seen elsewhere. The Joulters example is important because it shows that the shoal-generating physiography can be erased — the bar-and-channel topography has been extinguished and filled in, and clean ooid sands have been mixed with other sediments by burrowing to form a sand flat.

The relief of the Joulters ooid shoal over the surrounding sea floor is the result of the accumulation of ooid sands in several facies. A cross section (Fig. 43) reveals two basic parts to the shoal: (1) a narrow fringe of ooid grainstone, and (2) a more

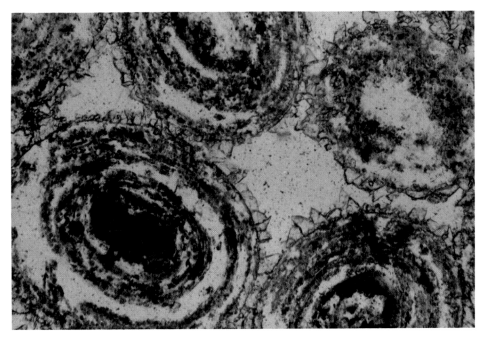

Figure 46—Photomicrograph of Pleistocene oolite from the outcrop shown in Figure 45. The ooids are cemented with rims of calcite crystals interpreted to have been cemented in a fresh-water phreatic environment. The ooids have undergone significant dissolution, so only remnants of the original aragonite remain. Width of photograph represents 1.5 mm.

widespread sequence comprised of lithoclast packstones and/or pellet wackestone at the base, peloidal packstone in the middle, and ooid packstone at the top. The ooid packstone forms a wedge that thickens in a seaward direction, while the peloidal packstone is the thicker part of an intraplatform sheet that thins in a seaward direction.

Shallowing-upward Sequence

The sequence of deposits recognized in cores from the Joulters ooid shoal and other Holocene examples, as well as in cores of Pleistocene limestones in the Bahamas, is one in which each succeeding unit is deposited in progressively shallow-water. This is largely because shallow-water sites provide optimum conditions for the physicochemical and biological production of carbonates, so platform carbonate sediments usually accumulate at greater rates than the rate of relative subsidence and repeatedly build up to sea level and above. This shallowing-upward sequence, the facies of which can be analyzed and related to an existing depositional environment in the Holocene, may be repeated numerous times in a thick succession of ancient platform carbonates and provides a valuable clue to their interpretation (Fig. 44).

Figure 47—Outcrop photograph of Pleistocene oolite from New Providence Island, Bahamas. The sedimentary structures indicate a change of depositional environment toward the top of the sequence. Complex cross-bedding of an active marine bar is dramatically displayed throughout most of the section; but toward the top, flat-lying laminae record the localized buildup to beach conditions.

The sequence of facies documented in the Holocene of the Bahamas is formed on a weathered surface of Pleistocene limestone. The surface is typically pitted and coated with a calcareous crust which forms a recognizable regional discontinuity surface that resulted from subaerial exposure during a sea level low. The Holocene shallowing-upward sequence results from depositional buildup during the subsequent flooding of this surface.

The initial deposits of the sequence are a transgressive basal lag of coarse coquina or lithoclast packstone. As increasing water depth creates a subtidal lagoonal environment, pelleted lime mudstone and wackestone are deposited. With improved circulation and more varied biota, the sediments gradually change to bioturbated wackestones, locally packstones, that may be rich in foraminifera and mollusk shells. Initiation of sufficient tidal flow across the platform may cause the localization of ooid formation on agitated bottoms. Ooid sands may be mixed with the lagoonal sediments that surround a shoal. If a shoal overlies the lagoonal deposits due to a transgression or progradation, the vertical sequence grades into laminated intertidal ooid grainstones. Burrowed, shallow-subtidal ooid sands may be present as well, the

Figure 48—Outcrop photograph showing closer view of cross-bedded marine deposits of Figure 54. Note the cut-and-fill structure of various cross-bed sets. The vertical holes are preserved burrows of marine worms and anemones. The scale bar is in centimeters.

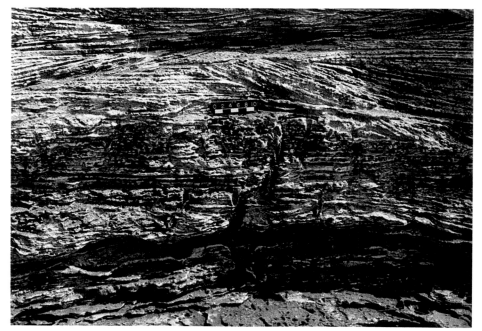

Figure 49—Outcrop photograph showing closer view of planar-bedded beach deposits of Figure 54. Note that there are gradual pinch-outs of the bed sets in both directions. A prominent set of climbing ripples is present immediately above the scale. The scale bar is in centimeters.

result of sand flat deposition. When the supply of sediment is large, the sequence may be capped by bedded supratidal ooid grainstone of beach ridge or dune deposition.

In the Bahamian model of deposition during the flooding of a carbonate platform, the sediment surface builds up to and exceeds sea level. The upward change of facies, forming in depositional environments that were superimposed through time, is a

shallowing-upward sequence. Clues to interpreting the water depth of such deposits lie in the grain types, sedimentary structures, depositional texture (percentage of mud), and diagenesis.

ANCIENT SAND BODIES

Sand body geometry and size, paleogeographic location and orientation, and internal structure are key

criteria to use in the interpretation of carbonate sand bodies. Usually, they cannot be determined from subsurface studies until late in the exploration of an area. As a result, the preliminary environmental interpretations are often based on grain type only. For instance, oolites may become ooid shoals, encrinites become crinoid banks, and caprinid sands become rudist thickets. Such interpretations are probably best withheld until more data are accumulated, but the interpretations may serve to inform initial exploration. Subsurface sand bodies, when thoroughly explored, may reveal the geometries of individual sand bars or back reef sands. For example, bar morphology has been documented for several sand bodies — the Mississippian McClosky sands, an informal unit, of the Illinois Basin by Carr (1973) and the Jurassic Smackover Formation of the Gulf Coast by Erwin and others (1979).

Pleistocene
Outcropping carbonates provide the most readily accessible and easily identifiable examples of ancient carbonate sand bodies. Pleistocene outcrops often retain much of their geometry left relict during exposure associated with sea level fall. In the

Figure 50—Outcrop photograph showing another view of the cross-bedded marine deposits of Figure 54. The layer of reworked limestone blocks (intraclasts of marine-cemented ooid sands) is typical of lithification that occurs on the flanks of marine bars and on beaches.

Figure 51—Stratigraphic cross section of the Upper Jurassic Smackover Formation and Buckner Member of the Haynesville Formation in Arkansas and Louisiana. Note that the Reynolds oolite, a thick accumulation of oil-bearing high-energy carbonate sands, changes updip to aggregates (grapestones) and carbonate mud. Shales and anhydrites of the Buckner Member act as seals for Smackover reservoirs.

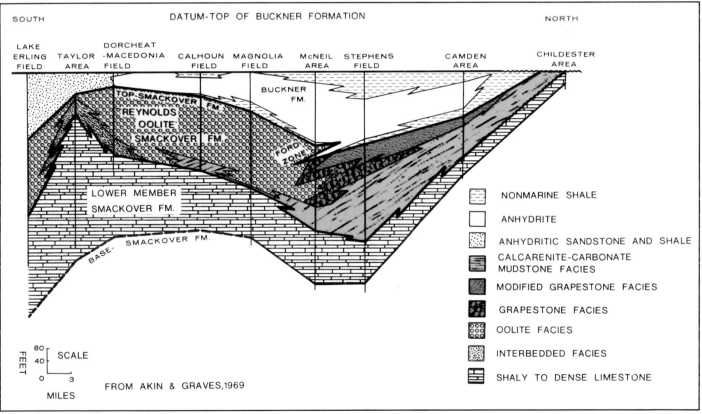

case of the Pleistocene Miami Oolite of south Florida, the geometry of the marine deposits is quite similar to that of analogous modern sand bodies (Hoffmeister et al, 1967; Halley et al, 1977).

The Pleistocene of south Florida is divisible into five marine units (termed from oldest to youngest Quaternary 1 through Quaternary 5), which are separated by regional discontinuity surfaces (Perkins, 1977). Paleotopography developed on the upper surface of the Q4 unit created a gently-dipping platform on which Q5 sediments were deposited. During deposition of Q5, tidal-bar deposits of the Miami Oolite formed along much of the eastern coast of south Florida. The sediments are ooid grainstones and packstones, locally

Figure 52—Photograph of core slab from the Upper Jurassic Smackover Formation in Mt. Vernon Field, Columbia County, Arkansas. Porous ooid grainstones have distinct centimeter-thick layering. The larger grains scattered within the oolite are algal balls. Porosity is 12% and permeability is 20 md.

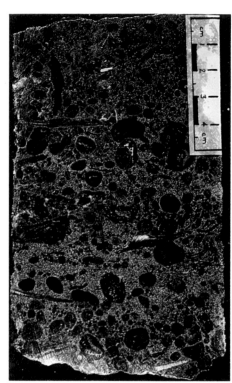

Figure 53—Photograph of core slab from the Upper Jurassic Smackover Formation in Walker Creek Field, Columbia County, Arkansas. These porous grainstones are a mixture of ooids, algal balls (both oncolites and rhodolites) and large pelecypod fragments. Porosity is 8% and permeability is 10 md.

Figure 54—Photograph of core slab showing Upper Jurassic Smackover oolite from Mt. Vernon Field, Columbia County, Arkansas. The distinctly cross-laminated ooid grainstone has significant oomoldic porosity.

arenaceous and often mixed with pellets. Extensively-burrowed pelletal sands accumulated in sheltered water behind the oolite bars. On outcrop, fresh-water cementation and dissolution have greatly enhanced recognition of the sedimentary structures in the Miami Oolite (Figs. 45, 46). Large cross bed sets typify the marine bar deposits.

Pleistocene outcrops in the Bahamas broaden our perspective of preserved depositional environments. Marine bars, beaches, and subaerial dunes are exposed on New Providence Island (Garrett, unpub. data). Ooids occur in ridges or irregular mounds that in outcrop may show a gradation upward from accretion and cross-rippled bar deposits to laminated beach deposits containing displaced beachrock blocks (Figs. 47 to 50). The dunes are composed of marine sands piled up from the beaches by wind and storms. They

are formed of coalescing sand lobes that are asymmetric toward the platform interior. Sedimentary structures consist of distinctive backset beds with low-angle bedding that change in a bankward direction through convex-upward spillovers to steeper foreset beds. The foresets change laterally into bioturbated sands, which were probably located in the lee of the developing dune.

Jurassic of the Gulf Coast

Shallow-water carbonates of the Upper Jurassic (Oxfordian) Smackover Formation of the Gulf Coast are one of the best documented ancient examples of hydrocarbon reservoirs in carbonate sand bodies. Imlay (1943), Dickinson (1968), Akin and Graves (1969), and Badon (1974) described the stratigraphy of the Gulf Coast Jurassic and summarized the environments and lithologies of the Smackover. Detailed field studies are

presented by Bishop (1968, 1971), Ottmann and others (1973), Chimene (1976), and Becher and Moore (1976).

The Smackover contains promising exploration targets and proven production (Newkirk, 1971). Structural traps related to salt movement have long been exploration targets, but now major objectives lie in more subtle stratigraphic traps. These may show little or no closure and may escape detection on seismic profiles. As a result, understanding the geology of modern carbonate sand bodies can play an important part in the exploration thinking for the Smackover trend.

The Smackover deposits in southern Arkansas and northern Louisiana were deposited in environments ranging from nearshore to basinal (Fig. 51). Quartz sands and fine-grained carbonates mark the up-dip limits of the Smackover; basinal limestones far downdip contain shales

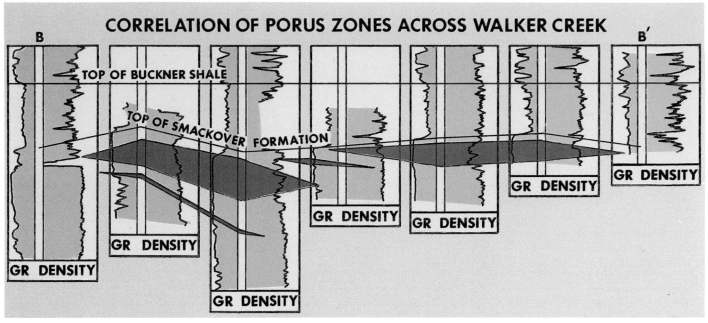

Figure 55—Density log cross section, Walker Creek Field. Orange areas have porosity greater than 4% and are continued to discrete horizons within the field (from Becher and Moore, 1976).

Figure 56—Inferred distribution of grainstones and mudstone resulting in porosity zonation within the Smackover Formation as illustrated in Figure 55 (from Moore, 1980).

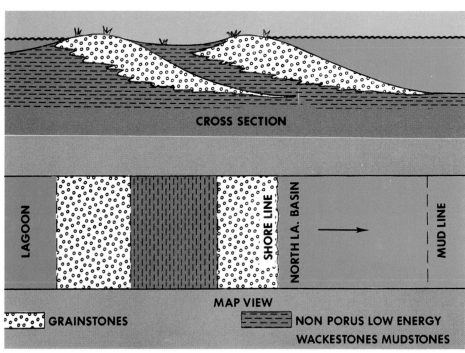

and sands. These basinal deposits extend to the north where they are overlain by a belt of shallower-water, high-energy carbonates that extends across southern Arkansas. The shelf deposits, locally termed the Reynolds oolite, are primarily ooid grainstones that were deposited in tidal bars and channels, beaches, and islands (Figs. 52 to 54). Porosity variations exist due to the presence of carbonate mud, and as a result of pore-filling cementation in clean grainstones.

Because the belt of high-energy shelf carbonates of the Smackover and the updip tidal flat equivalents of the Buckner Member of the Haynesville Formation prograded to the south through time, the position and continuity of reservoir-bearing grainstone bodies in the upper part of the Smackover are complex. In Walker Creek Field of southern Arkansas, pressure gradients measured on several wells define at least three separate reservoirs (Chimene, 1976). Several of these

porous zones have been delineated by Becher and Moore (1976; Fig. 55) and corresponding depositional environments inferred by Becher and Moore are illustrated in Figure 56. A north-south log correlation across Oaks Field, located in northern Louisiana to the southeast of Walker Creek Field, shows that the upper porous Smackover is made up of porous and tight zones (Fig. 57). The reservoirs, distinct lithic units stacked

on top of and to the south of one another, thin quickly downdip and pinchout updip within Buckner anhydritic shales (Erwin et al, 1979).

Understanding the sequence and timing of diagenesis is as important as understanding depositional environment relationships in predicting porosity in the Smackover Formation and many other ancient carbonate sands. The Smackover exhibits "early" cements that, based on their

Figure 57—Stratigraphic cross section of the Smackover Formation in northern Louisiana, based on log correlations and detailed petrographic observation. Three separated porosity zones each pinch out downdip within the Smackover and updip into the overlying tidal flat deposits of the Buckner Member of the Haynesville Formation.

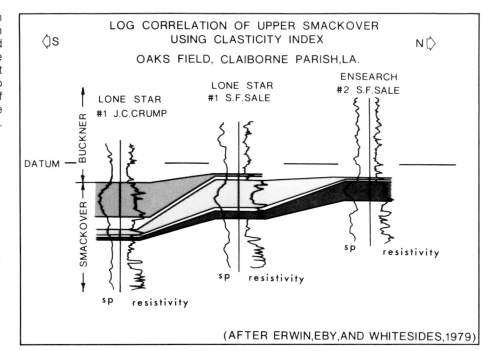

Figure 58—Photomicrograph of Upper Jurassic Smackover oolite showing oomoldic porosity (in white) and significant interparticle calcite cementation. The grains of the original sediment have been dissolved to form pore space and the porosity of the original sediment has been filled with cement. The resultant porosity of 30% is of reservoir quality, despite permeability of 1 md or less. Width of photograph represents 3 mm.

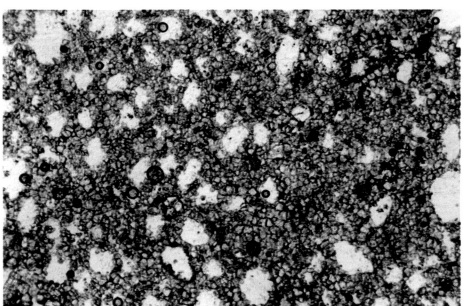

crystal morphology and distribution within the rock, resemble recent fresh-water cements (Harris, 1980). These blocky calcite cements may cause a significant reduction in primary intergranular porosity and locally produced seals within the upper part of the Smackover. Zones of oomoldic porosity may have also formed in these same fresh-water environments; the characteristically high porosities (as high as 35%) often have low associated permeabilities of only a few millidarcies (Fig. 58).

Although not as common in Smackover grainstones as fresh-water cements, cements that probably formed in the marine phreatic zone are present (see Fig. 59). Gypsum and dolomite also formed in shallow brine ponds and tidal flats where dissolved salts were concentrated by evaporation of sea water. The gypsum was converted to anhydrite during burial; the anhydrite may contain relics of ghosts of replaced grains and often pseudomorphs after gypsum.

In ancient carbonates, some of the primary porosity that was not destroyed by syndepositional fresh-water and marine cements is further decreased by "late" diagenesis. Late diagenetic processes are topics of active research and little agreement among carbonate petrologists. Even what constitutes "late" is a source of argument. Generally, late is meant to include diagenesis which occurs at depth and at elevated temperatures, that is in the mesogenetic zone of Choquette and Pray (1970). But some late processes, pressure solution for example, probably start at shallow depths and proceed continually or sporadically as burial continues. Diagenesis occurring during the burial of carbonate sands includes physical compaction, chemical compaction, cementation, and dissolution. Important variables involved in late diagenetic alteration are the vertical and lateral continuity of facies and diagenetic packages, the burial history, the nature of the pore fluids, early diagenetic history, geothermal history, and fluids from associated noncarbonate rocks including organic-rich sediments.

Perhaps one of the most significant late diagenetic events is the develop-

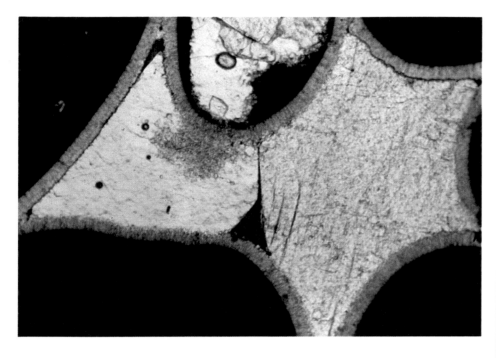

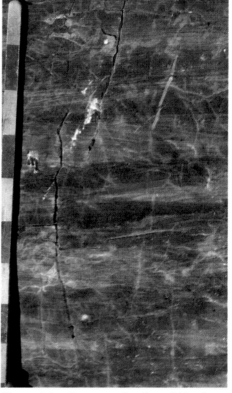

Figure 59—Photomicrograph of Upper Jurassic Smackover oolite with porosity filled by two stages of cementation — an early continuous, isopachous rim of marine cement, followed by a pore-filling coarse blocky cement. The remaining porosity is shown in blue. Width of photograph represents approximately 1.5 mm.

ment of secondary porosity, which has been documented recently in the Smackover Formation by Moore and Druckman (1980). Although incompletely understood, this process results in the dissolution of carbonate grains as well as early and late carbonate cements. The resulting vuggy porosity adds to the remnant interparticle porosity and results in ooid grainstones with enhanced reservoir properties.

Variations Among Ancient Examples

Although sand bodies share many of the attributes of the modern and Jurassic examples described previously, sand accumulations of differing age and geographic location present almost endless variety of internal detail. For example, cross-bedding typical of many sand accumulations may be absent if the original sediment was homogenized by burrowing organisms. Extremely well-sorted sand may not contain sufficient grain-size variation to produce prominent cross-bedding. Cementation may either enhance or disguise primary sedimentary features.

Subtle color variations may be produced by slight variations in grain size, grain composition, and cementation style (Figs. 60, 61). Other examples may retain little small-scale

evidence of cross-bedding, but large-scale evidence of sand body deposition may be preserved as bed forms in outcrop (Figs. 62, 63). Bedding features may survive extensive dolomitization, although such dolomitization may obliterate original grain textures (Figs. 64, 65).

Carbonate sand bodies occur in rocks of almost every geologic period, but are particularly abundant (and prolific hydrocarbon reservoirs) in rocks of Mississippian, Permian, Jurassic, and Cretaceous age. The range of geologic time over which sand bodies form results in compositional differences simply from the evolutionary changes which occur within the groups of organisms producing carbonate sand. For example, crinoidal sands are characteristic of the Mississippian, and molluscan or foraminiferal sands are typical of many Cretaceous sand bodies (Figs. 66 to 68); the reverse occurrences, Mississippian molluscan sands and Cretaceous crinoidal or foraminiferal sands, are extremely rare. Such evolutionary faunal changes and their effects on carbonate sediments are summarized by Wilson (1975).

Secondary porosity in carbonate sand bodies is common, as exemplified by Figure 58, but primary porosity is also often preserved at

Figure 60—Outcrop of oolite within the Cambrian Carrara Formation of Nevada. Rock exhibits small scale cross-bedding enhanced by color variations due to variable grain size and composition. Scale is in 10 cm intervals.

considerable depths (Figs. 69, 70). Often evidence of early, meteoric-water diagenesis is apparent in these porous rocks (Figs. 71, 72), and early diagenesis is frequently thought to be a factor in porosity preservation (Becher and Moore, 1976).

Despite the compositional and diagenetic variations among sand bodies of differing ages, their recognition is relatively straightforward bas-

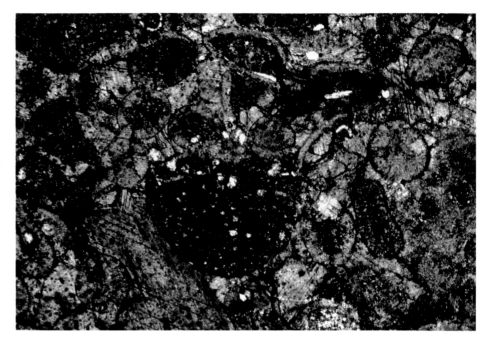

Figure 61—Photomicrograph of Cambrian ooid and skeletal grainstone from the outcrop shown in Figure 60. The thin section is artificially stained to reveal that the interparticle porosity is filled with one generation of iron-rich cement (blue). The grains are not stained blue because they are iron-poor. Width of photograph represents 3 mm.

Figure 62—Outcrop photograph of Mississippian oolite exposed in Kentucky. Individual bed sets are easily recognized, although the detailed cross-bedding has been obscured by thorough cementation and weathering. The outcrop is approximately 5 m high.

topographic relief on the structural feature. The predominant influences on sand production and the development of carbonate sand bodies are: (1) antecedent topography; (2) physical processes — tidal, wind, and wave-generated currents; (3) sea-level history; and (4) diagenesis. These influences do not act independently, but rather are interrelated and form a mosaic of facies at the surface which are even more complicated in three dimensions.

Nevertheless, some generalizations appear possible from the studies of modern Bahamian sand bodies. Sand bodies typically occur in shallowing, upward sequences with lagoonal deposits below, and supratidal, beach, or eolian deposits above. On windward margins, sand may migrate bankward, forming extensive sand sheets on the platform. On leeward margins, sands may build seaward, burying reefs and transporting sand to deeper water beyond the shelf margins.

Active portions of sand-shoal complexes are typically clean, rippled, cross-bedded sand, devoid of grasses and most burrowing organisms. Less active portions of sand shoals become colonized by plants and bioturbated by organisms within a few tens of years.

ed on paleogeographic position geometry, bedding, and textural characteristics described previously.

SUMMARY

Bank-margin sand bodies are characteristic of carbonate shelves and platforms and occur at a position just bankward of reefs; they may extend seaward to the bank margin or break in slope if reefs are absent.

They are typically composed of ooid and skeletal sands which form sand bodies parallel-to-flow if tidal currents are high (> 100 cm/sec) but normal-to-flow if currents are low (about 50 cm/sec). Gaps between bank-margin islands form significant sand accumulations as tidal deltas. Sand bodies also form on structural rises just seaward of bank margins or well into the basin center depending on the depth of the basin and

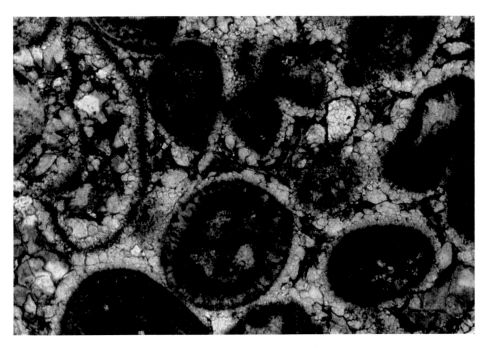

Figure 63—Photomicrograph of Mississippian oolite from the outcrop of Figure 62. Artificial staining of the thin section indicates that the porosity has been filled by two generations of cement — a rim of iron-poor crystals surrounding each grain, followed by iron-rich blocky crystals (blue). The earlier cement is similar to the fresh-water phreatic cement shown in Figure 32. Width of photograph represents 2.5 mm.

Despite the compositional variations of carbonate sands due to evolutionary processes and the diagenetic alterations produced during subaerial exposure and burial of carbonate sands, ancient sand bodies are relatively easily identified from characteristic textures, sedimentary structures, bedding styles, geometry, and paleogeographic location within a spectrum of carbonate facies.

ACKNOWLEDGMENTS

The authors thank the following persons for their contributions to this article: Mahlon Ball, Tom Bay, Donald Bebout, Jeff Dravis, David Eby, Paul Enos, Richard Inden, William Meyers, Clyde Moore, Kathy Nichols, Peter Scholle, and Gene Shinn. Their material and conceptual additions to early versions of this paper substantially improved the final product.

Figure 64—Outcrop of dolomitized Mississippian grainstone (Madison Group) in the Sawtooth Range, western Montana. Horizontal, thin bedding and low-angle cross-bedding are displayed despite complete dolomitization.

SELECTED REFERENCES

Adams, J. E., and M. L. Rhodes, 1960, Dolomitization by seepage refluxion: AAPG Bull., v. 44, p. 1912-1920.

Ahr, W. M., 1973, The carbonate ramp; an alternative to the shelf model: Trans., Gulf Coast Assoc. Geol. Socs., v. 23, p. 221-225.

Akin, R. H., and R. W. Graves, 1969, Reynolds Oolite of southern Arkansas:

AAPG Bull., v. 53, p. 1909-1922.

Badon, C., 1974, Petrology and reservoir potential of the Upper Member of the Smackover Formation, Clark County, Mississippi: Trans., Gulf Coast Assoc. Geol. Socs., v. 24, p. 163-174.

Ball, M. M., 1967, Carbonate sand bodies of Florida and the Bahamas: Jour. Sed. Petrology, v. 37, p. 556-591.

Basan, P. B., 1973, Aspects of sedimentation and the development of a carbonate bank in the Barracuda Keys, south Florida: Jour. Sed. Petrology, v. 43, p. 42-53.

Bathurst, R. G. C., 1966, Boring algae, micrite envelopes, and lithification of molluscan biosparites: Geol. Jour., v. 5, p. 15-32.

_____, 1975, Carbonate sediments and their diagenesis: New York, Elsevier Sci. Pub., 658 p.

Bay, T. A., 1977, Lower Cretaceous stratigraphic models from Texas and Mexico, *in* D. G. Bebout and R. G. Loucks, eds., Cretaceous carbonates of Texas and Mexico, application to subsurface exploration: Austin, Univ. Texas Bureau Econ. Geol. Rept. of Invest. No. 89, p. 12-30.

Becher, J. W., and C. H. Moore, 1976, The Walker Creek Field, a Smackover diagenetic trap: Trans., Gulf Coast Assoc. Geol. Socs., v. 26, p. 34-56.

Bishop, W. F., 1968, Petrology of Upper Smackover limestone in North Naynesville Field, Caliborne Parish, Louisiana: AAPG Bull., v. 52, p. 92-128.

_____, 1971, Geology of a Smackover stratigraphic trap: AAPG Bull., v. 55, p. 51-63.

Bliefnick, D. M., 1980, Sedimentology and diagenesis of bryozoan- and sponge-rich carbonate buildups, Great Bahama

Figure 65—Photomicrograph of dolomitized Madison Group showing moldic porosity partially filled with bitumen. Most grains have been altered so as to be unrecognizable. Width of photograph represents 3 mm.

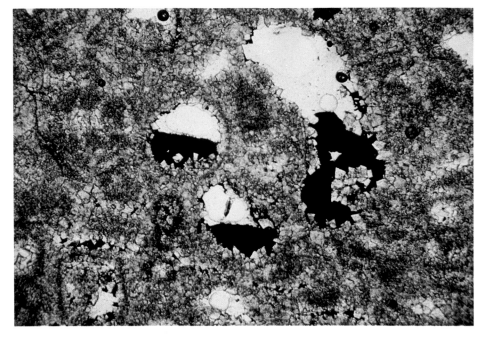

Figure 66—Photomicrograph of Mississippian crinoidal grainstone, New Mexico. The echinoderm fragments contain dark inclusions. They are overlain by two syntaxial cement generations, a clear early cement and an iron-rich (stained blue) later cement. Width of photograph represents 3 mm.

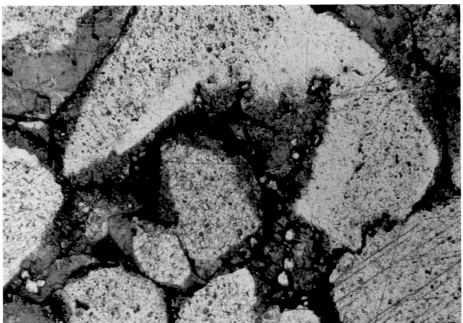

Bank: Santa Cruz, Univ. Calif., Ph.D. dissert., 289 p.

Boothroyd, J. C., and D. K. Hubbard, 1975, Genesis of bed forms in mesotidal estuaries, in L. E. Cronin, ed., Estuarine research, v. 2, geology and engineering: New York, Academic Press, p. 217-234.

Buchanan, H., 1970, Environmental stratigraphy of Holocene carbonate sediments near Frazer Hog Cay, British West Indies: New York, Columbia Univ., Ph.D. dissert., 229 p.

Carr, D. D., 1973, Geometry and origin of oolite bodies in the St. Genevieve Limestone (Mississippian) of the Illinois Basin: Ind. Dept. Natural Resources, Geol. Survey Bull. 48, 81 p.

Caston, V. N. D., 1972, Linear sand banks in the southern North Sea: Sedimentology, v. 18, p. 63-78.

Chimene, C. A., 1976, Upper Jurassic reservoirs, Walker Creek Field area, Lafayette and Columbia counties, Arkansas, in J. Braunstein, ed., North American oil and gas fields: AAPG Mem. 24, p. 117-204.

Choquette, P. W., and L. C. Pray, 1970, Geologic nomenclature and classification of porosity in sedimentary carbonates: AAPG Bull., v. 54, p. 207-250.

Davies, G. R., 1970, Carbonate bank sedimentation, eastern Shark Bay, Western Australia, in B. Logan, ed., Carbonate sedimentation and environments, Shark Bay, Western Australia: AAPG Mem. 13, p. 85-168.

Dickinson, K. A., 1968, Upper Jurassic stratigraphy of some adjacent parts of Texas, Louisiana, and Arkansas: U.S. Geol. Survey Prof. Paper 594, p. E1-E25.

Dravis, J. J., 1977, Holocene sedimentary depositional environments on Eleuthra Bank, Bahamas: Miami, Fla., Univ. Miami, M.S. thesis, 386 p.

_____, 1979, Rapid and widespread generation of Recent oolitic hardgrounds on a high energy Bahamian platform, Eleuthra Bank, Bahamas: Jour. Sed. Petrology, v. 49, p. 195-208.

Dunham, R. J., 1971, Meniscus cement, in O. P. Bricker, ed., Carbonate cements: Baltimore, Md., Johns Hopkins Univ., Stud. in Geology No. 19, p. 297-300.

Edwards, G. S., 1971, Geology of the West Flower Garden Bank: College Station, Tex., Texas A&M Univ., Sea Grant Program Pub. TAMU-SG-71-215, 199 p.

Enos, P., 1974, Surface sediment facies of the Florida-Bahamas Plateau: Geol. Soc. America Map, 5 p.

_____, 1977, Quaternary sedimentation in

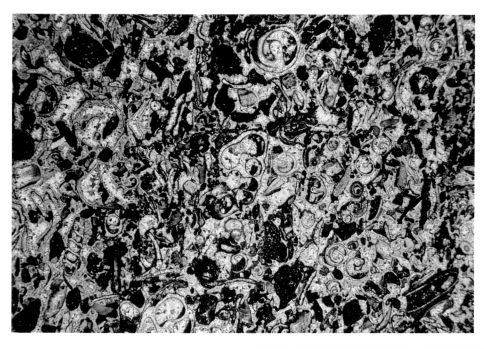

Figure 67—Photomicrograph of Lower Cretaceous, mollusk (gastropod), peloid grainstone, Louisiana. Molluscan sands such as this are rare in the Paleozoic, but occur commonly in Mesozoic and Tertiary carbonate rocks. Width of photograph represents 2 cm.

Figure 68—Photomicrograph of Lower Cretaceous, foraminiferal, peloidal grainstone, Texas. Photographed with polarizers crossed and gypsum plate inserted to emphasize two generations of cement. Width of photograph represents 3 mm.

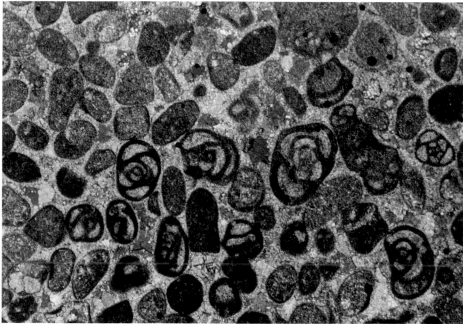

South Florida, Part I, Holocene sediment accumulations of the South Florida Shelf Margin: Geol. Soc. America Mem. 147, 198 p.

Erwin, C. R., D. E. Eby, and V. S. Whitesides, 1979, Clasticity index; a key to correlating depositional and diagenetic environments of Smackover reservoirs, Oaks Field, Claiborne Parish, Louisiana: Trans., Gulf Coast Assoc. Geol. Socs., v. 24, p. 52-62.

Garrett, P., and T. P. Scoffin, 1977, Sedimentation on Bermuda's atoll rim: Miami Fla., Univ. Miami, Proc., Third Internat. Coral Reef Symp., p. 87-96.

_____ and A. C. Hine, 1979, Probing Bermuda's lagoons and reefs (abs.): AAPG Bull., v. 63, p. 455-456.

Halley, R. B., and P. M. Harris, 1979, Fresh-water cementation of a 1,000-year-old oolite: Jour. Sed. Petrology, v. 49, p. 969-988.

_____ et al, 1977, Pleistocene barrier bar seaward of ooid shoal complex near Miami, Florida: AAPG Bull., v. 61, p. 519-526.

Hanshaw, B. B., W. Back, and R. G. Deike, 1971, A geochemical hypothesis for dolomitization by groundwater: Econ. Geology, v. 66, p. 710-724.

Harms, J. C., P. W. Choquette, and M. J. Brady, 1978, Carbonate sand waves, Isla Mujeres, Yucatan, in W. C. Ward and A. E. Weidie, eds., Geology and hydrology of northeastern Yucatan: New Orleans Geol. Soc., p. 60-84.

Harris, P. M., 1979, Facies anatomy and diagenesis of a Bahamian ooid shoal: Miami, Fla., Univ. Miami, Sedimenta

VII, Comparative Sedimentology Lab., 163 p.

_____, 1980, Freshwater cementation of Holocene and Jurassic grainstones (abs.): AAPG Bull., v. 64, p. 719-720.

_____ R. B. Halley, and K. J. Lukas, 1979, Endolith microborings and their preservation in Holocene-Pleistocene (Bahama-Florida) ooids: Geology, v. 7, p. 216-220.

Hayes, M. O., 1979, Barrier island morphology as a function of tidal and wave regime, in S. P. Leatherman, ed., Bar-

rier islands, from the Gulf of St. Lawrence to the Gulf of Mexico: New York, Academic Press, p. 1-28.

Heckel, P. H., 1972, Recognition of ancient shallow marine environments, in J. K. Rigby and W. K. Hamblin, eds., Recognition of ancient sediment environments: SEPM Spec. Pub. No. 16, p. 226-286.

Hine, A. C., 1977, Lily Bank, Bahamas; history of an active oolite sand shoal: Jour. Sed. Petrology, v. 47, p. 1554-1581.

Figure 69—Photomicrograph of core slab from producing zone of West Felda oil field, Lower Cretaceous, Florida. Zone of production is in mollusk, peloid grainstone at a depth of about 3800 m. Width of photograph is 3.5 cm.

Figure 70—Thin section photomicrograph of rock illustrated in Figure 69. Gastropod cast in center surrounded by and enclosing preserved primary porosity. Width of photograph represents 3 mm.

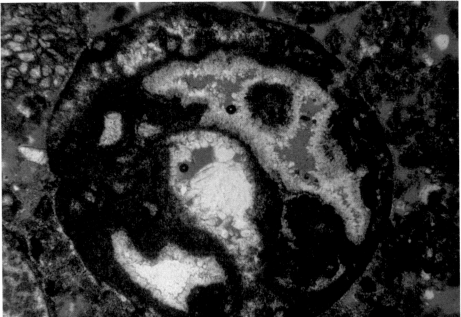

———— and A. C. Neumann, 1977, Shallow carbonate-bank-margin growth and structure, Little Bahama Bank, Bahamas: AAPG Bull., v. 61, p. 376-406.

———— R. J. Wilber, and A. C. Neumann, 1981, Carbonate sand bodies along contrasting shallow-bank margins facing open seaways; northern Bahamas: AAPG Bull., v. 65, p. 261-290.

Hoffmeister, J. E., K. W. Stockman, and H. G. Multer, 1967, Miami Limestone of Florida and its Recent Bahamian counterpart: Geol. Soc. America Bull., v. 78, p. 175-190.

Houboldt, J. J. H. C., 1968, Recent sediments in the southern bight of the North Sea: Geol. Minjbouw, v. 47, p. 245-273.

Hsu, K. J., and J. Schneider, 1973, Progress report on dolomitization-hydrology of Abu Dhabi sabkhas, Arabian Gulf, in B. H. Purser, ed., The Persian Gulf, Holocene carbonate sedimentation and diagenesis in a shallow epicontinental sea: New York, Springer-Verlag Pub., p. 409-422.

Illing, L. V., 1954, Bahamian calcareous sands: AAPG Bull., v. 38, p. 1-95.

Imbrie, J., and H. Buchanan, 1965, Sedimentary structures in modern carbonate sands of the Bahamas, in G. V. Middleton, ed., Primary sedimentary structures and their hydrodynamic interpretation: SEPM Spec. Pub. No. 12, p. 149-172.

Imlay, R. W., 1943, Jurassic formations of Gulf region: AAPG Bull., v. 27, p. 1407-1533.

Irwin, M. L., 1965, General theory of epeiric clear water sedimentation: AAPG Bull., v. 49, p. 445-459.

Jindrich, V., 1969, Recent carbonate sedimentation by tidal channels in the lower Florida Keys: Jour. Sed. Petrology, v. 39, p. 531-553.

Kobluk, D. R., 1977, Micritization and carbonate grain binding by endolithic algae: AAPG Bull., v. 61, p. 1069-1082.

Land, L. S., 1967, Diagenesis of skeletal carbonates: Jour. Sed. Petrology, v. 37, p. 914-930.

————, 1973, Contemporaneous dolomitization of Middle Pleistocene reefs by meteoric water, North Jamaica: Bull. Marine Sci., v. 23, p. 64-92.

Laporte, L. F., 1969, Recognition of a transgressive carbonate sequence within an epeiric sea; Helderberg Group (Lower Devonian) of New York State, in G. M. Friedman, ed., Depositional environments in carbonate rocks: SEPM Spec. Pub. No. 14, p. 98-118.

Lees, A., 1973, Les depots carbonates de

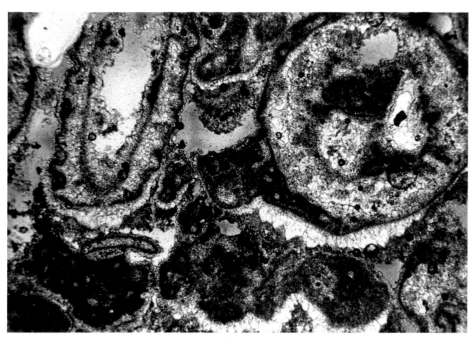

Figure 71—Photomicrograph of green-algal grainstone of Permian age, New Mexico. "Stalactitic" or gravitational cement below grains is evidence of meteoric-water cementation in the vadose zone. Note excellent preservation of primary porosity (blue). Width of photograph represents 3 mm.

Figure 72—Photomicrograph of mollusk grainstone of Cretaceous age, Florida. Fringing calcite cement is typical of early, meteoric, phreatic cements. Note excellent preservation of primary porosity (blue). Width of photograph represents 3 mm.

Plate-forme (Platform Carbonate Deposits): Bull. Centre Rech. Pau-SNPA, v. 7, p. 177-192.

Loreau, J. P., and B. H. Purser, 1973, Distribution and ultrastructure of Holocene ooids in the Persian Gulf, *in* B. H. Purser, ed., The Persian Gulf, Holocene carbonate sedimentation and diagenesis in a shallow epicontinental sea: New York, Springer-Verlag Pub., p. 279-328.

McKee, E. D., and R. C. Gutschick, 1969, History of Redwall Limestone of northern Arizona: Geol. Soc. America Mem. 114, 726 p.

Milliman, J. D., 1974, Marine carbonates: New York, Springer-Verlag Pub., 375 p.

Moore, C. H., and Y. Druckman, 1980, Burial diagenesis and porosity evolution, Upper Jurassic Smackover, Arkansas and Louisiana: La. State Univ., Dept. Geology, Applied Carbonate Research Program Tech. Ser. Cont. No. 5, 79 p.

Newell, N. D., 1955, Bahama platforms, *in* The crust of the earth, a symposium: Geol. Soc. America Spec. Paper 62, p. 303-315.

_____ and J. K. Rigby, 1957, Geological studies on the Great Bahama Bank; *in* R. LeBlanc and J. G. Breeding, eds., Regional aspects of carbonate sedimentation: SEPM Spec. Pub. 5, p. 15-72.

Newkirk, T. F., 1971, Possible future petroleum potential of Jurassic, western Gulf Basin, *in* Future petroleum provinces of the U.S., their geology and potential: AAPG Mem. 15, p. 927-953.

Off, T., 1963, Rhythmic linear sand bodies caused by tidal currents: AAPG Bull., v. 47, p. 324-341.

Ottmann, R. D., P. L. Keyes, and M. A. Ziegler, 1973, Jay Field—a Jurassic stratigraphic trap: Trans., Gulf Coast Assoc. Geol. Socs., v. 23, p. 146-157.

Palmer, M. S., 1979, Holocene facies geometry of the leeward bank margin, Tongue of the Ocean, Bahamas: Miami, Fla., Univ. Miami, M.S. thesis, 199 p.

Perkins, R. D., 1977, Quaternary sedimentation in south Florida, part 2, depositional framework of Pleistocene rocks in south Florida: Geol. Soc. America Mem. 147, 198 p.

_____ and P. Enos, 1968, Hurricane Betsy in the Florida-Bahamas area; geologic effects and comparison with Hurricane Donna: Jour. Geology, v. 76, p. 710-717.

Pfiel, R. W., and J. F. Read, 1980, Cambrian carbonate platform margin facies, Shady Dolomite, southwestern Virginia: U.S.A.: Jour. Sed. Petrology, v. 50, p.

91-116.

Plummer, L. N., et al, 1976, Hydrogeochemistry of Bermuda; a case history of ground water diagenesis of biocalcarenites: Geol. Soc. America Bull., v. 87, p. 1301-1316.

Purdy, E. G., 1961, Bahamian oolite shoals, *in* J. A. Peterson and J. C. Osmond, eds., Geochemistry of sandstone bodies: AAPG Spec. Vol., 232 p.

———, 1963, Recent calcium carbonate facies of the Great Bahama Bank; petrography and reaction groups, sedimentary facies: Jour. Geology, v. 71, p. 334-355, 472-497.

———, 1968, Carbonate diagenesis; an environmental survey: Geologica Romana, v. 7, p. 183-228.

Purser, B. H., 1973, Sedimentation around bathymetric highs in the southern Persian Gulf, *in* B. H. Purser, ed., The Persian Gulf, Holocene carbonate sedimentation and diagenesis in a shallow epicontinental sea: New York, Springer-Verlag Pub., p. 157-177.

——— and G. Evans, 1973, Regional sedimentation along the Trucial Coast, southeastern Persian Gulf, *in* B. H. Purser, ed., The Persian Gulf, Holocene carbonate sedimentation and diagenesis in a shallow epicontinental sea: New York, Springer-Verlag Pub., p. 211-231.

Rodriguez, R. W., 1979, Origin, evolution and morphology of the shoal, Escollo de Arenas, Viegras, Puerto Rico, and its potential as a sand source: Chapel Hill, Univ. N.C., M.S. thesis, 71 p.

Rowland, S. M., 1978, Environmental stratigraphy of the lower member of the Poleta Formation (Lower Cambrian), Esmeralda County, Nevada: Santa Cruz, Univ. Calif., Ph.D. dissert.

Runnells, D. D., 1969, Diagenesis, chemical sediments and the mixing of natural waters: Jour. Sed. Petrology, v. 39, p. 1188-1201.

Scholle, P. A., ed., 1978, A color illustrated guide to carbonate rock constituents, textures, cements, and porosities: AAPG Mem. 27, 241 p.

Shaw, A. B., 1964, Time in stratigraphy: New York, McGraw-Hill Pub., 365 p.

Shinn, E. A., 1969, Submarine lithification of Holocene carbonate sediments in the Persian Gulf: Sedimentology, v. 12, p. 109-144.

———, 1975, Polygonal cement sutures from the Holocene; a clue to recognition of submarine diagenesis (abs.): AAPG Ann. Mtg. Abs., v. 2, p. 68.

Southard, J. B., and D. J. Stanley, 1976, Shelf-break processes and sedimentation, *in* D. J. Stanley and D. J. P. Swift, eds., Marine sediment transport and environmental management: John Wiley and Sons Pub., p. 351-377.

Swift, D. J. P., 1972, Holocene evolution of the shelf surface, central and southern Atlantic Shelf of North America, *in* D. J. P. Swift, D. B. Duane, and O. H. Pilkey, eds., Shelf sediment transport; process and pattern: Stroudsburg, Pa., Dowden, Hutchinson, and Ross, p. 499-574.

Wilson, J. L., 1975, Carbonate facies in geologic history: New York, Springer-Verlag Pub., 469 p.

Winland, H. D., and R. K. Matthews, 1974, Origin and significance of grapestone, Bahama Islands: Jour. Sed. Petrology, v. 44, p. 921-927.

Young, G. M., and D. G. F. Long, 1977, Carbonate sedimentation in a Late Precambrian shelf sea, Victoria Island, Canadian Arctic Archipelago: Jour. Sed. Petrology, v. 47, p. 943-955.

10

Fore-reef Slope Environment

Paul Enos
Clyde H. Moore

Figure 1—Fore-reef slope deposits (left, slope is depositional) adjacent to Upper Devonian reefs (massive beds in the center) at Windjana Gorge, Napier Range, Canning basin, Western Australia. The bedded interval at the right is back-reef deposits. Height of the exposure is about 100 m (photograph courtesy of Phillip E. Playford).

*F*ore-reef slope deposits have been singled out from the broad spectrum of carbonate slope deposits (Cook and Mullins, 1982) because of distinctive characteristics associated with reef-dominated shelf margins and the proven economic importance of reef and perireef deposits. The skeletal framework and rapid vertical growth potential of ecologic reefs (Dunham, 1970) commonly, but not invariably, produce high-relief margins with steep slopes (Fig. 1). The shallow-water tropical setting and constructional framework of reefs favor rapid syndepositional submarine cementation that commonly extends into the fore-reef environment. Tropical settings and the typical highly developed ecologic niche structure of reefs favors intense bioerosion as well as framework construction, sediment production, and binding. The products of all these reef-related phenomena leave their imprint on perireefal sediments.

In an excellent summary of carbonate slope deposits, McIlreath and James (1978) categorized carbonate margins as by-pass and depositional margins. By-pass margins are generally steep slopes with a submarine escarpment of various origins across which sediment is transported from shallow to deep water without significant deposition on portions of the slope. Depositional margins are gentler, accreting slopes which merge gradually with basin floors. In either type of margin, the shallow-water portion may be formed by a reef or by carbonate-sand shoals (McIlreath and James, 1978). This emphasizes that reef margins may be either steep escarpments or relatively gentle slopes and that, conversely, not all steep shelf margins are the result of reef growth. As with all classifications, it does not emphasize the abundance of transitions between reef and non-reef margins which may be repeated in time and space. Modern margins present many examples of reef growth interspersed with carbonate sand shoals or even with terrigenous clastic deposits. Indeed, many reef tracts consist of interspersed lenticular pods rather than continuous reef walls, for example modern Florida reefs (Enos, 1977b), and mid-Cretaceous reefs of Mexico (Griffith, Pitcher, and Rice, 1969). Other transitions are produced by changes from windward to leeward locations, by sheltering effects of intermittent islands or barriers (Ginsburg and Shinn, 1964), by substrate variations, by sea-level fluctuations, and by the many other factors which govern reef growth. Viewed from the larger perspective of geologic time, variations in slope deposits occurred as the major reef-forming organism waxed and waned (Heckel, 1974; Wilson, 1975; James, 1978), resulting in significant periods when only mounds or banks (non-framework) existed or no reefs at all were formed (James, 1978, 1982).

Sediments of fore-reef slopes are a mixture of two end members, whether characterized by transport process or by origin of the particles. The major transport-process categories are directly gravity-induced deposits (resulting from talus fall, slump, and sediment gravity flow) and "gravity-defying" suspension sediments. The particles may originate from shallow-water carbonate environments, notably the reefs, or from elsewhere, primarily within the water column (pelagic), but also as deeper water benthos and from non-carbonate sources. For lack of better terms, the shallow-water carbonate environments will be referred to as neritic with the remaining environments described as oceanic or basinal sediments, following Laporte (1979). The groupings

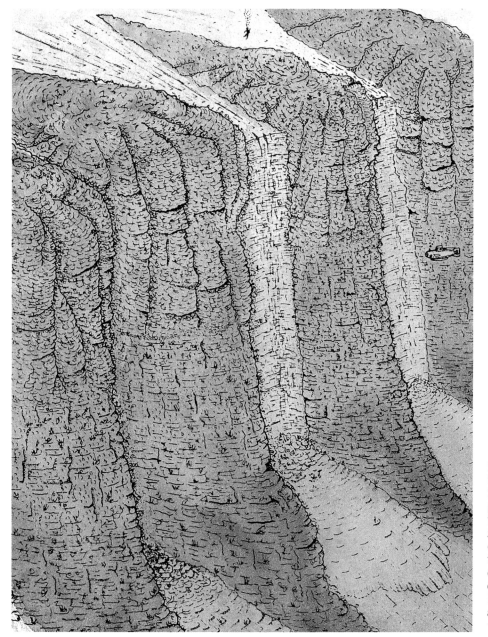

Figure 2—Sketch depicting four promontories and three reentrants along about 300 m horizontally of a near-vertical bypass slope in Jamaica. Promontories typically can be traced upward into zones of dense coral cover at the crest, extending downward and merging with the fore-reef slope. Talus cones accumulate in reentrants, with coarse debris characteristic only of apices. Height of the near-vertical portion is about 50 m (from Land and Moore, 1977).

resulting from these different perspectives are similar but not identical. Suspension sediments include carbonate mud derived from adjacent shelves, and gravity-induced deposits include pelagic sediment reworked from higher on the slope. Non-carbonate sediment may also arrive by either set of processes.

Definition

Fore-reef slope deposits include the spectrum of slope deposits located seaward of reef-rimmed shelves, platforms or atolls. They will here be considered to include all sediments with a significant input of gravity-induced neritic deposits, recognizing that the limit of such deposits may not be coincident with the limit of appreciable slope; they may extend beyond or stop short of the toe of the slope. In other words, the topographic slope may not be exclusively controlled by the accretion of a wedge of neritic sediment. Moreover, the seaward extent of the slope is extremely difficult to define in ancient examples and may even be somewhat arbitrary in modern examples where the slope merges gradually with basin floor. The "significant input" qualifier excludes essentially oceanic or basinal deposits with a few carbonate turbidites of shallow-water origin.

Diagnostic Characteristics

Slope deposits are characterized by a wide spectrum of lithologies representing transported reef and shelf debris mixed and interbedded with basinal-mud deposits. Talus blocks, slump deposits, debris-flow deposits, and turbidites all contribute. Perhaps

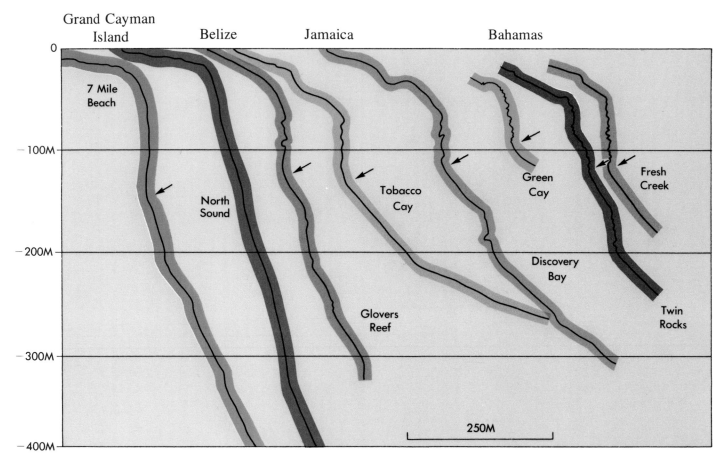

Figure 3—Representative profiles of reefal margins and upper fore-reef slopes from the Caribbean. Arrows indicate the top of the fore-reef deposits at the base of a by-pass slope. Vertical exaggeration is 1.75X (after Rigby and Roberts, 1976; James and Ginsburg, 1979).

Figure 4—Talus cone at a reentrant in the fore-reef wall, Discovery Bay, Jamaica (profile in Fig. 3). Water depth is 110 m. The rod-shaped sponge growing on the wall is approximately 50 cm high (from Land and Moore, 1977). Photograph by Noel P. James.

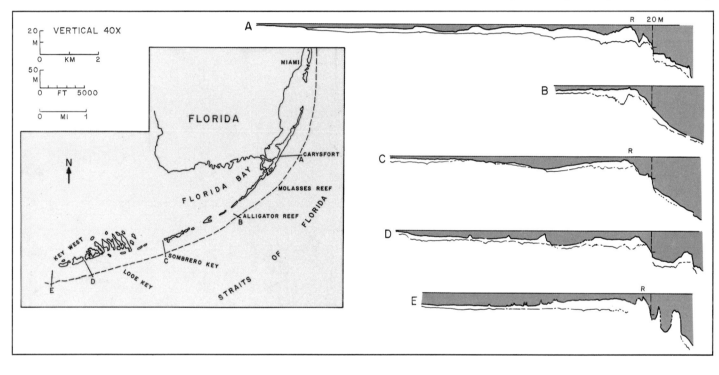

Figure 5—Block about 5 m in height fallen or slumped from the fore-reef wall, Tobacco Cay, Belize (profile in Fig. 3), 130 m depth (from James and Ginsburg, 1979).

Figure 6—Shelf and fore-reef slope reflection profiles from a depositional slope, south Florida shelf margin. Note vertical exaggeration; steepest slopes are 10°. Promontories on profiles **A, D,** and **E** are drowned Holocene reefs. The subsurface reflector is Pleistocene rock, also a reefal shelf margin, slightly modified by erosion. Pleistocene is exposed below 20 m in profiles **A** and **C.** "R" on profiles marks the position of a living shelf-edge reef (after Enos, 1977b).

the most diagnostic deposits are breccia sheets containing blocks of reef rock; the presence of actual reef blocks is critical to distinguishing fore-reef slope deposits from other types of slope deposits. Features indicative of slope are: synsedimentary folds formed by slump or creep; low-angle truncation surfaces, probably representing slump scars; and geopetal sediment deposits in primary voids, showing that enclosing bedding was deposited at an angle to horizontal.

SETTING

The location of fore-reef slopes is, by definition, dependent on the location of reefs, and the nature of the slope is directly influenced by the setting of the associated reef. The controls on reef location are discussed by James (1982). The global controls are oceanic temperature and circulation. With the steep latitudinal temperature gradients of the Quaternary, reefs are nearly confined to the tropics and occur preferentially on the western sides of oceans because of west-directed wind and current patterns in the tropics. Local controls that may be

important in predicting potential reservoir location are local circulation patterns, tectonic setting, subsidence rates and patterns, locations of suitable substrates, pre-existing constructional or erosional topography, and input and distribution patterns of terrigenous sediment.

The most extensively studied modern fore-reef slopes are on steep reef-bound margins of oceanic or intracratonic basins; some of these coincide with plate boundaries or tectonic zones. Examples are northern Jamaica (Goreau and Goreau, 1973; Goreau and Land, 1974; Moore, Graham, and Land, 1976; Land and

Figure 7—Seismic reflection profiles from fore-reef slopes southwest Florida (Uchupi and Emery, 1967; Enos, 1977b).

Figure 8—Diagrammatic representation of relative sediment contributions by different transport processes on fore-reef carbonate slopes with varying distance from a reef-bound basin margin. Suspension sediments (tan) are subdivided by origin. Individual accumulation rates and resultant types of deposit are too poorly known and too variable in different settings to attempt to quantify them. Talus deposits are confined to the top of the slope; turbidites typically extend many kilometers into the basin. A logarithmic distance scale would put the accumulated volumes (areas under the curves) into better perspective.

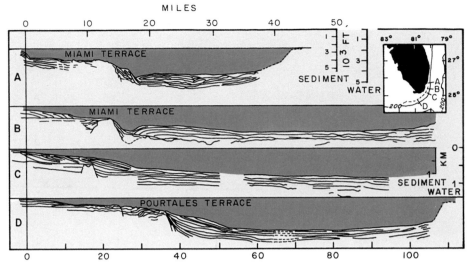

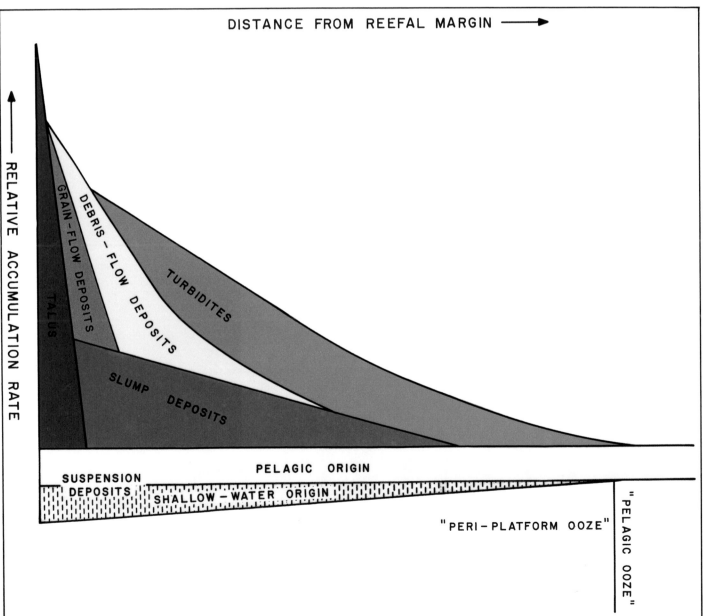

Figure 9A—"Matrix" of fore-reef talus. Perched cone of muddy *Halimeda* (green algal) sand in coral talus, Discovery Bay, Jamaica; 135 m depth.

Figure 9B—"Matrix" of fore-reef talus. Sponge with geopetal fill illustrating the initial dip (27°) of overlying layered deposits of the fore-reef slope, Sadler Limestone (Devonian) at McWhae Ridge, Canning basin, Western Australia (photo courtesy of Phillip E. Playford). Geopetal fills from shallower waters that have been rotated by resedimentation are also common in such deposits.

Figure 10—Intraformational truncation surface, Bone Springs Limestone, basinal equivalent of the Yeso reef (Permian, Leonard) Guadalupe Mountains, Texas. Base of the surface is at the observer's head; underlying and overlying beds are essentially horizontal (photograph courtesy of Peter A. Scholle).

Moore, 1977; Land, 1979, Land and Moore, 1980), Belize (James and Ginsburg, 1979), portions of the Bahama Banks (Mullins and Neumann, 1979; Schlager and Chermak, 1979), and Grand Cayman (Hannah and Moore, 1979). Unfortunately, no modern studies of slope sediments of atolls have been undertaken since the pioneering work of Emery, Tracy, and Ladd (1954). Studies of modern gentle depositional slopes are virtually non-existent; some data is available from study of adjacent shelves in south Florida

(Ginsburg, 1956; Enos, 1977b) and the Persian Gulf (Purser, 1973).

The modern slopes of northern Jamaica, Belize, Bahama Banks (eastern Little Bahama Bank and Tongue of the Ocean), and Grand Cayman are all characterized by a very steep by-pass zone flanked by a steep depositional slope (Figs. 2, 3) which merges seaward with a basin floor. Above the by-pass zone is the active reef with its spur-and-groove zone and steeper "fore-reef escarpment" (Jamaica; Moore et al, 1976) or "step" (Belize; James and Ginsburg, 1979). The reef is commonly bounded by a sand-covered slope ("fore-reef slope" of Moore et al, 1976) which steepens to a "brow" (Belize) or "drop-off" (Jamaica) at the top of the by-pass slope, termed the "wall" (Belize) or "deep fore-reef" (Jamaica). Sill reefs above the

by-pass slope locally trap shallow-water sediment or funnel it into canyons or corrugations in the by-pass slope (Goreau and Land, 1974).

The steepest part of the margin is a by-pass slope or wall, with slopes of 30° to near vertical or locally overhanging (Fig. 3). This configuration may extend from as shallow as 40 m to depths of 160 m. This zone supports some coral, calcisponge, and algal growth and receives a veneer of mud or skeletal sand, predominantly plates of the green alga, *Halimeda*. Small polished grooves within the generally corrugated front denote paths of sediment transport (Moore et al, 1976). Although virtually all sediment by-passes this zone, slow accretion does occur from skeletal growth and in situ cementation of skeletons and sediment.

To what extent the topographic

Figure 11—Allochthonous block of reef limestone in well-bedded slope deposits near McSherry's Gap, Canning basin, Western Australia (photograph courtesy of Phillip E. Playford).

features of these shallower, steeper parts of the margin are relicts of the rapid glacially-related fluctuations in sea level, and are thus likely to be atypical of much of the geologic record, is an open question. No ancient example of a by-pass slope or wall has been described (Schlager and Chermak, 1979). They likely exist in the middle Cretaceous of Mexico (Enos, 1974b), the Triassic of the Dolomites (Bosellini and Rossi, 1974), and elsewhere, but critical exposures or definitive studies are lacking.

The depositional slope below the wall is capped by a talus pile of reef-derived blocks (up to 10 m in diameter) with slopes up to 45°. Talus forms small cones in the corrugations of the wall (Figs. 2, 4). The talus zone is typically narrow with an abrupt lower boundary where it is overlapped by finer-grained sediment of the proximal slope. The proximal slope is as steep as 30°, but becomes gradually gentler and muddier until it merges, imperceptibly in some cases, with the basin floor or abyssal plain. The surface of the proximal slope is generally smooth (Fig. 3; Tobacco Cay, Discovery Bay), but it may be cut by canyons (Jamaica; Land, 1979) or riddled by small gullies (Tongue of the Ocean, Bahamas; Schlager and Chermak, 1979). These features signal

that, although the proximal slope is accreting, much sediment is being bypassed to the lower slope or basin. The gullies provide a line source of sediment resulting in laterally continuous sediment belts (Schlager and Chermak, 1979), whereas canyons suggest point sources. No sediment cones or submarine fan complexes, the expectable result of canyon point sources, have been described from either modern settings or the stratigraphic record. The slope may also be strewn with blocks (Fig. 5), some large enough (tens of meters) to act as local sediment dams (James and Ginsburg, 1979). Small escarpments of slump and/or fault origin may also trap or dam sediment (Land, 1979).

The seaward slope of the south Florida carbonate shelf with its intermittent shelf-edge reefs is a modern example of a gentle depositional slope. Local gradients are not more than 10° and overall slope is about 1° (Enos, 1977b). Relief is confined to the upper 35 m of the slope where linear buttresses of drowned Holocene reefs rise as much as 25 m above the general slope (Fig. 6). Modern coral growth has also encrusted exposed Pleistocene (reef) rock on the steepest portions of the slope, probably wave-cut nips formed

during the Holocene rise in sea level (Fig. 6A, C). Around areas of active reef growth, the shallow slope is strewn with coral rubble, mostly cylindrical sticks of branching *Acropora cervicornis,* although the bulk of reef rubble is transported landward (Ball, Shinn, and Stockman, 1967; Enos, 1977b). Most of the slope is a smooth accretionary wedge of muddy skeletal sediment. The slope is interrupted in two areas by extensive terraces (Fig. 7A, B, D) of outcropping Miocene(?) karstic limestone and phosphorite (Uchupi and Emery, 1967; Burnett and Gomberg, 1974; Enos, 1977b). Elsewhere post-Miocene sediment has draped over these terraces forming a smooth slope to the floor of the Straits of Florida (Fig. 7C).

Such relatively gentle depositional slopes are probably more typical of slopes of reef-fringed intracratonic basins and epeiric seas which comprise so much of the stratigraphic record. It is unfortunate that modern examples have not received more detailed study.

RELATION TO LATERAL FACIES

Lateral facies relationships of fore-reef carbonate slopes are as varied as the reef settings which produce them.

Figure 12A—Breccia beds of probably debris-flow origin. Large blocks of reef limestone floating in a fine-grained matrix, near Dingo Gap, Napier Range, Canning basin, Western Australia (photograph courtesy of Phillip E. Playford).

Figure 12B—Breccia beds of probably debris-flow origin. Thick debris bed in the Soyatal Formation (Upper Cretaceous) comprising blocks and bioclasts from time-equivalent and somewhat older (mid-Cretaceous) reefs that form the escarpment in the background, near El Doctor, Queretaro, Mexico.

Figure 13—Debris beds with channel form from the El Doctor (mid-Cretaceous) fore-reef slopes, Queretaro, Mexico. **A**. (top) Outcrop view of approximate strike section, 2 km from the reef escarpment; upper prominent ledges and lower ledge (arrow) are breccia beds with distinct channel forms truncated by a sharp faulted fold just to the left of the field of view. The ledge at lower left is probably a channel form too, but is intensely deformed. **B**. (middle) Diagrammatic reconstruction of the breccia beds in **A** from measured sections. Upper amalgamated beds are at 4x vertical exaggeration, lower bed is at 10x vertical exaggeration. The green diagram shows the same beds at no vertical exaggeration. **C**. (bottom) Grain size distribution in the upper amalgamated beds from field measurements.

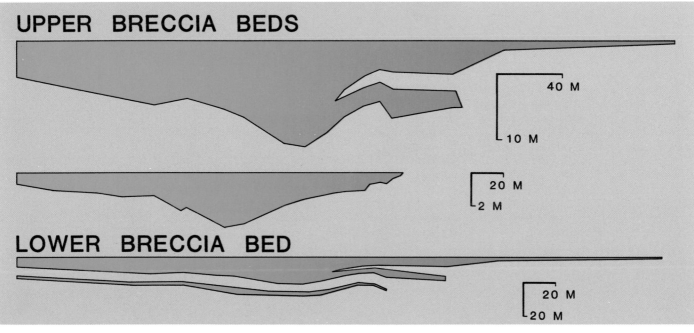

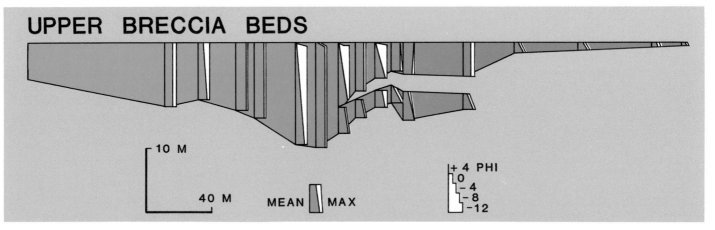

Figure 14—Base of breccia bed in lower slope facies at Dingo Gap, Napier Range, Canning basin, Western Australia (photograph courtesy Phillip E. Playford).

In areas marginal to the tropics, such as south Florida and the Bahamas, active reefs are interspersed with shelf-edge skeletal-sand or oolite shoals with their various slope topographies and sedimentary facies (Mullins and Neumann, 1979). The south Florida shelf passes laterally (northward) along both sides of the Florida plateau into areas of virtual non-deposition with a veneer of shells and even relict ooids from lower stands of sea level (Gould and Stewart, 1955; Enos, 1974a).

Exposed non-tectonic areas of the tropics and sub-tropics, such as the Florida and Yucatan peninsulas, are characterized by low-relief karst surfaces of young (Tertiary and Quaternary) limestones with little surface drainage. The lack of relief and surface drainage means that little detritus is delivered to the shelf. Where some terrigenous detritus is supplied by long-shore drift, as along the Florida coast, Holocene drowning of the coastline serves to confine detritus to the shoreline and inner shelf. The carbonate platforms are nearly devoid of submarine canyons, negating canyons as pathways for sediment to the slopes and basins. As a result, shelves and slopes in areas of active carbonate production receive nearly pure carbonate sediments, whereas areas where carbonate production is lacking are starved of sediment.

The Great Barrier Reef and the Belize barrier reef pass laterally (southward) into areas of terrigenous sediment input from hinterlands with considerable relief. The transition zones are characterized by terrigenous sediment in the shore zone and lagoons and by carbonate sediment at the shelf margin and on the slopes. Typical atolls are completely surrounded by fore-reef carbonate slopes.

LITHOFACIES AND SEDIMENTARY FACIES

The simplest scheme of fore-reef facies distribution from modern examples is that proposed by Emery and others (1954) for atolls of the northern Marshall Islands in the central Pacific Ocean. They found essentially depth-dependent boundaries with "coral sand" containing some coarser blocks (neritic sediments) in areas less than 550 m (300 fathoms) deep, grading into "*Globogerina* ooze" at depths down to 2500 m (1400 fathoms). *Globogerina* ooze (oceanic sediment) around the atoll pedestals passed into red clay below the carbonate compensation depth at 3600 to 5300 m (2000 to 2900 fathoms). Recent work with advanced techniques on steep fore-reef slopes has considerably refined the nature of the "coral sand" and emphasized the controls of biotic productivity and fore-reef topography on sediment types. Studies of ancient slope sequences have confirmed and further refined observations on sedimentary facies of modern slopes.

Each of the two major transport-process categories mentioned above includes several important members with contrasting sediment characteristics. Gravity-induced processes include talus fall, slump, grain flow, debris flow, and turbidity currents. These processes tend to produce discrete deposits with some distinguishing characteristics. In contrast, suspension sediments all settle particle-by-particle, but because of varied sources, produce a spectrum of intergradational sediment types. Pelagic biogenic products and mud derived from shallow water are the important carbonate sources, but these may be diluted or overwhelmed by terrigenous mud, volcanogenic sediment, siliceous biogenic products, or basinal evaporites. Figure 8 shows

Figure 15—Top of 15 m thick channel-form breccia bed (Fig. 13). El Doctor (mid-Cretaceous) fore-reef slope, Queretaro, Mexico. The top ½ m of the bed (at the hammer) is graded; large clasts from the ungraded massive part of the bed project up into the graded portion (just left of the hammer). The overlying thin beds are pelagic lime mudstone and chert.

qualitatively the spatial relationships of the major process-related sediment types.

Deposits of Gravity-induced Processes

Gravity-induced processes and the resultant deposits have been analyzed and described by Dott (1963), Cook and others (1972), and Middleton and Hampton (1973). Because these excellent references on process and recognition of deposits are available,

attention will be focused on criteria related to the fore-reef setting in the following discussion.

Talus is confined to a narrow zone at the top of the slope in all modern examples (James and Ginsburg, 1979; Moore et al, 1976; Mullins and Neumann, 1979; Schlager and Chermak, 1979). Talus is mentioned in many ancient sequences (Playford, 1980; Tyrrell, 1969) but no detailed accounts have been published. Blocks

are skeletal particles and fragments of the reef or peri-reefal sediment bound by submarine cement or encrusting organisms. Rocks which form the substrate of the reef and rocks or particles from back-reef environments may also be included. Thus, lithologies and cements produced in widely varying settings may be represented. Skeletal particles vary with geologic age of the deposit as the dominant reef-forming and bind-

Figure 16—Graded breccia bed, dipping 90°, from El Doctor (mid-Cretaceous) fore-reef slope, Puerto de Ortiga, Hidalgo, Mexico. Top of the bed (right) grades into pelagic lime mudstone.

Figure 17—Debris bed showing reverse grading and large clasts projecting into the overlying fine-grained rocks, Dingo Gap, Napier Range, Canning basin, Western Australia (photograph courtesy of Phillip E. Playford).

ing organisms varied. Arrangement of the blocks is chaotic, sorting is non-existent, and matrix is confined to mud and sand filtered into cavities. Such sediment commonly forms perched cones or layered geopetal fill (Fig. 9). Talus deposits may form linear prisms along the reef margin or may be isolated cones at reentrants (Figs. 2, 4).

Slumps may originate in the reef or reef-wall, but their deposits would be distinguished from talus only if other slope deposits were incorporated, or isolated "exotic blocks" were deposited. Slumps may originate anywhere on the slope (Fig. 8), but their formation is favored by steep slopes, rapid deposition, fine-grained sediment, and lack of intergranular support such as interlocking, bound, or cemented particles. The first two conditions are best met on the upper slope, but the last two are commonly confined to the lower slope.

Slump scars may be represented by concave up, low-angle "intraformational truncation surfaces," (Cook and Enos, 1977a) spectacularly represented in many fore-reef settings such as the Permo-Pennsylvanian of the Sverdrup Basin, Arctic Canada (Davies, 1977), the Triassic Steinplatte reef of Austria (Wilson, 1969), the Mississippian Lodgepole Formation of Montana (Smith, 1977), and the Permian Bone Spring Limestone of the Capitan escarpment, Texas-New Mexico (Fig. 10; Wilson, 1969). Whether these features result from slumps or some other type of submarine erosion (Yurewicz, 1977), or both, is still uncertain as suggested by Wilson's early designation as "cut and fill or slumping" (Wilson, 1969).

Slump deposits range from coherent but discordant blocks to

Figure 18—Delicate survivors of debris-flow transport. **A**. (left) Intact clusters of radiolitids (center, upper left), a reef-forming rudist bivalve, from the shallow El Abra (mid-Cretaceous) fore-reef slope deposits near El Lobo, Queretaro, Mexico. Radiolitids form rather cohesive clusters through interleaving shell flanges. **B**. (below) Platy clast of laminated lime mudstone, sandwiched between two intervals of fine, grainy limestone (outlined by dots where obscure). The grainy intervals were probably lightly cemented during transport to form a brittle clast. Note the more robust neighbors above and below, El Doctor (equivalent to El Abra) fore-reef, San Joaquin, Queretaro, Mexico.

masses highly contorted by soft-sediment folds or faults. By definition, slumping does not include loss of internal cohesion resulting in a fluid matrix (Dott, 1963). Examples of discordant exotic blocks are rudist-reef blocks in the basin-margin Tamabra Limestone of the mid-Cretaceous of northeastern Mexico (Carrasco, 1977; Enos, in press), algal-reef blocks as large as 1 km in the Lower Devonian Nubrigyn Formation of southeastern Australia (Conaghan et al, 1976), blocks up to 100 m across from the Devonian of the Canning basin, Western Australia (Fig. 11; Playford and Lowry, 1966; Playford, 1980), the "Cipit Limestone" of the Triassic in the Dolomites, northern Italy (Bosellini and Rossi, 1974), and the Permian Bell Canyon Formation of West Texas and New Mexico (Rigby, 1958). The "haystacks" of the North Jamaica island slope (Moore et al, 1976) may be modern examples; these are large, rounded, lithified blocks found at depths below 230 m subsea. Similar blocks litter the Belize fore-reef slope (Fig. 5; James and Ginsburg, 1979). Examples of folded slump deposits are described from the Permian Bell Canyon and Cherry Canyon Formations, west Texas and New Mexico, by Rigby (1958).

Lithologically, fore-reef slump deposits represent the entire spectrum from actual reef blocks to pelagic limestone or shale. Because slumps maintain internal cohesion, mixing of lithologies is largely confined to interlayering of the original deposits or piling up of successive allochthonous slump masses. Their distinguishing characteristics are discordance in attitude and, commonly, contrast in lithology with the surrounding rocks. Internal deformation may be slight to extreme. Adjacent beds may be

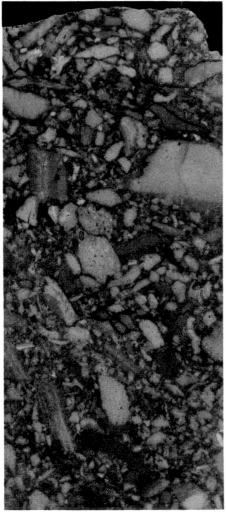

Figure 19—Randomly oriented clasts in breccia from Golden Lane fore-reef. Core of Tamabra Formation (mid-Cretaceous), Poza Rica field, Veracruz, Mexico. Core width is 9 cm.

Figure 20—Steeply inclined clasts (above and to right of hammer head; also at lower right) near lateral margin of channel-form breccia bed (Station I, Fig. 13B).

distorted by drag or "bulldozing" as the slump mass is emplaced. Superjacent beds drape or onlap the slump mass (Fig. 11).

Grain-flow deposits have been tentatively identified from modern fore-reef deposits of the Little Bahama Bank (Mullins and Neumann, 1979), and from Upper Devonian foreslope breccias adjacent to carbonate "buildups" in Alberta, Canada (Hopkins, 1977). The modern deposits, described from a single cored example, are inversely graded, contain floating clasts, and have injection structures at the base, features which Middleton and Hampton (1973) speculated might occur in grain-flow deposits. Fore-reef slopes are one of the few settings where the conditions for grain flow — slopes near the angle of repose and a supply of granular particles (Lowe, 1976) — may be maintained, but considerable doubt still exists as to whether grains flow is a significant depositional mechanism.

Debris flows form breccia beds that are the most spectacular elements of many fore-reef slope deposits (Fig. 12). They have been given various labels ranging from purely descriptive to highly genetic: breccia beds, megabreccia sheets, debris sheets, rudite sheets, olistostromes, mass breccia flows, submarine mass flows, debris avalanches, debris flow [deposits] (McIlreath and James, 1978). Ancient examples of fore-reef breccias of probable debris-flow origin include the Nubrigyn Formation (Lower Devonian), New South Wales, Australia (Conaghan et al, 1976); Napier Formation (Middle and Upper Devonian), Canning basin, Western Australia (Playford, 1980); Perdrix and Mount Hawk Formation (Middle Pennsylvanian-Lower Permian) Sverdrup Basin, Arctic Canada (Davies, 1977), Bone Spring Limestone (Leonardian, Permian), Delaware basin, Texas-New Mexico (Pray and Stehli, 1962); Cherry Canyon and Bell Canyon Formations (Guadalupian, Permian), Delaware basin, Texas (Rigby, 1958), and Tamabra Limestone (mid-Cretaceous), northeast Mexico (Carrasco, 1977; Enos, 1974b, 1977c). Few modern examples have been described, in part because of the difficulty of coring such materials, but Crevello (1978; Crevello and Schlager, 1980) reports a debris sheet that extends for 6400 sq m beneath the floor of Exuma Sound, a deep (2000 m) reentrant in Great Bahama Bank.

Debris beds range from less than one several tens of meters thick and may extend several tens of kilometers from the basin margin. It is rarely possible to reconstruct the depositional slope angle, but some beds were apparently deposited on slopes of no more than 1° (Cook et al, 1972). It is even more difficult to determine whether ancient flows moved onto the flat basin floor, but

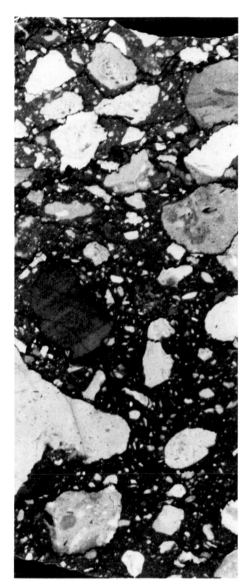

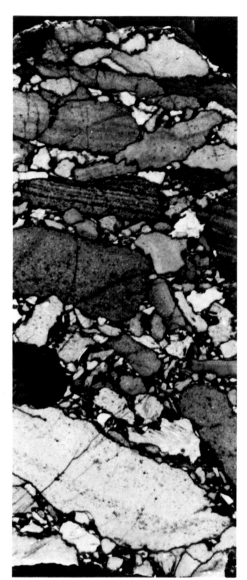

Figure 21—Breccia beds from core of Tamabra Formation, mid-Cretaceous, fore-reef of Golden Lane "atoll," Poza Rica field, Veracruz, Mexico (from Enos, 1977c). **A.** (left) Rudist fragment lime grainstone and microfossil lime wackestone clasts in matrix of pelagic lime muds, numerous stylolites. **B.** (right) Sparse matrix and sutured clasts including chert (dark clast left center); inclined fabric probably imbrication. Cores are 9 cm wide.

(Fig. 13). Sharp upper contacts are the norm (Fig. 15), but some sheets terminate by grading into pelagic deposits (Fig. 16). Hummocky tops formed by large projecting blocks are characteristic of debris-flow deposits (Fig. 17), but planar upper surfaces are more common (Fig. 15). Lateral and frontal margins are steep with abrupt termination or thinning.

Internally, sorting is extremely poor with mud and small clasts interspersed with blocks as large as 200 m in maximum dimension (Figs. 12 and 14; Conaghan et al, 1976; Playford, 1980). Despite such contrasts in size, delicate fossils and platy clasts are preserved (Fig. 18). Graded beds, both normal (Figs. 15 and 16) and reverse (Fig. 17), are reported but are not common. Particle orientation is commonly random ("chaotic," Fig. 19), but orientation parallel to sheet boundaries occurs near the base (Fig. 16) or less commonly at the steep sides or the top (Enos, 1977a). These are the regions of maximum flow whereas the interior of a flow may be rafted as a rigid plug (Hampton, 1972). Domains of parallel "packed" clasts within flows have orientations from bedding-parallel to nearly vertical (Fig. 20). Imbricate packing of clasts is also observed, and Hubert, Suchecki, and Callahan (1977) documented an inclined fabric shingled downslope, the reverse of normal imbrication, in the Cow Head Breccia (Cambro-Ordovician) of western Newfoundland.

Matrix content of breccia beds is characteristically high enough that clasts appear to float in matrix (Figs. 18 and 21A; "floatstone" of Embry and Klovan, 1973). Indeed the accepted definition of debris flow is support of clasts by matrix strength

Playford (1980) reports debris sheets that extend "several hundred meters over adjoining basin deposits." The modern debris sheet in Exuma Sound, Bahamas, extends perhaps 150 km beneath the basin floor (Crevello, 1978; Crevello and Schlager, 1980). Most debris beds are sheets in three dimensions, but lenticular to distinctly channelled forms occur interlayered in the same sequences (Fig. 13; Enos, 1973; Playford, 1980). Lower boundaries of debris beds are sharp and typically planar (Fig. 14). Sole marks are very rare; load casts are the most common. Underlying beds are commonly undeformed, but are locally deformed by shear fractures or dragfolds, or eroded by channels

during transport (cf. Middleton and Hampton, 1973). However, clast support is also common (Fig. 16, 21B, "rudstone" of Embry and Klovan, 1971). This need not imply that matrix support was lacking during transport, although some breccias that apparently had very little original matrix have been interpreted as debris-flow deposits. Apparent matrix volume can be reduced considerably by dewatering, as lime muds typically have about 75% original porosity (Enos and Sawatsky, 1981), somewhat higher than partially consolidated (dewatered) muddy intraclasts and at least 35% greater than partially lithified grainy clasts. Matrix is also susceptible to preferen-

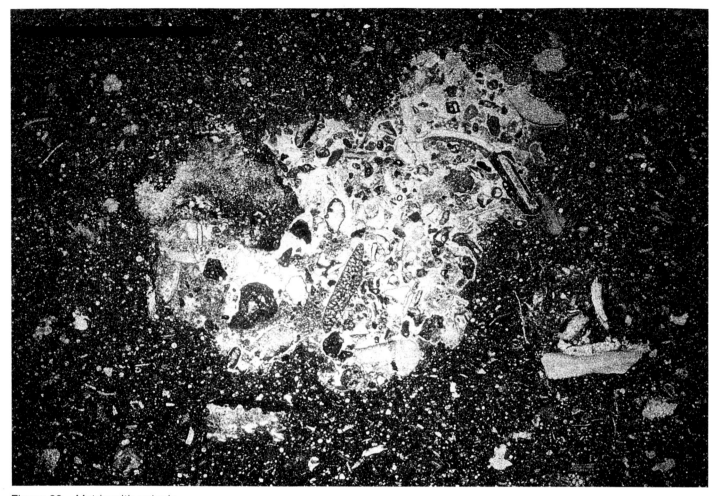

Figure 22—Matrix with pelagic microfossils (mostly calcispheres with some globular foraminifers) surrounding lightly cemented clast of shallow-water skeletal limestone; note thin rim of cement on projecting particles at right, Tamabra Formation (mid-Cretaceous) fore-reef of Golden Lane "atoll," Poza Rica field, Veracruz, Mexico. Scale bar is 5 mm (from Enos, 1977c).

tial removal during stylolite formation (Fig. 21).

Matrix comprises lime mud (silt and clay size) or a mixture of compositions, depending on slope and/or peri-reefal sediment composition. The common occurrence of pelagic fossils in the matrix (Fig. 22) points to deeper slope environments as a major source of matrix mud, either by erosion during transport or incorporation in the original flow, for example through slumping.

Composition of the clasts reflects the source and is the best means of establishing a reef source, independent of reconstructing regional geology. Clasts typically comprise blocks from: the reef, reef talus, or peri-reefal deposits; individual bioclasts or other carbonate grains; and muddy intraclasts. The percentage of muddy intraclasts, which are predominantly slope derived increases basinward (Enos, 1973, and in press). Grainy blocks must be bound or partially cemented to survive transport (Fig. 22) and reflect the organic binding or submarine cementation that is so common in reefs and upper forereef slopes. Blocks and bioclasts vary from angular (Figs. 14, 15, 22) to rounded (Figs. 16 and 21A), depending on degree of early lithification, original shape, and pre-transport history. Muddy intraclasts vary in composition depending on the composition of slope deposits and their locus of origin. Burrowed, or laminated lime mud and chert are common lithologies (Fig. 21). Although typical muddy clasts were probably unlithified, they are commonly angular or even platy (Figs. 15, 18, 21B). This is because they tend to separate along bedding planes and are protected from abrasion by matrix support during transport.

Deposits of turbidity currents, turbidites or allodapic limestones ("allodapische Kalke," Meischner, 1964), extend further into the basin than other types of gravity-induced deposits; they may be spread many kilometers across essentially flat basin floors. They can provide the most remote indication of the proximity of reefs, and, by paleocurrent reconstruction, point to the direction of the reef source. The difficulty lies in distinguishing reef-derived from non-reef carbonate sources. Again, the only clue to reef origin may be the composition of the clasts, unless regional paleogeography points to a

Figure 23—Detrital limestone bed showing Bouma divisions A-graded (note pebbles at base), B-planar laminated (?), and C-cross-laminated (with oversteepened ripple forms). The interval above the ripple forms may be a continuation of the same depositional unit with coarse-fine layering in division D, overlain by pelagic lime mudstone with chert (black, division E). From El Doctor (mid-Cretaceous) fore-reef, Puerta de Ortiga, Hidalgo, Mexico (from Enos, 1974b).

reef source or other, more obvious, fore-reef deposits (above) are interbedded. If the clasts are large enough to include lithoclasts, the diagnosis is considerably easier, although reef origin should not be interpreted based on small fragments alone. If only bioclasts are present, as is likely to be the case in more distal and therefore more critical areas, the task becomes one of reconstructing a typical reef assemblage for the particular age in question. In other respects, fore-reef carbonate turbidites do not differ from other carbonate turbidites. They are characterized by graded beds, partial or complete Bouma sequences (Fig. 23), sharp or erosive basal contacts, repetitious interbedding of grainy deposits containing shallow-water biota in pelagic deposits, and lack of shallow-water sedimentary structures (oscillation ripples, large-scale cross-bedding) or biogenic structures. Sole

marks are rarely observed in carbonate turbidites, in contrast to terrigenous turbidites. Whether this simply reflects relative scarcity of bedding-plane exposures in carbonate turbidites, particularly where the pelagic interbeds are also predominantly carbonate, or whether it reflects differences in rates of erosion between carbonate and terrigenous muds or even differences in characteristics of carbonate turbidite flow is at present an unanswered question (Ives, 1953).

Suspension Deposits

Suspension deposits adjacent to reefal or carbonate-sand shelf margins are mixtures of shelf-derived carbonate mud with pelagic carbonates ("peri-platform ooze," Schlager and James, 1978) and, commonly, with suspension material of other origins such as biogenic silica, terrigenous mud, or pyroclastic

material (Bosellini and Rossi, 1974). Studies of modern settings have documented progressive down-slope increase in mud content mixed with skeletal sands with various transport histories (James and Ginsburg, 1979; Moore et al, 1976), the potential contribution of carbonate-shelf mud to surrounding basins (Neumann and Land, 1975), and the composition of some of the particles (Moore et al, 1976; James and Ginsburg, 1979). The general characteristics and microfacies of ancient deep-water car-

Figure 25—Suspension deposits. **A**. (left) Laminated lime mudstone with sparse burrows filled with grainy carbonate reworked for interbedded fine-grained turbidites, shallow El Abra (mid-Cretaceous) fore-reef deposits, near El Lobo, Queretaro, Mexico. **B**. (below) Bedding-plane exposure of burrows shown in section **A**.

Figure 24—Finely laminated pelagic lime mudstone overlain by bioclastic-breccia bed of probable debris-flow origin, El Abra (mid-Cretaceous) fore-reef deposits near El Lobo, Queretaro, Mexico. Slab, 7 cm wide.

bonates have been treated in some detail by Wilson (1969, 1975), and Cook and Enos (1977b).

Carbonate mudstone, generally dark-colored, with fine lamination (Figs. 10, 15, 24), thick, structureless beds, or bioturbation (Fig. 25) are typical lithologies. Spiculitic (siliceous) limestone and nodular or interbedded chert are other common lithologies. The biota that dominates modern pelagic carbonates — planktonic foraminifers, pteropods, and calcareous nannofossils — first appeared in the Mesozoic, so older pelagic carbonates must be distinguished by other forms such as nautiloids, ammonoids, graptolites, planktonic crinoids, calcispheres (at least in the Mesozoic), tintinnids, and radiolarians.

Few criteria in peri-platform oozes distinguish the contribution of reefs or shelf-derived mud, although modern sediments offer a few leads which warrant further investigation. Distinctive, silt-sized, faceted chips are produced abundantly by boring activity of endolithic sponges, notably

Cliona (Futterer, 1974; Moore et al, 1976). Such particles, although not confined to them, are probably most abundant in tropical, shallow-water reefs (Futterer, 1974; Moore et al, 1976). Moore and others (1976) suggested that sponge chips with algal borings (''micrite envelopes'') may be useful shallow-water indicators. Distinctive silt-sized skeletal particles such as *Homotrema* (a red, encrusting foraminifer), tunicate spicules (a hemicordate), certain *Penicillus* needles (green alga; Perkins McKenzie and Blackwelder, 1972), and *Halimeda* plate fragments (green alga) are also of a size and density readily transported in suspension and, thus, are possible clues to tropical-shelf,

but not exclusively reefal, sediment contributions. Schlager and Chermak (1979) noted that high organic content and a green tint characterize sediment from Bahaman deep-water basins that was deposited during relatively high stands of sea level and thus received shelf-sediment contributions. Low-stand sediment, predominantly pelagic, is white with little organic matter.

The problem common to all such fine-grained tracers is preservation through the diagenetic curtain. The shallow-water skeletal components mentioned consist of metastable carbonate phases, aragonite and magnesian calcite, and are thus very susceptible to alteration during stabilization

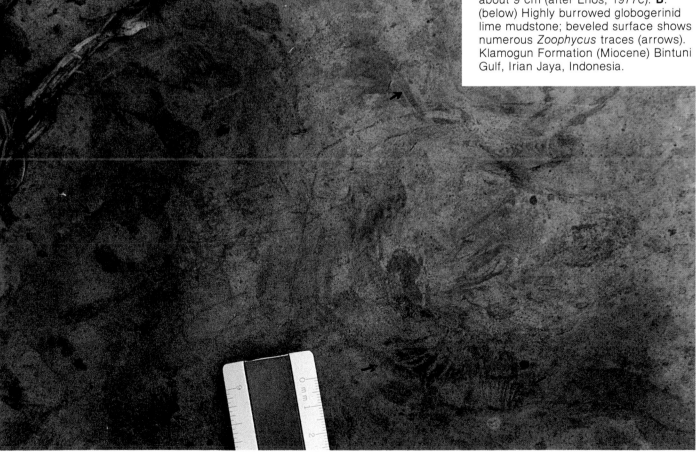

Figure 26—*Zoophycus* trace fossils, **A**. (above) Bioturbated pelagic microfossil lime mudstone from Tamabra Limestone (mid-Cretaceous), fore-reef of the Golden Lane Atoll; core from Zapotalillo field, Poza Rica trend, Veracruz, Mexico. Field is shown in Figure 30; width of view about 9 cm (after Enos, 1977c). **B**. (below) Highly burrowed globogerinid lime mudstone; beveled surface shows numerous *Zoophycus* traces (arrows). Klamogun Formation (Miocene) Bintuni Gulf, Irian Jaya, Indonesia.

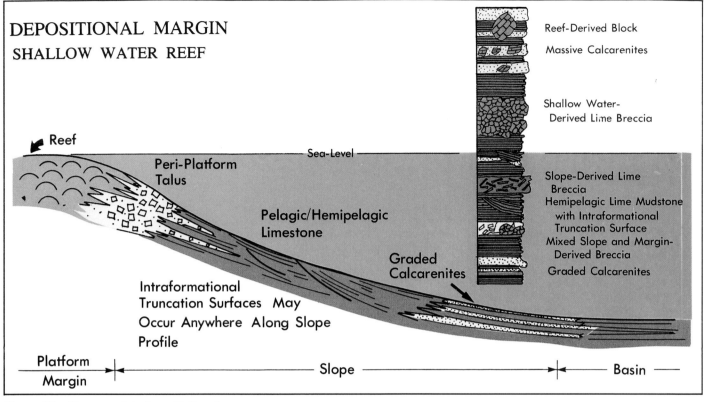

Figure 27—Slope models from McIlreath and James (1978). **A.** (above) Schematic model for a shallow-water, reef dominated, depositional carbonate margin, and illustration of a hypothetical sequence of deposits from slope accretion. **B.** (facing page) Schematic model for a shallow-water, reef dominated, by-pass type of carbonate margin.

to calcite. In contrast, pelagic carbonate particles are composed of calcite, except for aragonitic pteropods. If recognizable textures were destroyed, a tenuous clue to shallow-water input might remain in the form of abnormally high Mg or Sr (from aragonite) contents.

The paucity of bedding-plane exposures in carbonates, noted above, precludes observation of many types of biogenic structures potentially diagnostic of slope or basinal environments (Fig. 25B). Such structures as can be identified in vertical section, notably *Zoophycus*, are locally well-represented in carbonate rocks ranging in age from at least Devonian to Tertiary (Fig. 26).

Sedimentary structures indicative of depositional slope in suspension deposits are soft-sediment folds form-ed by down-slope creep, synsedimentary boudins (Hubert et al, 1977), and intraformational truncation surface (Fig. 10), discussed with slump deposits.

STRATIGRAPHIC SEQUENCE

The foregoing discussion illustrates the extreme variability of fore-reef slope deposits. For example, particle size ranges from nannofossils to the largest sedimentary particles known, with blocks exceeding 1 km in maximum dimension. This range of as much as 9 orders of magnitude in linear dimensions may occur within a single depositional unit. Although chaos seems the rule, the spatial distribution of sedimentary types shown in Figure 8 suggests that progradation of a fore-reef slope should produce a recognizable facies succession, despite extreme interdigitation. Suspension deposits should be succeeded by a preponderance of carbonate turbidites, then debris-flow deposits, slump deposits, possibly grain-flow deposits, reef talus, and finally the reef itself. McIlreath and James (1978) present schematic models of slope sequences showing these general trends with repeated interbeds of lime mud ("pelagic/hemipelagic limestone") reproduced here as Figure 27; Figure 28 is a diagrammatic representation of an actual prograding shelf margin and a transgressed margin from the Triassic of the Dolomites, northern Italy.

Distinguishing fore-reef slopes from slopes adjacent to carbonate sand margins is best accomplished using the petrology of the transported blocks which present large samples of the shelf margin. Indeed, many deposits eventually recognized as transported blocks from slopes were originally interpreted as patch reefs or bioherms (Pray and Stehli, 1962; Conaghan et al, 1976; summary by Mountjoy et al, 1972, Table 1). The composition of bioclasts may also be used to reconstruct a reef assemblage. James (1972, 1982) outlines major contributors to framework reefs at various times and emphasizes the times in which no framework reefs were formed, and thus no fore-reef slopes are to be expected. A more

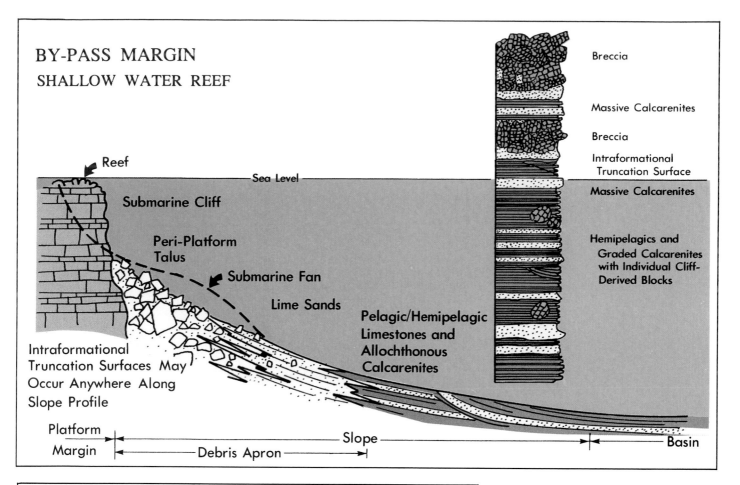

BY-PASS MARGIN
SHALLOW WATER REEF

Reef

Sea Level

Submarine Cliff

Peri-Platform
Talus

Submarine Fan

Lime Sands

Pelagic/Hemipelagic
Limestones and
Allochthonous
Calcarenites

Intraformational
Truncation Surfaces May
Occur Anywhere Along
Slope Profile

Platform
Margin

Debris Apron

Slope

Basin

Breccia

Massive Calcarenites

Breccia

Intraformational
Truncation Surface

Massive Calcarenites

Hemipelagics and
Graded Calcarenites
with Individual Cliff-
Derived Blocks

Figure 28—Accretionary profiles of carbonate bank margins from the Dolomites, South Tyrol, Italy, as interpreted from the fore-reef slope deposits. Left, an expanding, steadily prograding bank margin. Right, a bank margin that prograded and then shrank while continuing to accrete vertically. Green is massive dolomite, red is fore-reef talus (right) and exotic blocks (left), and blue is basinal pyroclastic rocks (after P. Leonardi, from Wilson, 1975).

detailed discussion is presented by Heckel (1974).

In a general way, it is predictable that the rigid framework of reefs resulting from organic binding and syndepositional cementation, would lead to a dominance of lithoclasts or blocks in fore-reef slope sequences. In contrast, the prevalence of detrital carbonate sands at a shelf margin should lead to a slope sequence dominated by calcarenite beds,

especially turbidites. While the geologic record seems to support this assertion, quantitative comparisons are lacking. It must also be noted that carbonate sands can be cemented by submarine mechanisms or during periodic exposure to fresh water, and therefore contribute copious blocks to the adjacent slope. Blocks alone do not indicate proximity to a reef; the petrology must be carefully examined.

LOG RESPONSE

Reconstruction of a low-relief fore-reef slope in the Permian of west Texas from log correlations is illustrated in Figure 29. Log expression of a wedge of fore-reef deposits within basinal limestones at the margin of a steep escarpment in the Poza Rica field, Mexico, is shown in Figure 30. Some of the prominent mechanical log markers were cor-

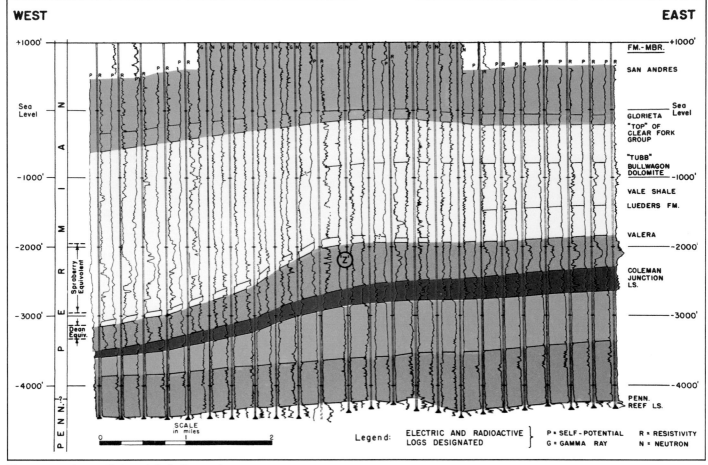

WEST   EAST

Figure 29—Low-relief reefal shelf margin
"Z" showing shelf, slope, and basinal
profile, Diamond-M field, Scurry County,
Texas. Vertical exaggeration about 4X
(from Van Siclen, 1958).

related throughout the Poza Rica
field by Barnetche and Illing (1956).
The markers represent intervals of
finer grained and non-porous rocks,
in part suspension deposits, within a
thick sequence of rudist grainstones
and breccias, subsequently interpreted
as sediment gravity-flow deposits
derived from the adjacent Golden
Lane reefs (see case histories to
follow; Enos, 1977c).

ECONOMIC CONSIDERATIONS

Some of the factors contributing to
the vast economic potential of reefs
are shared directly by the uppermost
part of the fore-reef slope. The
original texture of many reefs, coarse
skeletal grains with framework support, may be drastically modified by

diagenesis, beginning with submarine
cementation. The setting of typical
reefs as topographic highs at shelf
margins makes them particularly
susceptible to exposure to fresh water
which may enhance economic potential by enlarging and connecting pores
or may reduce porosity by cementation. Dolomitization also seems particularly prevalent in reefs. Finally,
settings at the edge of basins, wherein
most petroleum is generated, favors
charging of reservoirs.

These considerations apply directly,
although in reduced measure, to the
uppermost fore-reef slope. Most of
the particles are derived directly from
the reef and deposited with
framework support as, for example,
talus blocks. Submarine cementation
is particularly prevalent in upper fore-
reef settings in modern reefs of
Jamica (Land and Moore, 1977),
Belize (James and Ginsburg, 1979),
Cretaceous, Mexico (Aguayo, 1978),

and Devonian, Arctic Canada (Davies
and Krouse, 1975). The depths to
which submarine cementation is most
active are at present uncertain. Land
and Moore (1977) suggested that
cementation by magnesian calcite in
the Jamaican fore-reef is confined to
the warm, mixed surface layer above
the top of the thermocline at about
100 m. Evidence from Belize (James
and Ginsburg, 1979) indicates active
aragonite and Mg-calcite cementation
to at least 175 m subsea and Mg-
calcite precipitation to at least 600 m,
somewhat above the base of the local
thermocline layer. Later work in
Jamaica (Land and Moore, 1980) also
indicates important cementation
and/or alteration within the thermocline layer. Definitive statements
must await further study, particularly
within and below the thermocline
layer, but present indications are that
the generally deleterious effects of
submarine cementation are most pro-

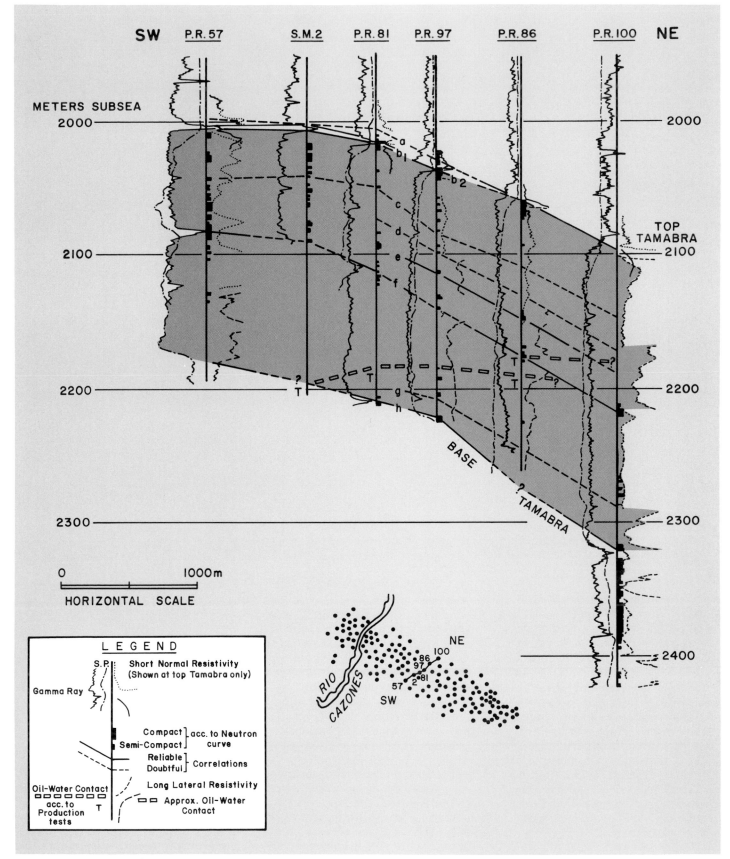

Figure 30—Representative cross section of Poza Rica field, Veracruz, Mexico, showing general character of mechanical-log "horizons," some of which are pelagic wackestones within fore-reef debris (from Barnetche and Illing, 1956).

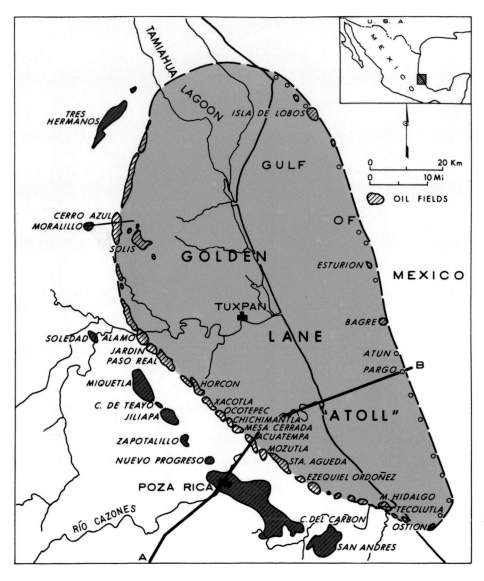

Figure 31—Location of the Poza Rica trend (red) and the adjacent Golden Lane fields, Veracruz, Mexico. The green area rimmed by oil fields is the mid-Cretaceous Golden Lane "atoll." Line A-B is location of Figure 33; short line at Cerro Azul is location of Figure 32 (after Guzman, 1967).

nounced in the surface mixed layer, significant within the thermocline layer, and much reduced below. It must be remembered, however, that submarine cementation does occur in the deep sea (Milliman, 1974).

Pleistocene fluctuations of sea level have exposed major portions of the fore-reef slope to subaerial diagenesis in both the vadose and phreatic zones (Land and Moore, 1977). Similar fluctuations have occurred in some past eras, for example, the Permian (Dunham, 1972) and Cretaceous (Ward, 1979).

Deeper fore-reef slopes may indirectly share the early diagenetic tendency toward change of shallower zones in proportion to the amount of

sediment derived from the reef and upper slope. In general, however, the prevalence of finer-grained sediment and reduced water temperatures and circulation mean lower initial porosities and less diagenetic alteration. Moreover, the mineralogy of post-Jurassic planktonic carbonate coccoliths and foraminifers is stable calcite, in contrast to the preponderantly metastable shallow-water skeletons. The deeper fore-reef is not devoid of economic potential, however, as case histories to follow will indicate. In fact, there might be more case histories if long-standing practice did not terminate exploration at the shelf edge. Historically, much of the economic interest in slope

deposits has focused on their use as "proximity indicators," pointing the way to nearby shelf margins.

The mineral potential of fore-reef slope deposits is also related to a high potential for porosity development in associated facies plus the probability of hydrothermal mineralization because of the association of reefs with continental margins and oceanic volcanic pedestals (atolls). Important hydrothermal copper-, silver-, and lead-sulfide deposits occur in basinal limestones and reef-derived breccias in the Zimapan area, Hidalgo, Mexico (P. Enos, personal observations) but no causal association with sedimentary facies has been demonstrated.

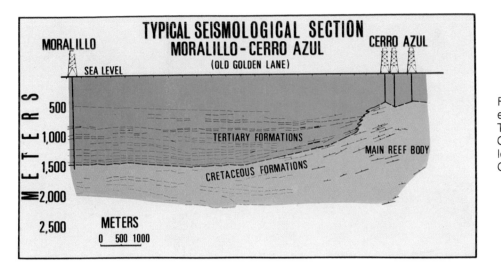

Figure 32—Typical seismic section from escarpment of Golden Lane (right) to a Tamabra Formation field, Moralillo, mid-Cretaceous, Veracruz, Mexico. Section location is shown in Figure 31 (from Guzman, 1967).

Figure 33—Regional cross section of the Tampico embayment showing mid-Cretaceous limestone formations and the location of the Poza Rica field. Acuatempa is a Golden Lane field. Production at Poza Rica is confined to the northeast limb of the gentle anticline. Initial slopes at the basin margin cannot be reconstructed; present dips on the Golden Lane escarpment, as steep 30°, are essentially depositional slope. Cross-section location shown in Figure 31 (from Enos, 1977c).

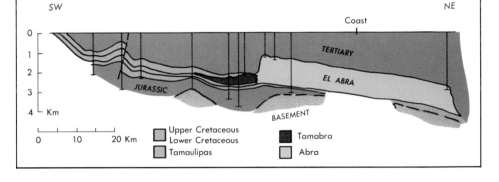

CASE HISTORIES OF PETROLEUM PRODUCTION

The best known example of fore-reef petroleum production is from the Poza Rica trend, Veracruz, Mexico (Barnetche and Illing, 1956; Enos, 1977c, and in press). The Poza Rica trend is a series of oil and gas fields located 10 to 20 km basinward of the subsurface escarpment of the mid-Cretaceous Golden Lane "atoll" (Figs. 31 and 32). Estimated ultimate recovery is more than two billion barrels, most of it from the greater Poza Rica field. Reservoir rock is the Tamabra Limestone, a wedge of detrital carbonate that thins from the base of the Golden Lane escarpment and interfingers with basinal lime mudstone and wackestone of the Upper Tamaulipas Formation (Fig. 33). Rudist-fragment lime packstone and grainstone (Fig. 34), probably resulting from sediment gravity flows, are the most important reservoir lithologies. Polymict breccias (Fig. 21), apparently debris-flow deposits, and dolomite are locally productive. Porosity includes some primary intergranular pores, but secondary molds of rudist skeletons are the most important type (Fig. 35). Deep down-dip circulation of fresh water from the exposed Golden Lane escarpment with its well-developed Cretaceous cave system is suggested as the process producing the extensive secondary porosity (Enos, 1977d). This indicates that even distal fore-reef zones may share in porosity enhancement resulting from the exposure of elevated reef complexes to an extensive freshwater system.

Trapping is provided by pinchout of the reservoir facies into dense basinal limestone coincident with draping in a gentle anticline (Figs. 33 and 30). Seals are provided by nonporous Upper Cretaceous pelagic lime mudstones overlying the Tamabra reservoir. Source rocks are uncertain but are thought to be Upper Jurassic shales rather than the adjacent Cretaceous basinal limestones which are low in organic content (Enos, in press).

Preliminary reports from the giant new Reforma trend in southern Mexico cite fore-reef deposits and talus as major contributors to the reservoirs (Fig. 36; Flores, 1978; Santiago and Mejia, 1980; Viniegra, 1981). Salt-flowage structures appear to control the accumulation sites in this highly dolomitized reservoir, however. In the offshore trend of the adjacent Campeche shelf, even more prolific production is obtained from carbonate breccias of Paleocene age, likewise ascribed to fore-reef deposition, although few details are as yet available (Santiago and Mejia, 1980; Viniegra, 1981).

Some of the production at the Golden Spike field, Upper Devonian, Alberta, Canada, is from coarse skeletal debris interbedded with dark lime muds as much as 1 km basinward from the reef margin. This is attributed to debris flows from the reef margin (Mountjoy et al, 1972). Many other examples of production or potential production from fore-reef

Figure 34—Rudist-fragment grainstone, Tamabra Formation. Primary intergranular porosity of 32% was reduced to 2% by cementation. Secondary moldic porosity 13.5%. Width of core slab is 10 cm (from Enos, 1977c).

slope facies can be expected with increased recognition of this facies and its economic potential.

SELECTED REFERENCES

Aguayo, C., 1978, Sedimentary environments and diagenesis of a Cretaceous reef complex, eastern Mexico: An. Centro Cienc. del Mar y. Limnol., Univ. Nacional Auton Mexico, v. 5, no. 1, p. 83-140.

Ball, M. M., E. A. Shinn, and K. W. Stockman, 1967, Geologic effects of Hurricane Donna in south Florida: Jour. Geology, v. 75, no. 5, p. 583-597.

Barnetche, A., and L. V. Illing, 1956, The Tamabra Limestone of the Poza Rica oil field, Veracruz, Mexico: Mexico D.F., 20th Internat. Geol. Cong., 38 p.

Bosellini, A., and D. Rossi, 1974, Triassic carbonate buildups of the dolomites, northern Italy, in L. F. Laporte, ed., Reefs in time and space: SEPM Spec. Pub. 18, p. 209-233.

Burnett, W. C., and D. N. Gomberg, 1974, Uranium in phosphate deposits from the Pourtales Terrace, Straits of Florida (abs.): Geol. Soc. America Abs. with Programs, v. 6, no. 7, p. 1028-1029.

Carrasco V., B., 1977, Albian sedimentation of submarine autochthonous and allochthonous carbonates, east edge of

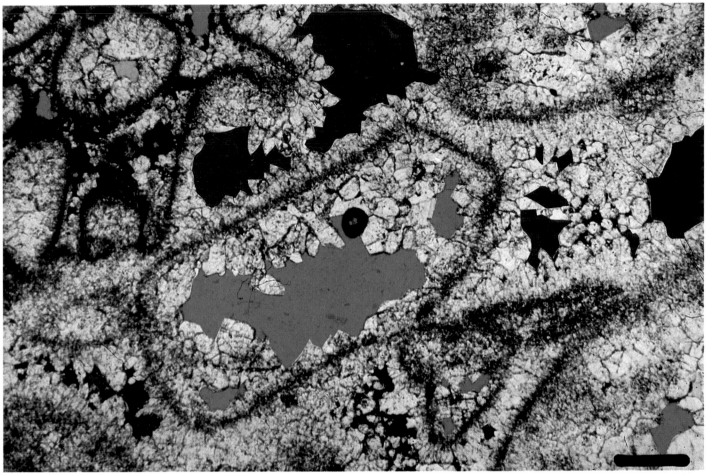

Figure 35—Thin section photo-micrograph of Tamabra Formation rudist-fragment grainstone in plane light. Primary intergranular porosity (red) and intragranular porosity (green) are reduced by cement. Moldic porosity (blue), also reduced by cement, lies within the preserved micrite rind. Note that all types of porosity contain residual oil stain. Scale bar (lower right) is 0.5 mm.

the Valles-San Luis Potosi platform, Mexico, *in* H. C. Cook and P. Enos, eds., Deep-water carbonate environments: SEPM Spec. Pub. 25, p. 263-272.

Conaghan, P. J., et al, 1976, Nubrigyn algal reefs (Devonian), eastern Australia; allochtonous blocks and mega-breccias: Geol. Soc. America Bull., v. 87, p. 515-530.

Cook, H. E., and P. Enos, 1977a, Deep-water carbonate environments—an introduction, *in* H. E. Cook and P. Enos, eds., Deep-water carbonate environments: SEPM Spec. Pub. 25, p. 1-3.

_____ and _____, eds., 1977b, Deep-water carbonate environments: SEPM Spec. Pub. 25, 336 p.

_____ and H. T. Mullins, 1982, Carbonate slopes and basin margins, *in* P. A. Scholle, D. G. Bebout and C. H. Moore, eds., Carbonate depositional environments: AAPG Mem. 33, this volume.

_____ et al, 1972, Allochthonous carbonate debris flows at Devonian bank (reef) margins, Alberta, Canada: Bull. Canadian Petroleum Geols., v. 20, no. 3, p. 439-497.

Crevello, P. D., 1978, Debris-flow deposits and turbidites in a modern carbonate basin, Exuma Sound, Bahamas: Coral Gables, Fla., Univ. Miami, M.S. thesis, 133 p.

_____ and W. Schlager, 1980, Carbonate debris sheets and turbidites, Exuma Sound, Bahamas: Jour. Sed. Petrology, v. 50, no. 4, p. 1121-1148.

Davies, G. R., 1977, Turbidites, debris sheets, and truncation structures in Upper Paleozoic deep-water carbonates of the Sverdrup Basin, Arctic Archipelago, *in* H. E. Cook and P. Enos, eds.,

Deep-water carbonate environments: SEPM Spec. Pub. 25, p. 221-247.

_____ and H. R. Krouse, 1975, Carbon and oxygen isotopic composition of late Paleozoic calcite cements, Canadian Arctic Archipelago—preliminary results and interpretation: Geol. Survey Canada, Paper 75-1B, p. 215-220.

Dott, R. H., Jr., 1963, Dynamics of subaqueous gravity depositional processes: AAPG Bull., v. 47, p. 104-128.

Dunham, R. J., 1970, Stratigraphic reefs versus ecologic reefs: AAPG Bull., v. 54, p. 1931-1932.

_____, 1972, Guide for study and discussion of individual reinterpretation of the sedimentation and diagenesis of the Permian Capitan geologic reef and associated rocks, New Mexico and Texas: Permian Basin Sec., SEPM Pub. 72-14, 235 p.

Embry, A. F., and J. E. Klovan, 1971, A Late Devonian reef trace on northeastern Banks Island, Northwest Territories: Canadian Petroleum Geols. Bull., v. 19, p. 730-781.

Emery, K. O., J. I. Tracey, Jr., and

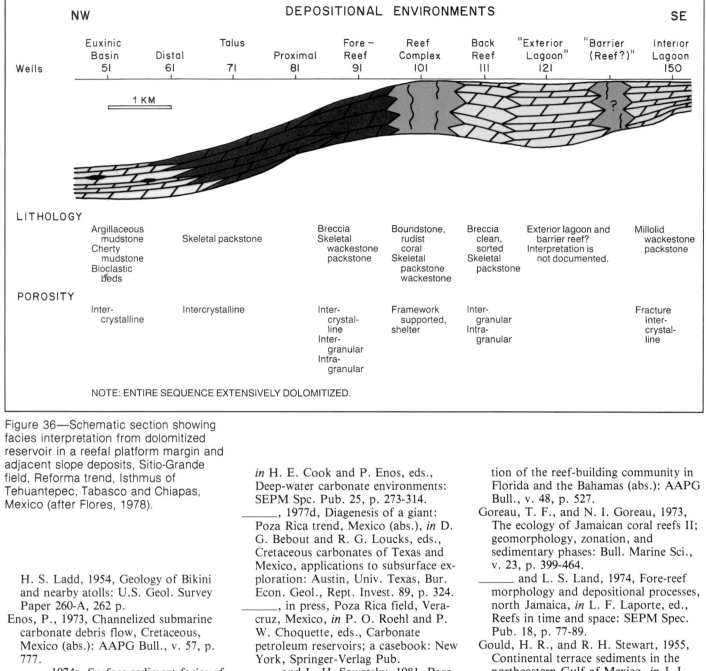

Figure 36—Schematic section showing facies interpretation from dolomitized reefal reservoir in a reefal platform margin and adjacent slope deposits, Sitio-Grande field, Reforma trend, Isthmus of Tehuantepec, Tabasco and Chiapas, Mexico (after Flores, 1978).

H. S. Ladd, 1954, Geology of Bikini and nearby atolls: U.S. Geol. Survey Paper 260-A, 262 p.

Enos, P., 1973, Channelized submarine carbonate debris flow, Cretaceous, Mexico (abs.): AAPG Bull., v. 57, p. 777.

_____, 1974a, Surface sediment facies of the Florida-Bahama Plateau: Geol. Soc. America, Map no. 5, 5 p.

_____, 1974b, Reefs, platforms, and basins of Middle Cretaceous in northeast Mexico: AAPG Bull., v. 58, p. 800-809.

_____, 1977a, Flow regimes in debris flow: Sedimentology, v. 24, p. 133-142.

_____, 1977b, Holocene sediment accumulations of the south Florida shelf margin, in P. Enos and R. D. Perkins, eds., Quaternary sedimentation in south Florida: Geol. Soc. America, Mem. 147, p. 1-130.

_____, 1977c, Tamabra limestone of the Poza Rica trend, Cretaceous, Mexico, in H. E. Cook and P. Enos, eds., Deep-water carbonate environments: SEPM Spc. Pub. 25, p. 273-314.

_____, 1977d, Diagenesis of a giant: Poza Rica trend, Mexico (abs.), in D. G. Bebout and R. G. Loucks, eds., Cretaceous carbonates of Texas and Mexico, applications to subsurface exploration: Austin, Univ. Texas, Bur. Econ. Geol., Rept. Invest. 89, p. 324.

_____, in press, Poza Rica field, Veracruz, Mexico, in P. O. Roehl and P. W. Choquette, eds., Carbonate petroleum reservoirs; a casebook: New York, Springer-Verlag Pub.

_____ and L. H. Sawatsky, 1981, Pore networks in carbonate sediments: Jour. Sed. Petrology, v. 51, p. 961-985.

Flores, V. Q., 1978, Paleosedimentologia en la zona de Sitio Grande–Sabancuy: Petroleo Internac., v. 36, p. 44-48.

Futterer, D. K., 1974, Significance of the boring sponge Cliona for the origin of fine grained material of carbonate sediments: Jour. Sed. Petrology, v. 44, no. 1, p. 79-84.

Ginsburg, R. N., 1956, Environmental relationships of grain size and constituent particles in some south Florida carbonate sediments: AAPG Bull., v. 40, no. 10, p. 2384-2427.

_____ and E. A. Shinn, 1964, Distribution of the reef-building community in Florida and the Bahamas (abs.): AAPG Bull., v. 48, p. 527.

Goreau, T. F., and N. I. Goreau, 1973, The ecology of Jamaican coral reefs II; geomorphology, zonation, and sedimentary phases: Bull. Marine Sci., v. 23, p. 399-464.

_____ and L. S. Land, 1974, Fore-reef morphology and depositional processes, north Jamaica, in L. F. Laporte, ed., Reefs in time and space: SEPM Spec. Pub. 18, p. 77-89.

Gould, H. R., and R. H. Stewart, 1955, Continental terrace sediments in the northeastern Gulf of Mexico, in J. L. Hough and H. W. Menard, eds., Finding ancient shorelines: SEPM Spec. Pub. 3, p. 2-19.

Griffith, L. S., M. G. Pitcher, and G. W. Rice, 1969, Quantitative environmental analysis of a Lower Cretaceous reef complex, in G. M. Friedman, ed., Depositional environments in carbonate rocks: SEPM Spec. Pub. 14, p. 120-138.

Guzman, E. J., 1967, Reef type stratigraphic traps in Mexico: Proc. 7th World Petroleum Cong., v. 2, p. 461-470.

Hampton, M. A., 1972, The role of subaqueous debris flow in generating turbidity currents: Jour. Sed. Petrology, v. 42, p. 775-793.

Hanna, J. C., and C. H. Moore, 1979, Quaternary temporal framework of reef to basin sedimentation, Grand Cayman Island, British West Indies (abs.): San Diego, Geol. Soc. America Ann. Mtg.

Heckel, P. H., 1974, Carbonate buildups in the geologic record; a review, *in* L. F. Laporte, ed., Reefs modern and ancient: SEPM Spec. Pub. 18, p. 90-154.

Hopkins, J. C., 1977, Production of fore-slope breccia by differential submarine cementation and downslope displacement of carbonate sands, Miette and Ancient Wall buildups, Devonia, Canada, *in* H. E. Cook and P. Enos, eds., Deep-water carbonate environments: SEPM Spec. Pub. 25, p. 155-170.

Hubert, J. F., R. K. Suchecki, and R. K. M. Callahan, 1977, The Cowhead Breccia; sedimentology of the Cambro-Ordovician continental margin, Newfoundland, *in* H. E. Cook and P. Enos, eds., Deep-water carbonate environments: SEPM Spec. Pub. 25, p. 125-154.

Ives, C., 1953, The unanswered question: New York, Southern Music Pub. Co., 8 p.

James, N. P., 1978, Facies models 10; reefs: Geosci. Canada, v. 5, p. 16-26.

_____, 1982, Reefs, *in* P. A. Scholle, G. Bebout and C. H. Moore, eds., Carbonate depositional environments: AAPG Mem. 33, this volume.

_____ and R. N. Ginsburg, 1979, The seaward margin of Belize barrier and atoll reefs: Internat. Assoc. Sedimentologists Spec. Pub. no. 3, 191 p.

Land, L. S., 1979, The fate of reef-derived sediment on the north Jamaican slope: Marine Geology, v. 29, p. 55-71.

_____ and C. H. Moore, Jr., 1977, Deep forereef and upper island slope, north Jamaica, *in* S. H. Frost, M. P. Weiss, and J. B. Saunders, eds., Reefs and related carbonates—ecology and sedimentology: AAPG Stud. Geology No. 4, p. 53-65.

_____ and _____, 1980, Lithification, micritization, and syndepositional diagenesis of biolithites on the Jamaican island slope: Jour. Sed. Petrology, v. 50, p. 357-369.

Laporte, L. F., 1979, Ancient environments: Englewood Cliffs, New Jersey, Prentice-Hall Pub., 2nd ed., 160 p.

Lowe, D. R., 1976, Grain flow and grain flow deposits: Jour. Sed. Petrology, v. 46, p. 188-199.

McIlreath, I. A., and N. P. James, 1978, Facies models 12; carbonate slopes: Geosci. Canada, v. 5, no. 4, p. 189-199. (also *in* R. G. Walker, ed.,

1979, Facies models: Geosci. Canada, Rept. Ser. 1, p. 133-149).

Meischner, K. D., 1964, Allodapische Kalke, turbidite in riff-nahen sedimentations-becken, *in* A. H. Bouma and A. Brouwer, eds., Turbidites: Amsterdam, Elsevier Sci. Pub., p. 156-191.

Middleton, G. V., and M. A. Hampton, 1973, Sediment gravity flows; mechanics of flow and deposition *in* G. V. Middleton and A. H. Bouma, eds., Turbidites and deep-water sedimentation: Los Angeles, Calif., SEPM Pacific Sec., p. 1-38.

Milliman, J. D., 1974, Marine carbonates: New York, Springer-Verlag Pub., 375 p.

Moore, C. H., E. A. Graham, and L. S. Land, 1976, Sediment transport and dispersal across the deep fore-reef and island slope (−55 m to −305 m), Discovery Bay, Jamaica: Jour. Sed. Petrology, v. 46, p. 174-187.

Mountjoy, E. W., et al, 1972, Allochthonous carbonate debris flows—worldwide indicators of reef complex, banks, or shelf margins: Montreal, 24th Internat. Geol. Cong., Sec. 6, p. 172-189.

Mullins, H. T., and A. C. Neumann, 1979, Deep carbonate bank margin structure and sedimentation in the northern Bahamas, *in* L. J. Doyle and D. H. Pilkey, eds., Geology of continental slopes: SEPM Spec. Pub. 27, p. 165-192.

Neumann, A. C., and L. S. land, 1975, Lime mud deposition in the Bight of Abaco, Bahamas, a budget: Jour. Sed. Petrology, v. 45, p. 763-786.

Perkins, R. D., M. D. McKenzie, and P. L. Blackwelder, 1972, Aragonite crystals within Codiacean algae; distinctive morphology and sedimentary implications: Science, v. 175, p. 624-626.

Playford, P. E., 1980, Devonian "Great Barrier Reef" of Canning basin, Western Australia: AAPG Bull., v. 64, p. 814-840.

_____ and D. C. Lowry, 1956, Devonian reef complexes of the Canning basin, Western Australia: Geol. Survey Western Australia, Bull. 118, 150 p.

Pray, L. C., and F. G. Stehli, 1962, Allochthonous origin, Bone Springs "patch reefs," west Texas (abs.): Geol. Soc. America Spec. Paper 73, p. 218-219.

Purser, B. H., ed., 1973, The Persian Gulf; Holocene carbonate sedimentation and diagenesis in a shallow epicontinental sea: Berlin, Springer-Verlag Pub., 471 p.

Rigby, J. K., 1958, Mass movements in Permian rocks at Trans-Pecos, Texas: Jour. Sed. Petrology, v. 28, p. 298-315.

_____ and H. H. Roberts, 1976, Geology reefs and marine communities of Grand Cayman Islands, British West Indies: Brigham Young Univ., Geol. Stud. Spec. Pub. A, p. 1-95.

Santiago, A. J., and D. O. Mejia, 1980, Giant fields in the southeast of Mexico: Trans., Gulf Coast Assoc. Geol. Socs., v. 30, p. 1-31.

Schlager, W., and A. Chermak, 1979, Sediment facies of platform-basin transition, Tongue of the Ocean, Bahamas, *in* L. L. Doyle and O. H. Pilkey, eds., Geology of continental slopes: SEPM Spec. Pub. 27, p. 193-208.

_____ and N. P. James, 1978, Low-magnesium calcite limestones, forming at the deep-sea floor, Tongue of the Ocean, Bahamas: Sedimentology, v. 25, p. 675-702.

Smith, D. L., 1977, Transition from deep-to shallow-water carbonates, Paine Member, Lodgepole Formation, central Montana, *in* H. E. Cook and P. Enos, eds., Deep-water carbonate environments: SEPM Spec. Pub. 25, p. 187-201.

Tyrrell, W. W., 1969, Criteria useful in interpreting environments of unlike but time-equivalent carbonate units (Tansill-Capitan-Lomar), Capitan Reef Complex, west Texas and New Mexico, *in* G. M. Friedman, ed., Depositional environments in carbonate rocks: SEPM Spec. Pub. 14, p. 80-97.

Uchupi, E., and K. O. Emery, 1967, Structure of continental margin off Atlantic coast of United States: AAPG Bull., v. 51, p. 223-234.

Van Siclen, D. C., 1958, Depositional topography—examples and theory: AAPG Bull., v. 42, p. 1897-1913.

Ward, J. A., 1979, Stratigraphy, depositional environments, and diagenesis of the El Doctor platform, Queretaro, Mexico: Binghamton, State Univ. of New York, Ph.D. dissert., 172 p.

Wilson, J. L., 1969, Microfacies and sedimentary structures in "deeper water" lime mudstone, *in* G. M. Friedman, ed., Depositional environments in carbonate rocks: SEPM Spec. Pub. 14, p. 4-19.

_____, 1975, Carbonate facies in geologic history: Berlin, Springer-Verlag Pub., 471 p.

Viniegra O. F., 1981, Great carbonate bank of Yucatan, southern Mexico: Jour. Petroleum Geology, v. 3, no. 3, p. 247-278.

Yurewicz, D. A., 1977, Sedimentology of Mississippian basin-facies carbonates, New Mexico and west Texas—the Rancheria Formation, *in* H. E. Cook and P. Enos, eds., Deep-water carbonate environments: SEPM Spec. Pub. 25, p. 203-219.

11

Basin Margin Environment

Harry E. Cook
Henry T. Mullins

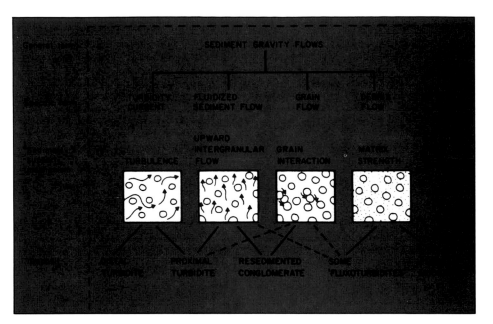

Figure 3—Classification of subaqueous sediment-gravity flows (from Middleton and Hampton, 1976).

Figure 4—SEM micrograph of soft peri-platform ooze. Note abundance of aragonite needles and low-Mg calcite cocoliths (*Emiliana huxleyi*). This mixture of shallow-water bank-derived aragonite needles plus planktonic debris such as coccoliths, foraminifers and pteropods is typical of modern peri-platform oozes on carbonate slopes in the Bahamas. Core 7507-020 from Tongue of the Ocean. Depth is 1000 fathoms; age is late Pleistocene or Holocene (from Schlager and James, 1978; courtesy W. Schlager).

tween these platform types is that the rimmed variety is characterized by a high energy rim of reefs, islands, or sand shoals found at the edge of the platform, whereas the former is an open, gently seaward dipping shelf devoid of any continuous energy barrier at the platform edge (Ahr, 1973).

Carbonate slopes surround all of these modern carbonate platforms. However, only the slopes in the northern Bahamas have been studied

in detail (Fig. 1; Neumann, 1974; Neumann et al, 1977; Mullins, 1978; Mullins and Neumann, 1979; Schlager and Chermak, 1979; Mullins et al, 1980a, b; Crevello and Schlager, 1980; Mullins et al, 1981, in prep.). Other regions of modern carbonate slopes have received only preliminary examination (Ewing et al, 1969; Worzel et al, 1973; Burne, 1974; Moore et al, 1976; Mitchum, 1978; Hana and Moore, 1979; James and

Ginsburg, 1979; Land, 1979). Thus, the examples from modern carbonate slopes in this paper will be strongly biased toward the northern Bahamas, as this is where most of the comprehensive studies have been completed.

In the northern Bahamas (Fig. 1), carbonate slopes are found along: (1) open oceans (an example being north of Little Bahama Bank) which are open to extremely large fetches; (2) open seaways, which are open to oceanic circulation at both ends (examples being the Straits of Florida and Providence Channels); or (3) closed seaways, which are open to ocean circulation at only one end (an example being Tongue of the Ocean; Exuma Sound). These slopes show great variability in morphology and sedimentary facies relationships that

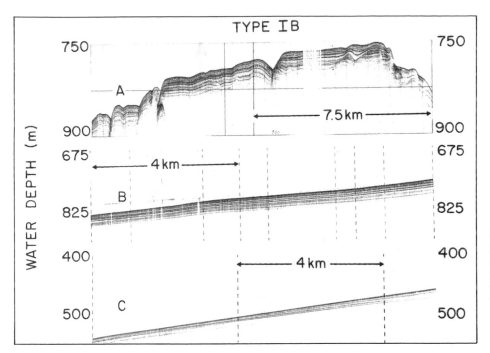

Figure 5—Examples of 3.5 KHz PDR (Precision Depth Record) profiles of peri-platform oozes in Northwest Providence Channel, Bahamas. Note the even, parallel continuous nature of the subbottom reflectors, indicative of uniform, widespread pelagic sedimentation (from Mullins et al, 1979).

Figure 6—Bottom photograph of outcrop of Pleistocene carbonate oozes in Great Abaco Canyon, Bahamas. Note the even, parallel laminations of these lithified deep-sea pelagic oozes. Holding cage at bottom of photograph is approximately 1 m across. Photo taken from DSRV ALVIN. Water depth is 3643 m (from Mullins et al, 1982).

are controlled by a combination of processes such as: (1) basement faulting, (2) direction and magnitude of off-bank sediment transport, (3) oceanic circulation, (4) gravity and pelagic sedimentation, (5) biological buildups, and (6) submarine cementation (Mullins and Neumann, 1979).

Modern carbonate slopes on the average tend to be steeper (5 to 15°) than terrigenous slopes (3 to 6°). However, average down-to-basin carbonate slopes are highly variable commonly ranging from as little as 1 to 60° with locally vertical to overhanging scarps (Mullins and Neumann, 1979). The widths and heights of modern carbonate slopes are also variable; widths ranging from a minimum of about 5 to 10 km along the Bahama Escarpment to more than 100 km off the northwest corner of Little Bahama Bank, and slope heights varying from a

minimum of 800 m in the Straits of Florida to a maximum of over 4,000 m in the Northeast Providence Channel.

Many of the carbonate slopes in the northern Bahamas are divisible into three morphologic parts. From the top of the slope to its base these are: (1) a marginal escarpment, which extends precipitously (up to 45° or more) from the shelf edge (30 to 50 m) to depths of 100 to 200 m (con-

tains the marginal facies in Fig. 1); (2) an upper gullied slope dipping seaward at slopes of 3 to 15° and dissected by numerous small canyons 20 to 150 m in relief (contains the slope facies of Fig. 1); and (3) a more gentle (1 to 5°), smooth to gently undulating lower slope or rise. The material at the base of the marginal escarpment commonly consists of mixtures of bank-derived and pelagic sediments plus talus blocks ("peri-

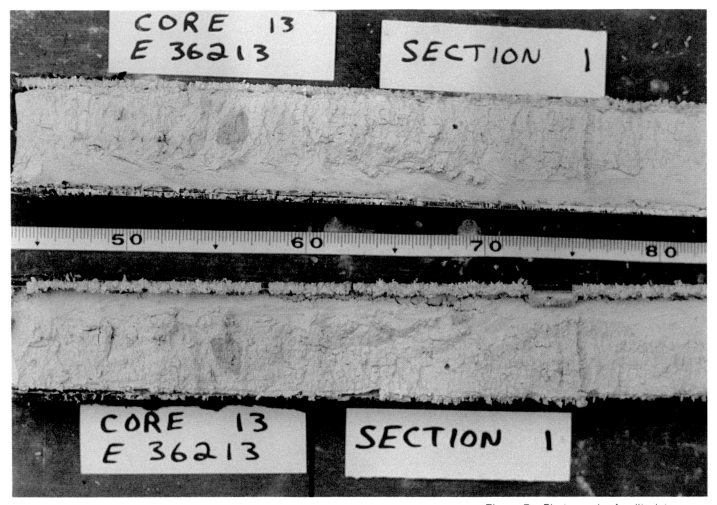

Figure 7—Photograph of split piston core of peri-platform oozes recovered from 850 m of water north of Little Bahama Bank. Note the lack of obvious layering in contrast to Figures 1 and 2. Also note the presence of a large circular burrow structure near the top of the core, as well as numerous thin, vertical burrows. Scale is in centimeters (photo by H. T. Mullins).

platform sand facies" of Mullins and Neumann, 1979); the upper gullied slope typically consists of silty, bioturbated carbonate ooze cut by channels that are floored with gravel and sand; the lower slope commonly consists of thick sediment gravity-flow deposits (turbidites, debris flows, grain flows) interbedded with carbonate ooze (Schlager and Chermak, 1979; Mullins et al, in prep.).

MASS TRANSPORT PROCESSES ON SLOPES

Mass transport is used here for the en masse downslope movement of material containing various amounts of water, for which gravity is the driving force (Dott, 1963, Cook et al, 1972). A selected list of papers that treat various aspects of mass transport includes: Bagnold (1954, 1956, 1966), Bouma (1962), Dott (1963), Dill (1966), Morgenstern (1967), Stauffer (1967), Middleton (1970), Fisher (1971), Cook and others (1972), Hampton (1972, 1975, 1979), Mountjoy and others (1972), Middleton and Hampton (1973, 1976), Walker and Mutti (1973), Carter (1975), Walker (1975), Wilson (1975), Lowe (1976a, b, 1979, 1982), Cook and Enos (1977a, b), Enos (1977c), Shanmugam and Benedict (1978), Stanley and Kelling (1978), Varnes (1978), Cook (1979a, b, c), Krause and Oldershaw (1979), Mullins and Van Buren (1979); and Nardin and others (1979). Table 1 and Figures 2 and 3 summarize the characteristics of the main types of mass transport giving the classification schemes that are currently most widely accepted.

Mass transport can be divided into three types — rockfalls, slides, and sediment gravity flows (Table 1).

Slides and sediment gravity flows can be further subdivided on the basis of their internal mechanical behavior and dominant sediment support mechanism (Table 1, Fig. 3).

Rockfalls, also referred to as talus accumulations, are only abundant in the marine environment at the base of steep slopes, canyon walls, or fault scarps. Deposits of this type accumulate by the rolling or freefall of individual clasts.

Slides can be divided into translational (glide) and rotational (slump) types (Varnes, 1978). The shear plane

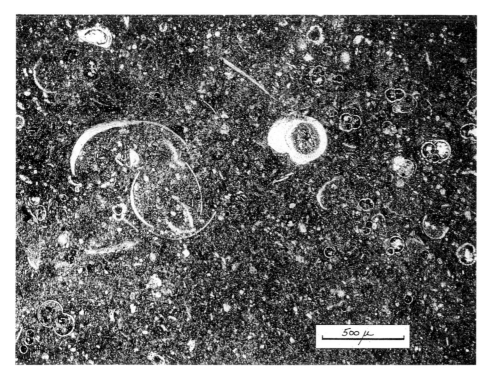

Figure 8—Thin section photomicrograph in plain polarized light of peri-platform chalk. Note the presence of pteropods (center left) as well as thick- and thin-shelled planktonic foraminifers (center right). Cement is micritic low-Mg calcite. Sample 555-25 from Tongue of the Ocean, Bahamas. Depth is 850 fathoms; age is middle Pleistocene. Scale bar = 500 millimicrons (from Schlager and James, 1978; courtesy of W. Schlager).

Figure 9—Fine-grained, laminated lime mudstone and wackestone interpreted as slope deposit from lower part of Hales Limestone, Upper Cambrian, Nevada (photo by H. E. Cook).

of a translational slide is predominantly along planar or gently undulatory surfaces parallel to the underlying beds. Slumps (rotational slides) exhibit concave-upward shear planes and usually a backward rotation of the slumped body. Slides can exhibit variable amounts of internal deformation. Some slides show purely elastic behavior, the original bedding virtually undisturbed except at the basal shear plane. Other slides behave in both an elastic and plastic manner, and semi-consolidated sediment is deformed into overfolds. Some slides become so internally deformed that they are remolded into debris flows (Cook and Taylor, 1977; Cook, 1979a, b, c).

Much of the literature on the mass transport of ancient submarine sediment fails to distinguish between deformed strata that have moved along discrete shear planes (slides) and deformed strata with no obvious basal shear plane. Also, literature on submarine slides often fails to differentiate between translational slides (glides) and rotational slides (slumps) (Table 1). Some authors commonly use the term "slump" for any type of feature that exhibits soft-sediment deformation but has no clear basal features. Thus, some "slumps" in the literature may be translational slides rather than rotational slides (slumps) and some "slumps" may simply be deformed strata with no sharp upper

TABLE 1

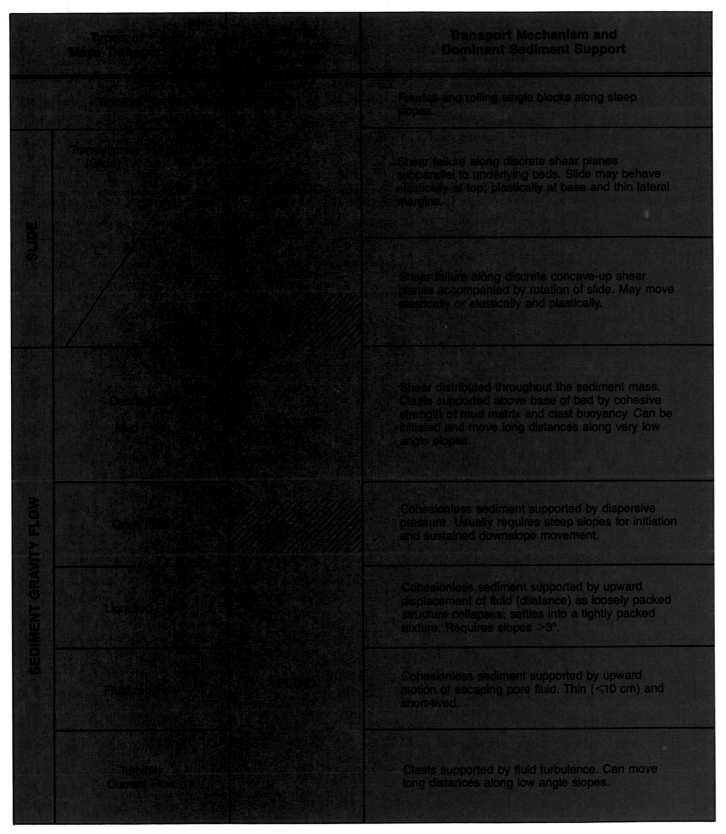

Types of Mass Transport			Transport Mechanism and Dominant Sediment Support
	Rockfall		Freefall and rolling single blocks along steep slopes.
SLIDE	Translational (Glide)		Shear failure along discrete shear planes subparallel to underlying beds. Slide may behave elastically at top; plastically at base and thin lateral margins.
			Shear failure along discrete concave-up shear planes accompanied by rotation of slide. May move elastically or elastically and plastically.
SEDIMENT GRAVITY FLOW	Debris Flow / Mud Flow		Shear distributed throughout the sediment mass. Clasts supported above base of bed by cohesive strength of mud matrix and clast buoyancy. Can be initiated and move long distances along very low angle slopes.
	Grain Flow		Cohesionless sediment supported by dispersive pressure. Usually requires steep slopes for initiation and sustained downslope movement.
	Liquefied Flow		Cohesionless sediment supported by upward displacement of fluid (dilatance) as loosely packed structure collapses; settles into a tightly packed texture. Requires slopes >3°.
	Fluidized Flow		Cohesionless sediment supported by upward motion of escaping pore fluid. Thin (<10 cm) and short-lived.
	Turbidity Current Flow		Clasts supported by fluid turbulence. Can move long distances along low angle slopes.

TABLE 1

Acoustic Record Characteristics	Sedimentary Structures and Bed Geometry	Reference To Figures
Strong hummocky bottom return, hyper-bolic and side echoes common. Weak, chaotic internal return; structureless.	Grain supported framework, variable matrix, disorganized. May be elongate parallel to slope and narrow perpendicular to slope.	15-17
Internal reflectors continuous and often undeformed; abrupt terminations. Strata of glide blocks may be unconformable or subparallel to underlying sediment.	Bedding may be undeformed and parallel to underlying beds or deformed especially at base and margins where debris flow conglomerate can be generated. Hummocky, slightly convex-up top, base subparallel to underlying beds; 10's to 1000's of meters wide and long.	18-41
Internal reflectors continuous and undeformed for short distances with deformation at toe and along base. Concave-up failure plane at head and subparallel to adjacent bedding at toe. Surface usually hummocky.	Bedding may be undeformed. Upper and lower contacts often deformed. Internal bedding at angular discordance to enclosing strata. Size variable.	
Sea floor reflectors may be hyperbolic, irregular, or smooth. Commonly acoustically transparent with few or no internal reflectors. Mounded or lens shaped with blunt termination at head. May be chaotic internally.	Clasts matrix supported; clasts may exhibit random fabric throughout the bed or oriented subparallel, especially at base and top of flow units; inverse grading possible. Clast size and matrix content variable. Occur as sheet to channel-shaped bodies cms to several 10's of meters thick and 100's to 1000's (?) of meters long; widths variable.	42-66; 126
	Massive; clast A-axis parallel to flow and imbricate upstream, inverse grading may occur near base.	67-71
Individual flow deposits very thin; may not be resolvable with present seismic reflection techniques. Repeated flows may produce a sequence of thin, even, reflectors.	Dewatering structures, sandstone dikes, flame and load structures, convolute bedding, homogenized sediment.	NONE
Thin, even, continuous, acoustically highly reflective units; onlaps slope or raised topography. Discontinuous, migrating and climbing in channel sequences.	Bouma sequences. Mms to several 10's of cm thick. 10's to 1000's of meters in length; widths variable.	72-89; 126-129; 131, 132

Table 1.—Major types of submarine mass transport on slopes and suggested criteria for their recognition (modified from Nardin et al, 1979.

Figure 10—Peri-platform ooze; evenly-bedded, gray lime mudstone with thin interbeds of argillaceous lime mudstone, Cooks Brook Formation, Middle Cambrian, Humber Arm, Western Newfoundland (from McIlreath and James, 1978; courtesy of N. P. James).

Figure 11—Thin even-bedded basin margin facies, Liberty Hall Formation, Middle Ordovician, Appalachians (from J. F. Read, 1980; courtesy of J. F. Read).

and lower boundaries.

Sediment gravity flows are defined by Middleton and Hampton (1976) as being "flows consisting of sediment moving downslope under the action of gravity. . .synonymous with mass flows. . . ." They distinguish four main types of these flows based on the forces that support the grains above the sediment-water interface during downslope transport due to gravity (Fig. 3): "(1) turbidity currents, in which the sediment is supported mainly by the upward component of fluid turbulence; (2) grain flows, in which the sediment is supported by direct grain-to-grain interactions (collisions or close approaches); (3) fluidized sediment flows, in which the sediment is supported by the upward flow of fluid escaping from between the grains as the grains are settled out by gravity; and (4) debris flows, in which the larger grains are supported by a "matrix," that is, by a mixture of interstitial fluid and fine sediment, which has a finite yield strength" (Middleton and Hampton, 1976).

Lowe (1976a) correctly draws the

Figure 12—Lower member of the Eocene Thebes Formation, Gebel Abu Had, Egypt. Slope facies consisting of thinly laminated chalks and fine-grained limestones with minor secondary chert (courtesy of P. D. Snavely, III, and R. E. Garrison).

Figure 13—Laminated lime mudstone and wackestone; pervasive sponge spicules (small light-colored blobs); larger light-colored spherules are authigenic pyrite. Bed is 4 cm thick. Lower part of Hales Limestone, Upper Cambrian, Nevada (from Cook and Taylor, 1977; photo by H. E. Cook).

distinction between fluidized sediment flow and liquefied sediment flow. In fluidized flows there is an upward movement of fluid between the grains which themselves are not moving downward. In liquefied flows the upward movement of water between grains is caused by the downward movement of grains which displaces the water upward. As Lowe (1976a) points out, except in volcanic vents and ignimbrites, where escaping gases fluidize the vitroclastic particles, fluidization as a sedimentary process under subaqueous environments probably does not occur. Biogenic gas from decaying organic matter or even gaseous hydrocarbons are unlikely to fluidize significant volumes of sediments, although they can significantly reduce shear strength and precipitate mass movements. Fluid escape structures in sediment gravity-flow deposits are probably the

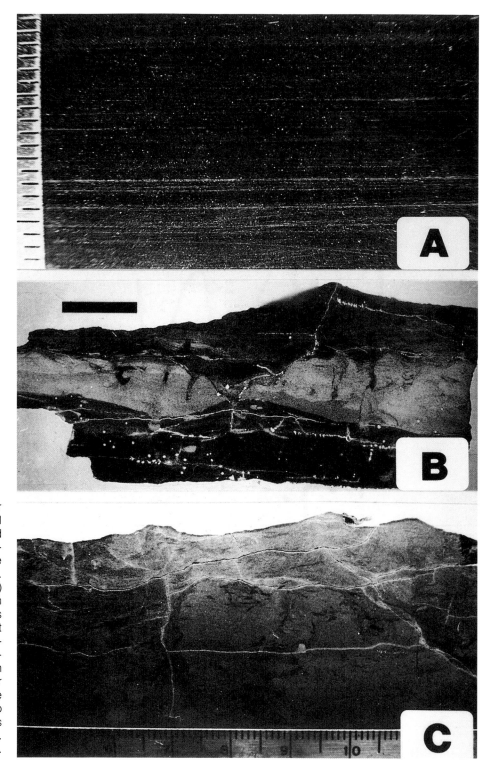

Figure 14—Lithologies in the Upper Devonian Middlesex Shale and Sawmill Creek Shale, New York. (**A**) Laminated Middlesex Shale from westernmost outcrop at Lake Erie. Clay-silt laminae couplets are undisturbed by bioturbation. Scale divisions are millimeters. (**B**) Bioturbate Sawmill Creek Shale from area east of Ithaca. Dark mudstone has been completely reworked; lighter silt layer is broken and mottled by burrowers. Bar scale equals 1 cm. (**C**) Totally bioturbated Sawmill Creek Shale from easternmost area of outcrop, near Sidney. Both silt and clay layers have been obliterated by reworking; almost no depositional structure remains. Outcrops in this facies are sparsely fossiliferous. Centimeter scale (from Byers, 1977).

result of liquefaction and not fluidization. Table 1 separates sediment gravity-flows into five types, drawing on the above distinctions as recognized by Lowe (1976b) and Nardin and others (1979).

In a grain flow, sediment is supported above the sediment-water interface by grain-to-grain interaction (that is, dispersive pressures; Middleton and Hampton, 1976). Because of these dispersive pressures, larger

grains are pushed to the zone of least shear stress near the top of the flow (Bagnold, 1954, 1956). Consequently when the grains are deposited, inverse grading theoretically develops, which is currently the main criterion for

Figure 15—(**A**, left) Large talus blocks resulting from defacement (joint-controlled rock fall) along walls of Great Abaco Canyon. Laminated material is Pleistocene ooze similar to that in Figure 6. Note vertical lamination indicating fallen blocks. Holding cage in foreground is about 1 m across, depth is 3663 m. (**B**, below) Partially cemented outcrop of manganese encrusted limestone boulders (Cretaceous in age) at base of vertical outcrop in Great Abaco Canyon. Note angular to rounded shapes of boulder set in a finer grained matrix. Holding cage is about 1 m across, depth is 3638 m (both photographs from Mullins et al, 1982).

recognizing grain-flow deposits. Middleton (1970) proposed that inverse grading is the result of a kinetic sieve mechanism whereby small grains fall downward between large grains during flow displacing the large grains upward. A kinetic sieve process may operate in sediments that have a low-density matrix. However, it is unlikely this process can account for inversely graded carbonate conglomerates that had a high density muddy matrix. Other criteria for grain flow include massive tops, grain orientation parallel to the flow direction, larger floating clasts near the top of the deposit, and injection structures at its base (Middleton and Hampton, 1976; Mullins and Van Buren, 1979).

Lowe (1976b) defines a "true" grain flow as the "gravity flow of cohesionless solids maintained in a dispersed state against the force of gravity by an intergranular dispersive pressure arising from grain interactions within the shearing sediments." Another very geologically significant part of his definition is the limitation that the "fluid interstitial to the dispersed grains is the same as the ambient fluid through which the flow is moving." Under these conditions, true grain flows require a steep slope of 18 to 30 + ° to sustain movement

(Bagnold, 1954; Lowe, 1976b; Middleton and Hampton, 1976). Lowe (1976b) further concludes that true grain flows of cohesionless sand-sized grains would produce deposits less than 5 cm thick.

Bagnold (1954) used spherical droplets composed of a lead stearate and paraffin mixture in his grain-flow experiments. To discuss the degree to which experimental data can be used to interpret field examples of possible

grain-flow deposits is beyond the scope of this paper. The reader is referred to Middleton (1970), Middleton and Hampton (1976), and Lowe (1976b) for aspects of this problem.

As pointed out by Lowe (1976b) "several processes may aid grain dispersive pressure in maintaining a dispersion against the force of gravity (Middleton, 1970): (1) the fluid interstitial to the grains may be denser

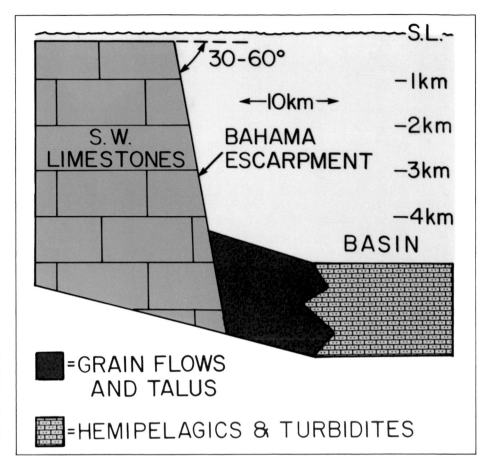

Figure 16—Schematic illustration of the facies relationships east of the modern Bahama Escarpment. Submersible observations at the base of the escarpment indicate the presence of talus blocks (see Fig. 15) and piston coring has recovered thick inversely graded grain-flow deposits. This base of escarpment facies changes rapidly seaward to hemipelagic sediments and thin distal turbidites (from Mullins and Van Buren, 1979).

than the ambient fluid; (2) shear may be transmitted downward to the flow from currents moving over its surface; (3) the interstitial fluid may become turbulent; and (4) escaping pore fluids may partially liquefy or fluidize the dispersed particles." A grain flow aided by any of the above processes is termed a "modified grain flow" by Lowe (1976b). Thus, a grain flow containing clay-sized material mixed with the interstitial fluid would be termed a modified grain flow. This type of density-modified grain flow could be mobile over slopes with gradients of 9 to 14° (Lowe, 1976b), considerably less than for true grain flows that require gradients of 18 to 30 + °.

The dominant internal mechanical behavior is plastic in the case of debris flows (the mixture of sediment and water has a finite strength). Liquefied flows, fluidized flows, and turbidity flows are considered to behave mainly as a fluid (the sediment-water mixture has no inter-

nal strength). Grain flows may behave either as a plastic or highly viscous fluid. The reader is referred to Dott (1963), Cook and others (1972), Hampton (1972, 1975), Walker (1975), Middleton and Hampton (1976), Lowe (1976a, b), Enos (1977c), and Nardin and others (1979) for a more detailed discussion of sediment gravity-flow processes.

We wish to stress that the classification shown in Table 1 and Figures 2 and 3 represents end-member concepts. Several processes can operate simultaneously during transport (Cook et al, 1972; Middleton and Hampton, 1976; Lowe, 1976b), and some of the photographs in this chapter clearly show that a single depositional unit can exhibit fabrics and sedimentary structures characteristic of more than one process. During mass transport of sediment, one process may dominate at any one point in time or space, even though several processes may operate before the sediment is deposited.

Also, keep in mind that the rock record is a picture of the final transportational, depositional, and compactional event(s). Compaction may modify clast fabric and increase the clast to matrix ratio and possibly influence interpretation of the transportational and depositional mechanism(s).

The terminology discussed above has been developed mainly from studies of ancient sediment and experimental studies. There are some differences between mass transport as observed in rocks and that observed on modern slopes where ephemeral or intermediate types of movement are recorded acoustically.

The subject of mass transport processes and classification schemes is an area of active research and is rapidly changing. As a result, some concepts discussed in this chapter and included in Table 1 and Figures 2 and 3 may become dated and should be applied prudently.

Figure 17—*Epiphyton-Renalcis* clasts in fore-reef slope, Devonian, Canning basin, western Australia (courtesy of J. F. Read).

0.8

1.0

1.2

Figure 18—10-cu-in air-gun seismic reflection profile from the base of the slope in the Florida Straits off the northwest corner of Great Bahama Bank. Note the presence of contorted-discordant reflectors overlying undisturbed, even, parallel reflectors. This superposition is indicative of the upper deformed unit having been deposited by a submarine slide. Note truncated reflectors at left indicating that slide was erosive along base. Numbers on right correspond to seconds of two-way traveltime. Horizontal, cross-photo scale is 20 km (from Mullins and Neumann, 1979).

MAJOR SEDIMENTARY UNITS ON SLOPES

In the discussion and illustrations below, slope sequences are divided into four major sedimentary units: (1) undisturbed in situ pelagic and hemipelagic sediment; (2) gravity induced mass-transport deposits; (3) bottom-current deposits; and (4) deep-water biologic buildups. The proportions of these units vary in different settings, but because slopes are especially susceptible to mass failure,

a high percentage of the section may contain allochthonous material. Upper slopes are typically characterized by pelagic-type sediment whereas lower slopes are usually dominated by sediment gravity-flow deposits and submarine slides.

Modern Pelagic Sediment

Fine-grained, undisturbed pelagic carbonate oozes may be present on any portion of a slope, but are volumetrically most important on the upper gullied part of the slope. Such

fine-grained carbonate sediments become volumetrically less important where they are winnowed by bottom currents, diluted by sediment gravity-flow deposits, or deposited below the calcite compensation depth (Mullins and Neumann, 1979).

Modern oozes surrounding carbonate platforms tend to be rich in bank-derived aragonitic and magnesian calcite muds (up to 80%; Boardman, 1978) as well as coccoliths and planktonic foraminifera which consist of low-Mg calcite (Fig. 4; Schlager

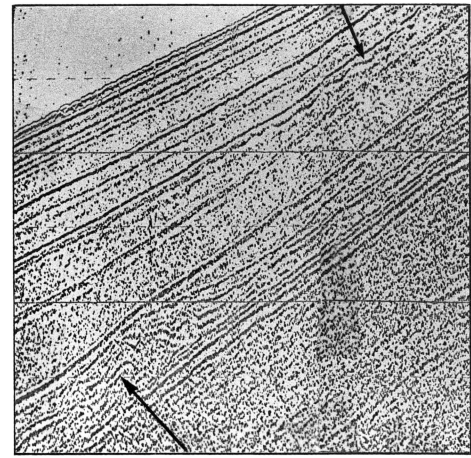

Figure 19—Uniboom seismic reflection profile from the slope north of Great Bahama Bank. Note the presence of regular, even reflectors in the upper sections of the profile which form part of an overall sigmoid terrace configuration (not seen in its entirety in this profile), as well as contorted-discordant reflectors (marked by arrows) formed by submarine sliding in the shallow subsurface (from Mullins and Neumann, 1979; courtesy of A. C. Hine).

Figure 20—3.5 KHz PDR profile illustrating a slump scar on the slope north of Little Bahama Bank. Note concave upward surface of failure indicating that the slide underwent some degree of downslope rotation. Cross photo scale is 5 km, water depth is 600 to 800 m. The numeral 6 corresponds to the location of a piston core (from Mullins, 1978).

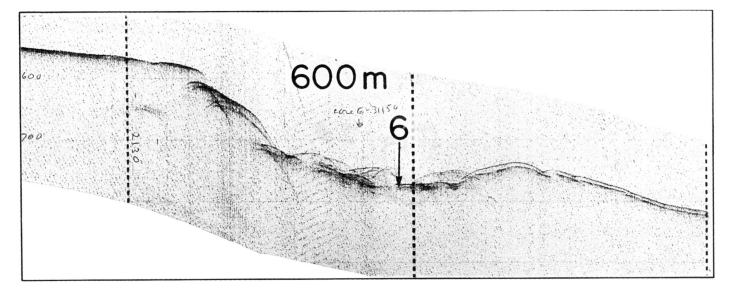

and James, 1978). Mineralogically these "peri-platform oozes" consist of aragonite, calcite and magnesian calcite in a ratio of about 3-to-2-to-1 (Schlager and James, 1978). However, the contribution of bank-derived sediments to slopes varies with the relative position of sea level with the greatest contribution of bank-derived sediment correlating with relative high stands of sea-level while the platforms are flooded (Kier and Pilkey, 1971; Hana and Moore, 1979).

Peri-platform oozes are deposited uniformly over broad areas commonly resulting in an even, parallel continuous seismic facies (Fig. 5), which

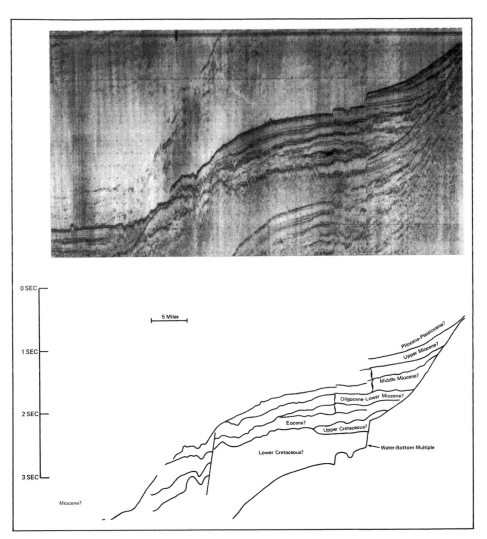

Figure 21—10 KJ sparker seismic reflection profile (top) from the west Florida slope illustrating large-scale slide scars and deposits. (Bottom) Interpretation of above profile (from Mitchum, 1978; courtesy of R. M. Mitchum).

Figure 22—Example of carbonate soft-sediment deformation in Liberty Hall Formation, Middle Ordovician, Appalachians (from J. F. Read, 1980; courtesy of J. F. Read).

Figure 23—Sequence of originally semiconsolidated slope limestone (ribbon limestone) that has deformed into a thin bed of breccia. Middle Cambrian, Cow Head Group, Western Newfoundland (courtesy of N. P. James).

Figure 24—Small-scale slide 50 cm thick within interchannel facies (see Fig. 125). Tape is 15 cm long. Lower part of Hales Limestone, Upper Cambrian, Nevada (modified from Cook, 1979a; photo by H. E. Cook).

Figure 25—Rotational slide (slump) in lower slope facies, 10 m thick which in turn is truncated by an overlying translational slide. Upper part of Hales Limestone, Lower Ordovician, Nevada (from Cook, 1979a; photo by H. E. Cook).

typically drapes antecedent topography on the sea floor. In outcrop, as viewed from research submersibles, lithified peri-platform chalks also display even, parallel continuous laminae, a few millimeters to centimeters thick (Fig. 6).

Modern peri-platform oozes are typically fine-grained (clayey silts), light gray to white in color, and show some evidence of bioturbation (Fig. 7). Where exposed near the sediment-water interface for extended periods, chalk hardgrounds may develop within 100,000 years; if buried, however, these oozes maintain their original composition for at least 200,000 to 400,000 years and may remain unlithified for tens of millions of years (Schlager and James, 1978). In thin section, chalk hardgrounds (having bored, stained, or encrusted upper surfaces) are planktonic foraminiferal-pteropod biomicrites (wackestone-packstone) cemented by micritic calcite (Fig. 8). In modern open marine settings in water depths

Figure 26—Translational slide in lower slope facies, 10 m thick and 400 m wide. Slump shown in Figure 25 occurs just above the right-hand margin of this slide. Downslope transport direction was southeast, obliquely out of the photo to the left. Hales Limestone, Upper Cambrian-Lower Ordovician, Nevada (from Cook, 1979a; photo by H. E. Cook).

Figure 27—Interior part of translational slide in Figure 26 showing large open overfolds developed in originally semiconsolidated hemipelagic limestone (from Cook, 1979a; photo by H. E. Cook).

Figure 28—Interior part of translational slide in Figure 26 showing the development of tabular clasts along the axis of an overfold. Clasts are dominantly tabular with either angular or subrounded terminations. Interclast space is filled with lime mud. Pencil is 8 cm long (from Cook, 1979a; photo by H. E. Cook).

less than 700 to 1,200 m (the boundary between intermediate and cold deep-water masses), the major cement type is magnesian calcite with up to 14 mole percent of $MgCO_3$; at greater water depths, the cement is typically low-Mg calcite of only 3.5 to 5.0 mole percent $MgCO_3$ (Fischer and Garrison, 1967; Wilber, 1976; Schlager and James, 1978; Mullins et al, 1980b). Exceptions to this are found in the Mediterranean and Red Seas with their exceptionally warm, saline bottom waters (Milliman et al, 1969; Milliman and Muller, 1973; Muller and Fabricius, 1974); magnesian calcite cements there occur at greater depths.

Ancient Pelagic Sediments

The characteristics of ancient deep-water carbonates are discussed in detail by Wilson (1969), Cook and Enos (1977b), and Scholle and others (1982). Throughout the geologic column, undisturbed slope sediment has numerous common features. Typical rock types are dark gray to black lime mudstones, calcisiltites, and wackestones. Variable amounts of insoluble residue are usually present as organic carbon, pyrite, silt size quartz grains, and clay minerals. Beds exhibit contacts ranging from planar and nearly parallel and continuous for tens of meters to more wavy and discontinuous (Figs. 9 to 12). Slope sediment is further characterized by its thin bedding to millimeter-thick laminae (Fig. 13). Preservation of laminae under quiet water conditions will depend mainly on whether the sediments formed in aerobic or anaerobic waters, and the influence these conditions exerted on burrowing organisms (Fig. 14; Byers, 1977). In silled basins only the upper part of the water column is well oxygenated, whereas at water depths below a few hundred meters the water is anoxic. In open ocean conditions, there is commonly a three-layer water system with the surface and deep waters being well oxygenated and water at intermediate depths on the slope having very low oxygen content (oxygen minimum zone). Thus, that part of the slope which is intersected by the oxygen minimum zone will have fewer burrowers and better preserved laminations than slope sediment which formed in well-oxygenated waters.

Modern Rock Falls

Carbonate platforms commonly form along the margins of rifted continents and as such, the edges of these platforms are thought to be controlled structurally at depth by faults (Ball, 1967; Sheridan, 1974; Mullins and Lynts, 1977). Such structural control, early lithification, and/or erosion of carbonate sediments can produce very steep escarpments (30 to 60°) with locally vertical to overhanging scarps (Paull and Dillon, 1980; Mullins et al, in press). Such scarps are highly unstable, and readily fail by rock fall processes (also termed "defacement" by Freeman-Lynde et

Figure 29—Base of a 3.5-m-thick translational slide in lower slope facies showing basal shear folds, developed in semiconsolidated sediment, breaking up into tabular clasts. Note that a range of clast sizes are in the process of forming. Tape is 45 cm long. Upper part of Hales Limestone, Lower Ordovician, Nevada (from Cook, 1979a; photo by H. E. Cook).

al, 1979). The bases of such escarpments are typically littered with pebble- to house-sized talus blocks (Fig. 15) which form a narrow facies belt paralleling the escarpment (Fig. 16). In the Bahamas, extensive rock fall deposits have been observed at the base of the marginal escarpments which ring the banks (depths 100 to 200 m; Neumann and Ball, 1970) and along the Blake-Bahama Escarpment in water depths of 1 to 4 km (Fig. 6; Freeman-Lynde et al, 1979; Mullins et al, in press).

Ancient Rock Falls

One example of rock fall material is illustrated in Figure 17 (see also McIlreath, 1977, and McIlreath and

James, 1978). In the absence of good stratigraphic relationships to an obvious steep scarp, suggested characteristics to distinguish rock fall deposits from other types of sedimentary conglomerates and breccias are discussed by Cook and others (1972) and Enos and Moore (1982).

Modern Slides, Slumps, and Intraformational Truncation Surfaces

Submarine slide deposits appear to be very common in ancient carbonate slope sequences. However, in the northern Bahamas very few slide masses have thus far been identified. Whether this discrepancy is due to an actual paucity of modern submarine slides or to technical limitations

(Cook, 1979a, b) is not known. However, one possible explanation may be that along current-swept deep-bank margins, submarine cementation stabilizes carbonate slope sediments by the formation of hardgrounds (Neumann, 1974). North of Little Bahama Bank, for example, evidence of submarine sliding has been found only where the near-surface sediments are not submarine lithified (Mullins et al, in prep.).

High-resolution, seismic-reflection profiles have, on occasion, defined the internal structure of modern carbonate submarine slides (Figs. 18 and 19). These deposits typically display a contorted-discordant chaotic seismic facies indicating that the slide

Figure 30—Basal shear zone in a 3.5-m-thick translational slide in lower slope facies. Here the shear plane is parallel to the underlying in situ hemipelagic beds and the shear zone is 30 to 40 cm thick. The shear folds are overturned in the direction of sliding (to the left). Pen is 10 cm long. Upper part of Hales Limestone, Lower Ordovician, Nevada (from Cook, 1979a; photo by H. E. Cook).

Figure 31—Completely conglomeratic texture developed at the base of a translational slide. Clasts are set within a pervasive lime-mud matrix. Note that the texture of the conglomerate is virtually identical to mass-flow deposits commonly inferred to be products of debris flows. Tape is 84 cm long. Lower slope facies, upper part of Hales Limestone, Lower Ordovician, Nevada (from Cook, 1979a; photo by H. E. Cook).

deformed plastically; they are commonly surrounded by undeformed, parallel reflectors. Unfortunately, there are no core data from these slide deposits in the northern Bahamas. Cores collected from the Campeche Escarpment in the Gulf of Mexico, however, clearly show the effects of submarine mass transport (Heezen and Hollister, 1971). The bedding in these cores is intensely deformed, suggesting plastic deformation during transport.

Whether the slide deposits in Figures 18 and 19 originated as translational slides (glides) or slumps is not known, as there are no data to show whether their basal shear planes were transitional or rotational. However, 3.5 KHz profiles from the slope north of Little Bahama Bank

have defined slump (rotational) scars (Fig. 20). These slump scars display concave-upward surfaces of failure suggesting that the slides underwent some degree of rotation during failure (Fig. 20). Such features may be analogous to concave-up, low-angle "intraformational truncation surfaces" (Cook and Enos, 1977a) that have been commonly observed in ancient deep-water limestone sequences (Wilson, 1969; Davies, 1977; Smith, 1977; Yurewicz, 1977).

Lower frequency seismic-reflection profiles from the west Florida slope have also revealed the presence of large scale submarine slides (Fig. 21; Mitchum, 1978). These large slides are recognized by their hummocky appearance and the presence of upslope scarps that abruptly truncate bedding (Fig. 21).

Ancient Slides, Slumps, and Intraformational Truncation Surfaces

Features described as slides and slumps range in thickness from a few centimeters to tens of meters or more (Figs. 22 to 26). Maximum three-dimensional geometries of ancient submarine slides are usually not ac-

curately known due to limited exposures.

The degree of internal deformation in slides ranges from only slight, to moderate, to complete disruption of bedding. Complete disruption of bedding occurs when the shear strength of the sediment is exceeded and the mass begins to deform plastically and move as a highly viscous debris flow (Cook and Taylor, 1977; Cook, 1979a, b, c). Not only are all gradations of intensity of internal deformation probably present in ancient slides but a single slide can exhibit various degrees of deformation. Well-exposed examples of the sequential stages of slides remolding into debris flows are found in Upper Cambrian and Lower Ordovician continental slope carbonates in the western United States (Cook, 1979a, b, c). Figures 27 to 33

Figure 32—Lateral margin of a 3.5-m-thick translational slide in lower slope facies. Figure 33 is a close-up view from the right side of photo. Upper part of Hales Limestone, Lower Ordovician, Nevada (from Cook, 1979a; photo by H. E. Cook).

Figure 33—Extreme lateral margin of a 3.5-m-thick translational slide, shown in Figure 32, showing the top of the slide where all original bedding has remolded into tabular clasts (from Cook, 1979a; photo by H. E. Cook).

illustrate various parts of these slides and Figures 34 to 37 diagrammatically show the inferred manner in which these slides underwent deformation with increased downslope movement. These data suggest that the basal and thin-tapering margins of slides are the first parts to lose their shear strength (Cook, 1979a, b, c).

Note the striking similarity between the deformation features in ancient slides (Figs. 27 to 30) and those for interpreted slide sediments on modern slopes (Figs. 18 and 19). We believe that some modern slides may have more internal deformation than is reported. Their "undeformed nature" may in some cases represent the problem of the limited resolution of conventional seismic-reflection systems (Cook and Taylor, 1977; Cook, 1979a, b).

Features that show intraformational truncation surfaces in carbonate slope sediment are illustrated in Figures 38 to 41. All three cases could represent slide scars, or alternatively they could have originated by

some type of abrasion process. Yurewicz (1977) prefers an abrasional origin for the surface in Figure 38. Figure 39, from the Permian in the Guadalupe Mountains of west Texas, exhibits a slight but distinct deformation of beds immediately at and above the truncation surface. This suggests that the beds immediately above the truncation surface have undergone soft-sediment deformation and these beds are part of the basal shear zone of a slide. This deforma-

tion is similar to, but not as well defined as, the basal shear zone in Figure 30. Wilson (1969) recognized similar "cut and fill" or slump structures in lime mudstones in Europe, Montana, and the Guadalupe Mountains of west Texas. Davies (1977) presents a lucid argument for the truncation surfaces in Figures 40 and 41 having formed by a gravity-slide mechanism rather than by some type of current scouring or other erosional process.

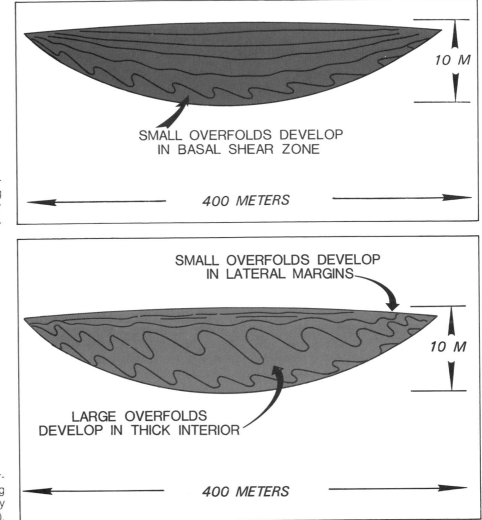

Figure 34—Model of progressive deformation of semiconsolidated slide moving downslope; stage 1 (drawing by H. E. Cook).

Figure 35—Model of progressive deformation of semiconsolidated slide moving downslope; stage 2 (drawing by H. E. Cook).

Sediment Gravity-Flow Deposits

Extensive piston coring of carbonate slopes in the northern Bahamas has documented the existence of sediments deposited by mass flows. However, turbidity currents and debris flows appear to be the dominant transport mechanisms for the downslope movement of coarse detritus on modern carbonate slopes.

In the ancient record, carbonate mass-flow deposits in slope and basinal settings are common throughout the geologic column on a world-wide basis (for example, Pray and Stehli, 1962; Thomson and Thomasson, 1969; Wilson, 1969, 1975; Cook et al, 1972; Mountjoy et al, 1972; Conaghan et al, 1976; Cook and Enos, 1977b; Keith and Friedman, 1977; Cook, 1979a, 1982;

Krause and Oldershaw, 1979; Pfeil and Read, 1980; Cook and Egbert, 1981a; Crawford, 1981; McGovney, 1981). Indeed, to find carbonate base-of-slope sequences with no hint of allochthonous sediment is most unusual. Of the five end-member types of sediment gravity-flow deposits, debris flows and turbidity-current flows are the best documented and appear to be the dominant processes for transporting large volumes of sediment-fluid mixtures downslope. We have no examples of liquefied or fluidized flows in modern or ancient carbonates.

Modern Debris-flow Deposits

In a debris flow, sediment is supported above the sediment-water interface by matrix strength and buoyancy (Cook et al, 1972; Mid-

dleton and Hampton, 1976; Hampton, 1975, 1979), which allows larger clasts to "float" along in a muddy matrix. Thus, the main criterion for recognizing debris flow deposits on modern carbonate slopes is massive, poorly-sorted, mud-supported debris commonly with floating clasts ranging from sand to gravel size (Figs. 42 to 44). Occasionally, modern debris flow deposits are capped by massive, normally graded carbonate sands (Figs. 42 and 43; Crevello and Schlager, 1980; Mullins et al, in prep.) suggesting that turbulence may have played an important role in supporting grains at the top of the flows (Hampton, 1972; Cook et al, 1972). Geometrically, debris flow deposits in Exuma Sound have been described as "sheets" that exhibit both lateral and downslope variations in grain size and

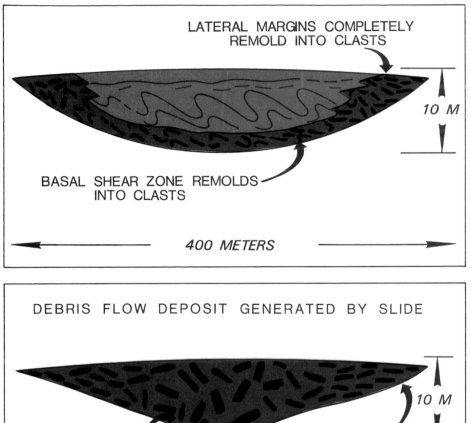

Figure 36—Model of progressive deformation of semiconsolidated slide moving downslope; stage 3 (drawing by H. E. Cook).

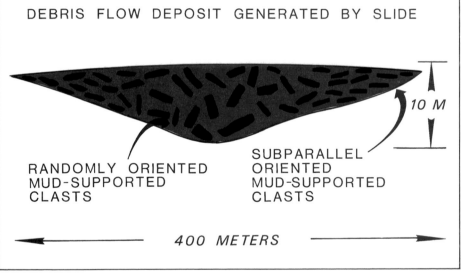

Figure 37—Model of progressive deformation of semiconsolidated slide moving downslope; stage 4. At this stage slide has completely remolded into clasts and mud and moves as a debris flow (drawing by H. E. Cook).

composition (Crevello, 1978; Crevello and Schlager, 1980). Such debris flow deposits appear to be most common on the lower slope (Mullins et al, in prep.) but have also been found to extend onto the adjacent basin floor over 100 km from their source (Crevello and Schlager, 1980).

Ancient Debris-flow Deposits

Coarse-textured debris-flow deposits occurring in both sheet and channel forms afford a striking contrast to the laminated dark lime mudstones of the enclosing pelagic and hemipelagic slope facies. This contrast is all the more evident because the debris-flow deposits have a resistant character and are usually

lighter colored than the enclosing host facies. What constitutes field evidence for debris flow (Table 1) is generally well accepted (Cook et al, 1972; Hampton, 1975; Walker, 1975; Middleton and Hampton, 1976; Enos and Moore, 1982). Figure 45 summarizes the main characteristics of Devonian debris-flow deposits in Canada that are composed of both shallow-water and deeper-water clasts (Cook et al, 1972). Many of these features are common to carbonate debris flows throughout the geologic column that originated at platform margins. Debris flows that originated in deeper water by the remolding of submarine slides can consist totally of dark-colored lime mudstone clasts (Cook,

1979a, b, c).

McIlreath and James (1978) proposed that shoal-water carbonate margins can be considered in terms of two end-member models — "depositional margins" with low depositional relief between the margin and slope, and "by-pass margins" where the margin-to-slope transition is separated by a cliff or submarine escarpment with up to 100 to 300 m or more of relief.

Debris flows can originate in areas of both low depositional relief ("depositional margins") as well as high depositional relief ("by-pass margins"). Field data at the well-exposed Devonian carbonate buildups of Alberta, Canada, demonstrate that

Figure 38—Intraformational truncation surface, Rancheria Formation, Mississippian, Sacramento Mountains, New Mexico. Width of outcrop about 15 m (from Yurewicz, 1977; photo by H. E. Cook).

Figure 39—Intraformational truncation surface, Bone Spring Formation, Permian, Guadalupe Mountains, west Texas. Width of outcrop about 5 m (photo by H. E. Cook).

impressive debris flows with clasts up to 25 × 50 m in cross section can be initiated in areas of low depositional relief and that, once initiated, the flows can transport very coarse-textured material 10 km or more across slope angles of 1° or less (Cook et al, 1972). Figures 46 to 52 illustrate these Devonian debris-flow sheet deposits initiated from "depositional margins."

The mobility of debris flows is further exemplified by a widespread debris-flow sheet (apron) from the Middle Devonian of the Yukon Territory, Canada. This ledge-forming sheet (Fig. 53) which lies within deep-water graptolitic lime mudstone is ac-

tually two or more debris-flow events separated by turbidites (Fig. 54). The debris originated at a low relief stromatoporoid bank margin 50 to 75 km away, yet the debris-flow sheet contains shoal-water clasts (pellet grainstone) up to 7 × 7 m in cross section (Figs. 54 and 55).

In this same area of the Yukon Territory, spectacular channels 100 to 200 m deep and 200 to 400 m wide are cut in deep-water, graptolitic lime muds of Middle Devonian age (Figs. 56 and 57). These channels are filled with shoal-water derived clasts that originated from a low relief bank margin 50 to 60 km from the channels.

High-relief "by-pass margins" are well illustrated in the Cambrian of Canada (McIlreath, 1977) and in the Cretaceous of Mexico (Enos, 1977a; Enos and Moore, 1982). Other examples of carbonate debris-flow deposits are shown in Figures 58 to 66.

Modern Grain-flow Deposits
Theoretically, very steep slopes (18 to 30 + °) are required to initiate and maintain grain-to-grain interactions. Thus, "true" unmodified grain flows are probably not significant mechanisms for long distance transport. Modified grain-flow deposits are found only locally at the

Figure 40—Large intraformational truncation surface in argillaceous and cherty limestones, Hare Fiord Formation, Permo-Pennsylvanian, Ellesmere Island, Arctic Archipelago. Note smooth curved concave-up (listric) geometry of the truncation surface and lack of obvious deformation of beds below or above truncation surface, downdip thickening of sedimentary fill, with highest beds parallel with beds below truncation surface. Shadow at lower center (arrow) just below truncation surface is of a helicopter. Width of view about 150 m (from Davies, 1977; courtesy of G. R. Davies).

Figure 41—Large truncation surfaces (arrows) from same area as in Figure 40. Width of view is about 500 m (from Davies, 1977; courtesy of G. R. Davies).

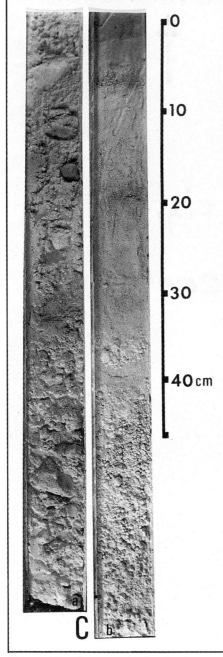

Figure 42—Photograph of split piston core recovered from the base of the slope at 850 fathoms of water in Exuma Sound, Bahamas. Note poorly sorted, mud-supported debris at the base of the core which grades upward into massive graded sands. The lower portions of this core are believed to have been deposited by a debris flow. Section "**b**" fits above section "**a**" (from Crevello, 1978; courtesy of P. D. Crevello).

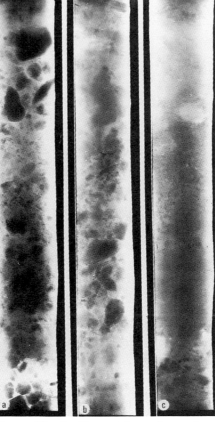

Figure 43—Positive print x-radiographs of a muddy debris-flow deposit which grades upward into massive carbonate sands. Core from 850 fathoms of water in Exuma Sound, Bahamas. Individual core sections are 37 cm long (from Crevello, 1978, courtesy of P. D. Crevello).

base of steep, carbonate escarpments in narrow facies belts that parallel the strike of the escarpment (Figs. 15, 16 and 67; Mullins and Van Buren, 1979). The top of the grain flow described by Mullins and Van Buren (1979) was turbulent as suggested by normal grading of the sediment. In addition, the grain flow contained lime mud. Both turbulence and a lime mud matrix probably aided dispersive pressure in supporting the grains above the base of the bed.

Ancient Grain-flow Deposits

In our experience, carbonate deposits that can be reasonably inferred to have resulted from true grain flows are rare in the ancient record. Perhaps this is to be expected due to the very high slopes required to sus-

Figure 44—Photograph of split piston core containing a debris-flow deposit. Note poorly sorted, mud-supported clasts derived from the upper slope. Core from 1000 m of water at the base of slope north of Little Bahama Bank. Total core length shown is 25 cm (photo by H. T. Mullins).

tain true grain flow, and the conclusion that true grain flows probably cannot form thick sedimentation units (Lowe, 1976a). Steep slopes will be areally restricted to special geological circumstances and this, combined with locating beds a few centimeters

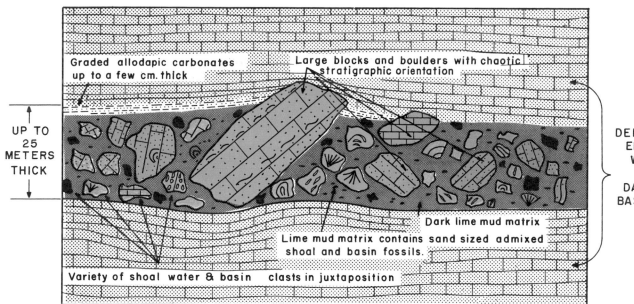

Figure 45—Generalized sketch of the major characteristcs of Upper Devonian carbonate debris-flow deposits, Rocky Mountains, Alberta, Canada (from Cook et al, 1972).

thick, limits the occurrence and geological importance of true grain flows.

Probable grain-flow deposits modified by the presence of a lime mud matrix are shown in Figures 68 and 69. Both of these examples are from the Upper Devonian Ancient Wall carbonate complex (Cook et al, 1972). Figure 68 shows a 50-cm-thick deposit that occurs 800 m from the margin of the Ancient Wall buildup. Reverse grading involves clasts up to 8 cm in maximum diameter. This bed is a mixture of clasts and interstitial lime mud matrix. Slopes on the bank margin were no more than 5 to 10° over a horizontal distance of 650 m (Mountjoy, 1967). These gradients decrease rapidly basinward to 1 or 2° (Fig. 46). Figure 69 is a 1-m-thick deposit with reverse grading of clasts ranging up to 5 cm in maximum diameter. This bed occurs about 4 km from the margin of the Ancient Wall carbonate buildup (Cook et al, 1972). It may have been initiated on slopes of 5 to 10° but, after a transport distance of less than 1 km, it moved across very low gradients, probably less than 2°. Its maximum transport distance is unknown.

Figure 70 shows some characteristics that may result from grain flow. This conglomerate exhibits probable reverse grading within the basal 10 cm of the bed overlain by massive(?) or normal grading(?). Parallel orientation of tabular clasts is clearly developed. A more pronounced example of reverse grading of tabular clasts is seen in Figure 71. Here the bed is capped by rippled, sand-sized carbonate grains. The clasts exhibit parallel orientation in the basal part of the bed and upslope imbrication in the upper half of the bed. Both of these conglomerates have a muddy matrix but contain a high concentration of clasts. These grain flows may have been modified by both a muddy matrix and some turbulence. Slope angles are not known for these two examples but their position on a continental slope and base of slope setting (Cook and Taylor, 1977; Cook, 1979a), and other field relations suggest that gradients were probably less than 5 to 10°.

Modern Turbidity-current Deposits
Sediments deposited from turbidity currents (turbidites) are obvious and volumetrically important components

of lower carbonate slope facies. In Exuma Sound, sediment gravity-flow deposits make up 25% of the upper 10 m of sediment found on the lower slope (Crevello and Schlager, 1980). These deposits typically exhibit sharp, erosional lower contacts, normal grading, and Bouma (1962) sequences (Figs. 72 to 75). Turbidites found on the lower slope tend to be thick (from 3 to 5 m) and coarse-grained deposits (commonly with gravel-sized clasts) (Fig. 72). In terms of Bouma (1962) sequences, lower slope turbidites commonly contain A, A-B, or A-C sequences (Fig. 74). Base-cut-out turbidites (B-E, C-E, D-E; Fig. 75) are uncommon on these lower slope environments. In the adjacent basins, however, thin (10 to 30 cm), fine-grained (calcisiltites and calcarenites), base-cut-out turbidites are much more common, comprising approximately 35% of the upper 10 m of basin fill (Bornhold and Pilkey, 1971; Mullins and Neumann, 1979).

Because the entire length of a shallow carbonate bank margin can supply sediment to the adjacent slope, sedimentation occurs from a "line" source (Schlager and Chermak, 1979; Crevello and Schlager, 1980; Mullins

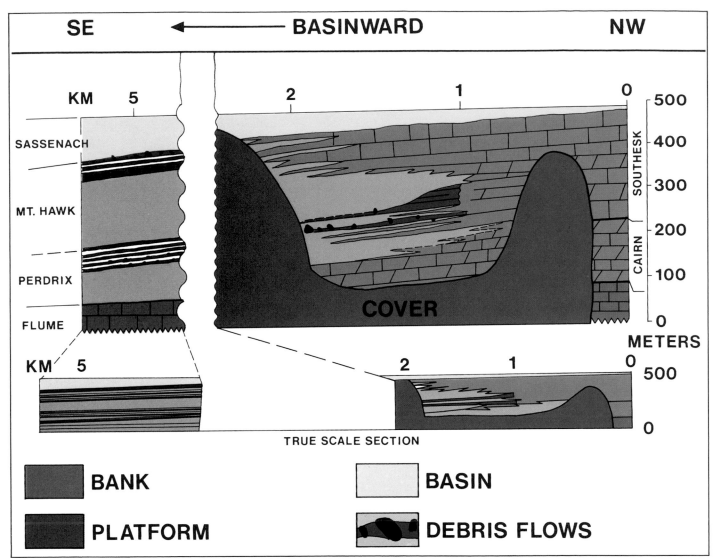

Figure 46—Stratigraphic cross-section at Ancient Wall carbonate complex, Upper Devonian, Alberta, Canada. Arrow closest to bank margin points to the same debris flow bed pointed to in Figures 47 and 48 and illustrated in Figures 49 and 50. Arrow farthest from the bank margin points to a series of sediment gravity flow deposits illustrated in Figure 52 (modified after Cook et al, 1972).

Figure 47—Southeast margin of the Ancient Wall carbonate complex, Upper Devonian, Alberta, Canada. Arrow points to debris flow bed shown in Figures 48 to 50. Skyline is at the 1 km mark on Figure 46 (modified after Cook et al, 1972; photo by H. E. Cook).

Figure 48—Southeast margin of the Ancient Wall carbonate complex, Upper Devonian, Alberta, Canada. Looking northwest, within the Perdrix basin facies, toward the buildup margin which begins at the skyline. Stratigraphic top to left. View shows the light-colored resistant nature of the debris sheet which is enclosed in dark, less resistant basin facies. The top arrow in photo points to a large knob at the stratigraphic top of the debris sheet which is a single clast about 10 × 30 m in cross section. The bottom arrow in photo points to a large knob which is a single clast 25 × 50 m in cross section (modified after Cook et al, 1972; photo by H. E. Cook).

et al, in prep.) rather than a "point" source that is commonly characteristic of terrigenous canyon-fan systems. Thus, the facies patterns that commonly develop on carbonate slopes are turbidite "aprons" or "sheets" (Fig. 76) on the lower slope that parallel the edge of the carbonate platform (Schlager and Chermak, 1979). Geometrically, these turbidite aprons or sheets are significantly different than clastic submarine fan geometries generated from point sources, in that they lack the fan morphology, are laterally much more

continuous, and do not develop systematic inner, mid, and outer fan facies.

Ancient Turbidity-current Deposits

Carbonate turbidites are very common on slope, base of slope, and more distal basinal settings (Fig. 77). As in clastic turbidites, carbonate turbidites are quite diverse in their sedimentary structures, textures, grain types, bed geometry, and origin.

Cobble-bearing carbonate turbidites are usually restricted to slope and near-slope settings where depositional

gradients are the highest (Figs. 78 to 81). There are exceptions to this, as seen in Figure 82 where a 15-cm-thick cobble-bearing turbidite containing shoal-water clasts was transported at least 75 km from a platform margin (see also Crevello and Schlager, 1980). Sand to pebble-sized carbonate turbidites can be found on slopes as well as in basinal settings (Figs. 83 to 87).

Some sand-sized turbidites appear to be genetically related to debris-flow deposits and represent the uppermost more dilute turbulent part of

Figure 49—Looking southeast (basin-ward) at the same portion of the debris sheet as in Figure 48. Stratigraphic top at top of photo. Arrow on right side of photo points to same 10 × 30 m clast in Figure 48; arrow on left side of photo points to same 25 × 50 m clast in Figure 48 (modified from Cook et al, 1972; photo by H. E. Cook).

Figure 50—Looking southeast (basin-ward) at the same portion of debris sheet in Figure 49. Large knob is the same 10 × 30 m clast seen in Figures 48 and 49. Note fairly planar base (modified from Cook et al, 1972; photo by H. E. Cook).

Figure 51—Textures in debris sheets shown in Figures 48 to 50. Note poorly sorted nature and variety of clast types. Large rectangular clast is a stromatoporoid. White circle is 2 cm wide (photo by H. E. Cook).

the debris flow (Figs. 45, 88, and 89; Cook et al, 1972; Krause and Oldershaw, 1979). This two-mechanism origin of debris-flow/turbidity-current-flow couplet is supported by experimental data on clastic debris flows (Hampton, 1972). The normally graded ripple-laminated upper part of the conglomerate in Figure 80 was probably the product of a turbidity flow, but the lower part may have formed under grain-flow and turbidity-flow conditions.

Carbonate turbidites that form part of a submarine fan are discussed later in this chapter under "Models" (Figs. 128 to 131).

Bottom-current Deposits

Modern

Along open-seaways in the northern Bahamas strong bottom (contour) currents (velocities of up to 60 cm/sec) associated with wind-driven surface circulation play an important role in slope sedimentation. Such bottom-currents are responsible for winnowing, mobilizing, redistributing, and depositing large volumes of carbonate sand. The bathymetric map of the northern Straits of Florida (Fig. 90) clearly indicates the presence of hemiconical-shaped sediment drifts up to 100 km long, 60 km wide, and 600 m thick off the northwest corners of both Little and Great Bahama Bank. Seismic reflection profiles of these sediment drifts reveal wedge-shaped geometries in longitudinal section (Fig. 91), and mound geometries in transverse section (Fig. 92). Internal reflectors are mostly oblique progradational and downlap on the underlying unconformity (Figs. 91 and 92). Core and surface grab sample data indicate that most of the sediment on these features is sand sized (Fig. 93), consisting of a mixture of pelagic and bank-derived material (Mullins et al, 1980a). The bank-derived sands are transported off the lee (west) side of the Bahama Banks by storm waves and currents (Hine and Neumann, 1977), onto the adjacent slope where they are remobilized into sand waves and ripple marks (Fig. 94), then transported along the slope and ultimately deposited on the sediment drifts. Rocks recovered from the drifts reflect the influence of strong bottom currents as they are typically coarse, grain-supported lithologies that commonly show evidence of crude cross-stratification (Fig. 95; Wilber, 1976; Mullins et al, 1980a).

Ancient

Well-documented examples of carbonate contourite deposits are sparse. In the absence of good paleocurrent data, clear facies associations, and regional trends in the slope, suspected contourites can often be ascribed to other origins.

Figure 96 is believed to represent thin-bedded carbonate contourites (Cook and Taylor, 1977; Cook and Egbert, 1981b). These calcarenites occur on the upper part of a north-

Figure 52—Arrow at left side of Figure 46 shows location of this photo. Looking southeast (basinward) at a series of stacked debris flow sheets, modified grain flow deposits, and turbidity-current deposits, Thornton Ridge southeast margin Ancient Wall carbonate complex. Stratigraphic top to right. Many of the sheets are separated by a few centimeters to meters of dark-colored basin facies. This stack of debris sheets totals about 50 m in thickness (from Cook et al, 1972; photo by H. E. Cook).

Figure 53—Looking northward at a ledge forming debris-flow sheet (arrows) which has a fairly uniform thickness of about 20 m over an area of at least 10 to 20 sq km. Shoal water clasts in debris flow sheet (Figs. 54 and 55) were transported at least 50 to 75 km from a bank margin to the east. Prongs Creek Formation, Middle Devonian, northern Mackenzie Mountains, Yukon Territory, Canada (photo by H. E. Cook).

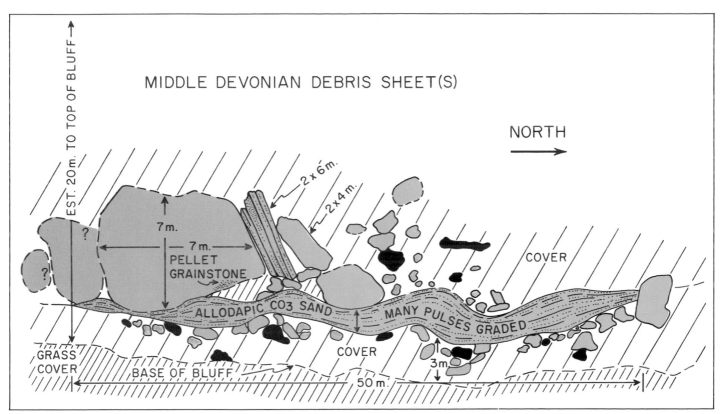

Figure 54—Sketch of ledge forming debris flow sheet(s) shown in Figure 53. Debris-flow deposits contain clasts of pellet grainstones up to 7 × 7 m in cross section (sketch drawn by P. N. McDaniel).

Figure 55—Texture of debris flow deposit shown in Figures 53 and 54. Most of the darker clasts are pellet grainstones. Light weathering matrix is lime mud. White circle is 2 cm wide (photo by H. E. Cook).

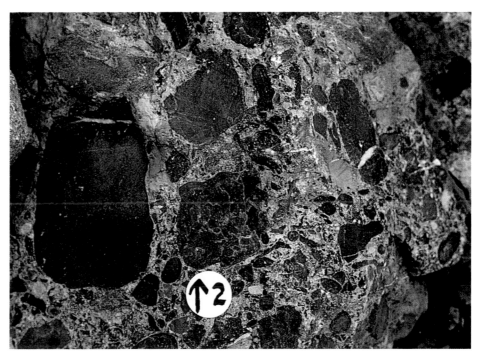

trending Paleozoic continental slope interbedded with pelagic and hemipelagic lime mudstone beds. The grains comprising these calcarenites are shoal-water-derived algae particles.

Paleocurrent data from these current-rippled calcarenites indicate a northerly current direction parallel to the paleoslope (that is, approximately perpendicular to the paleocurrent data on the carbonate mass-transport deposits). As pointed out by Cook and Taylor (1977) and Cook and Egbert (1981b), the rippled calcarenites do not appear to be the product of muddy turbidity currents. A different origin is indicated by: (a) the near perfect hydraulic sorting; (b) common lack of a mud matrix; (c) sharp lower and upper contacts; (d) laterally continuous evenly spaced current ripples; and (e) transport direction parallel to the slope. These sediments most likely are the result of winnowing of previously resedimented material by strong bottom-hugging contour currents (15 to 30 cm/sec). Similar limestone beds deposited on a Cretaceous continental slope have

Figure 56—Debris flow channel about 150 to 200 m deep and 300 to 400 m wide. Stratigraphic top at top of photo. Channel is cut in graptolite lime mudstones which are dipping about 30° into the photo. Note pronounced concave-up base. Channels have fairly flat tops. This large channel is one of at least 25 similar debris flow channel deposits that occur at the same stratigraphic horizon over a lateral distance of about 10 to 15 km. Channels occur about 50 to 60 km from bank margin. Prongs Creek Formation, Middle Devonian, northern Mackenzie Mountains, Yukon Territory, Canada (photo by H. E. Cook).

Figure 57—Debris flow channel deposits from channel shown in Figure 56. Cobble-sized clasts are mainly pellet grainstones and stromatoporoids set within a pervasive lime mudstone matrix (photo by H. E. Cook).

been ascribed a contourite origin (Bein and Weiler, 1976).

Modern Biological Buildups

Many geologists today still consider coral constructed mounds (bioherms) or reefs to be excellent indicators of warm, shallow, tropical environments despite the warnings of Teichert (1958) that ahermatypic corals are capable of constructing biohermal features in cold, deep waters. In the Florida Straits, numerous lithified, Pleistocene coral mounds, informally termed "lithoherms," have been discovered on the lower slope in water depths of 600 to 700 m west of Little Bahama Bank (Neumann et al, 1977). These lithoherms form a more or less continuous facies belt of current-oriented coral mounds over 200 km long, 10 to 15 km wide, and up to 70 m thick parallel to the edge of the platform making it one of the

Figure 58—Debris flow deposit. Inner submarine fan feeder channel about 10 to 15 m deep and 400 m wide that occurs at or near base of slope. The hand is on a single rectangular clast 3 × 15 m in cross section (dashed line) that lies subparallel to base of channel. Other clasts are randomly oriented and set within a lime mud matrix. Black solid line outlines base of channel. Top of channel is not visible in photo. Upper part of Hales Limestone, Lower Ordovician, Nevada (modified from Cook, 1979a; photo by W. W. Chamberlain).

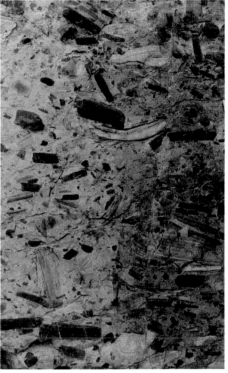

Figure 59—Thin lateral margin of debris flow channel deposit shown in Figure 58. Tabular clasts in lateral margins of channel are normally oriented subparallel to bedding but clast size distribution is still random. Thickness of bed shown is 60 cm.

largest coral "reefs" in the Bahamas (Fig. 1; Mullins and Neumann, 1979). Similar, although unlithified, coral mounds are also present on the lower slope north of Little Bahama Bank and on the Blake Plateau (Stetson et al, 1962; Mullins et al, 1981). Some deep-water coral buildups are also very diverse, containing as many as 11 genera and 16 species of ahermatypic corals many of which are solitary or weakly branched forms (Mullins et al, 1981).

On seismic reflection profiles these mounds appear as hummocky, chaotic masses that would classically be interpreted as submarine slide deposits (Fig. 97). However, detailed examination of such profiles indicates the mounds are discrete bodies separated by undisturbed, parallel reflectors (Fig. 98), suggesting that these mounds are in situ constructional features (Mullins and Neumann, 1979). Bottom photographs of lithoherms taken from the DSRV (Deep Submergence Research Vehicle) ALVIN indicate dense growths of ahermatypic coral

on the upcurrent ends of the mounds and an absence of coral on the downcurrent ends (Fig. 99). The corals form an organic framework (Wilber, 1976; Neumann et al, 1977) which baffles and traps other carbonate grains (Figs. 100 and 101). These materials are cemented by magnesian calcite which produces common geopetal structures (Fig. 102). Such sedimentological buildups intermittent or concurrent with submarine cementation also produce an "onion-skin" internal structure (Fig. 99A).

Ancient Biological Buildups

In contrast to the modern, which has numerous examples of deep-water ahermatypic coral buildups, analogous deposits in the rock record are very rare. In fact, only a handful of examples have appeared in the literature (for example, Squires, 1964; Coates and Kauffman, 1973; Stanley, 1979). Whether this disparity represents an actual paucity of ancient deep-water bioherms or simply their misinterpretation remains to be

seen. However, considering the commonality of modern examples, it is likely that more ancient examples of deep-water bioherms will soon be discovered (Mullins et al, 1981).

In those examples known from the rock record (ranging in age from Triassic to Pliocene), the deep-water

Figure 60—Areally very widespread 100 m thick megabreccia debris flow sheet(s). Arrow at left of photo is pointing to man standing at the top of this debris. Two large single clasts several meters across are shown just below man. Single clasts up to about 20 × 100 m in cross section have been found. Volume of this carbonate debris is estimated to be 60 to 140 cu km (see Johns et al, 1981). Hecho Group, Eocene, south-central Pyrenees, Spain (photo by H. E. Cook).

Figure 61—Shoal water foraminiferal grainstone clasts in debris flow sheet(s) shown in Figure 60. Matrix surrounding cobble-sized clasts is a mixture of sand-sized carbonate particles and peri-platform derived lime mud (photo by H. E. Cook).

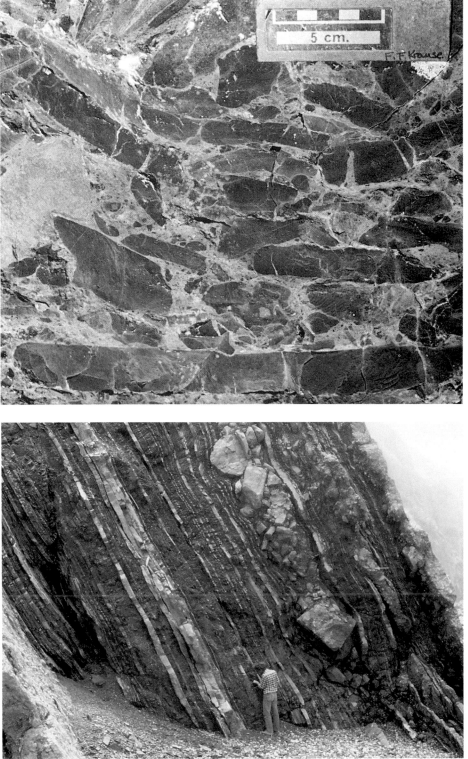

Figure 62—Tabular clasts with subparallel orientation set within a lime mud matrix. Sekwi Formation, Lower Cambrian, Mackenzie Mountains, Northwest Territories, Canada (from Krause and Oldershaw, 1979; courtesy of F. F. Krause).

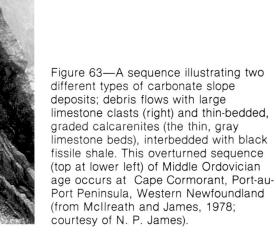

Figure 63—A sequence illustrating two different types of carbonate slope deposits; debris flows with large limestone clasts (right) and thin-bedded, graded calcarenites (the thin, gray limestone beds), interbedded with black fissile shale. This overturned sequence (top at lower left) of Middle Ordovician age occurs at Cape Cormorant, Port-au-Port Peninsula, Western Newfoundland (from McIlreath and James, 1978; courtesy of N. P. James).

bioherms appear as lenticular thickets with a framework usually constructed by a single species of coral (Squires, 1964; Coates and Kauffman, 1973). Large volumes of coral debris are also typical of such buildups, which appear to have developed in current-swept environments. The Miocene-Pliocene coral thickets of New Zealand are up to 3.4 m thick, 36.6 m long, and about 75 m in diameter (Squires, 1964). In addition to the corals themselves, a host of other calcareous invertebrates are commonly associated with these deposits (Stanley, 1979). Figure 103 is an example of an ancient biological buildup on a Paleozoic carbonate

Figure 64—Debris flow with large white boulders that are *Epiphyton* boundstones. Cow Head Group, Upper Cambrian, Western Newfoundland (courtesy of N. P. James).

Figure 65—Polymictic debris flow breccia, lower slope. Shady Dolomite Cambrian, Appalachians (courtesy of J. F. Read).

slope (see also Coates and Kauffman, 1973).

EARLY DIAGENETIC FEATURES

Modern

Because of the interaction of bottom currents, endolithic organisms, and the chemical instability and/or activity of carbonate minerals, early diagenetic features such as submarine cemented nodules, hardgrounds, and secondary bored porosity are common in some modern and ancient carbonate slope environments.

On the slope north of Great Bahama Bank piston coring has revealed the presence of in situ submarine cemented nodules that "float" in a sandy mud matrix (Fig. 104; Mullins et al, 1980b). Nodules are up to 6 cm across and irregular in shape with numerous projections of whole grains (Fig. 105). Constituent grains are entirely pelagic in origin, such as planktonic foraminifera and pteropods, or foram pteropod

Figure 66—Conglomeratic debris flow deposit, slope facies. Nodules composed of early cemented fine-grained carbonate, matrix composed of mud-supported fine-grained skeletal sand. Thebes Formation, Eocene, Egypt (courtesy of P. D. Snavely, III, and R. E. Garrison).

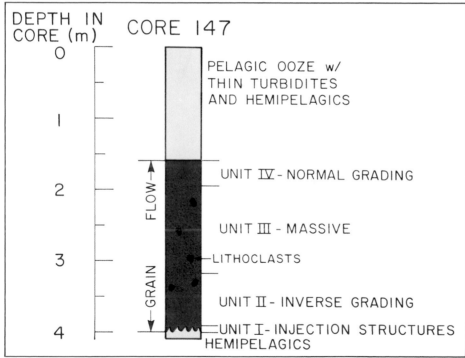

Figure 67—Schematic illustration of sedimentary structures in a modified carbonate grain flow deposit recovered by piston coring from 4000 m of water at the base of the Bahama Escarpment east of Little Bahama Bank. The basal part of the deposit exhibits inverse grading which becomes massive upward with "floating" lithoclasts. The top of this deposit exhibits normal grading which probably resulted from a transition from grain support by dispersive pressures at the basal part of the flow to grain support by turbulence at the top (from Mullins and Van Buren, 1979).

biomicrite intraclasts. The grains are typically cemented by amorphous to pelloidal, or occasionally bladed magnesian calcite (Fig. 106). The nodules appear to form on the middle portion of this gentle (1 to 2°) slope where bottom currents and bioturbation are active (Mullins et al, 1980b).

Facies relationships indicate gradual transitions downslope from hardgrounds (depths less than 375 m) to nodular sediment (375 to 500 m) to soft peri-platform oozes (depths greater than 500 m; Figure 107).

Submarine cemented hardgrounds that typically have intensely bored

and/or encrusted upper surfaces are widespread on carbonate slopes along the current-swept open-seaways in the northern Bahamas (Figs. 108 and 109). These hardgrounds are thin (less than about 20 cm), yet they are areally widespread cementation surfaces (Wilber, 1976). Texturally they are

Figure 68—A probable 0.5 m thick modified grain-flow deposit within the Perdrix basin facies about 800 m from the southeast margin on Ancient Wall buildup. The lower and upper contacts are shown in photo. Reverse grading involves clasts up to 8 cm in maximum diameter. Pencil is 15 cm long. Upper Devonian, Alberta, Canada (from Cook et al, 1972; photo by H. E. Cook).

Figure 69—One of the redeposited sheets shown in Figure 52, about 4 km from the southeast margin of Ancient Wall buildup. Probable modified grain-flow deposit 1 m thick. Reverse grading involves clasts ranging up to 5 cm in maximum diameter. The dark-colored resistant clasts are partially silicified fossil fragments. Ruler in inches (top), centimeters (bottom); the upper and lower contact is not seen in this photo (from Cook et al, 1972; photo by H. E. Cook).

commonly grain-supported lithologies cemented by pelloidal magnesian calcite (Fig. 110). Endolithic organisms bore into these submarine cemented surfaces producing both macro- (Figs. 111 and 112) and micro-borings (Fig. 113) that can produce up to 50% secondary porosity (Wilber, 1976; Wilber and Neumann, 1977; Zeff and Perkins, 1979). Boring sponges can also be important sediment producers. Moore and others (1976) report that more than 5% of the total volume of island slope sediments off Jamaica, and a mean of 24% of the silt fraction, consist of clinoid sponge chips.

Ancient

The degree of early marine cementation of slope material can range from the patchy development of pseudoclasts (Hopkins, 1977), to a dense network of nodules (Snavely, 1981), to a more uniform cementation, as suggested by the remolding of submarine slides into clasts (Cook, 1979a, b, c).

Nodular limestones that formed on an Eocene carbonate slope in Egypt have recently been reported by Snavely (1981). These early diagenetic nodules (Figs. 114 to 116) bear a striking resemblance in size and shape to nodules forming on modern slopes

(Mullins et al, 1980b; Figs. 104 and 105). Nodular limestone and pseudobreccias on slopes are probably more common than currently recognized (Hopkins, 1977). Early marine cementation is probably a fairly pervasive event on some slopes as indicated by the common occurrence of semiconsolidated pelagic and hemipelagic lime mudstones that are involved in submarine sliding (Figs. 22 to 30). These slides moved the uppermost 1 to 10 m of sediment, and

thus their semiconsolidated nature is unlikely to be the result of compaction alone.

The role that early marine cementation may play in the development of clasts and the initiation of conglomeratic mass-flows in the Devonian of Canada was discussed by Cook and others (1972) and Hopkins (1977). More recently Snavely (1981) has shown that nodular limestones and hardgrounds that formed on Eocene carbonate slopes were, in places, displaced downslope as debris flows.

MASS-TRANSPORT FACIES, FACIES SEQUENCES, FACIES ASSOCIATIONS, AND MODELS

Until the early 1970s, there were very few carbonates that had been interpreted as sediment gravity-flow deposits (Cook et al, 1972; Mountjoy

Figure 70—Possible modified grain-flow deposit (lower slope) with reverse grading (right side of photo) of clasts up to about the 10 in mark on tape. From the 10 in mark to the top of the tape the clasts may be slightly normally graded(?). Upper part of Hales Limestone, Lower Ordovician, Nevada (photo by H. E. Cook).

et al, 1972). The potential value of recognizing and correctly interpreting these types of deeper-water sediments was emphasized by Cook and others (1972) in their study of shoal-water derived mass-flow deposits at bank and reef margins of Devonian age in Canada. They stressed that these resedimented carbonates and the nature of their contained clasts are useful: (1) for establishing the existence of buildups or reefs in an area; (2) as proximity indicators for locating buildup or reef margins; (3) for genetic interpretation of the morphologic development and diagenesis

of the carbonate buildups (that is, the clasts are mostly limestone and are useful in establishing whether the buildup margin was a bank and/or reef and in determining time of diagenesis, particularly cementation and dolomitization, at the buildup margin); (4) for providing stratigraphic markers useful for correlation between carbonate buildups and the enclosing basin facies; and (5) for their reservoir potential in certain carbonate basins. The value of carbonate mass-flow deposits as petroleum exploration targets has recently been discussed by Cook and

Enos (1977a), Enos (1977a), Mullins and others (1978), Cook (1979a), Mullins and Neumann (1979), Enos (in press), and Cook (1982).

Mass-transport Facies
During the last ten years, the occurrence of mass-transport facies in carbonate basins throughout the geologic column on a worldwide basis has become clearly established (for example, Pray and Stehli, 1962; Garrison and Fischer, 1969; Thompson and Thomasson, 1969; Cook et al, 1972; Mountjoy et al, 1972; Cook and Enos, 1977b; McIlreath and

Figure 71—Possible modified grain-flow deposit. Bed exhibits reverse grading of tabular clasts, subparallel orientation of clasts, upslope imbrication of clasts in upper part of bed, and is capped by cross-bedded calcarenite sands. Bed is part of a mid-submarine fan distributary channel system (see Figs. 122 to 125). Lower part of Hales Limestone, Upper Cambrian, Nevada (photo by H. E. Cook).

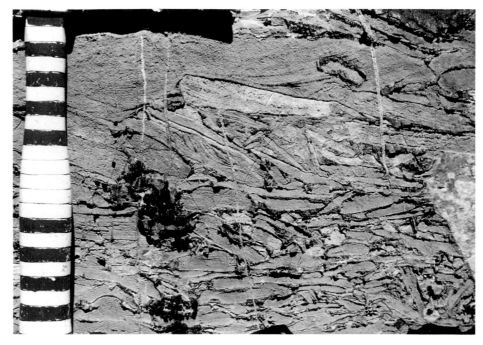

Figure 72—Photograph of split piston core containing carbonate turbidites. Both sections shown are from core E-36212 obtained from 1095 m of water on the base of slope north of Little Bahama Bank. Scales in centimeters. (A) Coarse-grained turbidite illustrating normal grading from gravels to medium sands. (B) Sandy turbidite exhibiting normal grading from coarse to medium sand. Note sharp erosional base. In both cases allochems are all slope derived (photos by H. T. Mullins).

James, 1978; Cook, 1979a; Krause and Oldershaw, 1979; Mullins and Neumann, 1979; Mullins and Van Buren, 1979, Crevello and Schlager, 1980; and Mullins et al, in prep.).

As illustrated in this paper, these facies occur as slides and slumps, megabreccias, conglomerates, calcarenites, and calcisiltites. All of these facies can be quite diverse in their clast types (shoal-water versus deep-water origin or mixtures of both), clast size (tens of meters across in some megabreccias), and clast shapes (spheroidal oolite and algal grains to plate-shaped clasts of slope or supratidal mudstones); some show massive bedding with randomly oriented clasts floating in a mud matrix, while others show distinct stratification with tabular clasts arranged in a parallel and/or imbricated manner. Bouma sequences and inverse grading are also common especially in the ancient. Their geometries vary from thick sheet-like bodies (aprons) covering hundreds of square kilometers to debris occurring in erosional channels up to 200 m deep.

Facies Sequences and Associations

Facies sequences and associations are used in the same sense given by Walker and Mutti (1973). Thus the

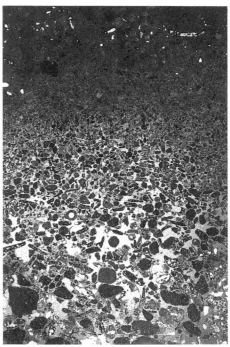

Figure 73—Thin section photomicrograph of normally graded carbonate turbidite. Sample from core C-32 obtained from 900 fathoms of water in Exuma Sound, Bahamas (from Crevello, 1978; courtesy of P. D. Crevello).

grouping of facies into preferred or commonly occurring sequences is purely descriptive without any environmental interpretations attached to the resulting facies sequences. The preferred field occurrence of sub-

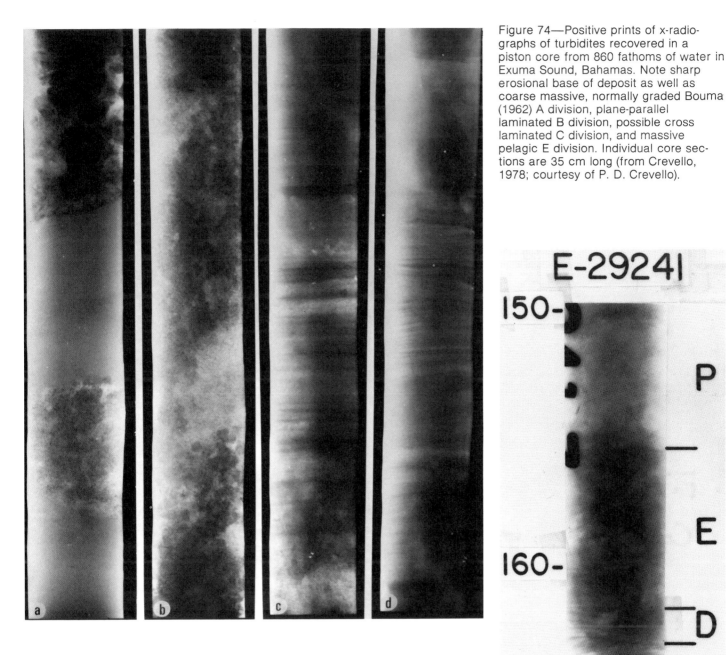

Figure 74—Positive prints of x-radiographs of turbidites recovered in a piston core from 860 fathoms of water in Exuma Sound, Bahamas. Note sharp erosional base of deposit as well as coarse massive, normally graded Bouma (1962) A division, plane-parallel laminated B division, possible cross laminated C division, and massive pelagic E division. Individual core sections are 35 cm long (from Crevello, 1978; courtesy of P. D. Crevello).

Figure 75—Positive print x-radiograph of base cut out carbonate turbidite displaying Bouma (1962) units B-C-D-E. Scale in centimeters on left. Piston core sample from 4000 m of water east of Little Bahama Bank (from Mullins, 1978).

marine slide and megabreccia facies would be a facies sequence. Assigning this descriptive facies sequence to a specific depositional environment is an interpretative step. This second level of interpretive facies organization identified facies associations. The same slide and megabreccia facies sequence may be assigned to a lower slope setting and be used to define a lower slope facies association.

Although carbonate mass-transport facies are now widely known from many basin-margin and basinal sequences, their recognition and organization into facies sequences, associations, and models is sparse.

This level of organization and interpretation is occurring but it has clearly developed at a slower pace than for redeposited clastic facies.

Various studies ranging from detailed lateral tracing of facies at specific bank and reef margins (Figs. 46 and 47) to the examination of thick, seaward prograding platform margin sequences (Figs. 117 to 120) allows some generalization to be made about facies sequences and associations (Thompson and Thomasson, 1969; Cook et al, 1972; Bosellini and Rossi, 1974; Keith and Friedman, 1977; Hubert et al, 1977; Cook and Taylor, 1977; Enos, 1977a;

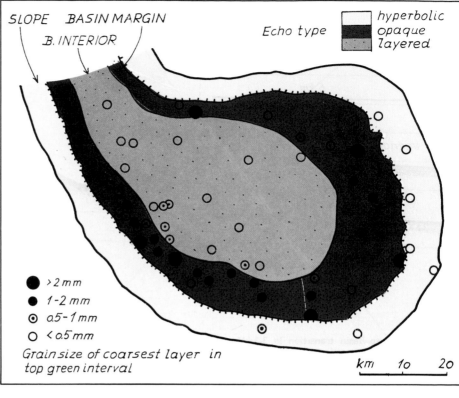

Figure 76—Distribution of near-surface (Late Pleistocene-Holocene) turbidite facies and 3.5 KHz PDR echo types in the cul-de-sac of Tongue of the Ocean, Bahamas. Grain size plotted is visual estimate of mode of coarsest part of coarsest turbidites at a particular location. Note good correlation between opaque reflection character and distribution of very coarse sand and rubble along basin margin (base of slope) (from Schlager and Chermak, 1979; courtesy of W. Schlager).

Figure 77—Sequence of 100 m thick carbonate turbidites and debris flow sheets (white) within pelagic and hemipelagic lime mudstones (dark). Section is about 30 km from a low-relief bank margin. Prongs Creek Formation, Middle Devonian, northern Mackenzie Mountains, Yukon Territory, Canada (photo by H. E. Cook).

McIlreath, 1977; McIlreath and James, 1978; Cook, 1979a; Cossey and Ehrlich, 1979; Cook and Egbert, 1981a; and others).

The overall downcurrent trend in mass-transport facies is from submarine slide and megabreccia facies succeeded by conglomerate and calcarenite facies and lastly by calcarenite and calcisiltite facies (Cook, 1982). There are notable exceptions, as for boulder-bearing debris-flow sheets that can be transported tens of kilometers into basin-plain settings (Figs. 53 to 57). McIlreath and James (1978), in an excellent summary of carbonate-slope sediments, present a series of depositional profiles that go from platform margin to slope to basin environments. On these profiles they

Figure 78—Normally graded limestone turbidite, 50 cm thick, contains both shelf- and slope-derived clasts. Tabular clasts exhibit subparallel orientation. Located in inner fan facies near base of slope. Lower part of Hales Limestone, Upper Cambrian, Nevada (modified from Cook and Taylor, 1977; photo by H. E. Cook).

Figure 79—Normally graded limestone turbidite, 30 cm thick, showing erosional base and flat top. Tabular clasts are oriented subparallel and exhibit upslope imbrication at right side of photo. This is a high clast-to-matrix ratio turbidite. Located in mid-fan distributary channel system. Lower part of Hales Limestone, Lower Cambrian, Nevada (modified from Cook, 1979a; photo by H. E. Cook).

Figure 80—Upper part of 1.5-m-thick channel deposit. Clasts are normally graded, imbricated in an upslope direction at the top of the channel and oriented subparallel below the top of the channel. Rippled calcarenites cap the bed. Located in mid-fan distributary channel system. Lower part of Hales Limestone, Upper Cambrian, Nevada (modified from Cook, 1979a; photo by H. E. Cook).

have schematically placed the inferred mass-transport facies sequences (Enos and Moore, 1982).

Models

One of the first depositional models for low-relief platform-margin facies was that by Cook and others (1972). This model (Fig. 121) was based on data from the Upper Devonian of Alberta, Canada, from surface and subsurface studies in the Or-dovician through Devonian of the Yukon Territory, Canada (Cook and McDaniel, unpub. data), from Pennsylvanian in the Marathon Basin of west Texas (see Thomson and Thomasson, 1969), from Permian in the Guadalupe Mountains of west Texas (Pray and Stehli, 1962; McDaniel and Pray, 1967), and from subsurface studies in the Permian of the Midland and Delaware Basin, west Texas (Cook, in prep.). The model in Figure 121 is meant to convey a variety of features. First, although there are abundant mass-flow deposits in the basin-margin facies, this debris is not organized into facies sequences that resemble clastic submarine-fan sequences. Rather, this material is more randomly distributed as thin-bedded calcarenite turbidites that occur in channel and sheet-form, and as volumetrically very large debris-flow

Figure 81—15-cm-thick conglomeratic turbidite. Clasts are normally graded. Note geopetal fabrics in gastropod shells indicating that shells were filled with lime mud after deposition of the turbidite. Turbidite occurs on slope 100 m from bank margin. White circle is 2 cm wide. Road River Formation, Siluro-Devonian, Wernecke Mountains, Yukon Territory, Canada (photo by H. E. Cook).

Figure 82—10-cm-thick normally graded conglomeratic turbidite. Note flat base and mounded surface. Bed occurs in graptolitic basinal lime mudstones 65 km from bank margin. Prongs Creek Formation, Middle Devonian, northern Mackenzie Mountains, Yukon Territory, Canada (photo by H. E. Cook).

sheets that represent major periodic events at bank and reef margins. The model also appears to be applicable to the carbonate debris sheets, wedges, and aprons in modern interplatform troughs of the Bahamas (Schlager and Chermak, 1979; Crevello and Schlager, 1980; Mullins et al, in prep.). Thus a carbonate slope and basin may contain a large amount of redeposited sediment of two end-member types. One type consists of thin-bedded calcarenite turbidites that are generated fairly continuously at a platform margin; the

Figure 83—Series of normally graded lime packstone turbidites, foreslope. Shady Dolomite, Cambrian, Appalachians (from Pfeil and Read, 1980; courtesy of J. F. Read).

Figure 84—Normally graded grainstone-packstone turbidite. Sekwi Formation, Lower Cambrian, Mackenzie Mountains, Northwest Territories, Canada. Scale in cm (courtesy of F. F. Krause).

Figure 85—Normally graded turbidite with Bouma division b at top of bed. White circle is 2 cm wide. Same locality as Figure 77, in Yukon Territory, Canada (photo by H. E. Cook).

second type will consist of thick, widespread boulder-bearing debris-flow deposits that are major, episodic events that affect large segments of platform margins. These volumetrically huge debris flows are capable of traveling more than 100 km down the slope into basin-plain settings and blanketing hundreds of square kilometers as a single event. This more or less random or episodic distribution of volumetrically large debris flows is not conducive to forming systematic submarine fan facies sequences as occurs in many clastic depositional settings (Cook and Egbert, 1981a; Cook, 1982).

Some workers have indiscriminately used the terms "fan" or "submarine fan" in their discussion of carbonate mass-flow facies without presenting evidence that demonstrates the existence of distinct fan facies sequences. Many of these carbonate mass-flow facies could as easily be a

series of simple sheet or apron deposits as in Figures 46 to 53, 77, and 121, rather than the systematic downcurrent development of inner, middle, and outer fan facies sequences. The term "submarine fan" and the defining criteria for recognizing submarine fan facies is firmly entrenched in the literature and should be prudently used. There is no reason to think that all carbonate mass-flow deposits form "submarine fans." A number of authors (Cook et al, 1972; McIlreath, 1977; Schlager and Chermak, 1979; Crevello and Schlager, 1980; Mullins et al, in prep.) have recognized that the mass-flow deposits they described are not organized into systematic submarine fan facies. These authors have referred to these deposits as sheets, aprons, wedges, etc., but not as submarine fans.

Until recently (Cook and Egbert, 1981a, b, c), the documentation of

carbonate mass-flow deposits that are organized into systematic vertical and lateral facies sequences resembling clastic submarine-fan sequences and fan models has been absent. During the early Paleozoic, the continental margin of the western United States was gradually prograding seaward (Fig. 117; Cook and Taylor, 1977). As Walker (1978) points out, "there are very few descriptions of complete ancient submarine fans. By complete, I imply a prograding sequence from basin-plain turbidites, upward through fan deposition into silty mudstones deposited on a prograding slope." The prograding continental margin sequence described by Cook and Egbert (1981a, b, c) is an unusually well-exposed 1500-m-thick carbonate section. This section includes basin plain, submarine fan, slope, and platform margin sequences (Fig. 117). Not only is there a continuous vertical record of these

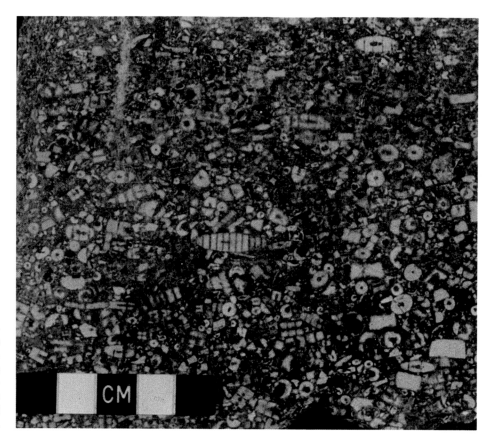

Figure 86—Coarse-grained crinoidal turbidite with load-casted base and crude size grading, with larger clasts irregularly distributed rather than confined to the base. Hare Fiord Formation, Permo-Pennsylvanian, Sverdrup Basin, Arctic Archipelago (from Davies, 1977; courtesy of G. R. Davies).

depositional environments, but the facies can be traced laterally along depositional strike for several kilometers (Fig. 117b). Figures 122 to 124 are preliminary local depositional models that summarize the main facies sequences and facies associations of this continental margin and carbonate submarine fan. Fan facies sequences consist of inner fan feeder channels (Fig. 58) associated with submarine slides (Fig. 27), upward thinning, braided distributary channels in mid-fan positions (Figs. 71, 125 to 128), upward thickening, outer fan-lobe sheets (Figs. 129 to 131), and fan fringe and basin plain facies (Fig. 132). These systematic facies sequences strongly contrast with the more episodic and randomly distributed debris-flow and calcarenite turbidites, sheets, and aprons that commonly occur in basin-margin sequences (Figs. 46 and 121).

Krause and Oldershaw (1979) have proposed a submarine sediment gravity-flow model for carbonate breccias based on field studies in the Cambrian of Canada (Figs. 133 and

134). One of the main differences between this model and models proposed by Walker (1975, 1978) for clastic conglomerates is the suggested downcurrent position for inversely graded conglomerates and breccias. In the Krause and Oldershaw model, inversely graded carbonates occupy a downcurrent position whereas Walker, based on theoretical grounds, suggests that inversely graded clastics occur in a more upcurrent position.

Submarine slides initiated in semiconsolidated lime mudstones on carbonate slopes can gradually remold into conglomeratic sediment gravity-flows (Cook and Taylor, 1977; Cook, 1979a, b). A model summarizing the stages of deformation and remolding that carbonate slides undergo is illustrated in Figures 34 to 37 and 135.

PETROLEUM CONSIDERATIONS

As discussed by Enos and Moore (1982), the best known example of petroleum production from carbonate slopes is from the Cretaceous of Mex-

ico (Enos, 1977a, in press). These reservoirs are mainly in carbonate sediment gravity-flow deposits.

Similar but less prolific petroleum reservoirs occur in Permian deep-water carbonate turbidites and debris-flow base-of-slope and basin margin deposits in the Delaware and Midland Basins, west Texas (Fig. 136; Cook et al, 1972; Cook, in prep.). These fields are 15 to 30 km from the shelf margin. Three basic types of shoal-water-derived debris make up these reservoirs: (1) carbonate megabreccia debris flows with clasts up to 6 m in one dimension; (2) pebble to cobble sized debris-flow and turbidity-current deposits; and (3) calcarenite turbidites. All three types of sediment gravity-flow deposits are porous, but the best permeability appears to occur in the calcarenite turbidites.

Most of the porosity in these Permian calcarenite turbidites is of a post-depositional origin. The dominant porosity type is solution interparticle as a result of the selective removal of lime mud between fusilinid and crinoid grains. Other

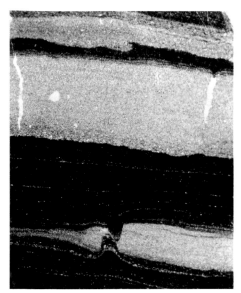

Figure 87—Normally graded grainstone turbidites interbedded with limey shale. Note erosional surface and flame structures at tops of shale layers, and boudin structures in lower grainstone layer (from Pfeil and Read, 1980; courtesy of J. F. Read).

Figure 89—Calcarenite turbidite comprised of a normally graded lower part, planar laminated middle part, and an upper portion having climbing ripples (Bouma divisions a, b, c), capping a debris flow breccia. Sekwi Formation, Lower Cambrian, Mackenzie Mountains, Northwest Territories, Canada (courtesy of F. F. Krause).

Figure 88—This normally graded turbidite sheet is 10 cm thick. It occurs discontinuously at the top of the megabreccia debris flow sheet at the southeast margin of Ancient Wall bank illustrated in Figures 48 to 50 and diagrammatically shown in Figure 45. Upper Devonian, Alberta, Canada (from Cook et al, 1972; photo by H. E. Cook).

types of porosity include intrabiotic, solution biomoldic, fracture, and solution fracture. In contrast to the calcarenite turbidite reservoirs, both the clasts and the dark lime mud matrix in the debris-flow conglomerate reservoirs are usually intensely dolomitized. Dolomitization of at least the mud matrix occurred after deposition. Porosity in the dolomitized muddy matrix is solution interparticle and solution fracture. Clasts have intercrystalline, vuggy, and fracture porosity.

As exploration continues into deeper water carbonate slope and basinal settings more reservoirs of these types will probably be sought and found. It will be increasingly important to understand the nature, origin, and facies associations of deeper-water carbonate sediment

gravity-flow deposits (Cook, 1982). As pointed out recently by Cook and Egbert (1981a) in their description of carbonate submarine fan facies, "mass-transport deposits, though common in carbonate basins, normally occur as widespread sheets or debris wedges. . ." (Figs. 46 and 121). Their recognition of carbonate submarine-fan sequences (Fig. 122) similar to clastic-fan sequences is new and raises several questions: "(1) What sedimentologic and tectonic conditions are conducive to fan development in carbonate provinces? (2) Do these conditions resemble those for clastic-fan development, or do carbonate provinces have unique requirements? By recognizing carbonate submarine fans and the geologic conditions that control their sediment dispersal patterns, areas of maximum sediment accumulation may be predicted as an aid in exploring for petroleum reservoirs of deeper water carbonate environments."

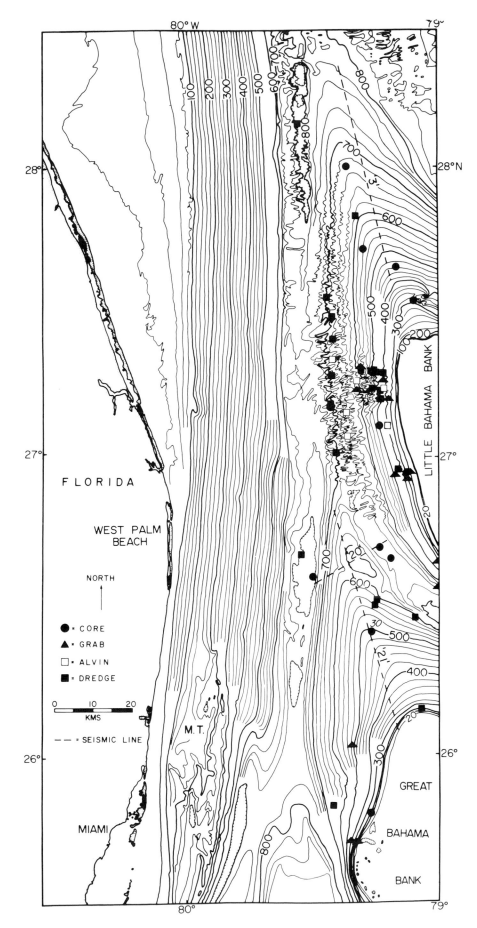

Figure 90—Bathymetry of the northern
Straits of Florida, based on Uchupi
(1969). Note the large hemi-conical
shaped slopes (sediment drifts) off the
northwest corners of both Little and
Great Bahama Bank. Contours in meters
(from Mullins et al, 1980a).

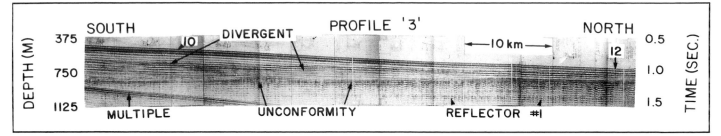

Figure 91—5-cu-in air-gun seismic reflection profile obtained along the crest of the Little Bahama Bank sediment drift, located in Figure 29. Note wedge-shaped body that has prograded north over an erosional/non-depositional unconformity of middle Miocene(?) age. Internal reflectors are oblique progradational, downlap on the underlying horizontal reflector and converge (pinch out) to the north, indicating a southerly source for the sediment (from Mullins et al, 1980a).

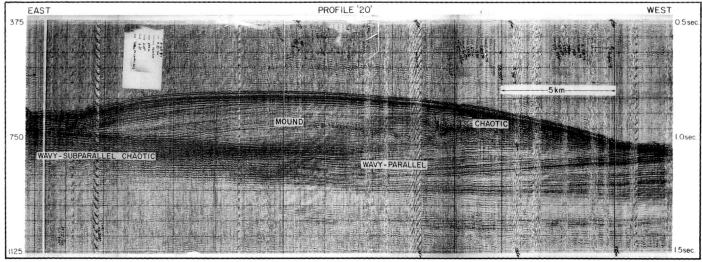

Figure 92—5 cu in air-gun seismic reflection profile "20" obtained along a track transverse to the strike of the Great Bahama Bank sediment drift located in Figure 29. Note the overall mound shape of this profile. Internal reflectors are oblique progradational. The chaotic seismic facies correlates with slide deposits (see also Fig. 8), the wavy-subparallel chaotic facies with turbidites, and the wavy parallel facies with pelagic sediments. The mound consists of contour, current-winnowed carbonate sands. Water depth in meters on left; seconds of two-way travel time on right (from Mullins and Neumann, 1979).

ACKNOWLEDGMENTS

We have assembled a collection of illustrations documenting the stratigraphic and geographic diversity and commonality of slope sequences. Illustrations were solicited from numerous people. We are greatly appreciative, because without their help a highly illustrated paper such as this would not have been possible. These people include: Alfonso Bossellini, Steven P. Crossey, Paul D. Crevello, Gerald M. Friedman, Robert E. Garrison, Albert C. Hine, Noel P. James, Jack Kofoed, Fed F. Krause, Wolfgang Krebs, Robert M. Mitchum, A. Conrad Neumann, Ronald D. Perkins, J. F. Read, Wolfgang Schlager, Parke D. Snavely, III, and R. Jude Wilber. The donor of each illustration is acknowledged in the figure captions.

Table 1, which provides a synthesis of this paper on carbonate slopes, is modified from Nardin and others (1979, Tables 1, 2, 3) with suggested modifications by M. A. Hampton. We have profited from discussions and reviews by

P. A. Scholle, M. A. Arthur, and W. C. Butler. The authors appreciate very much the clerical help of Terry Coit and Lillian Wood and the drafting services of Jeanne Blank.

SELECTED REFERENCES

Ahr, W. M., 1973, The carbonate ramp: An alternative to the shelf model: Trans., Gulf Coast Assoc. Geol. Socs., v. 23, p. 221-225.

Bagnold, R. A., 1954, Experiments in the gravity-free dispersion of large spheres in a Newtonian fluid under shear: Royal Soc. London Proc., Ser. A, v. 225, p. 49-53.

_____, 1956, The flow of cohesionless grains in fluid: Trans., Royal Soc. London Phil. Ser. A, v. 249, p. 235-297.

_____, 1966, An approach to the sediment transport problem from general physics: U.S. Geol. Survey, Prof. Paper 422-I, 37 p.

Ball, M. M., 1967, Tectonic control of the configuration of the Bahama Banks: Trans., Gulf Coast Assoc. Geol. Socs., v. 17, p. 265-267.

Barbat, W. F., 1958, The Los Angeles Basin area, California, in Habitat of Oil: AAPG Spec. Pub., p. 62-77.

Bein, A., and Y. Weiler, 1976, The Cretaceous Talme Yafe Formation; a contour current shaped sedimentary prism of calcareous detritus at the continental margin of the Arabian craton:

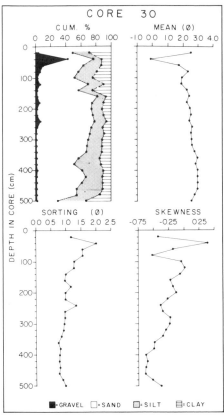

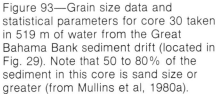

Figure 93—Grain size data and statistical parameters for core 30 taken in 519 m of water from the Great Bahama Bank sediment drift (located in Fig. 29). Note that 50 to 80% of the sediment in this core is sand size or greater (from Mullins et al, 1980a).

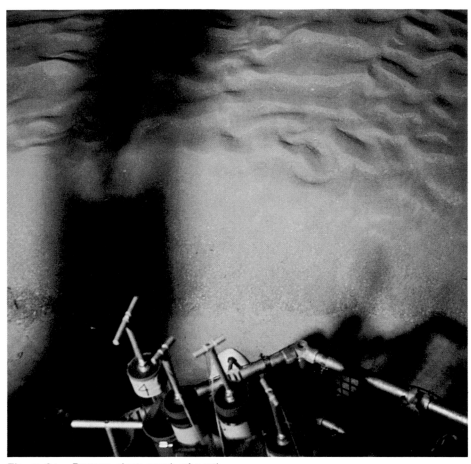

Figure 94—Bottom photograph of north-facing sand wave with superimposed north-facing ripple marks. These bedforms have formed in response to north-flowing bottom currents having velocities of up to 60 cm/sec. Sand wave slip face (center) is approximately 50 cm in height; holding cage in foreground is approximately 1 m across. Photo taken from DSRV ALVIN in 400 m of water on the slope west of Little Bahama Bank (from Mullins et al, 1980a; courtesy of A. C. Hine).

Sedimentology, v. 23, p. 511-532.

Bloomer, R. R., 1977, Depositional environments of a reservoir sandstone in west-central Texas: AAPG Bull., v. 61, p. 344-359.

Boardman, M. R., 1978, Holocene deposition in Northwest Providence Channel, Bahamas; a geochemical approach: Chapel Hill, Univ. of North Carolina, Ph. D. dissert., 155 p.

Bornhold, B. D., and O. H. Pilkey, 1971, Bioclastic turbidite sedimentation in Columbus Basin, Bahamas: Geol. Soc. America Bull., v. 82, p. 1341-1354.

Bosellini, A., and D. Rossi, 1974, Triassic carbonate buildups of the Dolomites, northern Italy, in L. F. Laporte, ed., Reefs in time and space: SEPM Spec. Pub. 18, p. 209-233.

Bouma, A. H., 1962, Sedimentology of some flysch deposits: Amsterdam, Elsevier Sci. Pub., 168 p.

—— et al, 1976, Gyre Basin, an intra-slope basin in northwest Gulf of Mex-ico, in Beyond the shelf break: AAPG Marine Geol. Comm. Short Course, v. 2, p. E-1 to E-28.

Burk, C. A., and C. L. Drake, 1974, Geologic significance of continental margins, in C. A. Burk and C. L. Drake, eds., The geology of continental margins: New York, Springer-Verlag Pub., p. 3-10.

Burne, R. V., 1974, The deposition of reef-derived sediment upon a bathyal slope; the deep off-reef environment, north of Discovery Bay, Jamaica: Marine Geol., v. 16, p. 1-19.

Byers, C. W., 1977, Biofacies patterns in euxinic basins; a general model, in H.

E. Cook and P. Enos, eds., Deep-water carbonate environments: SEPM Spec. Pub. No. 25, p. 5-17.

Carter, R. M., 1975, A discussion and classification of subaqueous mass-transport with particular application to grain-flow, slurry-flow, and fluxoturbidites: Earth Science Rev., v. 11, p. 145-177.

Coates, A. G., and E. G. Kauffman, 1973, Stratigraphy, paleontology, and paleoenvironment of a Cretaceous coral thicket, Lamy, New Mexico: Jour. Paleont., v. 47, no. 5, p. 953-968.

Conaghan, P. J., et al, 1976, Nubrigyn algal reefs (Devonian), eastern Australia; allochthonous blocks and mega-breccias: Geol. Soc. America Bull., v. 87, p. 515-530.

Cook, H. E., 1979a, Ancient continental slope sequences and their value in understanding modern slope development, in L. S. Doyle and O. H. Pilkey eds., Geology of continental slopes: SEPM Spec. Pub. No. 27, p. 287-305.

——, 1979b, Small-scale slides on inter-canyon continental slope areas, Paleozoic, Nevada (abs.): Geol. Soc. America Ann. Mtg., v. 11, p. 405.

——, 1979c, Generation of debris flows

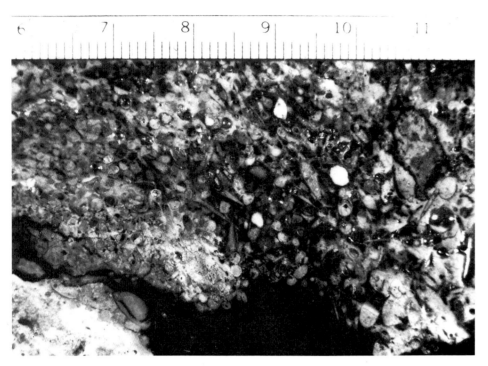

Figure 95—Polished slab section of biomicrudite (packstone) from the Great Bahama Bank sediment drift obtained by rock dredging in 595 m of water (located in Fig. 29). Coarse-grained pteropods, lithoclasts, planktonic foraminifera, and some shallow-water debris are cemented by magnesium calcite. Note crude cross stratification. Scale is in centimeters (from Mullins et al, 1980a; courtesy of R. J. Wilber).

Figure 96—Contourite grainstones occurring within upper slope facies. Composed of well-sorted silt to fine-grained, shallow-water derived alga grains. Matrix is virtually mud free and filled with sparry calcite. Ripple forms have periods of about 9 cm and 0.5 to 1.0 cm amplitudes. Both base and top have sharp contacts with enclosing hemipelagic slope mudstone. Upper part of Hales Limestone, Lower Ordovician, Nevada; scale in centimeters (photo by H. E. Cook).

and turbidity current flows from submarine slides (abs.): AAPG Bull., v. 63, p. 435.

_____ and R. M. Egbert, 1981a, Carbonate submarine fans along a Paleozoic prograding continental margin, western United States (abs.): AAPG Bull., v. 65, p. 913.

_____ and _____, 1981b, Late Cambrian-Early Ordovician continental margin sedimentation, central Nevada, *in* M. E. Taylor, ed., 2nd International Symposium on the Cambrian System Proceedings: U.S. Geol. Survey, Open-File Rept. 81-743, p.50-56.

_____, 1981c, Late Cambrian-Early Ordovician deep water carbonates, Hot Creek Range, central Nevada, *in* M. E.

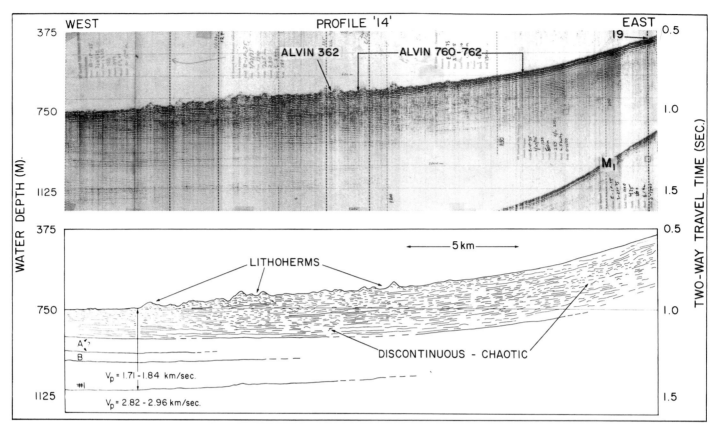

Figure 97—Photograph (top) and line drawing interpretation of 5 cu in air-gun seismic reflection profile in Straits of Florida west of Little Bahama Bank. Note the hummocky nature of reflections corresponding to lithoherms at the base of the slope. Also note the absence of similar features upslope. Locations of ALVIN dive sites are also shown (from Mullins and Neumann, 1979).

Figure 98—12 KHz PDR profile (top) of lithoherms west of Little Bahama Bank. Note the density of mounds plus their elevation above the sea floor (20 to 40 m). Water depths in meters on left and right. (Bottom) Blow-up of 5 cu in air-gun seismic reflection profile of lithoherms. Note that mounds (**L**) form discrete bodies of chaotic reflectors separated from undisturbed reflectors (**H**). Mounds appear to have built up vertically about 70 m from a horizontal reflector (**R**). Water depths in meters on left; seconds of two-way traveltime on right (photo by H. T. Mullins).

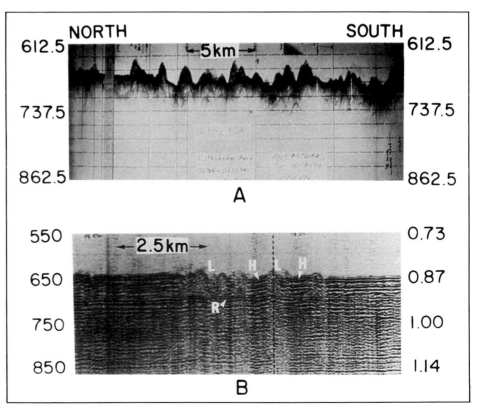

Figure 99A—Deep-water, ahermatypic coral (*Lophelia*) and sponges on the southern, upcurrent end of a lithoherm. Taken in the northern Straits of Florida, at water depths of 600 to 700 m from the DSRV ALVIN. Cross photo scale is approximately 3 m. Note steep sides.

Figure 99B—Close-up of ahermatypic coral and anemones in A. Cross photo scale is approximately 1.5 m.

Figure 99C,D—Rows of crinoids, left, situated on "micro-ledges" on the northern, downcurrent end of a lithoherm. Note steep sides and absence of coral. Cross photo scale is approximately 3 m. Outcrop, right, along the base of a lithoherm illustrating "onion-skin" internal structure. Cross photo scale is approximately 2.5 m (photos courtesy of A. C. Neumann).

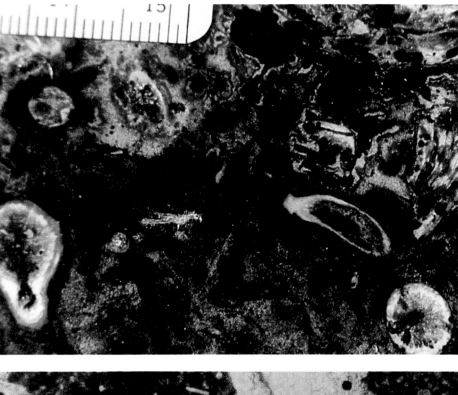

Figure 100—Polished slab section of lithoherm rock from 650 m of water in the northern Straits of Florida. Note organic framework produced by ahermatypic coral (white). Scale in centimeters (from Mullins et al, 1978; courtesy of R. J. Wilber).

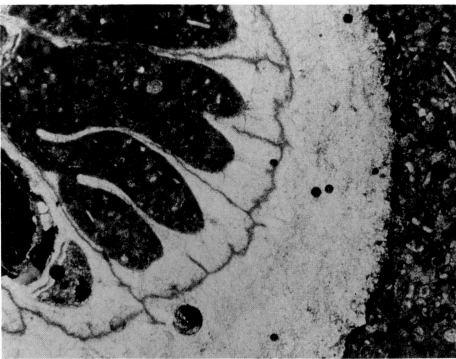

Figure 101—Thin section photomicrograph of lithoherm rock recovered from 950 m of water in the Northwest Providence Channel, south of Freeport, Bahamas. Note large coral skeleton that was bored (lower left) and later infilled by planktonic detritus (white) and cemented by micritic magnesian calcite (brown). Cross photo scale is 9 mm (from Mullins, 1978).

Taylor, ed., Cambrian stratigraphy and paleontology of the Great Basin and vicinity, western United States: 2nd Internat. Symp. on Cambrian System Field Trip Guidebook No. 1, p. 51-770.
_____ and P. Enos, 1977a, Deep-water carbonate environments—an introduction, in H. E. Cook and P. Enos, eds., Deep-water carbonate environments:

SEPM Spec. Pub. No. 25, p. 1-3.
_____ and _____, eds., 1977b, Deep-water carbonate environments: SEPM Spec. Pub. No. 25, 336 p.
_____ et al, 1972, Allochthonous carbonate debris flows at Devonian bank ("reef") margins, Alberta, Canada: Bull. Canadian Petroleum Geology, v. 20, p. 439-497.

_____ and M. E. Taylor, 1977, Comparison of continental slope and shelf environments in the Upper Cambrian and Lowest Ordovician of Nevada, in H. E. Cook and P. Enos, eds., Deep-water carbonate environments: SEPM Spec. Pub. No. 25, p. 51-82.
Cossey, S. P. J., and R. Ehrlich, 1979, A conglomeratic, carbonate flow deposit,

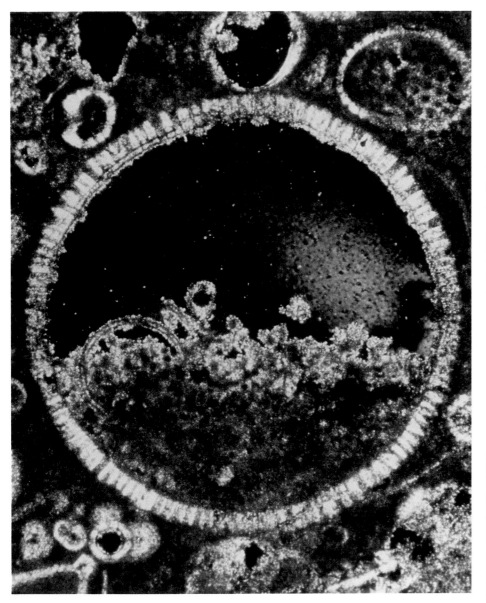

Figure 102—Thin section photomicrograph of geopetal structure in lithoherm rock from 600 to 700 m of water in the northeastern Straits of Florida. *Orbulina universa* test has been partially infilled by silt-size detritus. Cement is magnesian calcite of about 14 mole percent Mg. Geopetal fill in center is 0.74 mm in diameter (courtesy of A. C. Neumann).

Figure 103—*Epiphyton-Renalcis* bioherm (10 m high) in slope facies. Man at top of photo for scale. Shady Dolomite, Cambrian, Appalachians (from Pfeil and Read, 1980; courtesy of J. F. Read).

northern Tunisia; a link in the genesis of pebbly-mudstones: Jour. Sed. Petrology, v. 49, p. 11-22.

Crawford, G. A., 1981, Allochthonous carbonate rocks in toe-of-slope deposits (Permian, Guadalupian), Guadalupe Mountains, west Texas (abs.): AAPG Bull., v. 65, p. 914.

Crevello, P. D., 1978, Debris-flow deposits and turbidites in a modern carbonate basin, Exuma Sound, Bahamas: Coral Gables, Univ. of Miami, M.S. thesis, 133 p.

_____ and W. Schlager, 1980, Carbonate debris sheets and turbidites, Exuma Sound, Bahamas: Jour. Sed. Petrology, v. 50, p. 1121-1147.

Curran, J. F., K. B. Hall, and R. F. Herron, 1971, Geology, oil fields, and future petroleum potential of Santa Barbara Channel area, California, *in* Future petroleum provinces of the United States—their geology and potential: AAPG Mem. 15, p. 192-211.

Damuth, J. E., and D. E. Hayes, 1977, Echo-character of the east Brazilian continental margin and its relationship to sedimentary processes: Marine Geology, v. 24, p. 73-95.

Davies, G. R., 1977, Turbidites, debris sheets, and truncation structures in Upper Paleozoic deep-water carbonates of the Sverdrup Basin, Arctic Archipelago, *in* H. E. Cook and P. Enos, eds., Deep-water carbonate environments: SEPM Spec. Pub. No. 25, p. 221-247.

Dill, R. F., 1966, Sand flows and sand falls, *in* R. W. Fairbridge, ed., Encyclopedia

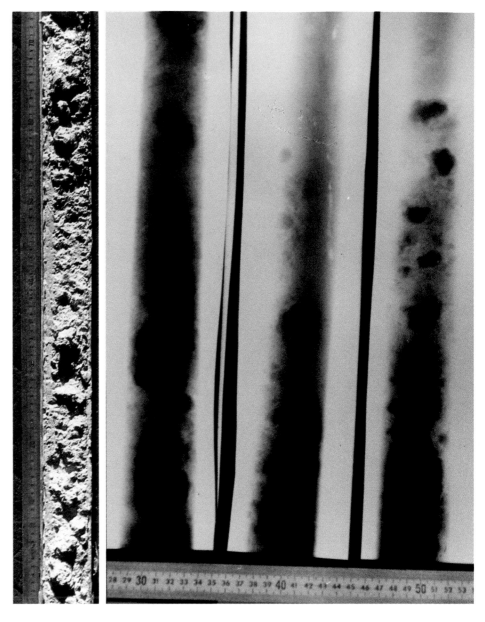

Figure 104— (**A**, left) Photograph of split piston core containing in situ submarine cemented nodules. Core from 450 m of water on the slope north of Great Bahama Bank (photo by H. T. Mullins). (**B**, right) Positive print x-radiograph of core containing in situ submarine cemented nodules. Note the subrounded nature of individual nodules which commonly float in the surrounding sandy mud matrix (from Mullins et al, 1980b).

of Oceanography: New York, Rheinhold, p. 763-765.

Dott, R. H., Jr., 1963, Dynamics of sub-aqueous gravity depositional processes: AAPG Bull., v. 47, p. 104-128.

_____ and K. J. Bird, 1979, Sand transport through channels across an Eocene shelf and slope in southwestern Oregon, *in* O. H. Pilkey and L. S. Doyle, eds., Geology of continental slopes: SEPM Spec. Pub. No. 27, p. 327-342.

Doyle, L., and O. H. Pilkey, eds., 1979, Geology of continental slopes: SEPM Spec. Pub. No. 27, 374 p.

Enos, P., 1977a, Tamabra Limestone of the Poza Rica Trend, Cretaceous, Mexico, *in* H. E. Cook and P. Enos, eds.,

Deep-water carbonate environments: SEPM Spec. Pub. No. 25, p. 273-314.

_____, 1977b, Diagenesis of a giant: Poza Rica Trend, Mexico (abs.), *in* D. G. Bebout and R. G. Loucks, eds., Cretaceous carbonates of Texas and Mexico, applications to subsurface exploration: Austin, Texas Bur. Econ. Geol., Rept. Inv. 89, p. 324.

_____, 1977c, Flow regimes in debris flow: Sedimentology, v. 24, p. 133-142.

_____, in press, Poza Rica field, Veracruz, Mexico, *in* P. O. Roehl and P. W. Choquette, eds., Carbonate petroleum reservoirs; a casebook: New York, Springer-Verlag Pub.

Ewing, M., et al, 1969, Initial reports of the deep-sea drilling project:

Washington D.C., U.S. Govt. Printing Office, v. 1, 672 p.

Fischer, A. G., and R. E. Garrison, 1967, Carbonate lithification on the sea floor: Jour. Geology, v. 75, p. 488-497.

Fisher, R. V., 1971, Features of coarse-grained, high-concentration fluids and their deposits: Jour. Sed. Petrology, v. 41, p. 916-927.

Flores, V. Q., 1978, Paleosedimentologia en la zona de Sitio Grande-Sabancuy: Petroleo Internacional, v. 26 (Nov. 1978), p. 44-48.

Freeman-Lynde, R. P., et al, 1979, Deface-ment of the Bahama Escarpment: EOS, v. 60, no. 18, p. 286.

Gardett, P. H., 1971, Petroleum potential of Los Angeles, California, *in* Future

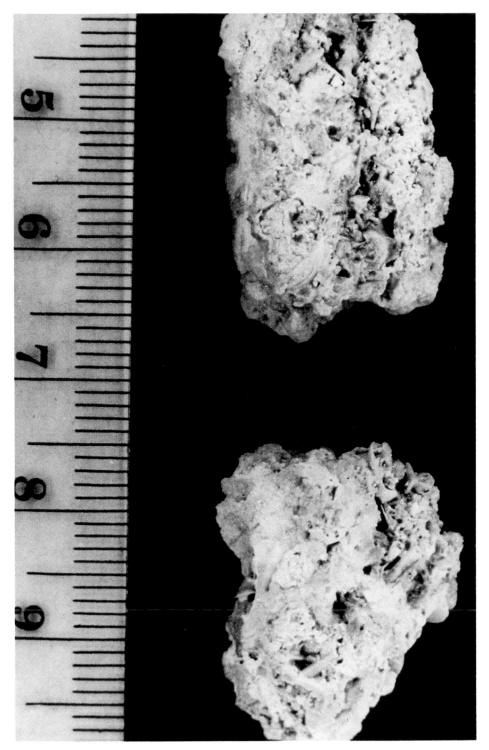

Figure 105—Photograph of in situ submarine cemented nodules. Note irregular outer surface of nodules and the projection of whole constituent grains indicating that the nodules have not been transported. Scale is in centimeters (from Mullins et al, 1980b).

petroleum provinces of the United States—their geology and potential: AAPG Mem. 15, p. 298-308.

Garrison, R. E., and A. G. Fischer, 1969, Deep-water limestones and radiolarites of the Alpine Jurassic, in G. M. Friedman, ed., Depositional environments in carbonate rocks: SEPM Spec. Pub. No. 14, p. 20-56.

Ginsburg, R. N., and N. P. James, 1974, Holocene carbonates of continental shelves, in C. A. Burk and C. L. Drakes, eds., Geology of continental margins: New York, Springer-Verlag Pub., p. 137-155.

Hampton, M. A., 1972, The role of sub-

aqueous debris flow in generating turbidity currents: Jour. Sed. Petrology, v. 42, p. 775-793.

_____, 1975, Competence of fine-grained debris flows: Jour. Sed. Petrology, v. 45, p. 834-844.

_____, 1979, Buoyancy in debris flows: Jour. Sed. Petrology, v. 49, p. 753-758.

Figure 106—Thin section photo-micrograph of a submarine cemented nodule illustrating intragranular cementation of a pteropod test by bladed magnesian calcite. Cross photo is 850 m (courtesy of R. J. Wilber).

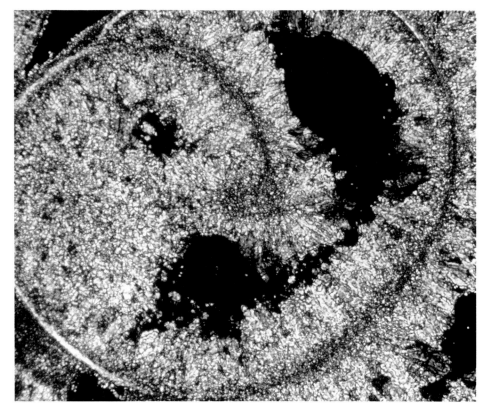

Figure 107—Facies relationships of nodular carbonates on the slope north of Great Bahama Bank. Core descriptions and percent mud in surface samples are superimposed on a 3.5 KHz PDR profile. Types IB, IIA, and IIb refer to echo types defined by Damuth and Hayes (1977). Note the downslope facies transitions from hardgrounds to nodular sediment interbedded with pelagic sediments to pure pelagics. These transitions correlate well with a downslope decrease of contour following bottom currents (from Mullins et al, 1979, 1980a).

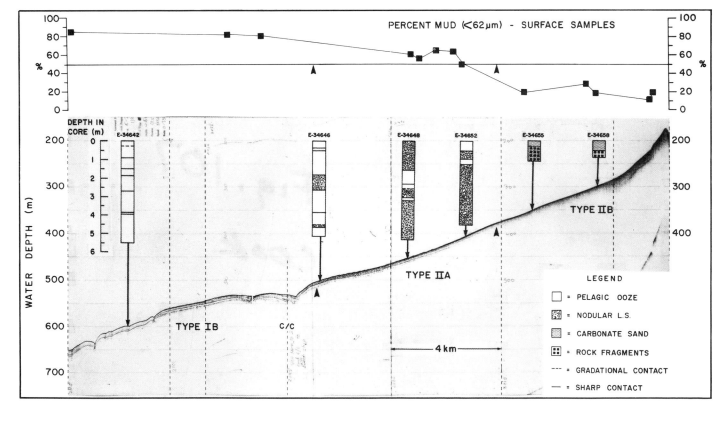

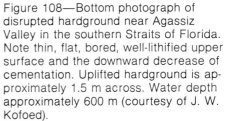

Figure 108—Bottom photograph of disrupted hardground near Agassiz Valley in the southern Straits of Florida. Note thin, flat, bored, well-lithified upper surface and the downward decrease of cementation. Uplifted hardground is approximately 1.5 m across. Water depth approximately 600 m (courtesy of J. W. Kofoed).

Figure 109—Close-up photograph of the upper surface of a hardground dredged from 230 m of water north of Great Bahama Bank. Note heavily bored, irregular surface. Scale is in centimeters (courtesy of A. C. Neumann).

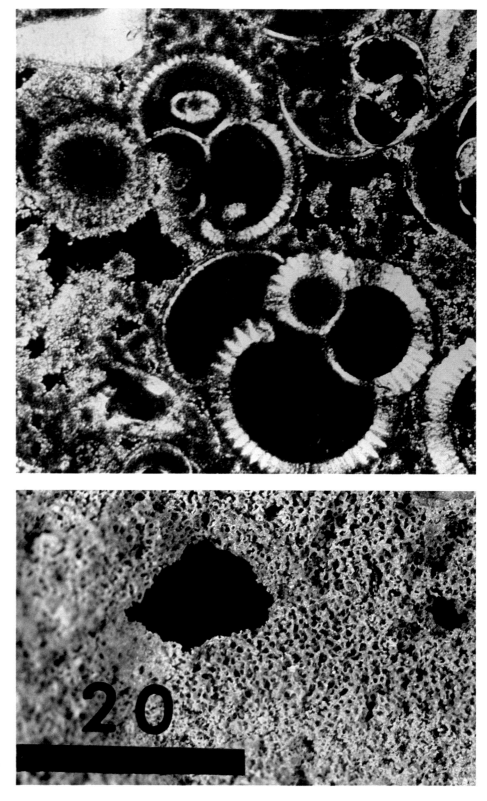

Figure 110—Thin section photomicrograph in polarized light of hardground dredged from 640 m of water in the northeastern Straits of Florida. Note the grain support produced by the planktonic forams. Also note the pelloidal nature of the intergranular magnesian calcite cement and the massive micritic texture of the intragranular cement (from Neumann et al, 1977; courtesy of A. C. Neumann).

Figure 111— Fresh surface of an intensively bored hardground from 620 m of water in the northeastern Straits of Florida. Note "frothy" texture. Such sponge borings are responsible for up to 50% secondary porosity in some carbonate slope hardgrounds. Scale is in millimeters (from Wilber, 1976; courtesy of R. J. Wilber).

Hana, J. C., and C. H. Moore, 1979, Quaternary temporal framework of reef to basin sedimentation, Grand Cayman, British West Indies: Geol. Soc. America Abs. with Programs, v. 11, p. 438.

Hedberg, H. D., 1970, Continental margins from the view point of the petroleum geologist: AAPG Bull., v. 54, p. 3-43.

Heezen, B. C., and C. D. Hollister, 1971, The face of the deep: New York, Oxford Univ. Press, 659 p.

Hine, A. C., and A. C. Neumann, 1977, Shallow carbonate bank margin growth and structure, Little Bahama Bank,

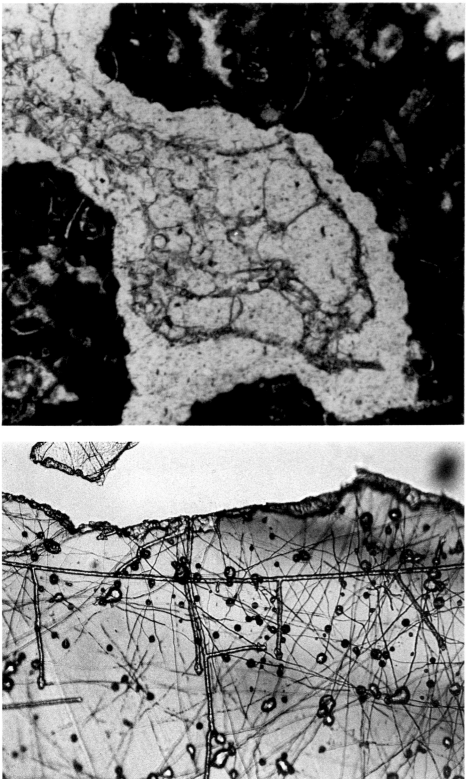

Figure 112—Thin section photomicrograph of in situ sponge borings. Note boring (white) that is occupied by a framework "skeleton" of sponge spicules; also note truncated allochems in host rock as well as hemispherical excavations. Cross photo scale is 2 mm (from Wilber, 1976; courtesy of R. J. Wilber).

Figure 113—Thin section photomicrograph in plane-polarized light of pteropod fragment with tubular, branching, fungal(?) microborings displaying perpendicular and nearly perpendicular side branches from straight tunnels. These right-angle branches terminate in conical swellings. Note the convergence of boring cavities without fusion. Finer filaments in background are fungal hyphae associated with sporangia. Cross photo vertical scale is 300 mm. Sample from 435 m in the northeastern Straits of Florida (from Zeff and Perkins, 1979; courtesy of R. D. Perkins).

Bahamas: AAPG Bull., v. 61, p. 376-406.

Hopkins, J. C., 1977, Production of foreslope breccia by differential submarine cementation and downslope displacement of carbonate sands, Miette and Ancient Wall buildups, Devonian, Canada, *in* H. E. Cook and P. Enos, eds., Deep-water carbonate environments: SEPM Spec. Pub. No. 25, p. 155-170.

Hubert, J. E., R. K. Suchecki, and R. K. M. Callahan, 1977, The Cowhead Breccia; sedimentology of the Cambro-Ordovician continental margin,

Figure 114—Bedding plane view of nodular hardground horizon, Middle Thebes Member, Eocene, Wadi Abu Hamadat, Red Sea coast, Egypt. Lens cap is scale. Hardground consists of framework of fused, early-cemented nodules of fine-grained limestone (courtesy of P. D. Snavely, III, and R. E. Garrison).

Figure 115—Close-up Figure 114. Early cementation and exposure of hardground at the sediment-water interface is evidenced by boring of nodules and common encrustation by oysters (courtesy of P. D. Snavely, III, and R. E. Garrison).

Figure 116—Uppermost portion of the Middle Thebes Member, Eocene, Wadi Saqia, Red Sea coast, Egypt. Upper slope facies consists of nodular chalk and nodular foram-rich sands. Evidence of strong bottom currents shown by large-scale, low-angle cross bedding. Vertical scale of photo approximately 8 m (courtesy of P. D. Snavely, III, and R. E. Garrison).

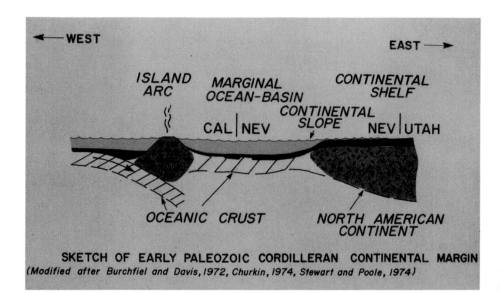

Figure 117A— Sketch of early Paleozoic Cordilleran continental margin.

Figure 117B— 1500-m-thick seaward prograding continental margin sequences. Width of photo about 2 km. Arrow points to base of 10-m-thick, 400-m-wide translational submarine slide shown in Figures 26 and 27 (photo by H. E. Cook).

Newfoundland, *in* H. E. Cook and P. Enos, eds., Deep-water carbonate environments: SEPM Spec. Pub. No. 25, p. 125-154.

James, N. P., and R. N. Ginsburg, 1979, The deep seaward margin of Belize barrier and atoll reefs: Internat. Assoc. Sedimentols. Spec. Pub. 3, 191 p.

Johns, D. R., et al, 1981, Origin of a thick, redeposited carbonate bed in Eocene turbidites of the Hecho Group, south-central Pyrenees, Spain: Geology, v. 9, p. 161-164.

Keith, B. D., and G. M. Friedman, 1977, A slope-fan-basin-plain model, Taconic sequences, New York and Vermont: Jour. Sed. Petrology, v. 47, p. 1220-1241.

Kier, J. S., and O. H. Pilkey, 1971, The influence of sea-level changes on sediment carbonate mineralogy—Tongue of the Ocean, Bahamas: Marine Geology, v. 11, p. 189-200.

Krause, F. F., and A. E. Oldershaw, 1979, Submarine carbonate breccia beds—a depositional model for two-layer, sediment gravity flows from the Sekwi Formation (Lower Cambrian(, Mackenzie Mountains, Northwest Territories, Canada: Canadian Jour. Earth Sci., v. 16, p. 189-199.

Land, L. S., 1979, Chert-chalk diagenesis; the Miocene island slope of north Jamaica: Jour. Sed. Petrology, v. 49, no. 1, p. 223-232.

Lowe, D. R., 1976a, Subaqueous liquefied and fluidized sediment flows and their deposits: Sedimentology, v. 23, p. 285-308.

_____ 1976b, Grain flow and grain flow deposits: Jour. Sed. Petrology, v. 46, p. 188-199.

_____, 1979, Sediment gravity flows; their classification and some problems of application to natural flows and deposits, *in* L. S. Doyle and O. H. Pilkey, eds., Geology of continental slopes: SEPM Spec. Pub. No. 27, p. 75-82.

_____, 1982, Sediment gravity flows; II. depositional models with special reference to the deposits of high-density turbidity currents: Jour. Sed. Petrology, v. 52, p.279-297.

Mattick, R. E., et al, 1978, Petroleum

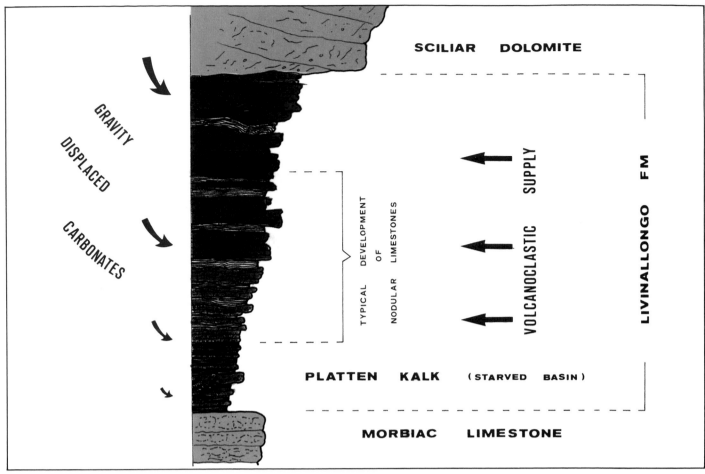

Figure 118—Sketch of Triassic seaward prograding platform margin carbonates, northern Italy (see Bosellini and Rossi, 1974; courtesy of Alfonso Bosellini).

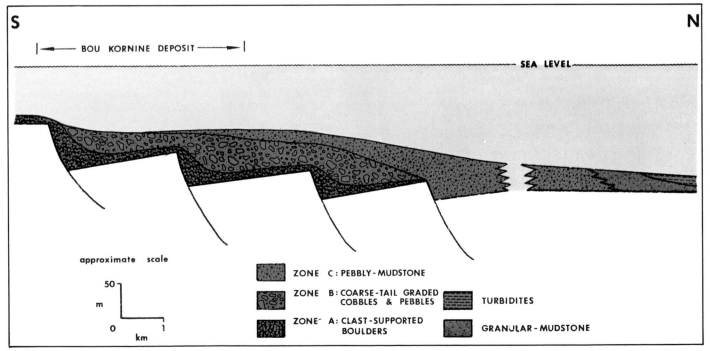

Figure 119—Summary of lithologic relationships of the Bathonian mass-flow deposit exposed on Bou kornine in Tunisia. Model is derived from measured sections and by combining several good exposures of smaller flow deposits, which appear more distal in nature (from Cossey and Ehrlich, 1979; courtesy of S. P. Cossey).

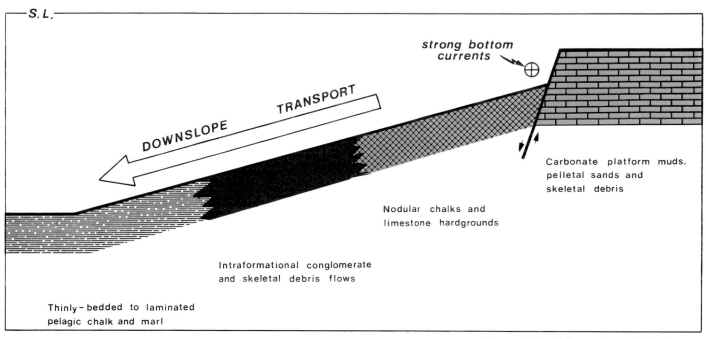

S.L.

strong bottom
currents

DOWNSLOPE TRANSPORT

Carbonate platform muds,
pelletal sands and
skeletal debris

Nodular chalks and
limestone hardgrounds

Intraformational conglomerate
and skeletal debris flows

Thinly-bedded to laminated
pelagic chalk and marl

Figure 120—Schematic model (after Mullins et al, 1980) showing the distribution of dominant lithofacies observed in the Thebes Formation, Eocene, Red Sea coast, Egypt (courtesy of P. D. Snavely, III, and R. E. Garrison).

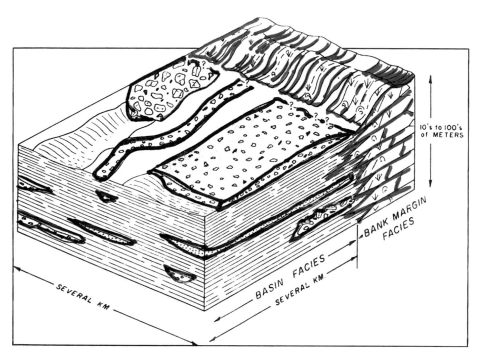

10's to 100's
of METERS

BANK MARGIN FACIES

BASIN FACIES
SEVERAL KM.

SEVERAL KM.

Figure 121—Depositional model of carbonate mass-flow deposits that are generated at bank and reef margins. Deposits commonly occur as widespread debris flow and turbidity-current flow "sheets," "aprons," and "debris wedges." Individual debris-flow sheets can cover 100 cu m and be transported more than 100 sq km into the basin. Note absence of lateral or vertical submarine fan sequences (from Cook et al, 1972).

potential of U.S. Atlantic slope, rise, and abyssal plain: AAPG Bull., v. 62, p. 592-608.

McDaniel, P. N., and L. C. Pray, 1967, Bank to basin transition in Permian (Leonardian) carbonates, Guadelupe Mountains, Texas (abs.): AAPG Bull., v. 51, p. 474.

McGovney, J. E., 1981, Resedimented deposits and evolution of Thornton (Niagran), northeastern Illinois (abs.): AAPG Bull., v. 65, p. 957.

McIlreath, I. A., 1977, Accumulation of a Middle Cambrian, deep-water limestone debris apron adjacent to a vertical, submarine carbonate escarpment, southern Rocky Mountains, Canada, in H. E. Cook and P. Enos, eds., Deep-water carbonate environments: SEPM Spec. Pub. No. 25, p. 113-124.

_____, and N. P. James, 1978, Facies models 12; carbonate slopes: Geoscience Canada, v. 5, no. 4, p. 189-199.

Meischner, K. D., 1964, Allodapische

Kalke, Turbidite in riff-nahen Sedimentatins-becken, in A. H. Bouma and A. Brouwer, eds., Turbidites: Amsterdam, Elsevier Pub., p. 156-191.

Middleton, G. V., 1970, Experimental studies related to problems of flysch sedimentation, in J. Lajoie, ed., Flysch sedimentology in North America: Geol. Assoc. Canada Spec. Paper 7, p. 253-272.

_____ and M. A. Hampton, 1973, Mechanics of flow and deposition, in

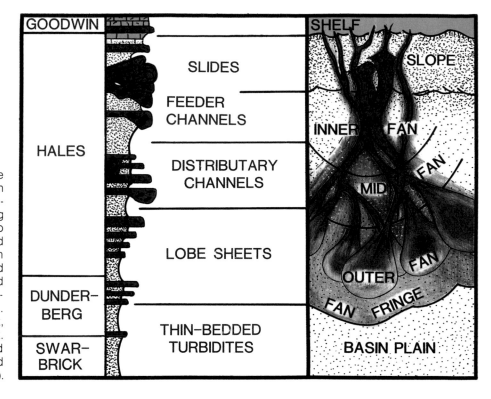

Figure 122—Preliminary local carbonate submarine-fan model showing that fan sediment is derived from both shoal-water shelf areas and by the remolding of deeper water slides and slumps into mass-flows, large slides and channelized conglomerates that occur in outer fan sites, calcarenites in non-channelized sheets in mid-fan sites, and thin-bedded silt to fine sand-sized carbonate turbidites in fan fringe and basin plain. Slope and fan facies about 500 m thick, basin plain facies about 1000 m thick. Model based on studies in Cambrian and Ordovician strata in Nevada (modified from Cook and Egbert, 1981b, c).

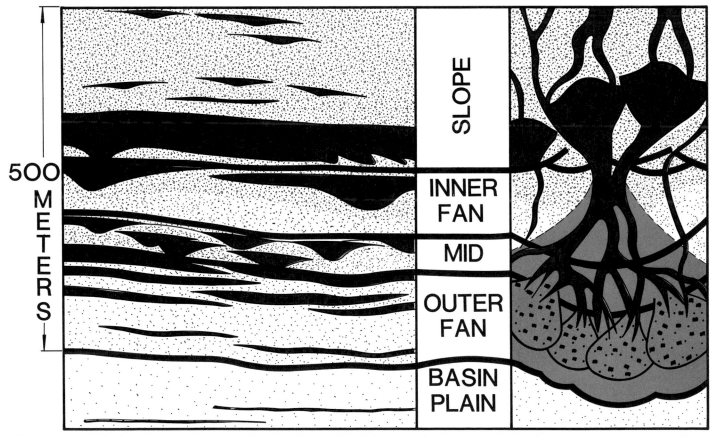

Figure 123—Preliminary local carbonate submarine fan model. Schematically shows vertical and lateral facies sequences that occur in prograding continental margin section illustrated in Figure 117b. Model based on studies in Cambrian and Ordovician strata in Nevada (from Cook and Egbert, 1981a).

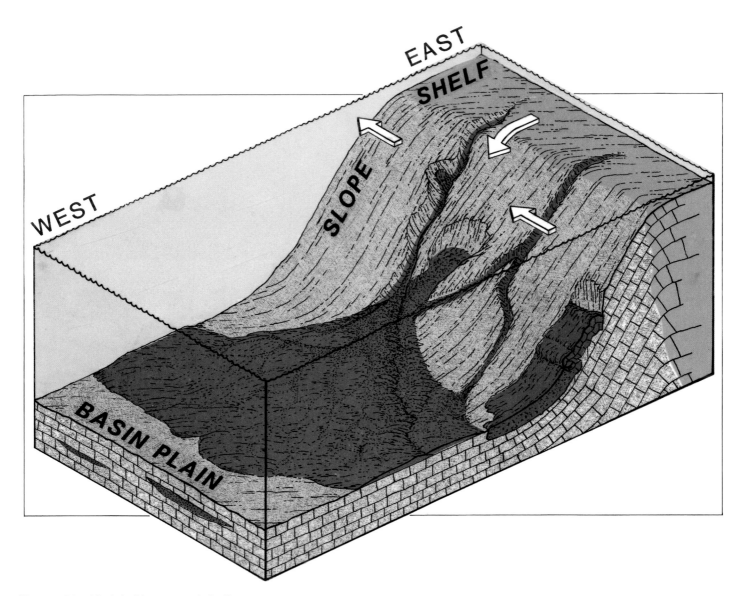

Figure 124—Model of interpreted shelf-slope-basin plain transition in the Late Cambrian and Early Ordovician of Nevada. Model shows slope is incised by numerous gullies but no major canyons; carbonate submarine fan develops at base of slope and basin plain; fan sediment is a mixture of shoal-water shelf carbonates and deeper water slide generated debris; contour currents flow northerly along upper slope (from Cook and Egbert, 1981a).

Figure 125—Model of some characteristics in mid-fan distributary channels shown in Figures 122 and 123. See Figures 71 and 126 for conglomerate in axis, Figure 127 for channel margin facies, and Figures 24 and 128 for inter-channel facies (drawing by H. E. Cook).

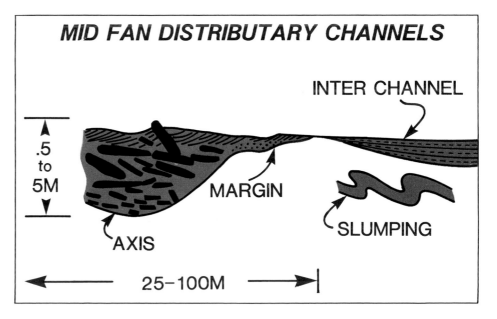

MID FAN DISTRIBUTARY CHANNELS

INTER CHANNEL

.5 to 5M

MARGIN

AXIS

SLUMPING

25–100M

Figure 127—Wavy, discontinuous calcarenite turbidites with climbing ripples occur near margins of channels where grain size is small yet deposition is rapid. White circle is 2 cm wide. Lower Hales Limestone, Upper Cambrian, Nevada (photo by H. E. Cook).

Figure 126—Thinning- and fining-upward sequences of mass-flow carbonates inferred to represent mid-fan distributary channels. Lower conglomeratic channel deposit is about 1.5 m thick. Lower Hales Limestone, Upper Cambrian, Nevada (photo by H. E. Cook).

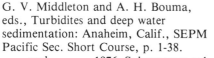

Figure 128—Calcarenite and calcisiltite thin-bedded, laterally continuous turbidites within interchannel facies. Lower Hales Limestone, Upper Cambrian, Nevada (photo by H. E. Cook).

G. V. Middleton and A. H. Bouma, eds., Turbidites and deep water sedimentation: Anaheim, Calif., SEPM Pacific Sec. Short Course, p. 1-38.

_____ and _____, 1976, Subaqueous sediment transport and deposition by sediment gravity flows, in D. J. Stanley and D. J. P. Swift, eds., Marine sediment transport and environmental management: New York, John Wiley and Sons, p. 197-218.

Milliman, J. D., and J. Muller, 1973, Precipitation and lithification of magnesian calcite in the deep-sea sediments of the eastern Mediterranean Sea: Sedimentology, v. 20, p. 29-46.

_____ D. A. Ross, and T. H. Ku, 1969, Precipitation and lithification of deep-sea carbonates in the Red Sea: Jour. Sed. Petrology, v. 39, p. 724-736.

Mitchum, R. M., Jr., 1978, Seismic stratigraphic investigation of West Florida Slope, Gulf of Mexico: AAPG Stud. in Geology No. 7, p. 193-223.

Moore, C. H., E. A. Graham, and L. S. Land, 1976, Sediment transport and dispersal across the deep fore-reef and island slope (¹55 m to ¹305 m), Discovery Bay, Jamaica: Jour. Sed. Petrology, v. 46, p. 174-187.

Morgenstern, N., 1967, Submarine slumping and the initiation of turbidity currents, in A. F. Richards, ed., Marine Geotechnique: Urbana, Univ. of Illinois Press, p. 189-220.

Mountjoy, E. W., 1967, Factors governing the development of the Frasnian, Miette and Ancient Wall, reef complexes (banks and biostromes), Alberta, in D. H. Osward, ed., International Symposium on the Devonian System: Calgary, Alberta Soc. Petroleum Geols., 1967, v. 2, p. 387-408.

_____ et al, 1972, Allochthonous carbonate debris flows—worldwide indicators of reef complexes, banks, or shelf margins: Montreal, 24th Internat. Geol. Cong., Sec. 6, p. 172-189.

Muller, J., and F. Fabricius, 1974, Magnesian-calcite nodules in the Ionian deep-sea; an actualistic model for the formation of some nodular limestones: Internat. Assoc. Sedimentol Spec. Pub. No. 1, p. 235-247.

Mullins, H. T., 1978, Deep carbonate bank margin structure and sedimentation in the Northern Bahamas: Chapel Hill,

Figure 129—Thickening and fining upward nonchannelized turbidite sheets. Note flat bases and tops of turbidite beds. Inferred to represent outer fan lobe sheets. Lower Hales Limestone, Upper Cambrian, Nevada (photo by H. E. Cook).

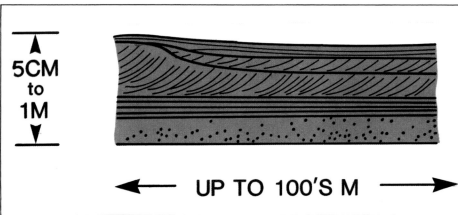

Figure 130—Sketch of sedimentary structures in individual outer fan lobe turbidite sheets shown in Figures 129 and 131 (drawing by H. E. Cook).

Figure 131—Outer fan lobe turbidite. Shows two turbidite events. Lower 14 cm consist of a Bouma A-C sequence overlain by a Bouma A sequence within upper 2 cm of photo. White circle is 2 cm wide (photo by H. E. Cook).

Figure 132—Thin-bedded turbidites interbedded with hemipelagic sediments in fan fringe and basin-plain facies. White circle is 2 cm wide. Dunderberg Shale, Upper Cambrian, Nevada (photo by H. E. Cook).

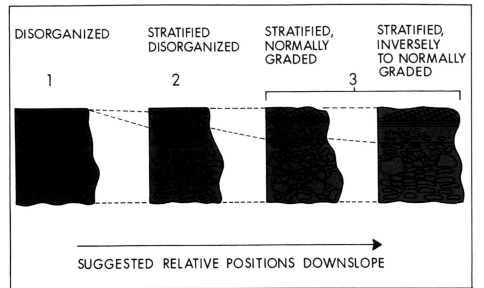

Figure 133—Breccia bed types observed in the field: (1) disorganized bed — breccia clasts are disordered and matrix or grain supported; (2) disorganized, stratified bed — disordered breccia clasts matrix or grain supported and overlain by massive and parallel-laminated grainstones of fragmented breccia material; (3) normally or inversely-normally graded, stratified beds — breccia clasts are normally or inversely to normally graded, matrix consists predominantly of comminuted breccia material, beds are capped by turbidite with A, B, and C intervals (from Krause and Oldershaw, 1979; courtesy of F. F. Krause).

Univ. of North Carolina, Ph.D. dissert., 166 p.

_____ and G. W. Lynts, 1977, Origin of the northwestern Bahama platform; review and reinterpretation: Geol. Soc. America Bull., v. 88, p. 1447-1461.

_____ and A. C. Neumann, 1979, Deep-carbonate bank margin structure and sedimentation in the northern Bahamas: SEPM Spec. Pub. No. 27, p. 165-192.

_____ and H. M. Van Buren, 1979, Modern modified carbonate grain flow deposit: Jour. Sed. Petrology, v. 48, p. 747-752.

_____ M. R. Boardman, and A. C. Neumann, 1979, Echo-character of off-platform carbonates: Marine Geology, v. 32, p. 251-268.

_____ et at, 1978, Characteristics of deep Bahama channels in relation to hydrocarbon potential: AAPG Bull., v. 62, p. 693-704.

_____ et al, 1980a, Carbonate sediment drifts in the northern Straits of Florida: AAPG Bull., v. 64, p. 1701-1717.

_____ et al, 1980b, Nodular carbonate sediment on Bahamian slopes; possible precursors to nodular limestones: Jour. Sed. Petrology, v. 50, no. 1, p. 171-131.

_____ et al, 1981, Modern deep-water coral mounds north of Little Bahama Bank: Criteria for the recognition of deep-water coral bioherms in the rock record: Jour. Sed. Petrology, v. 51, p. 999-1013.

_____ et al, 1982, Geology of Great Abaco Canyon; observations from the research submersible ALVIN: Marine Geology, v. 48, in press.

_____ et al, in prep., Anatomy of a modern open ocean carbonate slope; Northern Little Bahama Bank: Jour. Sed. Petrology.

Nagel, H. E., and E. S. Parker, 1971, Future oil and gas potential of onshore Ventura Basin, California, in Future petroleum provinces of the United States—their geology and potential: AAPG Mem. 15, p. 254-297.

Nardin, T. R., et al, 1979, A review of mass movement processes, sediment and acoustic characteristics, and contrasts in slope and base-of-slope systems versus canyon-fan-basin floor systems, in O. H. Pilkey and L. S. Doyle, eds., Geology of continental slopes: SEPM Spec. Pub. No. 27, p. 61-73.

Neumann, A. C., 1974, Cementation, sedimentation, and structure on the flanks of a carbonate platform, northwestern Bahamas, in Recent advances in carbonate studies (Abs. V): Fairleigh Dickinson Univ., West Indies Lab. Spec. Pub. 6, p. 26-30.

_____, 1977, Carbonate margins: NSA-NRC Rev. of Geology of Continental Margins, 13 p.

_____ and M. M. Ball, 1970, Submersible observations in the Straits of Florida; geology and bottom currents: Geol. Soc. America Bull., v. 81, p. 2861-2874.

_____ J. W. Kofoed, and G. H. Keller, 1977, Lithoherms in the Straits of Florida: Geology, v. 5, p. 4-10.

Paull, C. K., and W. P. Dillon, 1980, Erosional origin of the Blake Escarpment; an alternative hypothesis: Geology, v. 8, p. 538-542.

Pfeil, R. W., and J. F. Read, 1980, Cambrian carbonate platform margin facies, Shady Dolomite, southwestern Virginia, U.S.A.: Jour. Sed. Petrology, v. 50, p. 91-116.

Pray, L. C., and F. G. Stehli, 1962, Allochthonous origin, Bone Springs "patch reefs," west Texas (abs.): Geol. Soc. America Spec. Paper 73, p. 218-219.

Read, J. F., 1980, Carbonate ramp-to-basin transitions and foreland basin evolution, Middle Ordovician, Virginia Appalachians: AAPG Bull., v. 64, p. 1575-1612.

Schlager, W., and A. Chermak, 1979, Sediment facies of platform-basin transition, Tongue of the Ocean, Bahamas,

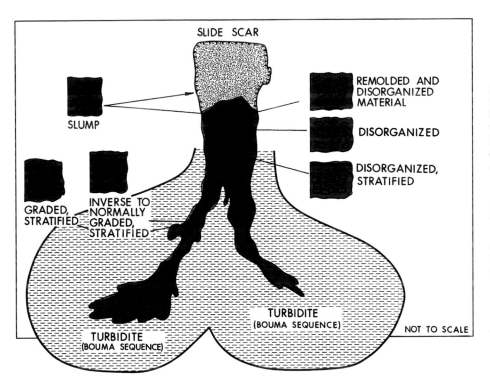

Figure 134—Suggested submarine sediment gravity-flow model for carbonate breccia bed deposits observed in the field. The process is initiated by slumping and sliding of slope deposits (calciturbidites). Material flows as a liquefied slurry, disaggregates and remolds; a deposit of disoriented clasts is produced. Continued downslope movement of the undrained sediment gravity flow and mixing with the overlying water mass produces a turbidity current on top of the flow; the deposits consist of a basal clast interval with a turbidite on top and a depositional sequence as suggested for Figure 133. The turbidity current races away from the basal portion of the flow as the latter comes to a stop. A turbidite lacking a breccia bed component will blanket the surrounding sea floor (from Krause and Oldershaw, 1979; courtesy of F. F. Krause).

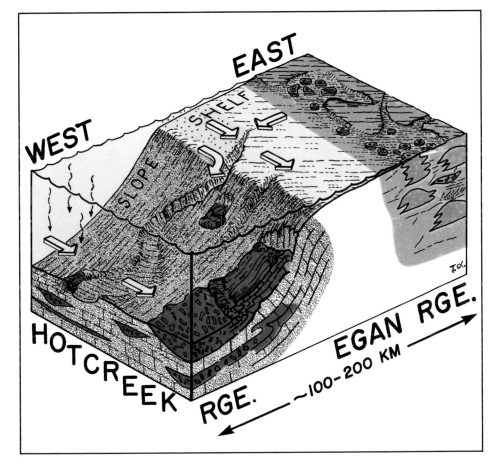

Figure 135—Depositional model showing continental margin in western United States during the early Paleozoic and the generation of mass-flow conglomerates and breccias from submarine slides on carbonate slopes (from Cook and Taylor, 1977).

Figure 136—Permian carbonate debris-flow deposit that forms a petroleum reservoir in the Midland Basin, west Texas. Consists mainly of shoal-water derived clasts and crinoid particles set within a dark lime mud matrix. Note post-depositional solution porosity in matrix (arrow); Permian, west Texas (photo by H. E. Cook).

in O. H. Pilkey and L. S. Doyle, eds., Geology of continental slopes: SEPM Spec. Pub. No. 27, p. 193-208.

————— and N. P. James, 1978, Low-magnesian calcite limestones forming at the deep-sea floor, Tongue of the Ocean, Bahamas: Sedimentology, v. 25, p. 675-702.

Schlanger, S. O., and J. Combs, 1975, Hydrocarbon potential of marginal basins bounded by an island arc: Geology, v. 3, p. 397-400.

Schlee, J., et al, 1977, Petroleum geology on the U.S. Atlantic Gulf of Mexico margins, in Exploration and economics of the petroleum industry: Proc., Southwestern Legal Found. v. 15, p. 47-93.

Scholle, P. A., 1977, Chalk diagenesis and its relation to petroleum exploration; oil from chalks, a modern miracle? AAPG Bull., v. 61, p. 982-1009.

Schanmugam, S., and G. L. Benedict, 1978, Fine-grained carbonate debris flow, Ordovician basin margin, southern Appalachians: Jour. Sed. Petrology, v. 48, p. 1233-1240.

Sheridan, R. E., 1974, Atlantic continental margin of North America, in C. A. Burk and C. L. Drake, eds., Geology of continental margins: New York, Springer-Verlag Pub., p. 391-407.

Smith, D. L., 1977, Transition from deep- to shallow-water carbonates, Paine Member, Lodgepole Formation, Central Montana, in H. E. Cook and P.

Enos, eds., Deep-water carbonate environments: SEPM Spec. Pub. No. 25, p. 187-201.

Snavely, P. D., III, 1981, Early diagenetic controls on allochthonous carbonate debris flows—examples from Egyptian lower Eocene platform-slope (abs.): AAPG Bull., v. 65, p. 995.

Squires, D. F., 1964, Fossil coral thickets in Wairarapa, New Zealand: Jour. Paleontology, v. 38, no. 5, p. 904-915.

Stanley, D. J., and E. Kelling, eds., 1978, Sedimentation in submarine canyons, fans, and trenches: Stroudsburg, Pa. Dowden, Hutchinson and Ross, 395 p.

Stanley, G. D., 1979, Paleoecology, structure, and distribution of Triassic coral buildups in western North America: Univ. Kansas Paleontol. Contrib. Art. 65, 58 p.

Stauffer, P. H., 1967, Grain flow deposits and their implications, Santa Ynez Mountains, California: Jour. Sed. Petrology, v. 37, p. 487-508.

Stetson, T. R., D. F. Squires, and R. M. Pratt, 1962, Coral banks occurring in deep water on the Blake Plateau: Am. Mus. Novitates, no. 2114, p. 1-39.

Teichert, C., 1958, Cold- and deep-water coral banks: AAPG Bull., v. 42, p. 1064-1082.

Thomsom, A. F., and M. R. Thomasson, 1969, Shallow to deep water facies development in the Dimple Limestone (lower Pennsylvanian), Marathon region, Texas, in G. M. Friedman, ed., Depositional environments in carbonate rocks: SEPM Spec. Pub. No. 14, p. 57-78.

Thompson, T. L., 1976, Plate tectonics in oil and gas exploration of continental margins: AAPG Bull., v. 60, p.

1463-1501.

Uchupi, E., 1969, Morphology of the continental margin off southeastern Florida: Southeastern Geology, v. 11, p. 129-134.

Varnes, D. J., 1978, Slope movement types and processes, in R. L. Schuster and R. J. Krizek, eds., Landslides; analysis and control: Transportation Research Board, Natl. Acad. Sci., Spec. Rept. 176, p. 11-33.

Walker, R. G., 1975, Generalized facies models for resedimented conglomerates of turbidite association: Geol. Soc. America Bull., v. 86, p. 737-748.

_____, 1978, Deep-water sandstone facies and ancient submarine fans; models for exploration for stratigraphic traps: AAPG Bull., v. 62, p. 932-966.

_____ and E. Mutti, 1973, Turbidite facies and facies associations, in G. V. Middleton and A. H. Bouma, eds., Turbidites and deep water sedimentation: Anaheim, Calif., SEPM Pac. Sec. Short Course, p. 119-158.

Wange, F. F. H., and V. E. McKelvey, 1976, Marine mineral resources, in G. J. S Govett and M. H. Govett, eds., World mineral supplies: New York, Elsevier Sci. Pub., p. 221-286.

Weeks, L. G., 1974, Petroleum resource potential of continental margins, in C. A. Burk and C. L. Drake, eds., The geology of continental margins: New York, Springer-Verlag Pub., p. 953-964.

Wilber, R. J., 1976, Petrology of submarine lithified hardgrounds and lithoherms from the deep flank environment of Little Bahama Bank (northeastern Straits of Florida): Durham, N.C., Duke Univ., M.S. thesis, 241 p.

_____ and A. C. Neumann, 1977, Porosity controls in subsea cemented rocks from deep-flank environment of Little Bahama Bank: AAPG Bull., v. 61, p. 841.

Wilde, P., W. R. Normark, and T. E. Chase, 1978, Channel sands and petroleum potential of Monterey deep-sea fan, California: AAPG Bull., v. 62, p. 976-983.

Wilson, J. L., 1969, Microfacies and sedimentary structures in "deeper water" lime mudstone, in G. M. Friedman, ed., Depositional environments in carbonate rocks: SEPM Spec. Pub. No. 14, p. 4-19.

_____, 1975, Carbonate facies in geological history: Berlin, Springer-Verlag Pub. 471 p.

Worzel, J. L., et al, 1973, Initial reports of the Deep-Sea Drilling Project: Washington, D.C., U.S. Govt. Printing Office, v. 10, 748 p.

Yarborough, H., 1971, Sedimentary environments and the occurrence of major hydrocarbon accumulations (abs.): Trans., Gulf Coast Assoc. Geol. Socs., v. 21, p. 82.

Yurewicz, D. A., 1977, Sedimentology of Mississippian basin-facies carbonates, New Mexico and west Texas—the Rancheria Formation, in H. E. Cook and P. Enos, eds., Deep-water carbonate environments: SEPM Spec. Pub. No. 25, p. 203-219.

Zeff, M. L., and R. D. Perkins, 1979, Microbial alteration of Bahamian deep-sea carbonates: Sedimentology, v. 26, p. 175-201.

Pelagic Environment

Peter A. Scholle
Michael A. Arthur
Allan A. Ekdale

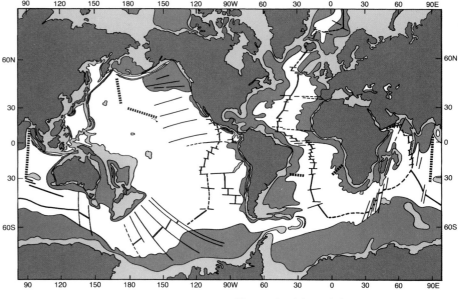

Figure 1—Map of the world showing major physiographic features and depositional provinces of the ocean basins. Brown, land areas; light green, shelf areas (less than 200 m water depth); dark green, continental slopes, rises and abyssal plains (dominated by turbidite and hemipelagic sedimentation); white, mid-ocean and aseismic ridge flanks, seamount provinces, and intervening basins (generally dominated by pelagic sedimentation); solid black, mid-ocean ridge crests and fracture zones; hachured black lines, aseismic ridge crests; blue lines, major deep-sea trenches.

The definition of "pelagic facies" is not a simple one because there are a wide variety of biogenic and non-biogenic components in most pelagic sediments and there can be no specific depth connotation to the word "pelagic." We follow Jenkyns (1978) in applying the term "pelagic" in a strictly descriptive sense to mean open-marine deposits, whether in shallow epicontinental seas and outer shelf areas or in the deep sea on oceanic crust in settings such as aseismic ridges, submerged plateaus, mid-ocean ridges and abyssal plains (Fig. 1). The term pelagic can also be applied to organisms that inhabit the open ocean and are excluded from marginal marine environments. In general, pelagic sedimentation implies a lack of significant influence of terrigenous sediment sources. In this chapter, we are primarily concerned with processes that influence the deposition of pelagic carbonate sediments and in recognizing depositional environments of pelagic carbonate facies. However, we also discuss other associated pelagic sediments (for example, biosiliceous, red clay, etc.) and hemipelagic sediments (for example, those containing a substantial amount of fine-grained terrigenous material). Pelagic sedimentation implies slow grain-by-grain settling of material biochemically produced in surface water, whereas hemipelagic sedimentation is characterized by redeposition of material, either downslope as in dilute-suspension turbidity currents or by settling out of bottom-currents or nepheloid layers.

Pelagic sediments are commonly simple in composition, consisting of various proportions of a few major biogenic components. However, the relative proportions of these components in pelagic facies change both in space, due mainly to various oceanographic factors and proximity to continents, and through time. Examples illustrated in this paper are mainly of late Mesozoic to Cenozoic age, a time when the pelagic flora was dominated by calcareous plankton, particularly coccolithophorids. It is estimated, for example, that at present about 67% of the calcium carbonate fixed by marine organisms is incorporated into tests of calcareous nannoplankton (Hay and Noel, 1976) and more than 50% of the present sea floor is covered by carbonate sediment (Fig. 2). This contrasts with most of Paleozoic and early Mesozoic time when shelf seas were dominated by carbonate platform deposits, carbonate-poor shales, or bedded cherts, while deeper basins were characterized by "starved-basin" facies. Although there may be a bias against preservation of pelagic facies because of ocean-floor subduction, there is little evidence of either widespread calcareous pelagic sediments or calcareous microplankton in Paleozoic rocks; much of the carbonate fraction found in deeper-water deposits represents comminuted shelf-derived carbonate detritus. During the Paleozoic, the plankton were primarily organic-walled, phosphatic, or siliceous varieties. Although ammonites, nautiloids, tentaculitids, styliolinids and other groups did contribute some skeletal carbonates to Paleozoic "pelagic" facies, this material was volumetrically and areally restricted. We might, therefore, consider the shelf cherts of certain times, the Ordovician and Devonian, for example, as the "chalks" of the Paleozoic. This contrast underscores the role of organic evolution in determining the composition of pelagic sediments through time. We will largely ignore this factor, however, and concentrate on the causes of variation in pelagic facies with water depth and changing sea level, climate, ocean circulation, and water chemistry.

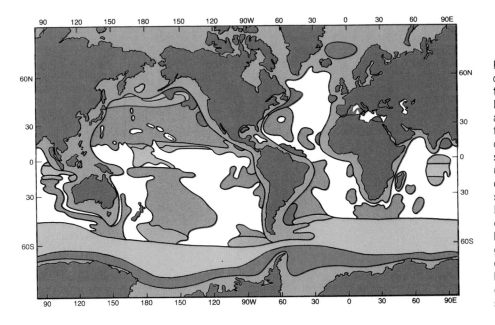

Figure 2—Map of the world showing distribution of major sediment types in the ocean basins (after Berger, 1974; Davies and Gorsline, 1976). Brown, land area; light blue, areas underlain by continental margin sediment dominated by clastic terrigenous debris or carbonate shelf deposits; dark blue, areas underlain by sediments of glacial-marine origin; light green, areas of primarily siliceous biogenic pelagic deposits; lavender, areas that contain pelagic red clay; pink, areas of overlap of siliceous biogenic sediment and red clay; dark green, deep sea fans and abyssal plains dominated by clastic terrigenous turbidite sediments; white, sea floor covered primarily by pelagic carbonate sediments.

CRITERIA FOR RECOGNITION

Sediment composition, especially the faunal and floral makeup of the sediment, is a key to the recognition of pelagic sediments. The dominant presence of planktonic foraminifers, coccoliths, pteropods, marine diatoms, radiolarians, or other groups of marine plankton or nekton, although not always easy to recognize after extensive diagenesis, is the most direct clue to a pelagic origin. Other criteria sometimes indicating a pelagic origin for sediments include: (a) condensed sections that were deposited at slow rates or are interspersed with numerous hiatuses; (b) presence of multiple hardground horizons that mark submarine lithification events associated with hiatuses; (c) fine-grained, well-bedded sediments of great lateral extent and with gradual lateral facies changes; (d) sediment characterized by presence of fecal pellets, small-scale lamination, or centimeter- to meter-thick bedding rhythms (chalk-marl cycles). Ripple marks and other small-scale bedding features may be present, but large-scale structures are typically absent or rare; or (e) burrow assemblages dominated by groups such as *Helminthoida, Paleodictyon, Zoophycos, "Skolithos,"* and *Chondrites* in deep-water settings and *Thalassinoides* and *Chondrites* in shallower-water settings.

Recognition and study of pelagic (and hemipelagic) facies are important for a number of reasons. These facies are commonly major hydrocarbon source rocks in continental margin settings, and recently we have begun to recognize their importance as reservoir rocks as well (for example, in the North Sea and U.S. Gulf of Mexico). Finally, their chemical composition and biota provide clues to patterns of circulation and chemical changes in ancient oceans.

PELAGIC SEDIMENT COMPOSITION AND TEXTURE

Pelagic sediments are simple chemical and mineralogic systems, in contrast with shallow-marine deposits; yet, commonly they have numerous biogenic and non-biogenic components. These components can include planktic, nektic, and benthic organisms, authigenic minerals, or detrital grains derived from terrigenous, volcanic, carbonate shelf, or other sources. The major components of modern and ancient pelagic sediments and their mineral composition are summarized in Tables 1 and 2, and some important biogenic constituents are illustrated in Figures 3 to 7.

For the past 100 to 150 m.y., pelagic carbonate sediments have been composed dominantly of planktic foraminifers, coccolithophores, and subordinate nannofossil groups. Mesozoic and Cenozoic pelagic sediments, therefore, have been uniformly fine grained except in areas winnowed by strong submarine current activity. Pelagic limestones from both shelf and deep-sea settings commonly have polymodal grain-size distributions with peaks corresponding to the average sizes of the dominant constituent grains and their breakdown products. Thus, many chalks and related limestones have grain-size peaks of up to about 0.5 micrometers (μm) (presumably individual calcitic laths from disintegrated coccoliths), 1 to 5 μm (corresponding to whole coccoliths), 5 to 20 μm (corresponding to larger coccoliths, coccospheres, and comminuted foraminifers and other skeletal fragments), 25 to 64 μm (corresponding to many whole foraminifers), and greater than 64 μm (foraminifers, *Inoceramus* prisms and other macrofossil fragments) (Hakansson et al, 1974; Black, 1980; Hancock, 1980).

Clearly, the relative proportions of these peaks will vary with the depositional setting of an individual sample, but analyses of grain size for

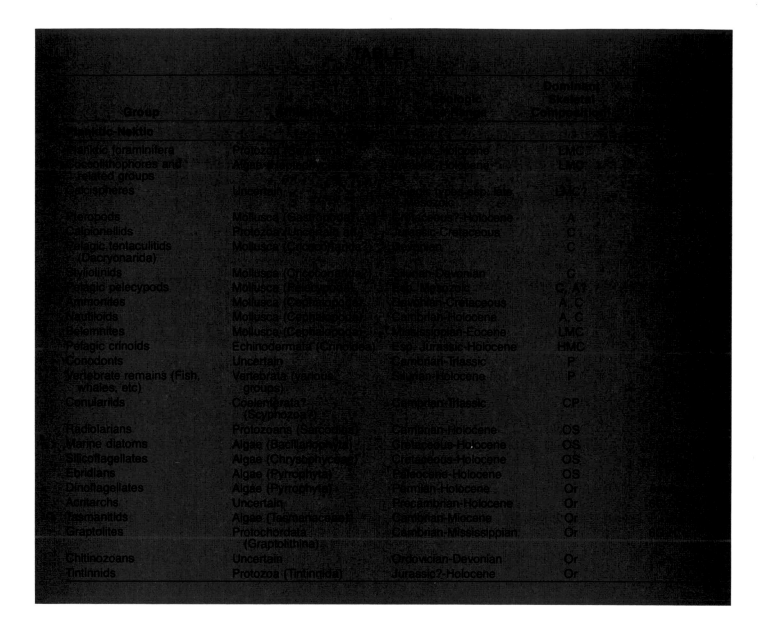

TABLE 1

Group	Biologic Affinities	Geologic Range	Dominant Skeletal Composition
Planktic-Nektic			
Planktic foraminifera	Protozoa	?-Holocene	LMC
Coccolithophores and related groups	Algae	?-Holocene	LMC
Calcispheres	Uncertain	?-Late	LMC?
Pteropods	Mollusca (Gastropoda)	Cretaceous?-Holocene	A
Calpionellids	Protozoa (Uncertain af...)	Jurassic-Cretaceous	C
Pelagic tentaculitids (Dacryonarida)	Mollusca (Cricoconarida)	Devonian	C
Styliolinids	Mollusca (Cricoconarida)	Silurian-Devonian	C
Pelagic pelecypods	Mollusca (Pelecypoda)	Esp. Mesozoic	C, A?
Ammonites	Mollusca (Cephalopoda)	Devonian-Cretaceous	A, C
Nautiloids	Mollusca (Cephalopoda)	Cambrian-Holocene	A, C
Belemnites	Mollusca (Cephalopoda)	Mississippian-Eocene	LMC
Pelagic crinoids	Echinodermata (Crinoidea)	Esp. Jurassic-Holocene	HMC
Conodonts	Uncertain	Cambrian-Triassic	P
Vertebrate remains (Fish, whales, etc)	Vertebrata (various groups)	Silurian-Holocene	P
Conularids	Coelenterata? (Scyphozoa?)	Cambrian-Triassic	CP
Radiolarians	Protozoans (Sarcodina)	Cambrian-Holocene	OS
Marine diatoms	Algae (Bacillariophyceae)	Cretaceous-Holocene	OS
Silicoflagellates	Algae (Chrysophyceae)	Cretaceous-Holocene	OS
Ebridians	Algae (Pyrrophyta)	Paleocene-Holocene	OS
Dinoflagellates	Algae (Pyrrophyta)	Permian-Holocene	Or
Acritarchs	Uncertain	Precambrian-Holocene	Or
Tasmanitids	Algae (Tasmanaceae)	Cambrian-Miocene	Or
Graptolites	Protochordata (Graptolithina)	Cambrian-Mississippian	Or
Chitinozoans	Uncertain	Ordovician-Devonian	Or
Tintinnids	Protozoa (Tintinnida)	Jurassic?-Holocene	Or

coccolith-rich, Cretaceous, pelagic shelf chalks yield an average size of 2 to 5 μm, with more than 90% (by weight) of the grains in size grades less than 64 μm (Hakansson et al, 1974). This small grain size implies that pore-throat diameters will be in the range of 0.1 to 1.0 μm (Price et al, 1976), and that matrix permeabilities for even highly porous chalks will be low. A typical chalk with 40% porosity will have approximately 5 millidarcys (md) permeability; with 20% porosity, permeability will average 0.4 md or less (Scholle, 1977a). Deep-marine pelagic limestones, because of their lower

average coccolith content and higher foraminifer content, generally have slightly coarser average grain sizes and correspondingly higher permeabilities.

The dominant constituents of Mesozoic and Cenozoic pelagic limestones have low-magnesium calcite compositions initially, with the exception of the aragonitic pteropods (see Table 1). Because low-magnesium calcite is the most stable form of calcium carbonate in most near-surface settings, the diagenetic behavior and reservoir properties (porosity and permeability) of Mesozoic and Cenozoic pelagic

limestones are relatively stable and predictable (Scholle, 1977a).

Composition, grain size, diagenetic history, and reservoir properties of pre-Jurassic pelagic limestones are far less predictable, however, because of the diverse origins of these units. Some pelagic carbonate-contributing organisms existed in pre-Jurassic time, but they were primarily macrofossils. Paleozoic fine-grained, deeper water carbonate rocks presumably consist almost entirely of carbonate debris reworked from shelf areas and transported into deeper water. Hence the abundance of Paleozoic starved-basin facies. The

TABLE 1

Group		Age Range	Composition[1]	Illustration number (this paper)
Benthic				
Agglutinated foraminifera		Cambrian-Holocene		
Calcareous benthic foraminifera		Devonian-Holocene	?, A	7A-B
Echinoderms (esp. ophiuroids and holothurians)	Echinodermata (various)	Cambrian-Holocene	HMC	7C
Ostracods	Crustacea (Ostracoda)	Cambrian-Holocene	LMC, HMC	
Inoceramids	Mollusca (Pelecypoda)	Cretaceous	LMC, A	7D-F
Benthic tentaculitids	Mollusca (Cricoconarida)	Ordovician-Devonian	C	
Other benthic macrofauna (Oysters, other molluscs, bryozoans, brachiopods, trilobites, ahermatypic corals, etc.)	Various	Various	C, A	7D, H
Siliceous sponges	Porifera (Hyalosponges and Demosponges)	Cambrian-Holocene	OS	5J-K
Detrital				
Reworked shelf limestone (coralgal sands, skeletal fragments, aragonite muds, etc.)	Various	Precambrian-Holocene		
Fecal pellets	Various	Precambrian-Holocene	HMC, LMC	6I-J
Vascular plant fragments	Various plant groups	Devonian-Holocene		
Spores	Lower plants	Silurian-Holocene	Or	6E
Pollen	Higher plants	Pennsylvanian-Holocene	Or	6F

[1]A, aragonite; C, calcite (Mg-content unknown); LMC, low-magnesium calcite; HMC, high-magnesium calcite; Mg, magnesium calcite; Or, organic matter; OS, opaline silica; P, phosphate (generally calcium fluorapatite); Q, quartz.

Table 1—Major biogenic components of modern and ancient pelagic sediments.

equally widespread occurrence of radiolarian chert and graptolitic shale in Paleozoic deeper marine sections is a further consequence of the absence of significant sources of pelagic carbonate during that era. Paleozoic basinal carbonates, then, commonly reflect the original skeletal composition of adjacent carbonate shelf areas, and their grain size is a function of transportation mechanisms and distance from the platform margins to the depositional site. Pelagic limestone deposited on isolated, pre-Jurassic submerged platforms ("Schwellen"), were dominated by benthic and/or nektic macrofossils (cephalopods, trilobites, brachiopods, etc.), and both grain size and original mineralogic composition are as dif-ficult to predict as for shallow-marine carbonate rocks.

CLASSIFICATION OF PELAGIC SEDIMENTS

The composition of pelagic sediments is a function of the relative rates of supply of biogenic carbonate and/or silica versus the influx of detrital, volcanic, or chemical contributions. The relative abundance of these components forms the basis for most classifications. Two of the more commonly used classification schemes are shown in Tables 3 and 4.

In addition to composition, the degree of lithification has traditionally played an important, if not always edifying, role in many classifications of pelagic sediments (as seen from Table 4). Largely unconsolidated pelagic carbonate sediment, for example, would be termed an ooze (foraminiferal or nannofossil ooze); the same sediment consolidated to the point of firmness would be termed a chalk; consolidated still further, it would be called a limestone or "chalkstone" (foraminiferal or nannofossil limestone). An analogous sequence exists for siliceous sediments that grade from oozes (radiolarian ooze or radiolarite) to porcelanites, and eventually to cherts as they undergo progressive diagenesis. The mixture of compositional terms with terms denoting the stage of diagenetic alteration can lead to considerable complexity and confusion. For

Figure 3—Calcareous planktic organisms. **A**. SEM (scanning electron microscope) photograph of modern globigerinid foraminifers; scale bar is 0.17 mm; photograph by Peter Roth. **B**. SEM photograph of modern globorotalid foraminifer; scale bar is 80 μm; photography by Peter Roth. **C.** Thin-section photomicrograph of planktic foraminifer in Upper Cretaceous chalk; plane-polarized light; scale bar is 80 μm. **D.** Thin-section photomicrograph of modern *Globorotalia* foraminifer in pelagic sediment; plane-polarized light; scale bar is 70 μm. **E.** SEM photograph of a modern coccosphere (*Emiliania huxleyi*); scale bar is 1.0 μm. **F.** SEM photograph of cross section of a coccosphere (*Emiliania huxleyi*) showing arrangement of individual plates (coccoliths); scale bar is 1.0 μm. **G.** SEM photograph showing diagenetically altered coccosphere from Oligocene oceanic chalk; scale bar is 2.3 μm. **H.** SEM photograph of single coccolith in Upper Cretaceous shelf chalk; scale bar is 0.7 μm. **I.** SEM photograph of mixed calcareous nannofossil debris including coccoliths, rhabdoliths, and prediscosphaerids from an Upper Cretaceous shelf chalk; scale bar is 3.4 μm. **J.** Light microscope view of coccoliths showing characteristic extinction patterns; smear mount of Eocene oceanic chalk (crossed polars); scale bar is 65 μm. **K.** SEM photograph of discoaster from Pliocene (?) oceanic chalk; scale bar is 0.8 μm.

SCALE OF EACH PHOTO

└─1 cm─┘

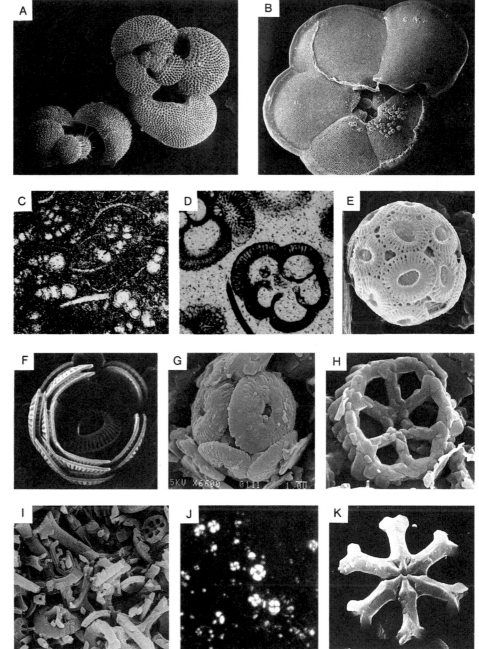

simplicity, the terms nannofossil chalk or foraminiferal chalk, as well as radiolarite and diatomite, should be used to denote sediment of dominant nannofossil, foraminiferal, radiolarian or diatom composition. Modifying terms such as ''lithified'' or ''unlithified'' should be used to denote the degree of diagenetic alteration.

CONTROLS ON PELAGIC SEDIMENT DISTRIBUTION

Biogenic Production of Pelagic Sediment

The presence and composition of pelagic sediments in modern oceans are controlled by a variety of primary factors. First, the production of pelagic microfossils in surface waters and supply to the sea floor are dependent on ecologic controls such as fertility of surface waters (nutrient sup-

ply from upwelling deeper water), water temperature, light, and salinity. These controls, in turn, are dependent on latitude, regional climate, and regional and local patterns of ocean circulation. Primary producers (phytoplankton) having mineralized tests are the coccolithophorids (calcareous) and diatoms (siliceous); the major consumers (zooplankton) are planktic foraminifers (calcareous) and radiolarians (siliceous).

Most calcareous plankton produc-

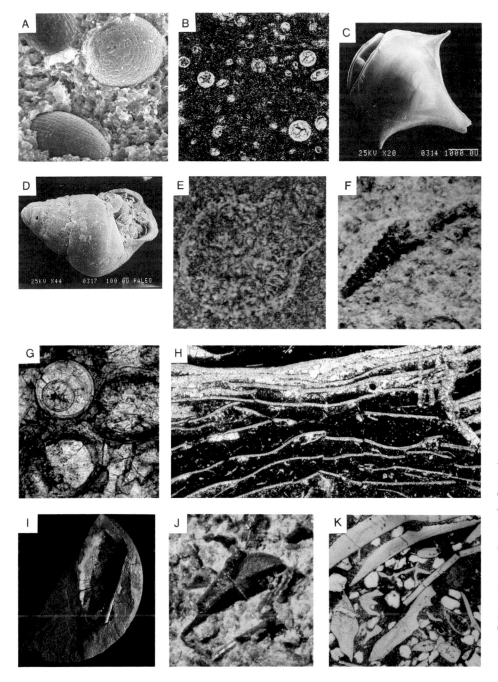

Figure 4—Planktic and nektic pelagic organisms. **A.** SEM photograph of calcispheres (*Pithonella ovalis*) from Upper Cretaceous chalk; scale bar is 4.9 μm. **B.** Thin-section photomicrograph of calcispheres filled with sparry calcite in Upper Cretaceous chalk; plane-polarized light; scale bar is 40 μm. **C.** SEM photograph of well-preserved modern aragonite pteropod, oceanic chalk; scale bar is 0.95 μm. **D.** SEM photograph of partly dissolved modern aragonitic pteropod from oceanic chalk; scale bar is 0.55 μm. **E.** Thin-section photomicrograph of calpionellid from Upper Jurassic limestone; plane-polarized light; scale bar is 55 μm. **F.** Tentaculitid from Silurian basinal limestone; scale bar is 2 cm. **G.** Thin-section photomicrograph showing cross section of sparry calcite-filled styliolinids of Devonian age; plane-polarized light; scale bar is 40 μm. **H.** Thin-section photomicrograph showing thin-shelled pelagic pelecypods (*Halobia*) from Triassic basinal limestone; plane-polarized light; scale bar is 0.25 mm. **I.** Calcitic ammonite aptychus of Late Jurassic age from oceanic limestone; scale bar is 1.2 cm; from DSDP Leg 11 report frontispiece. **J.** Phosphatic conodont (*Siphonodella*) from Mississippian limestone, scale bar is 0.44 mm. **K.** Thin-section photomicrograph showing cross section of numerous conodonts in Mississippian basinal phosphatic shale; plane-polarized light; scale bar is 0.15 mm.

SCALE OF EACH PHOTO
⊢ 1 cm ⊣

tion is in relatively warm surface waters that, at present, are at lower latitudes. Pelagic carbonate facies, therefore, are now confined to latitudes between 60° north and south (see Fig. 2). During climatic optima, such as in the Mesozoic and Early Cenozoic, pelagic carbonate facies probably were more widespread. Siliceous plankton are concentrated in, but are not confined to, areas of cooler surface waters and high nutrient supply (including dissolved silica). Such areas include high-latitude surface waters, cooler surface waters associated with upwelling along oceanic divergences (Antarctic and equatorial divergences), and along west-facing continental margins in the tradewind belts such as those adjacent to Peru and Chile, India and Pakistan, northwestern Africa, and southwestern Africa. The supply rate of pelagic carbonate and silica depends directly on the availability of nutrients. Productivity and supply are, therefore, highest in zones of upwelling along continental margins and oceanic divergences and lowest in the centers of large oceanic gyres. Seasonality and climate also control sea-surface productivity, especially at high latitudes where availability of light and sea-ice cover limit productivity during winter months. Useful reviews of the ecology and productivity of pelagic organisms can be found in Berger (1974, 1976), Berger and Roth (1975), Tappan and Loeblich

Table 2—Major non-biogenic components of modern and ancient pelagic sediments (modified from Berger, 1974).

(1971), and Zeitschel (1978).

Modification of Sediment Flux by Dissolution

The distribution and composition of pelagic sediments are modified by chemical and circulation factors within the oceanic water column and on the sea floor. The accumulation rate and distribution of carbonate sediments on the ocean floor are a function of the rate of supply versus dissolution. Deeper water masses are generally undersaturated with respect to calcite and aragonite, and thus, dissolution occurs both in the water column and on the sea floor. The degree of undersaturation is controlled by the CO_2 content of deep-water masses which is a function of the age of the water mass, the amount of organic matter oxidized, and temperature. Figure 8 illustrates the relationship between the degree of aragonite and calcite saturation in seawater and the occurrence of calcium carbonate in Atlantic and Pacific sediments. The observed level of greatly increased calcite dissolution in the oceans is termed the "lysocline," and the depth below which no pelagic carbonate sediment accumulates is the "carbonate compensation depth" (CCD). The CCD varies in different parts of the ocean basins (Figs. 8 and 9). The CCD is deeper at the equator, where the supply of carbonate material is greater. The CCD shoals toward the centers of the large oceanic gyres, where surface production of $CaCO_3$ is low, and along continental margins, where production of organic matter is high. The levels of the lysocline and the CCD also vary from one ocean basin to another (Berger and Winterer, 1974). At present, for example, the CCD is deepest in the North Atlantic and shallowest in the North Pacific. This is mainly due to the progressive increase in age and total CO_2 content of the deep waters from the North Atlantic to the Pacific (Broecker, 1974). Deep water loses oxygen and becomes more nutrient- and CO_2-rich with increasing age because of biologic consumption and organic-matter decomposition. Generally, an ocean basin with geologically "young" bottom water is a carbonate sink because the deep-water masses are not as aggressive in carbonate dissolution. Ocean basins with older bottom waters are more nutrient-rich and productive, but rates of carbonate dissolution are greater. This relationship has been termed "basin-basin fractionation" (Berger, 1970a).

Selective Dissolution

Carbonate dissolution takes place in the water column and on the sea floor. Aragonite is more soluble in sea water than calcite; therefore, the aragonite compensation depth (ACD) is shallower than the CCD. The major producers of aragonite in the pelagic realm today are the pteropods (Fig. 4C, D). In the Paleozoic and Mesozoic, ammonites and nautiloids were the major aragonite contributors to deeper water sediments. The calcite parts of these organisms are commonly preserved in sediments even when the aragonitic parts have been completely dissolved, yielding aptychus limestones (Fig. 4I).

Calcitic skeletons are inherently less soluble than aragonitic skeletons under most oceanic conditions. In addition, fine-grained carbonate particles, dominated by the calcitic coccolithophorids, may be shielded from

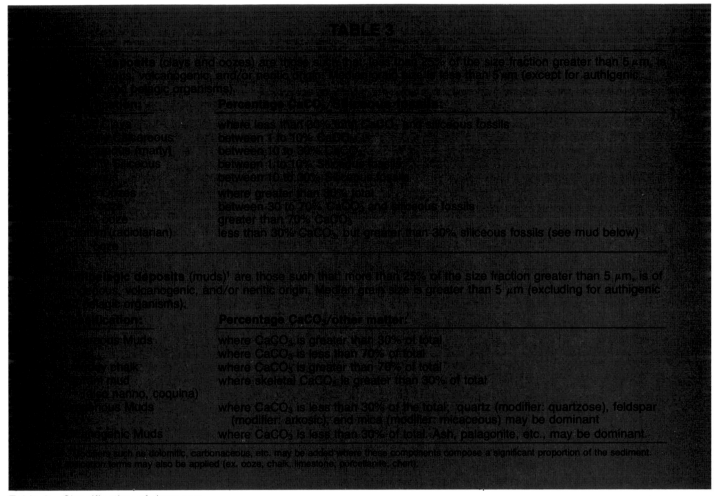

Table 3—Classification of deep-sea pelagic and hemipelagic sediments (after Berger, 1974).

dissolution by organic sheaths (as in fecal pellets) during their descent through the water column (Honjo, 1976). Such fecal pellets are abundant in some pelagic limestones (Fig. 6I, J). Thus, carbonate detritus is typically found in the surface sediment layer below the CCD (Adelseck and Berger, 1975). Most of this carbonate will be dissolved when the organic sheaths are oxidized. The amount of calcium carbonate dissolution that occurs in transit to the bottom depends on the degree of undersaturation in the water column (Peterson, 1966; Broecker and Takahashi, 1978), and on the shape, size, and density of the material (Lerman et al, 1974; Lal and Lerman, 1975), which determines its rate of fall to the sea floor. Pelagic

calcareous material can also be injected at rapid rates below the CCD by dilute-suspension turbidity currents (Hesse, 1975; Scholle, 1971). This material may then be buried and preserved because rates of dissolution are not high enough to remove the rapidly emplaced mass of carbonate (see also Kelts and Arthur, 1981).

Paleontologists can generally estimate the levels of past CCD's and lysoclines, as well as shorter term changes in rates of carbonate dissolution, by examining the composition and preservation of calcareous faunas and floras in the sediment (Berger, 1976; Adelseck, 1977; Thunell, 1976). The absence of more fragile, soluble forms, a high degree of test fragmentation, and low ratios of planktic to

benthic foraminifers are all signs of increasing carbonate dissolution. However, burial diagenesis can impose similar effects, and for most ancient pelagic sediments, the CCD is the only easily recognized chemical boundary.

Biogenic silica is also dissolved in the water column and on the sea floor. The patterns of silica accumulation and dissolution differ somewhat from those of the carbonate system. Berger (1976) argued convincingly that patterns of biogenic silica (opal) accumulation are directly related to productivity in surface waters. The sediments deposited under highly productive regions tend to be relatively rich in biogenic silica. The concept of basin-basin fractiona-

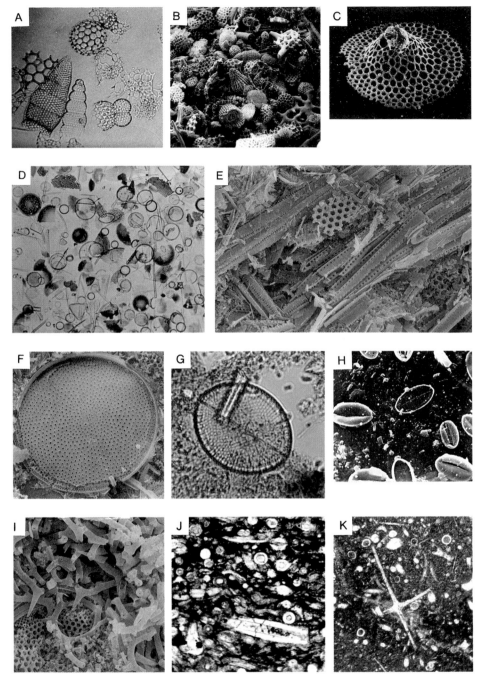

Figure 5—Siliceous planktic and benthic organisms. **A.** Light photomicrograph of coarse fraction of modern oceanic siliceous ooze containing a varied radiolarian assemblage; scale bar is 30 μm; photograph by C. R. Wenkam. **B.** SEM photograph of coarse fraction of modern oceanic siliceous ooze containing a fragmented but well preserved radiolarian assemblage; scale bar is 100 m (from Hein et al, 1979). **C** SEM photograph of single radiolarian that has been etched out of a Jurassic chert; scale bar is 32 μm; photograph by E. A. Pessagno. **D** Light photomicrograph of modern siliceous ooze containing a mixed assemblage of diatoms and sponge spicules; scale bar is 0.12 mm. **E.** SEM photograph of diatom-rich, Miocene siliceous ooze from the continental borderland off California; scale bar is 7.5 μm. **F.** SEM photograph of centric diatom from a Miocene continental borderland basin, California; scale bar is 12 μm. **G.** Light photomicrograph of a diatom from modern shallow-marine carbonate mudstone; plane-polarized light; scale bar is 7 μm. **H.** SEM photograph of diatoms from modern shallow-marine carbonate mudstone; scale bar is 3.6 μm. **I.** SEM photograph of probable silicoflagellates from Miocene siliceous ooze deposited in a continental borderland basin, California; scale bar is 10 μm. **J.** Thin-section photomicrograph of monaxon sponge spicules (with central canals) from a Pennsylvanian limestone; plane-polarized light; scale bar is 90 μm. **K.** Thin-section photomicrograph of a Cretaceous shelf chalk from England containing monaxon and multiaxon sponge spicules; plane-polarized light; scale bar is 0.1 mm.

SCALE OF EACH PHOTO

|— 1 cm —|

tion applies to silica as well. Relatively little biogenic silica accumulates in the North Atlantic basin today, whereas siliceous pelagic sediments are common under regions of fertile surface waters in the Pacific and Indian Ocean basins, where deep waters are older and more enriched in nutrients, including silica. Most dissolution of biogenic opal apparently occurs in warm, undersaturated surface waters (Edmond, 1974;

Heath, 1974). Post-depositional dissolution of siliceous tests in sediments is also a significant source of silica in pore waters; such dissolution also may contribute silica to oceanic deep waters by diffusion out of sediments.

Biogenic opal skeletons of different organisms, like calcareous tests, vary in resistance to dissolution. Modern forms range from least to most resistant as follows: (1) silicoflagellates,

(2) diatoms, (3) delicate radiolaria, (4) robust radiolaria, and (5) sponge spicules (Berger, 1976; Hurd and Theyer, 1977). Therefore, siliceous biogenic facies are clues to past rates of surface production, to possible past variations in ocean chemistry, and to ancient sites of upwelling and high productivity.

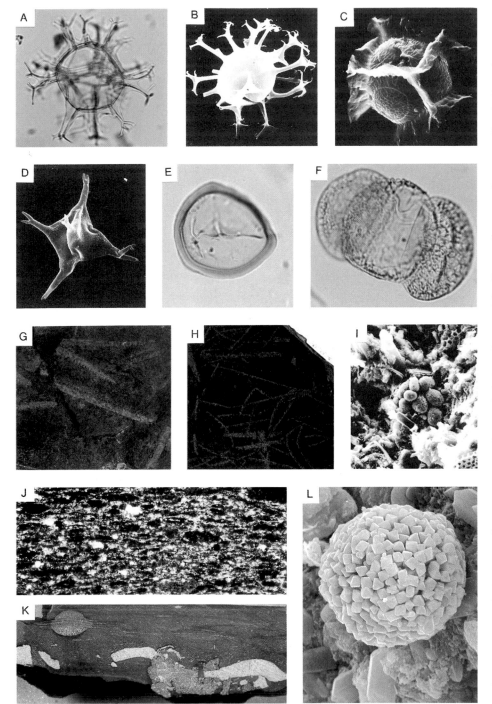

Figure 6—Organic-walled microfossils, fecal pellets and associated sulfide minerals. **A.** Light photomicrograph of stained preparation of Cretaceous shallow-marine dinoflagellate; scale bar is 9 μm. **B.** SEM photograph of Cretaceous(?) dinoflagellate; scale bar is 7.4 μm. **C.** SEM photograph of Paleozoic(?) herkomorph acritarch; scale bar is 12 μm. **D.** SEM photograph of another type of Paleozoic acritarch; scale bar is 13.4 μm. **E.** Light photomicrograph of stained palynological preparation showing single Cretaceous spore; scale bar is 7 μm. **F.** Light photomicrograph of stained palynological preparation showing single Cretaceous pollen grain; scale bar is 7 μm. **G.** Photograph of rock slab showing graptolites (*Monograptus*) on bedding plane of Silurian limestone, Nevada; scale bar is 1 cm. **H.** Photograph of rock slab showing graptolites on bedding plane in Silurian black shale, Alaska; scale bar is 1.4 cm. **I.** SEM photograph of pyritized fecal pellets from muddy diatomaceous ooze of Pliocene age; scale bar is 50 μm. **J.** Thin-section photomicrograph (plane-polarized light) of Upper Cretaceous shaly chalk showing fecal pellets (bluish-green) flattened in plane of bedding; fecal pellets retain some original porosity, and thus are stained by blue impregnation medium; scale bar is 0.1 mm. **K.** Photograph of slab of organic-carbon-rich·Upper Cretaceous chalk showing large pyrite concretions or crystal aggregates associated with inoceramid debris (note compaction of laminae around concretion in upper left); scale bar is 2 mm. **L.** SEM photograph of pyrite framboid from Upper Cretaceous chalk; these clusters of small isometric crystals are authigenic and form in association with organic matter (including fecal pellets); scale bar is 3 μm.

SCALE OF EACH PHOTO

└ 1 cm ┘

Sediment Modification by Submarine Currents

In addition to production and dissolution patterns, other processes influence the composition and rates of accumulation of pelagic sediments. Pelagic sediments can be winnowed and sorted by currents, both in shallow- and deeper-water environments. In pelagic shelf chalk settings, periodic winnowing or erosion can produce relatively pure calcarenites lacking a fine-grained fraction, or produce areas of nondeposition that may include chalk hardgrounds. Sequences of multiple hardgrounds or sequences in which carbonate beds relatively rich in clay-size detrital material alternate with beds of lower detrital content are probably produced, in part, by periodic changes in current velocity related to variations in climate or sea level (Kennedy and Garrison, 1975).

The abyss is not devoid of currents, either, contrary to earlier concepts (see Heezen et al, 1966; Heezen and Hollister, 1971). Measurements of current velocities in the deep sea, examination of bottom photographs,

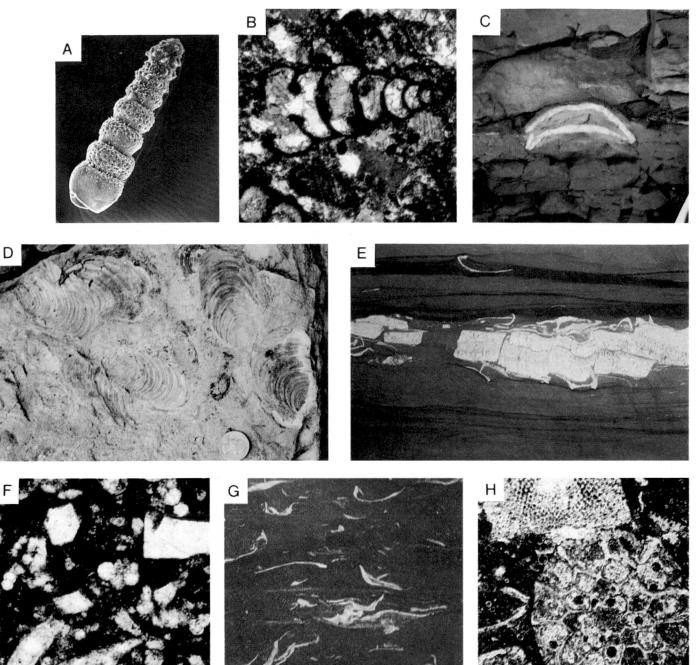

Figure 7—Calcareous benthic organisms commonly found in pelagic carbonates. **A.** SEM photograph of Holocene uniserial benthic foraminifer; scale bar is 36 μm. **B.** Thin-section photomicrograph (plane-polarized light) of uniserial benthic foraminifer of Pennsylvanian age; scale bar is 0.1 mm. **C.** Photograph of compacted Echinoid in Upper Cretaceous pelagic marlstone from Italy; scale bar is 2.5 mm. **D.** Photograph of slab of Cretaceous chalk showing articulated *Inoceramus labiatus* on bedding plane; scale bar is 3 cm. **E.** Photograph of both valves of a compacted inoceramid pelecypod, partly disaggregated and encrusted by oysters (*Pseudoperna congesta*), in partly laminated, organic-carbon-rich, Upper Cretaceous limestone; scale bar is 1.1 cm. **F.** Thin-section photomicrograph (plane-polarized light) of inoceramid fragments in Upper Cretaceous chalk; scale bar is 0.12 mm. **G.** Photograph of cross section of core slab of Upper Cretaceous chalk showing oysters deformed by compaction; scale bar is 1 cm. **H.** Thin-section photomicrograph (plane-polarized light) of bryozoan in Upper Cretaceous chalk; scale bar is 80 μm.

SCALE OF EACH PHOTO

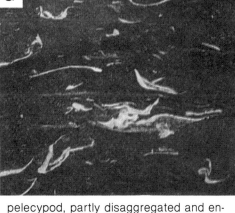

|___1 cm___|

TABLE 4

Sediment definitions assume only the names of those components present in quantities greater than 10%. Where more than one component is present, the most abundant component is cited farthest to the right in the string of descriptive modifiers, and other components are listed progressively to the left in order of decreasing abundance.

Induration is indicated by the sediment name. Although the determination of induration is highly subjective, the following criteria provide a usable guide:

A. Terrigenous sediments:

If the material is soft enough that the core can be split with a wire cutter, only the sediment name is used (e.g., silty clay, sand). If the core must be cut with a band or diamond saw, the suffix "stone" is added (e.g., silty claystone, sandstone).

B. Biogenic sediments:

Ooze — soft, with very little strength and readily deformed with a spatula blade.

Chalk — firm, partly indurated calcareous ooze, or friable limestone readily scratched with a fingernail or edge of spatula blade.

Limestone — hard, cemented or recrystallized calcareous rocks.

Radiolarite, diatomite, or spiculite — hard, cemented biogenic siliceous ooze.

The class limits of the sediment classification system are defined by percentages of components, as given below:

A. Terrigenous sediments:

greater than 30% terrigenous components, less than 30% calcareous microfossils, less than 10% siliceous microfossils, and less than 10% authigenic components. Sediments in this category are subdivided into textural groups on the basis of the relative proportions of sand, silt, and clay.

B. Volcanogenic sediments:

greater than 30% volcanogenic components; greater than 32 mm, volcanic breccia; less than 32 mm, volcanic lapilli; and less than 4 mm, volcanic ash (tuff, if indurated). Compositionally, these pyroclastic rocks are described as vitric (glass), crystalline, or lithic.

C. Pelagic clay:

greater than 10% authigenic components, less than 30% siliceous microfossils, less than 30% calcareous microfossils, and less than 30% terrigenous components.

D. Biogenic calcareous sediments:

greater than 30% calcareous microfossils, less than 30% terrigenous components, and less than 30% siliceous microfossils.

The principal components of biogenic calcareous sediments are nannofossils and foraminifers. Qualifiers are as follows (Foraminifera Percentage Name): less than 10% — *nannofossil ooze* (chalk, limestone); 10 to 25% — *foraminifer-nannofossil ooze* (chalk, limestone); 25 to 50% — *nannofossil-foraminifer ooze* (chalk, limestone); and greater than 50% — *foraminifer ooze* (chalk, limestone). The sediment is *calcareous ooze* if it contains more than 50% $CaCO_3$ of unknown origin. Calcareous sediments containing 10 to 30% siliceous fossils are qualified by "radiolarian," "diatomaceous," or "siliceous," depending upon the type of siliceous component.

E. Biogenic siliceous sediments:

greater than 30% siliceous microfossils, less than 30% calcareous microfossils, less than 30% terrigenous components.

Where radiolarians are greater in number than diatoms/sponge spicules, it is called a *radiolarian ooze (radiolarite)*. Where diatoms are greater in number than radiolarians/sponge spicules, it is called a *diatom ooze (diatomite)*. Where mixed, or when the siliceous microfossil source is unidentified, it is called a *siliceous ooze*. And where the silica is amorphous and lithified, it is called a *porcellanite*, or *chert*. Siliceous sediments containing 10 to 30% $CaCO_3$ are qualified by the terms "nannofossil," "foraminifer," "calcareous," "nannofossil-foraminifer," or "foraminifer-nannofossil," depending on the kind and quantity of the $CaCO_3$ component.

F. Transitional terrigenous-biogenic calcareous sediments:

greater than 30% $CaCO_3$, greater than 30% terrigenous components, and less than 30% siliceous microfossils.

"Marly" qualifies transitional sediments in the biogenic calcareous series (e.g., "marly nannofossil ooze"). If 10 to 30 % siliceous microfossils, the appropriate qualifier is used (e.g., "diatomaceous marly chalk").

G. Transitional terrigenous-biogenic siliceous sediments:

greater than 10% siliceous microfossils, less than 30% terrigenous components, and less than 30% $CaCO_3$.

Where 10 to 30% siliceous microfossils, it is called (name of siliceous fossil) mud or mudstone (e.g., 10 to 30% radiolarians = radiolarian mudstone). Where 30 to 70% siliceous microfossils, it is called muddy (name of siliceous fossil) ooze or (name of siliceous fossil) ite (e.g., 50% diatoms = muddy diatom ooze or diatomite).

Table 4—Classification of deep-sea sediments for general use in Deep Sea Drilling Project reports (modified from Benson, Sheridan et al, 1978; classification initially devised by JOIDES Panel on Sedimentary Petrology and Physical Properties).

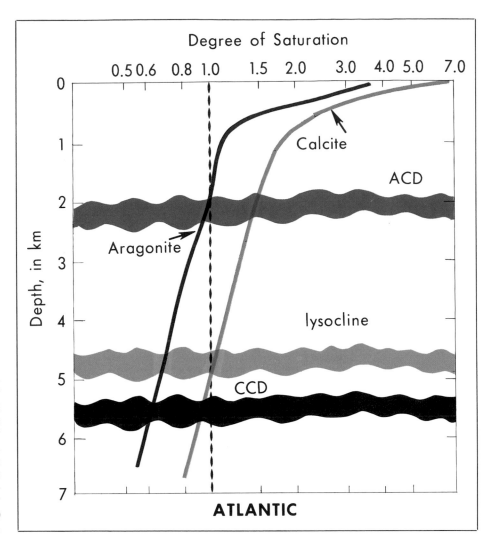

Figure 8A—Diagrammatic representation of the degree of saturation of seawater with respect to aragonite and calcite in the Atlantic Ocean. The lysocline marks the top of a zone of greatly increased rates of dissolution. The carbonate compensation depth (CCD) marks the depth below which no calcite is preserved. The aragonite compensation depth (ACD) is at much higher levels (approximately 2,000 m depth) than the CCD (after Broecker, 1974).

and inferences from textures of modern sediments have established that currents capable of transporting and/or resuspending and eroding most constituents of modern pelagic sediments do exist in certain parts of the ocean basins. The strongest deep currents (15 to 35 cm/sec and perhaps as high as 100 cm/sec) are commonly found along continental margins on the western side of ocean basins in the northern hemisphere. These geostrophic currents are termed "contour currents" because they flow along the "contours" of the continental slope and rise. They erode and transport large volumes of sediment, particularly along the east coast of North America (review by Stow and Lovell, 1979). Thick, rapidly deposited sediment drifts are generally at the down-current ends of con-

tour currents where velocities drop because of changes in topography, sediment supply, or water-mass stratification. Such deep-water currents can winnow, erode, or continually resuspend sediment creating coarse-grained lag deposits, and laminated or rippled sediment. Many hardgrounds in deep-sea deposits, including so-called lithoherms (Neumann et al, 1977; Mullins et al, 1980) may have ultimately resulted from current-induced erosion.

Dilution of Pelagic Sediment

The composition of pelagic sediments may be altered by dilution from terrigenous or other sources. Few pelagic sediments are entirely devoid of "contaminating" detritus. Even on the central Pacific deep-sea floor, pelagic red clays are deposited

below the CCD far from terrestrial sediment sources. They contain a small amount of wind-blown, fine-grained quartz and large volumes of clay minerals (Rex et al, 1969). The extent of dilution of pelagic biogenic constituents by terrigenous material depends mainly on the rate of supply of pelagic material, distance from continental margins, depth below sea level, local topography, and latitudinal position, particularly with reference to the prevailing climate and wind directions. Obviously, those locations away from the continents and on submarine highs, such as seamounts, oceanic rises, and plateaus, will contain relatively little terrigenous detritus. However, clay minerals can be transported large distances by prevailing winds, and nepheloid layers laden with very fine

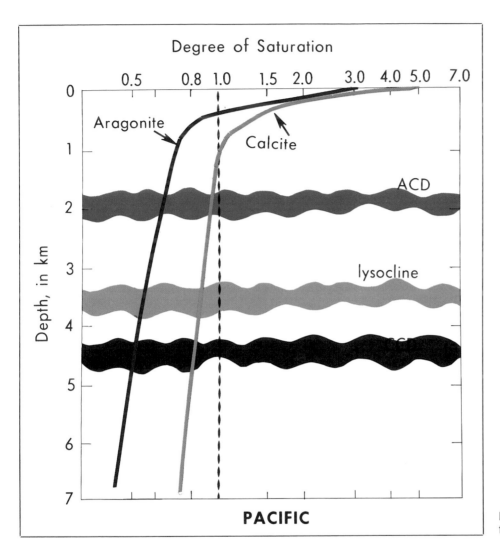

Figure 8B—Same as A but showing relations for the Pacific Ocean.

grained sediment (Eittreim et al, 1972, 1976) cover much of the ocean floor, particularly in regions swept by bottom currents and/or near continental margins. In basins off the arid coasts of west-facing continental margins situated in the latitude of the Tradewinds, such as the North Atlantic Basin off northwestern Africa, pelagic carbonates commonly contain significant amounts of eolian material, particularly clay to fine silt-size quartz. In more humid regions, rivers may dispatch large volumes of clay, silt, and sand across continental margins into the deep sea, building large cones or fans from the margin. An extreme example is the Bengal Fan in the Indian Ocean, which has as its source the Ganges-Brahmaputra rivers. Coarse-grained terrigenous material and clay are transported

across the Bengal Fan for more than 1,000 km into the adjacent ocean basin. In regions such as these, the pelagic biogenic flux to sediments is greatly diluted. Volcanic material also acts as a dilutent, especially near linear island chains, mid-ocean ridges and isolated seamounts.

Changes in Global Climate, Ocean Circulation and Pelagic Sedimentation Patterns

Studies of Pleistocene and older pelagic sequences, particularly those from piston cores and DSDP cores, have shown that pelagic sediments are fairly sensitive indicators of changes in ocean circulation, fertility, and temperature. The changes vary in rate and duration. For example, cycles or rhythmic bedding (see Figs. 18 to 21) occur in pelagic carbonate sequences

of all ages. Typical Pleistocene carbonate cycles have been extensively studied (Arrhenius, 1952; Berger, 1973; Adelseck, 1977; Adelseck and Anderson, 1978; Hays et al, 1976; Climap, 1976). The cycles consist of alternating carbonate-rich and carbonate-poor intervals that can be ascribed to fluctuations in carbonate production in surface waters, carbonate dissolution in deep waters, and/or terrigenous dilution. The relative importance of each parameter in any particular sequence varies, depending on the ocean basin and paleolatitude of deposition. The fluctuations in carbonate content are correlative and correspond to changes in climate (glacial-interglacial cycles) that, in turn, are partly determined by the Earth's orbit (the Milankovich hypothesis; Hays et al, 1976). The

Figure 9—Variations in the depth of the carbonate compensation depth (CCD) in the world ocean. Light blue, CCD deeper than 5,000 m; medium blue, CCD of between 4,000 and 5,000 m; and dark blue, CCD shallower than 4,000 m. Note that the CCD is deep in the Atlantic, and also is depressed in areas of high carbonate productivity, such as the Equatorial Pacific (modified from Berger and Winterer, 1974).

Figure 10—Global variations in sea level, average carbonate compensation depth (CCD), sediment accumulation rates, and hiatus abundance during the Cenozoic (after Arthur, 1979a). Note the coincident changes in all parameters, which are related to variations in climate, sea level, and rates of oceanic circulation.

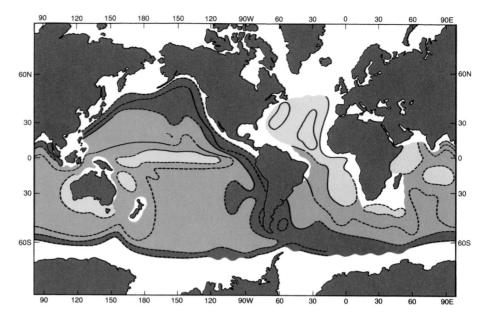

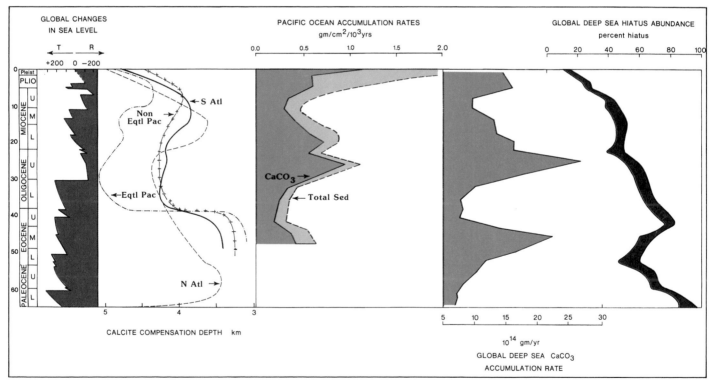

carbonate cycles have characteristic periods of from 20,000 to 100,000 years. This periodicity can be shown in older pelagic carbonate sequences as well (Dean et al, 1978; Arthur and Fischer, 1977; Arthur, 1979c; Fischer, 1980). Therefore, changes in global climate have induced variations in the chemistry and circulation of the oceans and in the input of sediment from land. These changes are imprinted on pelagic sediments as short-term variations in species compositions, in proportions of different faunal or floral groups, in the relative amounts and mineralogy of terrigenous material contained in pelagic sediments, and as shifts in pelagic facies patterns. These changes can be geologically rapid (less than 1 m.y.) or longer term (more than 30 m.y. or so; see Fig. 10).

Sea level plays an important role in pelagic sedimentation. First, relative sea level influences the amount of terrigenous material carried across continental shelves to the deep sea (Hay and Southam, 1977). Generally, low sea-level stands correspond to increases in detrital sediment reaching

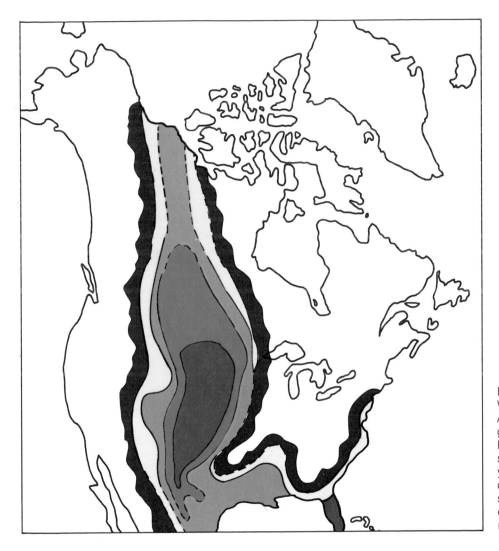

Figure 11—Facies patterns in the Western Interior seaway of North America during peak Greenhorn transgression (early Turonian) (modified from Kauffman, 1975). Brown, areas of erosion or local continental sedimentation; yellow, sand-dominated shoreline deposits; shades of green, areas of sedimentation of clay-rich to pelagic-carbonate-rich deposits, light to dark, respectively.

deep-sea basins (increases in average rate of deposition), whereas widespread flooding of continental shelves and epicontinental seas tends to trap terrigenous material near shore, causing lowered deep-sea accumulation rates.

Second, the area of shelf seas in combination with global climate probably determines balances in the carbonate system (Berger and Winterer, 1974). Other factors being equal, if the area of carbonate-shelf sedimentation increases significantly, less carbonate is available for sedimentation in the deep sea. Thus, the flux of carbonate material to deep-sea sediments will decline through a decrease in surface production and/or an increase in rates of dissolution. Changes in the rate of accumulation of carbonate sediment in

the deep sea and in the level of the CCD are probably related, in part, to this shelf-basin fractionation as well as to overall changes in ocean chemistry (Fig. 10).

The vertical continuity of pelagic sequences reflects a balance between supply and dissolution-erosion. There is a close correspondence between the incidence of hiatuses in deep-sea sediments and sea level, climate changes and variations in deep circulation (Moore et al, 1978). Changes in global climate, sea level, organic diversity, the composition of pelagic faunas, and pelagic sedimentation styles and rates are related to one another, as shown in Figure 10. Thus, there are significant patterns in pelagic sedimentation through time that aid in understanding the response of ocean chemistry, circulation, and

sedimentation to changes in continental positions, the opening of new ocean passages (Berggren and Hollister, 1977), and climate, sea level, and other parameters (Berger and Roth, 1975; Fischer and Arthur, 1977; Berger, 1979; Arthur, 1979a). This is currently an active field of research and a more complete discussion is beyond the scope of this paper.

PELAGIC FACIES MODELS

Shelf Seas

During times of eustatic high stands of sea level, significant volumes of pelagic sediment can accumulate on continental shelves or in epicontinental seaways. This accumulation is a consequence of several factors, including the raising

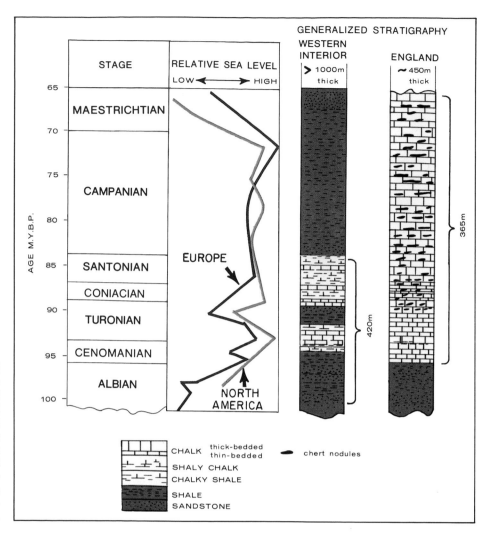

Figure 12—Generalized stratigraphy of the Cretaceous Western Interior seaway of North America (Kauffman, 1977) and the Cretaceous chalk seas of England (Gallois, 1965) compared to eustatic sea level changes (Hancock, 1975a). Note predominance of pure chalk in European sections versus argillaceous chalks and chalky shales in the Western Interior sequences. Relatively pure chalks are deposited only during transgressive maxima in the Western Interior.

of erosional base levels, decrease in relative elevation and area of continental land masses available for erosion, retreat of shorelines with consequent increases in the area of submerged shelves, and changes in climate and oceanic circulation patterns. The general effect of these variations is to decrease the supply of terrigenous debris to shelf areas and/or to trap the bulk of such material in nearshore areas. Thus, despite their relatively slow rate of accumulation, pelagic sediments can dominate large areas of shelves without being drastically diluted by terrigenous sediment.

The isolation of depositional sites from "extraneous" sediment influx is the key to the formation of pelagic sediments. Because of relatively narrow shelves and low sea-level stands, Holocene pelagic sediments are restricted largely to deep ocean basins far from the main influx of terrigenous debris. Even today, however, pelagic or hemipelagic sediments are accumulating in shallow water areas protected from non-pelagic sediment contributions. The Belize shelf lagoon, for example, in which terrigenous grains are trapped by a 30- to 50-m-deep trough, contains a significant percentage of coccoliths in the fine-grained, off-reef carbonate sediments deposited in 30 to 60 m of water (Scholle and Kling, 1972). Modern submerged seamounts, drowned platforms, aseismic ridges, and other isolated but relatively shallow-water depositional sites are also dominated by pelagic sediments provided they are in water sufficiently deep to exclude shallow-marine carbonate organisms (benthic algae, corals, etc.).

Throughout Phanerozoic time, pelagic sediments have dominated these types of settings. Lower Paleozoic graptolitic shales, radiolarian cherts, and subordinate limestones were deposited in "deeper water" continental margin or "oceanic" settings, as well as on isolated, submerged, paleotopographic high areas or "Schwellen" (see synthesis in Jenkyns, 1978). In Upper Paleozoic sections, these same settings are represented by cephalopod-(goniatite), Tentaculite-, or styliolinid-rich limestones, conodont-rich shales, or radiolarian cherts (Tucker, 1974, Jenkyns, 1978).

Likewise, the Mesozoic interval predating the rapid evolution of the modern groups of pelagic organisms was characterized by similar sedimentation patterns. Throughout pre-Middle Jurassic time, "Schwellen"

Figure 13—**A**. (left) Scanning electron microscope (SEM) photograph of epicontinental shelf chalk from Smokey Hill Shale Member of the Niobrara Formation. Sample shows well preserved, varied assemblage of cocoliths and rhabdoliths and porosity in the 35 to 40% range; scale bar is 10 μm. **B**. (below) SEM photograph of typical Tertiary deep-sea chalk from the Hatton-Rockall Basin. Note minor corrosion of coccolith margins. DSDP Leg 12, Site 116 at 462-m burial depth. Scale bar is 4.5 μm.

facies (or basinal areas that were isolated from shelf-derived turbidite sedimentation) were marked by low rates of sediment accumulation (starved basin facies). Those sediments that did accumulate were of three main types: shelf derived — terrigenous or carbonate muds transported by wind, nepheloid clouds, or other processes; siliceous pelagic materials — mainly radiolarian tests; or pelagic carbonate macro-fauna — such as ammonites or nautiloids. The total volume of pre-Jurassic pelagic carbonate is small indeed when compared with either shallow-marine carbonate sediment volumes of the same time interval or with carbonate pelagic sediments that post-date the evolution of modern pelagic groups.

The Jurassic, then, marks a watershed for pelagic sediments. The introduction of groups such as the coccolithophores (Jurassic), and then the planktic foraminifers (latest Jurassic to Early Cretaceous), meant that for the first time a true calcareous plankton existed. This, in turn, led to much higher rates of Jurassic to Holocene pelagic carbonate sedimentation than were possible earlier in the Phanerozoic. Pelagic carbonate sedimentation could "compete" on a

more equal footing with terrigenous input in many broad shelf, epicontinental sea, and isolated platform or "Schwellen" settings. Pelagic carbonates, therefore, are much more extensive in Upper Jurassic to Holocene sections than in earlier times, especially in relatively shallow marine shelf areas, although pelagic deposits are still excluded from areas of rapid terrigenous sedimentation.

Facies patterns of pelagic limestones are simpler than those of most carbonate sediments because of the relatively low topographic relief in outer shelf and epicontinental sea settings. Uniform water depths of 50 m or more in most shelf areas of pelagic sediment accumulation generally ensure the exclusion of strong wave and light influence and preclude the growth of most shallow-marine carbonate fauna and flora. Pelagic shelf limestones, therefore, commonly have gradual facies transitions and can have remarkably uniform lithology over areas of tens to hundreds of thousands of square kilometers.

Shelf chalk facies patterns have been summarized by Hancock (1975a) and can be illustrated using two of his examples from the Cretaceous of North America and Europe. In both

regions, the chalks represent pelagic sedimentation in broad shelf seas that were probably no deeper than a few hundred meters (see Hancock, 1975b, for discussion).

North American Chalks

The Upper Cretaceous of the Western Interior of North America (Fig. 11) contains two major intervals of pelagic limestone — the Cenomanian-Turonian Greenhorn Limestone and the Coniacian-Campanian Niobrara Formation (Kauffman, 1969). The chalky facies of both these units represent widespread pelagic carbonate deposition during transgressive maxima within an epicontinental seaway otherwise dominated by clastic terrigenous debris.

The seaway was bordered on the east by a relatively low-relief landmass that did not act as a major sediment source. To the west, however, an active orogenic belt supplied more than 5,000 m of largely nonmarine

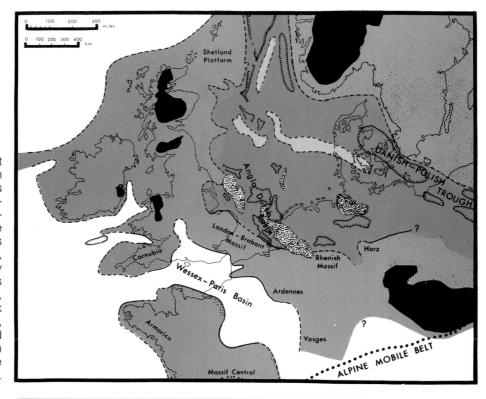

Figure 14—Distribution of chalks at maximum transgression in Western Europe during the Late Cretaceous (modified from Hancock, 1975b). Compare setting to that of the Western Interior of North America shown in Figure 11. Black, probable low-relief land areas on exposed massifs; yellow-green, areas of massifs periodically covered by seas in which thin sequences of chalks and transitional (marl, glauconitic marl, calcarenite) facies were deposited; dark green, chalk basins; hachured contours, troughs within chalk basins; stippled areas, sites of basinal chalk deposition that appear to have been uplifted by the Late Cretaceous.

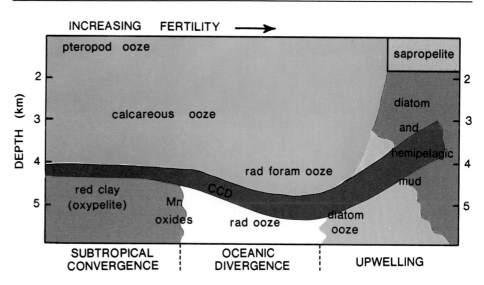

Figure 15—Generalized relationships of pelagic and hemipelagic facies versus depth, fertility and proximity to continental margins (after Berger, 1974).

terrigenous sediment along the western margin of the "geosyncline" (Reeside, 1944). The width of the seaway (nearly 1,500 km during transgressive maxima) allowed pelagic carbonate sediments to form at times of highest eustatic sea-level stands. During maximum transgression, coarse-grained terrigenous material was trapped in nonmarine and very shallow marine (inner shelf) settings. Clayey sediments were transported farther

from their source areas and were deposited in middle- to outer-shelf settings. Regions farthest from terrigenous sources were isolated from virtually all input of clays and silts (with the exception of volcanic ash). Substantial thicknesses (hundred of meters) of chalk and shaly chalk accumulated in these areas (Figs. 12 and 13A).

In the Western Interior, as elsewhere, the limiting factor for ac-

cumulation of pelagic limestones was the ratio of production of pelagic organisms to the influx of terrigenous material. The displacement of chalk deposition to the eastern margin of the trough is a consequence of the asymmetry of clastic sediment supply (dominantly from the west). The sheer distance to source areas at times of maximum transgression rather than any specific hydrographic barriers reduced the rates of supply of

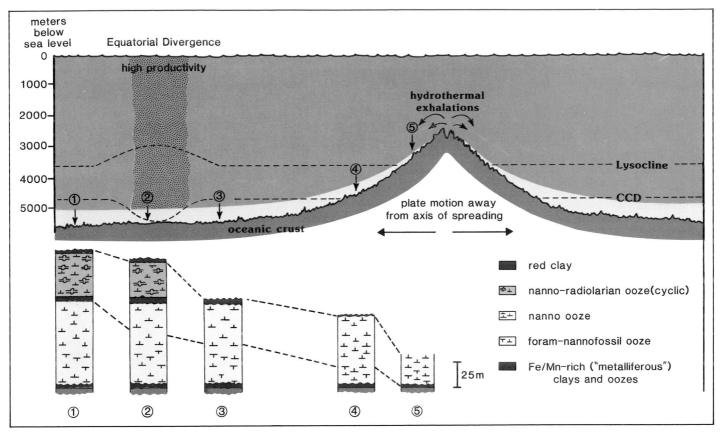

meters below sea level

Equatorial Divergence

high productivity

hydrothermal exhalations

⑤

Lysocline

④

CCD

① ② ③

oceanic crust

plate motion away from axis of spreading

red clay

nanno-radiolarian ooze(cyclic)

nanno ooze

foram-nannofossil ooze

Fe/Mn-rich ("metalliferous") clays and oozes

25m

① ② ③ ④ ⑤

Figure 16—Diagrammatic representation of subsidence of oceanic crust and succession of sediment facies related to changes in water depth and productivity of surface water.

terrigenous sediment to the chalk depositional areas. In this type of setting, chalk facies represent regions of maximum isolation, but not necessarily maximum water depth. Lateral facies changes are gradual and generally consist of a transition from chalk to marl or calcareous shale, to pure shale, and finally to siltstone and sandstone.

Because the Western Interior seaway was a relatively shallow body of water (probable maximum water depths of only a few hundred meters) that received extensive fresh-water runoff, water masses in the basin may have been salinity- and/or temperature-stratified. This stratification may have caused periodic abnormal bottom-water conditions that excluded most benthic, and especially infaunal, organisms. Indeed, the Western Interior chalks have a low diversity of benthic microfossils, as

well as macrofossils when compared with most European chalks of the same age. Inoceramids, oysters, and subordinate members of other molluscan groups constitute virtually the entire macrofauna in most of the Niobrara Formation, for example. High contents of organic carbon (2 to 3% average in the Smoky Hill Shale Member of the Niobrara Formation), widespread absence of burrowing organisms, preservation of fecal pellets of planktic and nektic organisms, and preservation of millimeter-scale lamination are also distinctive features of many of the Western Interior chalks (Hattin, 1975, 1981; also see Fig. 18).

European Chalks
The Upper Cretaceous strata of Europe contrast in many ways with their North American counterparts (see Hancock, 1975a; Wiedmann, 1979; Birkelund and Bromley, 1980; and references therein). During the Turonian to Maestrichtian in particular, pelagic carbonates were deposited over much of Europe (Fig.

14). Emergent landmasses during this period (Armorican Massif, Bohemian Massif, Baltic Shield) generally consisted of Hercynian and older basement complexes that had been eroded to low relief. The restricted terrigenous source terranes, their low relief, and the moderately arid climate that prevailed during much of the Late Cretaceous in Europe (Hancock, 1975a), combined with the high eustatic sea-level stands of that time (as much as 200 to 300 m higher than at present; Hays and Pitman, 1973) to yield an enormous shelf area in which pelagic carbonate deposition predominated because of the virtual absence of clastic terrigenous input.

Overall, the Upper Cretaceous strata of this region of Europe are much thinner than their North American equivalents. Chalks, however, are more widespread and considerably thicker in Europe than in equivalent Western Interior strata (Fig. 12). Thicknesses of Upper Cretaceous chalk in excess of 1,000 m are found in some of the basinal areas of the European chalk shelf,

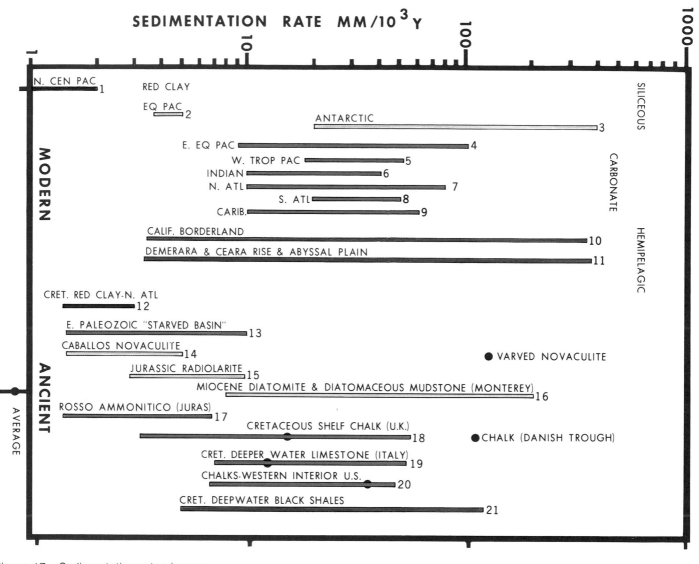

Figure 17—Sedimentation rates (uncorrected for compaction) of some modern and ancient pelagic and hemipelagic sediments. Reduction of modern rates by about 60 to 70% allows comparison with ancient rates. (Sources of data are Opdyke and Foster, 1971; Goldberg and Koide, 1963; van Andel et al, 1975; Hays et al, 1976; Hays, 1965; Swift, 1977; Luz, 1973; Berger et al, 1978; Geitzenaur et al, 1976; Hays et al, 1976; Olausson, 1967; Ericson et al, 1961; Prell and Hays, 1976; Prensky, 1973; Damuth, 1977; Jansa et al, 1979; Jenkyns, 1978; Folk and McBride, 1976; Lowe, 1976; Schlager, 1974; Garrison and Douglas, 1981; Hancock, 1975a, b; Arthur, 1979c; Kauffman, 1977; and numerous Deep Sea Drilling Project volumes).

especially within the central trough of the North Sea (Hancock and Scholle, 1975).

European chalk facies also have greater areal distribution than do their North American equivalents. Chalks were formed both in basinal areas and within close proximity to many massif shorelines. Because of the limited input of terrigenous detritus, chalk facies are common throughout the Upper Cretaceous, even in regressive intervals. Where chalk deposition was interrupted by lowered sea-level stands, the event is usually marked by thin hardground sequences in Europe rather than the thick regressive shales and sandstones of the United States Western Interior.

Most of these hardground units are concentrated on paleotopographic "highs" or "Schwellen," but units are also present in trough areas.

European chalks have a number of other characteristics that distinguish them from their North American equivalents. Whereas most North American chalks contain at least 10% clay, European chalks commonly have less than 1% total insoluble residue. The more open connections of the European shelf to the Atlantic and Tethyan oceans led to stronger water circulation and more normal salinity there than in the Western Interior seaway. As a consequence, European chalks have a more diverse benthic fauna including siliceous

Figure 18—Epicontinental shelf limestone deposited under anaerobic or dysaerobic conditions showing preservation of centimeter- and millimeter-scale laminations as well as decimeter-scale bedding cycles. Sediment contains primarily calcareous nannofossils and foraminifers with subordinate macrofauna of inoceramids and oysters. Upper Cretaceous Smoky Hill Chalk Member, Niobrara Formation, Kansas.

Figure 19—Unbioturbated, laminated, deep-water limestone of Early Cretaceous age from DSDP Site 367 (core 367-25-4, 30- to 44-cm interval) in the North Atlantic. The sediment contains 86.5% $CaCO_3$, 1.6% organic carbon, 0.6% total sulfur, and presumably reflects anaerobic conditions at the sea floor. Note progressive loss of lamination upward and presence of small burrows and lighter color near top indicating oxygenation.

Figure 20—Rhythmic alternation of burrowed limestone and laminated organic-carbon-rich marlstone shown in core. Transitions between each lithologic type are gradational. Limestone is composed primarily of calcareous nannofossils and planktic foraminifers. Benthic infauna was excluded by anaerobic or dysaerobic conditions on sea floor during deposition of laminated marls in a shallow epicontinental seaway. Upper Cretaceous Greenhorn Limestone, Pueblo, Colorado.

sponges, brachiopods, mollusks, bryozoans, echinoderms, arthropods, and foraminifers, among others. European chalks are also more extensively burrowed as a consequence of the large benthic population. The abundance of chert (derived mainly from siliceous sponges) in European chalks is a further result of this diversity of benthic organisms.

Lateral and vertical facies transitions in the European, normal marine, carbonate-dominated shelves are more varied than in the Western Interior. Sixteen facies types associated with European chalks have been summarized by Hancock (1975a). In addition to the transitions to shaly chalk, calcareous shale, shale, and sandstone noted in the Western Interior, transitions to greensands, shelly chalks, calcarenites, dolomitic chalks, phosphatic chalks

and condensed (hardground) sequences all are common in European sections.

Coastal areas were dominated either by a skeletal carbonate-secreting macrofauna or by glauconitic sandstones. Shoal areas without exposed land masses have hardground sequences, or, where current conditions led to high organic productivity, phosphatic chalks. As in the Western Interior, however, facies transitions are typically gradual, taking place over distances of up to tens of kilometers. Where tectonically active zones coincided with chalk deposition, such as in the compensatory troughs and grabens of the North Sea, great thicknesses of chalk accumulated, commonly as a result of

slumping of chalks from adjacent topographic highs (Hancock and Scholle, 1975; Watts et al, 1980). In such settings, facies transitions may be abrupt.

In summary, although the styles of pelagic shelf sedimentation are rather different in North America and Europe, they both represent similar responses to similar major controls. The widespread Cretaceous chalks reflect the high productivity of newly evolved calcareous plankton and reduced terrigenous influx caused by high sea-level stands. In the Western Interior, chalk deposition competed with orogenically supplied terrigenous detritus, and only occurred in small areas during maximum sea-level stands. In Europe, terrigenous influx was minimal and chalk deposition dominated an entire continent.

Although this discussion has fo-

Figure 21—Chalk-marl cycles in the Lower Chalk, Isle of Wight, England. Cycles average 30 to 50 cm thick and show alternation of clay-rich (as much as 60%) and clay-poor (less than 10%) layers.

Figure 22—Bedded chalks from the Isle of Wight, England. Bioturbation has completely homogenized most primary depositional fabrics within these beds, although primary bedding cycles are still weakly preserved, despite low average clay content (less than 5%).

cused on two examples of Cretaceous chalk deposition, similar factors apply to most pelagic sediments. The diatomites of the Miocene and Pliocene Monterey Formation in California represent deposition partly in isolated fault-bounded basins on a continental margin (Garrison and Douglas, 1981) similar to the Continental Borderland found off the coasts of California and northern Mexico today. Shoreward grabens trapped terrigenous debris, and seaward basins accumulated diatom-rich pelagic sediments. Likewise, the extensive Ordovician to Devonian radiolarian cherts deposited on continental crust in the United States interior and the Appalachian-Ouachita-Marathon belt (McBride, 1970; Folk and McBride, 1976) probably represent pelagic deposition in shelf to basin settings that were otherwise sediment-starved. As in the Cretaceous chalks, wide lateral extent, relatively slow sedimentation, and gradual facies transitions are characteristic.

Oceanic Areas

The distribution of deep-water pelagic sediments is fairly predictable within a framework of sea-surface biologic productivity and water depth. Figure 15 illustrates the general pattern of pelagic facies as a function of fertility and depth that can be compared with distribution of sediments in the modern ocean (Fig. 2). Little or no pelagic carbonate material is preserved below the CCD, whereas foraminifer-nannofossil ooze (Fig. 13b) accumulates in relatively low-fertility regions above the CCD. Slowly deposited "red" clay and manganese nodules form in relatively infertile regions of the oceans below the CCD. Manganese nodules are formed mainly where slow deposition, bioturbation and frequent current winnowing keep the nodules exposed at the sediment-water interface where growth occurs (Elderfield, 1977).

Where surface productivity is high, a significant part of the sediment may be composed of the siliceous (opal-A) skeletons of radiolaria and/or

Figure 23—Chalks with typical uniformity and lateral continuity of rhythmic bedding. Bedding cycles are accentuated by selective chertification along bedding planes. About 60 m of Upper Cretaceous chalk is exposed in cliff; section near Etretat, France.

Figure 24—Massive, organic-carbon-rich, basinal limestone that was rapidly deposited and shows few bedding features. These are inter-atoll, deep shelf-sea deposits that interfinger with reef-talus facies, and may be source rocks for part of the hydrocarbon accumulation in the Poza Rica and Golden Lane oilfields. Lower Cretaceous Tamaulipas Limestone, northeastern Mexico.

diatoms. At present, siliceous sediments occur along low-latitude continental margins and at high-latitude divergences. In the zone of equatorial upwelling, an area of high productivity, the CCD is somewhat depressed and there is a mixed facies of pelagic carbonate and biogenic siliceous ooze.

Near the continental margin, the flux of biogenic sediment produced in surface waters is diluted by land-derived sediment carried to the ocean by winds, in turbid nepheloid layers carried by geostrophic bottom currents (review by Stow and Lovell, 1979), and by turbidity currents and other mechanisms of downslope sediment redeposition. The extent of this terrigenous dilution depends on latitude, regional climate, geomorphology of the adjacent margin, width of the shelf, and the presence or absence of major rivers debouching onto the shelf.

The continental rise (Fig. 1) is dominated by redeposited sediment facies, including deep-sea fan facies and contourite drifts. Pelagic carbonates may be intercalated with terrigenous deposits in off-fan environments, but the pelagic constituents are substantially diluted. Seismic reflectors in the rise are discontinuous, and parts of the rise

prism are characterized by thick, seismically transparent zones (Tucholke and Mountain, 1979). The thick rise prism gives way seaward to a thinner sequence of dominantly pelagic facies in which seismic reflections are generally coherent and laterally continuous. Reflectors are commonly closely spaced and appear related to bulk-density changes due to variations in diagenetic alteration

(Schlanger and Douglas, 1974; Berger and Mayer, 1978). Contourite sediment drifts also characterize continental rises. These are thick sediment ridges composed of bottom-current transported sediment. They occur at the downcurrent end of intensified geostrophic currents, where current velocities decrease and sediment falls out of suspension. Contourite drifts may contain significant amounts of

transported pelagic carbonate (Mullins et al, 1980). Anomalously high (greater than 20 mm/1000 years) sedimentation rates and current-lamination are characteristic of pelagic carbonates deposited in contourite drifts.

Deep-water pelagic sediments may interfinger with redeposited shallow-water carbonate debris if they are deposited adjacent to continental shelf seas or island platforms that are dominated by carbonate buildups. Pelagic facies are intercalated with volcaniclastic deposits in the sediment aprons of island arcs, linear island chains, or other oceanic seamount provinces.

The pattern of deep-sea facies at any one time and the stratigraphic succession at various points is generally as shown in Figure 16. Newly formed oceanic crust subsides as it cools (Sclater et al, 1977, among others). The average ridge-crest elevation of a spreading center is about 2,700 m below sea level. Subsidence is rapid in the first 50 m.y., and the top of the crust reaches a depth of about 5,500 m below sea level at the end of this time. Assuming constant lysocline and CCD levels, typical oceanic crust begins its history of sedimentation above the lysocline and slowly subsides below the CCD within 50 to 60

Figure 25—Large-scale channels with minor slump features in chalks deposited in a relatively shallow epicontinental sea. These features may be associated with local biohermal buildups or they may represent current-related mounds. About 75 m of section is exposed, and bedding is accentuated by dark bands of chert nodules. Upper Cretaceous chalk, near Etretat, France.

Figure 26—Slumped and folded, dolomitic, pelagic limestone. Sediment was mixed and sheared during downslope transport on the flanks of an aseismic ridge. Organic carbon content is high and sediment above slumped unit is finely laminated. Deposition was probably in an anoxic basin or under an oxygen-minimum zone. Formation of dolomite in pelagic sediments is commonly associated with diagenesis in anoxic pore waters following sulfate reduction. Upper Cretaceous, DSDP Leg 40, Site 364, core 36-3, Walvis Ridge, South Atlantic. Scale in centimeters.

m.y. The first sediments deposited on oceanic crust will be either metalliferous clays of hydrothermal origin or interpillow pelagic limestones. Submarine weathering of pillow basalts may occur before they are covered by sediment. Subsequent nannofossil ooze is rich in planktonic foraminifers (deposited above the

Figure 27—Light-colored calcareous turbidites with hemipelagic black shale interbeds. Individual beds are graded and generally are less than 1 m thick, although beds as thick as 27 m have been reported (Scholle, 1971). Basal parts of beds contain some terrigenous sand and glauconite as well as coarser carbonate material. The bulk of the turbidites consists of reworked nannofossils and microfossils of contemporaneous age. Carbonate-free pelagic interbeds indicate deposition below the CCD. Upper Cretaceous Monte Antola Formation, near Genoa, Italy.

lysocline), then becomes more clay-rich upward, with "red" or brown pelagic clays (deposited below the CCD) sometimes accumulating. This is the general pattern both in lateral transects outward from the ridge crest and stratigraphically upward on older oceanic crust. More complicated stratigraphy may result if there are major variations in the depth of carbonate dissolution horizons, or if oceanic plate movement carries the crust under an oceanic divergence where sea-surface productivity is higher.

Thick sequences of pelagic carbonate may be deposited in local troughs or basins along the mid-ocean ridge crest, mainly by redeposition from submarine highs. Ponded pelagic turbidites are found along the present Mid-Atlantic Ridge, for example (van Andel and Komar, 1969).

In pre-Late Jurassic time, earlier than 140 to 150 m.y. ago, however, the CCD apparently was quite shallow (less than 2,700 m deep) and siliceous oozes accumulated directly on oceanic crust at the ridge crest (Garrison, 1974; Bosellini and Winterer, 1975). The CCD level apparently deepened abruptly about 140 m.y. ago, and deposition of pelagic limestone occurred on the sea floor above about 4,400 m. Therefore, the normal succession of oceanic crust to pelagic limestone to red clay with increasing age is inverted in Middle to Upper Jurassic sequences because of variations in paleoceanography, and perhaps as a consequence of the rapid evolution of calcareous pelagic fauna and flora at the time. This exemplifies the importance of paleoceanography which must be considered in the interpretation of pelagic sedimentary facies relations (see also Fig. 10).

Apparent CCD variations in the stratigraphic record, however, may also be due to large-scale regional uplift of parts of deep-sea basins, such as the 1,000-m uplift that occurred on the Cape Verde Rise in the early to middle Miocene (Lancelot and Seibold, 1978). Such uplift can raise large areas of ocean floor above the CCD with no shift in either the actual depth of the lysocline or the CCD.

Pelagic sediments may be relatively rich in organic matter when deposited at high sedimentation rate and/or in high productivity environments (Muller and Suess, 1979), under an oxygen-minimum zone (von Stackelberg, 1972), or in an oxygen-deficient basin such as the Black Sea (Degens and Ross, 1974). At times in the past, such as during Early to mid-Cretaceous time, large areas of the deeper ocean were oxygen-deficient, and organic-carbon-rich pelagic sediments were more widespread (see Arthur and Schlanger, 1979; Jenkyns, 1980, for reviews).

The oldest pelagic sediments in the present ocean basins are probably Early Jurassic in age; Upper Jurassic pelagic sediments have been penetrated by holes of the Deep Sea Drilling Project. However, most ancient pelagic sediments have been subducted. Knowledge of Jurassic and older pelagic facies has and will come from sedimentary terranes uplifted in mountain belts and in ancient subduction complexes in which oceanic crust and capping pelagic sediment were selectively preserved (Moore, 1975).

SEDIMENTATION RATES

Sediment accumulation in the deep sea is controlled by sediment supply, dissolution, and rates of erosion. Modern pelagic sedimentation rates (Fig. 17) range from near zero in the open ocean (far from land) to as much as 400 mm/1000 years in coastal basins where the pelagic material is diluted by terrigenous sediment.

Red or brown clay generally accumulates at rates of less than 2

Figure 28—Sole markings on base of calcareous turbidite. Flute casts predominate, but load casts and trace fossils are also visible. Current-cut sole markings are generally scarce in distal turbidite environments. Upper Cretaceous Monte Antola Formation, near Genoa, Italy.

Figure 29—Rhythmic intercalation of red marlstones and light-gray limestones deposited on a relatively deep sea floor or under an elevated CCD. Red marlstones may represent periodic clastic dilution of pelagic carbonate flux, or the carbonate beds may be pelagic turbidites deposited below the CCD. Upper Cretaceous, Zumaya, Spain.

mm/1000 years. Such sediments, which are common on sea floor areas below the CCD, are difficult to date because most microfossils are dissolved and numerous hiatuses may be present. Sediment erosion and reworking by currents is common in the deep sea, and as much as 80% of the stratigraphic record may be lost during certain time periods (Fischer and Arthur, 1977; Moore et al, 1978; Fig. 10). Major hiatuses are common

in areas of low sediment supply.

Biogenic silica is deposited at relatively high rates, ranging from 4 to 400 mm/1000 years. Biosiliceous pelagic sediments (greater than 30% biogenic silica) typically accumulate at high rates beneath fertile regions of the ocean. However, the highest rates in the modern ocean result from dilution of dominantly biosiliceous material by glacially derived sediment around the Antarctic divergences and

by fluvial and volcaniclastic material in coastal basins. In contrast, most ancient cherts were deposited slowly, even considering compaction, at rates comparable to those of modern deep-sea radiolarian oozes or radiolarian-rich nannofossil chalks. This relation suggests that most ancient radiolarites resulted from normal production of radiolarians with little or no dilution by carbonate or terrigenous clastic material.

Sedimentation rates of modern pelagic carbonates (uncompacted) vary by about an order of magnitude (Fig. 17), and average about 30 mm/1000 years. The variations are due largely to the degree of carbonate dissolution (for example, the depth of deposition) versus rate of supply. When corrected for differences in compaction, the range of rates and the average rate of accumulation of carbonate sediments in modern deep-water pelagic environments are similar to those of the Cretaceous deep sea (Fig. 17). However, the average accumulation rate of Cretaceous "shelf" chalks is higher than that of deep-sea chalks, partly because of the enhanced preservation of calcium carbonate in saturated surface waters. Deep-sea sedimentation rates may be locally high due to rapid redeposition of chalk from adjacent

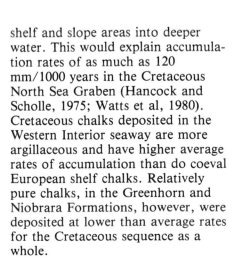

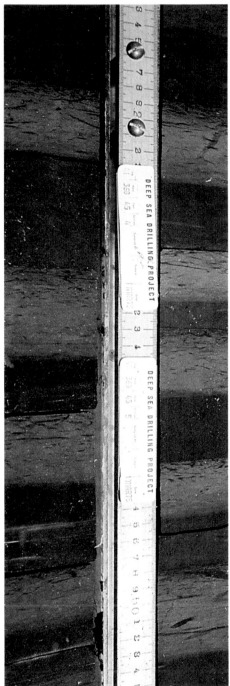

shelf and slope areas into deeper water. This would explain accumulation rates of as much as 120 mm/1000 years in the Cretaceous North Sea Graben (Hancock and Scholle, 1975; Watts et al, 1980). Cretaceous chalks deposited in the Western Interior seaway are more argillaceous and have higher average rates of accumulation than do coeval European shelf chalks. Relatively pure chalks, in the Greenhorn and Niobrara Formations, however, were deposited at lower than average rates for the Cretaceous sequence as a whole.

Figure 30—Thin (about 4 cm) hemipelagic dark green claystone overlying gray turbiditic marlstone. *Chondrites* burrows in uppermost 5 cm of marlstone (at 10-cm mark on scale) are filled with darker hemipelagic sediment from above. Flattening of the cross sections of individual *Chondrites* burrows is due to compaction. Note that *Chondrites* normally does not penetrate more than 10 cm (maximum of 20 cm) into the turbiditic marlstone from its upper surface, indicating rapid deposition of the marlstone. Absence of $CaCO_3$ in the dark green claystone suggests deposition below the CCD. Brown material at top of photograph is the base of next sandy turbidite. Upper Cretaceous Zementmergel, Bavaria, Germany (photograph by R. Hesse). Scale in centimeters.

Figure 31—Cycles in hemipelagic mudstone. Dark olive-gray parts are depleted in calcium carbonate and have sharp basal contacts with underlying light-colored parts of each couplet. Darker layers are slightly richer in terrigenous silt and organic carbon, and are interpreted as mud turbidites deposited along a continental rise. Note burrowing in upper part of light-colored interval and decrease in burrowing downward into the turbiditic portion. Eocene, DSDP, Leg 41, Site 368, core 45-4; off northwest Africa. Scale in centimeters.

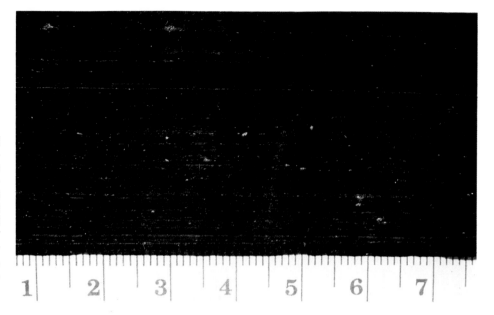

Figure 32—Polished slab of laminated basinal limestone containing approximately 2% organic matter. Although there is a minor siliceous planktic component, all the carbonate material is derived from adjacent shelf sources, as is typical for most Paleozoic basinal sediments. Upper Permian Lamar Limestone Member of Bell Canyon Formation, Delaware Basin, west Texas. Large-scale divisions in centimeters.

Figure 33—Uniformly thin-bedded (approximately 25 cm) limestones with thin, red shale partings. These rocks are highly indurated, which is typical of deeply buried pelagic limestone or turbidite sequences. Bedding may represent primary cyclic changes in deposition or turbidity current bedding that has been enhanced by transfer of carbonate during diagenesis. Lower Paleocene, Zumaya, Spain. Scale bar at center right is 15 cm long.

SEDIMENTARY STRUCTURES OF PELAGIC SEDIMENTS

Primary Sedimentary Structures of Carbonate Sequences

The dominant structure in most pelagic carbonate sequences is rhythmic bedding caused by fluctuations in depositional environment. Controlling factors include variations in terrigenous input, surface productivity and/or dissolution rates, current velocities, and water depth. These variations may be controlled by cyclic astronomic parameters, as discussed earlier. Cyclic characteristics occur on many scales (Figs. 18 to 23) ranging from millimeter-thick lamination, through centimeter and decimeter layering or bedding to sequences of beds up to tens of meters thick.

Millimeter-thick laminations are generally preserved only where a benthic infauna was excluded by anaerobic to dysaerobic conditions or by rapid sedimentation. Transitions between laminated and burrowed sediment commonly span a few centimeters and may represent rapid changes in oxygenation (Fig. 20). Larger-scale bedding cycles, tens of centimeters thick, are important features of nearly all pelagic carbonate sequences, even in highly bioturbated chalks. These bedding cycles generally consist of alternating clay-rich and clay-poor chalk (Fig. 21), but such cycles are also present in relatively pure chalk sequences (Fig. 22). In many sequences, cyclic bedding is accentuated by diagenetic features such as stylolitization of marl layers or selective chertification of more calcareous zones (Fig. 23). The periodicities of such bedding cycles are usually 20,000 to 100,000 years (Fischer, 1980). In some strata, the bedding is less pronounced (Fig. 24), possibly due to high sedimentation rates and intense bioturbation.

Bedding may be discontinuous,

Figure 34—Submarine-extruded pillow basalts in a Jurassic ophiolite sequence from Liguria, Italy. Shapes of pillows suggest that section is overturned. Scale bar in center is 15 cm long.

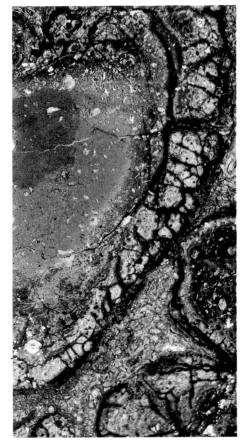

Figure 35—Pillow breccia deposited on submarine ridge crest showing alteration of original glassy margins of basalt pillows to smectitic clays. Also shown is some interpillow limestone. Lower Cretaceous, DSDP Leg 52A, Site 417A, core 30-2, western North Atlantic (photograph by K. Kelts). Area of core shown is about 5 cm wide.

disrupted or contorted due to slumping on gentle to steep slopes. Figure 25 illustrates lateral pinchout of chalk beds and other bedding relationships that suggest the filling of channels along the margins of large mounds in a shelf chalk. Some slumping occurred on the flanks of these structures. The channels and intervening high ridges have a vertical relief of tens of meters, and may be due to local biohermal buildups (Kennedy and Juignet, 1974), or to the action of strong submarine currents such as those affecting modern lithoherms (submarine-cemented mounds) in the Straits of Florida (Neumann et al, 1977). Hardgrounds (see section on diagenesis, Figs. 86 to 93), define many of the bedding planes in the chalks of Figure 25. Deeper-water pelagic sequences, especially those deposited along continental margins or other submarine topographic highs, may contain slumped intervals (Fig. 26). Carbonate dissolution along the base of escarpments or slopes may undermine slope-flank deposits causing such slumping, as on the flanks of the modern Ontong-Java Plateau (Berger and Johnson, 1976).

Bedding in pelagic carbonate sequences is not always caused by cyclic environmental changes. Many pelagic-type sediments have been largely

redeposited. Some of these redeposited beds are clearly identifiable as turbidites due to their internal structure, which may include features such as textural grading, current lamination, and sole marks. The thick beds of the Monte Antola Formation (Figs. 27 and 28) of northern Italy are an example of such a pelagic carbonate turbidite deposit (Scholle, 1971). The carbonate sediment apparently was redeposited from adjacent slopes or more local submarine highs, and consists almost entirely of pelagic organisms. Some redeposited pelagic carbonate beds do not have the recognizable primary sedimentary

structures of turbidites because of their uniformly fine grain size and/or later bioturbation. The beds shown in Figure 29 may be pelagic turbidites deposited below the CCD in a red clay sequence. Figure 30 illustrates another example of pelagic carbonate resedimented in an environment below the CCD. Exponential downward decrease in burrowing in such fine-grained turbidites is an important way of distinguishing them from normal pelagic carbonates. Figure 31 shows cycles that are possibly related to periodic terrigenous dilution of a clayey pelagic carbonate sequence, leading to

Figure 36—Cretaceous chaotic melange terrane in Anatolia, Turkey, with large, varicolored, allochthonous blocks. Red is red shale and radiolarian chert; white is pelagic limestone; and greenish gray is turbiditic sandstone and shale. These types of terranes are typical of uplifted mountain belts delineating ancient active continental margins (subduction zones).

Figure 37—Varved diatomaceous mud. Alternation of diatom-rich muds (light color) and silty-clay layers (dark color) represents blooms of diatoms under seasonal upwelling conditions. Deposition took place on slope under an oxygen-minimum zone. Benthic infauna was not present due to low-oxygen conditions, so laminations were not disturbed. Quaternary, DSDP Leg 64, Site 480, core 20-3, Gulf of California. Scale in centimeters.

lithology similar to that in Figure 30. Regular lamination in fine-grained pelagic carbonates may represent turbidite deposition rather than varving. In deep-water pelagic limestones of Paleozoic age (Fig. 32), from which pelagic carbonate microfauna and flora are not known, calcareous laminations may represent "microturbidites" of fine-grained carbonate derived from the shelves. Many younger, well-bedded pelagic carbonate sequences may also contain significant amounts of redeposited pelagic material that cannot be easily identified (Fig. 33).

Structures of Associated Sediments and Underlying Basement Rocks

Deep-water pelagic sediments generally overlie basaltic oceanic crust. The uppermost basaltic layer is typically composed of pillow basalts (Fig. 34), which indicate submarine extrusion of basaltic flows. The vesicularity of pillow lavas is an index of water depth; greater vesicularity characterizes shallower extrusions. Submarine "weathering" may cause

glassy margins of pillows to be devitrified and altered to smectitic clay minerals (Fig. 35). Interpillow sediments are typically metalliferous clays and/or pelagic limestone (Bostrom, 1973; Scott et al, 1974; Robertson, 1975). In uplifted foldbelts or subduction complexes, pelagic facies, and associated basement rocks may be exposed as chaotic melanges (Fig. 36). Lateral facies relations are virtually impossible to decipher in these rocks.

Siliceous biogenic sediments are common pelagic facies in highly productive regions of the ocean. Diatomaceous oozes are commonly laminated to bedded, particularly where deposited in coastal basins and slopes under the influence of low-oxygen conditions and strong seasonal productivity patterns. Laminations are due to seasonal influx of terrigenous clay superimposed on a seasonal pattern of upwelling and diatom productivity (Figs. 37 and 38). Laminations are preserved where there is a lack of burrowing infauna (Calvert, 1966; Schrader et al, 1980). Thicker, more massive intervals contain intercalated fine-grained turbidites that may consist largely of either terrigenous material or shallow-water to pelagic carbonate sediment that was redeposited from adjacent bank margins. Figures 38 and 39 illustrate an unaltered diatomaceous

ooze. With burial, diagenesis ensues, and the ooze is altered to brittle bedded porcellanite and chert (Fig. 40; see section on diagenesis).

Radiolarian-rich oozes (Fig. 41) are deposited in areas of equatorial upwelling and in association with diatomaceous oozes. Radiolaria are also subordinate constituents of red clay and pelagic carbonate oozes in regions of lower productivity. However, prior to the evolution of marine diatoms in the Late Cretaceous, radiolarian cherts were far more widespread in shelf and deep-basin settings. There are relatively few bedded cherts of Early Cretaceous and younger age, whereas Jurassic and earlier bedded

Figure 38—Laminated, organic-carbon-rich, diatomaceous mudstone intercalated with thin turbidites. Deposition was in nearshore but relatively deep-water block fault basins similar to those presently found on the California borderland. Environment was probably similar to that in the present-day Guaymas Basin, Gulf of California. Silica in this shallowly buried sample is still largely opal-A. The high content of marine organic matter makes these sediments excellent hydrocarbon source rocks where they are more deeply buried. Miocene to Pliocene Monterey Formation, near Santa Barbara, California. Pencil is about 16 cm long.

Figure 39—SEM photograph of sediment illustrated in Figure 38. Note abundance of marine diatoms (and sparse coccoliths), excellent preservation of primary texture, and absence of recrystallization or cementation. Scale bar is 10 μm.

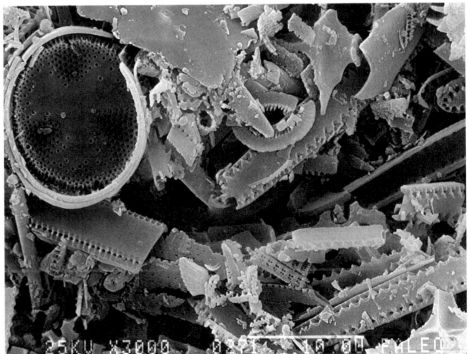

radiolarian cherts are more common (Fig. 42). These pre-Cretaceous bedded cherts have bedding cycles similar to those of pelagic carbonates, although the individual chert beds are thinner than their carbonate counterparts, reflecting the generally slower accumulation of radiolarian oozes. Some rhythmic bedding in cherts may be caused by siliceous turbidites intercalated with pelagic radiolarites (Nisbet and Price, 1974; Kalin et al, 1979). These turbidites may be cryptically graded (Fig. 43). In most cherts, radiolarians are poorly preserved (Fig. 44), and individual radiolarians may be pyritized or calcitized.

Pelagic facies rich in organic carbon are somewhat rare in the modern ocean, but were common at certain times in the past, such as during Early to mid-Cretaceous time (Arthur and Schlanger, 1979; Jenkyns, 1980). Millimeter-size laminae of pelagic carbonate sediments rich in organic car-

bon were widely deposited in Cretaceous epicontinental seas (Fig. 45), attesting to poorly oxygenated conditions at the sea floor. Lower to mid-Cretaceous deep-water pelagic limestone sections also commonly contain intercalations of organic-carbon-rich shale and/or siliceous ooze (Fig. 46), representing short-term increases in biological produc-

tivity or expansion of oxygen-depleted conditions in deeper water masses.

The color of normal pelagic sediment deposited in oxidizing environments ranges from white to tan to red with decreasing carbonate content (Fig. 47). These sediments, particularly the slowly deposited "red" clay (Figs. 47 and 48), are highly bioturbated and retain no primary

Figure 40—More deeply buried equivalent of previous example (Fig. 38). Silica has been altered to quartz through burial diagenesis. Diagenetic transformation has resulted in a significant decrease in matrix porosity. Concomitant increase in brittleness has led to extensive fracturing. A considerable amount of oil production from the Monterey Formation is from such fractured intervals. Miocene to Pliocene Monterey Formation, near Point Conception, California.

Figure 41—SEM photograph of coarse fraction of modern oceanic siliceous ooze containing a mixed assemblage that is dominated by radiolaria but contains minor sponge spicules and ebridian fragments (photograph from Hein et al, 1979). Scale markings at lower edge are 100 μm.

sedimentary structures. Color of pelagic sediments is controlled by the degree of bottom-water and pore-water oxygenation, trace constituents (pyrite, organic matter, iron oxides, volcanogenic grains), and subsequent weathering. The colors are ephemeral and may change during burial or with exposure on outcrop (Moberly and Klein, 1976). Ferromanganese nodules are commonly associated with red clay facies (Figs. 48 and 49). The nodules must remain at the sediment-water interface for long periods of time (millions of years; Glasby, 1977) in order to grow to appreciable size (10 to 15 cm in diameter). Such nodules are rarely preserved in ancient sediments because they are easi-

ly dissolved in anoxic pore waters on burial, but some ancient ferromanganese nodules are known from red (oxidized) sediments (Fig. 50) (Jenkyns, 1977).

Evaporite minerals are uncommon in pelagic sequences. However, in small, restricted ocean basins, such as the late Miocene Mediterranean (Hsu, 1972), and in areas of high evaporation, gypsum, anhydrite, and even halite (Fig. 51) may be deposited. In the late Miocene (Messinian) Mediterranean sequences, 1 to 2 km of evaporites are sandwiched between normal pelagic and hemipelagic facies. In the Permian Basin of west Texas-New Mexico, laminated evaporites, including anhydrite, halite, and sylvite, were deposited in water hundreds of meters deep, overlying basinal carbonate and clastic rocks rich in organic carbon.

Other "clastic" facies that may be associated with deeper water pelagic sediments are: ice-rafted detritus (Fig. 52), consisting of poorly-sorted con-

glomeratic debris; contourites (Fig. 53), which are current-laminated and winnowed sediment resulting from the action of geostrophic bottom currents; and graded, redeposited beds of shelf carbonate or volcaniclastic debris (Fig. 54).

BIOGENIC SEDIMENTARY STRUCTURES AND CHARACTERISTIC FAUNAS

Deep-Sea Pelagic Carbonates and Associated Facies

Benthic macro-organisms leave few if any body fossils in deep-sea environments, due to both calcium carbonate dissolution on the deep-ocean floor and the great preponderance of soft-bodied organisms in abyssal communities. Trace fossils (for example, biogenic sedimentary structures) provide virtually the only fossil evidence of the benthic macroinvertebrate community in ancient deep-sea environments. In fact, they often yield important ecologic and ethologic information, such as the relative importance of suspension feeders and deposit feeders in the bottom community, that body fossils do not provide. Burrowing organisms also strongly influence the preservation or destruction of primary sedimentary

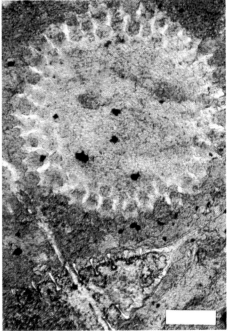

Figure 42—Interbedded chert (blackish-green) and hemipelagic (red) shale beds. Chert layers range from even and regular to nodular; variations in thickness are largely related to diagenetic history. Uncertainty exists as to whether the chert precursor was radiolarian ooze or silicified limestone. Jurassic, Lagonegro basin, Basilicata, Italy (photograph by E. F. McBride). Scale is 15 cm long.

Figure 43—Cross section of radiolarian in thin section. Relatively poor preservation is typical of radiolarians in ancient pelagic sediments and is a consequence of diagenetic alteration from original opal-A to quartz chert. Jurassic part of Franciscan Complex, Point Sal, California. Scale bar is 0.07 mm long.

structures in slowly deposited pelagic sediments.

Trace fossils, which include tracks and trails of epifauna as well as burrow systems of infauna, have been useful as paleodepth indicators in deep-water marine sequences in both outcrops and deep-sea cores (Seilacher, 1964, 1967). The depth limits for particular ichnofacies (true fossil assemblages) are variable, however, and can overlap with those of other ichnofacies. Trace fossils are bathymetrically controlled only in the sense that their creators, and/or the sedimentary facies in which they occur, are influenced by water depth. It is probably not depth alone, but rather water mass characteristics (temperature, salinity, nutrients, oxygen content, etc.) and substrate types (for example fluid versus hard bottoms) that control ichnofacies distribution.

Deep-sea organisms that produce trace fossils include forms that are both eurybathic (with broad depth tolerances) and stenobathic (with narrow depth tolerances). Among the eurybathic organisms are various genera of coelenterates, sipunculids, brachiopods, decapods, scaphopods, and ophiuroids. Stenobathic forms include various genera of sponges,

echiurans, pogonophorans, isopods, amphipods, and asteroids. Unfortunately, little information is available as to which kinds of animals make which kinds of burrows in the deep sea today.

The deep sea includes several general ichnofacies (Fig. 55) that are directly related to different depositional regimes in a relative depth sequence. Bathyal environments (200- to 2,000-m water depth) are typified by steep slopes and rapid rates of deposition, largely in redeposited sediment. Trace fossils in this setting are typically dominated by the deposit-feeding burrow system of *Zoophycos* (the *Zoophycos* Ichnofacies; Fig. 56). The distal turbidite facies, which may extend out onto the abyssal plain (2,000- to 6,000-m water depth), generally consists of finer grained sediment. The deposit-feeding burrow system known as *Chondrites* (Figs. 58 and 59) characterizes clay units, but a diverse assemblage of highly patterned, horizontal trace fossils (for example, *Helminthoida, Paleodictyon,* and *Nereites*) occurs on bedding planes and typified the distal turbidite facies (the *Nereites* Ichnofacies, for example; Figs. 60 to 63).

Seilacher's (1964, 1967) *Zoophycos* and *Nereites* ichnofacies have been

used as general models of deep-water trace fossil distribution in flysch basins and continental margin settings in both Europe (Ksiazkiewicz, 1970; Crimes, 1973, 1977; Roniewicz and Pienkowski, 1977; Kern, 1978) and North America (Chamberlain, 1971; Kern and Warme, 1974).

The "Deep-sea Ichnofacies" (Ekdale and Berger, 1978) is found in pelagic sediments and has a different character than does Seilacher's *Zoophycos* and *Nereites* ichnofacies of turbidites. Box-cores and ocean-bottom photographs of modern pela-

Figure 44—Redeposited bedded radiolarian cherts. These "ribbon-bedded" chert beds (light colors) contain 30 to 80% radiolarians, have graded bedding, and were presumably deposited by turbidity currents. Intercalated red shales (darker red layers) are presumably hemipelagic. Sequence overlies pillow basalts. Jurassic, Monte Roccagrande, Liguria, Italy (photograph by E. F. McBride). Scale in centimeters.

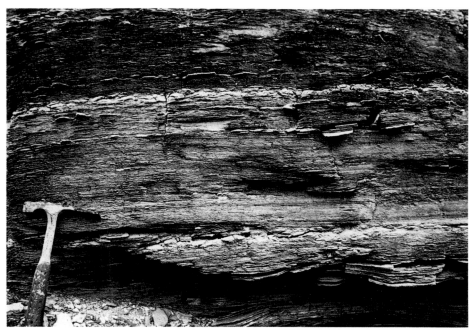

Figure 45—Millimeter-scale laminae preserved in shaly chalk (marl) deposited under anaerobic or dysaerobic conditions in a shallow epicontinental seaway. Preservation of large amounts of marine organic matter make this unit a potential hydrocarbon source rock. Upper Cretaceous Bridge Creek Limestone Member of Greenhorn Limestone, Kansas.

gic sediment reveal intense bioturbation of the abyssal sea floor. Surface tracks and trails, and even shallow burrows (less than 5 cm deep), are generally obliterated by the bioturbation of the deeper burrowing infauna. For example, although many surficial trails and burrows of the *Nereites* type have been photographed on the modern deep-sea bottom (Figs. 64 to 69; see also Ewing and Davis, 1967; Heezen and Hollister, 1971; Lemche et al, 1976; Kitchell et al, 1978) and even collected in box cores (Fig. 62B, see also Ekdale, 1980a), such traces are rarely preserved in the historical record of pelagic deposition (Ekdale, 1977; Ekdale and Berger, 1978; Berger et al, 1979). Only those structures produced beneath the top 5 to 8 cm of sediment are preserved. Thus, pelagic ichnofacies consist of infaunal trace fossils almost exclusively.

Analysis of deep-sea box cores reveals a three-tiered burrow stratigraphy in the surficial layers of pelagic sediment. The uppermost zone is the "mixed layer," a 5- to 8-cm-thick zone of virtually homogeneous, bioturbated sediment that grades downward (through the "mixed layer

transition") into the "transition zone." The "transition zone," a site of partial sediment mixing, which commonly extends 20 to 35 cm below the water-sediment interface, is the zone in which almost all preserved biogenic structures are produced. The bottom tier, termed the "historical layer," is the zone into which no burrowers penetrate, and is the level at which the processes of compaction and diagenesis take over.

Production and preservation of particular kinds of deep-sea burrows are dependent on the strength of the sediment, which in turn seems to be inversely correlated with the clay content (Berger and Johnson, 1976; Johnson et al, 1977). As the clay content of pelagic sediment increases, sediment shear strength decreases. This relationship is strongly dependent on water depth because of calcium carbonate dissolution below the lysocline in the ocean's stratified water column (Berger, 1970b, 1972,

1976). Below the CCD (Fig. 8), sediment strength is apparently too low to support several kinds of burrows that are common in carbonate-rich sediments from shallower depth. The burrows that are the most obvious reflectors of this trend are large (1 to 2 cm diameter), open, vertical burrows called *Skolithos* (Figs. 70 to 72; see also Ekdale and Berger, 1978). Although *Skolithos* is commonly considered an intertidal trace fossil, the name is properly applied to any simple, unbranching vertical tube. The abundance of vertical burrows in box cores or pelagic ooze is directly proportional to the percentage of calcium carbonate in the sediment; almost no *Skolithos* are found in clay-rich material with less than 60% calcium carbonate.

There are, therefore, two general ichnofacies for deep-sea pelagic sediment regimes. The first, for carbonate ooze above the CCD, contains abundant and diverse infaunal trace fossils, including *Skolithos* and *Planolites* with lesser numbers of *Chondrites, Teichichnus* and *Zoophycos* (Figs. 57 and 58, 70 to 75). Color contrasts among different burrows are high, and definition of individual burrows is good. This pelagic carbonate ichnofacies has been described in box cores of modern sediment (McMillen, 1974;

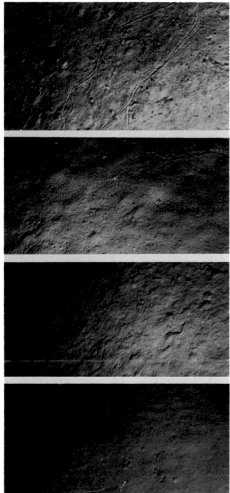

Figure 46—Deep-water pelagic limestone sequences, about 60 m thick, showing changes in oxidation state during deposition. Gray limestone near base is laminated; transition between gray and red to yellow-gray is near middle part of section in photograph. Dark layer (lower Turonian) within gray limestone sequence is a 1-m-thick radiolarian-rich argillite containing as much as 25% organic carbon. This layer represents sudden and short-term intensification of oxygen-deficits at the sea floor. Upper Cretaceous Scaglia Bianca Limestone-Scaglia Rossa Limestone contact, Umbria, Italy.

Ekdale and Berger, 1978; Berger et al, 1979; Novak, 1980) and in ancient pelagic carbonates recovered from DSDP cores (van der Lingen, 1973; Warme et al, 1973; Chamberlain, 1975; Kennedy, 1975; Ekdale, 1977, 1978, 1980b).

The second deep-sea ichnofacies, in pelagic clay below the CCD, contains abundant evidence of bioturbation, although the diversity and preservation state of burrows is low. No *Skolithos* and few *Chondrites, Teichichnus* or *Zoophycos* are found. The assemblage is dominated by *Planolites,* which commonly possess fuzzy and/or smeared burrow margins (Figs. 76 to 79). Trace fossils in pelagic clays have been described from box cores (McMillen, 1974) and DSDP cores (Ekdale, 1977, 1980b).

Trace fossils in DSDP cores have proved helpful in interpreting Cretaceous water depths in the Caribbean Sea (Warme et al, 1973), Tertiary water depths in the Mediterranean Sea (Ekdale, 1978), sedimentary facies changes in the Philippine Sea (Ekdale, 1980b), and mid-Cretaceous oceanic anoxic events in the north and south Atlantic (Arthur, 1979b).

Deep-sea trace fossils may be useful aids in interpreting paleoenvironmental aspects of pelagic deposits other than bathymetry and sediment types.

For example, oxygenation of bottom waters may have varied significantly in the past in pelagic depositional systems, and several modern and ancient deep-sea basins have undergone periods of stagnation (Demaison and Moore, 1980) (Figs. 19 and 20). Trace fossils can provide understanding of basinal stagnation histories by documenting the responses of benthic organisms to fluctuating levels of dissolved oxygen in the water at the sea floor. Most shelled organisms cannot tolerate dissolved oxygen concentrations of less than 1.0 milliliter/liter (ml/l) but soft-bodied organisms that do not require oxygen for shell construction may sustain themselves in environments in which dissolved oxygen concentrations are an order of magnitude less (as low as 0.1 ml/l) (Rhoads and Morse, 1971). The occurrence of burrows in a sediment indicates the presence of at least some oxygen in the bottom environment, and soft-bodied deposit feeders usually predominate under oxygen-poor conditions. The presence of the deposit-feeding burrow system *Chondrites* in an organic-rich layer in the absence of other biogenic structures or benthic shelled fossils probably indicates nearly (but not completely) anoxic conditions at the sea floor (Arthur, 1979b).

Figure 47—Progressive color change of fine-grained sediment with increasing water depth in the North Atlantic Basin. Note shift from cream-gray organic ooze (top photograph) to red abyssal clay (bottom photograph). Photographs (from top to bottom) are from water depths of 1,500 to 2,000, 3,462, and 4,914 m in the eastern part of the North Atlantic (photographs from Heezen and Hollister, 1971, plate 8).

Figure 48—Ferromanganese nodules on surface of pelagic red clay recovered in a box core taken in the North Pacific gyre region, north of Hawaii, in about 5,900 m of water. Note soupy nature of pelagic sediment.

Figure 49—Closely packed ferromanganese nodules formed on ocean bottom floor beneath the Antarctic circumpolar current in 3,924-m water depth on the eastern side of the Bellingshausen Basin, near entrance to Drake Passage (from Heezen and Hollister, 1971).

Figure 50—Ferromanganese nodules in Jurassic, red calcareous claystone, and yellow-red breccia. Some traces of stylolitic concentric structures are visible within the nodules. Jurassic Lower Ammonitico Rosso, Gelpach Quarry, northern Italian Alps (photography by J. G. Ogg). Nodules are 5 to 10 cm in diameter.

Bioturbation of Shelf-sea Pelagic Units

Most shelf-sea chalk is intensely bioturbated although individual burrows may be difficult to discern (Bromley, 1981). The diversity of identifiable trace fossils is great, commonly higher than in deep-sea deposits. For example, in addition to the *Chrondrites, Planolites, Skolithos, Teichichnus,* and *Zoophycos* that commonly abound in abyssal carbonates, shelf-sea chalk typically also contains *Asterosoma, Gyrolithes, Thalassinoides,* and several other ichnogenera (Fig. 80; Frey, 1970; Kennedy, 1970, 1975).

In both North America and northern Europe, there appear to be two major chalk ichnofacies (Frey, 1970; Kennedy, 1975; Bottjer, 1978). The first is a shallow-water assemblage dominated by dense *Thalassinoides* (Figs. 80 to 82) and other arthropod burrows (Seilacher's "*Cruziana* Ichnofacies"). The second assemblage is the *Chondrites-Planolites-Zoophycos* association. Transgressive sequences are commonly marked by the *Thalassinoides* assemblage grading upward to the *Chondrites-Planolites-Zoophycos* association.

Bored, encrusted, lithified surfaces in European chalk sections are commonly known as "hardgrounds" (Bathurst, 1971). Such surfaces are

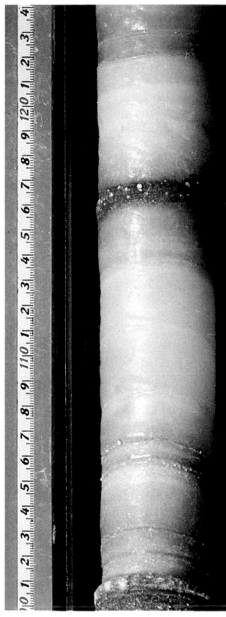

Figure 51—Interbedded thin, nodular gypsum and massive halite. These evaporite facies are used as evidence that the deep Mediterranean basin was isolated and desiccated during late Miocene time (Hsu et al, 1977). Miocene (Messinian); DSDP Leg 13, Site 134, core 10-2, Mediterranean Basin. Scale is in centimeters.

Figure 52—Ice-rafted debris on ocean bottom. Large, rounded and angular, ice-rafted boulders derived from Antarctica are scattered in finer grained sediments at 2,420-m depth. Holocene, Pacific continental slope, Antarctic Peninsula (photograph from Heezen and Hollister, 1971).

Figure 53—Contourite sediment from the North American Atlantic Continental Rise off New England in 4,746 m of water. Note intercalation of hemipelagic claystone and coarser, cross-laminated contourite siltstone (photograph from Heezen and Hollister, 1971). Large-scale divisions are in centimeters.

virtually absent from North American chalks (Bottjer, 1980) and most deep-sea carbonate deposits (Kennedy, 1975; but see Fischer and Garrison, 1967; Milliman, 1974a). Hardgrounds are paleoecologically interesting, because each contains a suite of post-lithification trace fossils (borings) superimposed upon a suite of pre-lithification trace fossils (burrows), each of which represents a succession of different ecologic conditions (Fig. 83). The pre-lithification trace-fossil assemblages are usually characterized by dense, anastomosing networks of the burrow system *Thalassinoides* (Figs. 81 and 82), created by burrowing crustaceans in a soft-sediment habitat (Bromley, 1967). Two or more sizes and geometries of *Thalassinoides* are present in some hardgrounds (Kennedy, 1967) and may have been produced at different times during the sequential stages of deposition and incipient lithification (Bromley, 1975). Post-lithification borings or borings of other hard substrates (Fig. 84; see also section on diagenesis) are typically diverse in size and shape, and most can be attributed to clionid sponges, mytilid and pholadid bivalves, acrothoracican barnacles, and worms of various types.

Distinct, parallel layers of siliceous nodules ("flints") typify Cretaceous chalk strata exposed in northern Europe (Fig. 23). Many of these horizons ostensibly represent silicified burrow networks, including *Thalassinoides* (fig. 81), *Gyrolithes, Zoophycos,* and *Bathichnus* associated with large, cylindrical "paramoudras" (Bromley et al, 1975). Indeed, Kennedy (1975, p. 392) declared that "most. . .of the flints in European chalks, and cherts in some Mesozoic limestones, are in fact silicified burrows; a transition from partly silicified burrow fill to perfect burrow pseudomorph can often be demonstrated." Silicified nodules and/or burrows are rare in Cretaceous shelf-sea chalks of North America.

Body Fossils
in Pelagic Carbonates

Body fossils of macro-organisms are very rare in deep-sea pelagic carbonate strata but may be moderately to extremely common in shelf-sea chalks. Many of the chalk megafossils

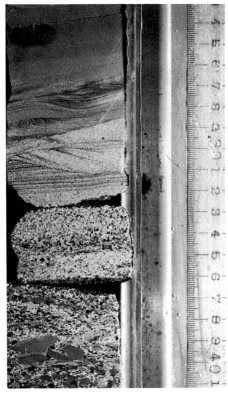

Figure 54—Graded volcaniclastic-carbonate turbidite. Carbonate constituents are reworked shallow-water material. Eocene of the Daito Basin, Philippine Sea; DSDP core 446-39-1, 132 to 142 cm interval. Large-scale divisions are in centimeters.

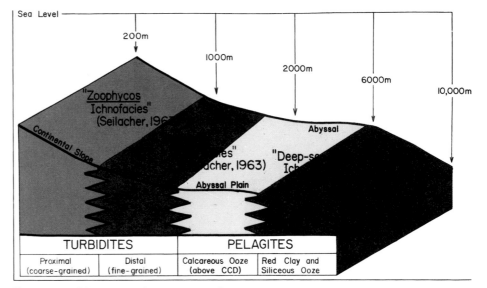

Figure 55—Diagrammatic representation of relationships among ichnofacies in deep-sea depositional environments.

are remains of nekton (ammonites and belemnites), but a significantly higher diversity of benthos (bivalves, echinoids and brachiopods) may also be represented. Oysters, inoceramids, echinoids and other calcite-shelled organisms are usually well-preserved. Ammonites, gastropods and other aragonite-shelled organisms, however, are generally more poorly preserved as internal or external molds or casts.

Kennedy (1978) outlined five general organism communities in the Cretaceous chalks of Great Britain containing pelagic faunas of cephalopods and fish as well as diverse benthic faunas of sponges, ahermatypic corals, brachiopods, bryozoans, bivalves, gastropods, serpulid worms, and echinoids. Body fossil associations in North American Cretaceous chalks are generally lower in diversity than those in European chalks. Molds of ammonites are com-

mon, but fossil nautiloids and belemnites are rare. Benthic communities inhabiting the Western Interior seaway of the United States were dominated by bivalves (mainly inoceramids and oysters). Other macrobenthos were uncommon, possibly due to poor oxygenation of some bottom waters (Hattin, 1971, 1975; Frey, 1972). Benthic communities preserved in the Gulf Coast chalks of Arkansas and Texas must have lived under more favorable conditions, because they include not only more diverse bivalve assemblages but also numerous species of echinoids, bryozoans and serpulid worms (Bottjer, 1978).

In contrast to the shelf-sea chalks, body fossils of macro-organisms are rare in deep-sea pelagic deposits. This is partly due to the fact that most studies of deep-sea sediment have been accomplished in small-diameter cores, so the chances of encountering megafossils are considerably lower than in outcrop sections. It is also due to the fact that many abyssal and hadal organisms lack durable hard parts and are thus geologically unpreservable. Even the calcite or aragonite shells of deep-sea mollusks and brachiopods cannot withstand the dissolution of calcium carbonate below the lysocline, making such fossils virtually absent from abyssal

pelagic clay and chert. Megafossils are occasionally found in DSDP cores, however. They include remains of nektic and pelagic organisms (ammonite shells and aptychi, fish bones and teeth and free-swimming crinoids) as well as benthic organisms (bivalves, serpulid tubes, echinoid spines and bryozoans).

Due to the dearth of macrofaunal remains in pelagic deep-sea deposits, most biostratigraphic correlations are founded upon foraminiferal, radiolarian and coccolith zonations. These are outlined in detail in the *Initial Reports of the Deep Sea Drilling Project*. On the other hand, biostratigraphy of the shelf-sea chalks commonly employs macrofaunal zonations (ammonites, belemnites, brachiopods, bivalves and/or echinoids) as well as microfossil zonations.

Prior to the appearance of calcareous plankton in the late Mesozoic, there was essentially no pelagic production of fine-grained carbonate. Pelagic cherts and hemipelagic mudstones of Paleozoic and early Mesozoic age typically contain fossil assemblages dominated by plankton and nekton with few benthic taxa represented. Terrigenous graptolitic shales in Ordovician and Silurian sections fall into this category. In deep-water facies of

Figure 56—Tightly meandering *Helminthoida* trails and a circular spreite of *Zoophycos* (in upper right) on a bedding plane of Jurassic to Cretaceous rocks in the Ligurian Apennines, Italy. Both trace fossils exhibit efficient foraging patterns produced by unidentified deposit-feeding animals (photograph by E. F. McBride). Area of photograph is about 50 cm wide.

Figure 57—Intensely bioturbated Eocene to Oligocene chalk from DSDP Site 192 (core 192A-1-3, 21 to 30 cm interval) in the northwest Pacific. The typical pelagic carbonate ichnofacies is represented by at least four kinds of burrows, and crosscutting relationships of burrows reveal a bioturbation sequence. *Zoophycos* is not cut by any other burrow, but it cuts across *Skolithos*, which itself cuts across *Chondrites*, which cuts across compacted specimens of *Planolites*. Thus, *Planolites* must have been produced first (and *Zoophycos* last) in the sequence. Core is 7 cm in diameter.

Devonian through Permian age, goniatite ammonoids, thin-shelled ostracods, free-swimming bivalves (pteriids and pectinids), fish and conodont-bearing animals are represented as fossils. Bedded cherts, which probably were pelagic in origin, are typically devoid of macro-organism remains.

Summary of Paleoecologic and Paleoenvironmental Implications

The major similarities between shelf-sea and deep-sea chalks include: (1) sediment composed primarily of calcareous nannoplankton (coccoliths) and microplankton (foraminifers); (2) sediment thoroughly bioturbated, so virtually no primary depositional structures are preserved; (3) at least some strata dominated by a characteristic trace fossil assemblage of *Chondrites, Planolites, Teichichnus* and *Zoophycos;* and (4) pelagic macrofaunal remains (for example, ammonites) rare to moderately common. Diagnostic differences be-tween shelf-sea and deep-sea chalks include the presence in shelf-sea deposits of hardgrounds, silicified burrow horizons, abundant benthic macrofauna, an additional *Thalassinoides*-dominated trace fossil assemblage occurring in shallower-water facies than the *Chondrites-Planolites-Zoophycos* assemblage, pelleted glauconitic marls, and moderately common terrigenous sand and silt. At sublysoclinal depths, abyssal carbonate strata become increasingly rich in pelagic clay, and the *Chondrites-Planolites-Zoophycos* assemblage of deep-sea chalks grades into a lower-diversity trace fossil assemblage dominated by smeared and deformed *Planolites,* which characterizes the abyssal pelagic clay facies.

In contrast to pelagic carbonate units, detrital carbonate strata deposited by turbidity currents or grain flows commonly exhibit: (1) primary depositional features (for example, graded bedding, soft-sediment deformation, sole marks and current features); (2) displaced benthic fossils; and (3) abundant horizontal trace fossils (for example, *Helminthoida* and *Paleodictyon*) occurring along bedding planes.

DIAGENESIS

Carbonate Sediments

Pelagic limestones normally undergo significant diagenetic alteration that may profoundly affect porosity, permeability, mineralogy, structural strength, and even grain or crystal size. Yet the diagenetic processes, and the rates at which they operate in pelagic units, are commonly quite different from those in shallow-water sediments. In Jurassic to Holocene pelagic carbonate deposits, the initial, stable, low-magnesium calcite composition and the improbability of early contact of "deeper-water" sediments with meteoric pore fluids minimizes the types of radical early diagenesis prevalent in shallow-marine carbonate settings. However, pelagic carbonate sediments are subject to several other

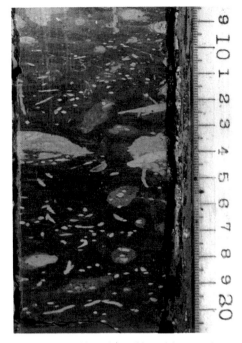

Figure 58—Abundant *Chondrites* and *Planolites* in Miocene silty-clayey chalk from DSDP Site 223 (core 223-12-5, 9 to 21 cm interval). Note that almost all the *Planolites* have been reburrowed by portions of *Chondrites* burrow systems. Large-scale divisions are in centimeters.

Figure 59—Large, evenly branched *Chondrites* burrow in Upper Cretaceous Monte Antola flysch in the Ligurian Apennines near Genoa, Italy. This highly organized burrow system was probably created by a deposit-feeding organism (perhaps an arthropod or worm-like animal) that used the structure as a dwelling as well as for feeding. Keys are about 4 cm long.

Figure 60—Abundant large plowmarks, known as *Scolicia*, covering a bedding plane of lower Tertiary flysch near Zumaya, Spain. Presumably these are locomotion trails made by a gastropod or echinoid crawling just beneath the sediment surface.

types of major alteration.

Sea floor (penecontemporaneous) lithification, or hardground formation, represents the earliest stage of diagenesis for some pelagic carbonates. Hardgrounds (Fig. 86) result from cementation of pelagic sediments by high-Mg calcite (Fig. 87) aragonite and, in some instances, glauconite and calcium phosphate (Figs. 88 and 89). Hardground formation is favored by long contact between sediment and seawater. Thus, areas of slow sediment accumulation, whether due to current removal of finer grained material or reduced primary sediment supply rate, are the most common sites of hardground formation. In many places, paleotopographic highs and restricted straits contain sections with abundant submarine cementation as a consequence of sediment winnowing in such areas, especially at times of lowered sea level (but not subaerial exposure). Such areas commonly show multiple hardgrounds (Fig. 89) with significant amounts of time

(probably tens to hundreds of thousands of years) representing each lithification surface.

Early diagenetic, sea-floor lithification in modern deeper water sediments has been described from a number of settings (Fischer and Garrison, 1967; Milliman, 1974a; Neumann et al, 1977). Early lithification may play a significant role in the preservation of carbonate sediment bodies (lithoherms and drifts) in areas of strong bottom currents.

The mechanisms and fabrics of hardground formation have been summarized by a number of authors, including Kennedy and Garrison (1975) and Bromley (1968, 1975). In many sediments, cementation first affects burrow fillings that may contain slightly coarser and more permeable sediment (Figs. 90 and 91). Eventually these hardened burrows may be exhumed by sea-floor winnowing and may coalesce partially or completely to form a continuous, lithified sur-

face. In other sediments, the process may cease short of hardground formation, yielding a discrete nodular fabric (Figs. 92 and 93).

Hardgrounds are ecologically im-

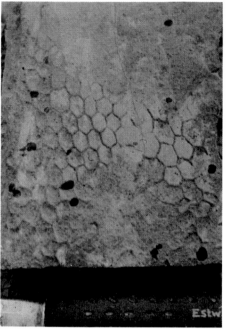

Figure 61—Meandering *Taphrhelmin-thopsis* trails on a bedding plane of lower Tertiary flysch near Zumaya, Spain. Trails of modern echinoids plowing across the deep-sea floor are common and appear to be identical to these ancient trace fossils.

Figure 62A—*Paleodictyon* is a calcareous turbidite of Jurassic age in the central High Atlas Mountains of Morocco. The "chickenwire" geometry of this type of burrow indicates that it was a rather permanent structure, because the builder must retrace at least part of its trail in order to complete construction of the total system. Structures such as *Paleodictyon* may serve as nets or farms by which the burrowers trap and collect benthic micro-organisms for food. Area of photograph is 15 cm across short axis.

Figure 62B—Mesh-like *Paleodictyon* burrow system in surficial calcareous ooze of a box core retrieved from 1,436 m of water in the southwestern Atlantic. Like *Cosmorhaphe, Paleodictyon* is a highly organized system of horizontal tunnels within only a millimeter or more of the sediment surface. The builder of the burrow system is unknown. Area of photograph is about 5 cm wide.

portant because they provide substrates for encrusting and boring organisms (Fig. 88) that would be unable to colonize the soft, even soupy, sediment-water interface of unlithified chalks or oozes. Thus, the specialized faunas present provide a key to hardground identification. Nodular fabric, reworked pebbles, and phosphate (or glauconite) replacement of carbonate grains are further criteria for hardground recognition (Figs. 88 and 89).

Hardground formation can drastically reduce the porosity of pelagic carbonate sediments; initial values of greater than 70% are lowered to less than 10% in many examples. This reduction in porosity has consequences for oil and gas production from reservoir rocks in these units. However, hardgrounds rarely affect a thick sediment section, and sea-floor cementation is not of major importance in the diagenetic history of pelagic limestones as a whole.

Compaction is of primary importance to the diagenesis of pelagic car-

bonates. Mechanical and chemical compaction accounts for the transformation of sea-floor oozes having greater than 70% porosity into fully lithified limestones (Figs. 94 and 95).

The relatively uniform starting compositions and grain sizes of Mesozoic and Cenozoic pelagic limestones lead to generally predictable patterns of burial diagenesis in these units. These patterns have been extensively discussed by Schlanger and Douglas (1974), Packham and van der Lingen (1973), Matter (1974), Neugebauer (1973, 1974), Scholle (1974, 1977a), and others, so they are only briefly summarized here. Mechanical compaction predominates during early stages of burial. Mechanical compaction includes simple dewatering as well as grain reorientation or breakage. Studies of

DSDP and other cores have shown that dewatering starts just below the sediment-water interface and commonly results in porosity reduction from original values of 60 to 80% to less than 50% within the first several hundred meters of burial (Schlanger and Douglas, 1974). As the constituent grains of the sediment are brought closer together, however, a stronger grain-supported fabric is formed. Thus, at porosity levels less than about 40%, mechanical compaction becomes less rapid and less important as a porosity-reducing mechanism.

As the surface area of grain contacts increases, chemical compaction (solution transfer) becomes the major mechanism of porosity reduction. Calcium carbonate dissolves at points of greatest differential stress and reprecipitates at points of minimal stress, generally forming overgrowth cements (Fig. 96). Three factors, in particular, enhance this process, they

Figure 63—Trace fossils distorted by juxtaposed load casts on the sole of a calcareous turbidite bed in lower Tertiary flysch near Zumaya, Spain. Scale in centimeters.

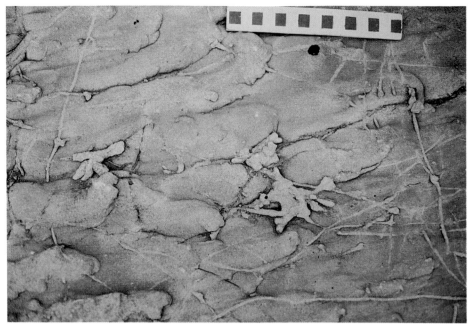

Figure 64—A "sea urchin" (irregular echinoid) creating a characteristic trail that closely resembles the trace fossil *Taphrhelminthopsis*. Photograph from continental rise off the Antarctic Peninsula; water depth, 4,153 m (from Heezen and Hollister, 1971). Area of narrowest part of photograph is about 1.2 m wide.

Figure 65—A trail resembling a tire tread, that was probably produced by a "sea cucumber" (Holothurian) crawling across the sea floor. Photograph from continental margin off southern Chile; water depth, 4,410 m (from Heezen and Hollister, 1971). Area of narrowest part of photograph is about 1.2 m wide.

are fresh (Mg-poor) pore fluids (Fig. 94), the presence of clay minerals, and significant overburden or tectonic stress (Neugebauer, 1973; Scholle, 1977a). Abnormally high pore-fluid pressures and/or hydrocarbon displacement of normal pore fluids can retard chemical compaction (Scholle, 1977a). Under normal circumstances, therefore, sediments that are clay-rich — those that have had fresh-water flushing of original

marine pore fluid, or that have been deeply buried — will be the most chemically compacted.

Variations in sediment composition and other factors on a bed-to-bed scale can lead to major diagenetic variations. For example, the primary depositional cyclicity in most pelagic sediments generally includes bed-to-bed or lamina-to-lamina variations in clay content (Figs. 18 to 23, and 97). The carbonate-poor (clay-rich) zones undergo significant dissolution of their carbonate grains. The dissolved carbonate commonly precipitates in adjacent carbonate-rich beds as overgrowth cement. Thus, a diagenetic "unmixing" accentuates primary compositional variations in the sediment, and leads to thinning of the clayey layers and cementation of the carbonate layers (Arthur, 1979c; Scholle, 1977a; Fig. 97).

The specific nature of the carbonate sediment can further modify these diagenetic patterns. Aragonite and high-Mg calcite are generally more soluble than low-Mg calcite; thus, layers rich in these minerals may lithify rapidly (Hattin, 1971). Nannofossils tend to resist dissolution and overgrowth cementation better than many other groups of organisms (Neugebauer, 1975). Even within a single group of organisms such as the

coccolithophores, for example, there are significant variations in resistance to both dissolution and overgrowth cementation from species to species (Adelseck et al, 1973).

Despite these variables, the diagenetic transformation of most pelagic carbonate sediments can be predicted with considerable confidence. Figure 94 summarizes the typical burial-related porosity (and permeability) changes of chalks under conditions of "normal" marine and Mg-poor fluids. Although porosity values of individual beds may differ significantly, the triangular envelope formed by the two curves and the baseline encompasses virtually all data on porosity versus burial depths for normally pressured European and North American chalks as well as DSDP cores.

Figure 95 illustrates some of the physical changes seen in chalks as they undergo burial diagenesis. Originally isolated and uncemented grains become progressively more welded by very finely crystalline overgrowth cements. The overgrowth cements, in addition to lithifying the grains, also gradually obscure the nannofossil origins of the sediment.

Accompanying these microscopic changes in the rock fabric are larger-scale changes visible in core, hand

Figure 66—An "acorn Worm" (enteropneust hemichordate) wanders across the sea floor, leaving behind a long, continuously coiled and meandering string of excrement. Photograph from the eastern side of the Kermadec Trench; water depth 4,871 m (from Heezen and Hollister, 1971). Area of narrowest part of photograph is about 1.2 m wide.

Figure 67—Furrowed trails produced by tube-dwelling annelid worms (Hyalinoecia) that drag their tubes along behind them as they wander across the sea floor. Photography from the continental slope off New England; water depth, 849 m (from Heezen and Hollister, 1971). Area of narrowest part of photograph is about 1.2 m wide.

Figure 68—A variety of trails, mounds and other surface features created by unknown organisms. Photograph of Canada Abyssal Plain; water depth, 3,790 m (from Heezen and Hollister, 1971). Area of narrowest part of photograph is about 1.2 m wide.

sample, or outcrop. Fissility or lamination may become more distinct with increasing burial as a consequence of rotation, grain breakage, or flattening of grains parallel to bedding (Fig. 6J). Solution seams and stylolites (Figs. 98 to 100) become more abundant with depth, and eventually become ubiquitous in the clayey parts of chalk-marl cycles (Garrison and Kennedy, 1977). Finally, the individual solution seams virtually fuse as the clayey zones approach 100% exportation of their original carbonate material (Fig. 100).

Changes in stable-isotope and trace-element values parallel the physical alteration of the rock during burial. These changes reflect the degree of diagenetic alteration, the site of alteration (near-surface versus burial diagenesis), the "openness" of the system (a general measure of degree of pore-fluid exchange with overlying and underlying rocks), the thermal regime, and other factors. Descriptions of the geochemistry of chalks have been summarized in Milliman (1974b), Scholle (1977a), Scholle and Arthur (1980), and many DSDP Leg Reports or summaries of DSDP studies (Anderson and Schneidermann, 1973; Matter et al,

1975; McKenzie et al, 1978). Chalk diagenesis has also been theoretically modeled (Land, 1980).

The diagenetic processes described thus far all involve destruction of primary porosity. Processes that generate secondary porosity in pelagic limestones are generally of minor importance. Although leaching of aragonite or other unstable minerals may contribute some secondary porosity, fracture development is generally the only significant source of secondary pore space. Fractures (Fig. 101) generally contribute greatly to the permeability of oil and gas fields in pelagic limestones, and, in some instances (for example, the Austin Group chalk reservoirs of the United States Gulf Coast) provide a major part of the effective porosity as well.

Although a number of complexities have been pointed out in the diagenesis of pelagic carbonate sediments, we should emphasize again that the diagenetic history of pelagic units is simple when compared with that of typical shallow-marine limestones. The latter contain grains of variable size and unstable mineralogy, have complex facies changes over short distances, and have a wide range of diagenetic pore-fluid compositions. Thus, the prediction of reservoir properties of pelagic

limestones is far more reliable than for shallow-marine equivalents. At the same time, however, these predictive models are applicable only to Upper Jurassic to Holocene pelagic units in which primary mineralogical composition and grain size are reasonably well known. Older pelagic limestones are not interpreted as easily because our knowledge of their initial properties is not as good.

Siliceous Sediments
The diagenesis of siliceous pelagic sediments is somewhat similar, but differs in many details from that of carbonate oozes. All biogenic siliceous sediments are originally composed of opaline silica (opal-A), and biogenic silica is by far the most important source of silica in pelagic sediments (Calvert, 1974). Yet unlike the calcite produced by most modern calcareous pelagic organisms, the opal-A contributed by diatoms, radiolarians, siliceous sponges, and silicoflagellates is quite unstable. As first shown by Bramlette (1946) and confirmed by studies of DSDP cores, silica undergoes relatively rapid and complete transformation during burial (Lancelot, 1973; Keene, 1975; Riech and von Rad, 1979).

The transformation of opal-A is generally a two-step process (Fig. 102). First, siliceous oozes are altered to porcellanite composed primarily of opal-CT (commonly termed

Figure 69—A variety of trails, mounds and holes created by unknown organisms. Abundant, small, twig-like structures may be macroscopic proto-zoans (zenophyophorids). Photograph of upper continental rise off New York State; water depth, 3,026 m (from Heezen and Hollister, 1971). Area of narrowest part of photograph is about 1.2 m wide.

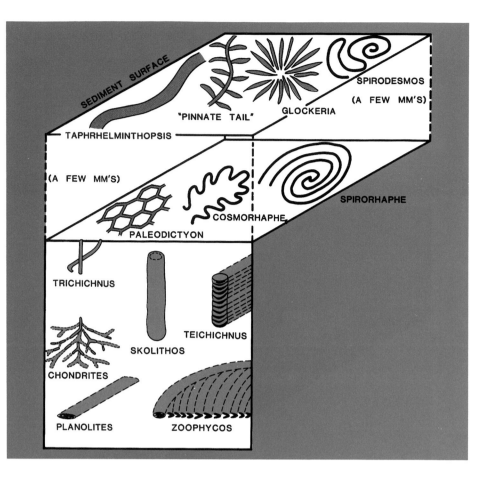

Figure 70—Diagrammatic representation of deep-sea ichnofacies in pelagic carbonates, depicting typical traces made on the sediment surface, just below the surface, and several centimeters below the surface.

cristobalite). Then, opal-CT-bearing porcellanite is converted to a chert containing mainly microquartz and chalcedony. Although temperature and pressure are the major controls on the rates of both of these solution-reprecipitation processes, other factors can retard or accelerate the reactions. Presence of clay minerals, for example, tends to retard the transformation, whereas carbonate minerals generally accelerate them (see summary in Kastner, 1979a). Although direct conversion from opal-A to quartz is possible, it appears to be relatively uncommon in marine sediments (Riech and von Rad, 1979).

The rates of conversion of the various silica polymorphs are difficult to quantify because of the conflicting effects of time, temperature, pressure, organic coatings, solution chemistry, sediment surface area, and presence of associated minerals (Kastner, 1979a). In general, however, the mineralogic changes noted above are generally accompanied by

recognizable fabric changes (Fig. 101). Opal-A found in sediments is mainly in the form of primary biogenic tests. During early diagenesis, these opaline frustules first begin to dissolve and appear to be frosted (Hurd et al, 1979). As dissolution progresses, opal-CT is precipitated both as a massive pore-filling or replacement mineral, and as clusters of bladed crystals (lepispheres) that are most commonly pore-filling (Fig. 102). Quartz is found as a replacement of opal-CT lepispheres, as pore-filling micro-quartz, and as vein-filling megaquartz (Fig. 102).

As with pelagic carbonate sediments, this progressive diagenesis is accompanied by a depth-related porosity loss. Unlike carbonate sediments, however, the porosity depth curve is not a smooth one but rather shows discrete steps at the points where the two stages of mineralogic transformation occur (Isaacs, 1981). Progressive loss of

matrix porosity is accompanied by increasing brittleness. Thus, fracture porosity is important in oil production from siliceous pelagic units such as the Monterey Formation of California.

Silica diagenesis also occurs within pelagic carbonate units whenever there is an admixture of siliceous tests in the original sediment. Most commonly, opaline silica is dissolved and reprecipitated as quartz chert in a variety of forms. Most cherts in European chalks, for example, occur as selective replacement of calcite in burrow fillings (Fig. 103), along bedding planes (Fig. 104), in limestone matrix (Fig. 105), and in the proximity of large siliceous sponges. In chalk sections rich in clay minerals, however, opal-CT lepispheres may predominate (Scholle, 1974).

Other Diagenetic Processes
Although largely beyond the scope of this paper, some mention should be made of several other diagenetic

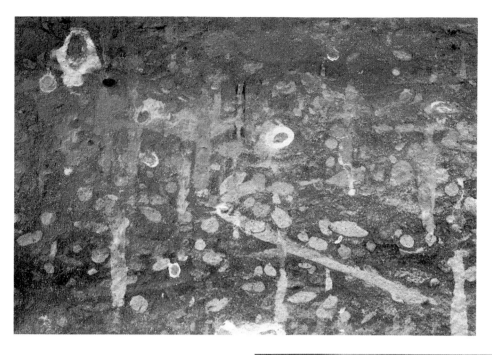

Figure 71—Vertical face (25 cm thick) of a box core from a water depth of 3,945 m in the equatorial Pacific. The typical pelagic carbonate ichnofacies is exemplified by abundant vertical burrows, called *Skolithos,* as well as by many low-angle, subhorizontal burrows, termed *Planolites.* White halos around several burrows are apparently caused by leaching of ferrous iron from the carbonate sediment, leaving unstained calcite to highlight the burrow walls.

Figure 72—X-ray radiograph (positive) of a 25-cm vertical slice from a box core of foraminiferal-coccolith ooze (about 76% $CaCO_3$ collected in the equatorial Pacific at a water depth of 3,945 m. Note grid-like pattern of prominent vertical burrows (*Skolithos*) superimposed on low-angle, subhorizontal burrows (*Planolites*) that characterize the pelagic carbonate ichnofacies.

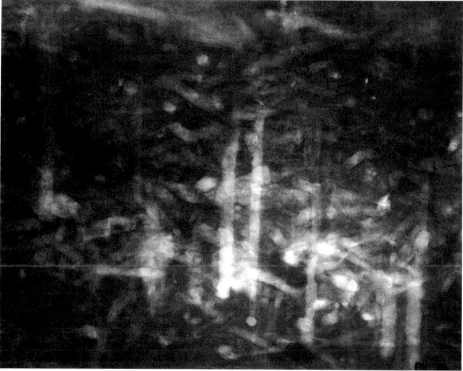

reactions that are common in pelagic sediments. Some are important because they are instrumental in the formation of economic mineral deposits, others because they are widespread, while still others are discussed only for the sake of completeness.

Pyrite — Generally of early diagenetic origin, pyrite forms as nodules of radiating crystals, as disseminated small blebs, or as framboidal fillings of foraminifers and other tests (Fig. 61). Pyrite (and occasionally marcasite or sphalerite) is nearly always associated with organic carbon-rich areas in the sediment (Figs. 6K, 106), most likely as a product of bacterial sulfate reduction (see review in Goldhaber and Kaplan, 1974). In organic carbon-rich pelagic sediments, such as the Niobrara Formation of the Western Interior, pyrite commonly constitutes 1% or more (by weight) of the total sediment.

Dolomite — An uncommon early-diagenetic replacement mineral of both shelf and deep-sea pelagic carbonates. It has been reported from a few European chalk localities (Fig. 82), mainly from areas near margins of massifs. Early fresh-water input and/or high primary Mg-calcite concentrations may have occurred in such areas, particularly in association

with overlying hardground surfaces (see summary in Hancock, 1975b). Dolomite in deep-sea carbonate sediment is related to high Mg^{2+} supply in pore waters derived from submarine alteration of basaltic basement rocks. Dolomite is also present in organic-carbon rich pelagic and hemipelagic sediments. Although the

exact mechanism of formation is not known, the dolomite rhombs precipitate after sulfate reduction in association with high pore water alkalinities (Baker and Kastner, 1980). Siderite (Fe^{+2} carbonate) may form in similar diagenetic environments following sulfate reduction in the presence of abundant

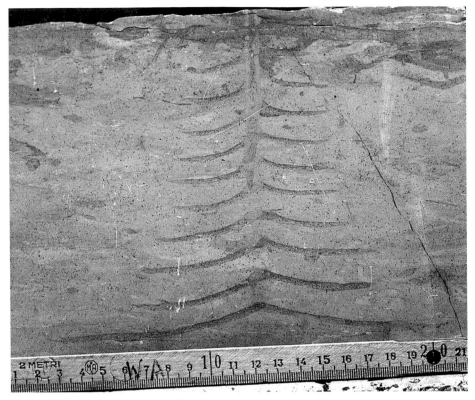

Figure 73—Vertical, cross-sectional view of a *Zoophycos* specimen that was split down its axis. Ten whorls of the helical spreite are visible. Sediment is pelagic limestone in the Scaglia Rossa (Eocene) from Umbria, central Italy. Large scale divisions in centimeters.

Figure 74—*Zoophycos* in silty-clayey-chalk of Miocene age from DSDP Site 223 (core 223-12-3, 118 to 132 cm interval) in the Arabian Sea. Note that *Zoophycos* spreiten (crescentic cross section) are pelleted, and also that the marginal tunnel of the burrow system is preserved at the edge of several spreite layers. One large scale division is 1 cm.

dissolved Fe^{+2}.

Phosphate (carbonate fluorapatite) — Generally a primary or early diagenetic (penecontemporaneous) replacement mineral in local areas, particularly in shelf-sea pelagic units and in current-scoured deeper water areas such as the Blake Plateau. Both nodular and granular phosphates are found in European Cretaceous chalks (Hancock, 1975b; Jarvis, 1980). Diagenetic phosphatization commonly takes place in association with submarine hardgrounds (Figs. 88 and 89), and in pebble beds at the base of chalk sequences. The phosphatization typically takes the form of brownish carbonate-apatite replacement of calcium carbonate in burrows or internal fillings of shells. Data on the occurrence, mineralogy, and economics of phosphorites have been summarized recently in Bentor (1980) and Manheim and Gulbrand-

sen (1979). Economic phosphate deposits in ancient sediments are commonly associated with organic-carbon-rich, siliceous pelagic facies deposited on shelves that were flooded during eustatic sea level rise (Arthur and Jenkyns, 1981).

Glauconite — Occurs as coatings on hardground surfaces (Figs. 86 and 89), as disseminated grains, or as fillings of foraminiferal tests in some pelagic sediments. Glauconite is a green, mica-like illitic mineral (K, Fe^{2+}/Fe^{3+}). It probably forms in diagenetic microenvironments characterized by low rates of sedimentation and generally low oxygen contents, in pore waters, or at the sediment-water interface (see reviews by McRae, 1972; Velde and Odin, 1975).

Barite — Marine barite generally occurs as 1- to 2-μm size euhedral crystals and can compose as much as 2% of pelagic sediment. The barium is probably derived from dissolution of biogenic detritus; at least some of the barite flux to sediments may come from precipitation in the seawater column (see Church, 1979, for review). Larger crystals and replace-

ment of gypsum or calcite may occur in volcanogenic sediment, in organic-rich pelagic sediments (Fig. 107; Dean and Schreiber, 1978), or in organic-rich pelagic sediments overlying evaporites in which pore water sulfate concentrations are elevated.

Zeolites — Common in pelagic sediments, particularly those rich in siliceous biogenic or volcanic material. Zeolites occur as isolated, stubby to elongate, euhedral crystals as much as 50 to 100 μm in size and form as pore-filling or test-filling cements in pelagic sediments (Figs. 108 and 109). Phillipsite and clinoptilolite are the most common forms in deep-sea sediments. Analcite, a sodium-aluminum-silicate, also occurs but is less common than phillipsite and clinoptilolite. Zeolites form authigenically in silica-enriched pore waters, usually as the result of alteration or solution of volcanic glass and/or biogenic opal. Clinoptilolite is generally the most common zeolite in siliceous biogenic sediment, whereas

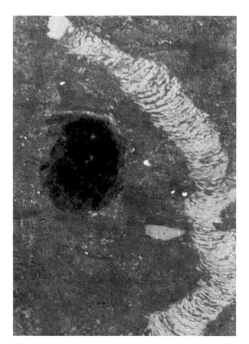

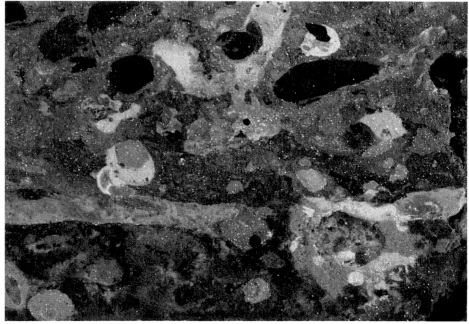

Figure 75—Large *Teichichnus* in Miocene silty-clayey chalk from DSDP Site 223 (core 223-14-3, 85 to 95 cm interval). Note that vertical spreite is pelleted in much the same way as are the *Zoophycos* specimens in Figure 74. Pellets may be fecal in origin. Area of photograph is about 6 cm wide.

Figure 76—Subhorizontal *Planolites* burrows of various colors in a box core of clay-rich calcareous ooze from water 4,850 m deep in the equatorial Pacific. Statistics of cross-cutting relationships of different colored burrows in box cores indicate that, as a general rule, the younger the burrow, the darker its hue. Thus, burrow color appears to fade slightly with age in the interval represented by a box core (the top 50 cm of sediment). Sediment shown is about 25 cm thick.

phillipsite and analcite are commonly related to alteration of volcanic material (see Stonecipher, 1978; Kastner, 1979b, for review).

ECONOMIC CONSIDERATIONS

Potential economic mineral deposits in pelagic sediments include ferromanganese nodules; copper, nickel, and other metals as hydrothermal deposits; phosphates; and oil and gas (Earney, 1980). In addition, pelagic rocks themselves have economic applications: chalk is used in marking materials, paper and paint additives, and mining; clays have applications in ceramics, paper production, and other fields; siliceous rocks are used as abrasives and filters; and all these rock types are used as construction materials (aggregate, cement, building stone).

The hydrocarbon reservoir potential of pelagic rocks, in particular, has been well defined in the past decade as exploration has expanded into deeper water settings and into "unconventional" and low-permeability reservoir rocks. Although pelagic limestone reservoirs are known from the Middle East and Libya, the most significant exploration successes in recent years have come from the North Sea, with subordinate discoveries in the United States Gulf Coast and Western Interior, and offshore eastern Canada. This exploration activity has been summarized by Scholle (1977b).

Production from chalk reservoirs falls into three basic groups: (1) reservoirs that have never been deeply buried (less than 1,000 m) and in which primary reservoir porosity and permeability are retained as a consequence of the absence of overburden-related compaction; (2) reservoirs that were buried to moderate or great depths (1,000 to 5,000 m) and that retain little or no primary matrix porosity or permeability but that may have exten-

sive secondary fractures; and (3) reservoirs that have been deeply buried but have retained high primary porosity as a consequence of abnormally high pore-fluid pressures (geopressuring or overpressuring) and/or early introduction of oil.

All three types of reservoirs have been exploited in the past decade. The Niobrara Formation, for example, is a shallowly buried, low permeability, natural gas reservoir in western Kansas, eastern Colorado, Nebraska and other areas of the United States Western Interior (Lockridge and Scholle, 1978; Smagala, 1981). The gas is of biogenic origin (Rice, 1980) and is generated from the organic-carbon-rich marls of the Smoky Hill Shale Member of the Niobrara Formation. Virtually all production is from units buried less than 1,000 m; such strata have generally retained 25 to 40% primary porosity and 0.5 to 2.0 md permeability because of shallow burial (Lockridge and Scholle, 1978).

Oil and gas production from Gulf Coast chalks generally falls into the second reservoir type. Most current exploration is in areas where the Austin Group has been buried to depths of from 2 to 3 km (Scholle, 1977b). Matrix porosity averages less than 10%, and matrix permeability is less than 0.5 md (Doyle, 1955). These

Figure 77—Smeared and deformed specimens of *Planolites* and *Chondrites* in calcareous clay of Miocene age from DSDP Site 445 (core 445-32-1, 48 to 53 cm interval) in the Philippine Sea. The characteristic ichnofacies for pelagic clay-rich sediment is typified by absence of vertical burrows and deformation of subhorizontal burrows. One large-scale division is 1 cm.

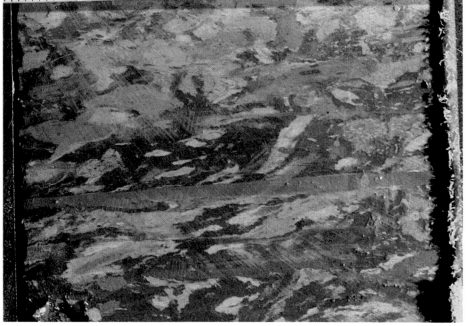

Figure 78—X-ray radiograph (positive) of 25-cm thick vertical slice from a box core of calcareous clay (about 35% CaCO$_3$) retrieved in the equatorial Pacific from a water depth of 4,850 m. Although the sediment is intensely bioturbated with abundant *Planolites,* well defined specimens of *Skolithos* are absent, as would be expected in the pelagic clay ichnofacies.

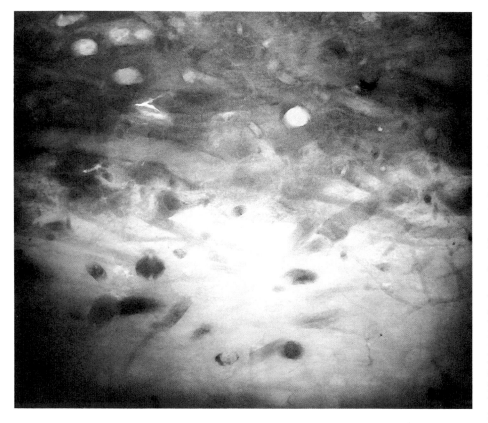

low-porosity values reflect the deep burial of the chalk with "normal" pore-fluid pressures (see Fig. 94). Fracturing is crucial for economic production of oil or gas in Gulf Coast chalks, and current seismic exploration utilizes sophisticated technology to find areas of maximum deformation and fracture development (Sumpter, 1981). Wells in Gulf Coast chalks, although economic successes in most cases, generally show a rapid decline of production from initial values of 200 to 400 b/d of oil (Stewart-Gordon, 1976; Sumpter, 1981), even where artificial stimulation techniques are used to enhance fracturing. Nevertheless, the widespread areal extent of Gulf Coast chalks and the high percentage of successful wells indicate that significant oil and gas will eventually be developed from Gulf Coast chalks.

The most prolific hydrocarbon production from chalk reservoirs is in the North Sea. Indeed, "the total cumulative hydrocarbon production from all North American chalk reservoirs equals less than 10% of the estimated recoverable reserves from North Sea chalks" (Scholle, 1977b). Production rates of 10,000 b/d of oil per well are typical in the Ekofisk area (Byrd, 1975), and total recoverable oil from chalk fields in the North Sea has been estimated at more than 6 billion barrels with an additional 8.1 Tcf gas (Tiratsoo, 1976; Norwegian Petroleum Directorate, 1980).

The oil and gas reserves from North Sea chalks are entirely in Norwegian and Danish waters, and are reservoired primarily in the eight fields shown in Figure 110. These fields all lie within the Central

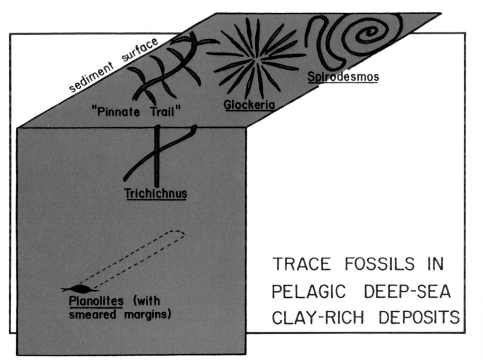

TRACE FOSSILS IN PELAGIC DEEP-SEA CLAY-RICH DEPOSITS

Figure 79—Diagrammatic representation of deep-sea ichnofacies in pelagic clay-rich deposits, depicting typical traces made on the sediment surface and several centimeters below the surface.

graben, a major structure that transects the North Sea basin, and that was active from at least Triassic through Tertiary time (Ziegler, 1978). All production is from uppermost Cretaceous (Maestrichtian) and lowermost Paleocene (Danian) chalks in areas where the Tertiary cover ranges from 2 to 3.5 km in thickness (Scholle 1977a). The Upper Cretaceous and Paleocene chalk section reaches a thickness of as much as 1,500 m within the Central graben, whereas thicknesses of 200 to 400 m are typical on the uplifted margins of the graben (Hancock and Scholle, 1975). The extremely high rates of carbonate sedimentation within the graben result from significant redeposition of chalks from adjacent uplifted areas. Slumps, debris flows, turbidites, and other bottom-traction deposits have been described from the Central graben chalks (Perch-Nielsen et al, 1979; Watts et al, 1980).

All fields producing hydrocarbons from North Sea chalks are related to structures formed by halokinetic movements of Permian salt that were, in part, synchronous with chalk sedimentation (Watts et al, 1980). These structures can be identified readily in seismic section (Fig. 111;

also see seismic section interpretations in Van den Bark and Thomas, 1981), and are equally evident in structure contour maps of the fields (Fig. 112). Salt movement created structural closure and fractures early in the history of chalk burial in the North Sea area, a fact that played a key role in determining the reservoir properties of these units.

Porosities in producing intervals in North Sea chalks are variable, but average between 25 and 35% (Childs and Reed, 1975; Rickards, 1974; Byrd, 1975). Values as high as 42% have been reported from some intervals in the Danian, and porosities as low as from 2 to 5% are common in the uppermost part of the Maestrichtian (Owen, 1972; Byrd, 1975; Childs and Reed, 1975). A typical lithology-porosity profile through a chalk field is shown in Figure 113. Important to note is that porosities average 30% or more over intervals of at least 100 m. Furthermore, as Harper and Shaw (1974, p. 7-8) stress, "most of this porosity is *effective porosity*. It is not the high but ineffective porosity often encountered in other fine-grained rocks such as shales."

Permeabilities measured on cores from North Sea chalks average less

than 3 md and range from 0 to 7 md (see Fig. 113). Large-scale permeability, measured by oil-flow rates, average 12 md (World Oil, 1971). This presumably demonstrates the important influence of microfracturing on reservoir permeability.

Such high porosity and permeability values at burial depths of 2 to 3.5 km are anomalous when compared with porosity-depth relations determined for other chalks (Fig. 94). Scanning electron microscope (SEM) examination of North Sea chalks (Fig. 114) shows that although some diagenetic alteration has taken place, primary porosity is preserved to a remarkable degree. The 30% porosity preserved in these samples is typical of chalks at burial depths of 1,000 to 1,200 m (Fig. 94), and the fabric visible in SEM is also typical of chalks buried to approximately 1 km rather than the 3 km burial depth from which the sample in Figure 114 was taken. Thus, North Sea chalks can be considered to have undergone abnormally slight diagenesis.

Many ideas have been proposed to explain the "abnormal" diagenesis of North Sea chalks (Harper and Shaw, 1974; Mapstone, 1975; Scholle, 1977a; Watts et al, 1980; Feazel et al,

Figure 80—Diagrammatic representation of trace fossil assemblage in shelf-sea chalk.

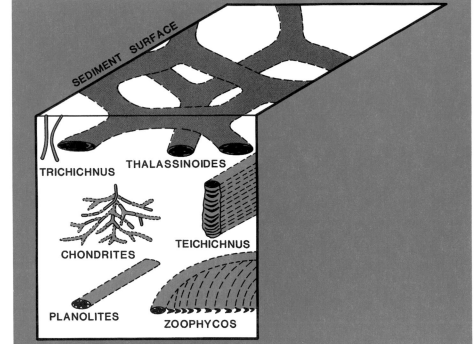

Figure 81—Bedding-plane exposure of partially exhumed, selectively silicified *Thalassinoides* burrow networks in Upper Cretaceous chalk near Etretat, France. *Thalassinoides,* which are the characteristic dwelling burrows of thalassinoidean decapod crustaceans such as the modern mud shrimp *Upogebia,* are common in shelf-sea chalks but are essentially unknown in deep-sea chalks.

1979; Van den Bark and Thomas, 1981). Primary sedimentation patterns appear to be important as an influence on porosity alteration. Resedimented beds, both in the Maestrichtian and the Danian, are generally high-porosity zones. Conversely, slowly deposited, burrowed, "primary" pelagic chalks are generally low-porosity intervals (Watts et al, 1980; W. J. Kennedy, personal com-

mun., 1979). Presumably, slowly deposited strata were subjected to more sea-floor cementation than were rapidly deposited debris-flow intervals. The influence of sea-floor cementation of pelagic strata is particularly apparent at the Cretaceus-Tertiary boundary in most Ekofisk area sections. Although no distinct hardground surfaces are present in that interval, sharply reduced porosity

values indicate extensive cementation of a 15 m thick section sandwiched between more porous Maestrichtian and Danian chalks (Van den Bark and Thomas, 1981). Differences in faunal and floral content between pelagic and redeposited units may also account for variations in post-depositional alteration.

The major factors influencing porosity values of North Sea chalks are pore-fluid pressures and oil-saturation values. There is a direct correlation between anomalously high porosity values in chalks and regions of overpressuring in the North Sea (Harper and Shaw, 1974; Scholle, 1977a). In the Ekofisk-Torfelt area, pore-fluid pressures of 7,100 psi have been reported from depths of 3,050 m (Harper and Shaw, 1974). Normal pore-fluid pressures at that depth would be about 4,300 psi, with a lithostatic load of nearly 9,000 psi. In the Ekofisk-Torfelt region, therefore, a major part of the lithostatic load is supported by the pore fluid, and net lithostatic pressure is reduced to about 1,900 psi. Thus, the rock is "sensing" a burial depth of only about 1,000 m in terms of the differential stresses leading to mechanical and chemical compaction.

Figure 82—Outcrop of dolomitized and chertified Turonian shelf-sea chalk near Etretat, France. Most of the chalk is intensely bioturbated and contains abundant *Thalassinoides* burrows. Note slumped zones occur at the top of the photograph.

Figure 83—Diagrammatic representation of trace fossil assemblage in chalk hardgrounds.

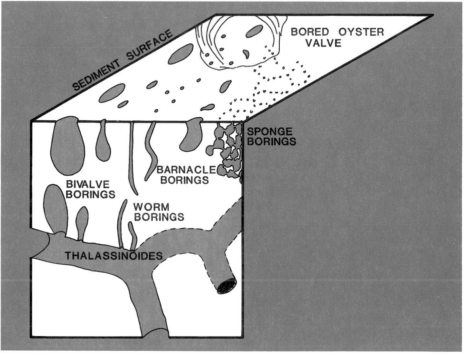

this is approximately the burial depth that would be estimated from the average porosity using the relations shown in Figure 94.

A typical pressure curve determined from drilling data and sonic-log shale travel time from an Ekofisk area well is shown in Figure 115. Clearly, the entire chalk section is strongly overpressured. This overpressuring is restricted to the Central graben area and accounts for anomalously high porosity, even in Central graben chalks that are not oil- or gas-bearing. The overpressures are presumably a consequence of a combination of rapid sedimentation throughout the Maestrichtian to Holocene interval, a low-permeability (chalk and shale) Cretaceous to Holocene section, high geothermal gradients, and lateral confinement by graben margins.

At least one other factor may be important in retarding the diagenesis of North Sea chalks. Oil-saturation values of 70 to 100% were reported by Harper and Shaw (1974) for Maestrichtian chalk at Torfelt field. Early formation of halokinetic structures and relatively early generation of overpressures implies probable early introduction of oil into North Sea chalk reservoirs. Deeply buried (Upper Jurassic Kimmeridge clay) source beds (Van den Bark and Thomas,

1981) and present-day geothermal gradients of 28 to 34°C/km (Harper, 1971) allow early oil generation in this region. Early oil saturation of North Sea chalks could have retarded diagenesis because dissolution and reprecipitation of calcium carbonate require at least films of water on grain surfaces. Indeed, Van den Bark and Thomas (1981) provided evidence that chalk porosity in Ekofisk wells is

directly related to the presence of oil. In areas lateral to the oil accumulation or below the oil-water contact, chalk porosity is sharply reduced in comparison with porosity within the oil reservoir.

Several other factors also contribute to the economic success of North Sea oil fields producing from chalks, although they may not have influenced diagenesis. Fracturing,

Figure 84—*Inoceramus* valves with extensive clinoid sponge borings. Upper part of the Austin Group (Coniacian-Santonian) near Waxahatchee, Texas. Coin is 21 mm in diameter.

Figure 85A—Ammonite "steinkerns" (casts of external molds) on a bedding plane of Ammonitico Rosso (Jurassic) in central Italy. These fossils represent syndepositional and post-depositional dissolution of the original aragonitic shell material, leaving an external mold of the animal which was later cast with fine-grained carbonate (photograph by J. G. Ogg). Lens cover is about 4 mm in diameter.

and gas reservoirs and are preferred exploration targets. Similar chalk or chalk-turbidite reservoirs may be discovered in future exploration in offshore areas, particularly in the deeper basins such as the Gulf Coast.

ACKNOWLEDGMENTS

We thank D. L. Gautier and W. E. Dean who provided thoughtful, constructive reviews of our manuscript. The assistance of C. Wenkam, J. Murphy, and T. Kostick in preparation of illustrations and of H. Colburn and M. Cunningham in typing the manuscript also are greatly appreciated.

SELECTED REFERENCES

Adelseck, C. G., Jr., 1977, Dissolution of deep-sea carbonate; preliminary calibration of preservational and morphologic aspects: Deep-Sea Research, v. 25, p. 1167-1185.

———G. W. Geehan, and P. H. Roth, 1973, Experimental evidence for the selective dissolution and overgrowth of calcareous nannofossils during diagenesis: Geol. Soc. America Bull., v. 84, p. 2755-2762.

——— and T. F. Anderson, 1978, The late Pleistocene record of productivity fluctuations in the Eastern Equatorial Pacific Ocean: Geology, v. 6, p. 388-391.

——— and W. H. Berger, 1975, On the

mentioned earlier, is important in increasing reservoir permeabilities. The low viscosity of the oil (36% gravity), high gas-oil ratio (GOR's of 1,000 to 2,500 were reported by Harper and Shaw, 1974), strong solution-gas drive and thick net pay sections (as much as 180 m), and high reservoir pressure all help to improve production.

Ekofisk-type fields, then, represent stratigraphic-structural traps. The high storage capacity of the chalk is a consequence of very high porosities retained by early oil input and over-pressuring. High production rates are related to the factors mentioned in the previous paragraph.

In summary, reservoirs that produce hydrocarbons from shallowly buried, high-porosity chalks or from deeply buried, low-porosity fractured chalks have relatively low reserves per well and low rates of oil or gas production. Production (or potential production) may extend over vast areas, however. Deeply buried, over-pressured, high-porosity chalks, on the other hand, can be prolific oil

Figure 85B—Example of carving on outcrop of bedding plane of basinal limestone facies of the Lower Tamaulipas Formation in the Sierra Madre Oriental, northern Mexico.

dissolution of planktonic and associated microfossils during settling and on the seafloor: Cushman Found. Foram. Research, Spec. Pub. 13, p. 70-81.

Anderson, T. F., and N. Schneidermann, 1973, Stable isotope relationships in pelagic limestones from the Central Caribbean; Leg 15, Deep Sea Drilling Project, *in* N. T. Edgar, J. B. Saunders, et al, eds., Initial reports of the Deep Sea Drilling Project, v. XV: Washington, D.C., U.S. Govt. Printing Office, p. 795-803.

Arrhenius, G., 1952, Sediment cores from the East Pacific: Rept. Swedish Deep Sea Exped. (1947-1948), Parts 1-4, v. 5, p. 1-288.

Arthur, M. A., 1979a, Paleoceanographic events; recognition, resolution and reconsiderations: Rev. of Geophysics and Space Physics, v. 17, p. 1474-1494.

_____, 1979b, North Atlantic Cretaceous black shales, the record at Site 398 and a brief comparison with other occurrences, *in* W. B. F. Ryan, J. C. Sibuet, et al, eds., Initial reports of the Deep Sea Drilling Project, v. XLVII, part II: Washington, D.C., U.S. Govt. Printing Office, p. 719-751.

_____, 1979c, Sedimentologic and geochemical studies of Cretaceous and Paleogene pelagic sedimentary rocks; the Gubbio sequence, pt. I: Princeton Univ., Ph.D. dissert., 173 p.

_____ and A. G. Fischer, 1977, Upper Cretaceous-Paleocene magnetic stratigraphy at Gubbio, Italy; lithostratigraphy and sedimentology: Geol. Soc. America Bull., v. 88, p. 367-371.

_____ and H. C. Jenkyns, in press, Phosphorites and paleoceanography, *in*

Figure 86—Hardground surface and cross section, exposed in fallen block on beach near Etretat, Normandy, France. Greenish cast on surface represents glauconitization that, in conjunction with phosphatization and carbonate cementation, is responsible for synsedimentary lithification of this horizon. Lighter colored chalk fills borings.

Figure 87—Peloidal and micritic magnesium-calcite cement and planktic foraminifers in modern submarine-cemented hardground surface from lithoherms (photograph by A. C. Neumann; from Neumann et al, 1977). Scale bar is about 75 μm long.

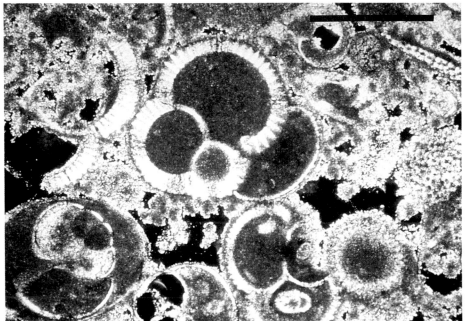

W. H. Berger, ed., Ocean chemical cycles: Oceanologica Acta, Spec. Issue.

_____ and S. O. Schlanger, 1979, Cretaceous "oceanic anoxic events" as causal factors in development of reef-reservoired giant oil fields: AAPG Bull., v. 63, p. 870-885.

Baker, P. A., and M. Kastner, 1980, The origin of dolomite in marine sediments (abs.): Geol. Soc. America Abs. with Programs, v. 12, no. 7, p. 381-382.

Bathurst, R. G. C., 1971, Carbonate sediments and their diagenesis: New York, Elsevier Sci. Pub., 620 p.

Benson, W. E., et al, 1978, Initial reports of the Deep Sea Drilling Project, v.

XLIV: Washington, D.C., U.S. Govt. Printing Office, 1005 p.

Bentor, Y. K., ed., 1980, Marine phosphorites: SEPM Spec. Pub. No. 29, 249 p.

Berger, W. H., 1970a, Biogenous deep-sea sediments; fractionation by deep-sea circulation: Geol. Soc. America Bull., v. 81, p. 1385-1402.

_____, 1970b, Planktonic foraminifera; selective solution and the lysocline: Marine Geology, v. 8, p. 111-138.

_____, 1972, Deep sea carbonates, dissolution facies and age-depth constancy: Nature, v. 126, no. 5347, p. 392-395.

_____, 1973, Deep-sea carbonates; Pleistocene dissolution cycles: Jour. Foram. Research, v. 3, p. 187-195.

_____, 1974, Deep-sea sedimentation, in C. A. Burk and C. L. Drake, eds., The geology of continental margins: New York, Springer-Verlag Pub., p. 213-241.

_____, 1976, Biogenous deep-sea sediments; production, preservation and interpretation, in J. P. Riley and R. Chester, eds., Treatise on chemical oceanography, v. 5: London, Academic Press, p. 265-388.

_____, 1979, Impact of deep sea drilling on paleoceanography, in M. Talwani, et al, eds., Deep drilling results in the Atlantic Ocean; continental margins and paleoenvironment: Washington, D.C., Amer. Geophys. Union, Maurice Ewing Ser. No. 3, p. 297-314.

_____ A. A. Ekdale, and P. F. Bryant, 1979, Selective preservation of burrows in deep-sea carbonates: Marine Geology, v. 32, p. 205-230.

_____ and T. C. Johnson, 1976, Deep-sea carbonates; dissolution and mass wasting on Ontong-Java Plateau: Science, v. 192, p. 785-787.

_____ J. S. Killingley, and E. Vincent, 1978, Stable isotopes in deep-sea carbonates; box core ERDC-92, west equatorial Pacific: Oceanologica Acta, v. 1, p. 203-216.

_____ and L. A. Mayer, 1978, Deep-sea carbonates; acoustic reflectors and lysocline fluctuations: Geology, v. 6, p. 11-15.

_____ and P. H. Roth, 1975, Oceanic micropaleontology; progress and prospect: Rev. of Geophysics and Space Physics, v. 13, no. 3, p. 561-585.

_____ and E. L. Winterer, 1974, Plate

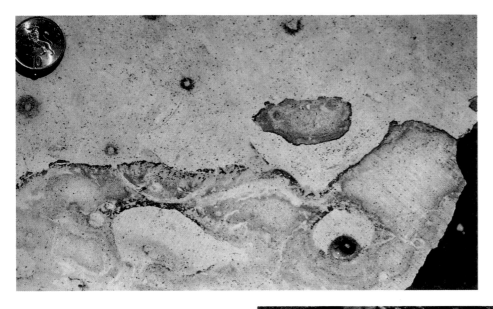

Figure 88—Close-up of hardground cross section. Note glauconitized and phosphatized pelagic carbonate sediment, multiple borings, and reworked hardground clasts, all of which provide evidence of penecontemporaneous formation. A specialized encrusting epifauna is also present on this surface although not readily visible in photograph. Upper Cretaceous Chalk Rock, southern England. Coin is 2.8 cm in diameter.

Figure 89—Polished slab of multiple hardgrounds. Note greenish color marking glauconitized areas and brownish tint denoting zones of phosphatization. Borings and reworked pebbles indicate that lithification was nearly contemporaneous with deposition. Multiple hardground surfaces are common in areas of high current velocities and low net sedimentation rate, especially on topographic high points. Upper Cretaceous Chalk Rock, southern England. Coin is about 2.8 cm in diameter.

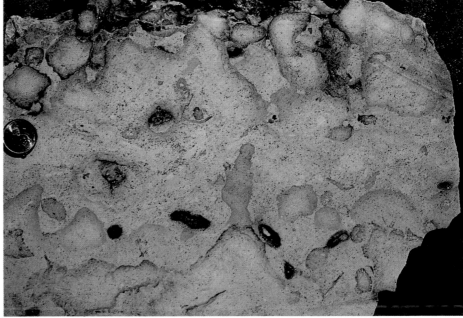

stratigraphy and the fluctuating carbonate line, *in* K. J. Hsu and H. Jenkyns, eds., Pelagic sediments; on land and under the sea: Internat. Assoc. Sedimentols. Spec. Pub. No. 1, p. 11-48.

Berggren, W. A., and C. D. Hollister, 1977, Plate tectonics and paleo-circulation—commotion in the ocean: Tectonophysics, v. 38, p. 11-48.

Birkelund, T., and R. G. Bromley, 1980, The Upper Cretaceous and Danian of NW Europe: Paris, 26th Internat. Geol. Cong., Guidebook No. 20, 31 p.

Black, M., 1980, On chalk, globigerina ooze and aragonite mud, *in* C. V. Jeans and P. F. Rawson, eds., Andros Island, chalk and oceanic oozes: Yorkshire Geol. Soc. Occasional Pub. No. 5, p. 54-85.

Bosellini, A., and E. L. Winterer, 1975, Pelagic limestone and radiolarites of the Tethyan Mesozoic; a genetic model: Geology, v. 3, p. 279-282.

Bostrom, K., 1973, The origin and fate of ferro-manganoan ridge sediments: Stockh. Contr. Geol., v. 27, p. 149-243.

Bottjer, D. J., 1978, Paleoecology, ichnology, and depositional environments of Upper Cretaceous chalks (Annona Formation; chalk member of Saratoga Formation), southwestern Arkansas: Indiana Univ., Ph.D. dissert., p. 424.

_____, 1980, The Arcola Limestone and the absence of extensive hardgrounds in North American Upper Cretaceous chalks (abs.): Geol. Soc. America Abs. with Programs, v. 12, no. 7, p. 390.

Bramlette, M. N., 1946, The Monterey Formation of California and the origin of its siliceous rocks: U.S. Geol. Survey Prof. Paper 212, 57 p.

Broecker, W. A., 1974, Chemical oceanography: New York, Harcourt, Brace, Jovanovich, 214 p.

_____ and T. Takahashi, 1978, The relationship between lysocline depth and in situ carbonate ion concentration: Deep-Sea Research, v. 25, p. 65-95.

Bromley, R. G., 1967, Some observations on burrows of thalassinidean Crustacea in chalk hardgrounds: Geol. Soc. London Quart. Jour., v. 123, p. 157-182.

_____, 1968, Burrows and borings in hardgrounds: Geol. Soc. Denmark Bull., v. 18, p. 247-250.

_____, 1975, Trace fossils at omission surfaces, *in* R. W. Frey, ed., The study

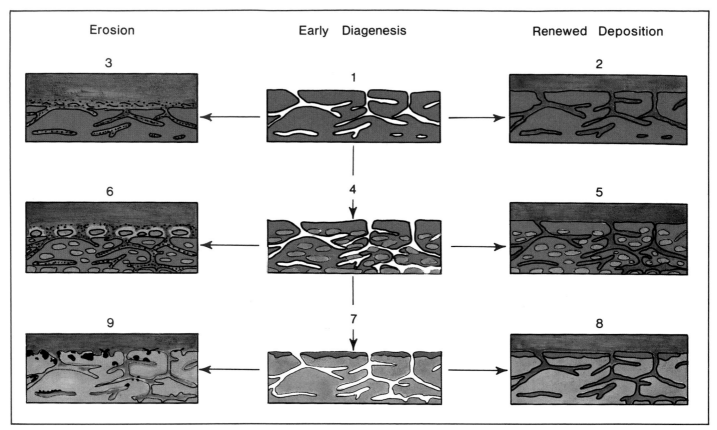

Erosion	Early Diagenesis	Renewed Deposition
3	1	2
6	4	5
9	7	8

Figure 90—Schematic diagram showing various stages and paths of shallow-water chalk diagenesis, hardground formation, and resulting morphologies (after Kennedy and Garrison, 1975): (**1**) open burrow (*Thalassinoides*) framework in initially deposited chalk; (**2**) morphology if sedimentation of later chalk follows immediately, filling burrow systems and allowing no formation of nodules or hardgrounds; (**3**) expected morphology if current winnowing occurs and renewed chalk deposition follows the hiatus — burrows are filled with calcarenite and a thin calcarenite layer forms between chalk layers; (**4**) if prolonged exposure occurs, calcite-cemented nodules form in chalk; (**5**) nodular chalk is formed and buried if renewed chalk deposition occurs soon after exposure; (**6**) if erosion and current winnowing follow nodule formation, nodules may be exhumed, phosphatized and glauconitized, and calcarenite may fill burrows; (**7**) nodules may coalesce to form fully cemented chalk layers; (**8**) indurated chalk layer is buried without hardground formation; (**9**) indurated layer forms hardground that is phosphatized, glauconitized, encrusted, and bored during prolonged exposure.

of trace fossils: Berlin, Springer-Verlag Pub., p. 399-428.

———— M. G. Schule, and N. B. Peake, 1975, Paramoudras; giant flints, long burrows and the early diagenesis of chalks: Biol. Skr. Dan. Vid. Selsk., v. 20, no. 10, p. 1-31.

————, 1981, Enhancement of visibility of structures in marly chalk; modification of the Bushinsky oil technique: Geol. Soc. Denmark Bull., v. 29, p. 111-118.

Byrd, W. D., 1975, Geology of the Ekofisk field offshore Norway, *in* A. W. Woodland, ed., Petroleum and the continental shelf of northwest Europe, v. 1: New York, John Wiley and Sons, p. 439-445.

Calvert, S. E., 1966, Accumulation of diatomaceous silica in the sediments of the Gulf of California: Geol. Soc. America Bull., v. 77, p. 569-596.

————, 1974, Deposition and diagenesis of silica in marine sediments, *in* K. J. Hsu and H. C. Jenkyns, eds., Pelagic sediments; on land and under the sea: Internat. Assoc. Sedimentols. Spec. Pub. No. 1, p. 273-300.

Chamberlain, C. K., 1971, Bathymetry and paleoecology of Ouachita geosyncline of southeastern Oklahoma as determined from trace fossils:

AAPG Bull., v. 55, p. 34-50.

————, 1975, Trace fossils in DSDP cores of the Pacific: Jour. Paleontology, v. 49, p. 1074-1096.

Childs, F. B., and P. E. C. Reed, 1975, Geology of the Dan Field and the Danish North Sea, *in* A. W. Woodland, eds., Petroleum and the continental shelf of northwest Europe, v. 1: New York, John Wiley and Sons, p. 429-438.

Church, T. M., 1979, Marine barite, *in* G. Burns, ed., Marine minerals: Mineralog. Soc. America Short Course Notes, v. 6, p. 175-209.

Climap, 1976, The surface of the ice-age Earth: Science, v. 191, p. 1138-1144.

Crimes, T. P., 1973, From limestones to distal turbidites; a facies and trace fossil analysis of the Zumaya flysch (Paleocene-Eocene), north Spain: Sedimentology, v. 20, p. 105-131.

————, 1977, Trace fossils of an Eocene deep-sea fan, northern Spain, *in* T. P. Crimes and J. C. Harper, eds., Trace fossils II: Liverpool, Seel House Press, p. 71-90.

Damuth, J. E., 1977, Late Quaternary sedimentation in the western equatorial Atlantic: Geol. Soc. America Bull., v. 88, p. 695-710.

Davies, T. A., and D. S. Gorsline, 1976,

Figure 91—Incipient early diagenetic cementation of chalks showing selective lithification of carbonate-rich *Thalassinoides* burrow-fills as seen in block fallen from cliff (bedding is vertical). Note rhythmic changes in burrowing intensity. This may be the first stage in the evolution of nodular hardgrounds. Upper Cretaceous chalk, near Etretat, France.

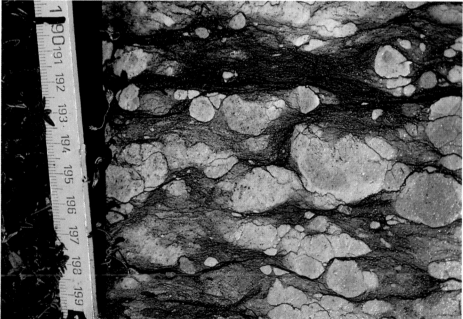

Figure 92—Isolated carbonate nodules floating in a matrix of stylolitized red calcareous claystone. Jurassic Ammonitico Rosso, northern Italian Alps (photograph by J. G. Ogg). Large scale divisions are in centimeters.

Oceanic sediments and sedimentary processes, *in* J. P. Riley and R. Chester, eds., Chemical oceanography, v. 5, 2nd ed.: London, Academic Press, p. 1-80.

Dean, W. E., et al, 1978, Cyclic sedimentation along the continental margin of northwest Africa, *in* Y. Lancelot et al, eds., Initial reports Deep Sea Drilling Project, v. XLI: Washington, D.C., U.S. Govt. Printing Office, p. 965-989.

_____ and B. C. Schreiber, 1978, Authigenic barite, leg 41 Deep Sea Drilling Project, *in* Y. Lancelot et al, eds., Initial reports of the Deep Sea Drilling Project, v. XLI: Washington, D.C., U.S. Govt. Printing Office, p. 915-931.

Degens, E. T., and D. A. Ross, eds., 1974, The Black Sea—geology, chemistry, and biology: AAPG Mem. 20, 633 p.

Demaison, G. J., and G. T. Moore, 1980, Anoxic environments and oil source bed genesis: AAPG Bull., v. 64, p. 1179-1209.

Doyle, W. M., Jr., 1955, Production and reservoir characteristics of the Austin Chalk in south Texas: Trans., Gulf Coast Assoc. Geol. Socs., v. 5, p. 3-10.

Earney, F. C. F., 1980, Petroleum and hard minerals from the sea: New York, John Wiley and Sons, 281 p.

Edmond, J. M., 1974, On the dissolution of carbonate and silicate in the deep ocean: Deep-Sea Research, v. 21, p. 455-480.

Eittreim, S., E. M. Thorndike, and L. Sullivan, 1976, Turbidity distribution in the Atlantic Ocean: Deep-Sea Research, v. 23, p. 1115-1128.

_____ et al, 1972, The nepheloid layer

Figure 93—Limestone with abundant nodules related to multiple hardground or incipient hardground surfaces. Nodular fabric has been accentuated by diagenesis; stylolites have formed by removal of calcite from internode areas and contributed to carbonate precipitation within the nodules (Jenkyns, 1974). Jurassic Upper Ammonitico Rosso, northern Italian Alps (photography by J. G. Ogg). Lens case is 8 cm in diameter.

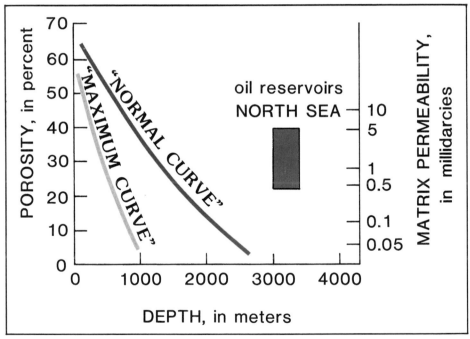

Figure 94—Trends with depth of burial for pelagic carbonate sediments (modified from Scholle, 1977a). "Maximum" curve represents pelagic carbonates that have undergone early fresh-water flushing; "normal" curve is for typical pelagic carbonates buried with marine or modified marine pore fluids; North Sea box represents rapidly deposited, over-pressured chalk reservoirs. Permeability decreases by more than two orders of magnitude with a 50% change in porosity.

and observed bottom currents in the Indian-Pacific, Antarctic Sea, *in* A. L. Gordon, ed., Studies in physical oceanography, v. 2: New York, Gordon and Breach, p. 19-35.

Ekdale, A. A., 1977, Abyssal trace fossils in worldwide Deep Sea Drilling Project cores, *in* T. P. Crimes and J. C. Harper, eds., Trace fossils, v. II: Liverpool, Seel House, p. 163-182.

_____, 1978, Trace fossils in leg 42A cores, *in* K. J. Hsu et al, eds., Initial reports of the Deep Sea Drilling Pro-

ject, v. XLII, Part I: Washington, D.C., U.S. Govt. Printing Office, p. 821-827.

_____, 1980a, Graphoglyptid burrows in modern deep-sea sediment: Science, v. 207, p. 304-306.

_____, 1980b, Trace fossils in leg 58 cores, *in* G. Dev Klein et al, eds., Initial reports of the Deep Sea Drilling Project, v. LVIII: Washington, D.C., U.S. Govt. Printing Office, p. 601-605.

_____ and W. H. Berger, 1978, Deep-sea inchofacies; modern organism traces

on and in pelagic carbonates of the western equatorial Pacific: Palaeogeography, Palaeoclimatology, Palaeoecology, v. 23, p. 263-278.

Elderfield, H., 1977, The form of manganese and iron, in marine sediments, *in* G. P. Glasby, ed., Marine manganese deposits: Amsterdam, Elsevier Sci. Pub., p. 269-290.

Ericson, D. B., et al, 1961, Atlantic deep-sea sediment cores: Geol. Soc. America Bull., v. 72, p. 193-206.

Ewing, M. E., and R. A. Davis, 1967, Lebensspuren photographed on the ocean floor, *in* J. B. Hersey, ed., Deep-sea photography: Baltimore, Johns Hopkins Univ., p. 259-294.

Feazel, C. T., J. Keany, and R. M. Peterson, 1979, Generation and occlusion of porosity in chalk reservoirs (abs.): AAPG Bull., v. 63, p. 448-449.

Fischer, A. G., 1980, Gilbert—bedding rhythms and geochronology, *in* E. L.

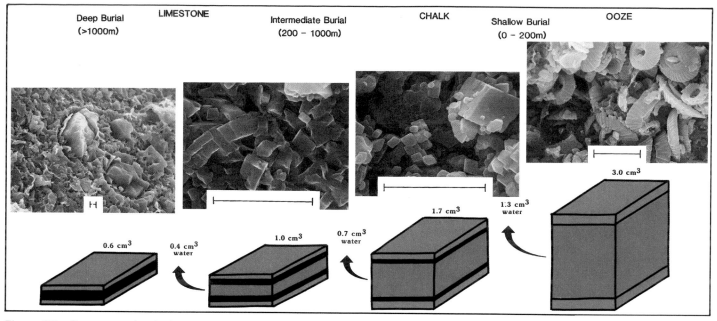

Figure 95—Sequence of pelagic carbonate diagenesis showing porosity reduction first by mechanical compaction and dewatering, then by chemical compaction and cementation (modified from Schlanger and Douglas, 1974). Green, volume of original carbonate constituents; light blue, interstitial water; dark blue, cement. SEM photographs (scale bar is 10 μm) show textures formed during progressive diagenesis.

Yochelson, ed., The scientific ideas of G. K. Gilbert: Geol. Soc. America Spec. Pub. No. 183, p. 93-104.

_____ and M. A. Arthur, 1977, Secular variations in the pelagic realm, *in* H. E. Cook and P. Enos, eds., Deep water carbonate environments: SEPM Spec. Pub. No. 25, p. 19-50.

_____ and R. E. Garrison, 1967, Carbonate lithification on the sea floor: Journal Geology, v. 75, p. 488-496.

Folk, R. L., and E. F. McBride, 1976, The Caballos Novaculite revisited, part 1; origin of Novaculite members: Jour. Sed. Petrology, v. 46, p. 659-669.

Frey, R. W., 1970, Trace fossils of Fort Hays limestone member of Niobrara chalk (Upper Cretaceous), west-central Kansas: Univ. Kansas Paleontol. Contribs., Art. 53, 41 p.

_____, 1972, Paleoecology and depositional environment of Fort Hays limestone member, Niobrara chalk (Upper Cretaceous), west-central Kansas: Univ. Kansas Paleontol. Contribs., Art. 58, 72 p.

Gallois, R. W., 1965, British regional geology: the Wealden district (4th ed.): London, Her Majesty's Stationary Office, 101 p.

Garrison, R. E., 1974, Radiolarian cherts, pelagic limestones and igneous rocks in eugeosynclinal assemblages, *in* K. J. Hsu and H. C. Jenkyns, eds., Pelagic sediments; on land and under the sea: Internat. Assoc. Sedimentols. Spec. Pub. No. 1, p. 367-400.

_____ and R. G. Douglas, eds., 1981, The Monterey Formation and related siliceous rocks of California: Pacific Sec., SEPM Spec. Pub., 332 p.

_____ and W. J. Kennedy, 1977, Origin of solution seams and flaser structure in Upper Cretaceous chalks of southern England: Sed. Geology, v. 19, p. 107-137.

Geitzenauer, K. R., M. R. Roche, and A. McIntyre, 1976, Modern Pacific coccolith assemblages; derivation and application to late Pleistocene paleotemperature analysis, *in* R. M. Cline and J. D. Hays, eds., Investigation of late Quaternary paleoceanography and paleoclimatology: Geol. Soc. America Mem. 145, p. 423-448.

Glasby, G. P., 1977, Marine manganese deposits: Amsterdam, Elsevier Sci. Pub., 523 p.

Goldberg, E. D., and M. Koide, 1962, Geochronological studies of deep sea sediments by the ionium/thorium method: Geochim. et Cosmochim. Acta, v. 26, p. 417-450.

_____ and _____, 1963, Rates of sediment accumulation in the Indian Ocean: Earth Sci. and Meteorology, no. 5, p. 90-102.

Goldhaber, M. A., and I. R. Kaplan,

Figure 96—SEM photograph of a coccolith with extensive overgrowth cementation. Note irregular length of individual elements of coccolith produced by varying amounts of overgrowth cement. Even one of the large, void-filling crystals is an overgrowth of a single coccolith element. Extensive overgrowth cementation can lead to a completely "welded" fabric as in the matrix shown in this photograph. Upper Cretaceous White Limestone, Northern Ireland. Scale bar is 1 μm long.

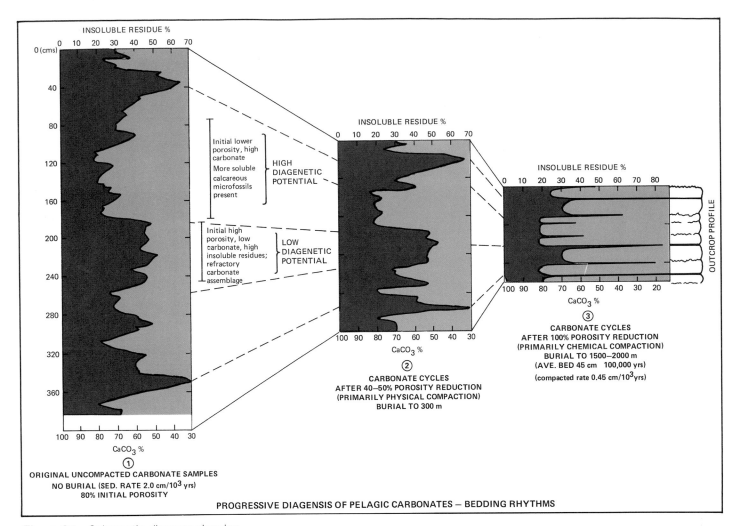

INSOLUBLE RESIDUE %

Initial lower porosity, high carbonate

More soluble calcareous microfossils present

HIGH DIAGENETIC POTENTIAL

Initial high porosity, low carbonate, high insoluble residues; refractory carbonate assemblage

LOW DIAGENETIC POTENTIAL

INSOLUBLE RESIDUE %

INSOLUBLE RESIDUE %

OUTCROP PROFILE

CaCO₃ %

③

CARBONATE CYCLES AFTER 100% POROSITY REDUCTION (PRIMARILY CHEMICAL COMPACTION) BURIAL TO 1500–2000 m (AVE. BED 45 cm 100,000 yrs) (compacted rate 0.45 cm/10³yrs)

CaCO₃ %

②

CARBONATE CYCLES AFTER 40–50% POROSITY REDUCTION (PRIMARILY PHYSICAL COMPACTION) BURIAL TO 300 m

CaCO₃ %

①

ORIGINAL UNCOMPACTED CARBONATE SAMPLES NO BURIAL (SED. RATE 2.0 cm/10³ yrs) 80% INITIAL POROSITY

PROGRESSIVE DIAGENSIS OF PELAGIC CARBONATES — BEDDING RHYTHMS

Figure 97—Schematic diagram showing compaction and solution transfer of carbonate during progressive burial, and the formation of limestone-shale rhythms in pelagic carbonate (after Arthur, 1979c). Blue, carbonate content; brown, clay content.

Figure 98—Toothed stylolites in pelagic limestone. Contrasts with previous examples in that individual, higher-amplitude stylolites tend to form in coarser grained limestones of lower insoluble-residue content, whereas wispy, closely spaced solution seams form in finer grained carbonates of higher initial insoluble-residue content. Note extent of dissolution along stylolite and offset of oblique burrow near top of slab (7 to 10 cm of original thickness is represented by offset of burrow and insoluble-residue content). Orange-red spots are planktonic foraminifers filled with iron oxides and sparry calcite. Upper Cretaceous Scaglia Rossa Formation, Gubbio, Italy. T is 1 cm high.

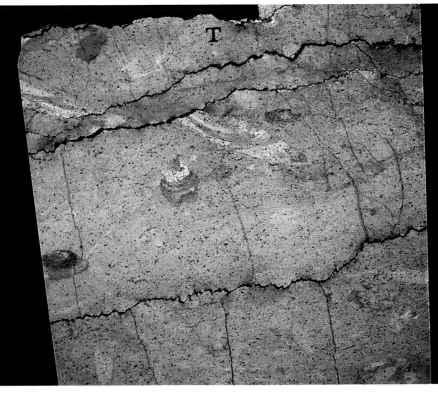

Figure 99—Pelagic limestone with extensive formation of wispy solution seams. Primary alternation of lithology has been accentuated by mechanical and chemical compaction with net export of carbonate from originally carbonate-poor zones to sites of cementation in more carbonate-rich zones. Much, but not all, dissolution has occurred along solution seams or stylolites that tend to be concentrated in initially carbonate-poor intervals. Upper Cretaceous limestone from 13,994.5 ft (4,266 m) depth, Chevron 1656-2 well, offshore Louisiana. Core width about 8 cm.

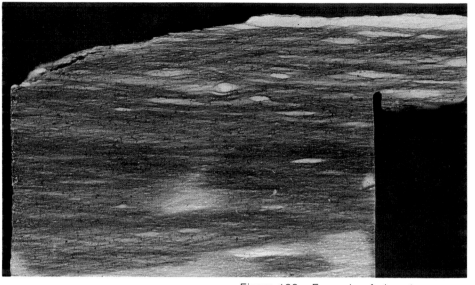

Figure 100—Example of almost complete dissolution of calcium carbonate along solution seams. Network of anastomosing solution seams has led to isolation of carbonate in zones that are probably burrow-fillings. Upper Cretaceous limestone from 14,028 ft (4,276 m) depth, Chevron 1656-2 well, offshore Louisiana. Core width about 8 cm.

Figure 101—Calcite filling of major fracture in the Upper Cretaceous Niobrara Formation, 2,973 ft (906 m) depth, Berthoud Field, north-central Colorado. Although matrix permeability in this unit is negligible, significant oil shows were recorded from fractures in this interval. Core width about 8 cm.

1974, The sulfur cycle, *in* E. D. Goldberg, ed., The sea, v. 5: New York, John Wiley and Sons, p. 569-655.

Hakansson, E., R. Bromley, and K. Perch-Nielsen, 1974, Maastrichtian chalk of northwest Europe—a pelagic shelf sediment, *in* K. J. Hsu and H. C. Jenkyns, eds., Pelagic sediments; on land and under the sea: Internat. Assoc. Sedimentols. Spec. Pub. No. 1, p. 211-234.

Hancock, J. M., 1975a, The sequence of facies in the Upper Cretaceous of northern Europe compared with that in the Western Interior: Geol. Assoc. Canada Spec. Paper No. 13, p. 84-118.

_____, 1975b, The petrology of the chalk: Proc., Geol. Assoc. London, v. 86, p. 499-535.

_____, 1980, The significance of Maurice Black's work on the chalk, *in* C. V. Jeans and P. F. Rawson, eds., Andros Island, chalk and oceanic oozes: Yorkshire Geol. Soc. Occasional Pub. No. 5, p. 86-98.

_____ and P. A. Scholle, 1975, Chalk of the North Sea, *in* A. W. Woodland, ed., Petroleum and the continental shelf of Europe: New York, John Wiley and Sons, v. 1, p. 413-425.

Harper, M. L., 1971, Approximate geothermal gradients in the North Sea: Nature, v. 230, p. 235-236.

_____ and B. B. Shaw, 1974, Cretaceous-Tertiary carbonate reservoirs in the North Sea: Stavanger, Norway, Offshore North Sea Technology Conf.

Paper GIV/4, 20 p.

Hattin, D. E., 1971, Widespread, synchronously deposited, burrow-mottled limestone beds in Greenhorn Limestone (Upper Cretaceous) of Kansas and southeastern Colorado: AAPG Bull., v. 55, p. 412-431.

_____, 1975, Stratigraphy and depositional environment of Greenhorn Limestone (Upper Cretaceous) of Kansas: Kansas Geol. Survey Bull. 209, 128 p.

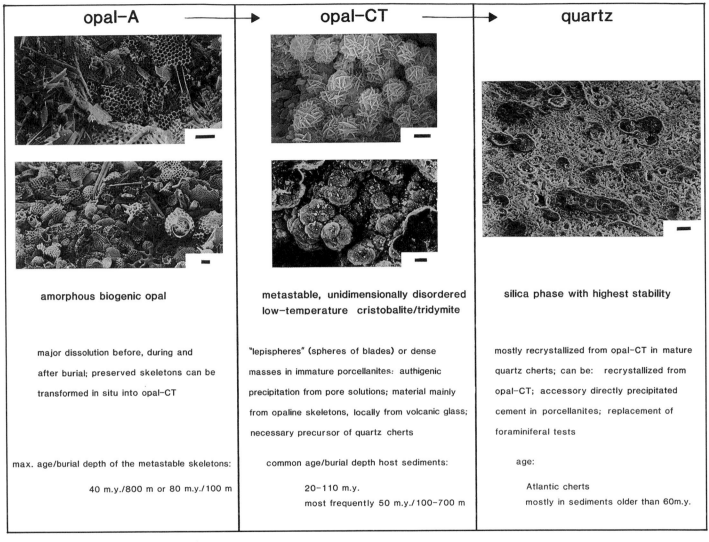

opal-A $\longrightarrow$	opal-CT $\longrightarrow$	quartz
amorphous biogenic opal	metastable, unidimensionally disordered low-temperature cristobalite/tridymite	silica phase with highest stability
major dissolution before, during and after burial; preserved skeletons can be transformed in situ into opal-CT	"lepispheres" (spheres of blades) or dense masses in immature porcellanites: authigenic precipitation from pore solutions; material mainly from opaline skeletons, locally from volcanic glass; necessary precursor of quartz cherts	mostly recrystallized from opal-CT in mature quartz cherts; can be: recrystallized from opal-CT; accessory directly precipitated cement in porcellanites; replacement of foraminiferal tests
max. age/burial depth of the metastable skeletons:	common age/burial depth host sediments:	age:
40 m.y./800 m or 80 m.y./100 m	20-110 m.y. most frequently 50 m.y./100-700 m	Atlantic cherts mostly in sediments older than 60m.y.

Figure 102—Sequence of diagenetic transformation of biogenic opal-A to chert (modified from Riech and von Rad, 1979; photographs by C. M. Isaacs and J. R. Hein). Scale bars are 10 μm long.

_____, 1981, Petrology of Smoky Hill member, Niobarar chalk (Upper Cretaceous), in type area, western Kansas: AAPG Bull., v. 65, p. 831-849.

Hay, W. W., and J. R. Southam, 1977, Modulation of marine sedimentation by the continental shelves, in N. R. Anderson and A. Malahoff, eds., The fate of fossil fuel CO2 in the oceans: New York, Plenum, p. 569-604.

_____ and M. R. Noel, 1976, Carbonate mass balance—cycling and deposition on shelves and in deep sea: AAPG Bull., v. 60, p. 678.

Hays, J. D., 1965, Radiolaria and late Tertiary and Quaternary history of Antarctic seas: Am. Geophys. Union Antarctic Research Ser., v. 5, p. 124-184.

_____ J. Imbrie, and N. J. Shackleton, 1976, Variations in the earth's orbit; pacemaker of the ice ages: Science, v. 194, p. 1121-1132.

_____ and W. C. Pitman, III, 1973, Lithospheric plate motion, sea level changes, and climatic and ecological consequences: Nature, v. 246, p. 18-22.

Heath, G. R., 1974, Dissolved silica and deep-sea sediments, in W. W. Hay, ed., Studies in paleo-oceanography: SEPM Spec. Pub. No. 20, p. 77-93.

Heezen, B. C., and C. D. Hollister, 1971, The face of the deep: Oxford, Oxford Univ. Press, 659 p.

_____ _____ and W. F. Ruddiman, 1966, Shaping of the Continental Rise by deep geostrophic contour currents:

Science, v. 152, p. 502-508.

Hein, J. R., et al, 1979, Mineralogy and diagenesis of surface sediments from DOMES Areas A, B, and C in J. L. Bischoff and D. Z. Piper, eds., Marine geology and oceanography of the Pacific manganese nodule province: New York, Plenum, p. 365-396.

Hesse, R., 1975, Turbiditic and non-turbiditic mudstone of Cretaceous flysch sections of the East Alps and other basins: Sedimentology, v. 22, p. 387-416.

Honjo, S., 1976, Coccoliths: production, transportation, and sedimentation: Marine Micropaleontology, v. 1, p. 65-79.

Hurd, D. C., and F. Theyer, 1977,

Figure 103—Chert nodules (flint) in pelagic limestone of Upper Cretaceous chalk, near Etretat, France. Chertification is the result of early diagenetic mobilization of biogenic silica. Shape of chert nodules is partially controlled by a pre-existing *Thalassinoides* burrow system. Dark areas are only partially silicified.

Figure 104—Tabular chert nodules along bedding plane in chalk. Distribution of chert is presumably controlled by variations in permeability and organic carbon content of replaced sediment. Bedding surfaces in this chalk sequence commonly represent zones of increased winnowing and improved permeability. Source of silica in this relatively shallow-water example was mainly from siliceous sponges; distribution of siliceous sponges, therefore, also can control sites of chertification. Upper Cretaceous Upper Chalk, southern England. Scale bar is 15 cm long.

Changes in the physical and chemical properties of biogenic silica from the central equatorial Pacific; part II; refractive index, density and water content of acid-cleaned samples: Am. Jour. Sci., v. 277, p. 1168-1202.

_____ et al, 1979, Variable porosity in siliceous skeletons; determination and importance: Science, v. 203, p. 1340-1343.

Hsu, K. J., 1972, Origin of saline giants; a critical review after the discovery of the Mediterranean evaporite: Earth-Science Reviews, v. 8, p. 371-396.

_____ et al, 1977, History of the Mediterranean salinity crisis: Nature, v. 267, no. 5610, p. 399-403.

Isaacs, C. M., 1981, Porosity reduction during diagenesis of the Monterey Formation, Santa Barbara coastal areas, California, *in* R. E. Garrison and R. G. Douglas, eds., The Monterey Formation and related siliceous rocks of California: Pacific Sec., SEPM Spec. Pub., p. 257-271.

Jansa, L. F., et al, 1979, Mesozoic-Cenozoic sedimentary formations of the North American basin; western North Atlantic, *in* M. Talwani et al, eds., Deep drilling results in the Atlantic Ocean; continental margins and paleoenvironments: Am. Geophys. Union, Maurice Ewing Series No. 3, p. 1-57.

Jarvis, I., 1980, The initiation of phosphatic chalk sedimentation—the Senonian (Cretaceous) of the Anglo-Paris Basin, *in* Y. K. Bentor, ed., Marine phosphorites—geochemistry,

Figure 105—Nodular chert in pelagic limestone illustrating incomplete silicification within the nodule, but with sharp boundaries to the silification front. Mottling within the nodule is probably due to burrowing prior to silicification. Silica was derived from dissolution of radiolaria, some of which are replaced by chert. Chert nodules are commonly darker colored than enclosing limestone as a result of abundant inclusions of organic matter, water, and other impurities in chert. Upper Cretaceous Scaglia Bianca, Umbria, Italy. T is 1 cm high.

Figure 106—Pyrite nodule in olive-gray hemipelagic mudstone. Pyrite nodules and disseminated framboids are typical of relatively rapidly deposited and/or organic-rich pelagic sediment sequences. Eocene, DSDP Leg 41, Site 368, core 30-1, off northwest Africa (photograph by W. E. Dean). Core piece is 8 cm long.

occurrence, genesis: SEPM Spec. Pub. No. 29, p. 167-192.

Jenkyns, H. C., 1974, Origin of red nodular limestones (Ammonitico Rosso, Knollenkalke) in the Mediterranean Jurassic; a diagenetic model, *in* K. J. Hsu and H. C. Jenkyns, eds., Pelagic sediments; on land and under the sea: Internat. Assoc. Sedimentols. Spec. Pub. No. 1, p. 249-271.

_____, 1977, Fossil nodules, *in* G. P. Glasby, ed., Marine manganese deposits: Amsterdam, Elsevier Sci. Pub., p. 85-108.

_____, 1978, Pelagic environments, *in* G. Reading, ed., Sedimentary environments and facies: New York, Elsevier Sci. Pub., p. 314-371.

_____, 1980, Cretaceous anoxic events; from continents to oceans: Jour. Geol. Soc. London, v. 137, p. 171-188.

Johnson, T. C., E. L. Hamilton, and W. H. Berger, 1977, Physical properties of calcareous ooze; control by dissolution at depth: Marine Geology, v. 24, p. 259-277.

Kalin, O., E. Patacca, and O. Rene, 1979, Jurassic pelagic deposits from southeastern Tuscany; aspects of sedimentation and new biostratigraphic data: Eclogae Geol. Helvet., v. 72, p. 715-762.

Kastner, M., 1979a, Silica polymorphs, *in* R. G. Burns, ed., Marine minerals: Mineralog. Soc. America Short Course Notes, v. 6, p. 99-110.

_____, 1979b, Zeolites, *in* R. G. Burns, ed., Marine minerals: Mineralog. Soc.

America Short Course Notes, v. 6, p. 111-122.

Kauffman, E. G., 1969, Cretaceous marine cycles of the Western Interior: Mtn. Geologist, v. 6, p. 227-245.

_____, 1975, The value of benthic Bivalvia in Cretaceous biostratigraphy of the Western Interior, *in* W. G. E. Caldwell, ed., The Cretaceous system in the Western Interior of North America; selected aspects: Geol. Assoc. Canada Spec. Paper 13, p. 163-194.

_____, 1977, Cretaceous facies, faunas, and paleoenvironments across the Western Interior basin: Mtn. Geologist, v. 14, p. 75-274.

Keene, J. B., 1975, Cherts and porcellanites from the north Pacific, DSDP leg 32, *in* R. L. Larson, et al, eds., Initial reports of the Deep Sea Drilling Project, v. XXXII: Washington, D.C., U.S. Govt. Printing Office, p. 429-507.

Kelts, K., and M. A. Arthur, 1981, Turbidites after ten years of deep-sea

Figure 107—Authigenic barite formed in rapidly deposited, organic-carbon-rich mud. Barite is common in sediments deposited under anoxic conditions and may mark changes in salinity and/or hiatuses in deposition. Lower Cretaceous, DSDP Leg 41, Site 369 off northwest Africa (photograph by W. E. Dean). Core shown is about 7 cm wide.

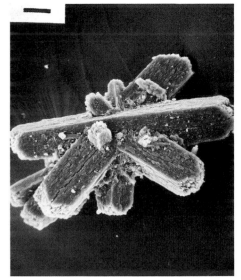

Figure 108—Twinned crystal of authigenic phillipsite in modern, pelagic, brown volcanogenic clay, west-central Pacific Ocean piston core (photograph by S. Stonecipher). Scale bar is 10 μm long.

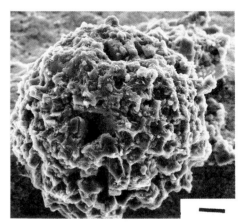

Figure 109—Internal mold of radiolarian test composed of clinoptilolite (a zeolite). Lower Cretaceous black zeolite claystone, western North Atlantic, DSDP Site 105, core 15-2 (photograph by S. Stonecipher). Scale bar is 10 μm long.

drilling—wringing out the mop?, in J. E. Warme, R. G. Douglas, and E. L. Winterer, eds., The Deep Sea Drilling Project; a decade of progress: SEPM Spec. Pub. No. 32, p. 91-127.

Kennedy, W. J., 1967, Burrows and surface traces from the Lower Chalk of southern England: Britian Mus. Nat. History Bull. (Geology), v. 15, p. 125-167.

————, 1970, Trace fossils in the chalk environment, in T. P. Crimes and J. C. Harper, eds., Trace fossils: Liverpool, Seel House, p. 263-282.

————, 1975, Trace fossils in carbonate rocks, in R. W. Frey, eds., The study of trace fossils: Berlin, Springer-Verlag Pub., p. 377-398.

————, 1978, Cretaceous, in W. S. McKerrow, ed., The ecology of fossils: Cambridge, Mass., M.I.T. Press, p. 280-322.

———— and R. E. Garrison, 1975, Morphology and genesis of nodular chalks and hardgrounds in the Upper Cretaceous of southern England: Sedimentology, v. 22, p. 311-386.

———— and P. Juignet, 1974, Carbonate banks and slump beds in the Upper Cretaceous (Upper Turonian-Santonian) of Haute Normandie, France: Sedimentology, v. 21, p. 1-42.

Kern, J. P., 1978, Trails from the Vienna Woods; paleoenvironments and trace fossils of Cretaceous to Eocene flysch, Vienna, Austria: Palaeogeography, Palaeoclimatology, Palaeoecology, v. 23, p. 230-262.

———— and J. E. Warme, 1974, Trace fossils and bathymetry of the Upper Cretaceous Point Loma Formation, San Diego, California: Geol. Soc. America Bull., v. 85, p. 893-900.

Kitchell, J. A., et al, 1978, Abyssal traces and megafauna; comparison of productivity, diversity and density in the Arctic and Antarctic: Paleobiology, v. 4, p. 171-180.

Ksiazkiewicz, M., 1970, Observations on the ichnofauna of the Polish Carpathians, in T. P. Crimes and J. C. Harper, eds., Trace fossils: Liverpool, Seel House, p. 283-322.

Lal, D., and A. Lerman, 1975, Size spectra of biogenic particles in ocean water and sediments: Jour. Geophys. Research, v. 80, p. 423-430.

Lancelot, Y., 1973, Chert and silica diagenesis in sediments from the central Pacific, in E. L. Winterer et al, eds., Initial reports of the Deep Sea Drilling Project, v. XVII: Washington, D.C., U.S. Govt. Printing Office, p. 377-405.

———— and E. Seibold, 1978, The evolution of the central northeastern Atlantic—summary of results of DSDP leg 41, in Y. Lancelot et al, eds., Initial reports of the Deep Sea Drilling Project, v. XLI: Washington, D.C., U.S.

Govt. Printing Office, p. 1215-1245.

Land, L. S., 1980, The isotopic and trace element geochemistry of dolomite; the state of the art, in D. H. Zenger, J. B. Dunham, and R. L. Ethington, eds., Concepts and models of dolomitization: SEPM Spec. Pub. No. 28, p. 87-110.

Lemche, H., et al, 1976, Hadal life as analyzed from photographs: Videnskabeuge Meddeluser fra Dansk Naturhistorisk Forening, v. 139, p. 262-336.

Lerman, A., D. Lal, and M. F. Dacey,

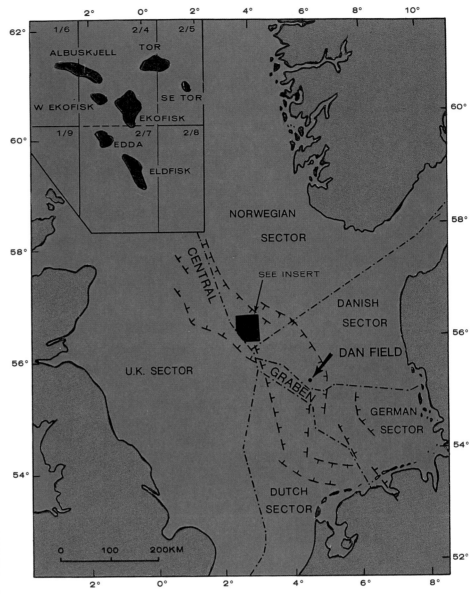

Figure 110—Index map of North Sea showing national boundaries, Central graben, and hydrocarbon accumulations in chalks in eight major fields (after Watts et al, 1980).

1974, Stokes settling and chemical reactivity of suspended particles in natural waters, *in* R. J. Gibbs, ed., Suspended solids in water: New York, Plenum Press, p. 17-47.

Lockridge, J. P., and P. A. Scholle, 1978, Niobrara gas in eastern Colorado and northwestern Kansas, *in* J. D. Pruit and P. E. Coffin, eds., Energy resources of the Denver Basin: Symp. Guidebook, Rocky Mtn. Assoc. Geologists, p. 35-49.

Lowe, D. R., 1976, Nonglacial varves in lower member of Arkansas Novaculite (Devonian), Arkansas and Oklahoma: AAPG Bull., v. 30, p. 2103-2116.

Luz, B., 1973, Stratigraphic and paleoclimatic analysis of late Pleistocene tropical southeast Pacific cores: Quaternary Research, v. 3, p. 56-72.

_____ and N. J. Shackleton, 1975, CaCO3 solution in the tropical east Pacific during the past 130,000 years, *in* W. V. Sliter, A. W. H. Be, and W. H. Berger, eds., Dissolution of deep-sea carbonates: Cushman Found. Foram. Research Spec. Pub. No. 13, p. 142-150.

Manheim, F. T., and R. A. Gulbrandsen, 1979, Marine phosphorites, *in* R. G. Burns, ed., Marine minerals: Mineralog. Soc. America Short Course Notes, v. 6, p. 151-174.

Mapstone, N. B., 1975, Diagenetic history of a North Sea chalk: Sedimentology, v. 22, p. 601-613.

Matter, A., 1974, Burial diagenesis of peletic and carbonate deep-sea sediments from the Arabian Sea, *in* R. B. Whitmarsh, et al, eds., Initial reports of the Deep Sea Drilling project, v. XXIII: Washington, D.C., U.S. Govt. Printing Office, p. 421-469.

_____ R. G. Douglas, and K. Perch-Nielsen, 1975, Fossil preservation, geochemistry, and diagenesis of pelagic carbonates from Shatsky Rise, northwest Pacific, *in* R. L. Larson et al, eds., Initial reports of the Deep Sea Drilling Project, v. XXXII: Washington, D.C., U.S. Govt. Printing Office, p. 891-921.

McBride, E. F., 1970, Stratigraphy and origin of Maravillas Formation (Upper

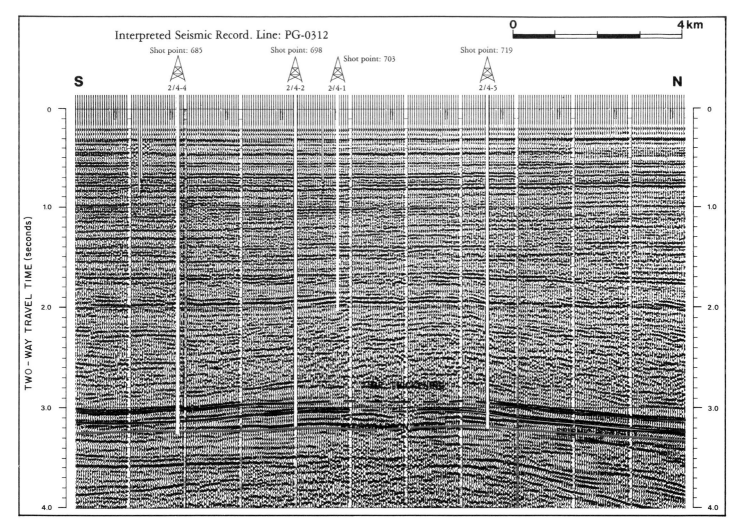

Interpreted Seismic Record. Line: PG-0312

Shot point: 685 Shot point: 698 Shot point: 703 Shot point: 719

2/4-4 2/4-2 2/4-1 2/4-5

0 4 km

S N

TWO – WAY TRAVEL TIME (seconds)

Figure 111—Multichannel seismic
reflection record (time section) across
the Ekofisk field (from Norwegian
Petroleum Directorate, 1980). Main
reservoir horizons are in thickened chalk
section between the lower red and
green lines. Green line, Cretaceous-
Tertiary boundary; lower red line, top of
the Danian; upper red line, top of the
Paleocene.

Ordovician), west Texas: AAPG Bull.,
v. 54, p. 1719-1745.

McKenzie, J., D. Bernoulli, and R. E.
Garrison, 1978, Lithification of pelagic-
hemipelagic sediments at DSDP site
372; oxygen isotope alteration with
diagenesis, in K. J. Hsu et al, eds., Initial reports of the Deep Sea Drilling
Project, v. XLII, pt. 1: Washington,
D.C., U.S. Govt. Printing Office, p.
473-478.

McMillen, K. J., 1974, Quaternary deep-
sea Lebensspuren and the relationship
to depositional environments in the
Caribbean Sea, Gulf of Mexico, and

the eastern and central North Pacific
Ocean: Houston, Tex., Rice Univ.,
M.A. thesis, 147 p.

McRae, S. G., 1972, Glauconite: Earth-
Sci. Reviews, v. 8, p. 397-440.

Muller, P. J., and E. Suess, 1979, Productivity, sedimentation rate and
sedimentary organic matter in the
oceans; I- organic carbon preservation:
Deep-Sea Research, v. 26, p. 1347.

Milliman, J. D., 1974a, Precipitation and
cementation of deep-sea carbonate
sediments, in A. L. Inderbitzen, ed.,
Deep sea sediments: New York,
Plenum, p. 463-476.

_____, 1974b, Marine carbonates: New
York, Springer-Verlag Pub., 375 p.

Moberly, R., and G. deVries Klein, 1976,
Ephemeral color in deep-sea cores:
Jour. Sed. Petrology, v. 46, p. 216-225.

Moore, J. C., 1975, Selective subduction: Geology, v. 3, p. 530-532.

Moore, T. C., Jr., et al, 1978, Cenozoic
hiatuses in pelagic sediments:
Micropaleontology, v. 24, p. 113-138.

Mullins, H. T., et al, 1980, Carbonate

sediment drifts in northern straits of
Florida: AAPG Bull., v. 64, p.
1701-1717.

Neugebauer, J., 1973, The diagenetic
problem of chalk—the role of pressure
solution and pore fluid: N. Jb. Geol.
Palaontol. Abhandl., v. 143, p.
223-245.

_____, 1974, Some aspects of cementa-
tion in chalk, in K. J. Hsu and H. C.
Jenkyns, eds., Pelagic sediments; on
land and under the sea: Internat.
Assoc. Sedimentols. Spec. Pub. No. 1,
p. 149-176.

_____, 1975, Fossil-diagenese in der
Schreibkreide: coccolithen: N. Jb.
Geol. Palaontol. Mh., v. 1975, p.
489-502.

Neumann, A. C., J. W. Kofoed, and
G. H. Keller, 1977, Lithoherms in the
Straits of Florida: Geology, v. 5, p.
4-10.

Nisbet, E. G., and I. Price, 1974, Silicified turbidites; graded cherts as
redeposited ocean-ridge derived
sediments, in K. J. Hsu and H. C.

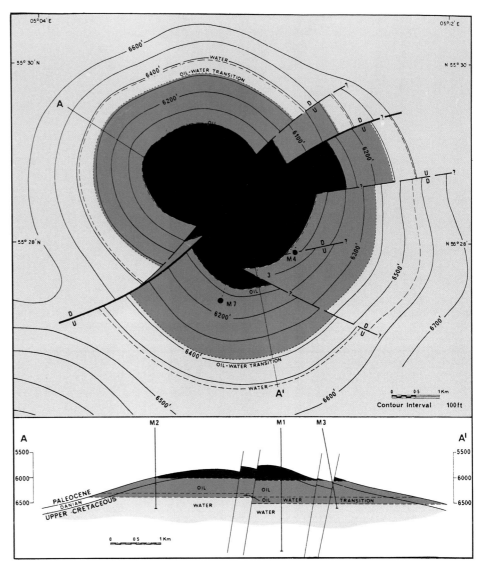

Figure 112—Structure contours drawn on the top of Danian chalk in the Dan field, Danish North Sea, coupled with a structural cross section of the field (from Childs and Reed, 1975).

Jenkyns, eds., Pelagic sediments; on land and under the sea: Internat. Assoc. Sedimentols. Spec. Pub. No. 1, p. 351-366.

Norwegian Petroleum Directorate, 1980, Lithology—wells 2/4-1, 2/4-2, 2/4-3, 2/4-4, and 2/4-5: Norwegian Petroleum Directorate Paper No. 25, 35 p.

Novak, M. T., 1980, Sedimentologic effects of bioturbation in deep-sea calcareous ooze: Univ. Utah, M.S. thesis, 97 p.

Olausson, E., 1967, Climatological, geochemical and paleo-oceanographical aspects of carbonate deposition: Progress in Oceanography, v. 5, p. 245-265.

Opdyke, N. D., and J. H. Foster, 1971, The paleomagnetism of cores in the North Pacific: Geol. Soc. America Mem. 126, 83 p.

Owen, J. D., 1972, A log analysis method for Ekofisk Field, Norway: 13th Ann. Logging Symp. Proc. Paper 10, Soc. Prof. Well Log Analysts, 22 p.

Packham, G. H., and G. J. van der Lingen, 1973, Progressive carbonate diagenesis at Deep Sea Drilling sites 206, 207, 208, and 210 in the southwest Pacific and its relationship to sediment physical properties and seismic reflectors, in R. E. Burns et al, eds., Initial reports of the Deep Sea Drilling Project, v. XXI: Washington, D.C., U.S. Govt. Printing Office, p. 495-521.

Perch-Nielsen, K., K. Ullenberg, and J. A. Evensen, 1979, Comments on "the terminal Cretaceous event; a geological problem with an oceanographic solution" (Gartner and Keany, 1978), in Proceedings of the Cretaceous-Tertiary boundary events

symposium: Copenhagen, Univ. Copenhagen, v. 2, p. 106-111.

Peterson, M. N. A., 1966, Calcite; rates of dissolution in a vertical profile in the central Pacific: Science, v. 154, p. 1542-1544.

Prell, W. L., and J. D. Hays, 1976, Late Pleistocene faunal and temperature patterns of the Columbia basin, Caribbean Sea, in R. M. Cline and J. D. Hays, eds., Investigation of Late Quaternary paleoceanography and paleoclimatology: Geol. Soc. America Mem. 145, p. 201-220.

Prenksy, S. E., 1973, Climatic and tectonic events recorded in late Pleistocene/Holocene deep-sea sediments, California borderland: Los Angeles, Univ. Southern Calif., M.S. thesis, 172 p.

Price, M., M. J. Bird, and S. S. D.

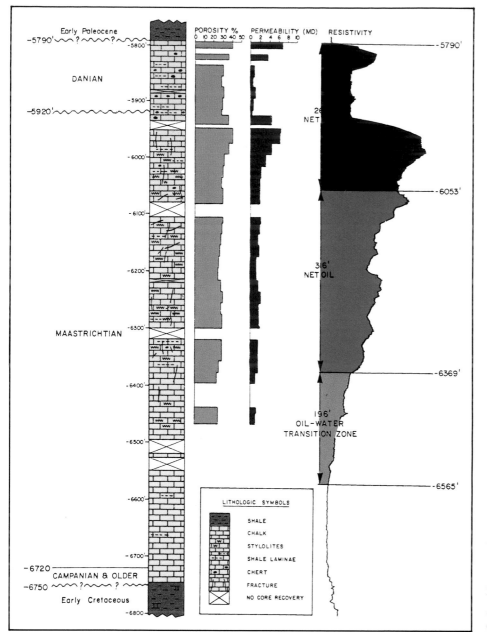

Figure 113—Plot of stratigraphy, lithology, porosity, permeability and resistivity in the M-1X well in the Dan field (from Childs and Reed, 1975). Porosity and permeability values are based on 486 plug analyses and were averaged over 10-ft intervals. Thicknesses are in feet, porosity in percent, and permeability in millidarcys (md).

Foster, 1976, Chalk pore-size measurements and their significance: Water Services, v. 80, No. 968, p. 596-600.

Reeside, J. B., Jr., 1944, Map showing thickness and general character of the Cretaceous deposits in the Western Interior of the United States: U.S. Geol. Survey Oil and Gas Inv. Prelim. Map. 10.

Rex, R. W., et al, 1969, Eolian origin of quartz in soils of Hawaiian Islands and in Pacific pelagic sediments: Science, v. 163, p. 277-279.

Rhoads, D. C., and J. W. Morse, 1971, Evolutionary and ecologic significance of oxygen-deficient marine basins: Lethaia, v. 4, p. 413-428.

Rice, D. D., 1980, Indigenous biogenic gas in Upper Cretaceous chalks, eastern Denver Basin (abs.): Geol. Soc. America Abs. with Programs, v. 12, no. 7, p. 509.

Rickards, L. M., 1974, The Ekofisk area, discovery to development: Stavanger, Norway, offshore North Sea Tech. Conf., Paper GIV/3, 16 p.

Riech, V., and U. von Rad, 1979, Silica diagenesis in the Atlantic Ocean; diagenetic potential and transformation, *in* M. Talwani, W. W. Hay, and W. B. F. Ryan, eds., Results of deep drilling in the Atlantic Ocean; continental margins and paleoenvironments: Am. Geophys. Union, Maurice Ewing Ser. No. 3, p. 315-340.

Robertson, A. H. F., 1975, Cyprus umbers; basalt-sediment relationships on a Mesozoic Ocean ridge: Jour. Geol. Soc. London, v. 131, p. 511-531.

Roniewicz, P., and G. Pienkowski, 1977, Trace fossils of the Podhale flysch basin, *in* T. P. Crimes and J. C. Harper, eds., Trace fossils II: Liverpool, Seel House Press, p. 273-288.

Schlager, W., 1974, Preservation of cephalopod skeletons and carbonate dissolution on ancient Tethyan seafloors, *in* K.

Figure 114—SEM photograph of typical Danian reservoir chalk in Ekofisk field. Sample is from core at 3,216 m depth in the 2/4-2x well, with approximately 35% porosity. Scale bar represents 4.5 μm.

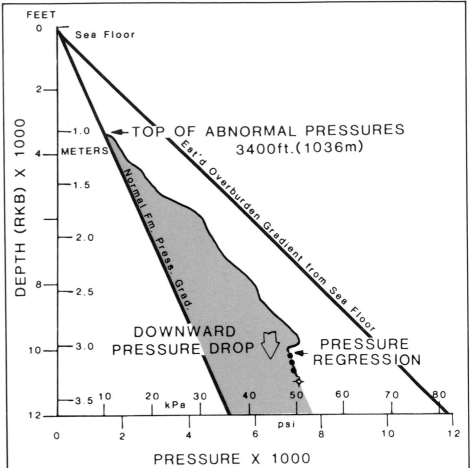

Figure 115—Generalized pressure-depth plot for typical Ekofisk well (from Van den Bark and Thomas, 1981). Variations in sonic travel-time (shale) and drilling factors indicate abnormal pore-fluid pressures in most of the lower Tertiary and Upper Cretaceous section including the chalk reservoirs at 3 to 3.5 km depth.

J. Hsu and H. C. Jenkyns, eds., Pelagic sediments; on land and under the sea: Internat. Assoc. Sedimentols. Spec. Pub. No. 1, p. 49-70.

Schlanger, S. O., and R. G. Douglas, 1974, Pelagic ooze-chalk limestone transition and its implications for marine stratigraphy, *in* K. J. Hsu and H. C. Jenkyns, eds., Pelagic sediments; on land and under the sea: Internat. Assoc. Sedimentols. Spec. Pub. No. 1, p. 117-148.

Scholle, P. A., 1971, Sedimentology of fine-grained deep-water carbonate turbidites, Monte Antola flysch (Upper Cretaceous), northern Apennines, Italy: Geol. Soc. America Bull., v. 82, p. 629-658.

_____, 1974, Diagenesis of Upper Cretaceous chalks from England, Northern Ireland and the North Sea, *in* K. J. Hsu and H. C. Jenkyns, eds., Pelagic sediments; on land and under the sea: Internat. Assoc. Sedimentols. Spec. Pub. No. 1, p. 177-210.

_____, 1977a, Chalk diagenesis and its relation to petroleum exploration—oil from chalks, a modern miracle? AAPG Bull., v. 61, p. 982-1009.

_____, 1977b, Current oil and gas production from North American Upper Cretaceous chalks: U.S. Geol. Survey Circ. 767, 51 p.

_____ and M. A. Arthur, 1980, Carbon isotopic fluctuations in pelagic limestones; potential stratigraphic and petroleum exploration tool: AAPG Bull., v. 64, p. 67-87.

_____ and S. A. Kling, 1972, Southern British Honduras; lagoonal coccolith ooze: Jour. Sed. Petrology, v. 42, p. 195-204.

Schrader, H., et al, 1980, Laminated diatomaceous sediments from the Guaymas Basin slope (central Gulf of California): 250,000 yr. climate record: Science, v. 207, p. 1207-1209.

Sclater, J. G., S. Hellinger, and C. Tapscott, 1977, The paleobathymetry of the Atlantic Ocean from the Jurassic to the present: Jour. Geology, v. 85, p. 509-522.

Scott, M. R., et al, 1974, Rapidly accumulating manganese deposit from the median valley of the mid-Atlantic Ridge: Geophys. Research Letters, v. 1, p. 355-358.

Seilacher, A., 1963, Lebensspuren und salinitatsfazies: Fortschr. Geol. Rheinld. und Westf., v. 10, p. 81-94.

_____, 1964, Biogenic sedimentary structures, *in* J. Imbrie and N. D. Newell, eds., Approaches to paleoecology: New York, John Wiley and Sons, p. 296-316.

_____, 1967, Bathymetry of trace fossils: Marine Geology, v. 5, p. 413-429.

Smagala, T., 1981, The Cretaceous Niobrara play: Oil and Gas Jour., v. 79, no. 10, p. 204-218.

Stewart-Gordon, T. J., 1976, High oil prices, technology support Austin chalk boom: World Oil, v. 183, no. 5, p. 123-126.

Stonecipher, S. A., 1978, Chemistry of deep-sea phillipsite, clinoptilolite, and host sediment, *in* L. B. Sand and F. A. Mumpton, eds., Natural zeolites, occurrence, properties, use: New York, Pergamon Press, p. 221-234.

Stow, D. A. V., and T. P. B. Lovell, 1979, Contourites; their recognition in modern and ancient sediments: Earth

Sci. Reviews, v. 14, p. 251-291.

Sumpter, R., 1981, Chalk play expands at fast clip in Texas: Oil and Gas Jour., v. 79, no. 12, p. 51-55.

Swift, S. A., 1977, Holocene rates of sediment accumulation in the Panama Basin, eastern equatorial Pacific; pelagic sedimentation and lateral transport: Jour. Geology, v. 85, p. 301-319.

Tappan, H., and A. R. Loeblich, 1971, Geobiologic implications of fossil phytoplankton evolution and time-space distribution, in R. Kosanke and A. T. Cross, eds., Symposium on palynology of the Late Cretaceous and early Tertiary: Geol. Soc. America Spec. Paper 127, p. 247-339.

Thunell, R. C., 1976, Optimum indices of calcium carbonate dissolution in deep-sea sediments: Geology, v. 4, p. 525-528.

Tiratsoo, E. N., 1976, Oil fields of the world, 2nd ed.: Beaconsfield, England, Sci. Press, 384 p.

Tucholke, B. E., and G. S. Mountain, 1979, Seismic stratigraphy, lithostratigraphy, and paleosedimentation patterns in the western North Atlantic, in M. Talwani, W. W. Hay, and W. B. F. Ryan, eds., Deep drilling results in the Atlantic Ocean; continental margins and paleoenvironment: Am. Geophys. Union, Maurice Ewing Series

No. 3, p. 58-86.

Tucker, M. E., 1974, Sedimentology of Paleozoic pelagic limestones; the Devonian Griotte (southern France) and Cephalopodenkalk (Germany), in K. J. Hsu and H. C. Jenkyns, eds., Pelagic sediments; on land and under the sea: Internat. Assoc. Sedimentols. Spec. Pub. No. 1, p. 71-92.

van Andel, T. H., G. R. Heath, and T. C. Moore, Jr., 1975, Cenozoic history and paleo-oceanography of the central equatorial Pacific Ocean: Geol. Soc. America Mem. 143, p. 1-134.

_____ and P. D. Komar, 1969, Ponded sediments of the mid-Atlantic Ridge between 22° and 23° North Latitude: Geol. Soc. America Bull., v. 80, p. 1163-1190.

Van den Bark, E., and O. D. Thomas, 1981, Ekofisk; first of the giant oil fields in western Europe, in M. T. Halbouty, ed., Giant oil and gas fields of the decade 1968-1978: AAPG Mem. 30, p. 195-224.

van der Lingen, G. J., 1973, Ichnofossils in deep sea oozes from the southwest Pacific, in R. E. Burns, et al, eds., Initial reports of the Deep Sea Drilling project, v. XXI: Washington, D.C., U.S. Govt. Printing Office, p. 693-700.

Velde, B., and G. S. Odin, 1975, Further information related to the origin of glauconites: Clays and Clay Mins., v.

23, p. 376-381.

von Stackelberg, U., 1972, Fazies verteilung in sedimenten des Indisch-Pakistanischen Kontinentalrandes (Arabisches Meer): "Meteor" Forsch.-Ergebn., v. E9, p. 1-106.

Warme, J. E., W. J. Kennedy, and N. Schneidermann, 1973, Biogenic sedimentary structures (trace fossils) in leg 15 cores, in N. T. Edgar et al, eds., Initial reports of the Deep Sea Drilling project, v. XV: Washington, D.C., U.S. Govt. Printing Office, p. 813-831.

Watts, N. L., et al, 1980, Upper Cretaceous and lower Tertiary chalks of the Albuskjell area, North Sea; deposition in a slope and a base-of-slope environment: Geology, v. 8, p. 217-221.

Wiedmann, J., ed., 1979, Aspekte der Kreide Europas; Stuttgart, E. Schweizerbart'sche Verlagsbuchhandlung: Internat. Union Geol. Sci. Ser. A, no. 6, 624 p.

World Oil, 1971, Massive Danian limestone key to Ekofisk success: World Oil, v. 172, no. 6, p. 51-52.

Zeitschel, B., 1978, Oceanographic factors influencing the distribution of plankton in space and time: Micropaleontology, v. 24, p. 139-159.

Ziegler, P. A., 1978, Northwestern Europe; tectonics and basin development: Geol. Mijnbouw, v. 57, p. 589-626.

Contributors

This final section contains biographical sketches of each contributor to this volume; they appear in alphabetical order. The biographies are written in the authors' own words, and have been edited only as necessary to blend them in general length and style.

Michael A. Arthur

*M*ichael is a native of Sacramento, and was raised in southern California. He developed an early interest in natural history, but majored in political science for two years in college before "seeing the light" and changing to geology. After receiving bachelor's and master's degrees (1971, 1974, respectively) in geology from the University of California, Riverside, he was drawn toward the east to Princeton University, where he obtained his doctorate in geology in 1979. The combined influence of S. O. Schlanger and A. G. Fischer directed his attention toward paleoceanography and patterns of marine sedimentation, which he pursued in his dissertation work on the sedimentology and geochemistry of Cretaceous-Paleogene pelagic sedimentary rocks.

His interest in deep-sea sedimentation and paleoceanography led him to a staff scientist position with the Deep Sea Drilling Project and post-graduate research geologist position at Scripps Institution of Oceanography, La Jolla, California, from 1977-1979. He has participated in DSDP legs in the North Atlantic and in the Japan Forearc Survey, Branch of Oil and Gas Resources, from 1979-1981. Michael is now assistant professor of geology at the University of South Carolina.

Michael has published extensively on the stratigraphy and geochemistry of deeper marine sediments, paleoceanography and the stratigraphic record, and on the burial and preservation of organic matter in marine strata through time. He is interested in chemical, stable isotopic and textural changes in carbonate rocks during diagenesis, in geochemical cycles (particularly changes in the cycling of carbon, phosphorous, and sulfur through the oceans through time), in the origin of and maturation of hydrocarbon source beds, and in sedimentary mass balances, tectonics, and sea level changes.

Don G. Bebout

A native of Monesson, Pennsylvania, Don earned his bachelor's degree in geology from Mount Union College in Alliance, Ohio, in 1952. He continued his education, receiving a master's degree (1954) and doctorate (1956) from the University of Wisconsin and University of Kansas, respectively.

He worked for Exxon Production Research Company in Houston for 12 years (1960-1972), doing research and teaching courses in carbonate facies, depositional environments, and diagenesis. In 1972, Don travelled to Austin where he began work for the Bureau of Economic Geology at the University of Texas. His work centered on carbonate research on the Lower Cretaceous of south Texas and subsurface mapping of the Eocene and Oligocene terrigenous systems along the Texas Gulf Coast. He also taught graduate courses in carbonate depositional environments and facies during this period at the university.

Beginning in 1979, Don moved to Baton Rouge, where he was director of research for the Louisiana Geological Survey and taught both undergraduate and graduate courses. He has now returned to work for the Bureau of Economic Geology in Austin.

Thilo Bechstadt

*T*hilo, though born in the eastern part of Germany, grew up in Austria. From 1964-1972 he studied geology at the University of Innsbruck. His 1972 thesis deals with Mid-Triassic carbonate-clastic sediments of the northern Dolomites of Italy.

Having worked in mineral exploration in Austria and Greece for Northgate of Canada from 1970-1971, Thilo joined the Geological-Paleontological Institute of Freiburg University as a scientific assistant in 1972. From 1974-1978 he worked in the same position at the Institute for General and Applied Geology of Munich University.

His habilitation paper at Munich University in 1978 deals with a facies analysis and palinspastic reconstruction of the Drau Range, eastern Alps. In 1978, Dr. Bechstadt received the Credner Award of the German Geological Society. He joined the petroleum industry in 1978 and worked for Deminex in Vietnam and Libya, his latest position being chief geologist.

Currently, Thilo is a professor at the Geological-Paleontological Institute of Freiburg University. His main topics of research interest are: facies analysis, especially of carbonates; stratabound ore deposits and their relation to facies developments; and facies reconstructions from well logs.

Philip W. Choquette

Harry E. Cook

Winterer.

Since 1974, Harry has been with the U. S. Geological Survey's Branch of Oil and Gas Resources in Menlo Park, California. His current work centers on the sedimentology of ancient and modern continental slopes, the geologic controls that promote carbonate submarine fans versus simple stacks of carbonate debris sheets and aprons, the similarities and differences between carbonate fans and clastic fans, and the petroleum geology of slope and rise settings.

*B*orn and raised in upstate New York, Philip did his undergraduate work at Allegheny College in Meadville, Pennsylvania, majoring in geology under William H. Parsons, with a year of study at the Institut Catholique in Paris. He earned master's (1954) and doctorate (1957) degrees at Johns Hopkins University, under the tutelage of a distinguished faculty which included Ernst Cloos, F. J. Pettijohn, Aaron Waters, and J. D. H. Donnay.

After two years with the U.S. Geological Survey (1956-1958) mapping low-grade metamorphic rocks in the North Carolina "Slate Belt" and clastic sedimentary rocks in the northwest flanks of the Black Hills, Phil joined the geological research staff of Marathon (then Ohio) Oil Company, newly organized by R. Dana Russell, in Littleton, Colorado. Happily transplanted to the Rocky Mountain region, Phil has remained with Marathon since 1958. Much of his work involved the assessment of frontier regions and studies of major carbonate reservoirs. His main field of interest is the stratigraphy/sedimentology of carbonate rocks, emphasizing diagenesis, with excursions into petrophysics, isotope geochemistry, thermal conductivity, and pore-system evolution of carbonates.

Philip has served as an associate editor of the *Geological Society of America Bulletin* and the *Journal of Sedimentary Petrology*. Currently he is SEPM councilor for mineralogy.

*H*arry was born in Fresno, California, near the western slopes of the Sierra Nevada. When faced with the option of college or an adventuresome interlude on the high seas, he chose the latter. It was during Harry's three years on a U. S. submarine that his attraction to marine geology was spawned.

In 1961 he received his bachelor's degree in geology from the University of California at Santa Barbara. Here his interests in field geology were inspired by Bob Webb, whose traditional field breakfasts of green jello and rice cakes are fondly remembered, albeit with mixed feelings. Harry went on to the University of California at Berkeley, where under the able tutelage of Charles Gilbert, Howell Williams, and Dick Hay he received his doctorate in 1966 in volcanology and low-temperature diagenesis.

While employed at Marathon Oil Company's Denver Research Center (1965-70) as a research geologist he learned, from Lloyd Pray, the merits of carefully deciphering ancient carbonates as an aid to understanding modern carbonates. At Marathon, Harry conducted studies on shoal-water and basinal carbonates with an emphasis on submarine mass-transport processes and their products. From 1970-1974, Harry taught at the University of California at Riverside and was a member of the Scripps Institution of Oceanography's Deep Sea Drilling Project. Here his attention shifted toward deeper marine environments focusing on the geologic history of the Equatorial Pacific in association with his enthusiastic colleagues and shipmates Seymour Schlanger and Jerry

Walter E. Dean

*W*alter was born and raised in Wilkes-Barre, Pennsylvania, the heart of the economically depressed northern anthricite coal fields of the 1940s. Thinking he wanted to be an engineer, Walt enrolled in civil engineering at Syracuse University. However, between his junior and senior years he took a summer geology field course with the University of Wisconsin—9,000 miles of regional geology over six weeks in South Dakota, Wyoming, Utah, Nevada, and Idaho. This marked the first time Walt had travelled west of Buffalo, and he went back to Syracuse, in gray upstate New York, finishing his senior year in geology. Walt began work for Limbaugh Aerial Surveys in Albuquerque, New Mexico, in 1961, doing survey data processing and highway earthwork computations. In 1963, Walt received a National Science Foundation-funded research assistantship with Roger Anderson at the University of New Mexico and completed a master's thesis on turbidites in the Marathon Basin of west Texas. As luck would have it, NASA offered Walt a new fellowship if he would stay for a doctorate, which he completed in 1967.

Walt had always wanted to know more about lakes, and Eville Gorham in the botany department at the University of Minnesota was looking for a postdoctoral candidate to study the geochemistry of lake sediments. In that one year at Minnesota, Walt really found his niche with lakes and lake sediments. Syracuse University had an opening in chemical sedimentology that sounded perfect, and for the next six years Walt taught courses in chemical sedimentology, general geology, limnology and oceanography, stratigraphic analysis, modern carbonates, geology of the algae, isotope geochemistry, ancient environments, and regional geology of the United States. While at Syracuse, he managed to continue some work on evaporates, continue Minnesota lake studies, and start some lake work in central New York, particularly on freshwater carbonates and manganese nodules.

In 1975, Walt joined the U. S. Geological Survey in Denver to work on geochemistry of oil shale. He was soon working on the outer continental shelf sediments in the Bering Sea while continuing his interest and activity in evaporites, carbonates, and lakes. All of these interests and activities continue today.

Barbara Dohler-Hirner

*B*arbara was born in Schwabisch Gmund, West Germany, and travelled to Munich to attend Ludwig-Maximilians-Universitat, studying biology and chemistry. In 1970, she directed her studies toward geology, doing field work in uranium exploration in the Bavarian Forest. She continued her education at the Albert-Ludwigs-Universitat in Freiburg, where field work centered in the Betic Cordilleras of southern Spain doing geological mapping and structure analysis on the western border of the Sierra Bermeja.

Barbara is currently working toward her doctorate, doing thesis research on the sedimentology and geochemistry of the Middle Triassic limestones in the Northern Limestone Alps and the Drau Range. In addition to field and laboratory work, she is conducting computer and statistical analysis of x-ray fluorescence.

Allan A. Ekdale

*A*llan A. (Tony) Ekdale is a specialist in animal-sediment interrelationships and ichnology, particularly with respect to bioturbation of pelagic deep-sea sediments. His research contributions include published work dealing with sediment-mixing by organisms and burrow construction and preservation in both modern and ancient deep-sea deposits. Currently, he is engaged in a study of bioturbation and its effects on early diagenesis in Upper Cretaceous chalk sequences in northern Europe. Other research interests have included quantitative synecology of coral reef-associated molluscan communities, tidal flat ichnology, and trace fossils in non-marine sediments.

Tony was born and raised in Burlington, Iowa. He received a bachelor's degree in geology from Augustana College (Rock Island, Illinois) in 1968, and earned his master's and doctoral degrees in geology from Rice University (Houston) in 1973 and 1974, respectively. In 1974, he joined the faculty of the University of Utah, where he now is associate professor of geology and associate chairman of the department of geology and geophysics.

Paul Enos

*T*he Pennsylvanian cyclothems of eastern Kansas frame the Enos family farm near Topeka where Paul Enos grew up. His father's interest in picking up fusulinids and horn corals from among the other organic remnants in the barnyard stimulated an early interest in things geologic that led to a bachelor's degree in geology from the University of Kansas in 1956. The next stop was again in limestone terrane (the Schwabische Jurassic) at a traditional paleo-stratigraphy school, the Universitat Tubingen, Germany, where Paul held a Fulbright scholarship during the last years of O. H. Schindewolf and the first years of Adolf Seilacher. Sands and sandstones followed; first as a pyschological warfare officer in the coastal plain of North Carolina, then as a Stanford graduate student working with Franciscan and younger clastic rocks in the Coast Ranges near Hollister, California (master's degree, 1961) and finally as a doctoral candidate under J. E. Sanders at Yale working in the Ordovician flysch of Gaspe peninsula through 1964. The sedimentologic portion of this work received the outstanding paper award of the *Journal of Sedimentary Petrology* in 1969.

Eventually, the opportunity to study carbonates came as a research geologist for Shell Development Company (1964-1970). Four years were spent working on modern carbonates, mainly of the Florida shelf margin and Florida Bay under the supervision of, first, R. N. Ginsburg, and later, R. J. Dunham. Two years at the Houston research center followed with a study of the Poza Rica trend, Veracruz, Mexico, as the main project.

In 1970, Paul joined State University of New York at Binghamton as an associate professor, and there he has stayed. Most of his research has been directed toward understanding depositional diagenetic processes that formed the mid-Cretaceous basinal, shelf margin, and slope facies carbonates in northeastern Mexico. Other projects include regional geology of Mexico; Kansas cyclothems; Tertiary carbonates in Irian Jaya, Indonesia; pore waters of Bermuda; and limestones of the Blake Plateau, sampled on the Deep Sea Drilling Project. A sabbatical year in 1976-1977 provided the opportunity to carry limestones to Liverpool, from Poza Rica, for petrographic study in R. G. C. Bathurst's laboratory. Paul leads several field trips each year to Florida and the Bahamas for various groups, including the AAPG, and helps instruct in the AAPG carbonate exploration school.

Mateu Esteban

B orn in Constanti, a province of Tarragona, Catalunya, Spain, Dr. Esteban studied geology at the University of Barcelona with grants from the Ministry of Education, the Research Council, and the Geological Museum of Barcelona. As an undergraduate, Mateu worked part-time in Aguas Industriales de Tarragona S.A. and in 1972 spent four months in the department of geology at the University of Liverpool. For his master's and doctorate theses (1969 and 1973, department of petrology, University of Barcelona) he worked on the stratigraphy, sedimentology and diagenesis of Cretaceous and Jurassic carbonates of the Catalan Coastal Ranges and obtained the Extraordinary Award of Doctoral Studies. During 1969-1974, he used a Fulbright post-doctoral grant in the Comparative Sedimentology Laboratory of the University of Miami to study Quaternary and Recent carbonates. Since 1972, Dr. Esteban has been a staff scientist in the unit of Marine and Regional Geology of the Institut Jaume Almera of the C.S.I.C. (Scientific Research Council) in Barcelona, and since 1979 has been involved in part-time consulting for ERICO and as an independent.

As a visiting scientist at the University of Wisconsin, Madison (1977), he worked on the pisolite facies of the Permian Capitan Reef Complex, and in 1980, was a lecturer in the AAPG continuing education program. He has done extensive field work in Spain, France, Italy, Sicily, Tunisia, and Morocco and has participated in research programs with the Universities of Liverpool, Wisconsin, Pisa, and Palermo. With students of the

University of Barcelona, Mateu created a research group interested in study of the sedimentology and diagenesis of the Mesozoic and Tertiary carbonates of the western Mediterranean as well as the petrology of the subaerial exposure in carbonates. His main interests are in Cretaceous and Tertiary reefs and platform margins, erosion surfaces, and the recognition of caliche and karst in the geologic record.

Thomas D. Fouch

T om was born in Portland, Oregon, but raised near the town of Milwaukie in the Willamette Valley. Tom left to attend Portland State University in 1960 where he majored in prelaw and economic studies in preparation for a career as an attorney. While waiting to enter law school, Tom took a geology course at Portland State as a lark. He graduated with a degree in earth sciences in 1966, and after switching the focus of his studies to geology, he became a geologic technician for the Oregon State Department of Geology and Mineral Industries, an engineering technician for the U. S. Department of Commerce, and a geologist for the Weyerhaeuser Timber Company exploring company lands in the Pacific Northwest for mineral deposits.

Tom's graduate work at the University of Oregon was designed to prepare him for the life of a mining geologist. His master's thesis in the Blue Mountains of Oregon focused on the stratigraphy and the petrology of igneous and metamorphosed sedimentary rocks. In spite of this, in 1968, Rufus LeBlanc of the Shell Oil Company offered Tom the chance to live in New Mexico while learning about sedimentology and petroleum geology. In his first days with Shell, he saw his first real limestone and more sandstones in a day than he had seen in his life prior to that time. While employed by Shell, Tom concentrated on the exploration of frontier provinces and unconventional hydrocarbon rocks. As a consequence, he had the opportunity to study most of the ancient lake basins of the western part of the United States and helped develop sedimentologic, mineralogic, and organic geochemical models of these units that could be applied to exploration efforts. He was part of a geologic team whose effort led to the discovery of the large Altamont-Bluebell field in the Uinta Basin.

He joined the J. M. Huber Corporation in 1973 as an explorationist for Montana, Colorado, and Utah. In 1974, Tom accepted an invitation from the U. S. Geological Survey to continue his work on the sedimentology and petroleum geology of unconventional hydrocarbon-bearing rocks.

Robert B. Halley

*B*ob Halley received a bachelor's degree in geology from Oberlin College (1969) and earned graduate degrees at Brown University (1971) and the State University of New York at Stony Brook (1974). At Brown, his thesis concerned environmental interpretations of Upper Cambrian limestones of the Hudson River Valley. At Stony Brook, he completed a dissertation describing the sedimentology and stratigraphy of a Cambrian formation in the Death Valley area of the southern Great Basin.

After graduate school, Bob con-

vinced State University of New York at Binghamton to support him in post-doctoral work for a year to study modern carbonate sediments in the Great Salt Lake, Utah. This limited experience with modern carbonate sediments helped him land his first job with the U. S. Geological Survey. His first "assignment" lasted almost six years at Fisher Island Station, Miami Beach, where his studies focused on modern carbonate sedimentation and early diagenesis. During the completion of his Miami tour, Bob became increasingly interested in applying what he had learned to ancient rocks. In 1980, this interest was fulfilled with a transfer to Denver and a dive into subsurface studies.

Paul M. Harris

*M*itch Harris is a carbonate sedimentologist who received a bachelor's degree in geology in 1971 from West Virginia University in Morgantown. He received a master's in geology from the same institution after completing thesis research on the stratigraphy of siliciclastic marshes along the coastline of Virginia. The focus of his doctoral studies at the University of Miami School of Marine and Atmospheric Science in Miami, Florida, was the subsurface facies relations and diagenesis of carbonate sand deposits in the Bahamas. He received a doctorate in marine geology and geophysics in 1977.

After working briefly for Getty Oil in Houston, Mitch became a project geologist for Gulf Research and Development Company in Houston.

He has worked the petroleum geology of carbonates from the Mesozoic of the U. S. Gulf Coast, the Permian in the Permian Basin and the Cretaceous of the Eastern Arabian Peninsula, and has organized industry field trips to study recent carbonates in Florida and the Bahamas and ancient carbonates in Texas and New Mexico.

Albert C. Hine

*A*l Hine is presently an assistant professor of marine science (geological oceanography) in the department of marine science at the University of South Florida. He received a bachelor's degree in geology at Dartmouth College (1967), a master's in geology from the University of Massachusetts (1973), and his doctorate in geology from the University of South Carolina (1975). He spent four years at the University of North Carolina in Chapel Hill as a research associate and as a research assistant professor of marine science.

Al's areas of specialization are: 1) modern, shallow marine depositional environments including coastal geology/sedimentation and carbonate bank sedimentation; and 2) high resolution structure and stratigraphy of the continental shelf. These studies have taken him to such diverse places as Iceland, Bermuda, and throughout the north-central Caribbean. He has also been principal or co-principal investigator on 24 research cruises. Results of these endeavors have been published in over 50 articles and abstracts. Some topics include: seismic stratigraphy and facies of reefs; carbonate bank margin sand accumulation; transport of carbonate sands; dynamics/sedimentological history of ooid shoals; structure and stratigraphy of carbonate bank margins; mechanisms of berm development on beaches; history of a glacial-outwash plain shoreline; bed form distribution on tidal deltas; and tidal inlet variability.

Current research involves deciphering paleo-environmental significance of buried channels within the continental shelf and the structure, stratigraphy, and sea-level cyclicity of phosphate-rich Miocene sediments onlapping the Carolina platform.

Richard F. Inden

*R*ichard was born in Milwaukee, Wisconsin, and received geology degrees from the University of Wisconsin at Milwaukee (bachelor's, 1965), University of Illinois (master's, 1968), and Louisiana State University in Baton Rouge (doctorate, 1972). During the next five years he did post-doctoral research and taught at the University of South Carolina, worked for the U. S. Geological Survey in Reston, Virginia, and taught at Kent State University in Canton, Ohio.

In 1977, he joined the Superior Oil Research Lab in Houston, and since 1980, has been with McAdams, Roux, O'Connor, and Associates in Denver. His interests center on carbonate and clastic stratigraphy, and the diagenesis of hydrocarbon-bearing carbonate sequences. During his career, he has been involved with numerous projects on the Lower Cretaceous of the Gulf

Coast, Liassic of Morocco, Cenozoic of the southeast Atlantic coastal plain, Gulf of Suez, and Dominican Republic; Mississippian of the Illinois Basin, Williston Basin, and Disturbed Belt of Montana; and the Lower Paleozoic of the Mid-Continent. Presently, he is involved in petroleum exploration in the Illinois Basin and Cincinnati Arch areas of Indiana and Kentucky.

Noel P. James

*N*oel is a native Newfoundlander who received his bachelor's degree from McGill University in 1967. Between advanced degrees he worked for Amoco Canada in Calgary, Alberta. He gained his doctorate in geology from McGill University in 1972, and subsequently joined the Comparative Sedimentology Laboratory, University of Miami. Since 1974, he has been at Memorial University of Newfoundland where he is presently professor of geology. In 1979, he was an Industrial Fellow at the Denver Research Center of Marathon Oil Company and this year is distinguished lecturer for the AAPG.

Noel's main field of research is the petrogenesis of carbonate rocks. He has studied modern carbonates and Pleistocene reefs throughout the Caribbean and has led many field seminars to Bermuda, Barbados, Florida, and the Bahamas. Together with Robert N. Ginsburg, he pioneered the use of research submersibles to study the deep zones of modern reefs and carbonates platforms. These studies of deep water carbonates continue, both on the sea floor, in the Mesozoic of Arabia, and in the Early Paleozoic of the Appalachians. In recent years, he has been particularly

active in detailing the paleontology, sedimentology, and diagenesis of the earliest paleozoic reefs.

Clif Jordan

*C*lif is a native of St. Louis, Missouri, where he attended Washington University and received his undergraduate training in geology. He later received master's and doctorate degrees in geology from Rice University in Houston. Most of his career has been spent in the petroleum industry, although he has also worked in mining geology as a geological consultant for various firms. His experience in the petroleum industry includes assignments in production geology, exploration geology, and exploration research. He is presently employed by Mobil Research and Development Corporation in Dallas, Texas.

Ramon Julia

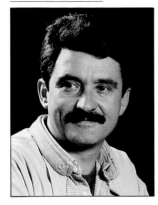

Colin F. Klappa

J. Richard Kyle

*D*r. Ramon Julia, born in Barcelona, Spain, studied geology at the University of Barcelona before becoming assistant professor in the department of geomorphology and tectonics there. The subject of his doctorate thesis was a study of Plio-Pleistocene and Recent continental carbonate deposits in Girona (northeastern Spain).

Since 1979 he has worked as a permanent research scientist at the Institute Jaime Almera of Geological Research belonging to the National Research Council of Spain (C.S.I.C.) The central thrust of his work involves sedimentological and geomorphological research on continental carbonates and some other aspects of the paleoclimatology and paleogeography of Quaternary deposits, focused mainly in northeastern Spain.

*C*olin is a native of Berkshire, England. He received his bachelor's degree in geology from the University of Liverpool after mapping, in the fog, Devonian-Carboniferous fluvial to shallow marine siliciclastics in southwest Ireland, and studying the stratigraphy, paleoecology, and diagenesis of Jurassic hardgrounds from southwest England for his thesis. In 1978, he earned his doctorate from the University of Liverpool, for his dissertation treating the origin and morphology of Recent and Pleistocene caliche from the coastal regions of the western Mediterranean.

These studies were followed by a westward shift to Canada where Colin carried out research, supported by a Memorial University of Newfoundland post-doctoral fellowship, on Cambrian and Ordovician limestones from southern Labrador and western Newfoundland. Continuing with a further westward shift, Colin moved to Calgary, Alberta, in 1980, where he is now employed in the Exploration Research & Services Division of Gulf Canada Resources Inc. He is currently involved in various projects on the petroleum geology of Devonian and Mississippian carbonates from the western Canada sedimentary basin.

*R*ichard was born in Athens, Tennessee. He received degrees in geology from Tennessee Technological University (bachelor's, 1970), the University of Tennessee (master's, 1973), and the University of Western Ontario (doctorate, 1977). He has worked as a minerals exploration geologist for a number of companies, including Cominco American, Inc., Pine Point Mines, Ltd., Exxon Minerals Company, U.S.A., Houston International Minerals Corporation, and Marathon Resources, Inc.

He is currently an assistant professor in the department of geological sciences at the University of Texas at Austin where he teaches courses in ore deposits geology and mineral resources and directs research on strata-controlled mineralization in volcanic and sedimentary environments. His current research interests include metallogeny, origin of stratiform sulfide deposits in sedimentary and volcanic rocks, geochemistry of alteration associated with ancient geothermal systems, diagenesis and formation of stratabound mineral deposits in carbonate rocks, application of fluid inclusion studies to geologic problems, and minerals exploration. He has published several papers on the diagenetic controls of lead-zinc mineralization in carbonate environments.

Robert G. Loucks

Edwin D. McKee

*B*ob received his bachelor's degree in geology from the State University of New York at Binghamton in 1967 and his doctorate from the University of Texas at Austin in 1976. He worked as an exploration and production geologist for three years with Texaco in Midland, Texas. From 1972-80 he was employed with the Bureau of Economic Geology, the University of Texas at Austin. There he completed studies on depositional systems and diagenesis of south Texas carbonates ranging in age from Jurassic through Lower Cretaceous, and on the Ellenburger dolomite in west Texas. He also participated in exploring for geopressured geothermal energy in the Tertiary section along the Texas Gulf Coast. With that project he did depositional systems analysis on the Frio and Vicksburg formations and extensive sandstone diagenetic studies of the complete Tertiary section.

After his employment with the Bureau, Bob worked on sandstone diagenesis of deep reservoirs with Mobil Research Lab in Dallas. He is now employed at Cities Service Research Lab in Tulsa where he is continuing research on carbonate depositional systems and diagenesis.

*B*orn in Washington, D.C., Edwin McKee's boyhood summers were spent in the hills of New Hampshire where he developed a lifelong interest in natural history. During the winters in the nation's capital, he came under the influence of Francois E. Matthes of the U. S. Geological Survey who was his Boy Scout leader and guided him into the fields of geology and earth sciences. His academic background included undergraduate work at Cornell University and graduate studies in geology at the University of Arizona, the University of California at Berkeley, and Yale University. He was granted an honorary Doctor of Sciences degree by Northern Arizona University.

Ed's professional career divides into four nearly equal parts. Beginning in 1929, he spent 12 years as park naturalist of Grand Canyon National Park, Arizona, following with 12 years teaching geology at the University of Arizona in Tucson, beginning as assistant professor and ending as professor and head of the department; concurrently, summers were spent as assistant director of the Museum of Northern Arizona in Flagstaff and as director of research at the institution. Work with the U. S. Geological Survey then began in Denver, first in developing the Paleotectonic Map Unit which produced geologic atlases for the Jurassic, Triassic, Permian, and Pennsylvanian systems, and later in field and laboratory studies of sedimentary structures, including considerable field work in foreign countries.

Ed is a strong believer in studying the present to interpret the past and sees great value in reproducing sedimentary features in laboratory ex-

periments. Subjects of special interest to him, as illustrated in his more than 150 publications, are environments of deposition represented by facies, both lithologic and faunal, transgressions and regressions, stratification and cross-stratification, key beds as media for tracing time planes, diastems and unconformities, paleogeography, cyclothems, and minor sedimentary structures.

Clyde H. Moore

*C*lyde was born in Jacksonville, Florida, and raised in New Orleans, Louisiana. He received his bachelor's degree at Louisiana State University in 1954 and then spent the next two years as an officer in the U. S. Army. He attended the University of Texas at Austin for graduate work and earned his doctorate in 1961.

After graduation Clyde joined Shell Development Company as a research geologist until 1966. During his tenure with Shell he investigated Lower Cretaceous shelf limestones under the direction of Frank E. Lozo, modern quartzose clastic tidal sands of the Atlantic Coast with Hugh Bernard, and finally the structural control of sedimentation in Mesozoic-Cenozoic sequences of the Pacific Coast. He joined the faculty at Louisiana State University as an assistant professor in 1966, where he is presently professor of geology and director of the Applied Carbonate Research Program. His research interests at L.S.U. have included intertidal cementation, fresh-marine water mixing diagenesis, reef related sedimentation and diagenesis, and most recently deep burial carbonate diagenesis.

Most of Clyde's work has been concentrated in the Gulf and Caribbean including south Florida and the Bahamas, Grand Cayman, Jamaica and St. Croix as well as the central Pacific. During his tenure at L.S.U. he has been an active consultant to the oil industry and a lecturer for the AAPG.

Henry T. Mullins

H ank was born in Ghent, New York, a small town in the Hudson Valley of upstate New York. During high school he showed little interest in academics, preferring basketballs to books, but a knee injury in 1969 dictated a change in professional aspirations and redirection of energy to more scholarly pursuits.

In 1973, Hank received his bachelor's degree in geology from State University of New York at Oneonta, where his professional interests in geology and oceanography were stimulated by the teachings of Jay Fleisher and Philo Wilson. It was here that he learned, from Jack Kepper, the potential of the Bahamas as a natural laboratory for the study of limestone genesis.

Hank received his master's degree in geology from Duke University in 1975, where his interests in modern carbonates were furthered by Ron Perkins and Orrin Pilkey. Numerous oceanographic research cruises to the northern Bahamas aboard Duke University's *R/V EASTWARD* led to his thesis

study of the "Stratigraphy and Structure of Northeast Providence Channel" under the direction of George Lynts. It was during this study that Hank recognized the need for more work on modern carbonate slopes and basins and, remembering his chilly, childhood days in upstate New York, the advantage of studying tropical carbonates.

His research on modern carbonate slopes continued with a doctoral dissertation on "Deep Carbonate Bank Margin Structure and Sedimentation in the Northern Bahamas," under the watchful eye of Conrad Neumann, at the University of North Carolina, where Hank received his doctorate in oceanography in 1978. As a part of the doctoral program, Hank spent a year studying marine geophysics with Bob Sheridan at the University of Delaware. From 1978 through 1982 Hank was an assistant professor of oceanography at the State of California's Moss Landing Marine Laboratories, where he initiated and developed a graduate program in marine geology. Although Hank's major research interests are in deep-water carbonate sedimentation and seismic stratigraphy, his interests have expanded to include genesis of marine phosphorites, sedimentation in oxygen minimum zones associated with coastal upwelling, and tectonics of the central California continental margin. Hank returned to New York on January 1, 1983, when he and his wife joined the geology faculty at Syracuse University.

W. Martinez del Olmo

W enceslao earned his master's degree in geology from the University of Madrid in 1965. He was employed by FINA IBERICA (Petrofina's Spanish branch) where he became chief exploration geologist, a position he held until 1975. He then joined HISPANOIL where he was actively involved in the exploration of Mediterranean areas of offshore Spain until 1980. He is presently employed by the Spanish National Oil Co. (ENIEPSA) as chief regional geologist for both onshore and offshore areas in Spain.

Peter A. Scholle

R aised in the scenic southern part of the Bronx, New York (now known as Spanish Harlem), Peter developed a deep and abiding love for the outdoors, rural areas, and other things un-Bronxish. After an undergraduate education in geology at Yale

(1965), a year at the University of Munich (on a DAAD/ Fulbright grant), and a year at the University of Texas at Austin (under Bob Folk), he received his master's degree (1969) and doctorate (1970) from Princeton University. The next two years were spent at Cities Service's research lab in Tulsa, followed by three years of graduate teaching at the University of Texas at Dallas.

Peter joined the U. S. Geological Survey in late 1974, where he currently holds the position of chief, Branch of Oil and Gas Resources, in Denver. Pete's research interests (which he indulges in only infrequently these days) include carbonate sedimentology and petrography, the diagenesis of deep water limestones (chalks), and sandstone petrography.

flat facies on northwest Andros, he transferred to Royal Dutch Shell in the mid-1960s to run a small field station in the Persian Gulf where he studied prograding arid tidal flats. In addition to publishing two papers on this work, he made a fundamental contribution by discovery and description of extensive submarine cementation, a process that results in unconformities near the base of prograding tidal flat cycles.

His chapter in this volume combines knowledge gained through extensive field studies and subsurface studies of ancient tidal flats with an extensive review of the works of others, and forms the basis of the new AAPG training film *Stratigraphic Traps — the Tidal Flat Model.*

Devonian reef complexes in Alberta, and the effects of cementation on reservoir quality in Devonian oil and gas fields in western Canada.

Additional published works include the paleoecology of Neocene echinoid assemblages of the Carolinas, experimental dissolution kinetics in carbonate skeletal material, and the isotopic composition of carbonate cements.

Dick was a lecturer of stratigraphy and sedimentology at McGill University (1975-1976), and since that time, has worked as an exploration geologist for Shell Canada Resources Ltd., and Canadian Hunter Exploration. He is currently vice president of Petrosec Exploration Inc., Houston.

Eugene A. Shinn

Richard A. Walls

William C. Ward

*G*ene Shinn is project chief of the U. S. Geological Survey's Fisher Island Field Station in Miami Beach, and former senior geologist with Shell Oil Company. Gene did much of the original research on Andros Island tidal flats in the early 1960s while employed by Shell Development Company and was a member of the Shell Field Research Station in Coral Gables, Florida, headed by R. N. Ginsburg. During that period, he discovered and described the environmental conditions and sedimentary structures associated with modern supratidal dolomite, providing much of the criteria used today for recognizing this form of dolomite in the geologic record. After Gene and his colleagues established the stratigraphy and sedimentary relationships of tidal

*D*ick was born in Fairborne, Ohio, and received his bachelor's in geology from Morehead State University in 1971. A master's degree in sedimentology and geochemistry from the University of North Carolina, and a doctorate (1977) in carbonate sedimentology from McGill University, Montreal, followed.

Dick's early geological experience included both field and subsurface mapping in the Appalachians. He has published papers on Devonian/ Mississippian deltaic facies in the subsurface of the southeastern Appalachian basin, and depositional and diagenetic models in Carboniferous carbonates of northeastern Kentucky. Recent studies and publications concern diagenetic models for

*B*ill Ward was born and raised on Cretaceous limestones in central Texas. He received his bachelor's and master's degrees in geology at the University of Texas in the quasi-Dark Ages of the mid-1950s. Attending Robert Folk's first carbonate petrology course gave him an itch to learn more about limestones. That itch someday had to be scratched, so in the mid-1960s Bill gave up a perfectly good job with Humble Oil and Refining Company to go to Yucatan — via Rice University and James Lee Wilson — to look at modern carbonates. Luckily, Bill arrived on the northeastern coast of Yucatan several years before the dunes were covered with jet-set hotels, and did his doctoral study on diagenesis of the eolianites.

Since 1970, he has been at the University of New Orleans, where he is a professor in the department of earth sciences. His research interests have centered around petrography and diagenesis of carbonates and sandstones.

James L. Wilson

*J*im Wilson was born in Waxahachie, Texas, and remained in the Lone Star state to attend Rice University and the University of Texas, where he received his bachelor's and master's degrees in 1942 and 1944. He earned his doctorate from Yale University in 1949 after serving two years in the U. S. Army. His geologic experience includes that of a field geologist in the Rocky Mountains, associate professor at the University of Texas at Austin (1949-1952), research geologist for Shell Development Company in Houston and for Shell International Research in Rijswijk, Netherlands, from 1952-1966. He became professor of geology at Rice University in 1966 and was chairman of the department from 1974-1977. In recent years he has also taught at the University of Calgary (1970) and the University of Munich, Germany (1972-1973). In 1979 he joined the faculty of the geology department at the University of Michigan where he currently teaches.

Jim was president of the SEPM in 1975, and became an Honorary Member in 1980. In 1975 he completed a book entitled *Carbonate Facies in Geologic History* (Springer-Verlag). He is a member of numerous geological societies and regularly participates in carbonate field and lecture courses with: the Laboratory of Comparative Sedimentology of Miami University, Florida; ERICO of London; the University of Houston; and the AAPG. Recent field experience includes work in Mexico, New Mexico, North Africa, the Rocky Mountains, and the Austroalpine area.

Index

A reference is indexed according to its important, or "key," words. Three columns are to the left of a keyword entry. The first column, a letter entry, represents the AAPG book series from which the reference originated. In this case, M stands for Memoir Series. Every five years, AAPG will merge all its indexes together, and the letter M will differentiate this reference from those of the AAPG Studies in Geology Series (S) or from the AAPG Bulletin (B).

The following number is the series number. In this case, 33 represents a reference from Memoir 33.

The last column entry is the page number in the volume where this reference will be found.

Note: This index is set up for single line entry. Where entries exceed one line of type, the line is terminated. (This is especially evident with manuscript titles, which tend to be long and descriptive). The reader must sometimes be able to realize keywords, although commonly taken out of context.